Springer Water

Series Editor

Andrey G. Kostianoy, Russian Academy of Sciences, P. P. Shirshov Institute of Oceanology, Moscow, Russia

The book series Springer Water comprises a broad portfolio of multi- and interdisciplinary scientific books, aiming at researchers, students, and everyone interested in water-related science. The series includes peer-reviewed monographs, edited volumes, textbooks, and conference proceedings. Its volumes combine all kinds of water-related research areas, such as: the movement, distribution and quality of freshwater; water resources; the quality and pollution of water and its influence on health; the water industry including drinking water, wastewater, and desalination services and technologies; water history; as well as water management and the governmental, political, developmental, and ethical aspects of water.

More information about this series at http://www.springer.com/series/13419

Mukand Babel · Andreas Haarstrick ·
Lars Ribbe · Victor R. Shinde ·
Norbert Dichtl
Editors

Water Security in Asia

Opportunities and Challenges in the Context of Climate Change

 Springer

Editors
Mukand Babel
Asian Institute of Technology
Bangkok, Thailand

Andreas Haarstrick
TU Braunschweig
Braunschweig, Germany

Lars Ribbe
TH Köln
Köln, Germany

Victor R. Shinde
Asian Institute of Technology
Bangkok, Thailand

Norbert Dichtl
TU Braunschweig
Braunschweig, Germany

ISSN 2364-6934   ISSN 2364-8198   (electronic)
Springer Water
ISBN 978-3-319-54611-7   ISBN 978-3-319-54612-4   (eBook)
https://doi.org/10.1007/978-3-319-54612-4

This Springer imprint is published by the registered company Springer Nature Switzerland AG
The registered company address is: Gewerbestrasse 11, 6330 Cham, Switzerland

Foreword

The relationship between water and climate change cannot be overemphasized. It is the primary medium through which the impacts of climate change are manifested. These impacts result in either too much water in the form of floods or too little water corresponding to water shortages. Already, many regions in the world are experiencing these impacts, and the trend will only continue, and perhaps intensity, unless corrective action is taken. Asian countries are particularly vulnerable for several reasons. Over the last few years, incidents related to floods and droughts have been on the rise in Asia, even in areas which traditionally have never faced such issues in the past.

Achieving water security is a prime mandate of governments across the continent. The last decade has seen several countries announce a flurry of policies, ambitious flagship programmes, and dedicated projects, all geared towards improving indigenous water security. Water is a strategic research and academic focus area of the Asian Institute of Technology (AIT). We are proud that AIT has been instrumental in conducting several evidence-based comprehensive scientific studies in some of the countries, that have paved the way for these initiatives.

Addressing climate change is a significant cog in the wheel in that journey towards water security. At the heart of an appropriate response to climate change is improved knowledge about the interplay of this phenomenon with the various societal dimensions that concern us. I am happy that this book seeks to contribute to this ever-growing body of knowledge on this subject. The editors, experts in their own right, have stringed together a compendium of high-class research studies that look at multiple aspects of climate change. This is important because it is now quite evident that combating climate change will require a multi-pronged approach involving diverse stakeholders and different sections of society.

I am optimistic that this book will serve as a useful reference media to a wide range of readers—academics, research scholars, practitioners, and students. I congratulate the editors and authors for their commendable work.

Dr. Eden Y. Woon
President
Asian Institute of Technology
Thailand

Preface

Water is at the heart of human security. It also serves as the vital link to a number of other human needs such as food, energy, and infrastructure. How water is managed will have repercussions on almost every aspect of human security. It is common knowledge that our planet is endowed with abundant water resources but the amount of that available in the form of usable water is very small. Over the years, the world's water resources have been increasingly facing severe pressure from many drivers: population growth, migration, industrialization, urbanization, socioeconomic development, among others. Climate change is exacerbating the crisis, particularly in Asia. Changing weather conditions are leading to increasing hydrologic variability and incidences of extreme events. Going forward, the Intergovernmental Panel on Climate Change (IPCC) suggests an increase in the number of years with above-normal monsoon rainfall or extremely low rainfall. Melting glaciers will affect water supplies, creating risks of glacial lake outbursts of floods and downstream flooding for some regions. This will lead to an overall reduction in water supplies from snow cover and glacial runoff in the long run. Droughts will also become an even more serious concern, particularly given the already strained water access issues.

Since the last five years, the Global Risk Landscape developed by the World Economic Forum has consistently highlighted water crises among the top global risks that the world faces. Enhancing water security, therefore, figures prominently in almost every global and national development agenda. The Sustainable Development Goal 6 (Ensure availability and sustainable management of water and sanitation for all) is an apt example for a global-level initiative. Similarly, almost all countries in Asia have prioritized water security and management in their policies and relevant development plans.

In order to truly achieve water security, it is important to have a holistic understanding of the various dimensions of water security, and for the stakeholders in the various sectors to refrain from working in "silos" and proliferate inter-sector cooperation and coordination. It is against this backdrop that in November 2016, the Asian Institute of Technology in association with Exceed centers CNRD (Cologne) and SWINDON (Braunschweig) in Germany organized an International

Scientific Conference on "Water Security and Climate Change: Challenges and Opportunities in Asia". The objective of this conference was to provide a platform for engaging leading experts in Asia and beyond in cross-sectoral discussions on water security issues in Asia to facilitate the path toward water-secure societies. The conference also sought to serve as a common interface for science, practice, and policy related work and trigger deliberations and dialog among the diverse stakeholders. The conference aimed to encompass as many dimensions of water security as possible and received papers related to sixteen thematic areas. These include: state of water resources and water security philosophies; assessing and quantifying water security; hydrological assessment and forecasting; groundwater security assessment and management; environmental water security assessment; water security for agriculture productivity; climate induced disaster risk assessment and impacts; climate resilient water supply management; enhancing ecosystem protection; climate change adaptation to reduce disaster risk; decision support tools for enhancing water security; water utility management; regional and national level water security enhancement interventions; water economics; wastewater reuse; stakeholder engagement and institution building. The conference, funded by the German Academic Exchange Service (DAAD), was attended by over one hundred and fifty experts from twenty countries.

We are pleased to present this book that comprises the highly appreciated and thought-provoking papers presented in the aforementioned international conference. The book is organized into different sections, each dwelling on a unique aspect of climate change and water security. The editors would like to thank all authors for their contribution to this publication. The editors would also like to acknowledge and thank Ms. Aakanchya Budhathoki Shah and Mr. Shashwat Sharma for their support in developing this publication. We hope this collection of papers will provide good insights into the holistic nature of water security and arouse more scientific interest in subsequent research and action that urgently needs to be taken.

Bangkok, Thailand	Mukand Babel
Braunschweig, Germany	Andreas Haarstrick
Köln, Germany	Lars Ribbe
Bangkok, Thailand	Victor R. Shinde
Braunschweig, Germany	Norbert Dichtl

About this Book

This book provides state-of-the-art scientific knowledge on the interlinkages between climate change and water, especially in the Asian context. Asia has been facing several water-related challenges for decades due to multiple factors such as increasing population, socioeconomic development, urbanization, and migration, and climate change now poses an additional threat. While significant efforts have been made by governments in Asia, much more work is needed to make Asian societies water secure. To do so will require improved knowledge about the interaction of climate change and water resources. This book attempts to meet this knowledge need through a multidisciplinary approach that looks at multiple perspectives and dimensions of water security. The book can serve as a beneficial resource for scientists and researchers involved in furthering the knowledge on water security and climate change and facilitate the translation of this knowledge into practice. In addition, it will provide policymakers, planners, and practitioners to gain useful insights for formulating sustainable water security enhancement strategies grounded in sound scientific evidence.

Contents

Part I   Water Security Landscape

1   Water, Food Security and Asian Transition: A New Perspective Within the Face of Climate Change   3
P. R. Khanal

2   Consolidating Drought Projections—Eastern Australia   17
Shahadat Chowdhury and Michael Sugiyanto

3   Planning for Climate Change and Mechanisms for Co-operation in Southeast Asia's Sesan, Sekong and Srepok Transboundary River Basin   31
N. J. Souter, D. Vollmer, K. Shaad, T. Farrell, H. Regan, M. E. Arias, T. A. Cochrane, and T. S. Andelman

4   Environmental Security Issues No Longer of Secondary Importance for Regional Cooperation or Conflict: The Case of the Mekong River   45
Christian Ploberger

5   Enhancing and Operationalizing Water Security: Present Landscape and Emerging Research Needs   61
Mukand S. Babel and Victor R. Shinde

Part II   Water Security Assessment and Planning

6   The Threats to Urban Water Security of Indonesian Cities   73
R. W. Triweko

7   River Basin Planning for Water Security in Sri Lanka   85
U. S. Imbulana

8   Water Quality Index: A Tool for Wetland Restoration   99
Nitin Bassi

9   **A Framework for Implementing Integrated Water Resources
    Management at River Basin Level in Indonesia**   111
    R. W. Triweko

10 **A Study on Developing Theoretical Framework, Dimensions,
    and Indicators of Urban Water Security in Indonesia**   123
    J. E. Wuysang and S. B. Soeryamassoeka

11 **Micro Level Vulnerability Assessment of Estuarine Islands:
    A Case Study from Indian Sundarban**   135
    R. Hajra, A. Ghosh, and T. Ghosh

12 **Assessing Water Security at District Level: A Case Study
    of Bangkok, Thailand**   155
    A. Onsomkrit, M. S. Babel, V. R. Shinde, and V. P. Pandey

13 **Assessment of Water Security in Indonesia Considering Future
    Trends in Land Use Change, and Climate Change**   167
    S. Tarigan and Y. Kristanto

Part III   Water Availability Assessment

14 **Application of Hydrological Study Methodologies Used
    in African Context for Water Security in Asian Countries**   181
    D. P. C. Laknath and T. A. J. G. Sirisena

15 **Implementation of Budyko Curves in Assessing Impacts
    of Climate Changes and Human Activities on Streamflow
    in the Upper Catchments of Dong Nai River Basin**   195
    Thi Van Thu Tran, Long Phi Ho, and Quang Phuoc Phung

16 **Contribution of Snow and Glacier in Hydropower Potential
    and Its Response to Climate Change**   207
    R. B. Chhetri, N. M. Shakya, and N. R. Sitoula

17 **Assessment of Variation of Streamflow Due to Projected Climate
    Change in a Water Security Context: A Study of the Chaliyar
    River Basin, India**   223
    S. Ansa Thasneem, Santosh G. Thampi, and N. R. Chithra

18 **Addressing Water Resources Shortfalls Due to Climate Change
    in Penang, Malaysia**   239
    N. W. Chan, A. A. Ghani, Narimah Samat, R. Roy, M. L. Tan,
    and Haliza Abdul Rahman

19 **Assessing the Physical Water Availability and Revenue Aspect
    of WUAs Under New Governance System: Case Studies
    of Pakistan**   251
    S. Ahmad, S. R. Perret, M. Imran, and S. A. Qaisrani

20 Simulation of Kathmandu Valley River Basin Hydrologic Process Using Coupled Ground and Surface Water Model 261
S. Basnet, N. M. Shakya, H. Ishidaira, and B. R. Thapa

21 Estimation of Water Availability in Rivers of Stung Sreng Basin, Cambodia, Using HEC-HMS 287
P. Hok, C. Oeurng, and S. Heng

Part IV Modelling Studies for Water Security Dimensions

22 Predicting Low Flow Thresholds of Halda-Karnafuli Confluence in Bangladesh 303
A. Akter and A. H. Tanim

23 Stream Flow Forecasting with One Day Lead Using υ-Support Vector Regression 315
Sameer Anipindiwar and Umamahesh V. Nanduri

24 Flood Discharge Estimation in Baddegama Using Pearson Type III and Gumbel Distributions 329
T. N. Wickramaarachchi

25 Impact of Climate Change in Rajshahi City Based on Marksim Weather Generator, Temperature Projections 339
M. Tauhid Ur Rahman, A. Habib, R. Tasnim, and M. Fida Khan

Part V Urban Water Security

26 Water Supply of Dhaka City: Present Context and Future Scenarios 351
Mohammed Abdul Baten, Kazi Sunzida Lisa,
and Ahmed Shahnewaz Chowdhury

27 Quantification of Municipal Water Supply and Role of Metering in Large Indian Cities 369
P. Sampat and Y. Alagh

28 Towards Holistic and Multifunctional Design of Green and Blue Infrastructure for Climate Change Adaptation in Cultural Heritage Areas 381
Zoran Vojinovic, Weeraya Keerakamolchai, Arlex Sanchez Torres,
Sutat Weesakul, Vorawit Meesuk, Alida Alves, and Mukand S. Babel

29 Adaptation to Flood Risk in Areas with Cultural Heritage 391
Zoran Vojinovic, Daria Golub, Weeraya Keerakamolchai,
Vorawit Meesuk, Arlex Sanchez Torres, Sutat Weesakul,
Alida Alves, and Mukand S. Babel

30 Rainwater for Domestic Use in Urban Area: A Simulation
    of Rainwater Harvesting System for Surabaya, Indonesia 401
    C. Kusumastuti and H. P. Chandra

31 Water Security and Climate Change: A Periurban Perspective . . . 413
    V. Narain

32 Selecting Multi-Functional Green Infrastructure to Enhance
    Resilience Against Urban Floods. 429
    A. Alves, A. Sanchez, B. Gersonius, and Z. Vojinovic

33 Climate Change and Sustainable Urbanisation: Building Urban
    Water Security in a Metro City of India. 443
    Shailendra K. Mandal and Gregg M. Garfin

34 An Analysis on Relationship Between Municipal Water Saving
    and Economic Development Based on Water Pricing Schemes 461
    W. C. Huang, B. Wu, X. Wang, and H. T. Wang

Part VI   Water Governance and Management

35 Improving Water Security to Mediate Impacts of Climate Change
    in the Ganges Basin 481
    B. Sharma, P. Pavelic, and U. Amarasinghe

36 Equitable Distrubution of Water in Upper Godavari Sub Basin:
    A Case Study from Maharashtra 493
    S. A. Kulkarni

37 Watershed Conservation for Ecosystem Services
    and its Implication for Green Growth Policies in the Context
    of Global Environmental Change: A Case of Bhutan 505
    Om Katel, Dhan B. Gurung, Kazuhiro Harada,
    and Dietrich Schmidt-Vogt

38 Testing Framing Effects on Subjective Wellbeing in Lao PDR 517
    J. Ward and A. Smajgl

39 Development Trade-Offs in the Mekong: Simulation-Based
    Assessment of Ecosystem Services and Livelihoods 533
    A. Smajgl, J. Ward, and T. Nuangnong

40 Groundwater Institutions and Governance in North India: Cases
    of Hoshiarpur and Jammu Districts 545
    Ishita Singh

41 The Effect of Cotton Management Practices on Water Use
    Efficiency and Water Security Challenges in Pakistan 557
    F. Zulfiqar and G. B. Thapa

42 Multi-stakeholder Negotiating Platforms for Effective Water Governance: A Case of Mashi Basin, Rajasthan, India 567
M. S. Rathore

43 Economic Benefits of EbA Measures to Assure Water Security: A Case Study on EbA for Sediment Trap Versus Dredging 581
R. Treitler, P. Kongapai, J. Ngamsing, and K. Sansud

Part VII Wastewater Engineering and Management

44 Coping with Salinity Intrusion in South-Western Bangladesh: The Continuing Struggle 597
Mokbul Morshed Ahmad and Muhammad Yaseen

45 Reclaimed Wastewater Reuse for Irrigation in Turkey 609
M. E. Aydin, S. Aydin, and F. Beduk

46 Domestic Wastewater for Climate Mitigation 619
K. Unwerawattana, T. Reinhardt, A. Michels, and C. Wongburana

47 Making the Case for Wastewater Irrigation in Bangladesh 631
S. T. Mahmood

48 Monitoring of Performance of Deammonification Process in Treating Wastewater in a Pilot Study, KTH, Sweden 643
M. T. Ur Rahman, U. R. Siddiqi, Md. A. Habib, and Md Rasheduzzaman

Part VIII Disaster Risk Assessment and Management

49 Assessment of Urban Flood Resilience for Water, Sanitation and Storm Water Drainage Sectors in Two Cities of India 659
S. Thakur and U. Bhonde

50 Vulnerability to Disaster in a Multi-hazard Coastal Environment in Bangladesh 675
M. M. Islam, M. Mostafiz, P. Begum, A. Talukder, and S. Ahamed

51 Analysis of Public Perceptions on Urban Flood in Phnom Penh, Cambodia 687
S. Heng, S. Ly, S. Chhem, and P. Kruy

52 Assessment of Climate Change Impact on Drought in the Central Highlands of Vietnam 703
Dao Nguyen Khoi and Pham Thi Thao Nhi

**53 Optimized Operation of Red-River Reservoirs System
     in the Context of Drought and Water Conflicts** 715
     L. X. Nguyen, T. D. Tran, S. T. Hoang, and P. T. Nguyen

Part IX Community Engagement for Water Security Enhancement

**54 Integrated Information Dissemination System for Coastal
     Agricultural Community** 735
     Aaron Firoz, Nazmul Huq, and Lars Ribbe

**55 Citizen Science on Water Resources Monitoring in the Nhue
     River, Vietnam** 749
     N. H. Tran, T. H. Nguyen, T. H. Luu, M. M. Rutten, and Q. N. Pham

**56 Remuneration for Conservation: An Ecosystem Service Changing
     Lives in the Hills of Bangladesh** 763
     F. K. Pushpa

**57 Role of Citizen Science in Safe Drinking Water in Nepal:
     Lessons on Water Quality Monitoring from Brazil** 775
     A. Gautam, J. Ramirez, L. Ribbe, K. Schneider, S. Panthi,
     and M. Bhattarai

**58 Community Mitigation Approaches to Combat Safe Water
     Scarcity in the Context of Salinity Intrusion in Coastal
     Bangladesh** 787
     M. Tauhid Ur Rahman, Md. Rasheduzzaman, Md. Arman Habib,
     Afzal Ahmed, Syed M. Tareq, and S. Md. Muniruzzaman

About the Editors

Mukand Babel is a Professor of Water Engineering and Management and the Chair of Climate Change Asia Initiative at the Asian Institute of Technology, Thailand. His professional experience in teaching, research, and consultancy spans over 35 years. He was a member of the Executive Board of International Water Resources Association (2016–2018) and a Board member of the Asia Water Council since 2016. He currently conducts interdisciplinary research relating hydrology and water resources with socioeconomic and environmental aspects of water to address diverse water problems including water security assessment, water-energy-food nexus, and climate change impact and adaptation in the water sector.

Andreas Haarstrick received his doctorate in 1992 in the field of biotechnological up- and down-streaming processes of biopolymers. Since 2006, he is Professor of Bioprocess Engineering at Technische Universität Braunschweig. His teaching and research interests cover modeling biological and chemical processes in heterogeneous systems, development of models predicting pollutant reduction in and emission behavior of landfills, growth kinetics at low substrate concentrations under changing environmental conditions, Advanced Oxidation Processes, and groundwater management. Since 2012, he is the managing director of the DAAD exceed-Swindon project dealing with the sustainable water management in developing countries (www.exceed-swindon.org).

Lars Ribbe is a Professor for integrated land and water resources management at TH Köln (University of Applied Sciences) and Dean of the Faculty of Spatial Development and Infrastructure Systems. Holding a Ph. D. in the field of Hydro-informatics, his current work in research and teaching is on spatio-temporal river basin assessment, modeling, and management, and he is specifically interested in developing knowledge systems—helping decision makers to cope with prevailing land and water resource challenges. Furthermore, he coordinates a global network of partner universities (www.cnrd.info) and an international network of MSc programs on "Integrated Water Resources Management".

Victor R. Shinde is a senior water management expert with the National Institute of Urban Affairs, the think tank of India's Central Ministry of Housing and Urban Affairs. He is involved with major projects of national importance in the country such as the developing frameworks for urban river management in the Ganga Basin, designing the national climate smart cities assessment, preparing the Master Plan for Delhi, among others. He has worked as a scientist and technical expert in nine countries in Asia and Africa-Bhutan, India, Japan, Mauritius, Philippines, Seychelles, South Korea, Thailand, and Vietnam.

Norbert Dichtl was Professor at the Technische Universität Braunschweig and head of the Institute of Sanitary and Environmental Engineering from 1994–2019. His specific work priorities lie in engineering processes that deal with the pursuing cleaning of municipal and industrial wastewaters and the digestion and dewatering of sludge and sewage sludge. Particular attention is also devoted to research on processes dealing with biological, physical, and chemical wastewater treatment (municipal, industrial), simulation, and optimization. Another key focus is the biogas production and reuse, anaerobic sludge treatment, wastewater reuse, and the sustainable environmental protection under the goals of Agenda 21. He is the author of more than 400 papers in peer-reviewed international and national journals.

Part I
Water Security Landscape

Chapter 1
Water, Food Security and Asian Transition: A New Perspective Within the Face of Climate Change

P. R. Khanal

Abstract Asia has witnessed a remarkable economic growth in recent times. The new economic environment offered by this growth helped millions of rural mass come out of poverty and starvation. Water has played a central role in this transformation supporting agricultural change and providing foundation for rural livelihood. While water is still at the center of food security and poverty reduction in Asia, the water management challenges are much more complex than before due to rapid economic growth, growing population, and changing rural urban linkages. Climate change will further compound the existing water management challenges changing the patterns of demand and supply of water for agriculture, livelihood and other economic activities.

This paper first analyzes the challenges facing water and food security within the context of ongoing Asian transition. It argues that food security is still a rural agenda, and future water and food security largely hinges on the prosperity of these rural population. Water management that facilitates these transitions and promote inclusive, equitable and greener growth will be a key to future water and food security agenda. It then looks at the broad and more specific impacts of climate change in major food production systems and explores how the extent and productivity of both rain fed and irrigated agriculture are expected to change in major Asian production system.

The paper shows that climate change will impact majority of the Asian production system covering some 70% of its population. It therefore calls for new practices in water management combining both hard and soft measures that boost water productivity, enhances water conservation, maintains water eco-systems and facilitates multiple water use services.

Keywords Water · Food security · Asian transition · Climate change

P. R. Khanal (✉)
FAO Regional Office for Asia and Pacific, Bangkok, Thailand
e-mail: PuspaRaj.Khanal@fao.org

© Springer Nature Switzerland AG 2021
M. Babel et al. (eds.), *Water Security in Asia*, Springer Water,
https://doi.org/10.1007/978-3-319-54612-4_1

1.1   Water Security in the Changing World

Since historical times, control over water has been pivotal for increasing agriculture production and productivity providing food and fiber for growing population. Access to irrigation was a key factor for success to the Green Revolution, which helped millions of poor farmers to come out of poverty and starvation improving stagnant agriculture sector. It also contributed to the fast socio-economic development in many parts of Asia. In recent times, innovative and affordable water management technologies are further revolutionizing agriculture sector with cheaper and efficient means for reliable access to water for crop production. The central role of water in meeting global food security is clearly illustrated by the fact that currently, about 40% of food production in the world comes from only 20% of irrigated land (FAO AQUASTAT).

But we are in a changing world and new challenges are emerging for water management. Despite remarkable economic growth, population continues to rise, and over one in eight in Asia still suffer from chronic hunger. Millions of people still lack basic sanitation facilities and access to drinking water especially in rural areas. Transferring the rural areas towards prosperity remains a daunting task in the twenty-first century. On the other hand, water demand continue to rise as a result of both population and economic growth. In the twentieth century, water withdrawals grew at almost twice the rate of population increase. Going forward in this century, it is anticipated that the world will need to produce 60% more food on average to feed a hungry world of more than nine billion people. The main challenge in our immediate future is to meet the increasing demand while also providing water for other economic, domestic and environmental uses—and all within the context of increasing water scarcity.

Asia is also experiencing rapid environmental degradation and widespread biodiversity loss despite progress in limited areas. Land cover and land use change continue with heavy consequences in the fragile eco-systems hindering the delivery of eco-system services. Soil erosion is already a major global threat and is expected to accelerate due to increased weather extremes like floods and drought resulting from climate change. Degradation of range land, deforestation and wet land destruction continue to intensify impacting the water supply regime and ultimately the food production system. All these will limit both land availability and water extraction in future with serious consequences to economic growth, food security and poverty reduction in coming days.

In future, climate change will exacerbate the pre-existing water challenges resulting from population and economic growth. Climate change is expected to intensify water scarcity worldwide and increase water related disasters. A warmer climate, with its increased climate variability, will increase the risk of both floods and droughts (IPCC 2007). In Asia, available reports suggests that by 2050, freshwater availability in Central South, East and South East Asia, particularly in large river basin will decrease. In contrast, the heavily populated mega deltas in the South, East and Southeast Asia will be at risk due to increased flooding from the sea and rivers. There is an urgent need for a concerted effort to incorporate water into climate policy and to reinforce the link between improved water security, sustainable food production and poverty reduction.

1.2 The Asian Transition

Asia is leading in economic growth worldwide. Between 2000 to 2012, its GDP (whole Asia and Pacific) increased from US$ 8.5 trillion to US$ 23.3 trillion at current prices. Likewise, its share in world GDP in terms of real US$ PPP increased from 23.2% to 38.8% between 1990 to 2014 and is expected to increase to 45% by 2025 (Barua 2015). This share is much larger than that of US and EU economy. Asia today is the fastest growing continent in the world and the region's ascendancy in the global economy is expected to continue for foreseeable future.

Together with the economic growth, it has made impressive progress in poverty reduction. Rural poverty in the developing world declined from 54% in 1988 to about 35% in 2008. This was largely due to massive reductions in poverty in East Asia, where rural poverty now stands at around 15% (IFAD 2011). Likewise, the proportion of hungry has been reduced by 12% points from the initial 24% between 1990/92–2014/16. In other words, while one in four people was undernourished some 25 years ago, today only one in eight is hungry (FAO 2015). Currently, Asia is urbanizing faster than any other regions in the world and is projected to be 64% urban by 2050 (UN 2014).

Despite striking economic success, over two third of the Asian population (excluding the advanced economies in Asia) still live in rural areas, and almost seventy percent of the poor are rural. It is the largest and most populous continent, and with 490 million people still suffering chronic hunger, the region is home to almost 62% of such people in the world and are mostly concentrated in South Asia. Eastern Asia and the South-Eastern Asia have shown great success in reducing the number of hungry people in the past decades. Southern Asia has not been able to reduce hunger as in other parts of Asia and continues to be the sub region with the largest number of hungry people in the world, a trend that is expected to continue for the foreseeable future.

Water, poverty and food security in Asia is therefore basically a rural phenomenon and rural agenda. The key challenge is to address issue of food insecurity and hunger in the region and bring prosperity to the rural mass who still are almost two thirds of the Asian population. A successful transition towards higher economic growth is essential in Asia to address absolute dimension of poverty and hunger and rapid rural transformation. However, given the large number of rural population and high number of rural poor, these growths should be inclusive and equitable. The issue is not only about producing more food, but more importantly about making food available to everyone's plate.

Available data shows that over two third of the rural poor still dependent on agriculture for their livelihoods and water is often the main limitation to improving agriculture productivity. A study on water and poverty in Asia (Khanal et al. 2014) shows that water related intervention can have high impacts on rural livelihoods and are a key point for alleviating rural poverty. The study further shows that almost half of the rural population will directly benefit from water intervention to support their livelihood. There is therefore clearly a great potential for poverty reduction through ensuring water security in rural Asia. The challenges therefore is how to achieve this security in a context of changing climate that is impacting the whole hydrological regime.

1.3 Ensuring Water Security Within the Context of Transition

Increased economic growth, reduced poverty and increasing urbanization has brought profound agricultural and dietary transformation in Asia. The intake of high quality food stuffs like meat, poultry and horticultural products has been rising whereas that of grains like wheat and rice has been stabilizing (Khanal et al. 2014). For example, between 1980 to 2010, consumption of poultry, fruits, vegetables, meat all increased by almost five times in China and many South East Asian nations followed the same trend. This, combined with population growth, will further increase agriculture water requirement in future. On the other hand changing lifestyles will increase domestic and industrial water demand.

The impact of economic and population growth in water use can be clearly seen from the Fig. 1.1 which shows the trends in water withdrawal in India and China.

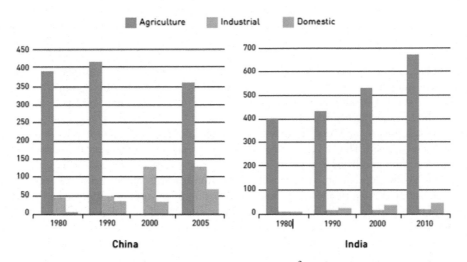

Fig. 1.1 Changes in water withdrawal in China and India (Km^3). Agriculture water use figure for 2000 for China is not available. *Source* Khanal et al. 2014

While agricultural water withdrawal has almost stabilized/ or decreased in China, it continues to rise in India. Demand for water for both domestic and industrial uses is rising in both countries at different pace. With the current trends of population and economic growth, global water demand is projected to grow by some 55% due to growing demand from manufacturing (+400%), thermal electricity generation (+140%) and domestic use (+130%) with Asia taking major share. In the face of these competing demands, there will be little scope for increasing water for agriculture and competition for water will be even more acute (OECD 2012).

The recent Asian water development outlook (ADB 2016) shows a strong link between economic growth and water security in the region. Countries with higher GDP are found to have higher water security. The issue however, is that countries with higher GDP have higher water security because they are rich or are they rich because they have higher water security. This is further explained by the GWP analysis which concludes that the relationship runs in both directions. Water-related investments can increase economic productivity and growth, while economic growth provides the resources to invest in institutions and capital-intensive water infrastructure. The conclusion is that water security increases with economic growth, but that also increases overall water use. It is however, difficult to quantify the share of each of these growth in total water withdrawal.

With its rising economic achievement, Asia has been reasonably able to maintain higher economic water security over the years, but has failed to maintain environmental water security as well as the security against water related disasters. This can be clearly seen from Fig. 1.2, which shows different dimensions of water security in Asia. The figure shows that economic achievements in Asia, in large part, has been obtained at the cost of environment. There is therefore an urgent need to reverse the past trends of unsustainable use of our limited water resources not only balancing supply and demand, but also preserving the eco systems of which water is the integral part. This is even important in the context of climate change described in next section.

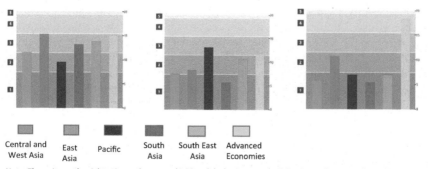

Note: The units on the right axis are the scores (1-20 scale); the ones on the left axis are the stages (1-5 scale from hazardous to model)

Fig. 1.2 Different dimensions of water security in Asia (from left to right): economic water security, environmental water security and security against water related disasters *Source* ADB 2016

Entwined between the economic growth and environmental degradation, Asia now faces two momentous and complex transitions. The first is the structural transformation linked to economic growth and would be main driver for future water and food security, poverty reduction and hunger elimination. The second is about transformation towards a greener growth for sustainable food production. These two transitions are rarely considered together, and in many instances they do not align well together. Future water and food security will depend on the way these two different transitions will be managed together. A key challenge is therefore how to adopt policies and strategies to successfully support and orient this double transitions in a sustainable and equitable manner.

1.4 Climate Change and Uncertainties: The Threat Multiplier

Climate change is superimposed on existing vulnerabilities and exacerbates current challenges in the management of water resources associated with ongoing global changes resulting from economic and population growth discussed earlier. With climate change, the whole water cycle is expect to undergo significant change impacting timing, quantity, quality and availability of water resources. Water is the vector by which most of climate change dynamics on societies and ecosystems are transmitted. Water management therefore plays crucial role in mitigating and adopting to effects of climate change. The importance of sustainable water management in climate change context is already illustrated by the fact that 82% of the nationally determined contributions (INDC) presented for COP 21 have an adaptation sector and among these, 92% mention water!

The combined impact of the ongoing economic transformation and the climate change can be presented as shown in Fig. 1.3. It is again the rural poor who would be disproportionally affected due to their high exposure, higher sensitivity and low coping capacity. In general, climate change will intensify the water stress, change land and water use pattern, increase incidences of water related disasters, cause sea level rise affecting the coastal and delta environment and accelerate loss of ecosystem and bio-diversity. These factors are not independent of one another in the wider climatic cycle, but they may impact agriculture, rural and urban assets, and other economic sectors differently. The net result will be increased production risk, diminishing agriculture, and reduced economic activities ultimately impacting the livelihood of the most vulnerable section of the society.

Climate Change is expected to alter both water availability and water consumption pattern. Rising temperature will result in higher evaporation which will impact both groundwater recharge and groundwater-surface water interactions limiting availability of both grey and blue water. This will have profound impact on both rainfed agriculture as well as in areas where agriculture depends on groundwater systems. Water consumption and demand will change as rising temperature will impact both

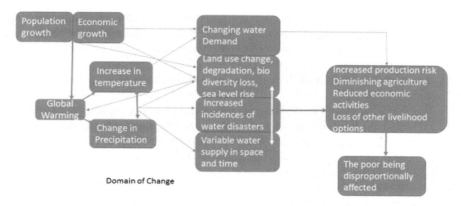

Fig. 1.3 A conceptual framework on impact of climate change and economic and population growth in water management and food production

crop water requirement and crop duration. In addition, it will also alter the whole snow based hydrological regime causing large scale impact to water storage and availability. Climate change also influences the precipitation pattern, and alters the hydrological regime impacting fresh water availability.

In many parts of Asia, mountain snowpack is the primary source of water supporting productive agriculture and other domestic and economic activities. Climate change will reduce snow storage and hence the water availability. Rising temperature means declining snow fall as snow will turn into rain before it hits ground or it will melt faster in the ground. It will therefor experience shrinking snow season, earlier snow melt and lengthening of the dry summer season. In addition, with rising temperature, the permanent snow line is expected to shift upwards hence losing large volume of snow storage. Total water availability for future will sharply decline in one hand and water induced disasters (floods and droughts) are expected to intensify on the other. The impact of this change in snow hydrology would be severely felt in Himalayan river systems of Asia that provides food and fiber in most areas of South Asia and China.

Though studies (Immerzel et al. 2013) show that Himalayan River may not experience water supply decline in near future, increasing weather extremes and temporal variation in precipitation will have profound impact in agricultural production and livelihood. The hydrological cycle is expected to accelerate as rising temperatures increase the rate of evaporation from land and sea. As a result, precipitation will increase in the tropics and humid regions while the arid and semi-arid zones will receive lesser precipitation. That means water scarcity will intensity in the drier areas while some humid regions may experience increasing flooding incidences.

The Asian continent has already experienced several large scale flood incidences in this decade, for example the Himalayan Tsunami of 2013, the floods in Thailand in 2011, Pakistan flood in 2011. On the other hand, drought intensity is increasing at an alarming rate in most of semi-arid zones of Asia. In fact in many semi-arid zones, drought has become a norm rather than an event threating the future of rainfed farming and range land production. Increasing drought risks have been a major production and livelihood challenges in countries like Pakistan, Afghanistan and India. Drought risk often ranks first in terms of weather related economic impact in many of the Asian countries. A future with expanding arid region and shrinking humid regions is eminent in future.

While large parts of South Asia will experience increased incidences of drought and flooding, the low lying deltas like the Mekong delta and the Ganga–Brahmaputra delta and many Asian coastal zones will experience degradation of land and water environment due to sea level rise, increased incidences of cyclones, inundation and flooding, salt water intrusion, and coastal erosion resulting in loss of agriculture land and bio-diversity. The frequency and intensity of cyclone have already increased in many coastal regions of Asia with heavy impact on lives and livelihoods. These are often areas with high population density, complex ecological environment and intensive agricultural production systems. The decline in eco-system services due to climate change could be devastating to local population.

Rising temperature and changing precipitation and water regimes described above will have direct consequences to rural production system as crops are very sensitive to climate. Climate change has affected crop yields in many places and is reducing production of staples such as rice and wheat. Nelson et al. (2009) predicts that by 2050, crop yield are expected to decline by about 1% in rainfed wheat and rice, 18% in irrigated wheat and 34% in irrigated rice due to climate change related impacts. The study further shows that in the absence of appropriate adaption measures, most of the developing countries will experience crop yield decline, particularly in South Asia ultimately leading to higher prices of agriculture commodities. Food insecure countries which have high rates of undernourishment, child stunting and wasting, and mortality among children will be impacted the most.

Changing climate and increasing climatic variability will increase agriculture production risk impacting farm income and food security in several ways. Erickson et al. (2011) warn that in many areas of global tropics including the Indo-Gangetic plain it may be too risky to pursue agriculture as a livelihood strategy in the long run. Many tropics in both South and South Asia will experience new growing condition and therefore will require adoption to agriculture systems. Risk of climate variability also affects livestock and fishery production, mainly arising from its impact on grassland and rangeland productivity as well as due to loss of bio-diversity and diminishing water quality. Results from several studies show that South Asia without sufficient adaptation measures, will likely suffer negative impacts on several crops that are important to large food-insecure human populations.

Climate change will not only affect the food production system (hence the food availability) discussed above, but also the other dimension of food security like food access, food utilization and food system stability. These other dimension of food security will be impacted mostly indirectly due to loss of livelihood assets, increasing migration, increase in water induced disasters, and conflict over available resource use. Further, studies have shown that climate change will also impact nutrition and nutrition value, which are also integral to food security. Most often those who already suffer from malnutrition and hunger are those most vulnerable to the impacts of climate change.

Climate change is the defining issue of our time and is expected to impact both economy and environment with consequences to future food security. As climate change manifests itself primarily through changes in the water cycle, adaption to climate change in large part directly translates to sustainable water management. Integrated and nature based approach to water management combined with enhanced local water management practices will play important role in future. Many countries, especially in South Asia are more vulnerable to impacts of climate change due to their low coping capacity of human system and their high dependency on agriculture. They need to be supported on effective adaption strategy thorough capacity development, awareness raising and on application of innovative water management technologies to minimize climate related production risk.

1.5 Response to the Climate Change in Various Production System in Asia

Discussions in previous paragraphs clearly show that climate change will significantly impact agriculture by increasing water demand, limiting crop productivity and by reducing water availability in areas where irrigation is most needed. Estimate of incremental water requirement to meet future demand for agriculture production under climate change may vary from 40–100% of the extra water needed without global warming (Turral et al. 2011). This clearly shows the importance of improved water management as the possibility of bringing new water to agriculture remains very limited for several reasons.

In order to identify water management options in the face of climate change, it is essential to know the systems in which the challenges are severe and climate change risks are high. Based on the importance of irrigation and other forms of agricultural water management, the following areas/systems are considered as high risk areas due to climate change impacts:

1. Glacier and snow fed large scale irrigation systems (those fed by Himalayan Rivers in India, Pakistan and China) may see diminishing water supply and increasing variability;
2. Coastal zones and deltas (Ganges–Brahmaputra in India and Bangladesh, Mekong delta in Vietnam, Irrawaddy delta in Myanmar) will experience sea level rise, increased incidences of cyclone induced flood and storms and salinity intrusion in fresh water system;
3. Surface and groundwater systems in arid and semi-arid areas, where rainfall will decrease and climatic variability will intensify; and
4. Humid tropics in monsoon regions will experience declining storage yield whereas intensity of peak flows are likely to increase resulting in net water loss for beneficial use.

The above four risk zones fall in the five major livelihood zones of Asia as seen in Fig. 1.4. These five zones include rice and rice–wheat-based irrigation systems, lowland rice based systems, rainfed (dry and humid tropic) and groundwater-based systems and they are also the major food production zones representing over 70% of Asian population. These areas, where agricultural water management is highly vulnerable to climate change require various grey, green and soft measures to minimize the future production risk and are summarized in Table 1.1. Grey measures refer to technological and engineering solutions. They mostly include hardware measures targeted at supply enhancement and are infrastructure-based, e.g. dams, water transfers, new tube wells and flood levees. Green measures are based on the ecosystem-based (or nature-based) approach and soft options include policy, legal, institutional, social, management and financial measures mostly targeting at demand management.

It can be seen from the Table 1.1 that the vulnerable regions have moderate level of adaption options in most cases and have wide variety of response options. Many of these response are generic, but will be applied in different combinations in specific contexts. These can also benefit from evolving approach to water management using ICT based technologies, innovative water accounting principle and consumption based water application. However, in many case it requires appropriate demand

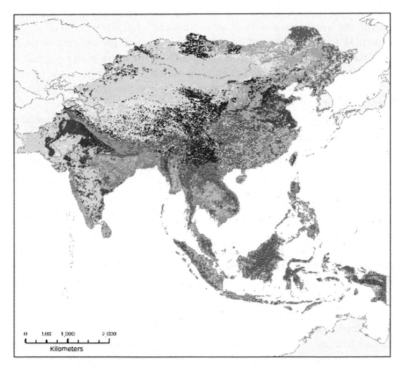

Fig. 1.4 Major livelihood zones in Asia *Source* Khanal et al. 2014

Table 1.1 Water management options for high climate change risk areas in Asia

Systems	River basin/geographic area	Climate change drivers	Adaptability	Response options
Rice or Rice—wheat based large irrigation	Glacier or snow based Himalayan river fed areas: Indo-Gangetic Basin, Mekong Basin and north China	Increased flow for a few decades followed by reduction in surface and groundwater, changes in runoff and peak flow	Low to medium	Demand management, increase water storage, drought and flood management, improved water management in rice
Low land rice-based systems	Rice based low land deltas and coastline (Ganga–Brahmaputra, Mekong, Brahmaputra)	Flooding, salinity (in coastal areas) damage to infrastructure, expected increase in groundwater recharge	Medium to high	Conjunctive use, atomistic irrigation, flood control small-scale irrigation
Rainfed dry tropics and subtropics; humid subtropics	Central Indian Plateau, northeast India, north central part of Southeast Asia	Increased drought and flooding	Medium	Use of micro-irrigation and other water conservation measures, Increase storage, groundwater recharge and atomistic irrigation
Cereal-based (rainfed)	Eastern China	Variable rainfall, drought, flooding	Medium	Drought management, improved water productivity, water harvesting
Groundwater based systems	Western India, northeast China	Increased incidence of drought	Medium	Groundwater interventions, replace high water requiring crops like rice with low water requiring high value crops

management, agricultural change and integrated management of land and water resources which are not easy to achieve in many socio-political settings. It requires strong commitment from government and support from people to achieve the desired results.

1.6   Way Forward: Managing Water for Future in Asia

The success of Asian Transition hinges heavily upon how it meets water and environment crises in a context of changing climate. There is no one-size-fits-all remedy given the very complex hydrological regime ranging from humid to arid climate, different topographical characteristics and varied social political system in the Asian Continent. Its domestic and agriculture water use continue to rise due to large population density and higher dependency on agriculture for livelihood. Its monsoon based hydrology yields very uneven spatial and temporal distribution of available water resources. Climate change will further intensify the water management challenges as the future will experience increasing water demand that needs to be satisfied with reduced and variable supply.

This needs to be addressed through new practices and investments applying both hard and soft measures described earlier. Innovation in technology and management will play a central role to enhance resiliency and sustainability of water system. In recent times, range of Remote Sensing and ICT based technologies have been successfully applied in effective water allocation, distribution and consumption and in establishing early warning systems to safeguard against water disasters. They enhance efficacy of both hard and soft measures and should be promoted widely. The choice of options depends on how different production systems are going to be affected and may involve combinations of several measures discussed in previous section.

It would require a multi-layered water management approach to ensure water security in future that boost water productivity, conserve water resources, preserve eco-system services, maintain water quality across agriculture, fishery and livestock and facilitates multiple water use and their supply chains and adopts bottom up participatory process in resource management. Water storage, both surface and subsurface will be critically important because of its importance in addressing variability. Equally important is to increase the use of green water and recycling and reuse of grey water while managing the demand in blue water. Such a multi-layered approach would also require changes in policy and institutions, and increased capacity of the local water management institutions.

References

ADB (2016) Asian water development outlook 2016. Strengthening water security in Asia and the Pacific, Asian Development Bank, Manila

Barua A (2015) Packing a mightier punch: Asia's economic growth among global market continues. Asia Pacific Economic Outlook Q1:2016

Ericksen P, Thornton P, Notenbaert A, Cramer L, Jones P, Herrero M (2011) Mapping hotspots of climate change and food insecurity in the global tropics. CCAFS Report No. 5. https://ccafs.cgiar.org/publications/mapping-hotspots-climate-change-and-food-insecurit yglobal-tropics#.UpjnYEwo7IU

FAO (2015) Regional overview of food insecurity, Asia and the Pacific: towards a food secure Asia and the Pacific. Food and Agriculture Organization of the United Nations, Rome

FAO AQUASTAT. https://www.fao.org/nr/water/aquastat/didyouknow/print3.stm. Food and Agriculture Organization of the United Nations, Rome

Hazell Peter BR (2011) The Asian Green Revolution, IFPRI Discussion Paper 00911, 2020 vision initiative

Immerzel WW, Pellicciotti F, Bierkens MFP (2013) Rising river flows throughout the twenty-first century in two Himalayan glacierized watersheds. Nat. Geoscience 6(9):742–745

International Fund for Agricultural Development (IFAD) (2011) Rural poverty report 2011. New realities, new challenges: New opportunities for tomorrow's generation. Rome: IFAD. https://www.ifad.org/rpr2011/index.htm.

IPCC 2007, IPCC Fourth Assessment Report: Climate Change 2007

Khanal PR, Saniti G, Merry DJ (2014) Water and the rural poor: Interventions to improve rural livelihood in Asia. Food and Agriculture Organization of the United Nations

Nelson GC, Rosegrant MW, Palazzo A, Gray I, Ingersoll C, Robertson R, Tokgoz S, Zhu T, Sulser TB, Ringler C, Msangi S, You L (2009) Food security, farming, and climate change to 2050: Scenarios, results, policy options. Washington: IFPRI. https://www.ifpri.org/sites/default/files/publications/rr172.pdf

OECD (2012) Meeting the water reform challenge

Turral H, Burke J, Jean-Marc F (2011) Climate change, water and food security. FAO Water Reports 36. Food and Agriculture Organization of the United Nations, Rome

UN (2014) World Urbanization Prospects. The 2014 Revisions. Highlights

Zhu T, Sulser TB, Ringler C, Msangi S, You L (2009) Food security, farming, and climate change to 2050: Scenarios, results, policy options. IFPRI, Washington. https://www.ifpri.org/sites/default/files/publications/rr172.pdf

Chapter 2
Consolidating Drought Projections—Eastern Australia

Shahadat Chowdhury and Michael Sugiyanto

Abstract Drought resilient water policy development relies on hydrologic analysis which traditionally assumes stationarity of the past climate. Climate change diminishes this assumption causing the need for projection of future climate. General Circulation Models (GCM) are the best tool to provide these projections. However, water managers are often overwhelmed by the wide choice of GCMs, the length of simulation period (year 1900 to 2100) and non-stationary CO_2 level. The choice of variables, GCM outputs, that may satisfactorily relate to drought severity is crucial. GCM accuracy varies on temporal, spatial and multivariate spaces. Different GCM may project different future. So how do we consolidate this GCM projection diversity at a level useful for policy development? This paper shares an insight on narrowing projections into measures or advices that are useful for planning water security.

The case study looks at future water security of Lachlan River at eastern Australia. Daily rainfall and evaporation from 15 GCMs are fed through a river system model to simulate inflow into major headwater storages, length of storage drawdowns, availability of water for irrigation and the river flow deficit at an environmentally significant site. The projection of 2020–2039 is compared against the simulation of twentieth century. No bias correction of GCM is attempted; instead the analysis has been limited to changes between the past simulation and future projection. Multi model consensus, rather than model average, is chosen for consolidated advice. The advice is further reduced into categories *WET*, *SIMILAR* and *DRY* for planning purpose. The case study forecasts less water for irrigation and environment without similar adverse conclusions for the headwater inflow or storage.

Keywords Drought · Agriculture · Climate change · Water policy · GCM · IQQM

S. Chowdhury (✉) · M. Sugiyanto
NSW Department of Planning, Industry and Environment, Parramatta, NSW, Australia
e-mail: shahadat.chowdhury@dpie.nsw.gov.au

© Springer Nature Switzerland AG 2021
M. Babel et al. (eds.), *Water Security in Asia*, Springer Water,
https://doi.org/10.1007/978-3-319-54612-4_2

2.1 Introduction

Reliable water supply is essential for human settlement followed by successful agricultural economy (Grafton and Hussey 2011). Water resources management and planning are evolved around managing demand and supply on a temporal and catchment scale. The demand is managed by implementing good policy and governance framework. The supply is managed through reservoirs to reduce temporal variability. The likelihood of supply deficit or drought severity determines the extent of water security. Robust water plans are designed to withstand any future droughts. Drought may be identified from rainfall deficit causing loss in agricultural production. The sensible prediction of this catchment specific possibility of drought is essential. How do hydrologists define drought characteristics useful to plan for water management?

Traditionally water managers develop their management plan based on historical inflow sequence inclusive of drought. The management plan assumes that the variability of the long term observed climate of the past (generally goes back to 1890 in Australia) gives the range of possible futures. The desired demand management option withstanding the historical sequence defines the extent of water security of the river system. The drought characteristics are implicitly reflected within historical data.

As described in the last paragraph, it has been customary to use past experience of drought to plan for the future dry spell. However, climate change is challenging this long held custom. By assuming that the past climate includes possible future ranges, we are stating that the long-term climate is stationary. Hydrologist can no longer rely on this assumption of stationarity (Matalas 2012; Milly 2008). The reconstruction of historical observation into future drought is becoming increasingly difficult (Kiem et al. 2016). Nonetheless, there is a strong community expectation to secure water supply against climate change (Bates et al. 2008; Neave et al. 2015). What are the practical challenges to consider climate change? What are the tools at our disposal to address this?

The current tool in assessing climate change impacts on water planning (Bhatt and Mall 2015; Chowdhury and Al-Zahrani 2013) is based on the General Circulation Models alternatively known as Global Climate Models (GCMs). Water managers are often overwhelmed by the sheer outputs generated by these GCMs. Firstly, there are multiple GCMs being available all around the world with different output resolution. Secondly the hind-cast accuracy (or bias) of multivariate outputs varies widely in space and time (Chen et al. 2011a, 2011b; Nahar et al. 2017; Ray and Brown 2015; Rocheta et al. 2014). The choice of variables that may satisfactorily relate to drought severity is crucial. This is often difficult to determine which variables are the most relevant indicators of drought and water security. Finally, the past performance of a hind casted variable is not necessarily indicative of forecast accuracy of the same variable. There are wide disagreements among GCMs' on future projections on a given greenhouse gas emission scenario. Among this backdrop of complexity, how to select a GCM for decision making and planning? There is additional temporal complication to this question that is discussed next.

GCMs are continuous simulation that extends beyond the current time, of up to the year 2100. The GCM prediction of future state is often referred to as 'projection' to differentiate it from real time forecasting. The GCM simulation injects increasing concentration of CO_2 level in the atmosphere over time causing a gradual change in climate. Unlike reproduction of past climate (either stochastic or observational), the GCM projection is not stationary. Selection of a wider time window amplifies the non-stationarity while smaller window reduces the chance of detecting extreme drought sequence. How to select simulation time window relevant for water planning purposes?

The remaining question, after selection of a projection within a time window, is how to detect change? As mentioned earlier there are wide disagreements among projections from different models. The change in future state (either worse or better) governs key policy response. Water resources plans are developed in response to prevailing climate variability of a catchment. Climate change generates a higher order statistic than the historical variability. The challenge is to separate 'the noise' of historical variability from 'the signal' of climate change.

This paper attempts to address the questions raised in this introductory section. This is done by using a case study of a rural catchment in eastern Australia. In summary, water security is about withstanding long dry spell or drought. Hydrologists need to revisit any change in severity of future drought due to climate change using available projections from GCM. This raises several challenges for practicing hydrologists. How to define drought for water management? How to deal with high number of GCM and their varying hind-cast skill and differing projections? How to detect change and simplify the findings useful for water security planning?

The paper is organised as followed. The methodology discusses the key drought characteristics, river system model to simulate those characteristics and how a change is detected. This is followed by algorithm for clearer presentation of computational steps which can easily be followed by the water managers. Finally, we present the case study findings.

2.2 Methodology

2.2.1 Defining Drought

There is no single way to define drought (Surendran et al. 2017; Wilhite and Glantz 1985). Unlike flood, the onset of drought is only detected in hindsight after months or years entering into the state. The impact of drought is felt differently from different perspective (Tate and Gustard 2000). Hydrologist defines drought very differently to economist or social scientists. The two popular choices are Palmer Drought Severity Index (PDSI) and Standardised Precipitation and Evapotranspiration Index (SPEI) or variants of these (Dai 2011; Palmer 1965). While the generic form of these indices is attractive there had been many attempts to redesign these indices to fit for purpose

(Hao et al. 2016; Heim 2002). This study analyses drought from the perspective of deficit in water resources, available mainly for agricultural production, regulated by large head water reservoirs.

This paper explores drought impact based on variables that are easily understood by the water planners. The following four variables are used to estimate the severity of drought which are commonly used for water planning (Murray 2012).

1. Drought inflow volume: This is identified as 75th percentile annual inflow to the major headwater storage. This is predominantly simulated through rainfall runoff model and water balance of the outflow and storage level.
2. Dry spell in days: Prolonged depletion of headwater reservoir level is an indicator of state of drought. This is calculated by counting number of consecutive days where the headwater storage remains less than half full.
3. Water allocated for irrigation: This is the share of storage water available to irrigators under the current water management policy and rules. The share is expressed in percentage allocation at the irrigation pump.
4. Low flow for environment: This is identified as a 75th percentile annual effluent flow to an environmentally significant site for example a wetland or fish habitat.

This study chose quartiles (25th and 75th) as an indicator of climate state. Use of quartiles is common in applied hydrology (Wang et al. 2009) and water management in the study region (Bureau 2020).

2.2.2 River System Model IQQM

This study uses a river system model to generate the four variables of interest listed above.

The case study utilises the simulations of a conceptual river system model known as Integrated Quantity Quality Model (IQQM). The hydrologic model IQQM was progressively developed in the 1990s by the hydrologists of the New South Wales (NSW) Government (Simons et al. 1996) in Australia. This is a conceptual deterministic model that mainly simulates (using a node link structure) daily rainfall runoff, river routing, reservoir operation, irrigation demand and associated extractions subject to legal compliance (Hameed and Podger 2001). The model has been extensively used for water resource planning of NSW and partly in Queensland (Murray 2012).

Various components of IQQM are first independently calibrated and assembled together using a node link structure. This is called 'the calibration version' of the model. For water resources planning reference, the various physical component of the calibrated model is frozen on a certain reference development level. The physical components may be the maximum irrigable land, population of the town, size of hydraulic structures on a particular year. Then the proposed planning rules are imposed on the model such as minimum flow requirement at a location, storage reserve, accounting methods of irrigation diversions, environmental watering regime and so on. This is called 'the planning version' of the model. The simulation of the planning version is carried out for a set of input time series (rainfall, evaporation, temperature etc.) for a period representative of climatic variability of the catchment (Murray 2012). Conventionally the time series includes long term historical observations (e.g. 1895 to 2009). The simulation provides probabilistic outcome of the proposed planning rules such as percentage of time the storage falls below critical level or reliability of irrigation allocation.

The key input time series for this study are daily rainfall and evaporation time series simulated by multiple GCMs'. Then we use IQQM to generate the four output variables of interest introduced in the earlier section. Note that the GCM hindcast accuracy of daily rainfall and evaporation affect the accuracy of the four IQQM generated variables. How do we navigate this accuracy question pragmatically? Should we amend the known biases, or choose the most accurate model, or combine the models, or seek the most agreed forecast? We explore it next.

2.2.3 Consolidating Multiple Projections

The hind-cast skills of GCM for hydrological variables are poor and vary considerably in space. The bias correction is onerous, may not hold for projection and bias correction research is ongoing (Johnson and Sharma 2015). Besides multiple GCMs' provide multiple answers to the same question. How to consolidate all these into a workable advice for water planning?

Frist of all, we limit our analysis on the relative change between the hind-casts (1900 to 1999) and the projections (forecast of 2030 climate) from the same model. This assumes that the relative change forecasted by the same GCM has better accuracy (i.e. the hind cast and projection errors are the same). The analysis provides multiple changes predicted by multiple GCMs. Secondly, we consolidate multiple answers into the most agreed prediction or on model consensus on change in variables. Past studies supported superiority of consensus forecast and various ways to determine consensus (Fritsch et al. 2000; Roebber 2010). The model consensus is defined here as the most agreed change (mode of change distribution) among the 15 GCM projections.

The paragraph above has described how the estimates of the four drought indicators are consolidated for the water planners. The next challenge is to determine whether any deviation within drought indicators is in excess of historical variability.

2.2.4 Climate Change Versus Historical Variability

GCM simulation forces increasing concentration of CO_2 level in the atmosphere from 2000 to 2100 simulation period subject to various emission scenarios. This CO_2 non stationarity prevents us from directly comparing the last century's hind-cast to this century's projection. We need a smaller time window around the planning period to satisfy the stationarity assumption. The planning time frame for this case study was 2030. We chose a 20-year time window of 2020 to 2039 for this study wrapping around the planning timeframe. This small-time window poses additional challenge of identifying signal from noise as explained next.

We first determine the most agreed change among all the projections as described in the earlier section. The changes are then flagged if they exceed the natural variability of a moving 20-year window within the twentieth century. We consider that the signal resolution is inadequate to estimate absolute quantum of change. Hence the future changes are categorised into three main classes according to its direction of change level: *DRY*, *SIMILAR* and *WET*.

2.3 Algorithm

The methodology can be explained by presenting it in a step by step execution algorithm.

1) Collate daily observed rainfall and evaporation time series. Estimate or generate to fill up any gaps within the study period. This case study used historical period of 1900 to 1999.

2) Simulate key drought variables (daily inflow, storage volume, irrigation etc.) using time series input of Step 1 into the IQQM setup.

3) Estimate selected statistics of each drought variable (e.g. mean annual inflow) from Step 2 for the period of 1900 to 1999.

4) Move a 20-year window throughout 1900 to 1999. Compute the departure of 20-year window statistics from 100-year statistics and determine the twentieth century variability. Determine the maximum range of variability.

5) Collate daily simulated rainfall and evaporation (or variables to compute evaporation) from available GCMs. This study used 15 GCM simulations over 1900 to 2100 period.

6) Simulate key drought variables (daily inflow, storage volume, irrigation etc.) using time series input of Step 5 into IQQM setup.

7) Compute selected statistics of each drought variable (e.g. mean annual inflow) from Step 6 for the period of 1900 to 1999. This is the notional replication 20th Century measure of the variables for a given GCM, inclusive of GCM inaccuracy.

8) Compute selected statistics of each drought variable (e.g. mean annual inflow) from Step 6 for the period of 2020 to 2039. This is the notional near future (here 2030) forecast post climate change state by the given GCM.

9) Subtract measures of Step 8 from Step 7. This is the change in drought variable (e.g. decrease in annual inflow) due to climate change for a given GCM.

10) Repeat Step 7 to Step 9 for all the GCMs (15 GCMs' in this study). Fit a kernel density and determine the mode of the distribution of the variable (e.g.

decrease in annual inflow). The mode provides the most agreed view among all the GCMs.

11) Compare the most agreed change found in Step 10 to the 20th Century variability estimated in Step 4. Decide if the future change is outside the variability and report accordingly.

2.4 Case Study

2.4.1 Catchment Description

The Lachlan River Valley located in eastern Australia, occupies around 85 000 km^2 which is half the size of Chao Phraya River Catchment of Thailand (Fig. 1). The eastern part of the catchment with higher elevation contributes most of the flow to the river. Average rainfall varies from 250 mm in the lower western reaches to 1 200 mm along the elevated eastern part of the catchment. Potential average annual evaporation exceeds average annual rainfall over the entire catchment with 1 210 mm in the east to 1 750 mm in the west. The river flows to westerly direction. Unlike most of the rivers, Lachlan drains to a swamp only and the swamp water rarely makes to the sea.

Lachlan is a long, arid and complex regulated river system. It has numerous anabranches, several regulating storages, wetlands, major irrigation developments and weirs. There are two head water dams, Wyangala and Carcoar, with a total

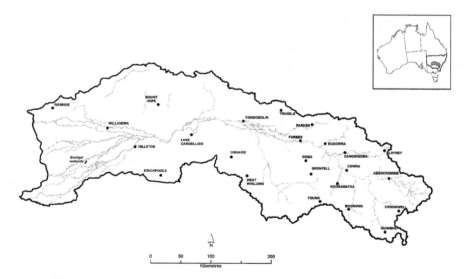

Fig. 1 Lachlan river catchment

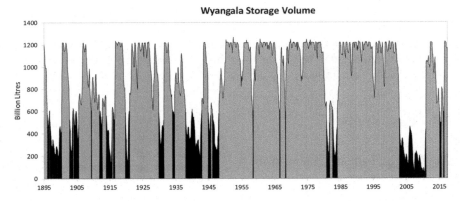

Fig. 2 Wyangala simulated storage volume. Storage volume is less than half during the dark shaded period indicating dry spells. One billion litre equals one million cubic meters

capacity of 1250×10^6 m^3. Two off river lakes in the midway, with a combined capacity of 215×10^6 m^3 are used to capture and re-regulate surplus water. Various weirs provide necessary head for diversion and short relieves during drought. Regulated release from the Wyangala dam may take three to four weeks to reach the furthest water user. Some 90 000 ha land is irrigated from the Lachlan River for the crops like cereals, lucerne (alfalfa) and cotton.

Summer crops are generally reliant on irrigation and hence on the headwater storage. The arid river system experiences long dry spells where the dam may not spill for up to 15 years (see Fig. 2). On average, every 1 in 10 years, no water is available for irrigation. There may be consecutive 8 summers when the storage cannot allocate any more than half of the irrigation entitlements. The poor reliability makes the knowledge of dry spell very important.

2.4.2 General Circulation Models

This paper attempts to use 15 readily available GCMs irrespective of their skill scores as a continuation of earlier work in the basin (CSIRO 2008). The different GCMs are assumed to be independent and equally as valid. The chosen GCMs (Table 1) simulate a future with A1B emission scenario. GCMs produce many output variables

Table 1 Names of GCM used in this study

CGCM3T47	CSIROMk3-0	GFDLCM2-0	INMCM3-0	MRICGCM2-3-2a
CGCM3T63	ECHAM	GISS-AOM	IPSLCM4	NCARCCSM3
CNRMCM3	ECHO-G	IAPFGOALSG1-0	MIROC3-2medres	NCARPCM

at continental scale on a monthly time interval. The variables of interest in this study were daily rainfall and evaporation.

Rainfall projections are downscaled into finer 5.5 km grids using weighted-inverse distance squared method (Liu and Zuo 2012). The monthly rainfall heights are disaggregated using stochastic generator WGEN (Richardson and Wright 1984). In contrast, the evaporations are not readily available from the GCMs. We derive evaporation through the ASCE Penman Monteith equation (Jensen et al. 1990) where the GCMs provide raw inputs like temperature, humidity, wind and solar radiation.

GCMs simulate climate from 1900 to 2100. The past period is taken to be the window of 1900–1999 (20th Century) as it has been a period that has been extensively used in Australian hydrology. It also captures the wide variability of Australian climate. The time for future planning is taken to be the period of 2020–2039 (centred on the year 2030).

2.4.3 Replication of 20th Century Variables

Lachlan River and its water resource use are simulated using daily input of observed rainfall and evaporation into IQQM for the period of 1890 to 2010. This provides the notional observed (or 20th Century behaviour) variable of interest. The following daily output are analysed for this study:

1. Inflow to Wyangala Dam—The annual inflow series is analysed to identify the number of years inflow is less than the dirtiest quartile (Fig. 3).

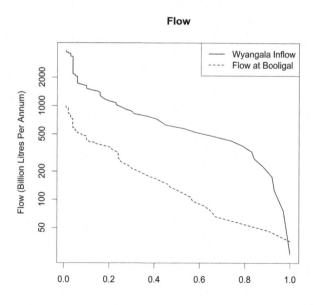

Fig. 3 Annual inflow rate into Wyangala Dam and at Booligal Wetland, the 75th percentile flow rate is showing the driest quartile

2.   Storage volume of Wyangala Dam—The length of consecutive days storage is less than half full is shown in Fig. 2 earlier. We study any change in this dry spell.
3.   Water allocated for irrigation at the start of irrigation season—The summer irrigation season in Lachlan starts at the beginning of September. We forecast any worsening of this allocation distribution in the future.
4.   River flow at Booligal for wetland health—We analyse any lengthening of the dry quartile inflow to Booligal Swamp.

The four outputs are further analysed to obtain 20th Century variability of a 20-year time window.

2.5   Results and Discussion

2.5.1   Poor Replication of 20th Century Variables

This study uses GCM sourced daily rainfall and evaporation for 1900 to 2100 as input variable to IQQM. This provides simulated time series of 4 variables × 15 GCM. The period of 1900 to 1999 is taken as notional reproduction of 20th Century behaviour.

Note that GCM simulation is not controlled by external forcing of temporal boundary conditions of the catchment. So we do not examine day by day reproduction of the time series. We rather compare cumulative distribution of the variables of interest. We find that the replication of 20th Century is poor for all 4 variables. Figure 4 shows replication of Wyangala Dam storage behaviour, for the sake of brevity this paper does not show other variables. The poor replication is unsurprising

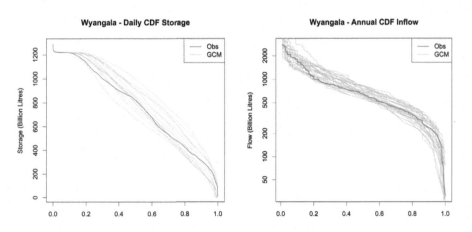

Fig. 4 Poor replication of 20th Century Wyangala Dam storage volume (left) and inflow (right)

Table 2 Detecting change in variables representing future drought

Variables	20th Century Variability (%)	2030 Change (%)	Conclusion
Dry quartile storage inflow	−28 to +66	−14	*SIMILAR*
Dry spell in days	−84 to +86	−4	*SIMILAR*
Irrigation Allocation in %	−54 to +121	−66	*DRY*
Dry quartile flow to wetland	−52 to +227	−56	*DRY*

given the fact that this study does not attempt to correct GCM sourced rainfall and evaporation of the last century prior to use.

Due to poor reproduction skill of 20th Century variable we limit our analysis into forecast of relative change only. The relative change is compared against the observed variability of 20-year window in 20th Century. The 20th Century variability is regarded as noise, any change outside this variability is considered as clear signal. Accordingly, this study detects future changes as shown in Table 2.

2.5.2 Similar Storage Inflow and Dry Spell

The GCM consensus is that climate change may reduce the dry quartile storage inflow by 14%. The estimate is based on analysis within a 20-year window of 2020 to 2039. However, a random selection of 20-year window in last century observed data exhibits a range of −28 to 66% variation of dry year inflows. The GCM projected decrease is well within this natural variability. Can this finding infer that the future dry year cumulative inflow will not be worse either? We look at another variable which is the length of dry spell next.

Readers are reminded that dry spell is defined as consecutive days when the storage is less than half full. The dry spell is found to be marginally shorter but occurring more often. Figure 5 is showing distribution of 15 model changes. The mode of the distribution (model consensus) is within natural variability. In terms of dry spell or storage inflow we conclude no worsening of future water resources volume. We next explore if more frequent dry spell leads to shortage of water available for irrigation.

2.5.3 Less Water Allocated for Irrigation

In NSW, water extraction from the river is regulated through perpetually owned water use right known as 'entitlement'. Farmers are allocated finite volumes of water for irrigation at the start of irrigation season. The allocation may be expressed at

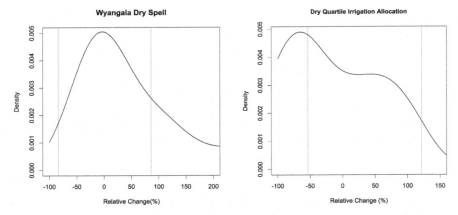

Fig. 5 Change in dry spell (left) is within the 20th Century variability, irrigation allocation (right) will be decreased beyond 20th Century variability. The vertical dash lines show 20th Century range

percentage where 100% mean that a farmer is allowed to use his or her full legal 'entitlement' during that irrigation season. Irrigation demand on storage is a function of prevailing rainfall, evapotranspiration and inflow from the intermediate tributaries. The study finds that (Table 2) there will be a significant reduction in available water for irrigation (at September) in the future (2030). This is counter intuitive as the storage behaviour does not display such decline. The explanation lies in the timing of the inflow which will be changed to the later part of the season making it less useful for irrigation. The timing of storage volume is important as stored irrigation water at the start of the season determines the cultivated land size for annual crops.

2.5.4 Less Flow to Wetland in Dry Years

Wetlands act as an environmental refuge, in the context of generally dry and highly variable climate of Eastern Australian catchments. They support various aquatic and terrestrial ecosystems. Lachlan River has many wetlands of high conservation significance. Booligal Swamp of the lower Lachlan River is a nationally signifi-cant wetland known for its capacity to support large water bird recruitment events within high-density lignum (*Muehlenbeckia florulenta*) dominated habitat (Driver et al. 2010). Booligal Swamp supports some of the largest recorded breeding popu-lations of ibis in Australia especially, straw-necked ibis (*Threskiornis spinicollis*), white Ibis (*Threskiornis molucca*) and glossy ibis (*Plegadis falcinellus*). The flow management rules in Booligal Swamp have focused on improving the aquatic envi-ronment favourable to bird breeding events (Chowdhury and Driver 2007). This study analysed flow into Booligal and found that dry quartile inflow will even be drier in the future (Table 2).

2.5.5 Conclusions

We have presented an example which simplifies the vast information in GCM future projections into a three-category forecast (*WET, SIMILAR, DRY*) useful for water planning. With the assumption that the model errors are structurally the same between the hind-cast and future period, we limit the analysis to the relative change of future and past period. The use of kernel density function fitted to the changes predicted by the available projections has given the most agreed view (mode of the distribution) on the future change. We compare the change (signal) to 20th Century variability (noise). The case study finds that while future drought may not be worse for the headwater inflow or storage, there will be less water for irrigation and environment.

Acknowledgements and Disclaimer The authors acknowledge the contribution of Tahir Hameed and Heping Zuo for IQQM model development and GCM simulations respectively. The original work was funded by the NSW Environmental Trust. The paper is intended to share knowledge on scientific methods and does not reflect the view or policy intend of the New South Wales Government.

References

Bates BC, Kundzewicz ZW, Wu S, Palutikof JP (eds) (2008) Linking climate change and water resources: impacts and responses, Chapter 3. In: Climate change and water. Technical paper of the intergovernmental panel on climate change, IPCC Secretariat, Geneva. 210 pp

Bhatt D, Mall RK (2015) Surface water resources, climate change and simulation modeling. Aquatic Procedia 4:730–738

Bureau of Meteorology (2020) Australia Water Information Dictionary, Streamflow Forecast, 15 August 2018. https://www.bom.gov.au/water/awid/product-water-forecasts-seasonal-stream flow.shtml

Chen J, Brissette FP, Poulin A, Leconte R (2011) Overall uncertainty study of the hydrological impacts of climate change for a Canadian watershed. Water Resour Res 47(12):W12509

Chen C, Haerter JO, Hagemann S, Piani C (2011) On the contribution of statistical bias correction to the uncertainty in the projected hydrological cycle. Geophys Res Lett 38(20):L20403

Chowdhury S, Driver P (2007) An ecohydrological model of waterbird nesting events to altered floodplain hydrology. In: Oxley L, Kulasiri D (eds) MODSIM 2007, International congress on modelling simulation, December, Christchurch, pp 2896–2902. ISBN 978-0-9758400-4-7

Chowdhury S, Al-Zahrani M (2013) Implications of climate change on water resources in Saudi Arabia. Arab J Sci Eng 38(8):1959–1971

CSIRO (2008) Water availability in the Murray-Darling Basin. A report to the Australian Government from the CSIRO Murray-Darling Basin Sustainable Yields Project. CSIRO, Australia. ISSN/ISBN 1835095X, 67pp

Dai A (2011) Characteristics and trends in various forms of the Palmer Drought Severity Index during 1900–2008. J Geophys Res 116:D12115. https://doi.org/10.1029/2010jd015541

Driver P, Chowdhury S, Hameed T, O'Rourke M, Shaikh M (2010) Ecosystem response models for lower Calare (Lachlan River) floodplain wetlands: managing wetland biota and climate change modelling. In: Saintilan N, Overton I (eds) Ecosystem response modelling in the Murray Darling Basin. CSIRO Publishing, April, pp. 183–196. ISBN 9780643096134

Fritsch J, Hilliker J, Ross J (2000) Model consensus. Weather Forecast 15:571–582

Grafton RQ, Hussey K (eds) (2011) Water resources planning and management. Cambridge University Press, pp xx–xxi

Hameed T, Podger G (2001) Use of IQQM simulation model for planning and management of a regulated river system. IAHS Red Book 2001:83–89

Hao Z, Hao F, Singh VP, Xia Y, Ouyag W, Shen X (2016) A theoretical drought classification method for the multivariate drought index based on distribution properties of standardized drougth indices. Adv Water Resour 92:240–247

Heim R (2002) A review of twentieth-century drought indices used in the United States. Bull Am Meteor Soc 83:1149–1165. https://doi.org/10.1175/1520-0477(2002)083%3c1149:arotdi%3e2.3.co;2

Jensen ME, Burman RD, Allen RG (ed.) (1990) Evapotranspiration and irrigation water requirement. ASCE manuals and reports on engineering practices. 70. New York, NY

Johnson F, Sharma A (2015) What are the impacts of bias correction on future drought projections? J Hydrol 525:472–485. https://doi.org/10.1016/j.jhydrol.2015.04.002

Kiem AS, Johnson F et al (2016) Natural hazards in Australia: droughts. Clim Change 139(1):37–54

Liu DL, Zuo H (2012) Spatial downscaling of daily climate variables for climate change impact assessment over New South Wales. Climatic Change, Australia. https://doi.org/10.1007/s10584-012-0464-y

Matalas N (2012) Comment on the announced death of stationarity. J Water Resour Plann Manage. https://doi.org/10.1061/(ASCE)WR.1943-5452.0000215,311-312

Milly PCD (2008) Stationarity is dead: whither water management? Science 319(5863):573–574. https://doi.org/10.1126/science.1151915

Murray Darling Basin Authority (2012) Hydrologic modelling to inform the proposed Basin Plan – methods and results, MDBA Publication No 17/12, Canberra. ISBN 978-1-922068-22-4

Nahar J, Johnson F, Sharma A (2017) Assessing the extent of non-stationary biases in GCMs. J Hydrol 549:148–162. https://doi.org/10.1016/j.jhydrol.2017.03.045

Neave I, McLeod A, Raisin G, Swirepik (2015) J. Managing water in the Murray-Darling Basin under a variable and changing climate. Water J Austr Water Assoc 42(2):102–107

Palmer WC (1965) Meteorological drought. In: U. S. D. o. Commerce (ed.) Washington, D.C., p 58

Ray PA, Brown CM (2015) Confronting climate uncertainty in water resources planning and project design: the decision tree framework. Washington, DC: World Bank. © World Bank. https://openknowledge.worldbank.org/handle/10986/22544. License: CC BY 3.0 IGO

Richardson CW, Wright DA (1984). WGEN: a model for generating daily weather variables. U.S. Department of Agriculture, Agricultural Research Service, ARS-8, 83 p

Rocheta E, Sugiyanto M, Johnson F, Evans J Sharma A (2014) How well do general circulation models represent low-frequency rainfall variability? Water Resour Res https://doi.org/10.1002/2012wr013085.

Roebber PJ (2010) Seeking consensus: a new approach. Mon Weather Rev 138:4402–4415. https://doi.org/10.1175/2010mwr3508.1

Surendran U, Kumar V, Ramasubramoniam S, Raja P (2017) Development of drought indices for semi-arid region using drought indices calculator (DrinC) – a case study from Madurai District, a semi-arid region in India. Water Resour Manage 31:3593–3605. https://doi.org/10.1007/s11269-017-1687-5

Simons M, Podger G, Cooke R (1996) IQQM-a hydrologic modelling tool for the water resources and salinity management. Environ Softw 11(1–3):185–192

Tate EL, Gustard A (2000) Drought definition: a hydrological perspective. Institute of Hydrology

Wang QJ, Robertson DE, Chiew FHS (2009) A Bayesian joint probability modeling approach for seasonal forecasting of streamflows at multiple sites. Water Resour Res 45(5). https://doi.org/10.1029/2008WR007355

Wilhite DA, Glantz MH (1985) Understanding: the drought phenomenon: the role of definitions. Water Int 10(3):111–120

Chapter 3
Planning for Climate Change and Mechanisms for Co-operation in Southeast Asia's Sesan, Sekong and Srepok Transboundary River Basin

N. J. Souter, D. Vollmer, K. Shaad, T. Farrell, H. Regan, M. E. Arias, T. A. Cochrane, and T. S. Andelman

Abstract Southeast Asia's 3S river basin, which comprises the Sesan, Srepok and Sekong rivers, is an important tributary basin of the Mekong River. The 3S rivers rise in Lao PDR and Vietnam, and flow through Cambodia, where they join before discharging to the Mekong. Vietnam has the highest forest loss and the most (46) dams. Lao PDR retains much of its forest, however, 15 hydropower dams are either being built or licensed for construction in its portion of the Sekong river basin. Cambodia is planning multiple dams on each of the three rivers' lower reaches. Cambodia's dams will have the greatest impact on system connectivity and thus the 3S's important migratory fishery. The flow regime of all three rivers has been altered as dam operations have increased dry season flow and reduced wet season flow. Increased Total Suspended Solids, Nitrate\Nitrite and Total Phosphorous have been observed in the Srepok River in Vietnam, whist there has been little change elsewhere. The three countries are at different stages of developing their national water governance frameworks, with Lao PDR the least developed. A range of climate and development scenarios have been generated for the 3S region. The 3S is highly vulnerable to climate change due to projected increases in temperature and the frequency of extreme floods and droughts. As the transboundary impacts of development in the 3S become clearer, so too does the need for cooperative management of the 3S basin and its water resources, to ensure a sustainable future as the climate changes.

N. J. Souter (✉) · T. Farrell
Conservation International, Greater Mekong Program, Phnom Penh, Cambodia
e-mail: nsouter@conservation.org

D. Vollmer · K. Shaad · T. S. Andelman
Conservation International, Betty and Gordon Moore Center for Science and Oceans, Arlington, USA

H. Regan
University of California, Riverside, USA

M. E. Arias
University of South Florida, Florida, USA

T. A. Cochrane
University of Canterbury, Christchurch, New Zealand

© Springer Nature Switzerland AG 2021
M. Babel et al. (eds.), *Water Security in Asia*, Springer Water,
https://doi.org/10.1007/978-3-319-54612-4_3

Keywords Mekong River Basin · 3S · Transboundary · Water governance · River regulation · Sesan · Srepok · Sekong · Climate change

3.1 Introduction

The lower Mekong's 3S River Basin comprises the catchments of three rivers, the Sekong, Sesan and Srepok. The 3S covers an area of 78,650 km^2 in Lao PDR (29%), Vietnam (38%) and Cambodia (33%). The rivers of the 3S rise in Lao PDR (Sekong) and Vietnam (Sesan and Srepok) before flowing through Cambodia to join the Mekong River. Although in area the 3S covers only 10% of the Mekong Basin, the 3S contributes 23% of the Mekong's mean annual volume (Adamson et al. 2009). Almost 15% of suspended sediment discharge in the Mekong (20 metric tons per year) originates in the 3S (Koehnken 2012). These sediments provide nutrients to both the Tonle Sap Great Lake and the Mekong Delta. The 3S is arguably the Mekong's most important catchment for maintaining migrating fish populations (Ziv et al. 2012) and with three hundred and twenty nine fish species is rich in biodiversity (Baran et al. 2014). Seventeen of these species are endemic to the 3S and it supports 14 critically endangered and endangered species (Baran et al. 2014). The 3S basin has a human population of around 3.4 million, the majority of whom (3 million) live in the Central Highlands of Vietnam (Asian Development Bank 2010). Here, ~24% of the population live in poverty, most of whom are minority groups living on marginal land (Giang et al. 2014). In Cambodia and Lao PDR, most people live close to rivers and are highly dependent upon natural resources for their daily survival (Asian Development Bank 2010).

The 3S river system remained largely undeveloped until the 1990's, when the first hydropower dams were commissioned on the Srepok (Vietnam) and Sekong (Lao PDR) rivers. Constructed to meet growing regional energy needs, more than 42 hydropower dams are being built or planned for the 3S river system (Cochrane et al. 2014). This pace of development is raising concerns about changing water flows. Full hydropower development in the 3S rivers of the 42 dams is anticipated to increase dry season flows by 63% and to reduce wet season flows by 22% (Piman et al. 2013). In the 3S, hydropower dams have a higher impact on daily and seasonal flows than climate change is anticipated to. Climate change may, however, increase long-term uncertainty in water flows required to generate electricity (Lauri et al. 2012). Sediment trapped behind dams could account for 92% of all 3S catchment yield (Kummu et al. 2010). And the Lower Sesan 2 and Lower Srepok and Sekong dam cascades are expected to have the most impact on fish migration and diversity (Ziv et al. 2012).

Future changes in water quality and nutrient fluxes can also result from hydropower development and land use change in the 3S basin. Hydropower dams will retain sediment and nutrients such as phosphorus, which are important to maintain downstream floodplain fertility and aquatic ecosystems. Land use change resulting in increases in agriculture, on the other hand, may increase nitrogen releases to the river system (Oeurng et al. 2016). Thus, water management is also important in regulating water quality.

The major future development activities in the 3S undertaken by the three countries will be linked to water. These decisions will have transboundary impacts that flow through the 3S system. It is well recognized that coordination is needed between the three nations to 'ensure sustainable management of land and water resources and equitable benefit sharing' within the 3S basin (Asian Development Bank 2010). A range of cooperative steps have already been taken. At the request of the three nations National Mekong Committees the Asian Development Bank undertook a 3S development study (Asian Development Bank 2010), whilst the IUCN is convening regional dialogue through its BRIDGE (Building River Dialogue and Governance) project (https://www.iucn.org/regions/asia/our-work/regional-projects/bridge-sekong-sesan-and-sre-pok-river-basins-bridge-3s). Despite the apparent willingness to cooperate in regional development transnational differences in policy and legislation, national development trajectories, local and regional pressures are likely to have a significant impact on what can be achieved.

The uncertainties of climate change coupled with the complexities of freshwater systems will further challenge water resources management in the 3S. Climate change presents three major areas of uncertainty: i) uncertainty in the climate projections due to the different general circulation models employed and the gas emission scenarios used as input for these models (IPCC 2014), ii) uncertainty in the impacts to freshwater ecosystems and hydrology under these projections, and iii) uncertainty in the effects of specific management and planning decisions on freshwater systems under climate and associated changes (Lawler et al. 2010). Irrespective of such uncertainties, management decisions will be made to meet society's freshwater needs. To minimize unintended consequences these decisions need to be based on current knowledge, without the benefit of full knowledge (Polasky et al. 2011). Scenario planning can assist in making informed management decisions.

Scenario planning is a framework for exploring options and for developing robust plans in the face of irreducible uncertainty (Peterson et al. 2003). Scenarios are highly uncertain yet plausible futures. Scenarios can represent plausible future states of a system such as different climate projections; different management or development plans, such as dam development or irrigation expansion; or the uncertain effects of a management or development plan. Scenario planning often uses an underlying conceptual model of the system, to which scenarios are applied. This provides a consistent and transparent basis for reliable comparison of system effects across the scenarios considered, allowing for robust decision making. Since the implementation of management plans can change the system, scenario planning frameworks should accommodate new data and highlight areas for monitoring to improve knowledge for better decision making. For this reason, scenario planning frameworks should be embedded in adaptive management frameworks that can adjust to updated information, changes in the system, behaviors of stakeholders, and new priorities.

In this paper, we discuss the mechanisms required for regional water management in the 3S basin in a changing climate by examining a range of factors. We will show that it is important to understand the differences in the current level of development between the three countries, their potential future development paths and likely consequences. We provide an assessment of the current status of the environmental vitality of the 3S system, and examine the legal framework within which the 3S system is managed. We review the use of scenarios in the 3S system, particularly as it relates to climate change. We then review the likely impacts of climate change on the 3S. And finally, we determine the transboundary impacts of development and provide recommendations on future resource planning in the 3S.

3.2 Material and Methods

3.2.1 Level of Development

Development within the three basin countries was assessed by examining dam construction and the extent of forest loss. A range of sources were consulted to determine the level of dam development within the three countries. The Mekong River Commission's Hydropower project database provided the main source of data. This was supplemented by information from Open Development Cambodia (2015), and WLE Mekong (2016). Information from these datasets was used to determine, which dams were operational, those under construction or licensed for future development, and potential dam sites as of December 2016. Both hydropower and irrigation dams were included in this analysis. Forest loss was determined using Hansen et al. (2013) global forest 2000–2014 data set. Analysis was conducted in Google Earth Engine and both the loss and gain in forest (trees >5 m in height) between 2000–2014 was calculated for the sections of the 3S basin in each of the three countries.

3.2.2 Environmental Vitality

Flow deviation and channel fragmentation indicators were calculated to assess the impact of dams using two scenarios: (Piman et al. 2013) Definite Future (DF) scenario, which includes all operational dams and those in advanced stages of construction in the 3S basin as of 2015 (Fig. 3.1); and Expected Dams on the Main Tributaries (EDMT) which were all current dams plus seven main tributary dams expected to be constructed in future (Fig. 3.1). The Amended Annual Proportion of Flow Deviation indicator, AAPFD (Gehrke et al. 1995; Gippel et al. 2011) was used to establish deviation of the DF scenario from modeled natural unregulated flow (Piman et al. 2013) baseline scenario). The AAPFD gives a score, the higher the number the greater the alteration. This score is assigned an Ecosystem Health Score (EHS). The EHS ranges of 0–1 with one being healthy. Four locations were assessed: (1) the 3S outlet to the Mekong; (2) the Sekong at the Cambodian/Lao PDR border; (3) the Sesan at the Vietnam/Cambodia border; and (4) the Srepok at the Vietnam/Cambodia border.

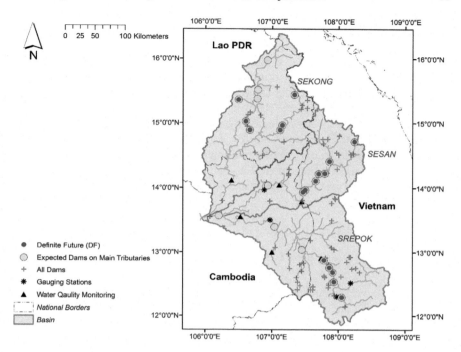

Fig. 3.1 Hydropower and irrigation dams in the 3S basin. All dams, proposed or potential dam sites

For channel fragmentation, the DF and EDMT scenarios were calculated using the Dendritic Connectivity Index DCI (Cote et al. 2009). DCI evaluates the hindrance to connectivity with the main stem with the Mekong caused by the dams. We assumed the 'passability' of the structures for fish in either direction to be zero.

Water quality was assessing from five Mekong River Commission (MRC) sampling sites. Three in Cambodia: Angdoung Meas (Sesan), Siempang (Sekong) and Lumphat (Srepok) and two in Vietnam: Pleicu (Sesan) and Ban Don (Srepok). There were no MRC sampling sites in Lao PDR. Total Suspended Solids TSS, Nitrate/Nitrite (NN) and Total Phosphorous (TP) were assessed over three periods: 2000–2006, when there was limited dam development; 2007–2011, when numerous dams were built and commissioned; and 2012–2014, representing current conditions. Mean ±95% confidence intervals for each period were calculated for the three parameters during both the dry (December-May) and wet (June-November) seasons. TP and NN were compared against the MRC water quality index for the protection of aquatic life (Ly et al. 2013) thresholds, which were 0.13 and 0.5 mg\L, respectively.

3.2.3   Legal and Management Framework

The governance framework was evaluated at two levels—national and transboundary. At each level, we assessed the institutional frameworks (e.g., rules and policies) along with their adherence to principles of "good governance", in particular, transparency, accountability, and participation. Relevant national and international policies, legislation and agreements were evaluated by undertaking a targeted literature review, and augmented with data from a survey administered in 2007 (UN-Water 2008). This survey assessed countries' progress towards implementing principles of integrated water resources management. National government agencies responded to the survey with results rated on a 5-point scale (0 being "not relevant" to 5 being "fully implemented).

3.2.4   Development and Climate Scenarios

The range of development and climate change scenarios developed for the 3S region were evaluated by undertaking a targeted literature review. The potential impacts of climate change were summarized.

3.3   Results and Discussion

3.3.1   Level of Development

The Vietnamese portion of the 3S basin is the most developed of the three countries. It has the largest number of operational dams and it appears that there is little further capacity for additional dam construction (Table 3.1). In 2000, Vietnam also had the lowest area of forest cover and the highest percentage of remaining forest lost in the following 14 years. Both Lao PDR and Cambodia had only one operational dam each,

Table 3.1 Level of dam (hydropower and irrigation) development as of December 2016 and percent 3S basin with forest cover in 2000, percent loss and gain from 2000–2014 for Cambodia, Lao PDR and Vietnam

Country	Operational dams	Dams under construction/licensed	Potential dams	% of total forested land area in 2000	% of forested area lost	% of forested area gained
Cambodia	1	1	21	54	15	2
Lao PDR	1	15	9	79	10	2
Vietnam	46	1	2	46	22	3

in the 3S catchment. Lao PDR is further advanced in dam construction/licensing than Cambodia. Conversely, Cambodia had less forest cover in 2000 and cleared more forest than Lao PDR from 2000–2014 (Table 3.1). This assessment confirms previous reports, which stated that the Central Highlands of Vietnam is undergoing intensive commercial and small scale agricultural development (Johnston et al. 2009; ICEM 2013a). One of the consequences of this is that the clearance of sloping land has led to flash floods and landslides.

3.3.2 Ecosystem Vitality

The Sesan River at the Cambodian border shows the largest degree of deviation in both high and low flows (AAPFD: 1.89; EHS: 0.4) (Fig. 3.2). The Sekong had the highest volume of the three rivers, and dam operations raise low flows (AAPFD: 1.67; EHS: 0.4). And the Srepok shows the least deviation under DF from unregulated conditions (AAPFD: 0.74; EHS: 0.6). Cumulatively, the 3S system at its outlet has an EHS (0.6) closer to that of the Srepok than to the other two rivers (AAPFD: 0.92). This is likely due to the lack of major dams in Cambodia, which mollifies the effect of those higher in the catchment. Modification in low flows can significantly alter riverine ecology and are highlighted by the AAPFD.

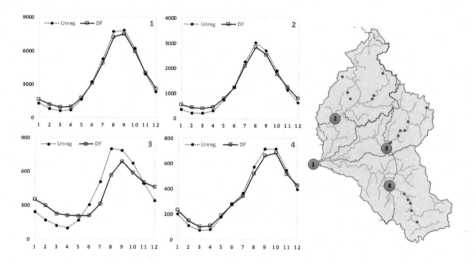

Fig. 3.2 Simulated flows for DF and natural unregulated conditions at the [1] outlet of the 3S, [2] Mekong at the Lao PDR /Cambodia border, [3] Sesan at the Vietnam/Cambodia border, and [4] Srepok at the Vietnam/Cambodia border. The plots compare the monthly averages of simulated flow over the period 1985–2008 (x-axis is month; and y-axis is m^3/s). The red stars represent dam locations

As the number of dams constructed in the 3S basin increases, connectivity to the Mekong decreases. Compared against a baseline connectivity (no dams) of 1, the current situation sees connectivity at 0.53 (DF scenario), whilst full dam development (EDMT) sees connectivity reduced to 0.12. The addition of dams on the main stems of the 3S and closer to their mouth i.e. those constructed in Cambodia, will have the greatest impact on migratory fish as the location of these dams decreases connectivity more than headwater dams. While dams with fish passage may allow for a degree of passability, improvements in connectivity will require most of the dams in the system to have effective fish passage. This is a challenging task given the number of dams involved, their remoteness and diversity of the fish species the passage system needs to cater for.

TSS changed little over time in the dry season at all sites except Dan Bon, where there was an increase (Fig. 3.3a). During the wet season, TSS was generally higher at all sites and more variable (Fig. 3.3b). An increase was most noticeable at Siempang, whilst a decrease was most noticeable at Dan Bon. Both Angdoung Meas and Lumphat exhibited increased TSS during the middle development period. In the dry season, there was little change in Nitrate/Nitrite at all sites except Ban Don, where its 95% confidence interval crossed the indicator threshold (Fig. 3.3c). A similar pattern was exhibited in the wet season, although the increase at Ban Don in the last period was much higher and the 95% CI exceeded the threshold (Fig. 3.3d). The 95% confidence intervals for TP at Dan Bon increased to above the threshold in the last two development periods of the dry season (Fig. 3.3e), whilst there was little change in the other parameters. In the wet season, TP at Dan Bon and Pleicu declined to below the threshold in the last period (Fig. 3.3f). Conversely, TP increased at Lumphat and Siem Pang.

Whilst the water quality parameters show high variability, the station at Dan Bon was consistently above the nutrient threshold values. This may be a result of extensive clearance and agriculture in this portion of the basin leading to increased run-off and input of fertilizer. With the exception of TP in the wet season, this did not tend to be mirrored downstream in Cambodia at Lumphat. The lower level of development in the Seasan and Srepok catchments is likely the cause of their generally lower and more consistent nutrient values.

Dam construction and operation is expected to reduce sediment flows in the 3S rivers (Kummu et al. 2010). Generally, there was little change in water quality at any of the sites except Dan Bon, which is downstream of five large dams on the Srepok River. A similar but smaller decline was also observed at Pleicu, which is downstream of six dams on the Sesan.

3.3.3 Legal and Management Framework

The four states of the Lower Mekong Basin (including Lao PDR, Vietnam and Cambodia) have engaged in transboundary cooperation since the 1950s, with the modern era of cooperation enshrined in the 1995 Agreement on the Cooperation for

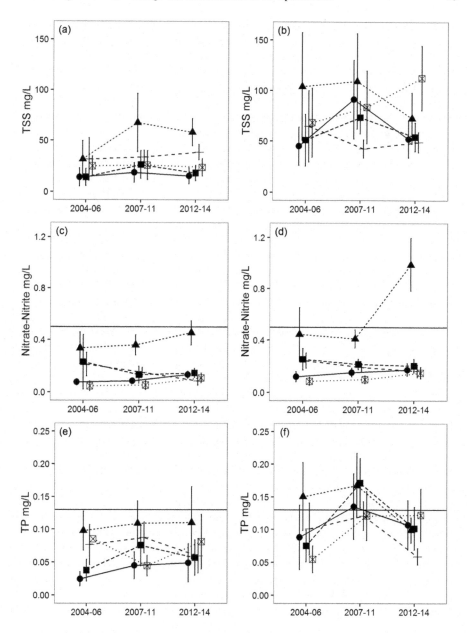

Fig. 3.3 3S basin mean (±95% CI) Total Suspended Solids in the **a** dry and **b** wet, Nitrate\Nitrite **c** dry and **d** wet, and Total Phosphorus **e** dry and **f** wet seasons for the three dam development periods. Angdoung Meas: circle, solid line; Ban Don; triangle, dashed line; Lumphat, square, dashed line; Pleicu: cross, dashed line; Siempang: cross in square, dotted line

the Sustainable Development of the Mekong (Mekong Agreement), which was subsequently held up as a model framework for transboundary basins (ESCAP 1997). The Agreement is meant to oversee river development between the four lower Mekong riparian states. It is a non-binding treaty intended to help countries address "altered hydrological flows that would arise as a consequence of intra and inter-basin diversions and of large storage dams" (ibidem, p 362 Mekong Agreement). Under the agreement, the Mekong River Commission (MRC) was established to foster cooperation between states. However, the Mekong Agreement is focused on the mainstream of the river and does not directly address activities within tributaries such as the 3S rivers. It does, however, include provisions for not causing significant harm within the territory of another state, and references the appropriate development of hydropower, fisheries, irrigation, timber floating, flood control, recreation and tourism. But these terms are not clearly defined, and the MRC has been criticized for its inability to regulate development or govern disputes in these matters.

The United Nations Convention on the Law of Non-Navigational Uses of International Watercourses (UNWC), entered into force in 2014 when Vietnam became the 35th global signatory. The UNWC provides an enforceable framework that is compatible with, but more specific than, the Mekong Agreement (IUCN 2016). It aims to ensure the development, management, conservation and promotion of optimal and sustainable use of water resources (Rieu-Clarke et al. 2012). The UNWC helps fill important gaps related to hydropower development for the Mekong and the 3S rivers, such as procedures relating to prior notification for states considering projects with potentially harmful impacts on other riparian nations, as well as binding dispute resolution mechanisms and procedures.

All three countries are at different stages of developing their national water governance frameworks, which influence not only how they manage resources within their borders, but also how they engage in transboundary matters. One enduring criticism of the current transboundary governance paradigm in the Mekong basin is that it is dominated by national interests, which marginalizes smaller interests as well as broader regional (i.e., transboundary) interests (Dore and Lebel 2010).

Although the results of the (UN-Water 2008) survey only reflect an "official" government response, essentially a brief self-assessment, they do reveal some interesting issues. At the time of assessment, Lao PDR was at a much earlier stage of developing its national water governance framework than the other two countries. None of the countries had established sub-national instruments (e.g., provincial water policies), something that could allow more flexibility in governing the 3S basin. There is also divergence in the degree of stakeholder participation, with Cambodia rating itself twice as high as Vietnam (whose score corresponds to "Under Development") (Fig. 3.4).

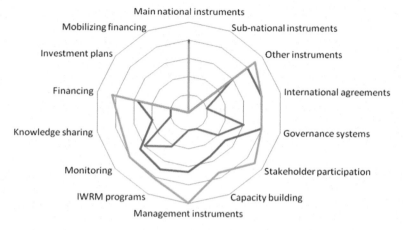

Fig. 3.4 Cross country comparison of implementation of integrated water resources management in 2007 in Lao PDR (blue line), Cambodia (green line) and Vietnam (blue line)

3.3.4 Development and Climate Scenarios

The likely impact on water and sediment flows in the 3S has been evaluated using a combination of four dam development scenarios, ranging from a baseline of no dams to development of all planned dams (Piman et al. 2013) and two IPCC emissions scenarios, A2 and B2 (Piman et al. 2015). In the Srepok basin in Vietnam Ty et al. (2012) developed four scenarios, which were compared against a baseline to examine changes in river flow. These scenarios examined various combinations of climate change (downscaled scenarios A2 and B2), future projections of land use change, human population change, hydropower dams, environmental flows, and water allocation rules.

The USAID Mekong ARCC project examined the likely effects of climate and hydrological change on a range of sectors in the lower Mekong basin including agriculture, livestock, natural systems, capture fisheries, aquaculture, health and rural infrastructure. Climate change threats were assessed using a range of scenarios based around the IPCC A1B emissions scenario and six downscaled general circulation models were used (ICEM 2013a).

The 3S basin was found to be a climate threat hotspot, being one of the top 5 areas most likely to be affected by either temperature and rainfall changes or flooding within the LMB. The largest increases in temperature in the lower Mekong basin will likely occur throughout the 3S basin, including a small portion of the Srepok catchment, which may see an increase of over 4 °C (ICEM 2013a). Climate change is predicted to cause a decline in annual rainfall across the basin, which is more pronounced in the dry season (Piman et al. 2015). Spatially, rainfall is likely to increase in the catchment headwaters (Lao PDR and Vietnam), whilst declining in the lowlands (Cambodia). Climate change will likely reduce river flow in the dry season at the 3S outlet, whilst total flow may either decline or increase depending on the emissions scenario. Larger and more frequent droughts and floods are also expected. Conditions for economically important crops including coffee and rubber will become less suitable in eastern Cambodia and the Vietnamese Central Highlands due to temperature increases, excessive rainfall, and periodic drought (ICEM 2013b). The subsistence crops, fisheries and non-timber forest products that are critical for the survival of many of the 3S basin's rural poor are likely be adversely affected (ICEM 2013a).

3.4   Conclusion

Whilst the 3S basin is highly vulnerable to climate change, managing its impact in this transboundary system is beset by numerous challenges. The rapid development and emerging transboundary environmental changes that are occurring are complex and interlinked. For example, whilst the largest changes to water quality occurred in the Srepok River downstream of Vietnam, likely due to extensive clearing and dam development, this river shows the lowest level of deviation from natural flows. This contrasts to the Sekong River, which has the greatest deviation, but shows little change in water quality. However, if the rate of forest loss in Lao PDR increases, and more dams are constructed, water quality may deteriorate, particularly downstream in Cambodia. Further development along the three rivers in Cambodia could also cause water quality to decline. Whilst this is not likely to impact upstream, if the dams constructed in Cambodia do not provide adequate fish passage, migratory routes will be blocked impacting fisheries and biodiversity. Climate change adds further complexity in trying to predict changes to the 3S basins water resources. The range of scenarios already used in the 3S have provided valuable insight into future resource management challenges. The current level of development in the 3S is already having transboundary impacts, and this will likely increase. Thus, cooperative transboundary management is required, and agreements such as the UNWC and dialogue is critical for managing the 3S in a future of climate change.

Acknowledgements This work was funded by grants from the Victor and William Fung Foundation and was undertaken as a part of the development of Conservation International's Freshwater Health Index.

References

Adamson PT, Rutherfurd ID, Peel MC, Conlan IA (2009) The Hydrology of the Mekong River. In: Campbell IC (ed) The Mekong: biophysical environment of an international river basin. Elsevier, Amsterdam, pp 53–76

Asian Development Bank (2010) Sesan, Sre Pok and Sekong River Basins Development Study in Kingdom of Cambodia, Lao People's Democratic Republic, and Socialist Republic of Viet Nam. Asian Development Bank, Bangkok

Baran E, Saray S, Teoh SJ, Tran TC (2014) Fish and fisheries in the Sesan, Sekong and Srepok River basins (Mekong watershed). In: On optimizing the management of cascades or system of reservoirs at catchment level. ICEM, Hanoi

Cochrane TA, Arias ME, Piman T (2014) Historical impact of water infrastructure on water levels of the Mekong River and the Tonle Sap system - supplement. Hydrol Earth Syst Sci 18(11):4529–4541

Cote D, Kehler DG, Bourne C, Wiersma YF (2009) A new measure of longitudinal connectivity for stream networks. Landscape Ecol 24:101–113

Dore J, Lebel L (2010) Deliberation and scale in Mekong region water governance. Environ Manage 46(1):60–80

ESCAP (1997) ESCAP and the Mekong cooperation. Water Resources Journal, September 1997, pp 1–6

Gehrke PC, Brown P, Schiller CB, Moffatt DB, Bruce AM (1995) River regulation and fish communities in the Murray-Darling river system, Australia. Regul Rivers: Res Manage 11:363–375

Giang TT, Wang G, Yan D (2014) Evaluation the factors leading to poverty issue in Central Highlands of Vietnam. Mod Econ 5(4):432–442

Gippel CJ, Zhang Y, Qu X, Kong W, Bond NR, Jiang X, Liu W (2011) River health assessment in China: comparison and development of indicators of hydrological health. ACEDP Australia-China Environment Development Partnership, River Health and Environmental Flow in China. The Chinese Research Academy of Environmental Sciences, the Pearl River Water Resources Commission and the International Water Centre, Brisbane

Hansen MC, Potapov PV, Moore R, Hancher M, Turubanova SA, Tyukavina A, Thau D, Stehman SV, Goetz SJ, Loveland TR, Kommareddy A, Egorov A, Chini L, Justice CO, Townshend JRG (2013) High-resolution global maps of 21st-century forest cover change. Science 342:850–853

ICEM (2013a) USAID Mekong ARCC climate change impact and adaptation study for the Lower Mekong Basin: main report. ICEM, Bangkok

ICEM (2013) USAID Mekong ARCC climate change impact and adaptation: summary. ICEM, Bangkok

IPCC: Climate Change (2014) Synthesis report. Contribution of working groups i, ii and iii to the fifth assessment report of the intergovernmental panel on climate change. [Core Writing Team, R.K. Pachauri and L.A. Meyer (Eds.)]. Intergovernmental Panel on Climate Change, Geneva (2014)

IUCN (2016) A window of opportunity for the Mekong Basin: The UN Watercourses Convention as a basis for cooperation (A legal analysis of how the UN Watercourses Convention complements the Mekong Agreement). IUCN, Bangkok

Johnston R, Hoanh CT, Lacombe G, Noble A, Smakhtin V, Suhardiman D, Kam SP, Choo PS (2009) Scoping study on natural resources and climate change in Southeast Asia with a focus on agriculture. IWMI, Vientiane

Koehnken L (2012) IKMP discharge and sediment monitoring programme review, recommendations and data analysis, parts 1 & 2. Mekong River Commission, Vientiane

Kummu M, Lu XX, Wang JJ, Varis O (2010) Basin-wide sediment trapping efficiency of emerging reservoirs along the Mekong. Geomorphology 119(3–4):181–197

Lauri H, de Moel H, Ward PJ, Räsänen TA, Keskinen M, Kummu M (2012) Future changes in Mekong River hydrology: impact of climate change and reservoir operation on discharge. Hydrol Earth Syst Sci 16(12):4603–4619

Lawler JJ, Tear TH, Pyke C, Shaw MR, Gonzalez P, Kareiva P, Hansen L, Hannah L, Klausmeyer K, Aldous A, Bienz C, Persall S (2010) Resource management in a changing and uncertain climate. Front Ecol Environ 8(1):35–43

Ly K, Larsen H, Duyen NV (2013) Lower Mekong Regional water quality monitoring report. Mekong River Commission, Vientiane

Oeurng C, Cochrane TA, Arias ME, Shrestha B, Piman T (2016) Assessment of changes in riverine nitrate in the Sesan, Srepok and Sekong tributaries of the Lower Mekong River Basin. J Hydrol Reg Stud 8:95–111

Open Development Cambodia: Hydropower dam (1993–2014) (2015). https://opendevelopment cambodia.net/dataset/?id=hydropower-2009-2014

Peterson GD, Cumming GS, Carpenter SR (2003) Scenario planning: a tool for conservation in an uncertain world. Conserv Biol 17(2):358–366

Piman T, Lennaerts T, Southalack P (2013) Assessment of hydrological changes in the lower Mekong Basin from Basin-Wide development scenarios. Hydrol Process 27(15):2115–2125

Piman T, Cochrane TA, Arias ME, Dat ND, Vonnarart O (2015) Managing hydropower under climate change in the Mekong Tributaries. In: Shrestha S et al (eds) Managing water resources under climate uncertainty. Springer, Cham, pp 223–248

Polasky S, Carpenter SR, Folke C, Keeler B (2011) Decision-making under great uncertainty: environmental management in an era of global change. Trends Ecol Evol 26(8):398–404

Rieu-Clarke A, Moynihan R, Magsig B-O (2012) UN Watercourses convention user's guide. IHP-HELP Centre for Water Law, Policy and Science (under the auspices of UNESCO), Dundee

Ty TV, Sunada K, Ichikawa Y, Oishi S (2012) Scenario-based impact assessment of land use/cover and climate changes on water resources and demand: a case study in the Srepok River Basin, Vietnam-Cambodia. . Water Resour Manage 26(5):1387–1407

UN-Water (2008) Status report on IWRM and water efficiency plans for CSD 16. UN-Water, Geneva

WLE Mekong (2016) Dam Maps. https://wle-mekong.cgiar.org/maps/

Ziv G, Baran E, Nam S, Rodriguez-Iturbe I, Levin SA (2012) Trading-off fish biodiversity, food security, and hydropower in the Mekong River Basin. Proc Natl Acad Sci 109(15):5609–5614

Chapter 4
Environmental Security Issues No Longer of Secondary Importance for Regional Cooperation or Conflict: The Case of the Mekong River

Christian Ploberger

Abstract When addressing environmental issues, a standard response is that environmental/ climate change issues are viewed with lesser sensitivity for regional cooperation and integration than topics related to national security, even though environmental/climate change related risks are identified as a potential 'threat multiplier'. Indeed, it is argued that cooperation on subjects like environmental security may contribute to cooperative behaviour between states even when confronted with serious issues of national security. Yet the paper challenges such a perception by arguing that non-traditional security threats have become more significant in their impact, with environmental security representing a prominent example. After all, the risks and development challenges related to environmental security are not only diverse, but increasing in their severity with regard to their impact on the development prospects of societies and the livelihood of the people. Not only can environmental/climate change related risks hamper development, but also undermine development goals that have already been reached. Consequently, we should no longer accept that environmental related risks are of lesser importance for national security and regional cooperation, indeed, environmental/climate change related risks indicate a new quality of security risks with potentially far-reaching consequences. It is also often emphasised that proximity represents a pre-condition for regional and subregional integration processes, yet, proximity with regard to environmental/climate change security issues takes on a different quality by harbouring potential serious negative implications for adjacent geographic locations with a particular regional context. The Mekong River, as a trans-national river system provides a specific case, since various countries having strong interests in utilising its resources for national development, and in doing so increases the prospect of a competitive development dynamic among the states within the Mekong region.

Keywords Environmental security · Security studies · development · Regional integration

C. Ploberger (✉)
Thammasat University, Bangkok, Thailand

© Springer Nature Switzerland AG 2021
M. Babel et al. (eds.), *Water Security in Asia*, Springer Water,
https://doi.org/10.1007/978-3-319-54612-4_4

45

4.1 Introduction

When considering the increasing negative impacts environmental/climate change related risks have on national development, and the livelihood of peoples, the implications are that such risks should no longer be viewed as a secondary security risk category compared to more traditional aspects of national security. With increasing environmental stress—steaming from various pollution issues, overuse of resources, and the impact of climate change—the relevance of environmental risks as a serious security challenge is bound to increase even further. Hence, environmental security is no longer just about the environment; it is increasingly connected with the overall development prospects of societies and countries and thus linked to political stability.

There exists a clear potential of environmental/climate change related risks undermining human development, consequently stimulating a process of redefining the meaning of security by contributing to a transformation in the awareness of security threats, in which human health, social welfare, and environmental degeneration are now identified as security concerns. Additionally, environmental/climate change related risks are trans-boundary in nature, which contribute to their potential as regional security issues. Hence, environmental/climate change risks not only represent complex global challenges, but a global challenge with varied specific local and regional impacts. After all, the impact of climate change as well as specific environmental issues occur in the context of a shared geographic location, hence environmental/climate change related risks are by nature trans-national and not constraint by administrative boundaries. Transnational geographic features, like river basins, mountain ranges, and coastal areas, represent some prominent examples, which will influence the distribution and dynamic of a specific environmental/climate change related impact.

Consequently, environmental risks should be treated as a serious regional and international relations challenge with potentially far-reaching implications for bilateral relations and regional co-operation. Indeed, despite environmental security/climate change related risks vary in their local, regional or global impacts, they may foster international competition and may even harbour the potential to stall regional integration processes. By adding to an already existing distribution challenge, the potential of environmental/climate change risks as 'threat multiplier' is already recognised. However, this should not suggest that environmental/climate change related risk would inevitably lead to armed conflicts, since such a direct link with armed conflicts is still rather the exception.

Even so, climate change and environmental related risk should not be evaluated in isolation, but in the context of other factors like an ongoing demand for development at local and national level as well as with the regional security context. After all, one can neither isolate environment/climate change related risks from the wider political dynamic within a specific region, nor from the national and sub-national political, economic, or social challenges, which contribute to the regional security dynamic. Hence, environmental/climate change related risk can carry critical implications for the stability of bilateral as well as multilateral relationships and thus can either support or undermine regional stability and integration.

The environmental security-development nexus and the implication of a shared geography, when addressing environmental/climate change related risks, can be made explicitly in the case of sharing the resources of the Mekong River Basin among the riparian countries. Competition over the Mekong's resources could generate a conflict dynamic since, the Mekong and its resources are of vital importance to some of those countries in fostering their national development processes. Indeed, if such a conflictual dynamic develops it may even generate a serious negative impact at the regional level, undermining regional political cooperation and development.

4.2 Environmental Issues as a Security Topic

Traditionally, interpreting security is based extensively on a realistic understanding of international relations. These relations are characterised by the following principal aspects: the nation-state as the key actor, international politics as a struggle for power in an anarchic world, nation-states relying on their individual capabilities to ensure their survival, states behave like rational-actors informed by national interests, power is the most important concept in explaining, and predicting state behaviour.

Even so, it is now widely acknowledged that what actually constitutes security underwent a critical re-evaluation, and a number of new subjects were added to the security discourse. This process of re-evaluating the meaning and focus of security was also facilitated by an issue-driven dynamic as non-traditional security issues like the prospect of development; environmental degeneration/climate change became increasingly recognised security topics.

4.3 Alternative Approaches to Security: Environmental Security

When investigating the process of re-evaluating the meaning of security, it is worth considering assertions made within the Critical Security Studies approach, which emphasises that an increasing complexity of security challenges can be identified, stating that the point of departure for conceptualising security lies in the real conditions of insecurity suffered by people and collectivists (Smith 2005). Insecurity, after all, is context-specific; an observation, which is especially applicable with regard to environmental security and the impact of climate change. This is also a vital consideration with regard to environmental security issues, since the link between security and development is already identified. Walker (1997), too supports these contentions, stating that the artificial distinction between security and development needs to be overcome.

As the damaging impact of excessive economic growth and economic modernisation on the environment became more recognised, environmental security rose to prominence. Another effort to integrate development with environmental security offers the UN Human Security concept. In re-interpreting the meaning of security, it highlights the complexity, variability and interdependence of different aspects of insecurity, like economic shocks (income, employment) or a deterioration of the people's livelihood through environmental degeneration (Human Security in Theory and Practice 2009). Once again, the link between security and development became more pronounced and recognised, after all, the prospect of development failure is a serious security and eminent political concern for many societies and countries worldwide.

However, Booth reminds us, that conceptualising security represents a particular political strategy, as different political and social actors interpret security challenges differently. Therefore, a political free definition of security does not exist (Booth 2005). Smith (2005) also states that the conceptualization of security is a product of different understanding of what politics is and should be about, emphasising the political nature of defining security, adding that security is something we choose to label as such. Alike, Buzan et al. (1998) to remind us that it is a political choice to securitize a particular issue, adding that securitization implies that a specific issue is presented as an existential threat. Hence, the implications are that security is not something objective, which just has to be uncovered.

Even so, we have to distinguish between securitization moves, the attempts to securitize an issue, and a successful securitization process, which is the acceptance of a specific issue as security subject. One can argue environmental concern underwent at least a partial securitization process. Partial, since not all feasible options to address environmental degeneration and the impact of climate change are neither comprehensively recognised nor implemented. Despite the recognition of the serious impact for human security and development, efforts to address environmental/climate change related risks fail quite short to address the underlining climate change dynamic, as current national mitigation and adaptation pledges are not sufficient for staying below the temperature limits agreed to in the Paris Agreement (Coninck et al. 2018, 315). An assessment supported by the recent UN Emission Gap Report (UNEP 2018), stating that the emission gap to limit global warming to $1.5°$ will not be achieved by the 2030 deadline, even we have the technology to do so. Consequently, the climate change dynamic and the impact this will generate will be continuing to increase. Recurrent extreme weather and climate events can be viewed as an indication of the challenges we will face in this respect, with the prospect that the character of climate change may change from a linear process to one characterised by abrupt changes, with the later representing an even more challenging challenge in term of human security.

Another challenge for a successful securitization process with regard to environmental/climate change related risks stems from the fact that traditional security challenges related to state security continue not only to be relevant but still receiving an overwhelmingly recognition from national and international actors. However, it is worth remembering that environmental degeneration and the impact of climate change is local specific and neither global nor universal in its impact, hence in many cases environmental related securitization moves are therefore local or regional in nature. The IPCC 5th Assessment Review already recognise the regional variability of climate change related impacts, like changes in regional precipitation, the impact a rising sea level can have on particular regions (Hewitson et al. 2014). What complicates the application of the environmental security agenda further is that we can distinguish between two types of disasters: concrete ones, with instantly felt implications (e.g. extreme weather events) and slowly developing ones (e.g. sea level rise) but often with alarming consequences for future development prospects for a particular population.

Recognising environmental security as a particular risk category as well as its uncertain character, one may consider applying a risk management approach in assessing the threat potential for a particular locality. After all, risk assessment does attempt to address the uncertainty of future events and consequently can support policy decision-making to account for various eventualities (Ploberger and Filho 2016). After all, the purpose of addressing environmental security is to help to reduce uncertainty regarding the potential impact on a particular location and to increase the resilience of the community affected. Yet, this in turn will require a more local and regionally focused risk assessment, one, in which local knowledge will form a critical part of the evaluation. This, may, in turn, also contribute to a closer relationship between academic analysis, local knowledge, and political decision-making. However, it should be mentioned that there exist potential limitations to the relevance of applying local knowledge, since further climate change related risks may not play out in a linear fashion, and consequently, lessons of adaptation may not be drawn from past experiences alone. However, this should not be interpreted as neglecting locally specific adaptation knowledge.

Even so, the wider regional context should not be ignored either, as context is imperative to identify environmental/climate change related threats, and the potential spill-over they may generate, after all, environmental issues and climate change related impacts are trans-national in nature, even when their impact is local specific. After all, geographical proximity is a relevant dimension for security in general and with regard to environmental security in particular. It is this interrelated process of environmental degradation and its impact on individuals and society alike, which can generate various feedback processes, thereby challenging the established political, economic, and social order.

4.4 Locating Environmental Security in the Context of Regional Security Challenges and Regional Processes of Integration

As stated before, environmental security issues and the impact of climate change not only need to be evaluated in the context of national development processes, but also at regional level as he potential impact is not restricted by administrative boundary, like national borders, but occurs within a shared geographic space, characterized by geographic features like river basins, mountain ranges, or coastal areas. Hence, proximity represent another critical aspect regarding the potential for increasing social, economic and political costs related to environmental/climate change related risks.

Proximity is usually interpreted in a positive way, when it comes to regional integration as it contributes to the formation of shared economic space, often transnational in character. Achieving infrastructure interconnectivity at the sub- or regional level constitute another vital aspect of the positive aspect of proximity within integration processes. Another underlining feature of integration dynamics is that there exists the expectation of an 'growth spill-over' effect. Yet, in the context of environmental/climate change related risk, proximity does take on a different meaning, one, which is no longer necessarily positive. A good example represents a statement made by India's PM, Narendra Modi that *"Blood and water cannot flow at the same time"*. Further, his announcement to suspend scheduled meeting between India and Pakistan on the sharing of the Indus River waters mirrors the conflict potential that is linked to the question how to share a transboundary water source (Why India's water dispute with Pakistan matters 2016). Even as it actually represents a political statement in response to a deadly attack on India's forces stationed in the disputed area of Kashmir, it still could lead to serious implications of how to share the water resources of the Indus and thus have the potential of further undermining their already stained, bilateral relationship.

In this context, it is worth considering the extent environmental/climate change related risks could be the source of international armed conflicts. Schleusser et al. (2016) present a critical evaluation of this subject. In their work on armed conflict risks and climate related disasters, their findings do not support a position, which would identify environmental/climate change risks as a single driver for armed conflicts as they recognise that poverty, income inequality, weak governance, natural resources exploitation, socioeconomic discrimination, and an ethnical fractionalized society are the main sources of armed conflicts. However, they maintain that environmental/climate change related events have the potential of amplifying already existing tensions contributing to instability.

Even so, environmental/climate change related risks should be identified as a particular risk category, but their actual or potential impacts need to be evaluated within a wider context, including economic, social, and political aspects, as well as with regard to the regional security dynamic. This will offer not only insight in assessing environmental/climate change related risks in general, but also in the wider context of bilateral and regional security dynamics.

In recognising the potential and actual risks environmental degeneration and climate change represent to the development prospects of countries, a new a critical risk category to the development framework is added, one, which may lead to a fundamental alteration of how the development challenge and regional co-operation is understood.

4.5 Shared Resources—Water, Development, and the Mekong River Basin Question

The Greater Mekong Sub-region (GMS) is an economic area bound together by the Mekong River and includes various parts of different countries, including: China, especially Yunnan province; parts of Myanmar; Laos, Thailand, Cambodia, and Viet Nam. The GMS cover 2.6 million square kilometres with a combined population of around 326 million people.

The political complexity of addressing the challenge to the Mekong River Basin is a rather serious topic and could lead to increasing political conflicts between the countries involved. Adding to the vulnerability of the river systems is the impact of climate change, characterized by a change in local and regional precipitation, and the occurrence of extreme weather events like drought and floods, each of those have the potential of interfering with the river's flow regime, and consequently on the amount and quality of the river water. Hence, the wider connection between development and the negative impact of climate change related risks are increasingly highlighted, thus climate change related risks are added to the wider development challenge the Mekong region is facing.

4.5.1 Addressing the Complexity of Regional Environmental Issues = Environmental-Development Nexus

A common goal for all states within the GMS is to facilitate economic development; addressing underdevelopment and marginalisation are of significant concerns. Yet, environmental degeneration/ climate change related risks have the potential of increasing those regional development challenges. Indeed, within the region, an intertwined link between economic development and modernisation, and environmental degeneration exists, not least because of a particular development paradigm, characterised by a 'growth first and clean up later' approach. This link between economic development and environmental degeneration is recognised in an Asian Development Bank report, that highlights that during the 1990–2010 period Southeast Asia was the region with the highest increase in carbon dioxide emissions globally and is on course of becoming a large emitter in the future. Consequently, an alternative model of economic development, which is low-carbon economy, should be applied (Raitzer et al. 2015).

Hence, as observable in other regions, the close link between the level of industrial and economic development and environmental degeneration becomes apparent again. In a 2014 IPCC evaluation it was stated that Southeast Asia could face a temperature increase of 3 °C, more extreme precipitations during the Monsoon season, and intensified droughts periods, adding that Southeast Asia faces a high degree of cumulative climate change risk impact (Hijioka et al. 2014). The official statement of the 5th GMS Summit (Joint Statement 2014) also states that the region is facing an increasing severity of natural disasters and environmental challenges at both the local and national level, hence addressing those threats would require a stronger regional cooperation.

At the same time, the significance of the Mekong for the development prospect of the people living along its banks was not only recognized early on and is still emphasized in recent regional summits and documents. Take for example the 1995 agreement on the Cooperation for Sustainable Development of the Mekong River Basin which states that the Mekong's natural resources and environment are of immense value to all the riparian countries, by supporting economic development and social wellbeing and therefore living standards of the people (Agreement on the Cooperation for the Sustainable Development of the Mekong river basin, Mekong River Commission 1995). Such an assessment is supported by other official statements like an assessment from the Mekong River Commissions' Ho Chi Minh City Declaration (April 2014) stating that while the exploitation of the Mekong River Basin's water sources largely contributes to the socio-economic development of the region, the further pressure on the water resources will add to the negative impact of climate change within the Mekong region (Ho Chi Min City Declaration 2014). Alike, the 5th GMS Summit (December 2014) affirmed that sustainable, environmentally friendly development was the way forward while ignoring the environmental aspect could undermine the development prospect within GMS (GMS e-Update 2014–2015).

Yet, all involved countries aim at utilizing the Mekong's resources for their own development, and the Mekong River Commission Strategic Plan 2016–2020 identifies some of the related challenges to the Mekong River water resources, which are linked with regional economic development prospects. Among them is tributary and mainstream hydropower development, expansion of irrigated agriculture, navigation, and water miss management (MRC Strategic Plan 2016–2020).

A specific challenge, which is identified in various reports, is to coordinate the development strategies of the riparian countries as the potential trans-national implications of national policies should not be ignored, as they may contribute to dynamics of regional conflicts. The MRC Environmental Program 2011–2015 already points out that transboundary effects of national projects are not sufficiently integrated in national assessments, and short-term national development benefits undermine long-term regional environmental protection (MRC Environmental Program 2011–2015). An Asian Development Bank report also states that a more selective approach in selecting specific targets should be applied, considering that broader regional integration represents an important agenda in the regional development agenda ('The Greater Mekong Sub-Region Economic Cooperation Program Strategic Framework 2012–2022). Alike argues Chayanis Krittasudthacheewa et al. (2019) by emphasising that national governments unabated exploit the Mekong's resources for short-term economic gains and competition over resources are increasingly transcending national borders without addressing the basin wide impact of this resource exploitation. As national development plans may increase the negative impact on neighbouring countries and thus are less optional from a basin wide perspective, the potential for resource distribution conflicts increases. Adding that a sense of urgency exists to take evaluate the impact of national development plans on basin wider development, especially with regard to food, water and energy security (MRC 2015 3).

Hence, once more, the transboundary nature of environmental/climate change related risks and their potential for generating interstate political conflicts should not be ignored in the context of national and local decision making when addressing specific environmental/climate change related risks. Consequently, bilateral and regional consultations regarding the potential impacts of national development strategies should be facilitated to address their potential regional impact. This in turn will add another layer of complexity for addressing particular environmental/climate change related risks and with it the potential for increasing political conflicts. However, addressing the trans-national implications of national development strategies successful may actually offer a positive incentive for bilateral or regional cooperation.

Regional development projects like the East–West Corridor offer additional insight into the complex development challenges the region faces. Though the implementation of such development projects will contribute to the regional development process, they will increase the pressure on the resources of the Mekong region and on the livelihood of the people living along its banks. Yet, it should not be neglected that other sections of the regional population will actually profit from such development projects. However, this may not offer much consolation for those who may lose their livelihood along the Mekong River, subsequently raising the potential for distributional conflicts at and below national level, which in turn could also trigger a more international assertive behaviour.

Taking together, this provides a background of the challenges of how to use the resources of the Mekong River to facilitate development and even to reach some of the UN Millennium Development goals. Thus, the challenge to the Mekong River Basin is a rather complex one, not only in the context of national development challenges, but also in the context of a transboundary perspective. This in turn could contribute to more tension between the involved countries.

4.5.2 Assessing the Potential for Regional Cooperation or Conflict Regarding Environmental Security Issues?

Assessing the potential of regional environmental and climate change challenges for either cooperation or conflict is not straightforward, and the pattern of interaction is unpredictable and could generate a dynamic in either direction.

Considering that such sub-regional integration processes are taking place in close proximity, and indeed proximity is highlighted as a driving force for cooperation, trans-border environmental issues could generate a quite negative impact on bilateral relations. Hence, considering the trans-national character, environmental/climate change related risks and that their distribution and impact are influenced by specific geographical locations and features instead of administrative borders, proximity can become a rather different connotation. This in turn can increase the challenge for both regional economic progress and regional wide integration. Consequently, cross-border environmental/climate change related risks can have fundamental, but varied, implications for interstate relationships, ranging from cooperation to conflict, depending on the intensity and how cross-border environmental/climate change risks will be addressed.

The building of hydropower plants along the Lancang/Mekong provides a good example, with China's up-stream hydropower plants are being the focus of criticism since the early 2000s. More recently, Laos hydropower plant expansion strategy along the middle section of the Mekong also became the focus of controversy. It latest, but not the last, dam project close to Luang Prabang, does generate increasing controversy among the other riparian countries. Thailand already complained about the test run of Lao's Xayaburi dam in 2019, claiming that this contributed to the Mekong's exceptionally low water levels during that year (Dam-building race threatens the Mekong River 2019). Thus, the countries within the middle and lower section of the Mekong request Lao to conduct rigorous transboundary impact assessments and enhance proposed measures to mitigate potential adverse impacts (MRC 2020b). It is worth take into account a recent water resource evaluation undertaken by the Mekong River Commission, which assessed the regional contributions of different catchment areas to the overall water flow of the Mekong river. Pointing out that 16% are contributed from upstream areas within China, 30% from within Laos, though another source of 23% is geographically shared between Laos and Cambodia, and 6% coming from Thailand (MRC 2020a). This assessment clearly indicates the significance of the middle section of Mekong's river basin for the flow regime of the Mekong itself and for the countries located further downstream, especially when considering that another 50 hydropower plants are planed within the middle section of the Mekong river. Hence, further developing the hydropower potential within its middle section does therefor provide a considerable challenge for resource distribution and the potential impact this may generates on whole river basin.

Yet, the exploitation of the Mekong's water resources does not stop here as extensive agricultural irrigation projects are planned by Thailand in support of development of its eastern regions along the Mekong, further reducing the amount of water and sediments reaching the down-stream areas of the Tongle Sap and the Mekong Delta. While it is estimated that the area suitable for agricultural irrigation is about 4.3 million ha, so far 3.6 million ha are actual covered by irrigation (FAO 2016). Hence, there exist a considerable potential for expansion, which would support local development but at the same time increase the pressure on the Mekong's water resources and on how to share this resource among the riparian countries. Vietnam (42%) and Thailand (30%) share the most parts of the irrigation area within the Mekong's basin and also the leading countries with regard to water withdrawal, with Vietnam accounting for 52%, Thailand for 29%, China 9%, Laos 5%, Cambodia 3%, and Myanmar 2% (FAO 2016). While in the case of Vietnam the high percentage is linked to the Mekong's delta which is often described as Vietnam's 'rice bowl' and its intensive use as agricultural area, whereas in the case of Thailand its high percentage is associated with considerable agricultural irrigation projects in its eastern provinces, located along and within the Mekong river. Indeed, Thailand's irrigation plans are extensive with planed increase of the land available for agriculture in its eastern provinces from 15,000 rai to 300,000 rai, covering 284 villages (Govt revives old plan to irrigate Isan 2019). Such extensive irrigation plans, by diverting the water resources of the Mekong, do not go done well with the downstream countries. Not least when considering that irrigation withdrawal accounts for 90.5% of the total water withdrawal from the Mekong river basin (FAO 2016). Once again geography counts, since location, either being located up- or down-stream within an IRB matters when it comes to resource competition.

Though, it its worth to mention that those four countries actually experienced an enduring drought situation and falling water levels along the Mekong at that time. However, since then, more hydropower plant projects are put forward not only by China, but especially by Laos, thus the concern with dam building along the Mekong and its tributaries is no longer limited to the upstream development in China, but also with dam building in its middle and lower sections. After all, dam building does generate an impact on the Mekong's flow regime and thus has the potential of impacting on the amount and quality of water available downstream. Considering that the general demand of water within the Lower Mekong Basin increases because of economic development, but also as more and more water of the Mekong is used for agricultural irrigation by Thailand, Laos and Cambodia. Thailand, for example, plans to use Mekong's water resources for national irrigation projects in its northeast region. Hence, the potential for political conflicts at country level is evident, this could even involve several ASEAN members, and consequently may even generate dissent within ASEAN. Both an increasing commercial and agricultural use of the Mekong's water resources can add to the vulnerability of the whole river systems.

Yet, the dam building in Laos represents an interesting case, since Lao emphasises that the revenue generated from its extensive planned increase in its hydropower potential will be used for poverty reduction and for facilitating domestic economic growth, indeed it should help Laos to reach the U.N. development goals. Even so, one can identify a potential distributional challenge within Laos, since the revenue generated from selling the electricity may not be directed at the affected communities, where the dams will be built. Even if one put aside the usual political competition over recourse distribution, one can find some good arguments for an alternative use of those founds, since Laos is facing general development challenges and Laos' complex topography, with strong variation in the livelihood of its people, just increase this challenge further.

Even as rural economy provides for two-thirds of the population, different perspectives and with it a different level of resilience to economic insecurity and climate change related risks could be identified within the rural sector. For example, on the one hand, communities near the Thai border, who have developed transborder trade relations, and such, which can take profit from tourist related activities, are among the most resilient ones, since those communities are able to diversify their economic incomes. On the other hand, remote communities in the highlands, not only suffering from missing connectivity to potential markets, but they are also severely affected by a changing climate, since changes in the weather pattern and extreme weather events could generate significant negative impacts on their livelihood with very limited option for generating alternative sources of income. Diversification of livelihood represents a fundamental task in increasing the durability of rural communities with generating connectivity and signifying a specific task.

Hence, Laos provides a good example for the complexity and the political conflict potential when addressing the question of how to share the resources of River Mekong. Not only the issue of interstate competition for resources is of relevance, but also the sensitive topic of limited resource distribution at domestic level, which in turn could increase the conflict potential at the international level.

With regard to dam building in general two more aspects need to be mentioned. One relates to the amount of sediments that have been transported downstream by the River and will be reduced because of dam building. This is of critical relevance for Cambodia and Viet Nam. In the first case, the Tongle Lap, Cambodia's huge inland lake, depend on the Mekong for its waters and nutritious sediments, as does the Mekong delta in southern Viet Nam. The latter is under pressure from seawater intrusion, which not only renders increasing parts of the delta unusable for agriculture, but also erodes the Delta itself. With a predicted sea-level rise based on global climate change, this aspect just becomes an increasingly serious topic. The other aspect is that dam building at the upper section of the Mekong may even help to address water shortage at the lower Mekong section during the dry season, consequently providing water year-round for agricultural irrigation (MRC Strategic Plan 2016–2020). A cooperative arrangement between up-stream and down-stream is a pre-condition for such an agreement, which in turn would contribute to a more cooperative relationship in sharing the Mekong recourses.

However, what further complicates the potential political conflict is that if neither Thailand nor Viet Nam would be prepared to purchase the power generated from the Lower Mekong River hydropower dams in Laos, these dams would very likely not be built at all (ICEM 2010, p 10). This in turn remains us on the complexity of the issue, as for example Viet Nam has continually protested about dam building in upstream countries, as it reduces water levels and withholds nutrition sediments, but on the other hand it finances some of the dam projects in Laos.

Within Southeast Asia, the territorial conflicts among various parties within the South China Sea represent a prominent issue, especially between China on the one hand and Viet Nam and the Philippines on the other. Though, there also exist other political and security challenges either within the region or with a wider geographical focus, namely within East Asia, which too could impede a more proactive cooperation regarding of how to share the resources of the Mekong. Two critical issues are which role an increasingly stronger China will play within the region, and to what extent the region get tangled into a U.S.-China-Japan competition at the regional level. Another, to what extent the ASEAN will be able to formulate a coherent position within its members and when confronted with challenges from outside. After all, ASEAN still adhere to its position of 'ASEAN Centrality', that it is the central regional actor within Southeast Asia. What's more all ASEAN members are still facing the challenge of development, which may limit their willingness for cooperation even with other member states, though it shall be mentioned that for the time being that is not a to prominent concern.

With regard to Southeast Asia's regional integration, one could argue that the region has indeed become more integrated over the last two decades with economic interests as the driving force behind it, even when formal regional integration agreements are missing. Yet, this year's launch of the ASEAN's Economic Community is a vital process in the regional integration dynamic. The recently set up Lancang-Mekong Cooperation (March 23, 2016) seems to indicate a more cooperative relationship among the countries sharing Mekong's resources, which includes all six countries along the Lancang-Mekong river basin, including China. This meeting produced a declaration, named "Sanya Declaration". The declaration states that development represents the top priority, but also mentions regional cooperation, connectivity, trans-border diseases, and climate change among its focus, and describes it as a new sub-regional cooperation.

It certainly will be an interesting subject to follow, to what extent this new cooperation framework will be instrumental in addressing the challenges of how to share Mekong's resources. If successful, it could lead to relationships that are more cooperative and thus could work against a potential development that could see heightened political conflicts among the involved countries. However, it should be pointed out that managing environmental issues successfully, especially across borders, might generate sufficient trust and dependencies, which aid to overcome other longstanding conflict structures. Consequently, helping to bridge a non-cooperative relationship between states on security concerns also matters national security.

Thus, so far, there seems to be no negative backlashes with regard to regional integration process and regarding inter-ASEAN rivalry. However, the Haze issues, involving Singapore and Malaysia on one side and Indonesia on the other, just indicate how fast this could change, even in the end they too managed to keep the political dispute of how best to address the Haze issue at a managerial level.

Once again, environmental/climate change related risks not only need to be evaluated in the national context, how they will be affected by national development strategies, but also in the regional context, in terms of the existing dynamic of cooperation and conflicts which represent political and economic challenges that will be addressed. Hence, the regional security context, the degree of amity and cooperation between the regional actors, is another prominent factor when considering the impacts of environmental security on bilateral and regional cooperation. Either way, we may have to re-consider to what extent environmental degeneration and the impact of climate change should be identified as security agendas. That is, we will likely move from what was once dubbed 'low politics' and instead include these dynamics to what is identified as 'high politics'.

4.6 Conclusion

The relevance of environmental security rose with the redefinition of what security refers to. Over time, we witnessed an issue driven process, in which the traditional, state-centric interpretation of security became increasingly challenged as the focus of security changed from the state to society and to individuals as the main reference for security. Consequently, the issue of underdevelopment raised to prominence and with it the concept of human security. Since environmental degeneration and climate change have an adverse impact on the development prospects of societies, environmental security has also become a more prominent issue.

When evaluating environmental/climate related risks, which are often transboundary, not only the national specific development strategy needs to be considered, but the regional development and security context should be taken into account as well. Not only is environmental degeneration linked with specific economic development strategies, like generating rapid economic growth or what is described as the 'growth first and clean up later' approach, but, in addition, the existing security interaction and dynamic within a region will also have an influence on how a trans-border environmental issue will be addressed.

In the case of trans-boundary river basins like the Lancang/Mekong, the complexity of addressing the challenges is apparent. Even as the specific impact of environmental/climate change related risks are felt locally, their impact carries across national borders. In addition, location also counts with regard to whether a country is located at the up- or downstream of an international river network. Alike, and as in the case of the Mekong River Basin, the economic development and modernisation strategies of the various countries will increase the pressure on the water resources of the Mekong River further. This in turn can lead to a stronger political conflict over the use and sharing the River's resources and depending on the situation, it may even have the potential of undermining regional co-operation and integration.

What's more, when assessing the potential of environmental/climate change related risks as either a source of co-operation or conflict, the bilateral and regional security dynamic need to be brought into perspective. After all, a regional dynamic of co-operation or conflict will form the framework, within which regional challenges, like environmental/climate change related risks, will be understood.

References

Agreement on the cooperation for the sustainable development of the Mekong River Basin, Mekong River Commission, 5 April 1995. https://www.mrcmekong.org/assets/Publications/policies/agreement-Apr95.pdf. Accessed 15 Apr 2017

Asian Development Bank (2011) The greater mekong subregion economic cooperation program strategic framework 2012–2022. Asian Development Bank, Mandaluyong City

Booth K (2005) Introduction to part I. In: Booth K (ed) Critical security studies and world politics. Lynne Rienner, Boulder, pp 21–25

Buzan B, Waever O, de Wilde J (1998) Security: a new framework for analysis. Lynne Rienner Pub, London

China denies dams have worsened drought in Mekong River Basin, 31/03/10, China Daily. https://www.chinadaily.com.cn/china/2010-03/31/content_9664697.htm. Accessed 08 Sept 2015

Krittasudthacheewa C, Navy H, Tinh BD, Voladet S (2019) Introduction: addressing development and climate challenges in the Mekong Region Development and Climate Change in the Mekong Region Case Studies. In: Krittasudthacheew CM, Navy H, Tinh BD, Voladet S (eds) Publisher: Stockholm Environment Institute, 10th Floor, Kasem Uttayanin Building, 254 Chulalongkorn University, Henri Dunant Road, Pathumwan, Bangkok, 10330 Thailand. Strategic and Information and Research Development Centre, No. 2, Jalan Bukit 11/2, 46200 Petaling Jaya, Selangor, Malaysia

Coninck de H, Revi A, Babiker M, Bertoldi P, Buckeridge M, Cartwright A, Dong W, Ford J, Fuss S, Hourcade JC, Ley D, Mechler R, Newman P, Revokatova A, Schultz S, Steg L, Sugiyama T (2018) Strengthening and implementing the global response. In: Masson- Delmotte V, Zhai P, Pörtner HO, Roberts D, Skea J, Shukla P R, Pirani A, Moufouma-Okia W, Péan C, Pidcock R, Connors S, Matthews JBR, Chen Y, Zhou X, Gomis MI, Lonnoy E, Maycock T, Tignor M, Waterfield T 2018 Global warming of 1.5°C, An IPCC Special Report on the impacts of global warming of 1.5°C above pre-industrial levels and related global greenhouse gas emission pathways, in the context of strengthening the global response to the threat of climate change, sustainable development, and efforts to eradicate poverty, pp 313–433. Viewed 02 December 2018. https://www.ipcc.ch/site/assets/uploads/sites/2/2018/11/SR15_Chapter4_Low_Res.pdf. Accessed 15 Apr 2019

Dam-building race threatens the Mekong River PUBLISHED: 19 AUG 2019 AT 20:05 https://www.bangkokpost.com/thailand/special-reports/1733071/dam-building-race-threatens-the-mekong-river. Accessed 10 May 2020

Five Features of Lancang/Mekong River Cooperation 2016/03/17, Ministry of Foreign Affairs of the People's Republic of China. https://www.fmprc.gov.cn/mfa_eng/zxxx_662805/t1349239.shtml. Accessed 22 May 2018

GMS e-Update 2014-2015:1-2 GMS e-Updated; September 2014 - February 2015, vol 8, Issue No 2. https://www.adb.org/sites/default/files/publication/154923/gms-e-updates-sep2014-feb2015.pdf. Accessed 16 Apr 2016

Hewitson BC, Janetos AC, Carter TR, Giorgi F, Jones RG, Kwon W-T, Mearns LO, Schipper ELF, van Aalst M (2014) Regional context. In: Barros VR, Field CB, Dokken DJ, Mastrandrea DM, Mach KJ (eds) Climate change 2014: impacts, adaptation, and vulnerability. Part B: regional aspects. Contribution of working group ii to the fifth assessment report of the intergovernmental panel on climate change, pp 1133–1197. Cambridge University Press, Cambridge. https://www.ipcc.ch/report/ar5/wg2/. Accessed 23 Aug 2015

Hijioka Y, Lin E, Pereira JJ, Corlett RT, Cui X, Insarov GE, Lasco RD, Lindgren E, Surjan A (2014) Asia. In: Climate change 2014: impacts, adaptation, and vulnerability. Part B: regional aspects. Contribution of working group ii to the fifth assessment report of the intergovernmental panel on climate change [Barros VR, Field CB, Dokken DJ, Mastrandrea MD, Mach KJ, Bilir TE, Chatterjee M, Ebi KL, Estrada YO, Genova RC, Girma B, Kissel ES, Levy AN, MacCracken S, Mastrandrea PR, White LL (eds)], pp 1327–1370. Cambridge University Press, Cambridge, New York

Ho Chi Minh City Declaration, Water, Energy and Food security in the context of climate change for the Mekong River Basin 5 April 2014. https://mrcsummit.org/download/HCMC-Declaration-V5-4Apr2014.pdf. Accessed 15 Jan 2016

Human Security Unit United Nation (2009): Human Security in Theory and practice. United nation Trust Fund for Human Security, 2009, Human Security Unit Office for the Coordination of Humanitarian Affairs, United Nations, New York. https://www.tr.undp.org/content/dam/turkey/docs/news-from-new-horizons/issue-41/UNDP-TR-HSHandbook_2009.pdf. Accessed 26 Oct 2014

ICEM (2010) MRC Strategic Environmental Assessment (SEA) of hydropower on the Mekong mainstream, Hanoi, Viet Nam. http://www.mrcmekong.org/assets/Publications/Consultations/SEA-Hydropower/SEA-FR-summary-13oct.pdf. Accessed 12 Mar 2016

IPCC (2014) Climate change 2014: impacts, adaptation, and vulnerability. Working group II contribution to the IPCC 5th assessment report - changes to the underlying scientific/technical assessment to ensure consistency with the approved summary for policymakers, March 2014. https://ipcc-wg2.gov/AR5/images/uploads/IPCC_WG2AR5_SPM_Approved.pdf. Accessed 08 June 2015

Joint Summit Declaration (2014) The Fifth Greater Mekong subregion (GMS) summit Bangkok, Thailand 19 – 20 December 2014. https://www.adb.org/sites/default/files/page/42450/5th-sum mit-joint-declaration-greater-mekong-subregion-gms.pdf. Accessed 22 Nov 2015

MRC (2020b) Laos urged to better assess impacts, provide effective mitigation measures, as Luang Prabang dam moves forwards, Vientiane, Lao PDR, 01 Jul 2020. https://www.mrcmekong.org/news-and-events/news/pr-luang-prabang-hpp-20200701/. Accessed Sept 2020

MRC (2020a) Understanding the Mekong River's hydrological conditions: a brief commentary note on the "Monitoring the Quantity of Water Flowing Through the Upper Mekong Basin Under Natural (Unimpeded) Conditions" study by Alan Basist and Claude Williams (2020) Vientiane: MRC Secretariat. Understanding-Mekong-River-hydrological-conditions_2020.pdf (mrcmekong.org). Accessed Sept 2020

MRC Environmental Program 2011–2015 (2011) Mekong River Commission. https://www.mrc mekong.org/assets/Publications/Programme-Documents/Environment-Programme-2011-2015-1-November.pdf. Accessed 25 Nov 2015

MRC Strategic Plan 2016–2020 (2016) Mekong River Commission.https://www.mrcmekong.org/assets/Publications/strategies-workprog/MRC-Stratigic-Plan-2016-2020.pdf. Accessed 25 Nov 2015

Raitzer DA, Bosello F, Tavoni M, Orecchia C, Marangoni G, Nuella J, Samson G (2015) Southeast Asia and the economics of global climate stabilization. Asian Development Bank, Mandaluyong City. https://www.adb.org/sites/default/files/publication/178615/sea-economics-global-climate-stabilization.pdf. Accessed 18 Apr 2016

Ploberger C, Leal Filho W (2016) Towards long-term resilience: the challenge of integrating climate change related risks into a risk analysis framework. In: Musa H, Cavan G, O'Hare P, Seixas J (eds) Walter Leal Filho W. Climate change adaptation, resilience and hazards. Springer, Cham, pp 369–380

Schleusser CF, Donges JF, Donner RV, Schnellhuber HJ (2016) Armed-conflict risks enhanced by climate-related disasters in ethnically fractionized countries. Proc Natl Acad Sci USA 113(33):9216–9221

Smith S (2005) The contested concept of security. In: Ken Booth K (ed) Critical security studies and world politics. Lynne Rienner, Boulder, pp 27–62

The Emissions Gap Report 2018, United Nations Environment Programme, Nairobi, viewed 02 December 2018. https://wedocs.unep.org/bitstream/handle/20.500.11822/26895/EGR2018_Full Report_EN.pdf?sequence=1&isAllowed=y. Accessed 08 May 2019

Walker RBJ (1997) The subject of security. In: Williams MC, Krause K (eds) Critical security studies. UCL Press, London, pp 61–82

Why India's water dispute with Pakistan matters, 28/09/16, BBC online. https://www.bbc.com/news/world-asia-india-37483359. Accessed 08 Oct 2016

Chapter 5
Enhancing and Operationalizing Water Security: Present Landscape and Emerging Research Needs

Mukand S. Babel and Victor R. Shinde

Abstract Water security is vital and crucial to human security. Water security, which was considered as an issue of water scarcity in the past, is a concept that is holistic and encompasses several aspects of water. Rapid advances have been made in water security related research in the last few years. These have contributed to an improved and an all-inclusive understanding of water security, along with unravelling the embedded complexities within the concept. The theoretical aspects of water security can be well explained through science; however, there is a wide scope of linking these theoretical aspects to come up with solutions to real–world problems. The philosophy of water security is robust but implementation on the ground is still feeble due to various challenges. Research in water security now needs to move into this solution space, where the theories are translated into practice. This paper provides an overview of the existing water security landscape and seeks to identify some key research needs necessary to operationalize this concept into reality. It recognizes that achieving water security may be a multi–faceted process, requiring the coordinated and concerted efforts from several stakeholders and diverse water use sectors. However, the creation of water–secure societies with an acceptable level of risk is very much possible, with science having an integral role in realizing this goal.

Keywords Climate change · Decision support systems · Integrated water resources management · Risk · Trade-off · Research · Water security

5.1 Introduction

The attention in academic world towards the term "water security" has been soaring in recent years. The research into this specific theme has progressed from a point

This article appeared in the J. Japan Soc. Hydrol. and Water Resour., Vol. 32, No.2, Mar. 2019 pp. 74–81. DOI: 10.3178/jjshwr.32.74. It is reproduced here with permission from the publisher.

M. S. Babel (✉) · V. R. Shinde
Water Engineering and Management, Asian Institute of Technology (AIT), Khlong Luang, Pathum Thani, Thailand
e-mail: msbabel@ait.ac.th

where it was classified merely as a water scarcity issue to a more holistic approach that comprises multiple water security dimensions. As understanding into water security deepened, it became evident that this concept is pertinent to several diverse sectors, with each of the sectors having their respective point of interest in the different aspects of water security. Table 5.1 provides a summary of how water security is described by these various disciplines/sectors.

Table 5.1 Water security definitions by disciplines/sectors (*Source* Cook and Bakker 2012)

Discipline/Sector	Water security focus or definition
Agriculture	• Input to agricultural production and food security
Engineering	• Protection against water related hazards (floods, droughts, contamination, and terrorism) • Supply security (percentage of demand satisfied)
Environmental science, environmental studies	• Access to water functions and services for humans and the environment • Water availability in terms of quality and quantity • Minimizing impacts of hydrological variability
Fisheries, geology/geosciences, hydrology	• Hydrologic (groundwater) variability • Security of the entire hydrological cycle
Public health	• Supply security and access to safe water • Prevention and assessment of contamination of water in distribution systems
Anthropology, economics, geography, history, law, management, political science	• Drinking water infrastructure security • Input to food production and human health/wellbeing • Armed/violent conflict (motivator for occupation or barrier to cooperation and/or peace) • Minimising (household) vulnerability to hydrological variability
Policy	• Interdisciplinary linkages (food, climate, energy, economy, and human security) • Sustainable development Protection against water-related hazards • Protection of water systems and against floods and droughts; sustainable development of water resources to ensure access to water functions and services
Water resources	• Water scarcity • Supply security (demand management) • "Green" (versus "blue") water security—the return flow of vapour

Table 5.1 shows the diverse array of dimensions of water security. It is therefore imperative to plan across diverse sectors and consider the perspectives of multiple stakeholders in order to achieve overall water security. A widely accepted definition is *"The capacity of a population to safeguard sustainable access to adequate quantities of and acceptable quality water for sustaining livelihoods, human well-being, and socio-economic development, for ensuring protection against water-borne pollution and water-related disasters, and for preserving ecosystems in a climate of peace and political stability"* (UN Water 2013). This definition thus necessitates water security to address the entire life spectrum and requires a holistic methodology that incorporates social, cultural, and economic aspects, while also considering scientific and technical solutions with a special emphasis on societal dynamics.

Translating the principles of water security into real–world practices has been a testing endeavour despite the considerable advances into the understanding the concept. The creation of water–secure societies is often hampered by several obstacles which include escalating population growth, degrading water quality, and recurring episodes of water-induced disasters such as floods and droughts, along with other hydrological effects of global changes. Globally, attaining a state of water security has been among the top challenges for quite some time. For example, water crisis has been a critical global risk in terms of impact (Fig. 5.1) as stated by the World Economic Forum (2018) in their Global Risk Report.

The global recognition of water security enhancement as a key challenge has given way for the issue to be deemed as an action item under the 2030 Developmental Agenda as well as having a dedicated Sustainable Development Goal (SDG) for water–SDG 6: Ensure availability and sustainable management of water and sanitation for all. It is under this umbrella that governments worldwide are making considerable efforts to create water secure communities. On a regional scale, the Asian Development Bank (2013) has also conducted a thorough assessment of water security for countries in Asia and the Pacific region.

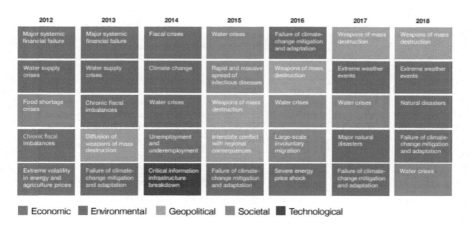

Fig. 5.1 Top 5 global risks in terms of impact (*Source* World Economic Forum 2018)

As governments are fulfilling their responsibilities in translating the water security philosophy into practice, the academic communities also need to intensify their efforts for the same cause by conducting research to fill the scientific gaps. As mentioned earlier, the theory and comprehension of water security has been on the rise in recent years, with researchers cognizant of the underlying intricacies and complex nature of issues related to water security. There have been many high-level publications on water security (e.g., Vörösmarty et al. 2010; Palmer et al. 2015) that have come into light in recent years. However, majority of the research on water security views it from a bird-eye angle hoping to encapsulate multiple perspectives and dimensions. The way forward in research should have the objective of identifying actions for holistic improvements in water security. This paper seeks to throw light on specific research needs necessary to bring about a tangible change in reality.

5.2   Emerging Research Needs

1.   **Identifying an acceptable level of water insecurity risk**
     Achieving a completely water secure society is close to impossible. There will always be some level of risk looming around. On a similar note, the OECD (2013) stated that the essence of water security is all about living with an acceptable degree of risk, which becomes more substantial in the face of climate change. Evident from Fig. 5.2, for the same probability, water risks are likely to be higher due to climate change.
     Imminent research needs to explore various avenues to identify acceptable risk levels for different settings and contexts. The appropriate level of water risk for a society should depend upon the balance between economic, social, and environmental consequences and the cost of improvement, the evaluation of which requires the consideration of limit of cost effective or practical water management.

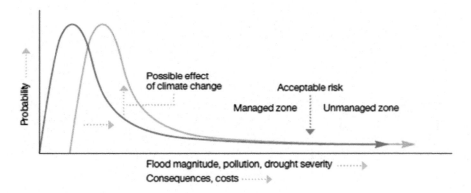

Fig. 5.2   Effect of climate change on the probability of extreme events

2. **Quantifying the impact of water security policies**

Improving water security has become a top priority for governments all over the world, and especially in developing countries. The Sustainable Development Goal Report (2018) states, "Too many people still lack access to safely managed water supplies and sanitation facilities. Water scarcity, flooding and lack of proper wastewater management also hinder social and economic development. Increasing water efficiency and improving water management are critical to balancing the competing and growing water demands from various sectors and users". Countries worldwide are therefore introducing new policies, laws, and strategies or consolidating existing ones in order to meet SDG-6. The development of a robust framework to measure the influence and effectiveness of various policies and strategies is crucial in enhancing water security.

3. **Operational water security assessments**

Examples of water security assessment at various scales is plentiful in literature. It is a crucial element undoubtedly, and is embodied through the adage, "you cannot manage what you cannot measure". However, it is now time for establishing connections between water security assessments and actual progress on the ground. In essence, water security assessments need to transition from an evaluation phase into a more operational phase. Essentially, these assessments need to focus on developing and implementing solutions rather than evaluating certain solutions after they have been implemented. Utilizing this fundamental shift in methodology leads to numerous research prospects in the area of designing water security solutions based on optimized assessment criteria and standards. This is imperative because optimization, in many cases, would be contextual and dependent on the stakeholders involved in the decision process. Reaching the optimal set of solutions will therefore require addressing multiple viewpoints and concerns. Seminal studies in this regard have been carried out in recent past. For instance, Babel and Shinde (2018) developed an operational water security index at a river basin scale with its sole purpose being the facilitation of monitoring and evaluating actions undertaken for water security enhancement. Similarly, a recent study by Assefa et al. (2019) dealt with the development of a generic domestic water security index and its application in Addis Ababa, Ethiopia. Hence, there is now a need to progress higher up based on the foundations of these research works.

4. **Addressing contentious trade-offs**

The holistic treatment of water security brings up challenges of addressing the concerns of multiple competitive users and sectors. There will always be trade-offs that will bring about both violent and non-violent conflicts. This will be notably seen between agriculture and other water use sectors, the former being the largest water user globally. Research is needed to reduce these trade-offs and develop synergies among competing users. Concepts like the Collaborative Risk Informed Decision Analysis (CRIDA) proposed by the Alliance for Global Water Adaptation (AGWA 2018) are already being advocated to tackle these pressing challenges. Research efforts urgently need to be intensified to overcome controversial trade-offs and enable the creation of an environment of shared understanding and management of water resources and associated risks.

5. **Climate change and water security**

   Impacts of climate change on water availability and water use sectors are the major areas of research currently being carried out related to water security and climate change. Components of water cycle such as snow and glaciers, river flows, groundwater, evapotranspiration have come under the deep scrutiny. The same goes for water use sectors such as agriculture, industry, water supply, hydropower, and aquatic life, among many others. The upcoming research endeavours need to focus more on climate change adaptation interventions. It is becoming evident that current mitigation strategies and policies are not likely to limit global warming to 2°C (Hulme 2016; Geden 2015) by the end of this century–a pact agreed upon by parties at the Climate Agreement in Paris in 2015. Therefore, climate change adaptation will be vital to sustainable development and ground-breaking research is required to expedite the adaptation process and actions for various water use sectors and diverse communities.

6. **Expanded decision support systems**

   Water systems have synergistic relationships with energy (e.g., water-energy nexus) and food systems (e.g., water-food nexus), or a combination of various systems (e.g., water-energy-food nexus). This necessitates inter- and multisectoral approach to management of resources. While this concept may look simpler in theory, it is a lot more complex in implementation in the real world. This may be due to inability or reluctance of various sectors to work collaboratively due to reasons such as mismatch of mandates, overlap of responsibilities, sharing of costs and benefits, and among others. The need for holistic management of the water sector is very much there but the means to achieve this is still very unclear. Therefore, future researchers have a great opportunity to incorporate various dimensions of water governance in order to tackle this clear research gap.

7. **Engaging the verticals of water governance for water solutions**

   Changing the present situation of water security in a particular context requires horizontal and vertical coordination among stakeholders. Vertical coordination is the point of interest here, with the horizontal coordination being discussed in the previous narrative. A key challenge in operationalizing water security in a holistic manner, especially in cities, is addressing the linkages of the cities with the basins in which they are situated. The multiple vertical tiers of water governance–which includes municipality, district, state, province, and any other relevant administrative boundary–must have coordination among them. Given that there is little to no interaction between these vertical levels in many geographic regions, policies and plans for holistic water management are often never translated into reality . There is therefore a need for researchers to build, explore, and test diverse models for ensuing vertical coordination and cooperation for water security solutions.

8.   **Creating a value for water**

Humans tend to cherish what they value most. The same idea fits perfectly with water, especially when considering environmental water security. Rivers in developing countries are exploited indiscriminately for consumptive uses and what used to be pristine conditions has now transformed into unimaginable polluted conditions without considering the repercussions. Environment has been neglected in the quest for economic prosperity. This needs to be changed, and this can happen only when people learn to value water environments such as lakes, streams, rivers, and oceans, and understand the services that they provide for our existence. These water environments fit between the spectrum of splendour and eye-sore, with the former transpiring when people attach economic, social, and aesthetic values to these environments and the latter when people lose connection with these environments. Contemporary research should explore various possibilities of bringing together citizens and these environmental assets to strike a balance of water use for human and nature.

9.   **Stepped-up citizen engagement**

Citizen support is vital to achieving water security and the need for effective communication in this respect cannot be stressed more. There are many instances where communication is considered secondary to a management plan. Additionally, orthodox communications tend to follow a top-down approach. There is therefore a need to step-up communication engagement strategies between citizens, which will encourage them to share the responsibility of action. The Government alone cannot fulfil all duties in order to achieve water security. Citizens have an integral role to play in enhancing water security, which can be done only when the research delivers joint and focused communication strategies to get them to play their individual and collective roles.

10. **Leveraging on the technological avalanche**

The universal growth of data provides value in almost every domain of science and society and the water field is not an exception. From preventing man-made disasters like overflowing rivers with toxic waste, to natural flooding, to raising public awareness in water savings and minimizing the impacts of drought in arid regions. All of these goals are possible with the effective use of Big Data technologies. Big Data has the promise to revolutionize not only research, but also science. Big data, if done responsibly, can deliver meaningful benefits and efficiencies in water resources analysis, scientific research, environment, and other specific areas.

5.3   Final Reflections

Realizing a water secure world is complex. It is similar to accomplishing Integrated Water Resources Management (IWRM) in that these both are multifaceted areas demanding a coordinated and collaborated effort from a wide range of stakeholders. Akin to IWRM, realization of water security requires interventions in three pillars–enabling environment (for policies and regulations), institutional role (for horizontal and vertical inter–sectoral co-ordination), and management instruments (for making a change in reality) of sustainable development. Assessment of current water security situations can deliver useful inputs in the IWRM planning cycle to improve water management as a whole for social equity, economic efficiency, and environmental sustainability.

The world will face overwhelming water management challenges as demand reaches the supply threshold leading to increased inter- and intra-sectoral competition, deteriorated water quality and threatened aquatic ecosystems. The concept of water security–tied up with the philosophy of IWRM—provides a way forward to help consider how we can best make social choices about water allocation and access as well as the sustainability of water resources and the infrastructure we use to manage those resources (Giordano and Shah 2014). Comprehensive and holistic water security solutions are vital in the present-day context. Furthermore, these solutions should be built upon sound scientific principles and should be agreeable to a wide range of stakeholders. Hence, research efforts now have to transform from an 'exploratory' domain to a 'solution' space, with science having an imperative role in bringing this transition to fruition.

References

Alliance for Global Water Adaptation (2018) Climate Risk Informed Decision Analysis. https://agwaguide.org/docs/CRIDA_Nov_2018.pdf.

Assefa YT, Babel MS, Susnik J, Shinde VR (2019) Development of a generic domestic water security index and its application in Addis Ababa, Ethiopia. Water 11:37. https://doi.org/10.3390/w11010037

Asian Development Bank (2013) Asian water development outlook: measuring water security in Asia and the Pacific. Philippines, Manila

Babel M, Shinde V (2018) A framework for water security assessment at basin scale. APN Sci Bull 8(1). https://doi.org/10.30852/sb.2018.342

Cook C, Bakker K (2012) Water security: debating an emerging paradigm. Glob Environ Chang 22(1):94–102

Geden O (2015) Policy: climate advisers must maintain integrity. Nature 521:27–28

Giordano M, Shah T (2014) From IWRM back to integrated water resources management. Int J Water Resour Dev 30(3):364–376. https://doi.org/10.1080/07900627.2013.851521

Global Risk Report (2018)

Hulme M (2016) The climate research agenda after Paris: should 1.5 degrees change anything? Nat Clim Chang 6:222–224

OECD (2013) Water and climate change adaptation: policies to navigate uncharted Waters. OECD Studies on Water. OECD Publishing. https://doi.org/10.1787/9789264200449-en

Palmer MA, Liu J, Matthews JH, Mumba M, D'Odorico P (2015) Manage water in a green way. Science 349(6248):584–585

Prosser I (2012) Governance to address risks of water shortage, excess and pollution. Paper presented at the OECD expert workshop on water security: managing risks and trade-offs in selected river basins, 1 June 2012, Paris

UN Water (2013) Water security and the global water agenda. A UN-Water Analytical Brief. https://www.unwater.org/publications/water-security-global-water-agenda/

Vörösmarty CJ, McIntyre PB, Gessner MO, Dudgeon D, Green P et al (2010) Global threats to human water security and river biodiversity. Nature 467:555–561

World Economic Forum (2018) The global risk report 2018. 13th edn. World Economic Forum Publishing. https://www3.weforum.org/docs/WEF_GRR18_Report.pdf

Part II
Water Security Assessment and Planning

Chapter 6
The Threats to Urban Water Security of Indonesian Cities

R. W. Triweko

Abstract Indonesian cities are facing complex urban water problems which are inter-related among the provision of water supply, water pollution, water related disasters, groundwater degradation, and poor solid waste management. The threats to urban water security are caused by population growth, climate change, and water conflict. Increasing population growth causes the increasing raw water demand which should be transferred from the rural areas due to heavy pollution of the water body in urban areas. Changing policy in water use from agricultural water to urban water supply causes water conflict between farmers in rural areas and water supply enterprise in urban areas, which could extend to inter-regional water conflict. In addition, over abstraction of the groundwater to fulfil water supply demand resulted degradation of groundwater table and land subsidence, which disturbance drainage system. In some coastal areas, those phenomena even increase the risk of the city to seawater flooding and intrusion. Finally, climate change which has changed the pattern of rainy season and rainfall intensity has been increasing the risk of the cities to flooding and water supply availability. To increase urban water security, each municipality should ensure the availability of water supply and sanitation system, mitigate water-related disasters, and maintain good collaboration with neighbouring regencies/cities. New paradigm of integrated urban water management should be developed to increase urban water security in Indonesian cities. The development of the Urban Water Security Index could be used to monitor and evaluate the progress of the cities in improving their urban water service.

Keywords Urban water security · Indonesian cities · Integrated urban water management

R. W. Triweko (✉)
Universitas Katolik Parahyangan, Bandung, Indonesia
e-mail: triweko@unpar.ac.id

© Springer Nature Switzerland AG 2021
M. Babel et al. (eds.), *Water Security in Asia*, Springer Water,
https://doi.org/10.1007/978-3-319-54612-4_6

6.1 Introduction

The Government of Indonesia in its National Strategic Development Plan 2015—2019 used water security as the goal of water resources development. Infrastructure development during that period is intended to ensure water security for supporting national security. Six objectives that will be achieved during that period are: (1) To fulfil raw water demand for domestic, municipality, and industry; (2) To fulfil the demand of irrigation water and raw water for urban areas; (3) To increase the performance of irrigation management; (4) To accelerate the utilization of water resources for electricity (hydropower); (5) To increase the prevention to water related disasters; and (6) To optimize water balance management.

The data from the Head Office of Statistic (BPS 2012) indicates that the number of urban population in Indonesia has exceeded 50% with the rate of growth 2.75%, while the national rate of population growth is 1.17% (BAPPENAS 2015). BPS (2016) estimated that the number of urban population in 2015 is 53.3% which will grow to 60,0% in 2025, and achieve 66.6% in 2035. This high rate of urbanization shows the critical condition of urban water service in Indonesia. Today, the level of water supply service by public water supply enterprise (PDAM), for example, is still under 50% of the total urban population. It means that this urbanization rate become a serious threat to urban water security. Up to now, the understanding and involvement of the decision makers to the issue of water security and especially urban water security is still very limited. Tjandraatmaja et al. (2013) developed a framework for water security assessment and capacity building and used the framework in assessing urban water security and climate change adaptation in Makassar, Indonesia. But so far, the developed framework has not been implemented nationally to measure the level of urban water security in other Indonesian cities.

Table 6.1 presented the number of regencies and cities in Indonesia. Cities are the area with a densely populated with main economic activities are public service, business, and industries. While regencies are the area in which their main economic activities are agriculture, including husbandry and horticulture. It can be said that the cities and the capital of regencies are urban areas. This table indicates that Java

Table 6.1 Number of regencies and cities in Indonesia

No.	Island (s)	# of Regencies	# of Cities	Total
1	Sumatera	120	34	154
2	Java	85	34	119
3	Kalimantan	47	9	56
4	Nusa Tenggara	37	5	42
5	Sulawesi	70	11	81
6	Maluku	17	4	21
7	Papua	40	2	42
Grand total		416	99	515

Island is the most urbanized area, followed by Sumatera, Sulawesi, and Kalimantan. Among those cities, 11 cities have population more than 1 million, 15 cities have population between 500,000 to 1,000,000, 20 cities have population between 200,000 to 500,000, 32 cities have population between 100,000 to 200,000, and 21 cities have population less than 100,000. It means that urban water problems in Indonesia varies from the metropolitan cities, big cities, to small cities with their own characteristics of geographical condition and water resources potential, social and economic activities, that will influence their water demand and water related problems.

This study is intended to attract urban water managers and decision makers in considering the important of using urban water security as a measure in managing urban water system. Implementing a framework of urban water security will help urban water managers in developing strategies for improving urban water service based on its own characteristics, water resources potential, and urban water problems.

6.2   Material and Methods

This paper is a result of literature study on urban water related problems in Indonesia. A simple analysis was done to the available secondary data related to water supply service in the eleven largest cities in Indonesia, which are assumed representing the general water supply condition in other cities. Analysis was also done from the available data of sewerage system constructed in Indonesia during the last 30 years. Comparing the results of the analysis with personal daily observation surface water pollution, solid waste management, and other information on groundwater degradation in some urban areas, resulting a comprehensive explanation on the complexity of urban water problems in Indonesian cities, in which the threats to urban water security are identified. Finally, a comprehensive program in implementing integrated urban water management in urban areas developed to improve urban water security in Indonesian cities.

6.3   Results and Discussion

Indonesian cities are facing complex urban water problems which are inter-related among the provision of water supply, water pollution, water related disasters, groundwater degradation, and poor solid waste management. Managing urban water system in Indonesian cities should be integrated with municipal solid waste management as well as groundwater management, due to its inter-related impacts to urban water security.

6.3.1   Water Supply Service

Table 6.2 presents the number of population and house connection from public water supply enterprise (PDAM). It is amazing that the number of populations in Bekasi is higher than Medan, while the number of populations in Tangerang and Depok are higher than that of Semarang, Palembang, and Makassar which are provincial capital. Those numbers demonstrate that the three neighbouring cities of Jakarta are growing very fast to support the existence of the Capital City of Jakarta with housing for the people who work in Jakarta, which form a huge metropolitan of the Greater Jakarta with total population almost 17 million, if we also include Tangerang Selatan. Administratively, each city has their own autonomy, but in practice their urban infrastructure system, included urban water system, will interact to each other.

Assuming that each connection serves 5 people in a household, the coverage indicates the percentage of the population in the city that enjoy water supply service from PDAM. This table shows that the largest coverage is own by Medan (94.6%), which is followed by Surabaya (73.6%). The smallest coverage of water supply service by PDAM is located in Tangerang (4.8%), which is followed by Depok (13.4%), and Bekasi (29.4%), whereas the three cities is part of the Jakarta Metropolitan with very rapid housing development to support Jakarta. In average, the level of water supply service in the eleven largest cities are 43%. It can be assumed that the level of service in other smaller cities is lower than 40%.

The reason of this still low level of service is the limited capacity of the public water supply enterprise to response the increasing demand from the high rate of population growth and fast economic development of the cities. Around 50% of the urban population in Indonesia depends on shallow groundwater, because of its easy access using dug wells, boreholes, hand pumps, or electric pumps. Social and business activities such as schools, offices, hospitals, malls, and hotels usually use deep groundwater for their water supply sources. New developed housings and apartments

Table 6.2 Public water supply service in the eleven largest cities in Indonesia

No	City	Population	# of PDAM Connection	% of Coverage*)
1	Jakarta	9,588,198	794,930	41.5
2	Surabaya	2,765,487	407,225	73.6
3	Bandung	2,394,873	146,247	30.5
4	Bekasi	2,334,871	137,474	29.4
5	Medan	2,097,610	397,065	94.6
6	Tangerang	1,798,601	17,243	4.8
7	Depok	1,738,570	46,716	13.4
8	Semarang	1,555,984	129,933	41.8
9	Palembang	1,455,284	142,651	49.0
10	Makassar	1,338,663	130,496	48.7
11	Tangerang Selatan	1,290,322	–	–

usually should find their own water supply service. In slum areas people usually get their water service from public hydrants, while in the area with high water scarcity like in North Jakarta, people should spend more money to water from the vendors. In peri-urban areas, sometime people use water springs, develop a distribution system, and manage it as community-based water supply management.

6.3.2   Waste Water Management System

In general, waste water management in Indonesian cities is still depended on the on-site system. Most households and public buildings rely on individual septic tanks. In some new simple housing area, they developed communal septic tanks, while in some real estate the developer installed sewerage system and advanced waste water treatment plant.

Conventional sewerage systems have been developed in some Indonesian cities since 1980s. However, only parts of the capacity are utilized, due to technical and financial problems. Table 6.3 shows the sewerage systems in Indonesia in the year 2012. In general, from the total capacity of 244,618 m³/day, it is only 114,847 m³/day or 46.9% has been used. With a total house connection of 170,178 the average waste water production for each connection is 675 L per day. The largest capacity of the sewerage system and wastewater treatment plant is located in Bandung. However,

Table 6.3   Sewerage systems in Indonesia in 2012

City	System	Total capacity (m³/day)	Used capacity (m³/day)	% Used capacity*	House connection
Medan	UASB	10,000	5,650	56.5	12,370
Prapat	Aerated Lagoon	2,000	115	5.7	253
DKI Jakarta	Aerated Lagoon	38,880	704	1.8	1,407
Bandung	Stabilization Pond	80,835	49,769	61.6	99,538
Cirebon	Stabilization Pond	20,547	9,667	47.0	13,165
Yogyakarta	Aerated Lagoon	15,500	7,314	47.1	11,000
Surakarta	Aer–Fac-Biofilter	9,504	6,325	66.6	11,978
Bali	Aerated Lagoon	51,000	31,185	61.1	8,647
Banjarmasin	RBC	10,000	2,568	25.7	8,968
Balikpapan	Extend. Aeration	800	800	100.0	1,452
Tangerang	Oxidation Ditch	2,700	600	22.2	1,200
Batam	Oxidation Ditch	2,852	150	5.2	300
Total*		244,618	114,847	46.9	170,178

Source The World Bank and Australian Aid (2013)
*Calculated by the author

among the smallest system, the best implementation of the sewerage system is in Balikpapan, where 100% of the capacity has been used. In contrast, only 1.8% of the available capacity in Jakarta is utilized.

Poor sanitation system and very limited capacity of the sewerage system increases water pollution in streams and rivers across the cities. Poor sanitation system in slum areas who depends on individual septic tanks which are not well constructed, results groundwater pollution, whereas the same groundwater is also used as water supply sources. Improving sanitation condition, Indonesian cities should develop innovative approach in sewerage and sanitation system. At least, new developed housing should implement a communal sewerage or small bored sewerage system combined with appropriate technology of wastewater treatment plant such as constructed wetlands or stabilization ponds.

6.3.3 Groundwater Degradation

Over abstraction of groundwater pumping, especially for industries and business activities resulted groundwater degradation in some urban areas. In coastal area of Semarang City, groundwater depletion followed by land subsidence has resulted negative impacts to drainage system which increasing the risks to sea water flooding. The problems of groundwater degradation are also experienced by the City of Jakarta, Bandung, Medan, Surabaya, etc. In Yogyakarta City, conflict heated up between local people and hotels investor due to the over-use of groundwater.

6.3.4 Flooding and Inundation

Most Indonesian cities are experiencing flooding and inundation. The City of Jakarta, Bandung, Semarang, Surabaya, and others are flooded every year. A small city of Garut in West Java Province experienced flash flood recently, in which around 26 people died, 19 people lost, and inundating public hospital and schools. Similar event was also experienced by the City of Manado several years ago.

6.3.5 Urban Water Problems in Indonesia

Figure 6.1 shows the complexity of urban water problems in Indonesian cities. The high rate of population growth in urban areas due to urbanization causes increasing water demand and land coverage. Increasing water consumption will increase wastewater production which in turn will cause pollution to the water body, due to the lack of wastewater management system. Increasing land coverage will increase direct runoff which results the problems of flooding and inundation.

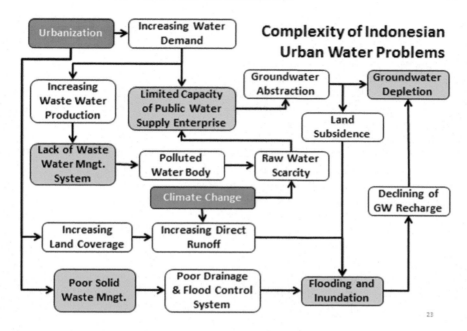

Fig. 6.1 Complexity of urban water problems in Indonesian cities

Climate change occurs in the form of increasing rainfall intensity which will greatly influence the increasing direct runoff and the shift of the season which will increase the scarcity of the water. Increasing urban population also influence solid waste management due to limited facilities, difficulties in finding final disposal site. Consequently, accumulation of garbage will accelerate drainage channel sedimentation, and dramatically reduce the capacity of the channel.

Pollution of the water body becomes a main constrain for the public water supply (PDAM) in increasing their capacity to response the increasing demand in the cities. Limited capacity of the PDAM causes most households and industries in Indonesian cities depend on the groundwater in fulfilling their water demand. Households usually use shallow groundwater for their domestic water demand, while institutions, business, and industries use deep water as a source of water supply.

Over abstraction of the groundwater, especially the deep groundwater, has resulted the decline in groundwater table, and even land subsidence, which disturbing urban drainage system and aggravate flooding and inundation problems. Beside the even worse flooding problems, saltwater intrusion also occurs in some coastal cities. Until now, all effort in developing artificial groundwater recharge has not been successful due to unbalance between the abstraction and the recharge, and might be also the decline of the aquifer capacity due to land subsidence.

6.3.6 The Threats to Urban Water Security

The concept of water security has been introduced by Global Water Partnership since 2000, who defined "Water security, at any level from the household to the global, means that every person has access to enough safe water at affordable cost to lead a clean, healthy, and productive life, while ensuring that the natural environment is protected and enhanced." This definition underlined the fulfillment of the basic demand of water supply for clean, healthy, and productive daily life and the sustainability of the environment. UN-Water Task Force on Water Security (2013) provides more comprehensive definition on water security as follows: "Water security is defined as the capacity of a population to safeguard sustainable access to adequate quantities of acceptable quality water for sustaining livelihoods, human well-being, and socio-economic development, for ensuring protection against water-borne pollution and water-related disasters, and for preserving ecosystems in a climate of peace and political stability." This recent definition covers not only the affordability of water supply as the basic demand, but also the role of water in socio-economic development, minimizing water-borne pollution, controlling water related disasters, preserving ecosystems, and maintaining peace and political stability.

Improving urban water security is an important part in achieving national water security. Strategic program in improving urban water security should include the following:

1. Maintaining sustainability of water supply service for households and supporting social and economic development;
2. Ensuring protection of the environment against water borne pollution;
3. Ensuring protection against water related disasters;
4. Increasing flexibility and adaptability to climate change;
5. Maintaining political stability at the city and regional level.

Asian Development Bank (2013) identified that the key dimensions of the National Water Security consist of five components, i.e. (1) household water security, (2) urban water security, (3) environmental water security, (4) resilience to water security disasters, and (5) economic water security. Household water security covers the fulfillment of water supply and sanitation for every households in the country, either they live in urban or in rural areas. Urban water security figure out the condition of water related problems in urban areas, which include the fulfillment of water supply for domestic, municipality, and industry, the availability of effective sanitation system preventing water pollution, and controlling water related disasters.

The threats to urban water security of Indonesian cities can be classified into three categories, i.e. (1) population growth, (2) climate change, and (3) water conflict. In general, the level of population growth in urban areas are higher than the average population growth of the country. Population growth increases water demand, land use changes, and water pollution. Land use changes increases direct runoff which increases the risk to water related disasters. Water pollution becomes a hindrance to the fulfillment of the increasing water demand. Usually people used groundwater to fulfill the demand, but over abstraction of the groundwater in the past and the declining groundwater recharge due to increasing land coverage has increased water scarcity.

Climate change phenomenon is increasingly understood, not only by experts in water resources and climatology, but also by lay people. Increasing rainfall intensity and duration, the shift of rainy and dry season, and frequent whirlwind indicates changes in climate. As a result, the risk related to water disasters increase in the form of flooding, land slide, and flashfloods. The shift of the rainy and dry season in turn will disturb water availability. Water related disasters will result a great impact to social and economic development.

Conflict of interest in water use due to increasing water demand in urban areas as a result of population growth, increasing quality of life, economic development, and industrialization needs availability of raw water. Polluted water body as a result of untreated wastewater from households, business, and industries becomes a constraint for cities in fulfilling their raw water demand. While the available groundwater resources also continuously decline. Consequently, they should find raw water sources from springs or other sources in remote rural areas which had been used for agriculture irrigation. As a result, conflict of interest in utilizing the only water sources, in turn will grow as a conflict between two water users, i.e. urban water supply enterprises and local farmers.

For Indonesian cities, increasing urban water security should be done in the following.

1. **Ensuring the availability of water supply and sanitation system** for all social and economic activities in the city, that is environmentally sustainable. To fulfill increasing water demand, new water sources should be found. The utilization of groundwater should be related to surface water. Water balance should be maintained, in terms of quantity as well as quality of water, both for groundwater and surface water. The implementation of appropriate technology in sanitation system is needed in improving water resources, both groundwater and surface water in urban areas.

2. **Mitigating water-related disaster.** Continuously growing urbanization in Indonesian cities, are increasing potential of water related disaster such as floods, droughts, landslide, and pollution. Inundation in some streets results traffic jams which in turn will disturb social and economic activities. Over abstraction of groundwater results water shortage, especially during the dry seasons. Polluted water becomes a hindrance to the effort in fulfilling water demand due to the scarcity of raw water resources. Mitigating those water related disasters, municipal government should control urban development.

3. **Maintaining good collaboration with neighboring cities and regencies** in water resources conservation, utilization, and controlling water related hazard. Water resources management in Indonesia is based on river basin territory as a management unit. Increasing water demand for agriculture, municipalities, and industries could create potential conflicts between cities and regencies in a river basin. In addition, increasing direct runoff due to urbanization could be

Table 6.4 Implementing integrated urban water management in Indonesian cities

	Social	Environmental	Economical
Water resources conservation	• Improving blue-green environment and more open space	• Improving water quality • Controlling groundwater abstraction	• Implementing water demand management
Water resources utilization	• Fulfilling basic water demand for all people • Developing public hydrants and sanitation facilities	• Maintaining environmental flow • Supporting community-based wastewater treatment	• Fulfilling water demand for industries and commerce • Replacing individual by public water supply system
Control of water related disaster	• Reducing number of people at risk to water related disasters • Restructuring of housing for people living in flood plain areas	• Preserving natural flood plain and wetlands • Increasing ground water recharge	• Implementing sustainable urban drainage system (SUDS) for storm water management

a potential flooding for other downstream areas. Neighboring cities and regencies, therefore, should maintain good collaboration in sharing resources and improving water and environment quality.

6.3.7 Integrated Urban Water Management

In the developed countries like the USA, the concept of integrated urban water management is understood as the integration of the three main functions of urban water system which includes water supply service, waste water management, and storm water management. Based on the above discussion, however, it is understood that the scope of integrated urban water management in Indonesia is not only covers water supply service, wastewater management, and storm water management, but also included ground water management as well as solid waste management. In addition, involvement of the stakeholders is very important in the efforts to improve quality of the urban water, because the success of those efforts always need a strong support from the stakeholders. Controlling groundwater abstraction, increasing groundwater recharge, implementing new sanitation or sewerage system, implementing the concept of reduce, reuse, recycle, and recovery (4R) in municipal solid waste management are urban water related programs that need large support from the government institutions, communities, industries, business, and other non-government organizations.

In Indonesia, it is understood that water resources have social function, environmental function, and economical function that should be managed in balance. It is also understood that water resources management should cover three main activities, i.e.

Water Resources Conservation, Water Resources Utilization, and Control of Water Related Hazards. Table 6.4 demonstrates a comprehensive program to implement integrated urban water management in Indonesian cities.

6.4 Conclusions

From the above discussion, it can be concluded that.

1. Indonesian cities are facing complex urban water problems which are inter-related among water supply, water pollution, water related disasters, groundwater degradation, and poor solid waste management.
2. The threats to urban water security of Indonesian cities come from high rate of population growth, climate change phenomena, and potential water conflict between agriculture and urban water demand.
3. Improving urban water security, Indonesian cities should implement the concept of integrated urban water management which integrate water supply service, wastewater management, storm water management, groundwater management, and municipal solid waste management into a municipal public utility office.
4. Urban water management has a mutual influence with water resources management in the river basin level. Consequently, coordination and cooperation among cities and regencies should be developed in river basin level.

Acknowledgements This paper is written based on the materials that had been presented in the Joint Seminar between Universitas Katolik Parahyangan and Hohai University (China) in Bandung on 18 July 2016.

References

Asian Development Bank (2013) Asian Water Development Outlook – Measuring Water Security in Asia and the Pacific, Asian Development Bank, Manila
BAPPENAS (2015) Rencana Pembangunan Jangka Menengah Nasional (RPJMN) 2015–2019
BPS (2012) Persentase Penduduk Daerah Perkotaan Menurut Provinsi, 2010 - 2035
GWP (2013) Integrated Urban Water Management (IUWM): Toward Diversification and Sustainability, Policy Brief, Stockholm, Sweden
Tjandraatmadja G, Kirono DGC, Neumann L, Larson S, Stone-Jovicich S, Barkey RA, Amran A, Selintung M (2013) Assessing urban water security and climate change adaptation in Makassar, Indonesia, International Congress on Modelling and Simulation, Adelaide, Australia, 1–6 December 2013
The World Bank and Australian Aid (2013) East Asia Pacific Urban Sanitation Review, Indonesia Country Report, September 2013
UN-Water Task Force on Water Security (2013) Water Security & the Global Water Agenda: A UN-Water Analytical Brief, United Nations University

Chapter 7
River Basin Planning for Water Security in Sri Lanka

U. S. Imbulana

Abstract In recent times, many actions were initiated in Sri Lanka to develop water resources and manage them in a sustainable manner. While these actions ranging from development projects to policy initiatives have addressed the most urgent issues, new challenges during the last few decades have made it necessary to critically examine future strategies. An analysis of recent water resources development projects in Sri Lanka shows that project formulation is increasingly adopting solutions within a river basin approach, specifically responding to climate change. However, there is a need to address sectoral issues in an integrated manner, supported by an enabling policy framework. While climate change is acknowledged as a threat to water security, water resources development planning and hydrological designs have to incorporate scientific predictions of future climatic conditions. River basin-based natural resources management, water quality management, community-based water resources management and building on traditional and local knowledge are among the subject areas where there is a potential to improve. The intricate dependencies and relationships among different water-uses such as irrigation and drinking water are important factors to be considered in addressing the current issues. The proposed approach focuses on filling the gaps and removing the barriers to water security and developing comprehensive river basin management plans.

Keywords IWRM · Water security · Barriers · Sri Lanka

7.1 Introduction

7.1.1 Background

Sri Lanka, having a total area of about 65,600 km^2, is located in close proximity to the equator. The estimated population of Sri Lanka was 21 million in 2015 (DCS 2016). The country experiences a tropical monsoon climate with a mean annual rainfall is 1,860 mm. The spatial distribution of rainfall defines the three climatic zones;

U. S. Imbulana (✉)
Climate Resilient Integrated Water Management Project, Colombo, Sri Lanka

© Springer Nature Switzerland AG 2021
M. Babel et al. (eds.), *Water Security in Asia*, Springer Water,
https://doi.org/10.1007/978-3-319-54612-4_7

Wet, Intermediate, and Dry. Dry Zone, which receives a rainfall less than 1,750 mm, extends over about 70% of the area in the northern, eastern and south-eastern parts.

There are 103 distinct river basins, out of which 20 rivers in Wet Zone carry nearly 50% of surface runoff (Arumugam 1969). Storage of water is enabled by about 250 large-scale reservoirs (T.J. Meegastenna, Personal Communication, December 10, 2015) managed by the Government. Most of them serve single sectors such as irrigation, hydropower and drinking water, though a few reservoirs are multi-purpose. There are also about 13,000 small reservoirs (DAD 2011) called "Village Tanks" that were traditionally managed by farmers, which now receive some degree of support from the Government. These reservoirs are arranged along the paths of small streams and form "cascades", which enable spill and drainage water from one reservoir to be utilized in the downstream reservoirs. They serve multiple needs of the community including domestic water supply, livestock needs, and irrigation requirements.

After the collapse of the famous "hydraulic civilization" of Sri Lanka in the thirteenth century, activities to resurrect irrigation infrastructure started during the latter part of the British colonial period. This gathered momentum after independence in 1948 (Imbulana and Neupane 2005). A Master Plan to harness the waters of the Mahaweli, Sri Lanka's largest river, was developed in the late 1960s. The "Accelerated Mahaweli Development Project" (AMDP) implemented from the late 1970s to mid-1980s made substantial positive impacts on agriculture and hydropower generation.

Displacement of people and adverse impacts on the environment due to deforestation influenced a period of emphasis on natural resources management and experimentation with institutional reforms that commenced towards the end of the AMDP (Imbulana and Neupane 2005). However, growing water demands necessitated another wave of water infrastructure development, and the Government implemented several medium and large-scale reservoir projects and inter-basin water diversions from the 1990s.

These interventions, which respond to public demands, show that water security is vital for the economy of the country and the livelihoods and social well-being of the individuals. Throughout history, investments have been made to ensure water security. However, the increased awareness of the importance of sustainable development and the environmental policies and regulations introduced since the 1980s demand novel strategies in water resources development now.

Impacts of climate change are becoming more pronounced in Sri Lanka in recent times. Consequences of climate change include increased frequency of flood and drought incidence, rapidly alternating in the same geographical locations. Such climate-induced hazards have resulted in the loss of livelihoods and decreased food security especially in the Dry Zone (UNDP, GEF/SGP 2016).

National policies framed for environmental conservation, drinking water, disaster management and climate change provide guidance for some aspects of water resources development. However, Sri Lanka's attempts to frame a comprehensive policy for the water sector during the last two decades were not successful due to socio-political reasons.

7.1.2   Rationale and Objectives of the Paper

Since the beginning of the new millennium, Sri Lanka appears to be trying to balance water resources development with improving water management using better technology and institutional arrangements (Imbulana and Neupane 2005) and environmental conservation. While an optimum balance of development and conservation is required for mutual sustainability, the constraints on water resources development and management due to technological, institutional and policy barriers need to be removed to ensure improved water security. Accordingly, this paper analyses the emerging water issues in Sri Lanka, barriers to addressing such issues and achieving water security, and explores the means to address the barriers.

7.2   Material and Methods

The analysis includes an assessment of emerging water issues and the responses of recent large-scale and multi-sector water resources development projects designed to address such issues. It identifies the trends of changing the focus of the new projects and concludes with an assessment of the opportunities for improving water security.

7.2.1   Emerging Water Issues

Climate Change: Recent studies indicate that temperatures in Sri Lanka show an increasing trend. Long-term trends of heavy rainfall events appear to be not statistically significant. However, an increasing trend of extreme positive rainfall anomalies during the South-West and North-East Monsoon seasons during the period 2010–2014 has been observed (Abeysekara et al. 2015). The period post-2010 coincides with a heavy incidence of breaching of village irrigation systems. Official records of the Department of Agrarian Development (DAD 2015) show that about 1,950 village irrigation reservoirs and diversions have been breached due to floods during 2012 and 2014.

The Sri Lanka Water Development Report suggested that Dry and Wet Zone boundaries may have shifted due to the changes in rainfall contours. The analysis compared rainfall isohyets of 1911–1940 with those of 1961–1990, which showed that some areas in the North-Western part of the country are becoming drier. Although the Report failed to highlight the importance of this shift from a water resources development point of view, the Government recently responded to growing water scarcity in the North Western Province, through Mahaweli water diversions planned under the proposed North Central Province Canal Project (NCPCP). Figure 7.1 shows the location of Mi Oya and Deduru Oya river basins in the North-Western Province,

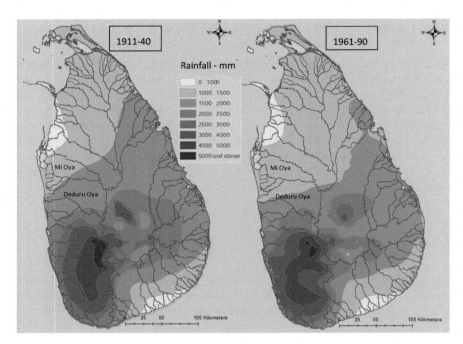

Fig. 7.1 Mi Oya and Deduru Oya river basins relative to rainfall isohyets of 1911–40 and 1961–90

which are to be benefitted by the NCPCP, relative to the rainfall isohyets of 1911–40 and 1961–90 periods (Imbulana et al. 2010).

Water-Related Disasters and Their Management: In recent years Sri Lanka, particularly the Dry Zone, has experienced a cycle of alternating droughts and floods. The drought of 2012 was one of the worst in past years, followed by floods in early 2013, droughts in early 2014 and again floods during late 2014 to early 2015 (DMC 2016, Nandy 2014). This cycle of water-related disasters resulted in large-scale food insecurity, drinking water shortages and the destruction of irrigation facilities.

Studies show that rainfall intensity in Sri Lanka, particularly in the north-central part of Dry Zone has increased (Ratnayake and Herath 2005) and increased rainfall intensity is a factor contributing to flash floods (Chen and Costa 2017). Recent experiences show that the Dry Zone is inadequately prepared in terms of institutional and technological capacity to deal with flash floods, which are observed to be increasing in frequency and magnitude. Even though Dry Zone villages have traditionally coped with drought and water scarcity by building small reservoirs and conserving forest cover, increasing pollution of water bodies and encroachment of forests have decreased their coping capacity. Moreover, the rapidly alternating occurrence of floods and droughts have adversely impacted the local economy and water infrastructure.

The Wet Zone is relatively better-prepared for floods. There is a higher concentration of rain and stream gauges in rivers which are traditionally designated as flood-prone. However, changes in monsoonal rain patterns and rainfall intensities, and some non-hydrological parameters have adversely affected the coping capacities in the Wet Zone. Observations show that inadequate urban drainage has contributed to the floods in the Western and Southern areas of Wet Zone in addition to rivers overflowing the river banks. Recent floods (2016) in the Western part of the country, particularly in Colombo and suburban areas are an example. Compared to a similar flood in 1989, the river water level was lower in 2016. But the number of affected people was much higher in 2016 (DMC 2016). While land use change is considered as a cause for this event, regular monitoring of such change is inadequate to use it as a parameter in disaster predictions.

Environmental Degradation and Pollution of Water Bodies: Pollution of water bodies is a growing concern in Sri Lanka. Inadequate waste disposal facilities and discharge of industrial waste and domestic waste to water bodies are sources of water pollution in urban areas (Imbulana et al. 2010). Excessive use of fertilizer, pesticides and other agrochemicals contribute to the pollution of rural water bodies. Saltwater intrusion, for which over-exploitation of groundwater has contributed (Panabokke and Perera 2005), is a concern in the coastal areas. Leaching of salts from agriculture to the groundwater table is identified as a cause of water pollution in northwestern aquifers (Kuruppuarachchi and Fernando 1999).

The recent emergence of health issues related to water has increased the attention of water quality in rural areas. One of the severe health issues in the Dry Zone is the increasing incidence of Chronic Kidney Disease of unknown aetiology (CKDu). This health issue affects the resilience to climate change impacts by decreasing the farmer's ability to engage in income-generating activities and increasing health expenditure. While the exact cause has not been identified, among the many suspected causes are heat stress-related dehydration (Tawatsupa et al. 2012), decreasing quality and availability of groundwater, and unsafe agricultural practices (Gunatilake et al. 2014) such as over-use and exposure to pesticides and fertilizers.

River basin development plans and studies point out that deterioration of the upper watersheds is a serious sustainability issue. Deforestation and improper land use have resulted in soil erosion, resulting in the silting-up of the reservoirs (SMEC 2010) and reducing the productivity of the land.

7.2.2 An Analysis of Selected Recent Water Resources Development Projects

Until the end of the twentieth century, a characteristic of water resources development projects was their sector-orientation. The main focus was on agriculture and hydropower development while flood management and drinking water provision were considered as secondary or indirect benefits. Four water resources development projects, which were carried out or planned since 2000, were selected as case studies and analyzed here to identify recent trends in water resources development and management.

Dam Safety and Water Resources Planning Project (DSWRPP): This project, funded by the World Bank, was implemented since 2008, after about three years of planning. The project development objectives are to establish long-term sustainable arrangements for the operation and maintenance of large dams and to improve water resources planning. Main project components include (i) improving the safety of the dams and their operational efficiency; (ii) upgrading and modernizing the existing hydro-meteorological information system; and (iii) preparing a national water use plan and integrated water resources plans for two selected river basins -Mahaweli and Mundeni Aru (The World Bank n.d.) (Fig. 7.2). These three project components were implemented in parallel. The improvement of dam safety and upgrading of hydro-meteorological information systems were responding to urgent needs at the time of project formulation and subsequently during implementation, rather than to a comprehensive river basin development plan.

Increased intensity and occurrence of extreme events induced by climate change emphasize the need for a technologically superior hydro-meteorological information system compared to the existing one and training for real-time analysis of flood situations. Increased intensity and frequency of extreme climatic events also increase the vulnerability of weak dams. Therefore, strengthening the dams would be vital to safeguard life and property. Although specific references to addressing climate change are not made in the project documents, the pressures that necessitated the interventions were later identified as results of climate change.

Climate Resilience Improvement Project: This project was initiated in 2014 and is planned to be completed in 2019. The development objective is to reduce the vulnerability of exposed people and assets to climate risk. The project aims at improving resilience to water and climate-related risk of flood, drought, and landslides (World Bank 2014). Accordingly, the interventions are expected to reduce the vulnerability of exposed people and assets to climate risk and to improve the Government's capacity to respond effectively to disasters (T.J. Meegastenne, Personal Communication, September 10, 2016).

The project has two major development components. The first component will produce river-basin investment plans in 10 selected basins including Mahaweli, Malwathu Oya, Maha Oya, Deduru Oya, Kala Oya, Kelani Ganga, Attanagalla Oya, Gin Ganga, Nilwala Ganga and Gal Oya (Fig. 7.2) using comprehensive flood and drought modeling. These models will take into account climate risks such as expected extremes of water scarcity and excess, and will produce reservoir operation rules and guidelines and risk communication rules. The second component deals with increasing the climate resilience of infrastructure by rehabilitating those damaged by recent floods or those at risk from future floods.

North Central Province (NCP) Canal Project: This Project, which is to be implemented from 2016–2030, will transfer water from Mahaweli River Basin to North

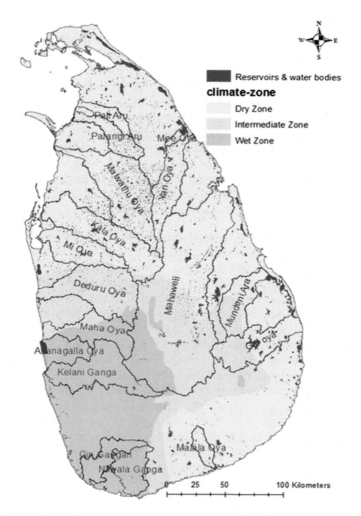

Fig. 7.2 Climatic Zones, distribution of reservoirs and river basins relevant to the case studies

Central, Northern, and North Western Province, to relieve water scarcity and improve productivity in those areas. The concept follows from the Mahaweli Master Plan of the 1960s but incorporates the new water demand areas in the NWP described in the preceding sections. The project, at completion, will comprise about 300 km of new and improved water transfer routes and would cost about US$1.64 billion (ADB 2014 and ADB 2016).

Phase 1 of the Project is currently named as Mahaweli Water Security Investment Program. It plans to divert Mahaweli water to selected irrigation systems in the North Western Province (NWP) and a part of the target area in the NCP, and will be funded by the Asian Development Bank (ADB) as the main financier. In the second phase, additional areas in the NCP and NP will be supplied with water. Water transfers envisaged under the project will benefit six river basins, namely Mi Oya, a tributary of Deduru Oya, Parangi Aru, Pali Aru, Kanakarayan Aru and Ma Oya (Fig. 7.2), in addition to the river basins currently serviced by the Mahaweli river flow (MCB 2014). It will augment the water storage in several large reservoirs and about 1,500 village irrigation reservoirs. The latter will be rehabilitated and farmers' capacity to manage them will be improved. Although drinking water supply schemes are not incorporated into the project scope, a definite quantity of water will be allocated for ongoing and planned water supply schemes. It is further planned to prepare strategies for improving the efficiencies of irrigation systems and their water productivity, which is expected to contribute to reducing the vulnerability to climate change. Another component will formulate recommendations and guidelines to institutionalize Integrated Water Resources Management (IWRM).

Climate Resilient Integrated Water Management Project (CRIWMP): This project was formulated based on the results of a feasibility study conducted with the technical support of the United Nations Development Programme (UNDP). The study aimed at finding the gaps in current approaches to water resources development, barriers to close the gaps and measures to remove the barriers. In the process, the study found that the river basin approach is the key to ensuring water security. It was noted that village irrigation systems were traditionally managed in an integrated manner, together with the immediate watershed home gardens and associated forests. However, most river basin-based development projects do not sufficiently deal with micro-level watershed management. As a result of the degradation of micro-watersheds, small reservoirs get silted up, evaporation increases and their capacity to mitigate both floods and droughts is reduced. Inadequate early warning of weather conditions and dissemination of climatic conditions affect the resilience of farmers. In addition, poor agricultural practices contribute to the pollution of these water bodies, which serve as a source for drinking water, both directly and through the recharge of groundwater aquifers. Summing up, the study notes that the deterioration of irrigation systems, degradation of watersheds, poor water quality, inefficient and inadequate weather and climate information are inter-linked, and when finding solutions, this connectivity within village tanks cascade systems and river basins should be given due consideration. Accordingly, the study deviates from the conventional approach of considering irrigation systems and drinking water supply schemes as mere "demand nodes" in a water balance equation.

The objective of the project, which is implemented with the funds from the Green Climate Fund (GCF) from 2017, is to strengthen the resilience of smallholder farmers, particularly women, in the Dry Zone through improved water management to enhance lives and livelihoods. The objective will be achieved through (i) upgrading village irrigation systems including their watersheds and scaling up climate-smart agricultural practices; (ii) enhancing safe drinking water supply and management; and (iii) strengthening early warning, forecasting, and water management systems to enhance the adaptive capacity of smallholder farmers to droughts and floods (GCF 2016). The project aims to revive the traditional methods of water management, which is feasible within the current socio-economic environment. Malwathu Oya, Yan Oya, and Mi Oya river basins have been selected for the interventions (Fig. 7.2).

7.3   Results and Discussion

7.3.1   Trends in Water Resources Development

A comparative assessment of the four development projects discussed above is presented in Table 7.1. It lists projects in the chronological order of implementation. There are certain noteworthy differences in the approaches such as the emphasis on interventions varying from infrastructure solutions to management solutions. However, the analysis indicates the following trends:

a.   An increasing acceptance of the river basin approach to water resources development and management
b.   Increasing recognition of climate change as a problem to be solved with specific targeting
c.   Increasing adoption of integrated solutions to address water-related issues

Table 7.1   A comparison of important features of recent water resources development projects

Project/commenced year	Positive features to be included in river basin planning
DSWRPP/2008	Assessment of the structural safety of major dams, hydrological modeling at river basin level and river basin plans (in selected basins), increasing the coverage of hydro-meteorological information systems
CRIP/2014	Flood and drought modeling in selected river basins with due consideration to climate risks, improving climate resilience, river basin development plans, improving risk communication management
NCPCP/2016	River basin based hydrological assessment including modeling, trans-basin diversion of water, cascade-based irrigated agriculture development, water allocation for drinking, improvement of irrigation efficiency and institutional arrangements for IWRM
CRIWMP/2017	Reviving IWRM at micro-level with Village Tank Cascade as the development unit, climate-smart agriculture, drinking water supply, irrigation water supply, improving hydro-meteorological information accessibility at the village level, community management of outputs generated by the project

7.3.2 Technological, Policy and Institutional Barriers Against Ensuring Water Security

Technological Barriers: The foregoing discussion points to heavy investments to enhance climate resilience, especially in the Dry Zone. Reviews have indicated that until recently, scientific knowledge about the future climate in Sri Lanka has not been conclusive, and there is no general consensus in the studies with regard to the temporal and spatial dimensions of climate change impacts (Imbulana et al. 2010; Eriaygama et al. 2010). There is a lack of comprehensive understanding of climate and disaster risk (World Bank 2014). In the absence of such knowledge, many of the interventions take the form of "no-regret solutions" such as adding water storages, adopting larger return periods for design floods, inter-basin water transfers, creating awareness, and improving preparedness etc. It is expected that ongoing activities including flood and drought modeling by CRIP and the research activities by The Regional Integrated Multi-Hazard Early Warning System for Africa and Asia (RIMES n.d.), of which Sri Lanka is a member country, could cover this knowledge gap in the future.

When the knowledge gaps on climate change are covered, they need to be incorporated into water resources development strategies. While the description of three climate zones (Wet, Intermediate and Dry) is useful for agricultural and natural resources development strategies, Manchanayake and Maddumabandara (1999) point out that it is rational to define the Dry Zone as the area where the average rainfall is less than the average annual potential evapotranspiration, for irrigation development strategies. As can be seen from Fig. 7.2, the concentration of reservoirs in the Dry Zone is comparatively higher. Considering the example of NCPCP having to cater for new water demand areas in the NWP, it will be useful to construct current and future Dry and Wet Zone boundaries based on both evapotranspiration and rainfall, for sustainable water resources development strategies.

Increased reports of small reservoir breaching suggest changes to the hydrological regime as one of the causes. There are several examples of using climate change knowledge in engineering designs. Ahmed and Tsanis (2016) identified the impact of climate change on design storm depths and its implications for storm-sewer network design, in a watershed in Canada. Devkota and Gyawali (2015) studied the changes of design flood flows under different climatic scenarios and their impacts on design parameters of hydraulic structures in the Koshi river of Nepal. While there is strong anecdotal evidence that rainfall intensities have increased in Sri Lanka as well, such changes need to be incorporated into hydraulic design criteria to ensure the sustainability of investments in water resources development.

Technology should also facilitate cost-effective monitoring of land use changes. Recent floods in the populous western region demonstrated that such changes alter the relationships between floods and hydrological parameters.

Policy Barriers: In the face of increasing acceptance of integrated solutions for water issues, sub-sector policies such as those for drinking water also need to be integrated. Among the issues to be addressed by such policies include the comprehensiveness of river basin development and management plans, including enabling the connectivity among macro-level inter-basin water transfers and micro-level water transfers at village irrigation systems. The disconnect between the 'bottom-up' approach adopted by projects such as CRIWMP and largely 'top-down' approaches adopted by large-scale river basin development projects needs to be addressed by the policies, as well. Another factor is the trade-off between different water uses (such as irrigation and drinking water) in the Dry Zone necessitated by drivers such as water pollution and climate change. While it is known that Dry Zone villagers traditionally adopted an integrated natural resources management system with water as the central focus but included the forests, pollution control, land use management and coping with extreme events, there should be policy support for such principles to be adopted, within the limits set by the current socio-economic environment. Water security and the sustainability of water investments substantially depend on the policy support to address such issues.

Institutional Barriers: While the ongoing and planned water resources development projects will result in strengthened water infrastructure, improved early warning systems and a higher capacity in inter-basin water transfers, there is a vagueness concerning how these tools and facilities are utilized in an integrated manner within a river basin. Another factor with increasing importance is the need to arrest the deterioration of watersheds including micro-level (such as in the case of village irrigation systems) and at the river basin level upper watersheds. While the management of different reservoirs is divided among several institutions based on their size, sectors served, agricultural area commanded etc., integrated management within a river basin requires a high degree of cooperation among these institutions. This need will be highlighted in the future when village irrigation schemes are connected to larger irrigation systems through trans-basin diversions such as NCPCP. Furthermore, as development and conservation activities are handled by different agencies whose area of authority often do not cover an entire river basin, Institutional arrangement for incorporating the conservation with development remains vague.

7.4 Conclusions

Water security depends on the sustainability of water resources development and management interventions, for which climate change remains a major challenge. It is noted that the adoption of IWRM principles with the river basin as a development unit is gaining acceptance. However, the sustainability and the effectiveness of such interventions will depend on the removal of several technical, policy and institutional barriers. It will enable the incorporation of increasing knowledge about climate change into development strategies.

The analysis indicates that several opportunities to enhance water security exist. The promotion of partnerships among the academics, government sector, and private sector is an opportunity to find solutions that could differ according to land use, demographic, and hydrological patterns in the river basins. The emerging issues in

Sri Lankan river basins highlight the need for integrated management of all the natural resources of a river basin while including the community and traditional knowledge in the management process. Policies on climate change and environmental conservation will be more effective with the existence of corresponding coherent development policies. A comprehensive river basin management and development plan enable the integration of such diverse policies for a common objective. Such a plan should be a "living document", frequently updated to accommodate climatic, policy, physical and technological changes.

Acknowledgements Review of this paper by Eng. (Ms.) T.J. Meegastenne, Director of Irrigation Drainage and Flood Systems) and Deputy Project Director (CRIP) is thankfully acknowledged.

References

Abeysekara AB, Punyawardena BVR, Premalal KHMS (2015) Recent trends of extreme positive rainfall anomalies in the Dry zone of Sri Lanka. Ann Sri Lanka Dept Agric 17:1–4

Ahmed S, Tsanis J (2016) (2016) Hydrologic and hydraulic impact of climate change on lake Ontario tributary. Am J Water Resour 4(1):1–15

Arumugam S (1969) Water Resources of Ceylon - Its Utilization and Development. Water Resource Board Publication, Colombo

Asian Development Bank (ADB) (2016) Facility Administration Manual, Democratic Socialist Republic of Sri Lanka: Mahaweli Water Security Investment Program. Project Number: 47381-001 June 2016. ADB, Manila

Asian Development Bank (ADB) (2014) Technical Assistance Consultant's Report. Sri Lanka: Water Resources Development Investment Program. Project Number: 47381 December 2014. ADB, Manila

Chen G, De Costa G (2017) Climate change impacts on water resources, case of Sri Lanka. Environ Ecol Res 5(5):347–356. https://www.hrpub.org

Department of Agrarian Development (DAD) 2015 Official records of the DAD on damages to irrigation infrastructure. Department of Agrarian Development, Colombo

Department of Agrarian Development (DAD) 2011.Department of Agrarian Development (DAD) (2011) Agro-Ecological and Watershed Handbook. Department of Agrarian Development, Colombo

Department of Census and Statistics (DCS) (2016) Department of Census and Statistics (DCS), 2016. Statistical Abstract 2016. DCS, Colombo

Devkota LP, Gyawali DR (2015) Impacts of climate change on hydrological regime and water resources management of the Koshi River Basin, Nepal. J Hydrol Reg Stud 4:502–515

Disaster Management Center (DMC) (2016) Disaster Management Information System Sri Lanka. https://www.desinventar.lk:8081/DesInventar/statistics.jsp

Eriyagama N, Smakhtin V, Chandrapala L, Fernando K (2010) Impacts of climate change on water resources and agriculture in Sri Lanka: a review and preliminary vulnerability mapping. (IWMI Research Report 135), International Water Management Institute, Colombo, Sri Lanka

Green Climate Fund (GCF) (2016) Consideration of funding proposals – Addendum, Funding proposal package for FP016. GCF/B.13/16/Add.08. 8 June 2016. https://www.greenclimate.fund/documents/20182/226888/GCF_B.13_16_Add.08_Funding_proposal_package_for_FP0 16.pdf/e0ef564a-e3fb-459d-bf01–1503a9760a0e

Gunatilake SK, Samaratunga SS, Rubasinghe RT (2014) Chronic kidney disease (CKD) in Sri Lanka - current research evidence justification: a review. Sabaragamuwa Univ J 13(2):31–58

Imbulana US, Neupana B (2005) Water, rural poverty and development options in Sri Lanka. In: Proceedings of the 12th world water congress, New Delhi, India, November 2005

Imbulana KAUS, Wijesekara NTS, Neupane BR, Aheeyar MMM, Nanayakkara VK (eds) (2010) Sri Lanka Water Development Report. MAD&AS, UN-WWAP, UNESCO, HARTI, and University of Moratuwa, Sri Lanka, Paris and New Delhi

Kuruppuarachchi DSP, Fernando WARN (1999) Impact of agriculture on groundwater quality: leaching of fertilizer to groundwater in Kalpitiya peninsula. J Soil Sci Soc Sri Lanka 11:9–16

Manchanayake P, Madduma Bandara CM (1999) Water Resources of Sri Lanka. National Science Foundation, Colombo

Mahaweli Consultancy Bureau (MCB) (2014) Environmental Impact Assessment. December 2014. SRI: Water Resources Development Investment Program Upper Elahera Canal (UEC) Prepared by Mahaweli Consultancy Bureau (Pvt) Ltd. for the Asian Development Bank. MCB. www.adb.org/sites/default/files/project-document/153180/47381-001-eia-01.pdf

Nandy S (2014) Resident/Humanitarian Coordinator Report on the Use of Cerf Funds Sri Lanka, Rapid Response, Drought. https://docs.unocha.org/sites/dms/CERF/HCRCReports/Sri%20Lanka%20RCHC%20Report%2014-RR-LKA-001.pdf

Panabokke CR, Perera APGRL (2005) Groundwater resources. In: Wijesekera NTS, Imbulana KAUS, Neupane B (eds) Proceedings, Workshop on Sri Lanka National Water Development Report, World Water Assessment Programme, Paris, France

Ratnayake U, Herath G (2005) Changes in water cycle: effect on natural disasters and ecosystems. In: Wijesekera S, Imbulana KAUS, Neupane B (eds) Proceedings of the preparatory workshop on srilanka national water development report. World Water Assessment Programme, Paris

Regional Integrated Multi-Hazard Early Warning System for Africa and Asia (RIMES) (n.d.) About RIMES. https://www.rimes.int/about_overview.php

SMEC International Pty Ltd. (in association with DHI Water and Environment (Denmark), Ocyana Consultants, Sri Lanka and Project Management Associates, Sri Lanka) (2010) DSWRP Project Component III, Initial Assessment Report, Mahaweli Basin. SMEC, Colombo

Tawatsupa B, Lim L, Kjellstrom T, Seubsman S, Sleigh A, The Thai Cohort Study Team (2012) Association between occupational heat stress and kidney disease among 37 816 workers in the thai cohort study (TCS). J Epidemiol 22(3):251–260

The World Bank (2014) Sri Lanka - Improving Climate Resilience Project. Washington DC; World Bank Group. https://documents.worldbank.org/curated/en/256411468334291922/Sri-Lanka-Improving-Climate-Resilience-Project

The World Bank (n.d.) LK Dam Safety and Water Resources Planning (DSWRPP). https://www.worldbank.org/projects/P093132/dam-safety-water-resources-planning-dswrpp?lang=en

UNDP Global Environmental Facility/Small Grants Programme (UNDP, GEF/SGP) (2016) Coping with climate change and variability: lessons from Sri Lankan communities. In: Proceedings of the National Workshop on Community Based Adaptation. Global Environmental Facility/Small Grants Programme Sri Lanka, UNDP, Colombo

Chapter 8
Water Quality Index: A Tool for Wetland Restoration

Nitin Bassi

Abstract Worldwide, wetlands which provide several benefits to human society and environment are subjected to increasing anthropogenic pressures resulting in loss of their hydrological and ecological functions. Such impacts are more prominent in case of wetlands in urban areas which are exposed to land use changes and developmental activities. In many Indian cities, natural water bodies such as lakes are heavily polluted due to runoff from farmlands in urban and peri-urban areas and discharge of untreated domestic and industrial wastewater. The major constraint for restoring such water bodies is difficulty in devising a concrete action plan for analyzing different set of water quality parameters. In order to address this challenge, a water quality index (WQI) which is a tool to summarize large amounts of water quality data, is computed for a natural urban lake in New Delhi metropolitan area. The mean WQI of the lake was estimated to be 46.27 which indicate a high level of water pollution. Season-wise assessment shows that water quality was worst during winters. Further, the analysis highlights the State failure in protecting water bodies in urban areas from large scale pollution and in regulating land use changes in their catchments. The paper discusses how these findings can be used by policy makers to formulate plans for wetland conservation. The paper also advocates the need for establishing a community based water quality monitoring system, backed by infrastructural support from the State, in order to restore the wetlands in urban areas.

Keywords India · Urban areas · Anthropogenic pressures · Water pollution · Water quality monitoring

8.1 Introduction

India, with its varying topography and climatic regimes, supports diverse and unique wetland habitats. As per the National Wetland Atlas 2011 estimates, which mapped wetlands on 1:50,000 scale, India has about 757 thousand wetlands with a total

N. Bassi (✉)
Institute for Resource Analysis and Policy [IRAP], Liaison Office, Delhi, India
e-mail: nitinbassi@irapindia.org

© Springer Nature Switzerland AG 2021
M. Babel et al. (eds.), *Water Security in Asia*, Springer Water,
https://doi.org/10.1007/978-3-319-54612-4_8

wetland area of 15.3 million hectares (m. ha), accounting for nearly 4.7% of the total geographical area of the country. Out of this, area under inland wetlands accounts for 69%, coastal wetlands 27%, and other wetlands (smaller than 2.25 ha) 4% (Space Applications Centre 2011). These wetlands are distributed in different geographical regions of the Country ranging from Himalayas to Deccan plateau.

Though wetlands provide many benefits which are categorized into cultural, supporting, provisioning and regulatory goods and services, they are continued to be ignored in the policy process (Turner et al. 2000; Ghermandi et al. 2008) and are threatened by increasing pollution as a manifestation of urbanization, population growth, land use changes and increased economic activities (Verma et al. 2001; Chan 2012). Urbanization exerts significant influences on the structure and function of natural wetlands, mainly through modifying the hydrological and sedimentation regimes, and the dynamics of nutrients and chemical pollutants (Faulkner 2004; Lee et al. 2006; Yuanbin et al. 2012).

In fact, water in most Asian wetlands (rivers, streams, tanks, lakes and ponds) is heavily degraded, mainly due to agricultural runoff containing pesticides and fertilizers, and industrial and municipal wastewater discharges, all of which causing widespread eutrophication (Prasad et al. 2002; Liu and Diamond 2005). Situation in India is no different. Particularly, the contribution of urban domestic and industrial effluent to this menace is noteworthy. In India, less than 31% of the domestic wastewater from urban centres is treated. In the 35 metropolitan cities (population size of 0.4 million or more) put together, treatment capacity is only 51%. Condition in smaller urban centres is even worse as treatment capacity exist for only about 18% of the sewage generated in Class I cities (population size of 100,000 or more but other than metropolitan cities) and 9% of the sewage generated in Class II towns (population between 50,000 and 100,000) (CPCB 2009). Actual sewage treatment is lower due to inadequacy of the sewage collection system and non-functional treatment plants.

Given the insufficient capacity to treat industrial and municipal wastewater, problem of water quality deterioration is alarming particularly in the case of water bodies such as lakes and ponds in small urban centres (Bassi et al. 2014). Results from Monitoring of Indian Aquatic Resources (MINARS) programme also show that water bodies such as rivers and lakes near to urban centres are becoming increasingly saprobic and eutrophicated due to contamination from partly treated or untreated wastewater (CPCB 2010).

Climate variability and change is further expected to affect wetland ecosystem adversely. Initial findings suggest that for Indian sub-continent, mean atmospheric temperature and frequency of occurrence of intense rainfall events has increased, while the number of rainy days and total annual amount of precipitation has decreased (Bates et al. 2008). Decreased precipitation will exacerbate problems associated with already growing demands for water and hence alter the freshwater inflows to wetland ecosystems (Bates et al. 2008; Erwin 2009), whereas, rise in temperature can aggravate the problem of eutrophication, leading to algal blooms, fish kills, and dead zones in the surface water (Gopal et al. 2010). Thus wetland water quality will be affected both on account of reduced freshwater inflows and eutrophication.

Considering the state of wetlands in urban areas, the foremost requirement is to determine the extent of water pollution and quality deterioration the wetlands have undergone. Thereafter, based on a correct assessment of water quality parameters, appropriate policy and management decision shall be taken for their restoration. However, to arrive at any strategy on wetland restoration through analysis of different set of water quality parameters is often difficult as it requires a multidisciplinary understanding of water pollution problems. The Water Quality Index (WQI) can be an important tool for decision and policy makers to formulate wetland restoration and management strategies based on water quality parameters.

Therefore, the main objective of this study is to compute the WQI for a lake situated in urban area and highlights its importance as a tool to assist policy makers, natural resource managers and conservationist to strategies actions which can lead to improvements in environmental conditions of the lakes, particularly quality of their water.

8.2 Material and Methods

8.2.1 Water Quality Index

Water quality indices are a convenient means to analyse large amounts of water quality data, using various groups of parameters (Wills and Irvine 1996; Bharti 2011). For this study, WQI developed by National Sanitation Foundation (NSF) was used as it provides best results for the indexation of the general water quality (Bharti 2011). In India, Central Pollution Control Board (CPCB) which is country's foremost pollution control body also uses NSF WQI, with a slight modification in weights, to classify some selected surface water bodies, mainly rivers, based on their pollution loads. NSF WQI is composed of nine parameters which include: Dissolved Oxygen (DO); Faecal Coliform (FC); pH; Biochemical Oxygen Demand (BOD); temperature change; total phosphate; nitrates; turbidity; and Total Solids (TS).

8.2.2 Study Area

WQI was computed for lake Bhalswa which is an important freshwater wetland of Yamuna river basin. It is a natural freshwater oxbow lake situated in the north-western district of Delhi metropolitan area. Most of the freshwater inflow in the lake occurs during the monsoon months (mid-June to September) when almost 90% of the rainfall is received. The average annual rainfall is around 750 mm. The water column (total depth) in lake varies from 89.7 m during monsoon to 72.3 m in summer.

Originally, the lake was shaped like a horseshoe. Over the years, due to large scale encroachments in its surroundings (as confirmed by satellite imageries of different time period and field survey), lake area has decreased by almost half. Presently, the lake is 1.5 kms long with water spread area of about 22 ha and is one of the three largest lakes in Delhi.

Along the west bank of the lake, there is a resettlement colony which is the main source of most of the wastewater being discharged into the lake. Majority of the households in this colony has dairy as their main occupation. One of the three major landfill sites of Delhi is also situated about 600 m from the western side of the lake. On the eastern side, Delhi Development Authority (DDA), an autonomous body under the Ministry of Urban Development of Government of India, has developed a golf course, water sports facilities and has also undertaken some plantation activities. Some plants (such as Prosopis sp. and grasses) suitable to semi-arid areas were found to be growing naturally on the western side of the lake.

Apart from its use as tourism and recreational site, the lake provides water for outdoor bathing, meeting the drinking and bathing needs of cattle and for the commercial fisheries activities. Further, it contributes to groundwater recharge which is indicated by shallow groundwater depth and greater rise in water levels post monsoon (in proportion to depth to groundwater level) in wells located near to lake than those located away from it but tapping the same aquifer (Fig. 8.1). Also, it attracts plenty of waterfowls (such as strokes, stilts, herons) in winter. In the past, the lake was connected to river Yamuna but due to large scale land use changes in the Yamuna flood plains and change of river course over the years that hydraulic inter-connectedness no longer exists. At present, the lake stands as an independent wetland providing the discussed benefits and services to the community and environment.

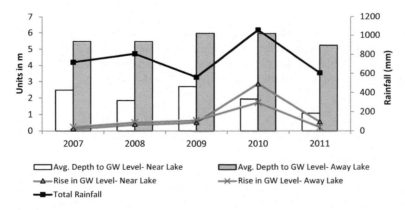

Fig. 8.1 Groundwater behaviour in wells located near and away from the lake

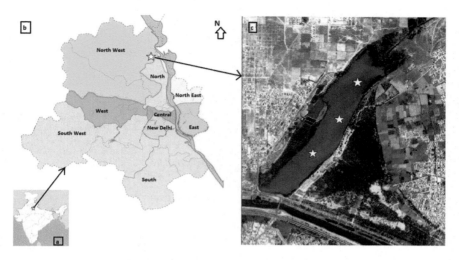

Fig. 8.2 a & b India map showing location of lake Bhalswa in Delhi, and **c** satellite image of lake with three sampling sites

8.2.3 Sampling and Analytical Procedure

Water samples were collected from a depth of 45 cm from three sites located near to northern end (site 1), middle (site 2) and southern end (site 3) of the lake during monsoon, winter and summer seasons, to capture both spatial and temporal variability in the water quality (Fig. 8.2). For bacteriological (FC) analysis, glass bottles were used to collect water sample. Parameters such as pH, DO, FC, turbidity and TDS were measured within 24 h of water sample collection. For analysis of BOD and FC, samples were stored at a temperature below 4 °C in the refrigerator and for nitrates, water samples were preserved below pH 2 by adding concentrated sulphuric acid (CPCB 2007).

Temperature was measured by dipping the thermometer at the selected sites in the lake. pH, DO, TDS and turbidity were measured by placing electrodes from the calibrated water testing kit in the water sample. Colorimetry method was followed for measuring total phosphates and nitrates in the water sample. For BOD, stored water sample were measured for DO and difference in DO values over three days. For measuring FC, Most Probable Number (MPN) method was followed (CPCB 2007).

8.2.4 WQI Computation

Results obtained from tests on the nine parameters of WQI are transferred to a weighting curve chart where a numerical value (Q-value) is obtained. Then for each

Table 8.1 NSF WQI parameters and assigned weights

Sr. No	Parameters considered	Desirable unit	Weight assigned
1	Dissolved Oxygen	% saturation	0.17
2	Faecal Coliform	colonies/100 ml	0.16
3	pH	Units	0.11
4	Biochemical oxygen demand	mg/L or ppm	0.11
5	Temperature Change	°C	0.10
6	Total Phosphates	mg/L or ppm	0.10
7	Nitrates	mg/L or ppm	0.10
8	Turbidity	NTU	0.08
9	Total Solids	mg/L or ppm	0.07
Overall Weight			1.00

Source Brown et al. 1970

parameter, the Q-value is multiplied by a weighting factor which is based on the significance of particular parameter in determining the water quality (Table 8.1). Sum of all weights used for determining the WQI is 1. The nine resulting values are then added to arrive at an overall WQI. The highest score a water body can receive is 100. Quality of water in surface water body is categorized from bad to excellent depending on the WQI score. Mathematically NSF WQI can be represented as:

$$NSFWQI = \sum_{i=1}^{p} Wi Ii \qquad (8.1)$$

where, Ii is the sub index (Q-value) for ith water quality parameter; Wi is weight (in terms of importance) associated with water quality parameter; and p is the total number of water quality parameters.

8.3   Results and Discussion

8.3.1   Assessment of Lake Water Quality

Results from the laboratory analysis of water samples are presented in Table 2a and 2b. No major intra-seasonal difference in temperature and pH was found across the three selected sites. Except for summer, pH was found to be within the permissible range (which is 6.5–8.5) as per the Indian Standard (IS) for the surface water bodies (IS: 2296). Higher alkalinity in summer can be due to increased concentration of carbonate salts as a consequence of greater evaporation and reduced inflow of freshwater in the lake. In fact, mean concentration of carbonate (52.43 mg/L) was found to be highest during the summer season.

Table 8.2a Variation in parameters within season

Parameters	Monsoon		Winter		Summer	
	Mean	SD	Mean	SD	Mean	SD
Temperature (°C)	32.0	0.0	10.3	0.6	31.0	0.0
pH	8.4	0.0	7.8	0.9	8.8	0.1
DO (mg/L)	4.8	0.1	4.7	0.2	4.7	0.1
Turbidity (NTU)	24.3	0.1	24.7	0.6	28.0	1.0
TDS (mg/L)	4318.7	60.3	4463.3	99.3	5259.7	59.5
Nitrates (NO3) (mg/L)	28.5	0.5	30.7	0.6	32.6	0.9
Total Phosphates (mg/L)	4.2	0.0	4.7	0.1	4.6	0.1
BOD (mg/L)	14.8	1.9	14.4	0.2	20.7	1.5
Fecal Coliform (No/100 ml)	28.3	4.0	37,666.7	15,176.7	7.0	8.7

Table 8.2b Variations in parameters across seasons

Parameters	Site 1		Site 2		Site 3	
	Mean	SD	Mean	SD	Mean	SD
Temperature (°C)	24.3	12.4	24.3	12.4	24.7	11.8
pH	8.4	0.4	8.4	0.4	8.2	0.6
DO (mg/L)	4.8	0.1	4.7	0.1	4.7	0.1
Turbidity (NTU)	25.3	2.3	25.3	1.5	26.3	2.3
TDS (mg/L)	4621.3	555.6	4680.0	455.0	4740.3	510.8
Nitrates (NO3) (mg/L)	30.3	2.4	30.4	1.6	31.1	2.2
Total Phosphates (mg/L)	4.5	0.2	4.5	0.2	4.5	0.3
BOD (mg/L)	17.4	3.4	15.8	2.7	16.7	4.6
Fecal Coliform (No/100 ml)	11,681.0	20,194.8	18,011.7	31,166.8	8009.3	13,848.0

Mean concentrations of DO in all seasons were observed to be above 4 mg/L which is considered to be minimum for fish culture and wildlife propagation but below the range of 5–6 mg/L which is considered ideal for drinking water and outdoor bathing use (as per IS: 2296). Similarly, mean concentration of BOD in all seasons were observed to be much above the permissible range of 2 to 3 mg/L (IS: 2296), thus making the lake water unsuitable for any kind of domestic use including that for outdoor bathing. High BOD concentration can mainly be attributed to high level of organic matter in the lake water which is due to discharge of untreated domestic wastewater from the habitations in its surrounding area. Large quantities of organic matter act as substrate for microorganisms and during its decomposition, dissolved oxygen in the receiving water is used up at a greater rate than it can be replenished causing oxygen depletion and increase in BOD concentration.

The lake water was also observed to have high turbidity. The major reason for high turbidity is due to high storm-water discharge from the surrounding built up area as it makes surface less impervious, reduces water retention and thereby increases runoff loaded with sediments. One of the major effects of high turbidity in water is that it reduces the light available to photosynthetic organisms which can have severe consequences on lake Biota. In fact, no major aquatic plants (submerged, floating or emergent) were seen in the lake.

TDS concentrations were found to be above the permissible range of 500–2100 mg/L as prescribed under IS: 2296 which makes the lake water not only unsuitable for domestic use but also for irrigation, industrial cooling and controlled waste disposal purposes. High TDS concentrations indicate presence of greater amount of dissolved solids which are due to high discharge of surface runoff containing sediments and untreated wastewater from the surrounding urban area.

Further, average concentration of total phosphates (soluble phosphorus and the phosphorus in plant and animal fragments suspended in lake water) was observed to be enough to promote algal blooms (Shaw 2004) which can interfere with the functioning of other productive systems in the lake. Though all the observed concentrations of NO_3 are below permissible limit of 50 mg/L which makes lake water suitable for drinking purpose with conventional treatment and disinfection, due to high BOD and TDS concentrations it is not suitable for domestic uses. The main source of nitrates is the animal waste which enters from the west bank of the lake.

FC was also found in all the seasons indicating unsuitability of water for drinking and other domestic uses. In winter season, FC count was found to be abnormally high. It may be due the fact that millions of people took a holy bath in the lake during a religious ceremony in winter and thereby may have contributed to its abnormally high FC count. Also, many migratory birds visit the lake during winters which might have also played a role in increasing the faecal content in the lake water.

With the results obtained from the water quality testing, it is difficult to arrive at any conclusion on the overall quality of the lake water and its suitability for various uses. Concentrations of some water quality parameters make use of lake water unsafe for municipal purposes, whereas others make it difficult for wildlife to sustain. For instance, concentrations of DO, BOD, TDS and FC were found to be above permissible limit for its use for drinking or outdoor bathing purpose and that of turbidity and phosphates makes it difficult for various aquatic flora and fauna to sustain. Thus, for better understanding of the overall health of the lake (in terms of its water quality) and take measures for its effective restoration, the results obtained from the water quality analysis need to be summarized. WQI plays an important role in achieving this.

8.3.2 Assessment of WQI for the Lake

Based on the observed concentrations of the water quality parameters, NSF WQI index was computed for lake Bhalswa using Eq. 8.1 (Fig. 8.3). Based on the index value, water quality is categorised as: excellent (90–100), good (70–90), medium (50–70), bad (25–50) and very bad (0–25). Highest WQI value was obtained for site

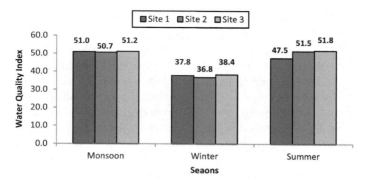

Fig. 8.3 Season and site-wise WQI of the selected lake

3 during summer season and lowest for site 2 during winter season. Subsequently, an independent sample t-test results showed that there is a significant difference between WQI in winter as compared to other two seasons (p value less than 0.05). However, WQI across sites does not vary significantly within a season. Nevertheless, in winter, the lake water quality falls in bad category and in summer and monsoon water quality was just above the range from being bad. Overall, water quality of the lake was found to be bad with a mean WQI value of 46.27. These results points to high level of pollution in lake Bhalswa.

8.3.3 Discussions

Wetlands in urban areas are under threat from the anthropogenic influences in their catchment and also from the climate induced environmental stresses. These influences have led to reduction in their areal extent and have also resulted in decline in the hydrological and ecological functions they used to perform (Bassi et al. 2014). As indicated by the WQI assessment of lake Bhalswa, the water quality has deteriorated to the extent that it is no longer useful for most of the domestic, economic and environmental uses. It was found that water quality at all the sampling locations was poor. Further, lack of concrete legislations and poor implementation of existing policies on wetland conservation and regulating land use changes has hastened their degradation. These factors are further discussed in the subsequent sub-sections.

Main reasons identified for the high degree of pollution in the lake water were discharge of storm-water loaded with sediments during monsoon months and untreated municipal wastewater from the surrounding habitations. As a result, the lake water was found to have high BOD, turbidity, TDS, and phosphates concentrations. The situation is not so different for wetlands in other urban areas. It is estimated that in India about 38,000 million litres per day (mld) of municipal wastewater is generated in the urban centres having population more than 50,000 (which encompass 70% of urban population). However, the treatment capacity exists for only about

29% (CPCB 2013). The untreated wastewater is discharged into water bodies such as rivers and lakes which increases their pollution load.

Other major reason for the poor state of wetlands in urban areas is the lack of government's effort in protection of their water quality. For instance, though lake Bhalswa was selected as one of the lakes for restoration by DDA, most of the works which were undertaken relate to creation of water sports facilities, de-silting and afforestation activities. No effort was made to restrict indiscriminate land use changes and real estate development in the catchment area of the lake. Neither any effort was made to control discharge of untreated wastewater into the lake. In fact, DDA has proposed buildings, hotels and high end housing on the eastern edge of the lake. Such proposal would further alter the lake catchment hydrological regime and lead to more pollution of this already degraded water body. The situation is similar for most of the wetlands in and around major metropolitan cities of India.

Also, climate variability influences the hydrological characteristics of the wetlands to a great extent. In India, there is a high degree of spatial and temporal variability in rainfall which greatly influences the inflows into wetlands. Such variability during years of low rainfall can affect the pollution assimilation capacity of those wetlands, like that of lake Bhalswa, which are fed by rainfall and runoff generated thereof. Further, in majority of wetlands, water inflow is mostly during monsoon months which significantly reduce their capacity to retain and treat pollutants during non-monsoon months. As a result, water quality deteriorates significantly in the non-monsoon months. Even in the case of lake Bhalswa, WQI was lowest during winter and summer months.

The other major factor leading to wetlands degradation is lack of proper legislation for wetland conservation and management. Though, there are some stand-alone instances of wetland restoration works, most of these efforts have only been on a few important wetlands and remained mostly in silos. Apart from a few States where wetlands offer tremendous tourism potential, they rarely figure in the government priority list. Further, ineffective implementation of pollution control programmes has led to discharge of untreated domestic wastewater and industrial effluent into wetlands, affecting their hydrological and ecological integrity (Bassi 2016).

Still, there is no legal instrument which directly relates to wetlands. However, they are indirectly influenced by a number of other legal instruments which include: Water (Prevention and Control of Pollution) Act, 1974; Forest (Conservation) Act, 1980; Environmental (Protection) Act, 1986; Wildlife (Protection) Amendment Act, 1991; Biodiversity Act, 2002; and Scheduled Tribes and Other Traditional Forest Dwellers (Recognition of Forest Rights) Act, 2006 (Prasad et al. 2002; Bassi et al. 2014).

Even the policy support for wetland conservation until 2010 was governed mainly by the programme initiated as per the international commitment such as National Wetland Conservation Programme (NWCP), National Lake Conservation Plan (NLCP) and National River Conservation Plan (NRCP). Post 2010, the Central Government has notified the Wetlands (Conservation and Management) Rules, 2010. But these rules seem to have narrow objectives as they regulate only some selected wetlands (such as those under Ramsar Convention and UNESCO World Heritage site) and completely ignore large number of smaller wetlands such as lakes, tanks and ponds which play an important role in urban and peri-urban landscapes (Bassi 2016). This clearly shows the lack of interdisciplinary expertise and understanding in dealing with issues related to wetland conservation. Further, policies governing land use planning is also flawed as in many cases wetland catchment area has been altered leading to reduced or no water inflows.

8.4   Conclusions

Wetlands in urban areas are undergoing large scale deterioration in terms of both water availability and quality. Climate variability and change is further going to exacerbate the wetland water quality through alteration in their hydrological regime due to changes in precipitation, runoff, temperature and evapo-transpiration. As indicated by WQI assessment for lake Bhalswa, main reasons for the high pollution loads in the wetlands are discharge of storm-water loaded with sediments and untreated municipal wastewater. It has been also found that the Governments have failed in protecting wetlands water quality from large scale pollution and land use changes in their catchments across many river basins in India. Further, inadequate legislative support and poor design and implementation of policies on wetland conservation and management have resulted in their further degradation. Given the state of affairs, WQI can be an important tool for policy makers, natural resource managers and concerned agencies to develop plans which can reflect the extent of effort required to restore the wetlands in urban areas. However, at present, water quality assessments are undertaken randomly and only for a few large wetlands. Thus, an effective and proper water quality monitoring strategy needs to be developed for all wetlands taking into consideration, the ecological, economic and social benefits they generate.

Further, community is often dependent on the wetlands for various domestic and economic purposes. Therefore including them in the water quality monitoring exercise as an important stakeholder can bring about remarkable results. Selected community members can be given responsibility of water sample collection and its transfer to designated scientific laboratory. In fact, community can also be given portable water quality testing kit which can give instant results on temperature, pH, turbidity, TDS and OD. However, relevant trainings and capacity building activities need be designed and undertaken for these selected community members. Necessary government support in terms of laboratory infrastructure and scientific manpower will also be required. Based on the results of water quality monitoring (in terms of concentrations of parameters and WQI values over different time periods), appropriate measures for regulating land use changes and treating wastewater can be adopted. This will also generate a reliable data on the present condition of wetlands water quality which can be used for further planning.

Acknowledgements Reprinted from *Water Policy* vol 19, issue 3, pp 390-403 (2017) with permission from the copyright holders, IWA Publishing.

References

Bassi N, Kumar MD, Sharma A, Pardha-Saradhi P (2014) Status of wetlands in India: a review of extent, ecosystem benefits, threats and management strategies. J Hydrol Reg Stud 2(1):1–19

Bassi N (2016) Implications of institutional vacuum in wetland conservation for water management. IIM Kozhikode Soc Manag Rev 5(1):1–10

Bates BC, Kundzewicz ZW, Wu S, Palutikof JP (eds) (2008) Climate change and water, Technical Paper VI. Inter-governmental Panel on Climate Change, Geneva

Bharti NK (2011) Water quality indices used for surface water vulnerability assessment. Int J Environ Sci 2(1):154–173

Brown RM, McClleland NI, Deininger RA, Tozer R (1970) A water quality index – do we dare? Water Sewage Works 117(10):339–343

Central Pollution Control Board (CPCB) (2009) Status of water supply, wastewater generation and treatment in Class-I cities and Class-II towns of India. Central Pollution Control Board, Ministry of Environment and Forests, Government of India, New Delhi

Central Pollution Control Board (CPCB) (2010) Status of water quality in India 2009. Central Pollution Control Board, Ministry of Environment and Forests, Government of India, New Delhi

Central Pollution Control Board (CPCB) (2007) Guidelines for water quality monitoring. Central Pollution Control Board, Ministry of Environment and Forests, Government of India, New Delhi

Central Pollution Control Board (CPCB) (2013) Status of water quality in India 2011. Central Pollution Control Board, Ministry of Environment and Forests, Government of India, New Delhi

Chan NW (2012) Managing urban rivers and water quality in Malaysia for sustainable water resources. Int J Water Resour Dev 28(2):343–354

Erwin KL (2009) Wetlands and global climate change: the role of wetland restoration in a changing world. Wetlands Ecol Manag 17(1):71–84

Faulkner S (2004) Urbanization impacts on the structure and function of forested wetlands. Urban Ecosyst 7(2):89–106

Ghermandi A, van den Bergh JCJM, Brander LM, Nunes PALD (2008) The economic value of wetland conservation and creation: a meta-analysis. Fondazione Eni Enrico Mattei Working Paper No. 79.2008, Milan, Italy

Gopal B, Shilpakar R, Sharma E (2010) Functions and services of wetlands in the eastern Himalayas: Impacts of climate change. Technical Report 3. International Centre for Integrated Mountain Development, Kathmandu, Nepal

Lee SY, Dunn RJK, Young RA, Connolly RM, Dale PER, Dehayr R, Lemckert CJ, McKinnon S, Powell B, Teasdale PR, Welsh DT (2006) Impact of urbanization on coastal wetland structure and function. Austral Ecol 31(2):149–163

Liu J, Diamond J (2005) China's environment in a globalizing world. Nature 435(7046):1179–1186

Prasad SN, Ramachandra TV, Ahalya N, Sengupta T, Kumar A, Tiwari AK, Vijayan VS, Vijayan L (2002) Conservation of wetlands of India-a review. Trop Ecol 43(1):173–186

Shaw B, Mechenich C, Klessig L (2004) Understanding Lake Data. University of Wisconsin-Extension, Cooperative Extension, WI, USA

Turner RK, Van Den Bergh JC, Söderqvist T, Barendregt A, van der Straaten J, Maltby E, van Ierland EC (2000) Ecological-economic analysis of wetlands: scientific integration for management and policy. Ecol Econ 35(1):7–23

Verma M, Bakshi N, Nair RP (2001) Economic valuation of Bhoj wetland for sustainable use. Project report for World Bank assistance to Government of India, Environmental Management Capacity-Building. Indian Institute of Forest Management, Bhopal

Wills M, Irvine KN (1996) Application of the national sanitation foundation water quality index in the Cazenovia Creek, NY, pilot watershed management project. Middle States Geographer 29(1):95–104

Yuanbin C, Hao Z, Wenbin P, Yanhong C, Xiangrong W (2012) Urban expansion and its influencing factors in Natural Wetland Distribution Area in Fuzhou City, China. Chin Geogr Sci 22(5):568–577

Chapter 9
A Framework for Implementing Integrated Water Resources Management at River Basin Level in Indonesia

R. W. Triweko

Abstract As an archipelagic country, water resources management in Indonesia is based on river basin territory (RBT) which is defined as a unit area of water resources management in one or more catchment area and/or small islands with a total area of less than or equal to 2,000 km^2. In total, Indonesia has 128 RBTs which are classified into five categories, i.e. international, inter-provincial, national strategic, inter-regency/city, and in-one-regency/city river basin territories. The authority in managing each river basin territory is distributed among Central Government, provincial governments, and regency/city governments. The uneven distribution of population that also influences the gap in socio-economic development between Java Island and other islands has resulted unique problems for each river basin territory. Moreover, variability in human resources capacity and financial capacity among those provincial and regency/city government becomes constraints in solving the problems. Facing with this variability in water resources management problems, this paper proposed a framework in implementing integrated water resources management at river basin level to increase water security in Indonesia.

Keywords Water resources problems · River basin territory · Integrated river basin management

9.1 Introduction

As a large archipelagic country, developing water resources management in Indonesia is not an easy task. Uneven distribution of the population between Java and other islands, which also influence the distribution of social-economic activities results an unbalanced water demand. Unfortunately, water availability in Java is very limited even though annual precipitation in this island is relatively high. Water resources infrastructure which had been built since the Dutch period, are continuously developed in Java in order to respond the increasing demand, and the need to utilize

R. W. Triweko (✉)
Universitas Katolik Parahyangan, Bandung, Indonesia
e-mail: triweko@unpar.ac.id

© Springer Nature Switzerland AG 2021
M. Babel et al. (eds.), *Water Security in Asia*, Springer Water,
https://doi.org/10.1007/978-3-319-54612-4_9

water resources potential, including hydropower development. Although many water resources development projects have been implemented in other island such as Sumatra, Sulawesi, and Nusa Tenggara, most water resources potential in Indonesia have not been fully utilized. On the other hand, the fulfilment of domestic water supply, which is the basic water demand and part of the human rights, is still a big challenge, especially in villages and remote areas. Water resources development is also needed in other islands as the basic infrastructure to support social and economic development of the region.

Water resources management in Indonesia, has been gradually developed and periodically adjusted to the new challenges and the changes in the political system. The role of the government in water resources management had been already initiated during the Dutch period, based on the *Algemeen Waterreglement Year 1936,* which was used as the basic regulation on water management until Indonesia enacted Law No. 11 Year 1974 on Watering. In the past, water resources management has centralized its authority and responsibility in the hand of the central government. The changes in the political system towards more democratic, opened, accountable, and participatory have been influencing the institutional arrangements in water resources management, at the national level as well as at lower level.

Since the implementation of reforms in the political structure in 1998, democratic environment in Indonesia indicates a positive development which is implemented in a new structure of the House of Representatives, political parties system, general election, direct presidential election, and head of region election which are continuously developed. The implementation of the regional autonomy since the year 2000 has also gradually modified to find the most favourable format for Indonesia. This political context influences the pattern of water resources management in Indonesia. Law No. 7 Year 2004 on Water Resources is a result of the reforms in water resources management which is influenced by the spirit of democratization and regional autonomy. Unfortunately, in February 2015, the Constitution Supreme Court annulated the law, and decided to use the previous Law No. 11 (1974), which is more centralistic. Fulazzaky (2014) presented the changing paradigms in water resources management in Indonesia, which compared the old (1971–1998) and the new (1998–present) paradigms and their implications in water resources management. His analysis considered the elements of government regime, administration rule, decision making process, budgetary system, role of water in regional development, management responsibility, and role of government in construction, project setting up orientation, project interest, and project preparation.

The changes in political system, governmental structure, policies, and strategies, are influencing water resources management. The challenge of water resources development, however, is still similar on how fulfilling water demand, utilizing water potential, and solving water related problems. Therefore, those changes should not influence the level of service and sustainability of water resources management, because the role of water resources is very important in supporting social, economic, and environmental sustainability.

This paper aims to present a framework of the water resources management at river basin level in Indonesia. This framework is intended as a guideline in the development of water resources management at the river basin level, in order to gradually increase the level of service in each river basin, in line with its water potential, problems, and demand.

9.2   Material and Methods

This study is based on a literature study related to water resources management in Indonesia. The first stage of this study is intended to understand the challenge of water resources management in Indonesia. An analysis on water resources potential was done to compare per capita water availability among the main islands. In addition, the concept of flow thickness that indicates the wet or dry condition of the island was introduced by calculating the annual water resources potential divided by the area of the island. The second stage is a chronological analysis of river basin territory division as a consequence of the changes in the water resources law in Indonesia, that indicates the development of the concept in river basin classification. The third stage is an analysis of the dynamics of water resources management problems in a river basin territory to explain the complexity of the problems and its impacts on water security. As a large archipelagic country, each river basin in Indonesia has different condition of its potential, problems, and utilization of water resources which demands a unique approach in its policy, strategy, and management. Finally, a framework for water resources management in a river basin is developed as guideline principles to develop its level of service and improving its sustainability.

9.3   Results and Discussion

9.3.1   The Challenge of Water Resources Development and Management

Table 9.1 demonstrates the water resources potential and per capita water availability for the main islands of Indonesia. The table demonstrates an uneven population distribution. In Java, this is the smallest among the large islands, live 57.5% of the Indonesian population. The second most populated island is Sumatra, while other islands are still relatively still under populated. The largest water resources potential however, is owned by Kalimantan which has 1,314.0 billion m^3 (BCM) annually (or 33.6% of the national water resources), followed by Papua with 1,062.1 BCM annually (or 27.2% of the national water resources potential). If the amount of the water resources potential is divided by the total population in each island, there will be a comparison on per capita water availability. It is clear that Java has the lowest number 1,210 m^3 annually, while Papua has the largest number of almost 300,000 m^3 annually.

Table 9.1 Water resources potential and per capita availability in the main islands of Indonesia

No	Island(s)	Total population in 2012	%	Water resources potential (billion m³/year)	%	Per capita water availability (m³/year)	Flow thickness (mm/day)
1	Sumatera	50,630,931	21.3	840.7	21.5	16,730	4.56
2	Java	136,610,590	57.5	164.0	4.2	1,210	3.62
3	Kalimantan	13,787,831	5.8	1,314.0	33.6	95,900	6.74
4	Sulawesi	17,371,782	7.3	299.2	7.7	17,510	4.33
5	Bali and NusaTenggara	13,074,796	5.5	49.6	1.3	3,880	1.86
6	Maluku	2,571,593	1.1	176.7	4.5	71,100	5.43
7	Papua	3,593,803	1.5	1,062.1	27.2	299,500	7.07
	Indonesia	237,641,326	100.0	3,906.5	100.0	16,600	5.56

Flow thickness is calculated from annual water resources potential divided by the area of the island. Therefore, flow thickness figures out the average daily water availability in the island, or indicates the wet or dry condition of the island. Papua and Kalimantan are the wettest islands in Indonesia with almost 7 mm/day of flow thickness, while Nusa Tenggara is the driest islands with only 2 mm/day of flow thickness. Abundant water from the wettest islands, however, cannot be easily transferred to the driest islands. This information indicates that social economic development, and especially agricultural policies, should be adjusted to water potential in each island. At the national level, the above water balance becomes an important input for the national strategic economic development, which in turn is implemented as virtual water policies.

The study of water balance should be done for each river basin as the basis for the strategic water resources development in fulfilling the present and future water demands. Those studies even should be done in more detail for each catchment area or small island. For the neighboring catchment areas which are parts of a river basins, understanding of a comprehensive water balance can be used as the basis for the possibility of inter-catchment, or even inter-basin water transfer. For the individual small island, knowledge of flow thickness, potential water availability, and water demand is needed in developing strategic water resources development, because they should be autonomous in their water management.

9.3.2 The Development of River Basin Territory Division

Water resources management in Indonesia has been based on river basin territory (RBT) since 1982 with the implementation of the Law No. 11 (1974) on Watering.

Article 1 f of the Government Regulation No. 22 (1982) on Water Management System defined River Basin as a unit of water management according to Article 1 No. 7 Law No. 11 (1974) which consists of one or more catchment areas. The implementation of the GR No. 22 (1982) is Minister of Public Works Regulation No. 39/PRT (1989) on River Basin Division which divided the whole country into 90 RBTs. The criteria for the RBT division is based on the approach of hydrology, government administration and planning, as it was mentioned in the GR No. 22 (1982). Minister of Public Works Regulation No. 48/PRT (1990) on Water and Water Sources Management in River Basin divided the authority in managing river basin into three categories, i.e. 73 RBTs managed by local governments, 15 RBTs managed directly by MPW, and 2 RBTs managed by a certain legal body.

Minister of Public Works Regulation No. 11A/PRT (2006) on Criteria and River Basin Decision is an implementation of Law No. 7 (2004) on Water Resources, which divided RBTs into five categories, i.e. (1) international, (2) inter-province, (3) national-strategic, (4) inter-regency/city, and (5) in-one regency river basin territory. In Law No. 7 (2004), river basin territory is defined as a unit area of water resources management in one or more catchment area and/or small islands with a total area of less than or equal to 2,000 km^2.

Presidential Decree No. 12 (2012) on River Basin Decree as an implementation of the Law No. 7 (2004) on Water Resources and Government Regulation No. 42 (2008) on Water Resources Management. Divided into 131 RBTs which consists of 5 international RBTs, 29 inter-province RBTS, 29 national-strategic RBTs, 53 inter-regency/city RBTs, and 15 in-one regency/city RBTs.

Minister of Public Works and Housing Regulation No. 04/PRT (2015) on Criteria and River Basin Decision as a follow up of the re-enactment of Law No. 11 (1974) on Watering by the Constitution Supreme Court, which are divided into 128 RBTs, i.e. 5 international RBTs, 31 inter-province RBTs, 28 national-strategic RBTs, 52 inter-regency RBTs, and 12 in-one-regency RBTs.

The changes of the classification and authority in river basin management indicates the changes of the policy and strategy in water resources management at the national level, which directly or indirectly influence the condition of water resources management in the related regions.

9.3.3 Water Resources Management in Indonesia

Water resources management in Indonesia is based on the Article 33 (3) of the Constitution of the Republic of Indonesia which stated: "The land, the waters and the natural resources within shall be under the powers of the State and shall be used to the greatest benefit of the people." This idea is then implemented in three pillars in managing water resources in Indonesia, i.e. (1) water resources should be managed based on the basic principles of conservation, balance, social benefit, integration, harmony, fairness, transparency, and accountability; (2) water resources are managed comprehensively, integrated, and environmentally sound, with a goal to materialize sustainable benefit for the maximum people welfare; and (3) water resources have social, environmental, and economical functions which are organized and implemented harmoniously. A comprehensive water resources management covers all aspects of water resources management which includes conservation, utilization, and control of the water hazard, and covers of the whole management territory which includes all process of planning, implementation, monitoring and evaluation. Integrated water resources management is implemented by involving all stakeholders, inter-sector as well as inter-region. Water resources management with environment insight will consider and maintain the balance of the ecosystem and its environment capacity. Sustainable water resources management is intended to consider water demand, not only for present generation, but also future generation (Team of Indonesia Country Paper 2009).

9.3.4 River Basin Organization and Management

The classification of river basin in Indonesia into five categories resulted some alternatives in the form of river basin organization and management. International river basins, inter-province river basins, and national strategic river basins which are under the central government authority are managed by River Basin Office *(Balai Wilayah Sungai*/BWS) or River Basin Head Office *(Balai Besar Wilayah Sungai/BBWS)* depends on its complexity of the water resources problems and the level of water resources infrastructure development. BWS or BBWS represents the central government in implementing water resources infrastructure projects in the related river basins. The constructed infrastructures are then handed over to State-own Enterprise *(Perum Jasa Tirta/PJT)* who operate and maintain the infrastructure. Inter-regency river basins which are under the provincial governments authority usually managed by Water Resources Management Office *(Balai Pengelolaan Sumber Daya Air)*, while In-one-regency river basins managed by Department of Water Resources *(Dinas Pengelolaan Sumber Daya Air)* of the local governments. Financing of the river basin management becomes the responsibility of the related authority, but in case of financial burdens, local governments can ask financial support from the central government. Beside the government budget, in principle beneficiaries should also contribute to the cost of water resources management service *(Biaya Jasa Pengelolaan Sumber Daya Air/BJPSDA)*, with an exception to the use of water for community irrigation. This stipulation, however, has not been implemented smoothly in all river basins, due to the variety in legal system as the basis for implementation (Team of Indonesia Country Paper 2009).

It should be noted that water resources management in a river basin are involving sectors and stakeholders with their own interests, such as agriculture, forestry, mining, health, environment, industries, and municipalities. Coordination among sectors and stakeholders of water resources management in river basin level is implemented in a coordination body called Coordination Team of Water Resources Management *(Tim Koordinasi Pengelolaan Sumber Daya Air / TKPSDA)*. This Coordination Team responsible to extend recommendation of the Water Resources Management Pattern (*Pola Pengelolaan Sumber Daya Air*) to the Minister of Public Works and Housing. It is a master plan of water resources management in river basin level which provide long term direction and guideline in water resources development and management. This master plan is then implemented into Water Resources Management Plan (*Rencana Pengelolaan Sumber Daya Air*), which is a five-year strategic plan of water resources development and management in the related river basin (Team of Indonesia Country Paper 2009). Bhat (2008) elaborates extensively the pattern of water resources management in Brantas River Basin (Indonesia) in comparison with The Murray Darling River Basin (Australia).

9.3.5 The Dynamics of Water Resources Management Problems

Indonesia has 128 river basin territory (RBT) which has various problems, depends on its location, geological condition, hydrological characteristic, water resources infrastructure, population distribution and growth, and social and economic activities in the river basin. Figure 9.1 elaborates typical complexity of water resources management problems in a river basin territory. Basically, population growth is the main reason for the dynamics of water resources management problems in a river basin.

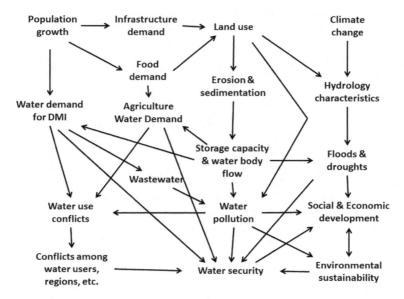

Fig. 9.1 The dynamics of water resources management problems in a river basin territory

Population growth increases water and infrastructure demands. Increasing water demands for domestic, municipalities, industries, and agriculture some time resulting water use conflicts which can encourage conflicts among water users, or inter-region water conflicts. The fulfillment of infrastructure demand has the consequences of land use changes which in turn increases erosion and sedimentation, which results the decline of storage water capacity as well as the flow capacity of the water body. The decline of storage capacity will influence fulfillment of the water demand, while the decline of flow capacity will have an impact to water quality. The phenomenon of climate change which is also happening in Indonesia, have greatly influence to hydrological characteristics, which in turn resulting more floods and droughts. Water shortage due to increasing water demands, social conflicts among water users and regions, water pollution, and increasing floods and droughts are becoming the threats to water security, which is a very critical for the social and economic development.

Understanding the complexity of the problems is very important in developing a comprehensive water resources management in a river basin. It is a requirement that each river basin develop its own schematic diagram explaining the complexity of the water resources management in the related river basin based on its water resources potential, increasing water demands, and water related problems in the area.

9.3.6   A Framework of River Basin Management

Global Water Partnership (2000) defined integrated water resources management (IWRM) as a process which promotes the coordinated development and management of water, land and related resources in order to maximise the resultant economic and social welfare in an equitable manner without compromising the sustainability of vital ecosystems. Implementing the concept of IWRM in a river basin means that water resources management should consider the inter-relationship of water resources potential, social and economic activities which influence water demand and water related problems in the basin, and involving all relevant stakeholders in the decision making process to maximize the benefit and maintain its sustainability.

Figure 9.2 demonstrates a framework of water resources management in an Indonesian river basin. The uniqueness of each river basin depends on its physical conditions, social-economic-cultural conditions, and institutional arrangements. Different condition of each river basin demands different water resources policies, strategies, which in turn will be implemented in the daily water resources management system. The final goal of the water resources management, however, is to achieve water resources sustainability.

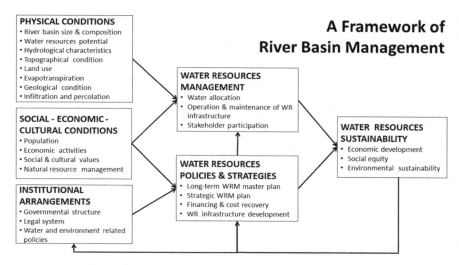

Fig. 9.2 A framework of river basin management

9.3.7 *Physical Conditions*

- River basin size & composition. The area and composition of a river basin territory will influence the development of water resources infrastructure, water resources utilization. Some river basin territory consists of a major river basin and the surrounding smaller catchment areas, like RBT Musi, RBT Kapuas, or RBT Citarum. Other RBT consists of a large number of small catchment areas, such as RBT Flores or RBT Nias. Other RBT consists of small islands, such as RBT Kepulauan Seribu. Other RBT consists of a large river basin in the main island, but includes the surrounding small islands, like the RBT Noelmina.
- Water resources potential includes rainfall, surface water, and groundwater. Potential of water resources in a river basin, catchment areas, and small islands will influence the possibility its utilization. This potential includes annual and seasonal rainfall, surface water availability in rivers, lakes, and reservoirs, and water availability in groundwater basin. Water resources potential also includes its potential for water transportation, generating electricity, or its potential for water related sport and recreation.
- Topographical condition will influence characteristics hydrology and water related infrastructure needed to utilize its potential.
- Evapotranspiration will influence hydrological cycle in the river basin.
- Geological condition will influence the capacity of the aquifer in storing groundwater, and also influencing the feasibility of dam construction.
- Infiltration and percolation indicates the potential of the catchment area in groundwater recharge, but in other cases also indicates the level of water loss in the irrigation system.

- Land use will influence hydrological cycle, both quantitatively and qualitatively. Water quality in rivers depends on social-economic activities in the catchment areas.

9.3.8 Social—Economic - Cultural Conditions

- Population. Number of the population in a river basin indicates water demand not only for the domestic water supply, but also indirectly for food production as irrigation water demand.
- Economics activities in the river basin such as business, industries, agriculture, tourism, etc. result water demand in line with its own characteristics of activities. Water use in economic activities also has an impact to quality of water body.
- Social and cultural values related to people attitude and behavior to water.
- Natural resources management such as mining, agriculture, forestry, and environmental management should be considered, because it's directly or indirectly impacts to water resources management in a river basin.

9.3.9 Institutional Arrangements

- Governmental structure will influence institutional arrangement of water resource management. In the past, governmental structure in Indonesia is more centralized, but since the year 2000 it is more democratic and decentralized by the implementation of new law on regional autonomy.
- Legal system at the national, provincial, and local government level will influence the institutional arrangement of water resource management in a river basin level.
- Water and environment related policies at the national, provincial, and local level will influence water resources management at the river basin level.

9.3.10 Water Resources Policies and Strategies

- Water resources management master plan is needed as a guideline to maintain the long-term direction of water resources development in a river basin. This master plan should be developed based on the uniqueness of each river basin according to its physical conditions, potentials, problems, and increasing and changing water demand. The process of the water resources master plan in a river basin should involve the relevant stakeholders, and approved by the authorized government ensuring that the plan is in line with the national water resources policy.
- Strategic water resources management plan is a medium term (say in 5 year period) to provide a real action in implementing the master plan. The achievement of the strategic plan implementation should be measured quantitatively in terms of key

performance indicators. The process of the strategic plan should also involve the relevant stakeholders and approved by the authorized government to ensure that the plan in line with the water resources master plan and recent social-economic development policies.

- Financing and cost recovery. Maintaining sustainability of water resources management in a river basin need a strong financial support, which is basically becomes the responsibility of the central, provincial, and local government in providing this basic infrastructure. However, for the beneficiaries who get economic benefit from the water service it is rational if they also contribute to the cost of the service. Implementation of the financing and cost recovery policy is also intended to control water use and water quality.

9.3.11   Water Resources Management

- Water allocation is the implementation of water resources policy and strategy, so that all water demand in the river basin can be equally fulfilled.
- Operation and maintenance of water resources infrastructure is the daily activities that should be done by the water resources manager to ensure that all water demand in the basin will be fulfilled according to the agreement of water availability.
- Stakeholder participation should be considered in water resources management in a river basin level. Different interest and perception of stakeholders related to water resources utilization and management should be harmonized in a coordination body.

9.3.12   Water Resources Sustainability

The final goal of water resources management in a river basin is to maintain its sustainability. It means that water resources utilization will support economic activities in the basin which resulted economic development. But this economic development will be sustainable if social equity can be enjoyed by all people in the basin, and the environment maintained.

9.4   Conclusions

From the above discussion, it can be concluded that

Water resources management in Indonesia is based on river basin territory, which is categorized as international, inter-province, national-strategic river basin territory which are under the authority of the Central Government, inter-regency/city river basin territory which are under the authority of the provincial governments, and in-one regency river basin territory which are under the authority of the regency governments.

Differences in water resources potential and social economics in each river basin resulted different complexity of the water resources management problems. Developing a comprehensive, integrated, and environmentally sound water resources management, the complexity of the problems should be clearly understood as the basis for developing water resources management pattern in a river basin territory.

This paper has developed a framework of water resources management in river basin level, in which the development of water resources policies and strategies should be based on physical conditions, social-economic-cultural conditions, and institutional arrangements in each river basin. However, the final goal of water resources management is the sustainability of water resources.

Acknowledgements This study is a result of the author's involvement through different role as a resources person, invited speaker, team member, and coordinator in several water resources management study in Indonesia.

Reference

Bhat A (2008) The politics of model maintenance: the murray darling and brantas river basins compared. Water Altern 1(2):201–218

Fulazzaky MA (2014) Challenges of integrated water resources management in Indonesia. Water 6(7):2000–2020

Global Water Partnership (2000) Integrated Water Resources Management, Technical Advisory Committee Background Papers, No. 4, Stockholm, Sweden

Law No. 11 (1974) on Watering

Law No. 7 (2004) on Water Resources

Minister of Public Works Regulation No. 39/PRT (1989) on River Basin Division

Minister of Public Works Regulation No. 11A/PRT (2006) on Criteria and River Basin Decision

Presidential Decree No. 12 (2012) on River Basin Division

Team of Indonesia Country Paper (2009) River Basin Organization on Managing Water Resources in Indonesia. Regional Workshop on River Basin Organization and Management, Indonesia Water Partnership, Directorate General of Water Resources, NARBO and GWP-SEA. Yogyakarta, 17–19 June 2009

Triweko RW (2014) Water security for Indonesia: academician perspective (in Indonesian). Indonesia Water Learning Week (IWLW), Water Security for Indonesia: Examining the Water-Energy-Food Nexus, Jakarta, 24–26 November 2014

Chapter 10
A Study on Developing Theoretical Framework, Dimensions, and Indicators of Urban Water Security in Indonesia

J. E. Wuysang and S. B. Soeryamassoeka

Abstract Increasing economic activity in urban areas is driving Indonesia's economic growth. This causes massive rural to urban migration. With this rapid urbanization some cities currently facing the critical condition of clean water and the services of urban wastewater. The significant increase in water demand has resulted in water demand exceeding the natural availability of the supply. Similarly, the problems of pollution, land subsidence, and damage to groundwater are growing. Due the reasons, Indonesian cities should implement the concept of Integrated Urban Water Management as a part of Integrated Water Resources Management that integrates water and sanitation services to improve urban water security. By analyzing the definition of water security from previous studies and literature and examining the complexity of urban water problems in Indonesian cities, further develop the theoretical framework of urban water security and determine the dimensions and indicators that affect urban water services in Indonesia. The theoretical framework of urban water security that are considered to describe urban water security in Indonesia consist of six dimensions and fifteen indicators, such as: Household Water Security (indicators: access to piped water, access to non-piped water, access to sanitation), Environmental Water Security (indicators: access to urban wastewater treatment, urban water drainage), Economic Water Security (indicators: access to water resources, access to improve sanitation), Water Security Independency (indicators: access to government water reserved storage, access to household water reserved storage), Resilience to Water Related Disaster (indicators: access to updated weather information, level of understanding of technology, urban condition), and Institutions (indicators: institution, regulation, community involvement).

J. E. Wuysang (✉) · S. B. Soeryamassoeka
Parahyangan Catholic University, Bandung, Indonesia
e-mail: humkoler@unpar.ac.id; mail@upb.ac.id

S. B. Soeryamassoeka
e-mail: untan_59@untan.ac.id

J. E. Wuysang
Panca Bhakti University, Pontianak, Indonesia

S. B. Soeryamassoeka
Tanjungpura University, Pontianak, Indonesia

© Springer Nature Switzerland AG 2021   123
M. Babel et al. (eds.), *Water Security in Asia*, Springer Water,
https://doi.org/10.1007/978-3-319-54612-4_10

Keywords IWRM · IUWM · Water security · Theoretical framework · Urban water security

10.1 Introduction

Indonesia is an archipelago with abundant water resources besides being on the equator with quite high rainfall throughout the year, with the average annual rainfall reaches about 2,500 mm. However, by the rapid population and the economy growth, massive land reform, some regions in Indonesia have experienced water supply shortage. That should be a main concern is a city with some central of activities such as economy and governance, industry, and a region with high density of population. The growing city will automatically be followed by the increasing of water demand, water pollution, surface run off and ground water as a consequence of the majority of the population and industries could not have access to the clean water. In 2015, Directorate General of Cipta Karya (2015) mentioned that only 67.7% urban population in Indonesia which have adequate water supply, the rest of it experienced so limited in acquiring clean water supply, while the wastewater treatment is still 77.15%. Due to the limitation of clean water availability therefore the people and the industry that could not have the access from water pipe supply, conducting ground water exploration with boreholes. Over abstraction of ground water will have impact to declining of groundwater table and land subsidence. The Center of Groundwater and Environmental Technology of The Ministry of Energy and Mineral Resources mentioned that land subsidence of some cities in Indonesia such Jakarta, Yogyakarta and Semarang are around 5 to 30 cm each year.

Due the reasons, Indonesian cities should implement the concept of Integrated Urban Water Management that integrates water and sanitation services to improve urban water security. It is necessary to acquire some dimensions and indicators that could explain the urban water security where the main goal of urban water security is improving urban water services for urban people. Water security has many parameters, including supplies of sufficient good quality water to users and the environment, mitigating risks hazards and avoiding conflicts over shared waters. Therefore, the dimensions and indicators selected will reflect of urban water security in Indonesia.

10.2   Material and Methods

10.2.1   Literature Study

10.2.1.1   Water Security, Definitions and Dimensions

Water security defined as every person has access to enough safe water at affordable cost to lead a clean, healthy and productive life, while ensuring the environment is protected and enhanced (Global Water Partnership 2014). Lautze and Manthrithilake (2012) stated that a conceptual of water security is proposed that contains four components: basic needs, agricultural production, the environment, and risk management and Grey and Sadoff (2007) stated a component that contain some issues related to water for national security or independence component. Table 10.1 illustrates the degree of differentiation in scope and variables of analysis used by different disciplines and organisations (Cook and Bakker 2012).

Table 10.1 Narrow disciplinary framing of water security (Cook and Bakker 2012)

Discipline	Water security focus or definition
Agriculture	Input to agricultural production and food security
Engineering	Protection against water-related hazards (floods, droughts, contamination, and terrorism)
	Supply security (percentage of demand satisfied)
Environmental	science Access to water functions and services for humans and the environment
	Water availability in terms of quality and quantity
	Minimising impacts of hydrological variability
Fisheries, geology/ geosciences, hydrology	Hydrologic (groundwater) variability
	Security of the entire hydrological cycle
Public health	Supply security and access to safe water
	Prevention and assessment of contamination of water in distribution systems
Anthropology, economics, geography, history, law, management, political science	Drinking water infrastructure security
	Input to food production and human health/wellbeing
	Armed/violent conflict (motivator for occupation or barrier to cooperation and/or peace)
	Minimising (household) vulnerability to hydrological variability
Policy	Interdisciplinary linkages (food, climate, energy, economy, and human security)
	Sustainable development
	Protection against water-related hazards
	Protection of water systems and against floods and droughts; sustainable development of water resources to ensure access to water functions and services
Water resources	Water scarcity
	Supply security (demand management)
	Green (versus blue) water security

Indonesia Ministry of Public Work (2012) stated that, water security draws upon keys dimension from national water legislation. Water security as defined in Indonesia's Water Resources Laws: conservation of water resources, utilisation of water resources, and control of damage from water. The indicators are: water utilisation (sufficient, stressed, shortage), flood management (no of people, share of people affected, hindrance for traffic), erosion and sedimentation (low, medium, high impacts), and water quality (quality of river water compared to standard).

Asian Water Development Outlook (2013), developed a comprehensive framework for national water security (Fig. 10.1). It was an outcome-based approach and crafts a comprehensive vision of water security recognising the need for security in households, economies, cities, the environment, and resilient communities. The framework transforms the vision of water security into a quantitative assessment in five key dimensions. The key dimensions of water security are related, interdependent, and should not be treated in isolation. Measuring water security by aggregating indicators in these key dimensions recognises their interdependencies. Increasing water security in one dimension may simultaneously increase or decrease security in another dimension and affect overall national water security.

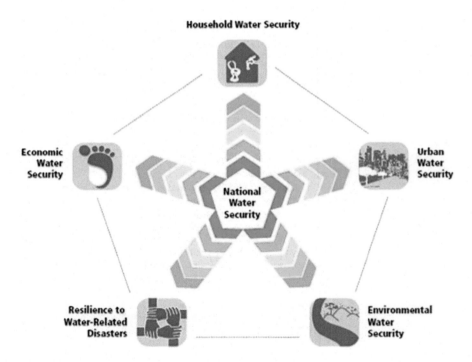

Fig. 10.1 Water security framework of five interdependent dimensions (AWDO 2013)

10.2.2 Water Security at Scale

Water security relies on effectively integrating water resources management at various scales, in particular at national, river basin, and local scales and includes the essential elements of economic efficiency, social equity, and environmental sustainability. People assign meaning to the concept of water security depending on the scale at which it is applied. Most reports to date address water security at a national scale. This, together with food and energy security, underlines the critical importance of water security to countries sustainable development. Most commonly, water security is addressed at country, river basin, city, and community scales. In some cases, water security may be considered for a specific region or unit, such as a large metropolitan area, a delta, or an island (Beek and Arriens 2014).

10.2.3 Potential Indicator for Urban Water Security Key Dimensions

The selection of potential indicators and/or sub-indicators to develop an urban water security key dimension should meet the criteria of relevance, availability, credibility, universality, and statistical independence. Additionally, they should consider the water quantity and quality aspects (*state*) of the channel, the main driving processes (climate and human pressures) affecting them, and existing (or lacking) societal responses. Once the urban water security indicators are developed, they must go through sensitivity analysis and validation processes, using a wide range scenarios, before their final application (Global Water Partnership 2014).

10.2.4 Integrated Water Resources Management and Water Security

Integrated water resources management (IWRM) is a well-established framework in the water sector which has been adopted by governments in all regions and at all levels of economic development. Defined by the Global Water Partnership as "a process which promotes the coordinated development and management of water, land and related resources, in order to maximize the resultant economic and social welfare in an equitable manner without compromising the sustainability of vital ecosystems," it takes into account both human and ecological needs (Global Water Partnership 2000).

IWRM is an approach that can be adopted at multiple scales. It is often associated with river basin-level management, but its principles can be applied at all spatial scales, from the local community to international level, (Biswas 2008). The city is a promising level for the adoption of IWRM both theoretically and practically as it

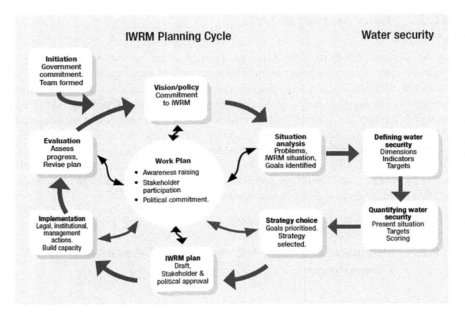

Fig. 10.2 Factoring water security into the IWRM planning cycle (Beek and Arriens 2014)

corresponds to existing administrative and political units and to the spatial reach of much existing water infrastructure (Van de Menee et al. 2011).

Integrated Water Resources Management (IWRM) and water security have the same general objective – improving the conditions related to water for human well-being. Water security can never be achieved because conditions will change, demand for water will continue to grow, and limited financial resources will constrain. IWRM will help improve water security although improvements will depend largely on the amount and quality of resources invested in the effort. IWRM and water security are symbiotic, and this is factored into the continuous IWRM planning cycle (Fig. 10.2). An important step in the planning cycle is the situation analysis in which the problems are identified, and goals set. Water security quantifies those goals by identifying the dimensions of water security and specifying indicators to measure this, preferably including clear targets. In practice, IWRM offers a framework for addressing water-related problems and issues (Beek and Arriens 2014).

Indonesian cities are facing complex urban water problems which are inter-related among water supply, water pollution, water related disasters, groundwater degradation, and poor solid waste management. Improving urban water security, Indonesian cities should implement the concept of Integrated Urban Water Management as a part of Integrated Water Resources Management which integrate water supply service, wastewater management, stormwater management, groundwater management, and municipal solid waste management into a municipal public utility office. Urban water management has a mutual influence with water resources management in the river basin level. Consequently, coordination and cooperation among cities and regencies should be developed in river basin level (Triweko 2014, 2016).

10.2.5   Methodology of Study

This research is mainly focused on literature study of urban water related problems and determine some parameters that affect to urban water security in Indonesia. Analysis was done to the literature related to water supply service in the cities in Indonesia, which are assumed representing the general water supply condition in other cities.

First step is collecting empirical material, this step involves the collection of relevant empirical material for the study, all literature and official documentation on the urban water or water security is gathered in a specific sequence in an attempt to draw a systematic explanation leading to best outcome.

Second step is building the complexity scheme of urban water problems in Indonesian cities. Urban water problems originate from economic activities, this causes urbanization, resulting in urban population growth, land use change, clean water needs and sanitation services, if infrastructure is inadequate then there will be floods, inundation, accumulation of solid waste, lack of clean water, land subsidence, groundwater degradation and sea water infiltration, beside the climate change, overlapping responsibilities of institutions, low awareness of compliance with regulations and low community involvement in implementing regulations also affect.

Third step is building the theoretical framework urban water security in Indonesia which provided the underlying basis for dimension selection.

Fourth step is to determine the relevant dimensions of urban water security in Indonesia. Examples include household water security, economic water security, and environmental water security. The dimensions selected will depend on the situation, and the specific objectives of stakeholders and decision-makers.

Fifth step is to determine the indicators that reflect the key dimensions. Indicators selection involved the selection of appropriate indicators for the field of research given their relevance to current issues, their appropriateness to the area in question, their scientific and analytical basis plus their ability to effectively represent the issues they are designed for (measurability). This involved an investigation of indices such as the Water Poverty Index (Lawrence et al. 2003; Sullivan et al. 2003; Garriga and Foguet 2010), Sustainability Index (Attari and Mojahedi 2009; Carvalho et al. 2009; Sondoval-Solis et al. 2011; Linhos and Ballweber 2015), Water Stress Index (Falkenmark and Lundqvist 1989), Human Development Index (Central Bereau Statistic of Indonesia 2015), Water Security Index (Lautze and Mathrithilake 2012; Asian Water Development Outlook 2013). For example, indicators and sub-indicators in access to piped water, access to non-piped water and access to sanitation could measure household water security.

10.3   Results and Discussion

Figure 10.3 shows the problems that give affects to urban water security with their phenomena in Indonesian cities. For this reason, a theoretical framework for urban water security was developed that can describe the situation. From the explanation, the theoretical framework of urban water security in Indonesia can be explained as follows. The theoretical framework of urban water security in Indonesia (Fig. 10.4) consist of six key dimensions, such as: 1) Household Water Security (indicators: access to piped water, access to non piped water, access to sanitation); 2) Environmental Water Security (indicators: access to urban wastewater treatment, urban water drainage); 3) Economic Water Security (indicators: access to water resources,

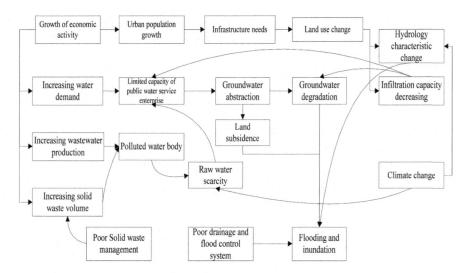

Fig. 10.3 Complexity of urban water problems in Indonesian cities

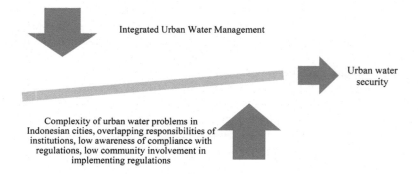

Fig. 10.4 The implementation plan for integrated urban water management in Indonesia

access to improve sanitation); 4) Water Security Independency (indicators: access to government water reserved storage, access to household water reserved storage); 5) Resilience to Water Related Disaster (indicators: access to updated weather information, level of understanding of technology, urban condition), and 6) Institutions (indicators: institution, regulation, community involvement). The purpose of indicator in every key dimension explains what will be measured and all of substantial indicators urban water security.

A purpose of the first key dimension is measuring the satisfaction of household water and sanitation need, the second key dimension is measuring how well drainage channel developed and managed to sustain ecosystem service, the third key dimension is measuring how the city use water to sustain industry and energy, the fourth key dimension is measuring the urban independency and potential raw water availability to support urban development that sensitive of water demand, the fifth key dimension is measuring how the city encounter and recover from the impact of water related disaster, and the sixth key dimension is measuring how the institutions and regulations sustain water continuity, wastewater treatment, drainage system and water consumption. The following figure are the complexity of urban water problems in Indonesian cities (Fig. 10.3), the implementation plan for Integrated Urban Water Management in Indonesia (Fig. 10.4), the theoretical framework of urban water security (Fig. 10.5) and a matrix of urban water security key dimensions and indicators (Table 10.2).

Fig. 10.5 Theoretical framework of urban water security in Indonesia

Table 10.2 Indonesia urban water security key dimensions and Indicators

No	Key Dimension	Indicator	
1	Household Water Security	Access to piped and non-piped water supply, access to improved sanitation and hygiene	*Access to piped water*
			Access to non-piped water
			Access to sanitation
2	Environmental Water Security	Wastewater treatment, drainage system, channel health, pressure/treat to channel system from disturbance and pollution, resilience to alterations to natural flows by infrastructures development and biological factors	*Access to urban wastewater treatment*
			Urban water drainage
3	Economic Water Security	industry and energy	*Access to water resources*
			Access to improved sanitation
4	Independency Water Security	Available raw water source to fulfill present and future water demand	*Access to government water reserved storage*
			Access to household water reserved storage
5	Resilience to Water Related Disaster	Consist the type of hazard, measuring exposure, basic population vulnerability, infrastructure and non infrastructure approach	*Access to updated weather information*
			Level of understanding of technology
			Level of understanding of urban society
6	Institutions	Compliance levels of society of the regulations and the cooperative of society to keep the implementation of the regulations	*Institutions*
			Regulations
			Communities involvement

10.4 Conclusions

From the above discussion, it can be concluded that.

1. Indonesian cities are facing complex urban water problems which are inter-related among water supply, sanitation, water related disasters, overlapping

responsibilities of institutions, low awareness of compliance with regulations and low community involvement in implementing regulations.

2. Indonesian cities should implement the concept of Integrated Urban Water Management as a part of Integrated Water Resources Management which integrates water supply and sanitation services to improve urban water security.

3. Urban water security theoretical framework, dimensions and indicators can be used to monitor and evaluate the progress of a city in improving its urban water services.

4. Six key dimensions those are acquired from the main dimension in theoretical framework are connected with the level of urban water security in Indonesia, and every key dimension has certain relation with one another and could be integrated in the interconnected framework.

5. The selected of Indonesia urban water security key dimensions, consist of six key dimensions and fifteen indi cators, such as: Household Water Security (indicators: access to piped water, access to non piped water, access to sanitation), Environmental Water Security (indicators: access to urban wastewater treatment, urban water drainage), Economic Water Security (indicators: access to water resources, access to improve sanitation), Water Security Independency (indicators: access to government water reserved storage, access to household water reserved storage), Resilience to Water Related Disaster (indicators: access to updated weather information, level of understanding of technology, urban condition), and Institutions (indicators: institution, regulation, community involvement).

Acknowledgements The authors would like to thank Robertus Wahyudi Triweko and Doddi Yudianto from Civil Engineering Department, Parahyangan Catholic University Bandung, Indonesia for their support and assistance conducting this study.

References

Asian Water Development Outlook (2013) Measuring Water Security in Asia and the Pacific. Asia Development Bank. Manila

Attari J, Mojahedi SA (2009) Water Sustainability Index: Application of CWSI for Ahwaz County. World Environmental and Water Resources Congress 2009: Great Rivers © 2009 ASCE

van Beek E, Arriens WL (2014) Water Security, Putting the Concept into Practice. Tec Background Paper. Global Water Partnership, Stockholm

Biswas AK (2008) Integrated water resources management: is it working? Int J Water Resour Dev 24:5–22

De Carvalho SCP, Carden KJ, Armitage NP (2009) Application of a sustainability index for integrated urban water management in Southern African cities: case study comparison – Maputo and Hermanus. ISSN 0378-4738 = Water SA, vol 35, no 2 (Special WISA 2008 edition) 2009 ISSN 1816-7950 = Water SA (on-line)

Central Bereau Statistic of Indonesia (Badan Pusat Statistik Indonesia) (2015) Statistik Indonesia. Badan Pusat Statistik Indonesia, Jakarta

Cook C, Bakker K (2012) Water security: debating an emerging paradigm. Glob Environ Chang 22:94–102

Directorate General of Cipta Karya (2015) Kebijakan Pembinaan dan Pengembangan Infrastruktur Permukiman. Kementerian Pekerjaan Umum dan Perumahan Rakyat Republik Indonesia, Jakarta

Falkenmark M, Lundqvist J (1989) Toward water security; political determination and human adaptation crucial. Nat Resour Forum 21(1):37–51

Garriga RG, Foguet AP (2010) Improved method to calculate a water poverty index at local scale. J Environ Eng 136(11):1287–1298

Global Water Partnership (2000) Towards Water Security: A Framework for Action. GWP. Stockholm, Sweden, pp 1–18

Global Water Partnership (2014) Assessing Water Security with Appropriate Indicators. Proceedings from the GWP Workshop. Global Water Partnership, Stockholm

Lautze J, Manthrithilake H (2012) Water security: old concepts, new package, what value. Nat Res Forum 36(2):76–87

Lawrence P, Meigh J, Sullivan C (2003) The water poverty index: an international comparison. United Nations J 27(3):189–199

Ministry of Public Works (DGWRD) (2012) Java Water Resources Strategic Study. Kementerian Pekerjaan Umum/Bappenas, Jakarta

Sandoval-Solis S, McKinney DC, Loucks DP (2011) Sustainability index for water resources planning and management. J Water Resour Plan Manag 137:381–390

Sullivan CA, Meigh JR, Giacomello AM, Fediw T, Lawrence P, Samad M, Mlote S, Hutton C, Allan JA, Schulze RE, Dlamini DJM, Cosgrove W, Priscoli DJ, Gleick P, Smout I, Cobbing J, Calow R, Hunt C, Hussain A, Acreman MC, King J, Malomo S, Tate EL, O'Regan D, Milner S, Steyl I (2003) The water poverty index, development and application at the community scale. Nat Res Forum 27(2003):189–199

Triweko RW (2014) Ketahanan Air Untuk Indonesia: Pandangan Akademisi. Indonesia Water Learning Week (IWLW), Water Security for Indonesia: Examining the Water-Energy-Food Nexus, Jakarta, 24–26 November 2014

Triweko RW (2016) Urban Water Security for Indonesian Cities. Joint Seminar between Parahyangan Chatolic University and Hohai University, Bandung, 18 July 2016

Van de Meene SJ, Brown RR, Farrelly MA (2011) Towards understanding governance for sustainable urban water management. Glob Environ Chang 21:1117–1127

Chapter 11
Micro Level Vulnerability Assessment of Estuarine Islands: A Case Study from Indian Sundarban

R. Hajra, A. Ghosh, and T. Ghosh

Abstract The estuarine islands of Indian Sundarban are highly vulnerable due to erosion, flooding, and population pressure. Quantitative assessment of vulnerability at micro-region level in this delta is still limited, which acts as a barrier to formulate effective implementable management strategies for increasing resilience. This study quantified the level of vulnerability using 783 household survey data from twenty-seven sampled 'Mouza' (lowest administrative boundary; village) through cluster random sampling, from the Sagar, Ghoramara, and Mousuni islands. Mouza level analysis has been carried out following the 'Composite Vulnerability Index' (CVI) considering physical and social variables like erosion, housing condition, electrification, population density, adult education, sanitation, and economic status as the major percentage of people 'Below Poverty Line' (BPL). Result suggests that all these Mouzas are within the rank of 'moderate to high' vulnerability and susceptible to both socio-economic and environmental changes. The 'hot spot' Mouzas identified are Sapkhali, Ghoramara, Bankimnagar, Shibpur, and Baliara, while 66% of remaining Mouzas are at the edge of vulnerability, and further deterioration of socio-ecological conditions may convert those into highly vulnerable. Considering the ecological importance of this delta, identification of thrust areas and action research to increase social resilience should be a priority to minimize the existing vulnerable conditions in this region.

Keywords CVI · BPL · Estuarine Islands · Vulnerability · Resilience · Indian Sundarban

A. Ghosh · T. Ghosh (✉)
School of Oceanographic Studies, Jadavpur University, Kolkata, West Bengal, India

R. Hajra
Department of Geography, Polba Mahavidyalaya, Hooghly, West Bengal, India

© Springer Nature Switzerland AG 2021   135
M. Babel et al. (eds.), *Water Security in Asia*, Springer Water,
https://doi.org/10.1007/978-3-319-54612-4_11

11.1   Introduction

Coastal regions are always being the favourite destination for human from ancient civilizations due to rich biodiversity and multiple livelihood options. Marine ecosystems play a vital role in regulating climate as they are a major carbon sink and oxygen source (MOEF 2009). According to the estimates of the United Nations in 1992 more than half of the world population lives within 60 km of the coast, which has been projected to rise to almost three quarters by 2020 (MOEF 2009). Natural factors and climate change, including aggravated cyclone, erosion, and sea level rise along with anthropogenic factors including human adaptation processes are making these coastal areas vulnerable. Vulnerability can be defined according to the Intergovernmental Panel on Climate Change (IPCC) Second Assessment Report (IPCC 1996) as *'the extent to which climate change may damage or harm a system; it depends not only on system sensitivity but also the ability to adapt to new climatic conditions'*. In addition to these broad-scale influences, local factors have also been shown to affect vulnerability at household level (Eriksen et al. 2005). Household level vulnerability is also associated with social vulnerability, which is the exposure of groups or individuals to stress as a result of social and environmental change, where stress refers to unexpected changes and disruption to livelihoods (Adger 1999). This study tries to assess the composite vulnerability analysis to focus not only on the environmental factors but also socio-economic pressures. However, being the vulnerable region, coastal areas have got immense importance in such type of studies. The similar situation is also observed in the coastal West Bengal and more specifically, the areas of the Indian Sundarban Delta (ISD) system within a dynamic deltaic environment. Land loss due to submergence and increases in soil salinity and land fragmentation all make life of the islanders of ISD difficult. Around 34% of the 4.6 million people residing on different islands of the ISD are under extreme poverty (Hazra et al. 2014). To get a more complete scenario Mouza level vulnerability assessment of islands of ISD are the focus of this study.

The western boundary of ISD is the major focus of this study, Sagar, Ghoramara and Mousuni Island in specific (Fig. 11.1). Administratively, Sagar Island is classified as a "CD-Block", a collection of Mouzas or villages, which is the largest island in ISD with 206,844 total populations (Census 2011). It is bounded by the Hooghly River to the north and west, the Muriganga River to the east, and the Bay of Bengal to the south. Overall, there are 42 villages on Sagar Island. Ghoramara Island located to the north of Sagar Island covers an area of 4.43 km^2 (Ghosh et al. 2014) with a total population of 5,193 (Census 2011). The major villages on this island include Khasimara, Baishnabpara, Hathkola, Baghpara, Raipara, Mandirtala, Chunpuri, Lakshmi Narayanpur, and Khasimara Char. Out of these, Khasimara Char, Lakshmi Narayanpur, Khasimara, and Baishnabpara have already been disappeared (Jana et al. 2012; Ghosh et al. 2003; Ghosh et al. 2014). Mousuni Island covers 24 km^2, and, according to 2011 census figures, is home to 3,340 families and 22,073 people (Census 2011). This island is encircled by the Muriganga/Bartala River to the west and northwest, Pitt's Creek/Chenayer River to the east, and the Bay of Bengal

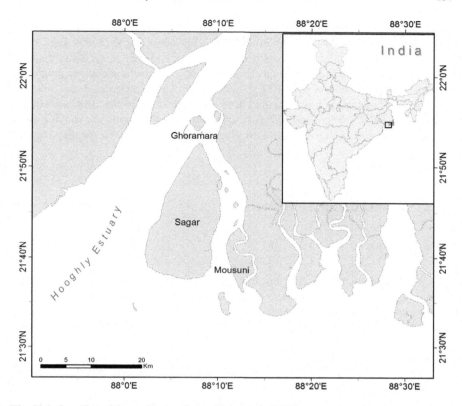

Fig. 11.1 Location of the study area. *Source* Hajra et al. (2017)

to the south (WWF 2010). Mousuni Island is a single Gram Panchayat (GP) unit, with four Mouzas—Bagdanga, Kusumtala, Baliara and Mousuni—under Namkhana CD Block.

The deltaic islands of the Indian Sundarban are highly vulnerable to frequent embankment failures, submergence and flooding, beach erosion, cyclone and storm surges (Hazra et al. 2002, 2014; Hajra et al. 2017). The study area is characterised by a flat alluvial plain with a ground elevation varying from 2.10 to 2.75 m above mean sea level (Purkait 2009). The Hooghly estuarine is classified as a mixed macro tidal estuary with tidal ranges 4.64 m (Ghosh et al. 2001). The rate of erosion estimated from 1990 to 2015 was 0.2, 0.02, and 0.08 km^2 in Sagar, Ghoramara and Mousuni Island, respectively (Hajra et al. 2016). The surface air temperature increase rate has been reported about 0.019 °C with a projected 1 °C by the year 2050 (Hazra et al. 2002) and rainfall increased 2.88 mm per year (Hajra and Ghosh 2016). Sagar and Mousuni are densely populated areas with decadal growth rate of 1.5 and 1.03% at 2011. Ghoramara has experienced negative growth rate of −0.08% from 2001 to 2011, which is mainly attributed to out-migration. Population density in these islands is 872, 1,172, 786 persons per km^2, respectively.

About 89% of the total population of the region is dependent on mono-crop cultivation (Aman paddy) (Hazra et al. 2002; Hajra and Ghosh 2016). Less profit in paddy cultivation, high degree of occupational dependence on agriculture and its rapidly declining income share is an indication of a higher incidence of poverty in these islands. Overall less than 30% of houses are of pakka structure and only 258 persons per thousand populations had access to electricity in the Blocks of Indian Sundarban (HDR 2009). Both climatic and non-climatic events are making these islands vulnerable. This study aims to assess the vulnerability to identify the 'hot spot' Mouzas.

11.2 Material and Methods

11.2.1 Data

This study uses data collected through direct interviews with households within the study area and secondary data including demography, economic patterns, and housing conditions. The basic demographic data on total population, rate of literacy, and workers status have been gathered from the Census Report (2011) of Government of India. Socio-economic data are obtained from economic and statistical report published by India and State Government.

To assess actual socio-economic conditions of the area, a thorough household-level socio-economic survey was conducted in the sampled villages of the study area from 2012 to 2013. A two-stage cluster random sampling was used for this study. In the first stage, Mouzas were chosen randomly from all three islands; in the second stage, households within these areas were selected randomly for the survey. The survey was carried out through direct interviews in 52% of the inhabited Mouzas of Sagar Block, including Ghoramara (22 Mouzas out of 42) and 100% of the inhabited Mouzas (four Mouzas) of Mousuni Gram Panchayat of Namkhana Block. One-to-one direct interviews were conducted with members of 783 households from 27 villages of the study area, consisting of a total number of surveyed populations of 4,500.

11.2.2 Analytical Method

This study aims to assess village/Mouza level vulnerability of Sagar, Ghoramara and Mousani Island considering both physical and socio-economic criteria. Physical parameters like erosion, house types, electrification, social parameters like population density, adult education achievement level, sanitation, and economic parameters like percentage of people below poverty line have been considered. It has been found that there are lots of vulnerability assessment methodologies used, substantiated and documented through different literatures. However, taking cues from those literatures (Gornitz and White 1992; Gornitz et al. 1997; Klein and Nicholls 1999; Pethick and Crooks 2000; Thieler and Hammar-Klose 1999; Thieler et al. 2002; Hazra et al. 2014) Composite Vulnerability Index (CVI) formulation based on the square root of geometric mean of the ranked variables has been adopted for this study.

Rankings have been done following the modified equal step method and the guidelines of International Federation of Red Cross and Red Crescent Societies (2007). The difference of maximum and minimum values is divided by three to get class interval and then values are classified accordingly. Each parameter was normalized with percentage value and then ranked as high-medium–low (or 3, 2, 1), and finally resolved in a GIS platform. The Mouza level vulnerability ranks are assessed based on predetermined criteria. Few examples are given below in Tables 11.1, 11.2 and 11.3.

Similarly, seven of such components of Physical, Social and Economic vulnerabilities have been assessed at the Mouza level from primary survey data. The geometric mean of different vulnerability classes with rank has been derived to assess Mouza level Vulnerability Rank (VR).

$$VR = \sqrt[7]{(V_1 * V_2 * V_3 \ldots \ldots V_7)} \tag{11.1}$$

The different vulnerability maps of study islands have been prepared separately in interactive GIS platform. The Composite Vulnerability Ranks of the families have plotted on risk zones I, II, and III.

11.3   Results and Discussion

11.3.1   Vulnerability Assessment

The major principle used for vulnerability assessment in this study is the composite vulnerability– which is a product of physical and socio-economic vulnerability. The chosen seven parameters have been assigned with certain ranks based on their data values. The coastal vulnerability index is calculated by geometric mean of seven

Table 11.1 Example I - Dwelling vulnerability

Vulnerability rank	Vulnerability class	Criteria
1	Less vulnerable	% of pakka[1] houses in each Mouza
2	Moderately vulnerable	% of semi-pakka houses in each Mouza
3	Highly vulnerable	% of kachcha[1] houses

(1) Definition of "pakka" and "kachcha": see paragraph below "*Vulnerability of housings*".

Table 11.2 Example II - Sanitation vulnerability

Vulnerability rank	Vulnerability class	Criteria
1	Less vulnerable	More than 75% of families having good sanitation
2	Moderately vulnerable	55–75% of families having good sanitation
3	Highly vulnerable	Less than 55% of families having good sanitation

Table 11.3 Example III - Economic vulnerability

Vulnerability rank	Vulnerability class	Criteria
1	Less Vulnerable	Less than 50% of families under poverty line
2	Moderately vulnerable	50–65% of families under poverty line
3	Highly vulnerable	More than 65% of families under poverty line

variables. Composite Vulnerability Index is commonly used to integrate and assess the situation of whole ongoing system.

11.3.2 Vulnerability of Erosion

Erosion has been identified as a major threat, which brings changes to the ongoing system. All three islands have experienced land loss. The rate of erosion estimated from 1990 to 2015 was 0.2, 0.02, and 0.08 km^2 in Sagar, Ghoramara, and Mousuni Island, respectively. Mouazs having landloss more than 0.081 km^2 per year are identified as the most vulnerable Mouzas, and Mouzas having landloss less than 0.04 km^2 are considered as less vulnerable Mouzas. Rest of Mouzas have been taken as moderately vulnerable (ranges between 0.041 to 0.08 km^2 land losses). It is evident from the erosion vulnerability map that coastal parts such as Sibpur, Doblat, Beguakhali, Sumatinagar, Kachuberia, Muriganga, and Ghoramara of Sagar Block, and Baliara of Mousuni Block are worst affected due to erosion (Fig. 11.2).

11.3.3 Vulnerability of Dwellings

House types have been classified as Kachcha (both wall and roof made of natural materials like mud, sand, thatch and other low-quality materials), Semi-pakka (either wall or roof made of finished and/or rudimentary materials like raw wood planks, palm/ bamboo, stone, unpolished cement) and Pakka (both wall and roof made of finished material like polished wood, marble, ceramic tiles, polished cement). This classification has been followed by Ministry of Rural Development, Govt. of India (2006) and National Family Health survey by International Institute for Population Sciences (IIPS) (2007). Mouzas with more than 30% Kachcha surveyed houses

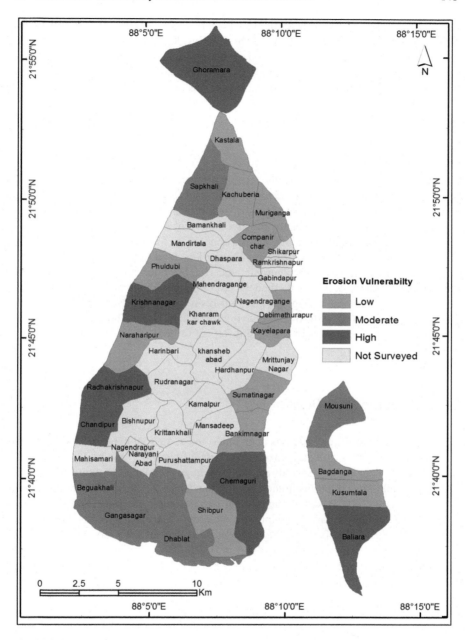

Fig. 11.2 Erosion vulnerability map

are considered as highly vulnerable, less than 15% Kachcha houses are considered less vulnerable, and others with 15–29% are considered as moderately vulnerable. Sapkhali (50%), Kusumtala (44%), Ghoramara (43%), Mousuni (38%), Bagdanga (31%), and Krittankhali (31%) are the most vulnerable with respect to house types that are most susceptible to damage during storms and coastal flooding (Fig. 11.3).

11.3.4 Electrification

Traditionally stable electric supply has been regarded as the engine of economic progress. Electricity is a key factor behind socio-economic development and in providing vital services that improve the quality of life. Hence rural electrification could be an influencing driver for the assessment of socio-economic vulnerability. To assess the present situation of electric connectivity Mouza level vulnerability has been measured. The surveyed Mouzas having less than 10% houses without electricity are considered as low vulnerable, and 10–30% houses without electricity are moderately vulnerable and more than 30% of houses without electricity are considered as highly vulnerable (Fig. 11.4). It has been found that 20 out of 27 surveyed Mouzas i.e. almost 75% of total are having more than 30% of houses without electricity (Fig. 11.4). This condition would impact on the systems sustainability in a negative way.

11.3.5 Population Density

Population density (Fig. 11.5) has been classified as less vulnerable (<600 persons per km^2), moderately vulnerable (601–900 persons per km^2), and highly vulnerable (>901 persons per km^2). The study islands have huge population pressure. Among 27 Mouzas Bankimnagar has the highest population density (1,263) and Ramkrishnapur the lowest one (260). Based on the ranking Bankimnagar, Kochuberia, Rudranagar, Phuldubi, Company Char, Chandipur are the most vulnerable Mouzas. On striking feature has been found from density analysis. The total area of Ghoramara recorded in the census as 10 km^2 throughout 40 years (1971–2011). But this study reveals that Ghoramara has area of 4.4 km^2. Ghoramara has density of around 1,180 persons per km^2 based on estimated 4.4 km^2 area, and it falls under highly vulnerable zone.

11.3.6 Education Level

Social vulnerability has been assessed by adult education achievement level. Education vulnerability (Fig. 11.6) has been classified as per ranking based on percentage of adults achieve secondary level education. Level of education is directly related to the

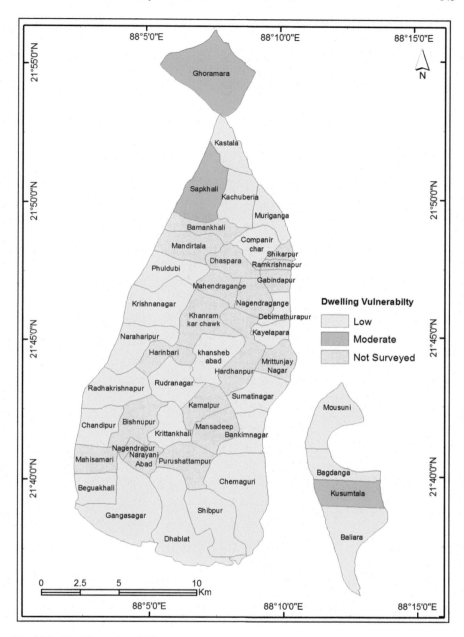

Fig. 11.3 Dwelling vulnerability map

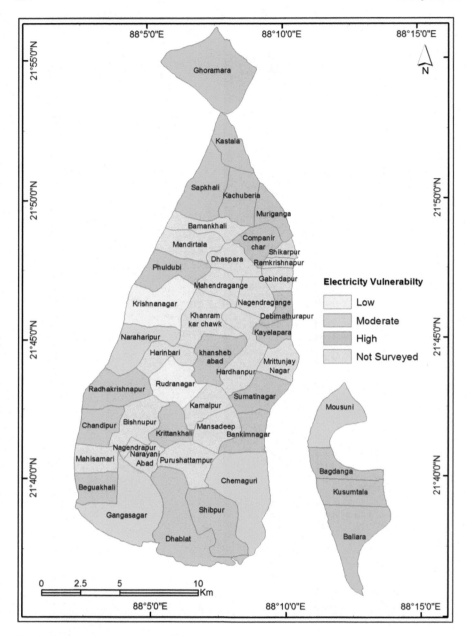

Fig. 11.4 Electrification vulnerability map

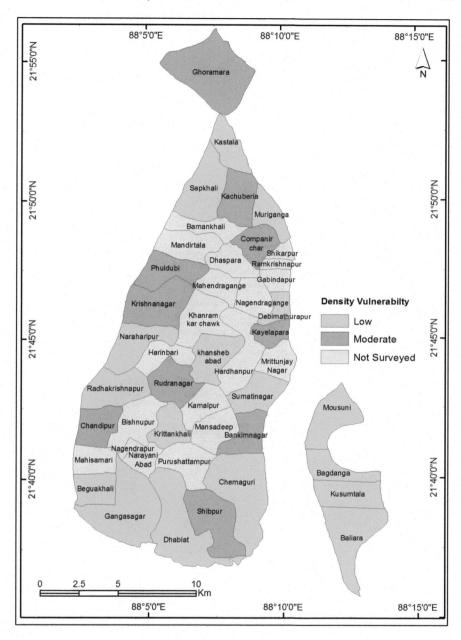

Fig. 11.5 Population density vulnerability map

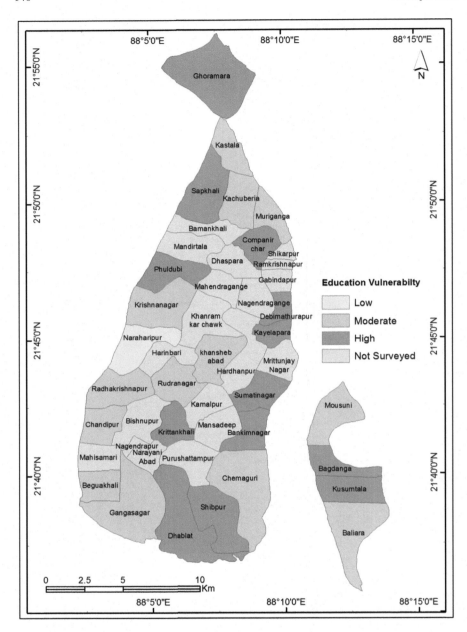

Fig. 11.6 Secondary education achievement vulnerability map

social and economic structure of a society. Education vulnerability has been classed as low (more than 40% adults achieve secondary education), moderately (20–40% adults achieve secondary education), and highly vulnerable (less than 20% adults have secondary education). Sapkhali (7%), Krittankhali (11%), Debimathurapur (12%), Company char (18%), Phuldubi (18%), Sumatinagar (18%), and Shibpur (18%) are the most vulnerable Mouzas.

11.3.7 Sanitation Vulnerability

Sanitation can be considered as one of the several parameters for determination of the health status of a population. The type of sanitation in Sundarban is classified as permanent, semi-permanent, and open space (i.e. households having no sanitary toilet within/close to the premises). Mouzas having less than 55% sanitation facilities are considered as highly vulnerable (Fig. 11.7). Those with sanitation facilities between 55–75% are considered as moderately vulnerable, and Mouzas with more than 75% sanitation facilities are evaluated to be less vulnerable. Bagdanga, Baliara, Bankimnagar, Begualkhali, Chemaguri, Companychar, and Chandipur are the most vulnerable Mouzas.

11.3.8 Economic Vulnerability

Economic vulnerability of the studied Mouzas is calculated on the basis of household survey. Families living below poverty line have been estimated in reference to Indian standard poverty line value. Vulnerability based on percentage of below poverty families has been classified into three classes: low (<50% families under poverty line), moderate (50 to 64% under poverty line), and high vulnerability (>65% under poverty line). Bagdanga (81%), Baliara (76%), Bankimnagar (74%), Beguakhali (69%), Chandipur (69%), and Chemaguri (67%) are most poverty-stricken Mouzas and highly vulnerable (Fig. 11.8).

11.3.9 Composite Vulnerability Index

Depending on the evaluations of the multi-dimensional aspects of vulnerability of the 27 Mouzas of study area, a composite vulnerability map has been generated. Ghoramara Island has been identified as the most vulnerable one. But the composite vulnerability rank is showing Sapkhali, Shibpur, Dhablat, Gnagasagar, and Phuldubi Mouza of Sagar Island, and Baliara, and kusumtala of Mousuni Island are at the threshold of highly vulnerable condition. Any degradation of condition may

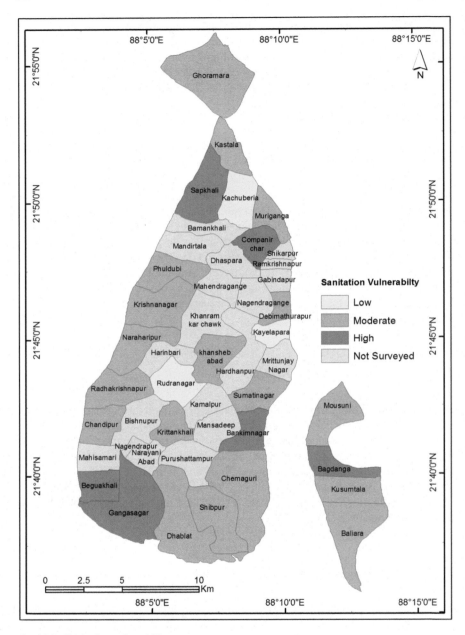

Fig. 11.7 Sanitation vulnerability map

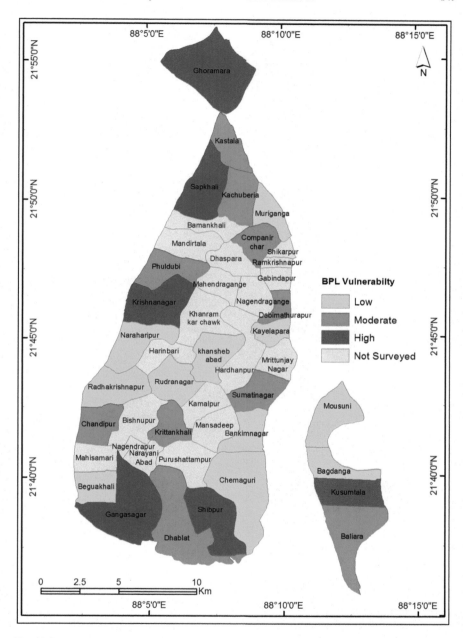

Fig. 11.8   Economic vulnerability map

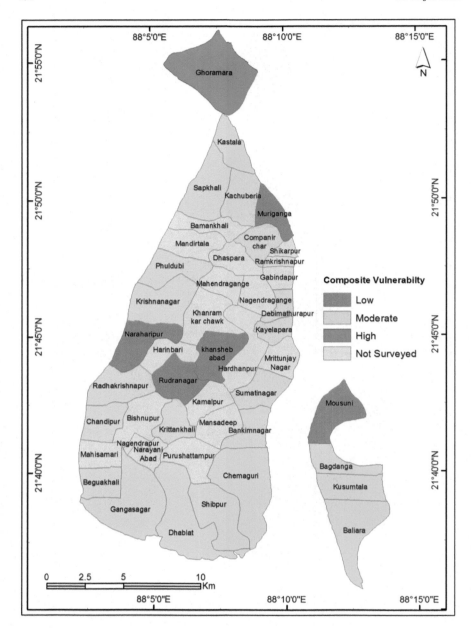

Fig. 11.9 Composite vulnerability map

worsen their condition (Fig. 11.9). Another finding of this study is that the vulnerability ranking of villages is declining with the distance from the coast. This may be attributed to erosional effect, storm surge, and saline water inundation.

11.4 Conclusions and Recommendations

These deltaic islands are highly vulnerable to the impacts of climate change and natural hazards. In addition, large proportions of the households in the study area are poor and thus have limited access to resources and facilities. The main objective of this paper was to assess Mouza level vulnerability using socio-ecological variables. The results of the study confirmed that surveyed Mouzas are vulnerable to specific socio-ecological conditions, and many of them are very close to highly vulnerable value. This issue needs prior attention. In this context, several policy recommendations for improving system condition have been given here.

First, the intensity of disaster impact depends on societal resilience, and with proper developmental and management plans the adverse effect of disaster on households could be controlled. It is evident from the field survey that the embankments are only maintained by mud pilling, become unstable and often collapse during equinox tide in September. The situation is worst at Ghoramara, Sibpur, Beguakhali, Sumatinagar, and Muriganga Mouza of Sagar, and Baliara Mouza of Mousuni. Construction of stable embankment along the sea and tidal rivers, and regular repair and maintenance of embankments of utmost importance in these islands.

Second, 22% of the surveyed Mouzas including Ghoramara, Sapkhali, Gangasagar, Sibpur, and Kusumtala are under the level of 'severely vulnerable' mostly due to poverty. These Mouzas need immediate economic incentives from authority. There should have focus on adaption strategies in income generation sector. For example, more people need to be trained and engaged in cultivation of salt resistant varieties of paddy. New Self Help Groups can be set up or existing ones need to be strengthened through different training programs preferably at Mouza level to reduce the poverty level through income generation. Planned rehabilitation could be a useful measure to reduce the disaster risk and consequently the poverty.

Third, 75% of surveyed population does not have secondary education. Upgrading of education facilities, night schools, vocational trainings, and technical schools for adults could be a useful measure for this problem. Elimination of gender disparities and equal access to educational facilities should be ensured by proper policy implementation.

Fourth, planned housing structure is needed especially in Sapkhali, Kusumtala, and Ghoramara Mouzas. Sapkhali, Compani Char, Beguakhali, and Bagdanga have less than 50% of houses with no proper sanitation facility. Monitoring and upgrading sanitation condition from Gram Panchayat (base level administrative unit) and awareness among locals about sanitation and healthy life can be raised by organizing special classes at schools, clubs, etc. Meeting the energy requirement is one of the prime ways to achieve an economic stability, as 81% of the surveyed Mouzas are deficient of energy. Under Rajiv Gandhi Grameen Vidyutikaran Yojana, West Bengal Renewable Energy Development Agency (WBREDA) is working in these islands for alternate energy resource using a combination of fossil fuel with a renewable source like wind power or biomass or entirely solar photovoltaic. But establishment of solar panel at household level or access to solar energy is often beyond reach for poor households. This issue should be addressed by giving subsidies, expanding the distribution network, and emphasis on supplying industrial consumers as well as domestic consumers. Given the vulnerability assessment and socio-ecological issues, the lessons learned from the study area could be adopted in other delta regions and beyond.

References

Adger WN (1999) Social vulnerability to climate change and extremes in coastal Vietnam. World Dev 27(2):249–269

Census Report of India (2011) Government of India

Eriksen S, Brown K, Kelly PM (2005) The dynamics of vulnerability: locating coping strategies in Kenya and Tanzania. Geogr J 141:287–305

Ghosh T, Bhandari G, Hazra S (2001) Application of a 'bio- engineering' technique to protect Ghoramara Island (Bay of Bengal) from severe erosion. J Coast Conserv 9:171–178

Ghosh T, Bhandari G, Hazra S (2001) Assessment of land use/ land cover dynamics and shoreline changes of Sagar Island through remote sensing. Paper presented at the 22nd Asian Conference on Remote Sensing, Singapore, 5–9 November 2001

Ghosh T, Hajra R, Mukhopadhyay A (2014) Island erosion and afflicted population: crisis and policies to handle climate change. In Leal Filho W, et al (eds) International perspectives on climate change, climate change management, pp 217–225. Springer, Switzerland. https://doi.org/10.1007/978-3-319-04489-7_15

Gornitz VM, White TW (1992) A coastal hazards database for the U.S. East coast. ORNL/CDIAC-45, NDP-043 A. Oak Ridge National Laboratory, Oak Ridge, Tennessee, U.S

Gornitz VM, Beaty TW, Daniels RC (1997) A coastal hazards data base for the U.S. West coast. ORNL/CDIAC-81 NDP-043 C. Oak Ridge National Laboratory, Oak Ridge, Tennessee, U.S. https://doi.org/10.3334/CDIAC/ssr.ndp043c

Hajra R, Ghosh A, Ghosh T (2016) Comparative Assessment of Morphological and Landuse/Landcover Change Pattern of Sagar, Ghoramara, and Mousani Island of Indian Sundarban Delta Through Remote Sensing. In Hazra et al (ed) Environment and Earth Observation: case studies in India. Springer. https://doi.org/10.1007/978-3-319-46010-9_11

Hajra R, Ghosh T (2016) Migration pattern of Ghoramara Island of Indian Sundarban-identification of push and pull factors. Asian Acad Res J Soc Sci Humanit 3(6):186–195

Hajra R, Szabo S, Tessler Z, Ghosh T, Matthews Z, Foufoula-Georgiou E (2017) Unravelling the association between the impact of natural hazards and household poverty: evidence from the Indian Sundarban delta. Sustain Sci 1–12. Springer

Hazra S, Das I, Samanta K, Bhadra T (2014) Impact of climate change in Sundarban Area West Bengal, India. School of Oceanographic Studies. Earth Science and Climate Book- 9326/17.02.00. Report submitted to Caritus India, SCiAF

Hazra S, Ghosh T, Das Gupta R, Sen G (2002) Sea Level and associated changes in the Sundarbans. Sci Cult 68(9–12):309–321

HDR (2009) Human Development Report, South 24 Parganas District. Development and Planning Department. Government of India

IIPS (2007) National Family Health Survey (NFHS-3), 2005–06: India: Volume I. Mumbai: International Institute for Population Sciences and Macro International

International Federation of Red Cross and Red Crescent Societies (2007) Vulnerability and capacity Assessment Guideline. Geneva

IPCC (1996) Second Assessment Report: the Science of Climate Change. Cambridge University Press, Cambridge, p 564

Jana A, Sheena S, Biswas A (2012) Morphological change study of Ghoramara Island, Eastern India using multi temporal satellite data. Res J Recent Sci 1(10):72–81

Klein RJT, Nicholls RJ (1999) Assessment of coastal vulnerability to climate change. Ambio 28(2):182–187

MOEF (2009) World Bank Assisted Integrated Coastal Zone Management Project. Environmental and Social Assessment. MOEF-ICZM Project Report. Govt of India, Centre for Environment and Development, Thiruvantapuram

Pethick J, Crooks S (2000) Development of a coastal vulnerability index: a geomorphological perspective. Environ Conserv 27:359–367

Purkait B (2009) Coastal erosion in response to wave dynamics operative in Sagar Island, Sundarban delta. India, Front. Earth Sci. China 3(1):21–33

Thieler ER, Hammar-Klose E (1999) National assessment of coastal vulnerability to sea-level rise. Preliminary results for U.S. Atlantic Coast. Open-file report 99-593. U.S. Geological Survey, Reston, VA, 1 sheet. https://pubs.usgs.gov/of/1999/of99-593/. Accessed 20 May 2015

Thieler ER, Williams SJ, Beavers R (2002) Vulnerability of U.S. National Parks to sea-level rise and coastal change. U.S. Geological Survey fact sheet FS 095-02. U.S. Geological Survey, Reston, VA, 2 p. https://pubs.usgs.gov/fs/fs095-02/. Accessed 20 May 2015

WWF (2010) Sundarbans: Future Imperfect Climate Adaptation Report (ed.) Danda, A. WWF.

Chapter 12
Assessing Water Security at District Level: A Case Study of Bangkok, Thailand

A. Onsomkrit, M. S. Babel, V. R. Shinde, and V. P. Pandey

Abstract This study applies the water security framework in Bangkok city, Thailand at different spatial (five districts: Sathon, Lat Phrao, Nong Chok, Bangkok Noi and Nong Kheam) and temporal (2007–2014) scales. The framework consists of an index (water security index, WSI), five dimensions, ten parameters and twelve indicators. The five dimensions cover following aspects for a water-secure city: (1) Every person at household level can access easy piped water supply with sufficient quantity to meet basic needs and be of acceptable safe quality; (2) Water productivity of economic activities in the city is reasonably high; (3) Water-bodies in the city are not affected from pollution and contamination generated in the city; (4) Acceptable level of water-related disaster to people in the city that consider urban flood damage and rainfall variation; and (5) Water governance is effective for resource use, management, and capacity development. Results showed that the overall status of water security in the study districts are "moderate" level. We also found that some of the indicators and parameters were found inappropriate at district (or sub-city) level due to lack of data. This application demonstrates suitability of the framework at a city-scale rather than sub-city (or district), as data of a finer scale are lacking at sub-city scale and most of the actions to secure water are taken at the city level.

Keywords Bangkok · Framework · Index · Indicators · Water security

12.1 Introduction

Early societies arose along rivers and lakes because these natural assets provided significant water security for domestic use, irrigation, transport, fisheries, and power (from water wheels to hydropower). However, as population and water demands have grown, man-made infrastructure became necessary to supplement natural assets to maintain water security. There is evidence of dams built over 4,000 years ago to store water in ephemeral rivers (Fahhlbusch 2001 cited in Grey and Sadoff 2007).

A. Onsomkrit · M. S. Babel (✉) · V. R. Shinde · V. P. Pandey
Asian Institute of Technology, Khlong Luang, Pathum Thani, Thailand
e-mail: msbabel@ait.ac.th

© Springer Nature Switzerland AG 2021
M. Babel et al. (eds.), *Water Security in Asia*, Springer Water,
https://doi.org/10.1007/978-3-319-54612-4_12

In countries with adequate wealth and technology, dams, wells, canals, pipelines, and municipal water supply systems have been built to provide storage and delivery functions much similar to the lakes, rivers and springs and treatment plants that provide the cleansing functions to wetlands and aquifers. From natural to man-made and from small-scale to large, a continuum of options has evolved to meet the challenges of water security (Grey and Sadoff 2007).

Awareness is growing that water is a scarce and precious resource, which must be carefully managed if frightening future water crises are to be avoided (GWP 2000). The world has recognized that secure livelihoods, strong economies, and sustainable ecological systems depend on the availability of water, and principles for its management have been internationally agreed (GWP 2000). The urgent challenge that remains is to translate these agreed principles into practice. In 1997 at the first World Water Forum, professionals from around the world agreed that a mass mobilization and awareness campaign was needed to alert people and politicians to the fragile status of the world's water resources (GWP 2000).

The concept of water security emerged in the 1990s and has evolved significantly since then (Cook and Bakker 2012). Held in 2000, the 2nd World Water Forum conceptualized the first definition focused to tackle the global water crisis by directing the need to work towards "water security" as an overarching goal. Therefore, the Global Water Partnership (GWP) introduced an integrative definition of water security which gave the definition as "water security at any level from the household to the global means that every person has access to enough safe water at affordable cost to lead a clean, healthy and productive life, while ensuring that the natural environment is protected and enhanced." On the other hand, Grey and Sadoff (2007) have defined water security as "the availability of an acceptable quantity and quality of water for health, livelihoods, ecosystems and production, coupled with an acceptable level of water-related risks to people, environments and economies". Also, Cook and Bakker (2012) showed that framings of water security have become more diverse, expanding from an initial focus on quantity and availability of water for human uses to water quality, human health, and ecological concerns. Thus, they proposed four interrelated themes that dominated the published research on water security: water availability; human vulnerability to hazards; human needs (development-related, with an emphasis on food security); and sustainability. However, the concept of water security remains largely unquantified (Lautze and Manthrithilake 2012 cited in GWP 2014), due to which developing and managing water resources to achieve water security remains at the heart of the struggle for growth, sustainable development, and poverty reduction (Grey and Sadoff 2007).

Furthermore, scale is also critical in assessing water security because of the scalar variability of hydrology, as illustrated by a study (Vorosmarty et al. 2010 cited in Cook and Bakker 2012). Cook and Bakker (2012) argued that different disciplines tend to focus on different scales. Development studies tend to use national scales, hydrologists often focus on watershed scales from the regional to the national, and social scientists regularly work at the community scale (Cook and Bakker 2012). Moreover, water security assessment at the national scale can mask significant variations in security at the local scale (Vorosmarty et al. 2010 cited in Cook and Bakker

2012). Dun et al. (2009) also stated that indicators are often site-specific or framed for a specific scale that may not be transferable to other scales (e.g. national or international level indicators may not be sensitive enough to identify water issues at a local level) (GWP 2014). Thus, this study aims to apply the water security framework to assess status of water security at the district scale of Bangkok city, Thailand.

12.2 Water Security Assessment Framework

From a previous study, Onsomkrit (2015) established the water security framework at city scale that defines water security as–*Every person at household level can access easy piped water supply with sufficient quantity to meet basic needs and be of acceptable safe quality; Water productivity of economic activities in the city is reasonably high; Water-bodies in the city are not affected from pollution and contamination generated in the city; Acceptable level of water-related disaster to people in the city that consider urban flood damage and rainfall variation; and Water governance is effective for resource use, management, and capacity development.* The framework consists of an index (Water Security Index, WSI), five dimensions (reflecting the definition), ten parameters and twelve indicators to measure the dimensions that are showed in Fig. 12.1. This framework was developed by using DPSIR (drivers, pressures, state, impact, and response) framework and SMART (specific, measurable, actionable, relevant, and time-bound) criteria. Moreover, a scoring system from 1 (water insecurity) to 5 (very high-water security) was employed to represent and interpret the water security situation. Equal weight was also applied for this framework. Different weights to dimensions, parameters, and indicators can be given based on their importance in a particular city. Multi-criteria decision analysis techniques such as Analytical Hierarchy process can be applied to define the weights to the indicators, parameters, and dimensions. The methodology of this study is based on water security framework from Onsomkrit (2015), which has been applied to assess water security status at district level.

12.3 Study Area

This study considers a small spatial scale (district level) in Bangkok, Thailand. The districts include Sathon, Lat Phrao and Nong Chok (east of Chao Phraya river), and Bangkok Noi and Nong Kheam (west of Chao Phraya river) as shown below in Fig. 12.2. The summary of characteristics of the selected districts is given in Table 12.1.

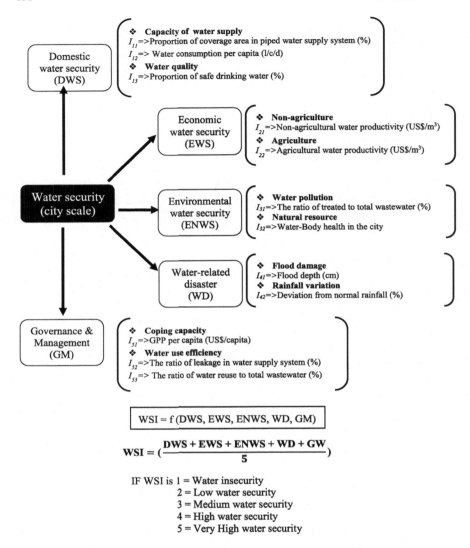

Fig. 12.1 The overall water security framework at city scale (*Source* Onsomkrit 2015)

12.4   Results and Discussion

The water security index is calculated for the five selected districts of Bangkok. The temporal scale of the study was eight years, from 2007 to 2014. The values of each indicator have been presented in Table 12.2. Some of the data was available at provincial level and hence such values were assumed to be same for the districts within a province. The results of application of the framework at district level, as

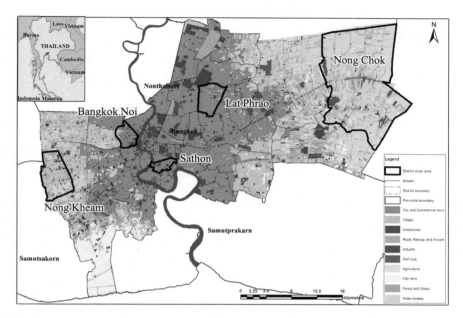

Fig. 12.2 Selected districts for assessing water security status

shown in Fig. 12.3, indicates that the level of water security of selected districts is of moderate level over the 8–year period. Sathon, Lat Phrao, Bangkok Noi and Nong Kheam have fluctuating trends over the years while for the Nong Chok district, there is a slight increase in water security status compared to the other four districts.

Firstly, the *domestic water security dimension* in each district shows a high level of water security. The indicators of coverage area in water supply system and proportion of safe drinking water in each district has the same trend over the years. This dimension varied according to water consumption per capita per year. In Sathon, Lat Phrao and Bangkok Noi, which has the most urban area coverage relative to its total, people may use more water than in rural areas. Nong Chok and Nong Kheam districts have lesser extent of urban areas which might lead to people using less water from piped systems, but more from other sources such as groundwater and rainwater.

Secondly, for the analysis of *economic water security dimension*, the study was used with the same value in different scales because of data availability. This study assumed that each district has same status of economic water security. Sathon and Bangkok Noi has a very high score of water security because these districts only have non-agricultural economic activity. Lat Phrao, Nong Chok and Nong Kheam have non-agricultural and agricultural economic activities, due to which these districts experience fluctuations according to agricultural water productivity.

Thirdly, the *environmental water security dimension* in selected districts shows low water security levels, except for some years in Sathon. The indicator of the ratio of treated to total wastewater at Sathon is the highest over the years because this

Table 12.1 Summary of characteristics of selected districts of Bangkok, Thailand

Characteristic	Sathon	Lat Phrao	Nong Chok	Bangkok noi	Nong Kheam
Location within Bangkok	Inner area of Eastern part	Middle area of Eastern part	Outer area of Eastern part	Inner area of Western part	Outer area of Western part
Area (km^2) (% of total area in Bangkok)	9.33 (0.59%)	21.86 (1.39%)	236.26 (15.06%)	11.94 (0.76%)	35.82 (2.28%)
Population (persons) (% of total population in Bangkok) (2013)	83,898 (1.48%)	122,441 (2.15%)	159,962 (2.81%)	117,503 (2.07%)	151,877(2.67%)
Density (person/km^2) (2013)	8,996	5,601	677	9,837	4,239
BOD (mg/L)/DO (mg/L) of canals in districts (average value in 2013)	18.36/0.14	7.63/0.58	3.55/3.54	7.36/2.96	4.97/2.69
WWTP (% of coverage area/Capacity of treated (m^3/day)	1 plant (100%/200,000)	–	–	–	1 plant(around 30%/157,000)
Economic activities	Non-agriculture	Non-agriculture and Agriculture	Non-agriculture and Agriculture	Non-agriculture	Non-agriculture and Agriculture
Urban flood occurrences (Maximum flood depth (cm)) (2013)	2 (20 cm)	3 (15 cm)	–	1 (15 cm)	1 (20 cm)

BOD = Biochemical oxygen demand; DO = Dissolved oxygen; WWTP = Wastewater treatment plant

Table 12.2 Values of Water Security Indicator in selected districts of Bangkok, Thailand

No.	Indicators	Unit	Spatial scale	Year							
				2007	2008	2009	2010	2011	2012	2013	2014
1	Proportion of coverage area in water supply system (I_{11})	%	Sathon [E]	96	100	100	100	100	100	100	100
			Lat Phrao [E]	96	100	100	100	100	100	100	100
			Nong Chok [E]	96	100	100	100	100	100	100	100
			Bangkok Noi [W]	96	100	100	100	100	100	100	100
			Nong Kheam [W]	96	100	100	100	100	100	100	100
2	Water consumption per capita (I_{12})	l/c/d	Sathon [E]	211	210	204	205	198	205	205	206
			Lat Phrao [E]	256	254	255	257	244	253	253	256
			Nong Chok [E]	129	136	141	149	149	160	165	177
			Bangkok Noi [W]	253	252	255	263	270	286	282	291
			Nong Kheam [W]	151	150	149	152	155	163	166	170
3	Proportion of safe drinking water (I_{13})	%	Sathon [E]	100	100	100	100	100	100	100	100
			Lat Phrao [E]	100	100	100	100	100	100	100	100
			Nong Chok [E]	100	100	100	100	100	100	100	100
			Bangkok Noi [W]	100	100	100	100	100	100	100	100
			Nong Kheam [W]	100	100	100	100	100	100	100	100
4	Non-agricultural water productivity (I_{21})	US$/m³	Sathon [E]	240.0	260.4	254.4	289.3	341.6	354.0	381.8	362.2
			Lat Phrao [E]	240.0	260.4	254.4	289.3	341.6	354.0	381.8	362.2
			Nong Chok [E]	240.0	260.4	254.4	289.3	341.6	354.0	381.8	362.2
			Bangkok Noi [W]	240.0	260.4	254.4	289.3	341.6	354.0	381.8	362.2
			Nong Kheam [W]	240.0	260.4	254.4	289.3	341.6	354.0	381.8	362.2

(continued)

Table 12.2 (continued)

No.	Indicators	Unit	Spatial scale	Year							
				2007	2008	2009	2010	2011	2012	2013	2014
5	Agricultural water productivity (I_{22})	US$/m³	Sathon [E]	NA	NA	NA	NA	NA	NA	NA	NA
			Lat Phrao [E]	0.20	0.31	0.37	0.46	0.34	0.24	0.24	0.43
			Nong Chok [E]	0.20	0.31	0.37	0.46	0.34	0.24	0.24	0.43
			Bangkok Noi[W]	NA	NA	NA	NA	NA	NA	NA	NA
			Nong Kheam[W]	0.20	0.31	0.37	0.46	0.34	0.24	0.24	0.43
6	The ratio of treated to total wastewater (I_{31})	%	Sathon[E]	65	60	43	68	100	95	66	62
			Lat Phrao[E]	0	0	0	0	0	0	0	0
			Nong Chok[E]	0	0	0	0	0	0	0	0
			Bangkok Noi [W]	0	0	0	0	0	0	0	0
			Nong Kheam [W]	21	22	22	20	20	20	21	18
7	Water-Body health in the city (I_{32})	0–100	Sathon [E]	8.5	11.7	3.8	4.4	0.6	4.6	0.6	4.5
			Lat Phrao [E]	7.7	16.8	6.8	8.5	6.8	9.3	5.2	3.3
			Nong Chok [E]	46.5	47.1	49.4	54.6	47.9	53.3	49.0	41.1
			Bangkok Noi [W]	40.6	38.5	40.7	29.9	41.5	34.0	33.5	32.4
			Nong Kheam [W]	33.3	29.9	27.9	31.1	35.2	33.8	29.2	22.9
8	Flood depth (I_{41})	cm	Sathon [E]	20	30	30	30	30	20	20	20
			Lat Phrao [E]	10	–	–	–	15	–	15	–
			Nong Chok [E]	–	–	–	–	–	–	–	–
			Bangkok Noi [W]	15	–	20	20	20	15	15	20
			Nong Kheam [W]	–	–	18	25	–	–	20	20

(continued)

Table 12.2 (continued)

No.	Indicators	Unit	Spatial scale	Year 2007	2008	2009	2010	2011	2012	2013	2014
9	Deviation from normal rainfall (I_{42})	mm	Sathon [E]	17.2	25.5	55.6	42.6	52.4	17.8	16.3	−12.9
			Lat Phrao [E]	17.2	25.5	55.6	42.6	52.4	17.8	16.3	−12.9
			Nong Chok [E]	17.2	25.5	55.6	42.6	52.4	17.8	16.3	−12.9
			Bangkok Noi [W]	−19.3	4.9	3.9	−10.9	7.4	0.4	−19.2	7.7
			Nong Kheam [W]	−19.3	4.9	3.9	−10.9	7.4	0.4	−19.2	7.7
10	GPP per capita (I_{51})	US$/capita	Sathon [E]	10,517	11,074	10,507	11,818	13,073	14,042	15,191	13,922
			Lat Phrao [E]	10,517	11,074	10,507	11,818	13,073	14,042	15,191	13,922
			Nong Chok [E]	10,517	11,074	10,507	11,818	13,073	14,042	15,191	13,922
			Bangkok Noi [W]	10,517	11,074	10,507	11,818	13,073	14,042	15,191	13,922
			Nong Kheam [W]	10,517	11,074	10,507	11,818	13,073	14,042	15,191	13,922
11	The ratio of leakage in water supply system (I_{52})	%	Sathon [E]	31.6	29.1	28.6	27.3	25.5	24.5	21.8	21.1
			Lat Phrao [E]	23.0	19.4	19.5	19.5	19.2	18.2	16.9	16.3
			Nong Chok [E]	23.3	21.2	17.4	16.5	17.4	15.0	13.3	12.5
			Bangkok Noi [W]	35.1	36.1	34.7	30.6	33.2	30.4	31.5	28.9
			Nong Kheam [W]	34.4	31.5	30.1	29.7	32.4	31.7	34.9	31.1
12	The ratio of water reuse to total wastewater (I_{53})	%	Sathon [E]	9.5	8.5	11.6	7.8	11.8	13.2	10.5	10.3
			Lat Phrao [E]	0.0	0.0	0.0	0.0	0.0	0.0	0.0	0.0
			Nong Chok [E]	0.0	0.0	0.0	0.0	0.0	0.0	0.0	0.0
			Bangkok Noi [W]	0.0	0.0	0.0	0.0	0.0	0.0	0.0	0.0
			Nong Kheam [W]	0.2	0.3	0.3	0.3	0.3	0.3	0.4	0.3

Note [E] = District is in eastern part and [W] is District is in western part of Bangkok

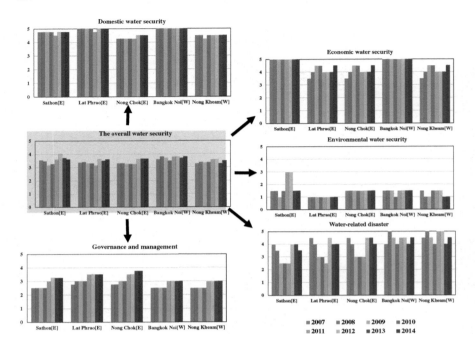

Fig. 12.3 Water Security Index of five selected districts in Bangkok

district has a Wastewater Treatment Plant (WWTP) covering the entire area. The indicator of water-body health in the city at Nong Chok has the highest value over the years because this district has less urban area and less population density that affects the wastewater generated in the area. However, water security in each district is still low because water quality of natural water sources in each district has seen an increased deterioration due to population growth and economic development and inadequacy of resources to treat wastewater.

Fourthly, for the *water-related disaster dimension*, the value of each indicator in selected districts has a fluctuating trend over the years. The level of water security varied from low to high. The score of this dimension in Sathon is the lowest among the five selected districts. Also, districts in western area (Bangkok Noi and Nong Kheam) have a higher score than districts in eastern area.

Finally, the fifth dimension–*Governance and Management*–includes three indicators to reflect management capacity. The score of GPP per capita, as the first indicator, shows high level of water security over the years. The leakage in water supply system of Sathon, Lat Phrao and Nong Chok districts increased slightly over the years while the leakage in Bangkok Noi and Nong Kheam has the same trend over the years. The score of the water use indicator in each district is same as water security/insecurity level over the years. Although water reuse in Sathon is of the highest magnitude, this value also ranks in the level of water insecurity. Hence, this dimension varied according to the leakage in water supply system and the ratio of water reuse in the respective district.

12.5 Conclusions

The overall status of water security in the selected districts has been at a moderate level over 8 years. Furthermore, the result of the study shows that the domestic water security dimension is of the highest level while the environment water security dimension is of the lowest level in each district. By applying the framework at different spatial scales, it was found out that the city scale can mask significant variations in water security situation at the district scale. The study also found that some of the indicators and parameters were found inappropriate at district (or sub-city) scale due to lack of data. Hence, this framework could be better to apply at a city scale and not at a sub-city (or district) scale because there is a lack of finer–scale data and most of the interventions of water security are implemented at the city scale.

Acknowledgements The authors would like to acknowledge the financial support provided by the Asia-Pacific Network for Global Change Research (APN) and the Asian Institute of Technology, Thailand, to conduct this research. The authors would also like to thank the Metropolitan Waterworks Authority, the Department of Drainage and Sewerage, the Department of Groundwater Resources, the Royal Irrigation Department, the Thai Meteorological Department, the National Statistical Office and the Office of the National Economic and Social Development Board for providing important and valuable data for the research.

References

Ait-kadi M, Arriens WL (2012) Increasing Water Security-A Development Imperative. Retrieved August 2014, from Global Water Partnership Website. https://www.gwp.org/ToolBox/PUBLIC ATIONS/Perspectives-Papers/
Asian Development Bank (2013) Asian Water Development Outlook 2013: Measuring Water Security in Asia and the Pacific, Manila, Philippines
Cook C, Bakker K (2012) Water security: debating an emerging paradigm. Glob Environ Chang 22(1):94–102
Global Water Partnership. (2000). Towards Water Security: A Framework for Action. Stockholm, Sweden
Global Water Partnership (2014) Proceedings from the GWP workshop: Assessing water security with appropriate indicators, Stockholm, Sweden
Grey D, Sadoff CW (2007) Sink or Swim? Water security for growth and development. Water Policy 9(6):545–571
Onsomkrit A (2015) Development of a Water Security Index at City Scale: A Case Study of Bangkok. Master research study, Asian Institute of Technology (2015) Asian Institute of Technology, Thailand
Sinyolo S, Mudhara M, Wale E (2014) Water security and rural household food security: empirical evidence from the Mzinyathi district in South Africa. Food Secur 6(4):483–499

Chapter 13
Assessment of Water Security in Indonesia Considering Future Trends in Land Use Change, and Climate Change

S. Tarigan and Y. Kristanto

Abstract Java Island as the heart of Indonesia economic and where 60% of Indonesian population live are facing serious water security. Existing average index of water use in Java Island has exceeded 100%. Both land use and climate change factors contribute further to the water security problems in Indonesia. But, there is lack of investigation on the relative contribution of both factors to the future water security in Indonesia. Knowledge on their relative contribution enables us to select the most appropriate mitigation or adaptation options. In this study we used SWAT hydrological model to investigate the relative contribution of both factors on a water quantity as one component of water security in Ciliwung watershed. Like all watersheds in Java Island, the Ciliwung watershed is experiencing rapid development of residential area which affect the water cycle in the watershed. Result of the SWAT simulation showed that the relative impact of a climate change on the future water security is more pronounced than the impact of land use change. This result is mostly consistent with another study with different typology of watershed in Sumatra Island where the land use change is dominated by rapid development of monoculture plantations instead of rapid development of residential area.

Keywords Climate change · Land use change · SWAT model · Water security

13.1 Introduction

Situated at a humid tropical region, Indonesia has rich water resources. Annual quantity of available water exceeds demands in most regions in Indonesia. Nevertheless, water scarcity phenomenon in Indonesia occurs in almost all regions during dry season. Seasonal variability of water availability is very high in most part of Indonesia. Into a certain extent, reservoir construction can buffer the seasonal fluctuation. But, the storage capacity of existing reservoir in Indonesia is still lacking and it's getting worse due to the sedimentation problems.

S. Tarigan (✉) · Y. Kristanto
IPB University, Bogor, Indonesia

© Springer Nature Switzerland AG 2021
M. Babel et al. (eds.), *Water Security in Asia*, Springer Water,
https://doi.org/10.1007/978-3-319-54612-4_13

The water security in Indonesia is affected considerably by land use and climate change (Boer and Faqih 2004; Naylor et al. 2007; Tarigan et al. 2016; Tarigan. 2016a, 2016b). Land use change and population growth intensify the water scarcity problem during dry season in the future. Land use chan0ges enhance seasonal flow fluctuation and population growth trigger the deterioration of watersheds leading to sedimentation problem and reducing reservoir capacities. It is said that climate change will increase evaporation due to the warming effect. According to The Department for International Development (DFID) and World Bank (2007) in 2020, it was projected that mean temperature increased somewhere between 0.36 to 0.47 °C compared to 2000. Higher evapotranspiration reduces water availability in dry season (Mcintyre 2007). Climate change alters rainfall, evaporation, and river discharge. To a certain extent, temperature rise will have effects on agriculture and thus food security. The largest concern for Indonesia with regards to the impacts of climate change could be the risk of decreased food security.

The objective of this study are; (a) to assess streamflow water quantity as one component of a water security in Indonesia considering future trends in land use change and climate change and (b) to differentiate relative contribution of land use and climate change factors on the streamflow water quantity using SWAT model.

13.2   Material and Methods

The study was carried out in Java, Indonesia. The study area has two distinct seasons, namely wet and dry seasons. The wet season occurs from October to March and the dry season from April to September. The temperature ranges from 22 to 32 °C.

The relative impact of the land-use and climate change on a water quantity was simulated using SWAT model in Ciliwung watershed, West Java province, Indonesia. A water quantity is a component of a water security. The SWAT models quantify the water balance of a watershed on a daily basis, which can be used for the assessment of ecosystem services such as freshwater for agricultural uses, instream flows, flood risk, climate change simulation and other water resource infrastructure (Arnold et al. 2012). The application of the SWAT model has been tested in many part of the South East Asia with good performance (Marhaento et al. 2017; Wangpimool et al. 2017; Tarigan et al. 2018; Tarigan et al. 2020).

Daily meteorological datasets, baseline data or climate change projection data are required for the simulation model. For calibration and validation, daily streamflow datasets of Upper Ciliwung River at the Katulampa outlet are required. In this study, SWAT simulation requires a minimum of three years of data, first year for warming up model, second year for calibration, and third year for validation. Model input data sources for the SWAT model include soils, land use, weather, and streamflow data (Table 13.1).

After calibration procedures, the land use and climate change impact on the water quantity were simulated using the SWAT model based on the scenarios in Table 13.2.

Table 13.1 SWAT model input data

No.	Data	Year	Source
1	Daily precipitation, maximum and minimum air temperature, relative humidity, wind speed, and solar radiation	2015–2018	Meteorological Station of Citeko-Bogor (BMKG)
2	Daily streamflow data at Katulampa outlet	2016–2017	Ciliwung-Cisadane River Basin Agency (BBWS Ciliwung-Cisadane)
3	Land use/land cover	1996, 2017	Ministry of Environment and Forestry of the Republic of Indonesia (KLHK)
4	Digital elevation model 8 × 8 m	–	Geospatial Information Agency (BIG)
5	Soil map scale 1:50.000	–	Indonesian Center for Agricultural Land Resource Research and Development (BBSDLP)
7	Distance from river, main road, settlement, and forest area	–	Geospatial Information Agency (BIG)

Table 13.2 Land use and climate change scenarios for the SWAT model simulations

Scenario	Land use data	Climate data	Remarks
Baseline	2017	2017	
Land use change	2030	2017	The land use in 2030 was predicted using multilayer perceptron-neural network by QGIS Land use Change Modeler (LCM)
Climate change	2017	2030	Climate change scenario was adapted from Naylor (2007), Hulme and Sheard (1999), Tarigan and Faqih (2019)

13.3 Results and Discussion

Water for irrigation and hydropower have the biggest proportion in the total water demand in Indonesia. Total paddy field in Indonesia is 13.1 million ha, some 4.4 million ha of which are irrigated paddy fields. Total national irrigation water demand is about 6,8 $m^3 s^{-1}$ (Table 13.3). From these demand, 86% is supplied by the run-of-river diversion, 13% is supplied from reservoirs, and about 1% from groundwater.

Based on the 80% reliable flow, the average index of water use in Java is 110 and 115% in year 2015 and 2030 respectively (Fig. 13.1). Despite over supply shown by low water use index in other islands, there is drought risk during dry season due to the high river flow fluctuation and lack of reservoir storage.

Table 13.3 National water demand and supply

Island	Supply (m³ s⁻¹)	Consumptive demand (m³ s⁻¹)		Water use index
		Total	Irrigation	
Java	2,819	3,342	3,065	1.10
Sumatra	18,129	2,108	1,883	0.12
Kalimantan	28,551	593	412	0.02
Sulawesi	5,850	814	668	0.14
Papua, NTB, NTT, Bali, Maluku	30,500	901	870	0.03

Source Puslitbang Air (2011)

Fig. 13.1 Water use index in Indonesia based on district boundaries (Puslitbang Air 2011)

13.3.1 Simulation of a Land Use Change Impact on a Streamflow Water Quantity

Rapid population growth forces land use conversion into agriculture and residential area. Land use conversion into agriculture and build urea reduce infiltration rate which in turn reduce base flow in dry season.

The SWAT model showed good performance with NSE value of 0.80 and 0.76 for calibration and validation, respectively (Fig. 13.3). Compared to baseline (2017), residential area increased almost double in 2030. The streamflow in Ciliwung watershed was simulated based on the predicted land use change in 2030 (Fig. 13.2).

As it is obvious from the Fig. 13.4, the land use change in the Ciliwung watershed increased peak flow during wet season (January–March) and decreased baseflow during dry season (May–August). The increased peak flow was mainly caused by lower soil infiltration under rapid development of the residential area.

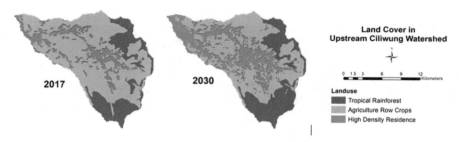

Fig. 13.2 Land use change map in Ciliwung watershed used for SWAT model simulation

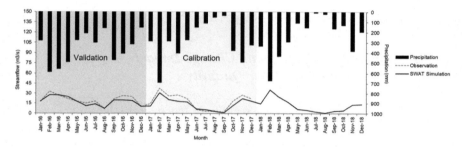

Fig. 13.3 SWAT model validation and calibration

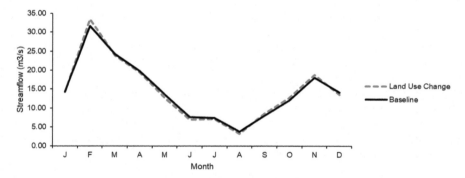

Fig. 13.4 Land-use change impact on streamflow shown by the red line representing land use in 2030 in comparison with land use in 2017 (baseline)

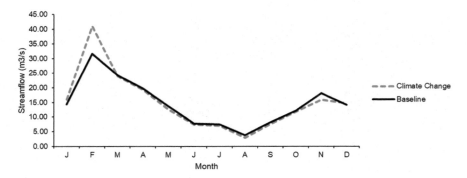

Fig. 13.5 Climate change impact on streamflow in 2030

13.3.2 Simulation of a Climate Change Impact on a Streamflow Water Quantity

The second factor that affects water security in Indonesia is climate change. The Indonesian climate is controlled primarily by the monsoon circulation system with two distinct seasons of wet and dry seasons. The climate change alters temperature and the precipitation pattern. The temperature is projected to increase in the future over all of Southeast Asia, including Indonesia. The increased temperature alters proportion of evapotranspiration of the water cycle. The higher the temperature, the higher the evapotranspiration and the lower the annual runoff volume is. There is rather consistent projection in the literatures on the temperature rise. The climate change will cause increased temperature up to 2 °C in 2050–2080 (Met Office Hadley Centre, Table 13.4). In contrast, change of precipitation pattern is more varied depending on the geographical conditions. It was projected that rainfall patterns will change considerably, with the wet season ending earlier and the length of the wet season becomes shorter (Naylor et al. 2007; Hulme and Sheard 1999). According to Naylor et al (2007), the seasonal pattern of rainfall has changed with up to 75% decrease in rainfall in the dry season (July–September). According to Hulme and Sheard (1999), during the wet season (December-February), parts of Sumatra and Kalimantan become 10 to 30% wetter by the 2080's. In contrast, rainfall pattern during the dry season (June–August) are becoming drier. Boer and Faqih (2004) reported similar projection of precipitation pattern where southern region of Indonesia the wet season rainfall has increased while the dry season rainfall has decreased, meanwhile the contrast pattern was observed in the northern region of Indonesia. Decrease of rainfall in the dry season will reduce water security sharply.

To see the impact of a climate change on water security, the SWAT model was used to simulate the streamflow using a land use map of 2017 (baseline) and climate change scenario. Climate change scenario was adapted from Naylor (2007), Hulme and Sheard (1999), Tarigan and Faqih (2019), where rainfall during wet season peak (December-February) increases 20% and during dry season peak (July–September)

Table 13.4 Impact of climate change on water resources in Indonesia

Source	Climate impact on water resources
ICCSR—Scientific Basis: Analysis and Projection of Temperature and Rainfall (2010)	The projection of changes of rainfall pattern in general is more varied than the pattern of temperature increase. There is the tendency of increasing rainfall in wet months and the decline of rainfall in dry months compared to the baseline condition (1961–1990). According to GCM output analysis, this condition will be enhanced near 2080s period. In regard to the decline of rainfall in dry months, Western Java area is the area with higher potential hazard compared to the other areas
Boer and Faqih (2004)	Analyzing rainfall data from 210 stations and compared period of 1931–1960 and 1961–1990 and reported that annual precipitation has decreased by two to three percent across all of Indonesia over the last century. But, there is significant spatial variability, where there has been a decline in annual rainfall in the southern regions of Indonesia (e.g., Java, Lampung, South Sumatra, South Sulawesi, and Nusa Tenggara) and an increase in precipitation in the northern regions of Indonesia (e.g., most of Kalimantan, North Sulawesi)
Boer and Faqih (2004)	Analyzing some GCM results including those of CSIRO Mark 2 (Australia), HadCM3 (England), and ECHAM-4 (Germany). They found a varied projection particularly for rainfall
Climate: Observations, projections and impacts (Met Office Hadley Centre)	For the A1B emissions scenario projected temperature increases over Indonesia are generally in the range of 2–2.5 °C. There is good agreement between the CMIP3 models over all of Indonesia. Precipitation changes show quite low agreement between CMIP3 models over Indonesia
Naylor et al. (2007)	There is a significant 30-day delay in the onset of monsoon season and a substantial decrease in precipitation later in the dry season
Ibrahim and Ginting (2009)	Impact of climate and land use changes on hydrologic characteristics and water resources in Indonesia: Increased temperature and increased evapotranspiration due to the climate change
Setiawan et al. (2009)	Impact of climate changes on water resources in Cidanau watershed, Indonesia: Increased temperature and evapotranspiration

(continued)

Table 13.4 (continued)

Source	Climate impact on water resources
Santoso and Forner (2006)	Climate change projections for Indonesia: Temperature increase 2.15 °C and rainfall increase 1.0–4.6% in South East Asia in 2050 but they found that GCM projections for rainfall in Indonesia have large uncertainties
Mulyantari (2009)	Flooding and rainfall trend in West Java during 1993–2005: No indication of increased flooding and rainfall intensity and frequency
Pawitan (2009)	Impact of climate change on water availability in Indonesia 2030–2050: Decreased river discharge 8.3% in 2030 and 13.8% in 2050 respectively. Contribution of climate change couldn't be determined
Babel et al. (2014)	Climate change impact on water resources and selected water use sectors (Case of Bagmati River Basin, Nepal): T_{max} is predicted to increase by 2.1 °C under A2 scenario and by 1.5 °C under B2 scenario in 2080

decreases 10 during. As it is obvious from the Fig. 13.5, the climate change in the Ciliwung watershed increases peak flow and decreases baseflow.

Based on the SWAT model simulation, both land use and climate change tend to increase peak flow during wet season. But, the climate change has more pronounced impact than land use change on the increase of the peak flow and water yield. The impact of climate change is also greater than the impact of land use change on the decrease of dry season discharge (Table 13.5).

Rather similar findings of this study was obtained from another study in different typology of a watershed in Sumatra Island where land use changed has been dominated by rapid development of monoculture plantations instead of residential area development (Tarigan and Faqih 2019). In this watershed typology, the climate change showed greater impact on the increase of peak flow than land use change

Table 13.5 Comparison of land use and climate change impact on the water quantity

	Baseline	Land use change	Climate change
Annual rainfall (mm)	3500	3500	3600
Water yield (mm)	3000	3000	3100
Peak flow (m^3 s^{-1})	49.7	56.8	75.8
Dry season discharge (m^3 s^{-1})	0.80	0.48	0.45

impact as well. But, the impact of the land use change on the dry season discharge is more significant compared to that of the climate change impact. The different finding of the study regarding the dry season discharge is due to the different watershed typology between Sumatra and Java Island. In Sumatra island a land use change is trigged by rapid development of monoculture plantations. Meanwhile, in Java island a land use change is dominated by rapid development of residential areas.

Knowing relative contribution of a land use and climate change on future water security enables government and communities to plan for appropriate combination of mitigation and adaptation measure in Indonesia. To reduce seasonal fluctuation as a result of land use change, we need to implement proper land use management such as determining minimal protection forest areas in a watershed. Forest land use has been identified by many researchers as the most effective land use in increasing water flow regulation of a watershed (Bruijnzeel, 1989; 2004). In addition to enough forest area in a watershed, effective soil and water conservation measures should be implemented in agricultural area in a watershed. The soil and water conservation measures will increase water infiltration and also reduce sediment flowing to the downstream reservoir. Both measures will greatly enhance water security in the future. Impact of climate change on water resources, especially change of precipitation pattern and extreme event, are relative difficult to predict accurately. In this respect, in addition to land use management described above, adaptation measures should be prepared for more intense flooding and drought in the future.

13.4 Conclusions

Among five big islands in Indonesia, water security in Java Island is substantially affected by land use and climate change in Indonesia. The reason is that the Java Island has already high water use index and the Java Island is continuing as the heart of economic activity and home to almost 60% of Indonesian population. Despite low water use index, other islands are still facing drought risk during dry season and more frequent flooding during wet season.

There is still lack of researches on how to differentiate the relative contribution of both factors to the water security condition in Indonesia. Knowledge on their relative contribution enables us to focus on the appropriate mitigation or adaptation options. Based on our simulation using SWAT model, the climate change had more pronounced impact on the water security compared to that of land use change impact. The result of this study is mostly consistent with another study with different typology of watershed in Sumatra Island where the land use change is dominated by rapid development of monoculture plantations instead of rapid development of residential area. In general, both climate and land use impacts reduce dry season flow and increase peak flow during wet season. Researches in this issue should be intensified in the future.

Acknowledgements The water resources data and water use index were provided by Directorat General of Water Resource and Puslitbang Sumber Daya Air, Ministry of Public Work Indonesia

References

Arnold JG, Moriasi DN, Gassman PW, Abbaspour KC, White MJ, Srinivasan R, Santhi RC, Harmel RD, van Griensven A, Van Liew MW, Kannan N, Jha MK (2012) SWAT: model use, calibration, and validation. T ASABE 55:1491–1508

Babel MS, Agarwal A, Shinde VR (2014) Climate change impacts on water resources and selected water use sectors. In Shrestha S, Babel MS, Pandey VP (eds.) Climate change and water resources. CRC Press

Boer R, Faqih A (2004) An integrated assessment of climate change impacts, adaptation and vulnerability in water shed areas and communities in Southeast Asia. Report from AIACC Project No. AS21 (Annex C, 95-126) International START Secretariat, Washington, District of Columbia

Bruijnzeel LA (1989) (De)forestation and dry season flow in the tropics: a closer look. J Tropical Forest Sci 1(3):229–243

Bruijnzeel LA (2004) Hydrological functions of tropical forests: not seeing the soil for the trees? Agr Ecosyst Environ 104(1):185–228

DFID and World Bank (2007) Executive Summary: Indonesia and Climate Change. Working Paper on Current Status and Policies

Mulyantari F (2009) Analysis of flooding and rainfall trend in West Java Indonesia. In Identification of Climate Change Impact on Water Resources on Indonesia. ISBN 978-979-8801-36-5. Ministry of Research and Technology, pp 105–120

Hulme M, Sheard N (1999) Climate change scenarios for Indonesia. Climatic Research Unit, Norwich , p 6

ICCSR (Indonesian Climate Change Sectoral Roadmap) (2010) Scientific Basis: Analysis and Projection of Temperature and Rainfall. In Triastuti UH, Mintzer I, Thamrin S, von Luepke H Dieter B (eds)

Ibrahim AB, Ginting S (2009) Impact of climate and land use changes on hydrologic characteristics and water resources in Indonesia. In Identification of Climate Change Impact on Water Resources on Indonesia ISBN 978-979-8801-36-5. Ministry of Research and Technology, pp 55–72

Mcintyre N (2007) How will climate change impact on fresh water security? Dept. of Civil and Environmental Engineering and Grantham Institute for Climate Change, Imperial College London https://www.theguardian.com/environment/2012/nov/30/climate-change-water. Accessed 15 Sept 2016

Naylor RL, Battisti DS, Vimont DJ, Falcon WP, Burke MS (2007) Assessing risks of climate variability and climate change for Indonesian rice agriculture. Proc Natl Acad Sci USA 104(19):7752–7757

Pawitan H (2009) Climate change impact on water availability in Indonesia. In Identification of Climate Change Impact on Water Resources on Indonesia. ISBN 978-979-8801-36-5. Ministry of Research and Technology, pp 105–120

Puslitbang Air (2011) Penelitian Kebutuhan Air pada Wilayah Sungai di Indonesia

Setiawan BI, Satyanto A, Kusmayadi S (2009) Impact of climate change on water resources in Cidanau Watershed, Banten Province, Indonesia. In Identification of Climate Change Impact on Water Resources on Indonesia, pp 73–84. Ministry of Research and Technology. ISBN 978-979-8801-36-5

Tarigan SD (2016a) Land cover change and its impact on flooding frequency of batanghari watershed Jambi Province Indonesia. Procedia Environ Sci 33(2016):386–392

Tarigan SD (2016b). Modeling effectiveness of management practices for flood mitigation using GIS spatial analysis functions in Upper Cilliwung watershed. IOP Conf Ser Earth Environ Sci, vol 31, p 012030. https://doi.org/10.1088/1755-1315/31/1/012030.

Tarigan SD, et al (2016) Mitigation options for improving the ecosystem function of water flow regulation in awatershed with rapid expansion of oil palm plantations. Sustain Water Qual Ecol https://doi.org/10.1016/j.swaqe.2016.05.001

Tarigan S, Faqih A (2019) Impact of changes in climate and land use on the future streamflow fluctuation: case study Merangin Tembesi watershed, Jambi province, Indonesia. Jurnal Pengelolaan Sumberdaya Alam dan Lingkungan 9(1):181–189

Tarigan S, Wiegand K, Sunarti Slamet B (2018) Minimum forest cover required for sustainable water flow regulation of a watershed: a case study in Jambi Province, Indonesia. Hydrol Earth Syst Sci 22:581–594

Tarigan S, Stiegler C, Wiegand K, Murtilaksono K (2020) Relative contribution of evapotranspiration and soil compaction to the fluctuation of catchment discharge: case study from a plantation landscape. Hydrol Sci J 65(7):1239–1248

Wangpimool W, Pongput K, Tangtham N, Prachansri S, Gassman PW (2017) The impact of para rubber expansion on streamflflow and other water balance components of the Nam Loei River Basin, Thailand. Water 9:1–20. https://doi.org/10.3390/w9010001

Marhaento H, Booij MJ, Rientjes THM, Hoekstra AY (2017) Attribution of changes in the water balance of a tropical catchment to land use change using the SWAT model. Hydrol Process 31:2029–2040. https://doi.org/10.1002/hyp.11167

Part III
Water Availability Assessment

Chapter 14
Application of Hydrological Study Methodologies Used in African Context for Water Security in Asian Countries

D. P. C. Laknath and T. A. J. G. Sirisena

Abstract Climate change has threatened the water resources of the African and Asian continents, emphasising the importance of climate resilient water resources management. In this context, hydrological studies are prerequisite components of water resource management plans, including water conservation and mitigation of disasters related risks due to floods and drought. However, factors such as (i) *data inadequacy* and (ii) *unaffordable costs of the already accepted methodologies* are major constraints inherent in hydrological studies of developing countries. The main objective of this study is to address these two constraints related to the African country and to propose similar solutions for Asian countries, which are being experienced similar constraints. A hydrological study was carried out for the tributaries of the Nile River (*Nyabarongo—Akanyaru*) in Rwanda, focusing on a potential dam site. As incompleteness of recorded rainfall and discharge data is a major issue identified during the analysis, suitable scientific and systematic measures have been taken. For the selected minimum of 30 years of analysis period, flow characteristics were simulated by 3 methods: (i) Specific Discharge Method (SDM), (ii) MIKE 11 NAM, and (iii) HEC—HMS simulations. From our analysis, it has been identified that all 3 methods have given approximately similar results with the same accuracy level. However, compared to NAM and HEC-HMS, SDM was recognized as easy handling, fast and cost effective method. Thus, as an outcome of this research, *(i) suitable adjustments for incomplete data and (ii) cost effective, fast and easy handing methodology for flow characteristic simulations* were introduced for Asian countries which are being constrained with the similar barriers.

Keywords Data inadequacy · Water conservation · Flow characteristics simulation · Cost effectiveness

D. P. C. Laknath (✉)
Lanka Hydraulic Institute, Moratuwa, Sri Lanka

T. A. J. G. Sirisena
IHE Delft Institute for Water Education, Delft, The Netherlands

© Springer Nature Switzerland AG 2021
M. Babel et al. (eds.), *Water Security in Asia*, Springer Water,
https://doi.org/10.1007/978-3-319-54612-4_14

14.1 Introduction

As a result of global climate change and subsequent effects on potentially highly sensitive hydrological systems, water resources in the world have been severely affected (Minville et al. 2008). In a global context, changes of precipitation patterns have been emerged in various locations such as increasing precipitation in high latitudes and reduction of precipitation in China (Mohammed and Dore 2005). Moreover, climate change has adversely affected the water resources of the African and Asian continents, emphasising the importance of climate resilient water resources management to ensure the water security. According to Lauri et al. (2014), several large basins in the Asian region have sparse surface observation networks of the hydro-meteorological parameters, which are a prerequisite for hydrological model simulations. Hydrological models are often used in water resources management plans, including water conservation and mitigation of disasters related risks due to floods and droughts, which set strict requirements for model accuracy and reliability. Generally, in case of larger river basins, substantial amount of meteorological data is required to drive the hydrological model. However, in some regions in Africa and Asia, available surface observations may be sparse, and the quality of the historical measurement data can often be questionable. Many large river basins in Asia, such as the Ganges–Brahmaputra, Irrawaddy, Salween, and Mekong (excluding the Thai part of the basin), suffer from rather poor data coverage. At the same time, mainly due to the large hydropower potential with the rapidly growing energy needs of the region, water resources–related developments are rapid in the region (Grumbine and Pandit 2013; Johnston and Kummu 2012). Apart from the data inadequacy and lack of quality, unaffordable costs of the already accepted methodologies such as software licence costs and costs associated with computer resources are major constraints that are inherent in hydrological studies of both African and Asian countries. Accordingly, the main objective of this study is to address identified hydrological data constraints (i.e. *unavailability, poor quality, and expensive simulation methodologies*) related to an African country as a case study and to propose similar solutions for Asian countries, which are being experienced similar constraints.

14.2 Material and Methods

14.2.1 Study Area

A hydrological study was carried out for the tributaries of the Nile River (*Nyabarongo—Akanyaru*) in Rwanda, focusing on a potential dam site at a suitable location across the *Nyabarongo River* (Fig. 14.1). This dam site is situated approximately 30 km north-east of Kigali, which is the capital and largest city of Rwanda. Rwanda has two principal basins called eastern Nile basin (85% of the entire basin) and the western Congo basin (15% of the entire basin). These basins

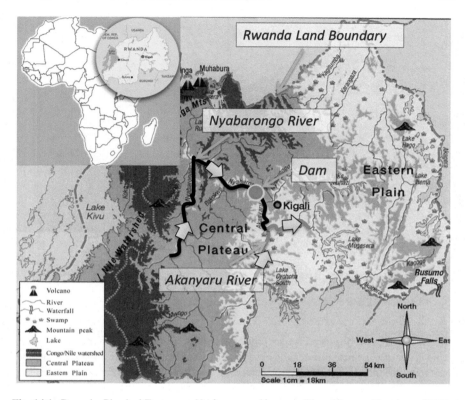

Fig. 14.1 Rwanda: Physical Features—*Nyabarongo -Akanyaru* River (*Source* Henninger 2013)

are divided by Congo-Nile alpine regions. The *Nyabarongo* and *Akanyaru* rivers of this study area belonged to the Nile basin. *Nyabarongo* and *Akanyaru* tributaries consist of catchment area of over 16,000 km². The proposed dam could aim to cater for municipal and irrigation water and to power needs of the country and stabilization of *Nyabarongo River* for flood control.

14.2.2 Current Status of Hydrological Database

For the hydrological study of the study area (i.e. area belongs to *Nyabarongo—Akanyaru* River Basins), status of the available hydrological database was assessed. Accordingly, sources, completeness and consistency of rainfall and discharge data belonging to the study area were evaluated. In case of rainfall gauging stations (GS), there are considerable numbers of station records for rainfall than any other meteorological variable. Thus, rainfall data at 133 stations obtained from Rwanda Meteorology Agency was analysed. After the analysis, incompleteness of recorded data at

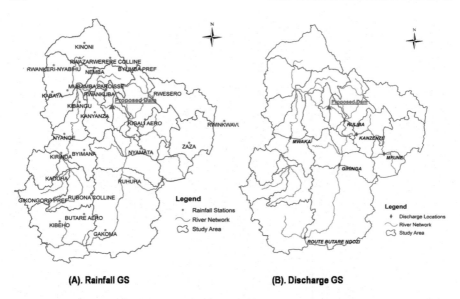

(A). Rainfall GS   **(B). Discharge GS**

Fig. 14.2 Rainfall and discharge gauging stations (GS) in the study area

many stations was observed for a significant length of period during the last 30 years. This is mainly because most of the hydrometric observations in Rwanda were stopped as a result of the civil war in 1994. Thus, non-availability of rainfall data is dominant between for the period of 1990–2010. Hence, rainfall data analysis was carried out for 1971–1991 (period 1) and 2010–2013 (period 2). Finally, 26 rainfall stations were selected considering the data recorded in terms of duration and spatial distribution (Fig. 14.2-A).

Discharge data (i.e. stage gauge data) were available at 6 GSs (*Kanzenze, Mfune, Mwaka, Route Butare Ngozi, Ruliba, and Gihinga*) for this study. However, rating curve data were not available at one GS (i.e. Gihinga). Hence, discharge data at 5 GSs were selected for the hydrological study (Fig. 14.2—B). Thus, identified 3 main problems, as they were (1) *inadequacy*, (2) lack of *quality*, and (3) lack of *accuracy* of both rainfall and discharge data, were minimized as follows.

14.2.3 Rainfall and Discharge Data Gap Filling

Hydrological study was initially expected to carry out for the last 30 years (i.e. 1983–2013). However, alternative periods were considered (i.e. 1971–2013) due to the non-availability of continuous hydrological data in the study area for past 30 years. After analysing rainfall data at 133 stations, incompleteness of recorded data at many stations was observed. Generally, gap filling of rainfall data for a long period (approximately 10 years) is quite difficult since identification of trend, cyclic effect and randomness is difficult. To fill rainfall data gaps at daily time scale at a station, recorded rainfall data at other few stations around the missing data station are needed. But such data are not available for very long periods and thus filling missing data at daily time steps is not possible for a long period such as 1992–2009. Thus, a decision was made not to use that period for the study, for which no data exist at each station in the catchment. In case of discharge data, data is not available at all GSs for the period between 1991 and 1994 (approximately). Further, after 1994, data are not continuously available at other stations. To apply the gap filling method (e.g. double mass analysis) for missing discharge data at a gaging station, discharge data should be available at another gaging station for the same data missing period. Considering these difficulties, gap filling was carried out focusing on alternative periods. After selecting 26 numbers of suitable rainfall stations on the basis of data availability, gap filling was carried out by using best method out of 3 gap filling methods (i.e. simple arithmetic averaging method, normal ratio method, inverse distance method). Three gap filling methods were evaluated using statistical parameters such as maximum, average, median, standard deviation (SD), and root mean square error (RMSE). Then the best rainfall gap filling method was selected for each station and gap filling is carried out accordingly. To check any trends and discontinuities, tests for trends, stability of variance and mean were carried out for the gap filled rainfall data. Trend analysis was carried out by using non-parametric Mann–Kendall test (Mann 1945). The tests for stability of variance and mean verify not only the stationarity of a time series, but also its consistency and homogeneity. Tests for stability of variance and mean were carried out with F-test and t-test, respectively.

Figure 14.3 illustrates the availability of discharge data between 1971–2013 periods. After 1990, data are not available for significant periods. Therefore, the data missing period between 1971–1991 (Period-1) and 2010–2013 (Period-2) were filled by implementing "double mass analysis". Accordingly, it was attempted to fill the missing discharge data at Ruliba GS by using discharge at Route Butare Ngozi GS. For this purpose, discharge data were only available at Route Butare Ngozi GS for the data missing periods. Similarly, for Mwaka, Kanzenze and Mfune GSs, discharge

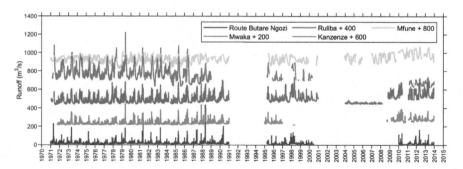

Fig. 14.3 Discharge Data Availability and Missing Periods at 5 Discharge Gauging Stations. For illustration purposes, a fixed value: 200, 400, 600, and 800 m³/s was added to daily runoff data at four stations: Mwaka, Ruliba, Kanzenze, and Mfune, respectively (shown in legend)

data were only available at Ruliba GS for the data missing periods. For all cases, double mass curve follows a straight line, and same trend was observed. Thus, it can be assumed that conditions have not changed at locations, which were used for double mass analysis. However, Route Butare Ngozi gauging station was not considered for gap filling at Ruliba GS since its rating curve has been developed in 1955. Discharges at Mwaka, Kanzenze and Mfune GSs were filled by using discharge at Ruliba GS. It is notable that there is a significant drop of discharge values at Ruliba GS between 2003 and 2008 than in previous years. Further, discharge values at Kanzenze GS have been significantly dropped after 2010. Hence, these data were not considered for the analysis. Further, double-mass curve technique is used to check the consistency of the discharge data records. This technique is based on the principle that when each recorded data come from the same parent population, they are consistent. Accordingly, annual discharges of the selected station and the neighbourhood station were analysed. Thus, accumulated discharge for the period 1971–2013 at each station was calculated. Values of accumulated discharge between Mwaka—Ruliba, Kanzenze—Ruliba and Mfune—Ruliba GSs were plotted. Coefficient of Determination (R^2) value for all double-mass curves was found approximately equal to 1. Thus, on the basis of the obtained linear trend line, consistency of discharge data was verified at Mwaka, Ruliba, Kanzenze and Mfune GSs. After improving both rainfall and discharge data, their applicability is explained in following sections.

14.2.4 Rainfall Data Analysis

The areal rainfall was estimated through the application of Thiessen Coefficients based on the observed (gap filled) rainfall data for the period 1971–2013. This method assigns an area called a Thiessen Polygon to each gauge. In effect, the precipitation surface is assumed to be constant and equal to the gauge value throughout the region. Based on Thiessen method, estimated average monthly areal rainfall for the period 1971–2013 (excluding the period 1992–2010) at proposed dam site is given in Table 14.1. As it can be seen there, for the periods February–May and September–December, significantly more rainfall has occurred than in the other months. Highest rainfall has occurred in April. Estimated average annual rainfall (*water year basis*) was 1331 mm for the considered period.

Table 14.1. Average monthly and annual areal rainfall data at the proposed dam site (*Unit: mm*)

Period	Jan	Feb	Mar	Apr	May	Jun	Jul	Aug	Sep	Oct	Nov	Dec	Annual
1971–2013	93	111	146	208	147	43	16	46	110	134	158	120	1331

14.2.5   Discharge Data Analysis

After enhancing the quality of the historical measurement data by statistical adjustment (e.g. check for outliers), flow characteristics were identified at each GS and given in Table 14.2. The highest and lowest average discharges have been recorded at Ruliba and Route Butare Ngozi gauging stations, respectively.

For the selected minimum 30 years of analysis period, flow characteristics were simulated by 3 methods as follows:

1. Flow discharge calculation using "Specific Discharge Method (SDM)"
2. Rainfall runoff analysis using "MIKE 11 NAM model"
3. Rainfall runoff analysis using "HEC-HMS model"

These methods were opted to represent factors such as easiness of operation, degree of data requirement, and cost factor. Accordingly, trade off among the accuracy of outcome, cost and easiness were compared by considering the applicability of the best method for Asian countries, who are being constrained with limited data availability and state-of-the art computer simulation methodologies. Focusing on the necessity of water security and the conservation purposes, discharge data at the potential dam site were considered as the main outcome from alternative methodologies.

(1). Flow discharge estimation using Specific Discharge Method (SDM)
Based on *"Specific Discharge Method"*, river discharge at the proposed dam site and required locations were computed. Specific discharge method is based on "flow" and "area" relationship. Thus, to calculate discharge at a particular point, annual mean flow discharges at other gauge stations and relevant catchment area of each gauging station were considered. Accordingly, *"Specific Discharge Curve"* for the drainage basin of the *Nyabarongo* River was developed. Flow-Area relationship is illustrated in Fig. 14.4.

The developed liner relationship between annual mean flow discharge and catchment area is given by Eq. (14.1).

$$Q_{Annual} = 0.288 \times C_{Area} + 411.7 \qquad (14.1)$$

Table 14.2 Flow characteristics at 5 main gauging stations

Items	Unit	Route Butare Ngozi	Mwaka	Ruliba	Kanzenze	Mfune
Catchment area	km^2	1,479	2,427	8,355	13,923	15,534
Max. discharge	m^3/s	227.0	240.7	331.9	325.0	265.0
Min. discharge	m^3/s	2.8	13.0	33.8	24.2	28.7
Average discharge	m^3/s	16.1	41.3	99.7	130.3	140.0
Median discharge	m^3/s	13.2	32.9	88.1	123.6	140.6

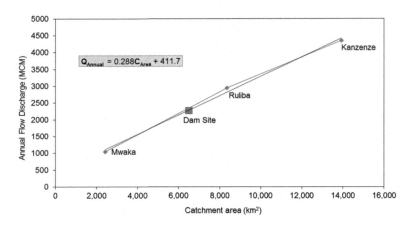

Fig. 14.4 Flow-area relationship

where, Q_{Annual} is Annual mean flow discharge (MCM) and C_{Area} is Catchment Area (km^2).

Using Eq. (14.1) and catchment area at the dam site, mean annual discharge at dam site was calculated as 2281 MCM (million cubic meter). Thus, ratio of annual mean flow between dam site and nearest GS of the dam site (Ruliba) was calculated. Mfune gauging station was not considered for the analysis since it was observed that there was a considerable water abstraction from the river between Kansenze and Mfune gauging stations. Route Butare Ngozi gauging station is also not considered since its rating curve has been developed in 1955. For the calculation of daily discharge, observed discharge data (gap filled) was used.

$$\frac{\text{Annual Mean Flow Discharge at Dam Site}}{\text{Estimated Annual Mean Flow Discharge at a Nearest Station to Dam Site}} = 0.7726$$

Based on above ratio and available measured (gap filled) daily discharge data at Ruliba GS, daily discharge at the dam site was calculated for the desired period using Eq. (14.2).

$$\text{Daily Discharges at Dam site (Approx.)} = \text{Daily Discharge at Ruliba} \times 0.7726$$
$$(14.2)$$

Basically, reproduction of discharge data was carried out without rainfall data in this method (i.e. SDM).

(2). Rainfall Runoff Analysis Using MIKE 11 NAM Model

MIKE 11 NAM ("Nedber-Afstremnings-Model") is the hydrological model, which is capable of simulating the rainfall-runoff processes at the catchment scale or a large river basin scale. Thus, rainfall runoff processes in the study area was simulated by using MIKE 11 NAM hydrological model as an alternative approach to SDM. As a main outcome, discharge at proposed dam site was simulated. MIKE 11 NAM model estimate the runoff volume generated from the rainfall of corresponding basin. The reliability of the MIKE11 NAM is evaluated based on the Efficiency Index (EI) as described by Nash and Sutcliffe (1970) for the long term basis. The EI was developed to evaluate the percentage of accuracy or goodness of the simulated values with respect to their observed values. To setup the NAM model, total catchment area, which belongs to Nybarongo and Akanyaru River system, was sub divided into 22 sub catchments in order to incorporate the rainfall runoff processes within the study area. Combined catchment option of MIKE 11 NAM model was used to determine the runoff at the dam site. The main and sub catchments were delineated using DEM data. Rainfall stations in the study area were used to calculate the weighted areal rainfall catchments. Theissen option in NAM model was used to determine the weighted rainfall of catchments. As the meteorological data input of NAM model, rainfall and evaporation data of the simulation period were used. Observed discharge values at Ruliba GS were used for the calibration and validation purpose of the model. Calibration and verifications were carried out for the periods 1972–1973 and 1985–1986, respectively. The rainfall runoff model established for Ruliba catchment was calibrated first. The calculated and acceptable goodness of fit parameters are given in Table 14.3.

Calibration and validation plots for Ruliba catchment are shown in Figs. 14.5 and 14.6, respectively. It was observed that there is no any considerable discrepancy between the observed and simulated values. The calibrated and validated rainfall-runoff model was simulated for the period 1971–1991 in order to obtain the required runoff time series at the proposed dam site across the *Nyabarongo* River.

Table 14.3 Goodness of fit parameters MIKE 11 NAM

Gauging station	Calibration			Validation		
	EI	Volume error (%)	RMSE (m³/s)	EI	Volume error (%)	RMSE (m³/s)
Ruliba	0.87	2.84	12.86	0.88	−1.86	14.80

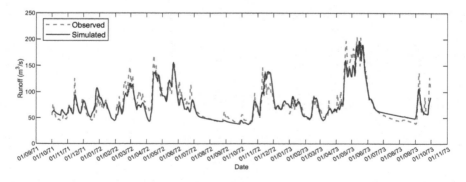

Fig. 14.5 Calibration Results at Ruliba: Simulated vs. Observed Flow Comparison (1972–1973)

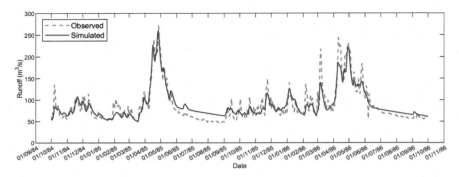

Fig. 14.6 Validation Results at Ruliba: Simulated vs. Observed Flow Comparison (1985–1986)

(3). Rainfall Runoff Analysis Using HEC-HMS Model

As the next alternative method to establish the rainfall runoff relationship and simulate discharges at the proposed dam site, HEC-HMS numerical modeling system was used. The Hydrologic Modeling System (HEC-HMS) is designed to simulate the rainfall-runoff processes of dendritic drainage basins. The HEC-HMS model components are used to simulate the hydrological response in a watershed. The model includes basin models, meteorological models, control specifications, and input data. It simulates the rainfall-runoff response in the basin model for the given input from the meteorological model. The control specifications define the time period and time step of the simulation run. Input data components, such as time series data, paired data, and gridded data are often required as parameters or boundary conditions in basin and meteorological models. Junction at Ruliba discharge location was used as the model calibration point. Fifteen sub basins, 9 reaches and 10 junctions were created for the Ruliba catchment model. In meteorological model, rainfall is spatially distributed by Theissen weight method. Under the Control specification, calibration was carried out for the period 1971–1973 and validation was done for the period 1984–1986. The performance of the HEC-HMS model results were evaluated similar to NAM model and presented in Table 14.4.

As seen in Table 14.4, calculated error parameters and validation results have demonstrated good model performance. Calibration and validation plots for Ruliba catchment are shown in Figs. 14.7 and 14.8, respectively. The calibrated and validated model was used to simulate for the period of 1971–1990 in order to obtain the required runoff time series at dam site along the *Nyabarongo* River.

Table 14.4 Goodness of fit parameters HEC-HMS

Gauging station	Calibration			Validation		
	EI	Volume error (%)	RMSE (m³/s)	EI	Volume error (%) (%)	RMSE (m³/s)
Ruliba	0.84	3.1	13.4	0.75	12.6	20.6

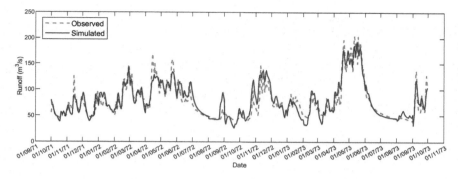

Fig. 14.7 Calibration Results at Ruliba: Simulated vs. Observed Flow Comparison (1972–1973)

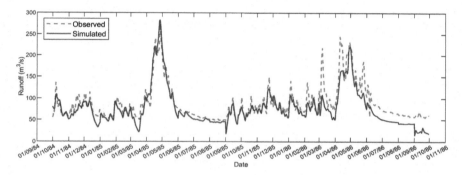

Fig. 14.8 Validation Results at Ruliba: Simulated vs. Observed Flow Comparison (1985–1986)

14.3   Results and Discussion

After improving existing hydrological data base, flow characteristics were simulated at the potential dam site. To assess and to identify the applicability of the most efficient and effective methodology, same outcome was reproduced with three different methodologies. Then, the identified flow characteristics are tabulated in Table 14.5. Comparing simulated values in the period 1971–2013, significant differences were not identified among the outcomes of each methodology. It was identified that 3

Table 14.5 Discharge characteristics at dam site

Flow characteristics (m³/s)	Method		
	SDM	MIKE11 NAM	HEC-HMS
Max. discharge	257	248	235
Min. discharge	26	1	1
Average discharge	77	78	48
Median discharge	68	70	41

methods have given approximately similar results with the same accuracy level. Considering the maximum and average discharge values, SDM and MIKE 11-NAM have shown closeness, while HEC-HMS exhibited lower values. However, it seems that SDM could not capture low flow values, while other methodologies have done it. Generally, it is clear that SDM has given more conservative values for hydraulic structure design and flood protection works comparing with MIKE 11-NAM and HEC-HMS simulation values.

14.3.1 Probabilistic Flood

For all 3 methods, the value of annual maximum flood data in the period 1971–2013 were used to analyse and to predict flood for different return periods. Predicted flood values are important for flood protection works. By using three goodness of fit tests (i.e. *Kolmogorov Smirnov Statistic, Anderson Darling Statistic, and Chi-Squared Statistic*) and best fit distributions (e.g., *Weibull, Gumbel Max, Normal, Johnson SB, Log-Pearson 3, Pearson 5 (3P), Lognormal (3P), and Frechet*) for the predicted discharges were identified. Thus, identified probabilistic flood for best fit distributions are summarized in Table 14.6.

Comparing magnitudes of floods of different return periods, except 2 years of return period, SDM has given higher values than values obtained from the other methods. HEC-HMS simulated flood frequency values are lesser than of both SDM and MIKE 11-NAM values. Hence, on the basis of estimated maximum flood values, suitable values are selected from Table 14.6 for the flood protection and design works.

Table 14.6 Predicted probabilistic flood by different methods and estimated flood values for flood protection works, *Units (m^3/s)*

Return Period (Year)	Method			Estimated Flood	Selected Method	Best Fit Probabilistic Distribution
	SDM	MIKE11 NAM	HEC-HMS			
2	182.7	184.6	170.1	184.6	NAM	Normal
5	214.3	212.4	205.1	214.3	SDM	Johnson SB
10	233.5	232.0	222.0	233.5		Johnson SB
20	260.5	258.4	239.8	260.5		Frechet
25	270.0	267.4	248.9	270.0		Frechet
30	278.0	274.9	256.6	278.0		Frechet
50	301.6	297.1	279.3	301.6		Frechet
100	336.7	329.9	313.2	336.7		Frechet
200	375.6	366.1	351.0	375.6		Frechet
1000	484.0	466.0	457.0	484.0		Frechet
10000	695.5	658.0	666.4	695.5		Frechet

MIKE11-NAM is commercial software. Hence, cost component is charged as the value of product and their service. HEC-HMS is made available to the public by the developer. However, compared to SDM, demand for several input data is still there. Thus, SDM can be used as a conservative approach instead of highly input data demanding and computationally expensive methodologies. Hence, while selecting suitable methodology to simulate discharge values at a potential dam site, SDM has demonstrated promising results in this study.

14.4   Conclusions

In this study, flow characteristics simulation was carried out for a potential dam site across the *Nyabarongo* River in Rwanda. Inadequacy of essential hydrological data, low quality of existing data and accuracy issues were identified as challenges to be addressed. Rainfall and river discharge data were mainly considered as important data from simulation and hydraulic structure design perspectives. Selection of appropriate number of rainfall stations considering the recorded duration (i.e. availability) and spatial distribution, use of different gap filling methods based on statistical tests for different rainfall stations, ensuring consistency and homogeneity of dataset were some important measures have been taken to enhance the quality and accuracy of the existing database.

Though many large river basins in Africa and Asia have poor data coverage, water resources—related developments such as dams are rapid in the region at the same time, mainly due to water conservation, water supply, and rapidly growing energy needs in the region. Hence, a hydrological study was carried out focusing the simulation of discharges at a potential dam site. Accordingly, flow characteristics were simulated by 3 methods: SDM (*Specific Discharge Method*), MIKE 11-NAM, and HEC–HMS modeling for an African country, where available surface observations were sparse, and the quality of the historical measurement data are often questionable. These 3 methods were opted to represent factors such as easiness of operation, degree of data requirement, and cost factor. Accordingly, trade off among the accuracy of outcome, cost and easiness were compared by considering the applicability of the best method for Asian countries. Focusing on the necessity of water security and the conservation purposes, discharge data at the potential dam site were considered as the main outcome from alternative methodologies.

Comparing simulated maximum and average discharge values in the period 1971–2013, significant differences of the simulated discharge values were not identified among each methodology. They have given approximately similar results with the same accuracy level. Considering the maximum and average discharge values, SDM and MIKE 11-NAM have shown approximately close values while HEC-HMS has exhibited low values. Generally, SDM has given more conservative values than MIKE 11-NAM and HEC-HMS simulation values for hydraulic structure design and flood protection works.

For SDM, rainfall data is not required to reproduce the discharge data at the interested point. Also, it is a cost effective, fast and easy handing methodology. MIKE 11-NAM is commercial software; hence it is an expensive option. HEC-HMS is made available to the public by the developer. However, compared to SDM, several input data, including scarce rainfall data is prerequisite for later two methodologies. Thus, considering all these, SDM can be used as a conservative approach to simulate discharge values for hydraulic structure designs and flood protection works. Therefore, SDM can be introduced as a cost effective, fast and easy handing methodology for flow characteristic simulations for Asian countries, which are being constrained with the similar barriers in African context such as data inadequacy and unaffordable computer resource cost.

Acknowledgements The authors wish to acknowledge to *Rwanda Meteorology Agency* for providing data for the success of this research.

References

Dore MHI (2005) Climate change and changes in global precipitation patterns: what do we know? Environ Int 31:1167–1181
Grumbine RE, Pandit MK (2013) Threats from India's Himalaya dams. Science 339:36–37
Henninger S (2013) Does the global warming modify the local Rwandan climate? Nat Sci 5:124–129
Johnston RM, Kummu M (2012) Water resource models in the Mekong basin: a review. Water Resour Manag 26:429–455
Lauri H, Rasanen, TA, Kummu M (2014) Using reanalysis and remotely sensed temperature and precipitation data for hydrological modeling in monsoon climate: mekong river case study. J Hydrometeorol 17:1532–1545
Mann HB (1945) Non-parametric test against trend. Econometrica 13:245–259
Minville M, Brissette F, Leconte R (2008) Uncertainty of the impact of climate change on the hydrology of a nordic watershed. J Hydrol 358:70–83
Nash IE, Sutcliffe IV (1970) River flow forecasting through conceptual models Part I. J Hydrol 10:282–290

Chapter 15
Implementation of Budyko Curves in Assessing Impacts of Climate Changes and Human Activities on Streamflow in the Upper Catchments of Dong Nai River Basin

Thi Van Thu Tran, Long Phi Ho, and Quang Phuoc Phung

Abstract Climate changes (CC) and human activities (HA) have been evidently identified as the primary factors for short- and long-term annual streamflow variations, which directly affect water resources management in river basins. In this study, Ta Pao and Ta Lai, the two upper sub-basin of Dong Nai river basin in the South of Vietnam, are analysed in order to assess the influences of climate variability and human impacts on river discharges in the period of 27 years, from 1987 to 2013. While the Budyko curves are modified to evaluate empirical climatic data as precipitation, potential evapotranspiration, and actual evapotranspiration, unsupervised classification of Landsat images is adopted to identify land covers in comparing periods – from 1987 to 1999 (1990s) and from 2000 to 2013 (2000s). Consequently, in Ta Pao sub-basin, the discharge of the latter period has increased by 30% approximately compared with the former, in which CC accounted for 92.2% and HA 7.8%. On the contrary, in Ta Lai sub-basin, while CC makes streamflow increased by 65.3%, HA makes it decreased by 34.7%. Therefore, the discharge in 2000s was only 4.5% more than in 1990s. Furthermore, as results of land cover classifications of the two sub-basins in the two periods, urban area in Ta Lai region is 4 times larger than in Ta Pao. In conclusion, climate variability has the tendency to raise the average annual discharge in upper catchments of Dong Nai river basin but the impacts of human activities in urbanization has a decreasing trend to the discharge that expressed significantly in Ta Lai sub-basin comparing 1990s and 2000s.

Keywords Budyko curves · Streamflow · Climate changes · Human activities · Landsat images

T. V. T. Tran (✉) · L. P. Ho · Q. P. Phung
Center of Water Management and Climate Change, Vietnam National University Ho Chi Minh City, Ho Chi Minh City, Vietnam

© Springer Nature Switzerland AG 2021
M. Babel et al. (eds.), *Water Security in Asia*, Springer Water,
https://doi.org/10.1007/978-3-319-54612-4_15

195

15.1 Introduction

Water resource is the crucial factor in the sustainable existence and development of ecosystems and environment in any region. The total volume of global water is approximately 1.4 billion km^3, in which the majority (97%) is salt water in oceans, whereas the rest (3%) is fresh water in river systems, ground layers and glaciers. However, it is important to acknowledge that the problems of fresh water shortage will increase globally (Oki and Kanae 2006), due to the fact that climate change is getting worse with severe droughts or devastating floods while water demands for human activities are getting larger. Thus, the integrated research approach of key effects of the above causes is essential to reach a harmonious socio-economic development and protect the environment in the present crisis of water.

The objective of this study is to assess the long-term effects of CC and HA on streamflow variations in a catchment through analyzing the available time-series data of hydrology, meteorology and land covers. The study focuses on the upper Dong Nai river basin in the South of Vietnam with two specific sub-basins, Ta Pao and Ta Lai. The time-series data is observed in the period of 27 years from 1987 to 2013, divided into two small periods for comparison—the 1990s (from 1987 to 1999) represents the under-developed stage and the 2000s (from 2000 to 2013) represents the urbanized one. The methodology of the study is to implement Budyko hypothesis to create Budyko curves based on meteorological data such as precipitation, potential evapotranspiration, actual evapotranspiration… that expresses the impacts of CC and HA. Meanwhile, the unsupervised classification of Landsat images is exploited to identify land cover changes driven by HA. The results are desired to determine influences of CC and HA on increasing or decreasing streamflow tendencies in the above sub-basins over two developing stages. Such a research study would contribute to an integrated water resource management framework within diversified watershed development programmes in a sustainable way.

15.2 Material and Methods

15.2.1 Study Area

The upper Dong Nai river basin (Fig. 15.1) is located at the southeastern part of Sai Gon – Dong Nai, the second largest river system in the South of Vietnam, with the basin area of 14,640 km^2 and the river length of about 280 km. There are two main sub-basins in this catchment, namely Ta Pao (2,075 km^2) and Ta Lai (9,506 km^2), which form of the two major rivers, La Nga and Da Nhim respectively. The study watershed originates from Lang Biang highland (Lam Dong Province) at an attitude of 1000 m and above with the average slope of about 30%. With the tropical monsoon climate type and the near-equatorial region, the average annual precipitation is observed from 2,200 ÷ 2,900 mm/year that generates the genetic diversity of vegetational covers,

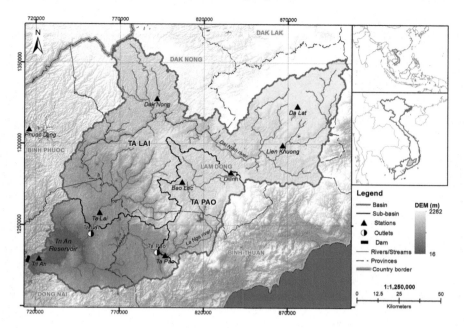

Fig. 15.1 The upper Dong Nai river basin

particularly industrial plants (rubber, tea, coffee…) and food plants (maize, peanuts, potato…). Moreover, the lower part of this river basin lies the multi-purpose Tri An reservoir, spreading over a surface area of 323 km², that has an important function of electricity production and water supplies to the key economic southern region at the downstream, Ho Chi Minh City specifically. Therefore, the water resource of the upper Dong Nai catchment plays the primary role in the sustainable socio-economic development of the regional provinces, also affects routine operations of Tri An reservoir in dry and rainy seasons significantly.

15.2.2 Data

The daily data of the upper Dong Nai river basin consist of precipitation (P) from gauging stations and daily streamflow (Q) measured at two outlets as Fig. 15.1 during the period from 1987 to 2013. Figure 15.2 diagrams the time series data of annual P and Q in the two study sub-basins, where the annual P in Ta Pao is larger than in Ta Lai, and both of them are slightly increasing; while Ta Pao has a rising trend in the annual Q, Ta Lai does the opposite. Besides, the available terrain elevation was exploited with grid revolution 100 × 100 m (Fig. 15.1). Furthermore, monthly average meteorological data such as maximum and minimum air temperatures, relative humidity, and sunshine hours from seven stations in the basin in the period of

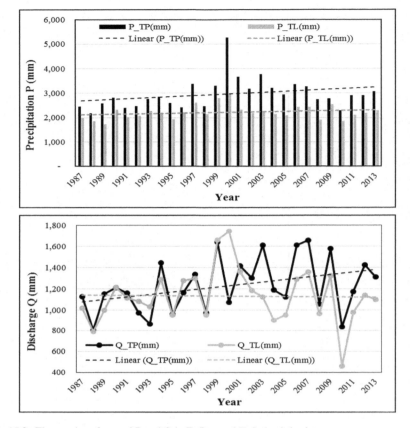

Fig. 15.2 Time series of annual P and Q in Ta Pao and Ta Lai sub-basin

27 years were applied to calculate potential evapotranspiration (PET) using Morton's model (Morton 1983; McMahon et al. 2013). Actual evapotranspiration (AET) is figured out by the mass balance law if considering that average P is equal to average Q plus average AET in a watershed within a long-term hydrological cycle.

15.2.3 Budyko Hypothesis

Based on empirical data from several catchments, Budyko (1974) proposed a hypothesis of a balance between available energy and water to assess water resources in historical climatic conditions. The Budyko curve is derived from a combination of the average annual actual evapotranspiration index (E/P) and the average annual potential evapotranspiration index (E_p/P) or the aridity index as follow:

$$\frac{E}{P} = \left[\left(1 - exp\left(-\frac{E_p}{P} \right) \right) \frac{E_p}{P} tanh\left(\frac{E_p}{P} \right)^{-1} \right]^{0.5} \tag{15.1}$$

Various researches have been developed and formulated in order to consolidate other factors in the average annual water balance at the watershed scale. In the limit of this paper, the equation conducted by Fu (1981) is applied to evaluate the average annual streamflow in the climatological water balance, whereas, the parameter ω is determined so as to adjust observed data to the Budyko curve of Eq. (15.2).

$$\frac{E}{P} = 1 + \frac{E_p}{P} - \left[1 + \left(\frac{E_p}{P} \right)^{\omega} \right]^{1/\omega} \tag{15.2}$$

From the Budyko curves, the decomposition method (Wang and Hejazi 2011) is proposed to quantify the direct human and climate contributions to the mean annual streamflow change as Fig. 15.3. Considering the post- and pre-change periods, point A (pre-change period) is denoted as E_{p1}/P_1 and E_1/P_1 and point B (post-change period) is denoted as E_{p2}/P_2 and E_2/P_2, thus, A would be shifted to B due to both CC and HA interferences. Under CC only, point A would be transformed into point C $(E_{p2}/P_2, E_2'/P_2)$ along the Budyko curve. Therefore, the difference of B and C would be affected by HA influences. If water storage change is negligible, the discharge Q is calculated by $P - E$ or $P(1 - E/P)$. Then, the ΔQ total streamflow change is computed by $Q_2 - Q_1$, in which $Q_2 = P_2 - E_2$ and $Q_1 = P_1 - E_1$. Similarly, the HA impact on discharge change can be solved by $\Delta Q^h = P_2\left(E_2'/P_2 - E_2/P_2 \right)$, and discharge change due to CC is $\Delta Q^c = \Delta Q - \Delta Q^h$.

Fig. 15.3 Human and climate contributions to the mean annual streamflow change. (*Source:* Wang and Hejazi (2011))

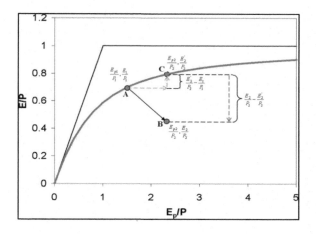

15.2.4 Landsat Images

By reason of distant information acquisition, remote sensing can be benefited in various domains including monitoring and assessing environmental impacts. One of available remote sensing data used in this research is from Landsat satellite. These data is retrieved from the source of U.S. Geological Survey (https://earthexplorer.usgs.gov). LANDSAT-4 and LANDSAT-5 Thematic Mapper images (Batson et al. 1985) are used as an object of remote sensing in land cover identification of the upper Dong Nai river basin. After geometric calibration and subsetting, the unsupervised classification, particularly ISODATA classification, is applied to the Landsat images of 1995 and 2011 that expressed as land cover changes of 1990s and 2000s of the study area. Cloud removal is also taken part in as pixels where clouds are would be assigned as no data. The outcome is maps of upper catchment of Dong Nai river basin in five major types of land covers: open water, forest, agriculture, urban area, and rangeland in both the comparing stages – 1995 and 2011.

15.3 Results and Discussion

The average annual values of precipitation (P), discharge (Q), potential (PET) and actual (AET) evapotranspiration in two comparing stages are expressed in Table 15.1. In particular, there has been a marked increase in the mean annual P to both Ta Pao and Ta Lai sub-basin, however, considering the annual streamflow, Q in Ta Pao is rising greatly than in Ta Lai by differences of 627 and 5 respectively. In the other hand, pairs of PET and AET values of each sub-basin have happened in the contrary way, namely, while the average annual evapotranspiration in Ta Pao had an increase trend to PET and a decrease one to AET, it did the opposite in Ta Lai.

Figure 15.4 presents the comparison of the annual ratios of potential and actual evapotranspiration to precipitation (PET/P and AET/P) between observed data and the Budyko curves during the period of 27 years in two sub-basins. The parameter w of Fu (1981) equation is calculated to fix the Fu (1981) curve on observed values, namely 2.11 in Ta Pao and 2.44 in Ta Lai. It can be clearly seen in the graph that 27 grey observed dots are plotted in the right side of the limit line and less than 1.0 due to the tropical monsoon climate and the high relative humidity of the Dong Nai river

Table 15.1 Average annual values of P, Q, PET and AET in two comparing stages

Sub-basin	The under-developed stage (1990s) (mm)				The urbanized stage (2000s) (mm)				Discharge change (mm)
	P	Q	PET	AET	P	Q	PET	AET	ΔQ
Ta Pao	2669	1392	1233	1277	3241	2018	1313	1222	627
Ta Lai	2174	1129	1342	1045	2272	1133	1316	1139	05

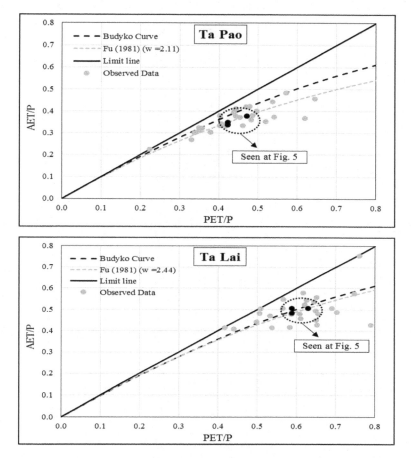

Fig. 15.4 Comparison of annual ratios of PET/P and AET/P between observations and the Budyko curves

basin, the aridity index PET/P is located in the energy-limited region which has the water supply larger than the evapotranspiration demand (P > PET).

In order to analyze streamflow changes in the catchments in two periods due to CC and HA, discharges Q solved out from Eq. (15.2) and the observed values are tabulated in Table 15.2. Combinations of PET/P and AET/P between equation and observation indicated as A, B and C in Fig. 15.5 demonstrates that while C is in the lower position than A on the Fu (1981) curve in both sub-basins, to make it clearer, there has been a wetter tendency (P > PET) to the climate in the region to make the streamflow increase. In contrast, in the actual condition of the evapotranspiration, differences of B and C positions implement impacts of HA on streamflow changes in the post and pre-change stage, whereas, there are a slight increase (B lower than C) in Ta Pao and a remarkable decrease (B higher than C) in Ta Lai. As a result, in Ta

Table 15.2 Percentages of discharge changes due to CC and HA in two comparing stages

Sub-basin	The under-developed stage (1990s)			The urbanized stage (2000s)				Discharge change		Changes in discharge due to			
	PET/P	AET/P Fu Eq.	Q_1 (mm)	PET/P	AET/P Fu Eq.	AET/P Observed	Q_2 (mm)	ΔQ (mm)	$\Delta Q/Q_1$ (%)	CC (mm)	HA (mm)	CC (%)	HA (%)
Ta Pao	0.47	0.38	1658	0.42	0.35	0.34	2147	489	29.5	451	38	92.2	07.8
Ta Lai	0.63	0.51	1071	0.59	0.48	0.51	1119	48	04.5	102	54	65.3	34.7

Fig. 15.5 CC and HA influences on streamflow changes

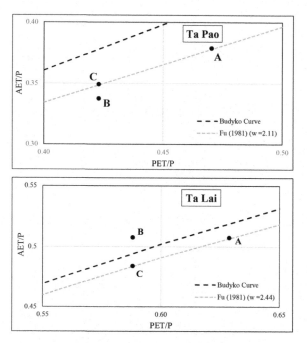

Pao sub-basin, the discharge of the post period has increased by 30% approximately compared with the pre-change period, in which CC accounted for 92.2% and HA 7.8%. On the contrary, in Ta Lai sub-basin, while CC makes streamflow increased by 65.3%, HA makes it decreased by 34.7%, thus, the discharge in the urbanized stage was only 4.5% more than in the under-developed one.

Results of land cover changes from LANDSAT images described in Fig. 15.6 and Table 15.3 are declared that although Forest and Rangeland have a huge majority in the whole sub-basin area in both the considering periods in the upper Dong Nai watershed, changing paces of Urban and Agriculture are faster than others. While, in Ta Pao sub-basin, the percentage of Forest is a constant, Rangeland is transferred to small parts to Open Water and Urban, and the larger rest to Agriculture. Otherwise, in Ta Lai sub-basin, Rangeland is almost stable, Forest is shifted to Open Water and Agriculture, and Urban for the rest. Moreover, the urbanization speed of Ta Lai area is four times faster than in Ta Pao area that makes the decrease trend to annual Ta Lai streamflow due to HA become clearer.

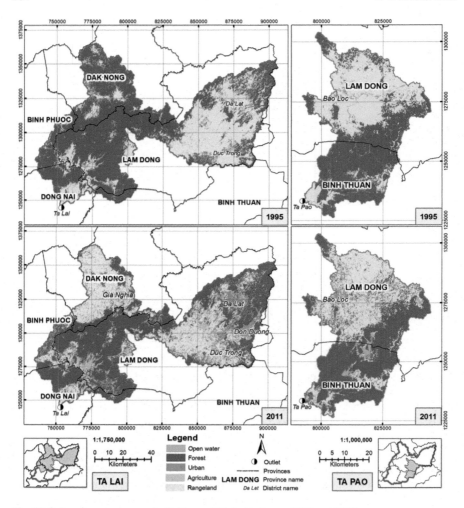

Fig. 15.6 Land cover changes in Ta Pao and Ta Lai between 1995 and 2011

Table 15.3 Percentages of land cover types in two comparing periods of two sub-basin

Land cover type	Ta Pao sub-basin			Ta Lai sub-basin		
	% of land cover changes		Changing pace (Post/Pre)	% of land cover changes		Changing pace (Post/Pre)
	1995	2011		1995	2011	
Open water	0.99	2.10	2.12	1.65	2.17	1.32
Forest	49.98	49.89	1.00	58.43	48.86	0.84
Urban	0.59	2.39	4.06	0.23	4.16	18.11
Agriculture	0.02	1.23	54.16	0.68	2.33	3.43
Rangeland	48.42	44.40	0.92	39.01	42.49	1.09
Total	100.00	100.00		100.00	100.00	

15.4   Conclusions

The study was to carry out the Budyko hypothesis in order to determine the influences of CC and HA on the average annual streamflow in Ta Pao and Ta Lai, the two sub-basins of the upper Dong Nai watershed. As a result, while the climate in the basin with a wetness trend that las increased the discharge due to the down ratio of potential evapotranspiration and precipitation PET/P, the impacts of HA caused the discharge to decrease due to the up ratio of actual evapotranspiration and precipitation AET/P, particularly in Ta Lai sub-basin, in the comparison between the urbanized stage (from 2000 to 2013) and the under-developed one (from 1987 to 1999). Moreover, when identifying land cover changes from LANDSAT images of the two sub-basins in the two periods, the urbanization in Ta Lai region was four times more spacious than in Ta Pao, a clear evidence of water shortage driven by HA. These results provided a broader perspective on the integrated water resources management not only in the study area but also the downstream Dong Nai river basin, especially the direct and considerable effects on the routine operations of Tri An reservoir to produce electricity and distribute water demands.

Acknowledgements This research is funded by Vietnam National University Ho Chi Minh City (VNU-HCM) under grant number A2013-48-01.

References

Batson RM, Kieffer HH, Borgeson WT (1985) Geometric accuracy of LANDSAT-4 and LANDSAT-5 Thematic Mapper images
Budyko MI (1974) Climate and life: English. In Miller DH (ed) Academic Press
Fu BP (1981) On the calculation of the evaporation from land surface. Sci Atmos Sin 5(1):23–31
McMahon TA, Peel MC, Lowe L, Srikanthan R, McVicar TR (2013) Estimating actual, potential, reference crop and pan evaporation using standard meteorological data: a pragmatic synthesis. Hydrol Earth Syst Sci 17(4):1331–1363
Morton FI (1983) Operational estimates of areal evapotranspiration and their significance to the science and practice of hydrology. J Hydrol 66(1):1–76
Oki T, Kanae S (2006) Global hydrological cycles and world water resources. Science 313(5790):1068–1072
Wang D, Hejazi M (2011) Quantifying the relative contribution of the climate and direct human impacts on mean annual streamflow in the contiguous United States. Water Resour Res 47(10)

Chapter 16
Contribution of Snow and Glacier in Hydropower Potential and Its Response to Climate Change

R. B. Chhetri, N. M. Shakya, and N. R. Sitoula

Abstract Hydropower development is considered as backbone of economic water security in context of Nepal. Most of the rivers originate from high mountains and are considered as snow fed rivers. Hence the impact of climate change on snow and glacier affects such rivers. In this study, snow and glacier are delineated from LANDSAT TM7 image by ERDAS IMAGINE and melt discharge is estimated by Energy Budget Method (EBM) along with the conceptual TANK model. Runoff generated from the watershed along with precipitation in non-glacerised area are given as inputs to HEC-HMS for simulating discharge of the whole basin. The significance of snow and glacier melt contribution at different downstream stations is assessed. Finally, future precipitation and temperature data obtained using AR5 scenario is used to assess the significance of climate change in river flow. From the analysis it is found that Koshi basin (1.62%) is covered by larger areas of glacier than that of the Mahakali basin (1.16%). The maximum contribution of snow and glacier in Koshi at Chatara and Mahakali at Chameliya are 12.02% and 17.35% respectively. While analysing the effect of climate change on the hydropower potential for storage project, it is observed that most of the sub-basins will have decreased volume of annual flow in 2030 A.D., but that will increase in 2050 A.D. There will be an increase in flow volume in Mahakali in the year 2030 A.D. and 2050 A.D. But for RoR project designed at Q40 flow volume will increase in both 2030 A.D. and 2050 A.D. in Koshi and Mahakali basin.

Keywords AR5 · EBM · ERDAS IMAGINE · Glacier · HEC-HMS · Hydropower · Snow

R. B. Chhetri · N. M. Shakya (✉) · N. R. Sitoula
Department of Civil Engineering, Institute of Engineering, Tribhuvan University, Lalitpur, Nepal
e-mail: nms@ioe.edu.np

© Springer Nature Switzerland AG 2021
M. Babel et al. (eds.), *Water Security in Asia*, Springer Water,
https://doi.org/10.1007/978-3-319-54612-4_16

207

16.1 Introduction

Glaciers are the world's water towers and the origin and lifeline of the major river systems, which are made up of fallen snow that, over many years, compresses into large, thickened ice masses. The observed widespread glacier retreat is considered as the consequences of global temperature increase and is rated as one of the most reliable natural indicators of ingoing warming trends (IPCC 2001). Variations in glacier mass are generally determined by the balance between incoming (mass gain) and outgoing (mass loss) terms. Due to the variety of involved time scales, the state of a glacier bears valuable information both on climatic changes in the past and on changes occurring at present (Singh and Singh 2001). The link between glaciers and climate is not a simple, one-directional relationship though. Via a number of feedback mechanisms, changes in ice cover can potentially influence the ambient air and thus modify the local climate. The unique surface characteristics of ice, namely the comparatively high reflectivity in the solar spectrum and a surface temperature never exceeding 0 °C, affect the components of the surface energy balance and near surface exchange processes. In addition to direct physical interactions between a surface ice cover and the overlying air mass a further feedback mechanism consists in freshwater originating from glacier melt entering the oceans and thereby modifying oceanic circulations and deep water formation and contributing to sea level rise (IPCC 2001).

Nepal's climate influenced by elevation as well as by its location in subtropical latitude ranges from subtropical monsoon conditions in Terai, through a warm temperature climate in the mid-mountains to an alpine climate along the lower slopes of the Himalaya Mountains. In the recent years, Nepal is witnessing continuous disturbances in its ecology due to climate change resulting floods, severe landslides, and soil erosion and so on. About 93% of the total annual precipitation occurs during the monsoon period.

So, it is imperative to understand the current status and behaviour of glacier for proper planning of the water resource. Direct field observations are very difficult to carry out because of rugged and remote mountain terrain. For better understanding of the behaviour, different models have been developed around the world; which vary from simple methods based on the temperature measurement to more physical methods. Melt models generally falls into two categories: energy balance models, attempting to quantify melt as residue in the heat balance equation, and temperature-index models assuming an empirical relationship between air temperature and melt rates (Shrestha 2008). Hence, for better simulation of the watershed the contribution of snow and glaciers should be included in suitable model.

Therefore, a proper understanding of the status, environment and physical dynamics of these glaciers is very important for long-term planning in the use of water resources. These models are generally developed for estimation of snow and ice melt from debris free clean glaciers and need further study for debris covers since the ablation area of glaciers in the Nepalese Himalayan are mostly covered by debris layer. In this study, an attempt has been made to assess the impact of climate change on the seasonal variability of stream flow in the Koshi and Mahakali basins.

Himalayan glaciers have been retreating in the accelerating face of global climate warming resulting increase in the number and size of glacier lakes escalating the threat of (Glacier Lake Outburst Flood) GLOF (WWF 2005). The glaciers in the Nepal Himalayas are already melting so rapidly that there will be no glaciers left in 2180 even without any further warming (Chaulagain 2006). Referring to such statements, it can be understood that the effect of climate changes specially the global warming has an important role in deciding the fate of not only the glaciers but the whole water resources system of Nepal.

The local hydrology of every river in the world is likely to be affected by climate change in some ways. It affects different aspects of the local hydrology of a river such as the timing of water availability, quantity and quality. These changes in the river hydrology will induce risks to water resources facilities in the form of flooding, landslides, sedimentation from more intense precipitation events (particularly during the monsoon) and greater unreliability of dry season flows.

The latter poses potentially serious risks to water and energy supplies in the lean season. For the long-term planning and management of water resources, the future changes in the pattern of land use, water demand and water availability should be analyzed well in advance. It entails understanding the manner in which a water resources system responds to changing trends and variability, the manner in which it is affected by these conditions today and how it might respond if these conditions undergo change.

The objective of this study is to determine the distribution pattern of snow and glacier over the basin by analysing the satellite image & Digital Elevation Model (DEM) and further determine the contribution of snow and glacier melt on the river basins by using Energy Balance Model (EBM), TANK and Hydrologic Engineering Centre-Geospatial Hydrologic Modeling System (HEC-HMS) models. In this study, impact of climate change on river hydrology is carried out using the Fifth Assessment Report (AR5) scenario in order to estimate the change in hydropower potential of the river under climate change scenarios.

16.2   Material and Methods

16.2.1   Description of Study Area

The Koshi river basin is located at the eastern part of the country from latitude 26°52'0" to 29°6'41" N and longitude 85°44'51" to 89°14'53" E while the Mahakali river basin is located at the far western part of the country from latitude 29° 49' 48" to 30° 34' 48" N and longitude 80.00° to 81° 6' 36" E. These basins cover three distinct latitudinal physiography zones: Mountain, Siwalik and Terai of Nepal.

The Koshi basin has an effective watershed area of 60,400 km² at Chatara out of which 31,940 km² (53%) lies within Nepal (WECS 2011). Koshi River basin has seven major sub-basins: the Tamor, Arun, Dudhkoshi, Likhu, Tamakoshi, Balephi and Indrawati. The Koshi Basin is selected as it has significant coverage of active glaciers and glacier lakes. The tributaries of Koshi River encircle Mt. Everest from all sides and are fed by the world's highest glaciers. The Koshi River Basin comprises 779 glaciers with an area of 1,409.84 km² and has an estimated ice reserve of 152.06 km³. (Mool et al. 2001). Some major running and proposed hydropower plants of Nepal are in Koshi basin. The Koshi river basin contributes 22,350 MW out of the total hydropower potential of Nepal which is 83,290 MW. And the economically exploitable potential is assessed as 10,860 MW out of 42,140 MW for whole Nepal (Shrestha 1985).

The Mahakali River Basin has a total catchment area of 15,260 km² at Banbasa barrage of which 5,410 km² (35%) lies within Nepal (WECS 2011). Mahakali basin has four administrative districts- Baitadi, Dadeldhura, Darchula and Kanchanpur. The Mahakali River is about 223 km long and originates from ApiHimal. This western boarder river has two main tributaries on the Nepalese side. There are the Chamelia River and Surnagad River. In this study Chamelia river is considered because a major portion of the flow in this river is due to the glacier melt at higher altitude. The estimated runoff from this basin is 698 m³/s out of which 247 m³/s flows from Nepal. There are altogether 758 numbers of glacier in the Mahakali basin out of which only 87 lie in Nepalese territory. The total glacier area of the Mahakali basin within the Nepalese territory is 143.23 km² with an estimated ice reserve of 10.06 km³ (WECS 2011). The location map of Koshi and Mahakali basin is shown in Fig. 16.1.

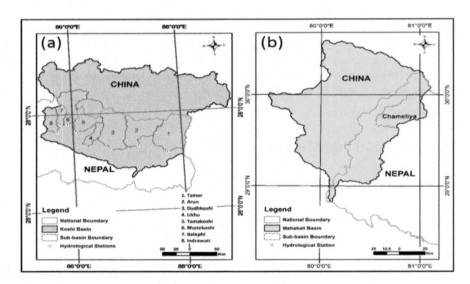

Fig. 16.1 Location map of the Koshi and Mahakali Basin

16.2.2   Theoretical Description of Model

A simplified energy balance model is used for the simulation of snow accumulation and snow/glacier melt, based on a lumped representation of the snow pack as a single layer. Energy content and water equivalence are the main state variables that describe the snow pack. The energy exchange at the snow surface is simulated using physical descriptions, and is balanced by a net heat flux into the snow pack. TANK model is selected for snow and glacier melt runoff simulation as it is very simple and flexible mathematical model. It can be applied in mountainous catchment or flatter terrain, in humid climates or in arid climates as the structure of the model can easily be changed to suit the particular situation. But, TANK model itself is not sufficient to represent the hydrological process in the non-glacerised part of the study area due to the watershed area in these portions is large enough. So, for this portion semi-distributed model HEC-HMS is chosen because with little additional effort semi-distributed model can give better result.

The input to the conceptual TANK model is the sum of precipitation as rainfall; snowmelt and glacier melt from debris-free and debris-covered glaciers. Whereas, the input to HEC-HMS model is: stream discharge from glacerised watershed and precipitation (with reduction of evaporation loss) from non-glacerised area.

16.2.3   Model Input Data

The 30 m resolution DEM and Image needed for delineation for snow and glacier are obtained freely from the site of Advanced Space Borne Thermal Emission and Reflection Radiometer GDEM (ASTER GDEM) and Glovis.usgs.gov respectively. Images for the required month and year are selected and analysed for cloud cover not more than 15%. The eight bands of Land Resources Satellite Thematic Mapper (LANDSAT TM7) are stacked into single multispectral image in Earth Resource Data Analysis System IMAGINE (ERDAS IMAGINE) software. Scan Line Correction (SLC) is applied for the satellite imagery imaged after 2030 AD to remove black strips and then multiple adjacent images are mosaicked to cover the whole area of the basin.

In this study, different layers are formed by using different indices in Geographical Information System (ArcGIS) and ERDAS IMAGINE. The elevation and slope layers are made using DEM in ArcGIS. Normalized Difference Snow Index (NDSI), Normalized Difference Vegetation Index (NDVI), Land and Water Mask (LWM) and supervised class layers are made using satellite image in ERDAS IMAGINE. Once all the layers are formed, filtration is carried out on the basis of fixed criteria. The filtration of NDSI and NDVI layers are done in ERDAS IMAGINE whereas filtration of LWM, slope and elevation layers are done in ArcGIS. The supervised layer are intersected with filtered layer in ArcGIS to obtain more refined image of snow and glacier Estimation of snow and glacier melt is done by using surface energy balance method. In EBM, aerial altitude distribution of watershed area and glacerised area are required, which are extracted from ArcGIS. For the runoff simulation of glacerised watershed a conceptual continuous lumped hydrological TANK model is used. The runoff thus generated from whole glacerised watershed is then used as an input to HEC-HMS for simulation of discharge at non-glacerised area. 12 hydrological and 21 meteorological stations are used in this study.

For the assessment of climate change in the Koshi and Mahakali River basin, bias corrected Regional Climate Model (RCM) data of temperature and precipitation is used as the input. In climate change, AR5 scenario is used. Precipitation data are extracted from CANESM2 Model whereas Temperature data is extracted from RegCM Model. Once the change in river flow is obtained from climate change analysis, river hydrograph before and after climate change is analysed to determine the change in potential of river due to climate change. Change in total volume of water in a year due to climate change is determined followed by preparation of flow duration curve for base year 2030 and 2050 AD, and determination of Q40 for each year.

16.2.4 Model Setup

The watershed is divided into a number of altitude bands and the snow and debris cover are distributed in each altitude band. The energy-balance model coupled with mass balance of snow is used to calculate the snow accumulation, snow and glacier ablation from all bare areas, debris-free and debris-covered areas in each altitude band of the watershed and averaged specific melt and rainfall is calculated. The energy flux estimations are driven by meteorological parameters and surface conditions. The surface condition is classified as bare area, debris-covered glacier, debris-free glacier areas over which the new snow accumulation occurs. Accumulation of new snow in each altitude band is calculated with the help of air temperature. The energy fluxes for all the surfaces is computed and positive energy balance is deemed as the energy used to melt the new snow and glaciers. Thus obtained snow and glacier melt from energy balance and precipitation as rainfall are used as the inputs for the TANK hydrological model for runoff simulation.

Using the HEC-HMS models, runoff is calibrated and validated at different points of basins. Glacerised watershed is simulated by TANK model and non-glacerised watershed is simulated by the HEC-HMS model. Finally, climate change impact of river discharge due to changes in temperature and precipitation is carried out using the new climate change scenario AR5. In this study, the effect of climate change at different locations of Koshi and Mahakali basins are estimated.

16.3   Results and Discussion

16.3.1   Areal coverage of Snow Cover, Debris Covered Glacier and Clear Ice Glacier

After the delineation of glacerised watershed, percentage of snow and glacier to the total catchment area for selected downstream is calculated. Looking on overall basin, around 1.62% area of the basin is covered by glacier (DC and CI). Looking at this data it can be seen from the Table 16.1 that the percentage of glacier to total watershed is greater in the Koshi basin than other basins of Nepal but Mahakali basin has a less percentage of glacier than other basins of Nepal. It is clearly seen from the Table 16.2 that there is decreasing trend of the snow cover area with time. This may be due to climate change effect. Figure 16.2 represents the distribution of snow and glacier at different time of Koshi and Mahakali Basin.

16.3.2   Calibration and Validation of TANK Model

For the hydrological modeling of glacerised watershed, three snow melt gauging stations are selected. The station of DHM at Humla (Halji, field) at an elevation of 4214 amsl is used for modelling Chamelia of Mahakali Basin. Station at Kyangjing at an elevation of 3655 amsl is used to simulate the western part of the Koshi Basin. Similarly, the station at Dingboche at an elevation of 4375 masl is used to simulate the eastern part of the Koshi Basin.

The model is calibrated at the basin outlet at the hydrological station set by DHM for the year 1992 for Halji, 1988 for Kyanging and in 1989 for Dingboche by using the observed discharge. The manual calibration technique is used to estimate the model parameter of the model with care to limit the parameter values within acceptable limits. In the EBM model the glacier area obtained from the processing of satellite image of 2001 is used considering the glacier area does not change much in short interval of time. The model simulated the discharge at Halji field, Kyangjing and Dingboche with Nash–Sutcliffe Efficiency (NSE) of 92%, 98% and 96% respectively. Similarly, the volume difference during calibration in these three stations are −1.14%, −0.41% and 0.49% respectively.For validation of the model, flow is simulated for the year 1991 for Dingboche and Kyangjing. The model is well validated with NSE of 95.32% and the volume difference of −0.42% for Dingboche and NSE of 78% and the volume difference of −1.63% for Kyangjing. But, due to lack of hydro-meteorological data, validation could not be done in Halji field. Figure 16.3 shows the calibration and validation of TANK at Kyangjin.

Table 16.1 Aerial Distribution of Glacier Cover in Koshi and Mahakali basin

S.N.	Sub-basin	Basin area (Km²)	Debris Covered (DC) Glacier				Clear Ice (CI) Glacier			
			Summer, 2001		Summer, 2013		Summer, 2001		Summer, 2013	
			Area (Km²)	%	Area (Km²)	%	Area (Km²)	%	Area (Km²)	%
1	Tamor at Majhitar	4,046	71	1.75	67	2.00	88	2.17	69	1.71
2	Arun at Turkeghat	2928	28	0.96	21	1.00	18	0.61	31	1.06
3	Dudhkoshi at Rabuwa Rabuwa bazar	3747	184	4.91	154	4.00	47	1.25	63	1.68
4	Likhu at Sangutar	851	9	1.06	5	0.59	2	0.24	4	0.47
5	Tamakoshi at Busti	1493	23	1.54	14	0.94	21	1.41	12	0.80
6	Bhotekoshi at Barabise	373	0	0.00	0	0.00	0	0.00	0	0.00
7	Balephi at Jalbire	623	10	1.61	4	0.64	11	1.77	7	1.12
8	Indrawati at Dolalghat	1231	1	0.08	0	0.00	3	0.24	2	0.16
9	Mahakali except Chamelia	3687	25	0.68	12	0.33	16	0.43	14	0.38
10	Chamelia at karkaleGaun	1163	14	1.20	5	0.43	8	0.69	8	0.69

Table 16.2 Aerial Distribution of Snow Cover in Koshi and Mahakali basin

S.N.	Sub-basin	Basin area (Km²)	Snow Cover							
			Summer, 2001		Summer, 2013		Winter, 2001		Winter, 2013	
			Area (Km²)	%	Area (Km²)	%	Area (Km²)	%	Area (Km²)	%
1	Tamor at Majhitar Majhitar	4,046	466	11.52	328	8.11	1200	29.66	1166	28.82
2	Arun at Turkeghat	2928	339	11.58	219	7.48	865	29.54	611	20.87
3	Dudhkoshi at Rabuwa Bazar	3747	550	14.68	342	9.13	1349	36.00	814	21.72
4	Likhu at Sangutar	851	35	4.11	22	2.59	136	15.98	118	13.87
5	Tamakoshi at Busti	1493	121	8.10	79	5.29	507	33.96	405	27.13
6	Bhotekoshi at Barabise	373	9	2.41	7	1.88	83	22.25	58	15.55
7	Balephi at Jalbire	623	84	13.48	74	11.88	239	38.36	206	33.07
8	Indrawati at Dolalghat	1231	56	4.55	26	2.11	198	16.08	135	10.97
9	Mahakali except Chamelia	3687	130	3.53	91	2.47	567	15.38	559	15.16
10	Chamelia at karkaleGaun	1163	44	3.78	41	3.53	327	28.12	292	25.11

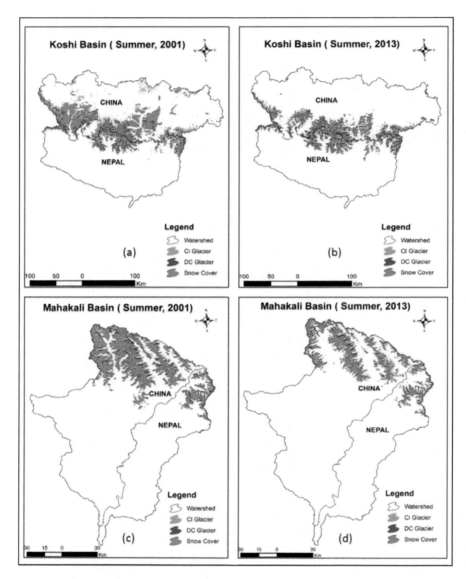

Fig. 16.2 Distribution of Snow and Glacier in Koshi and Mahakali Basin

16.3.3 Calibration and Validation of HEC-HMS Model

For the calibration of HEC-HMS, the model is run for year 1989 A.D. for Tamor, Dudhkoshi and Likhu& for the year 1988 for Balephi & Indrawati. In Koshi Basin, calibration of Arun, Tamakoshi and Bhotekoshi could not be done due to lack of

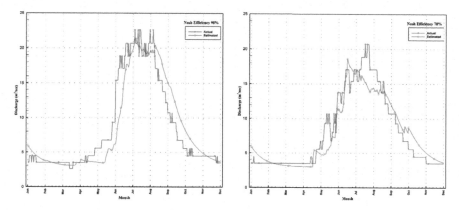

Fig. 16.3   Calibration and Validation of TANK at Kyangjing Station

sufficient meteorological data as the watershed of these basins extended upto China. Similarly, validation of HEC-HMS is done for the year 1991 for all basins. In case of Mahakali there is only one glacerised sub-basin with proper hydrological station i.e. Chamelia. The calibration of Chamelia is done for year 1992 but due to lack of hydro-meteorological data validation could not be done. Calibration and Validation of Koshi at Chatara is done in 1989 and 1991 respectively (Fig. 16.4). Nash–Sutcliffe Efficiencies of different sub-basins for calibration and validation is shown in the Table 16.3.

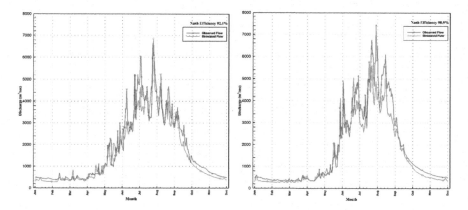

Fig. 16.4   HEC-HMS Calibration and Validation of Koshi Basin at Chatara

Table 16.3 NSE for Calibration and Validation of HEC-HMS

S.N.	Name of Sub-Basin	NSE		S.N.	Name of Sub-Basin	NSE	
		Calibration	Validation			Calibration	Validation
1	Tamor at Majhitar	75.0	71.0	5	Indrawati at Dolalghat	74.1	71.0
2	Dudhkoshi at Rabuwa bazar	78.1	74.0	6	Chamelia at KarkaleGaun	70.4	–
3	Likhu at Sangutar	70.0	71.0	7	Koshi at Chatara	92.1	90.9
4	Balephi at Jalbire	72.7	72.0				

16.3.4   Snow and Glacier Melt Contribution

After the calibration and validation of HEC-HMS, the model is run without precipitation to determine the contribution of snow and glacier in each of the sub-basin as well as in the basin outlet as a whole. Also, the contribution is determined annually as well as in driest month. Here, month March, April and May are taken as dry month.

The result shows that contribution of snow and glacier melt discharge to the annual flow at Koshi at Chatara is about 7.87% with dry month contribution of 12.02%. While determining the contribution at different sub-basins, it is seen that snow and glacier melt contribution is maximum in Dudhkoshi at Rabuwa Bazar (20.83% on annual basis and 59.03% during dry periods).

In case of Mahakali basin, the total contribution of snow and glacier to the Mahakali River could not be determined because there is no hydrological station for the record of river discharge. But, while analyzing the river discharge of Chamelia River at KarkaleGaun it is seen that about 11.74% of the total flow is contributed by snow and glacier melt (annual basis). But, this contribution is significantly increases in dry periods (17.35%). Figure 16.5 represents the snow and glacier melt contribution of Dudhkoshi at Rabuwa Bazar and Chameliya at Karkale Gaun. The percentage contribution of snow and glacier to flow at outlet of different sub-basins of Koshi and Mahakali basins in both annual basis and during dry period is shown in Table 16.4.

16.3.5   Assessment of the Impact of Climate Change

For climate change impact analysis, future data of precipitation and rainfall of year 2030 A.D. and 2050 A.D. are prepared by using the Intergovernmental Panel on Climate Change (IPCC) 5th assessment (AR5) scenario. This data is given as input to the optimized hydrological model. For this year 1990 A.D. is taken as the base year.

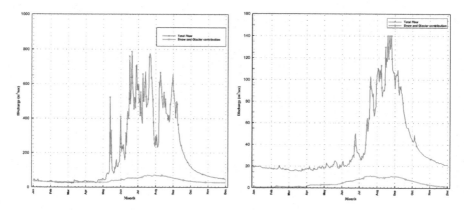

Fig. 16.5 Snow and Glacier melt Contribution of Dudhkoshi at Rabuwa Bazar & Chameliya at KarkaleGaun

Table 16.4 Contribution of Snow and Glacier in Koshi and Makali Basin

S.N.	Station Name	Dry period (%)	Annually (%)	S.N.	Station Name	Dry period (%)	Annually (%)
1	Tamor at Majhitar	32.12	17.37	6	Bhote-koshi at Barabesi	-	-
2	Arun at Turkeghat	6.28	5.48	7	Balephi at Jalbire	26.45	18.11
3	Dudhkoshi at Rabuwa Bazar	59.03	20.83	8	Indrawati at Dolalghat	17.84	7.05
4	Likhu at Sangutar	31.92	9.47	9	Koshi at Chatara	12.02	7.87
5	Tamakoshi at Busti	12.52	5.26	10	Chamelia at KarkaleGaun	17.35	11.74

During climate change analysis, the optimized parameter of the model is kept same and the model is run for the period 2030 A.D. and 2050 A.D. in order to obtain their respective runoff volume. The analysis of the effect of climate change on river flow of Arun, Tamakoshi and Bhotekoshi of Koshi basin and the Mahakali river of Mahakali basin could not be carried out due to lack of future climate data of these basins. Due to which impact of climate change effect could not be determined in Chatara of Koshi basin. Similarly, the effect of climate change could not be determined in the Mahakali river of Mahakali basin due to lack of hydrological station. From the result, it is seen that in general the annual flow volume will decrease in 2030 A.D. but will again increase in 2050 A.D. The percentage change in flow volume of different sub-basins of Koshi and Mahakali basins in the year 2030 A.D. and 2050 A.D. due to climate change is shown in Table 16.5.

Table 16.5 Change in Available Discharge in Koshi and Mahakali Basin

S.N	Station Name	Annually (%)		Q_{40} (%)	
		Simulated (2030)	Simulated (2050)	Simulated (2030)	Simulated (2050)
1	Tamor at Majhitar	−9.93	1.69	4.30	48.30
2	Dudhkoshi at Rabuwa Bazar	20.21	38.06	59.16	100.92
3	Likhu at Sangutar	−6.31	9.79	0.56	103.09
4	Balephi at Jalbire	−0.58	29.63	72.95	106.56
5	Indrawati at Dolalghat	8.88	28.40	34.13	148.34
6	Chamelia at Karkale Gaun	12.42	36.41	69.29	116.85

Finally, change in river flow due to climate change is analyzed to determine the effect of climate change on hydropower potential of river. For determination of climate change impact on the hydropower potential for storage project, change in total annual volume is analysed. It is observed that in Koshi basin that there will be decrease in volume for three sub-basins and increase in volume for remaining sub-basins in 2030 A.D. But, the volume will increase in all sub-basins of Koshi basin in 2050 A.D. Similarly, there will be increase in volume in Chamelia sub-basin of Mahakali basin in year 2030 A.D. and 2050 A.D. It is seen from the result that climate change effect is favouring storage project in 2050 A.D. than in 2030 A.D. Table 16.5 shows the result of climate change on hydropower potential for storage project in these basins.

A similar process is used for determination of climate change impact for RoR project. Design discharge is fixed at forty percentile of time and the change of available discharge in 2030 A.D. and 2050 A.D. is determined. The result shows that

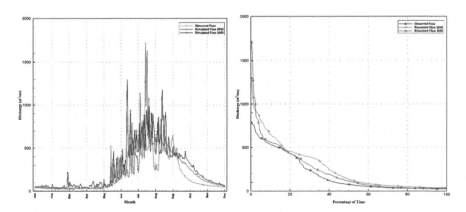

Fig. 16.6 Available Discharge and FDC of Dudhkoshi at Rabuwa Bazar

the available flow of RoR project will increase in both 2030 A.D. and 2050 A.D. of Koshi and Mahakali basin. The effect of climate change on potential of river for year 2030 A.D. and 2050 A.D. is shown in Table 16.5. The available discharge and Flow Duration Curve (FDC) for Dudhkoshi at Rabuwa Bazar is shown in Fig. 16.6.

16.4 Conclusions

This study is carried out to assess the climate change impact on river discharge of various sub-basins of Koshi and Mahakali basin. During this study, the altitude wise area of snow and glacier along with precipitation data in glacerised watershed is given as input to Energy Budget model to calculate melt volume from the watershed. TANK model is used for runoff generation from these watersheds for which melt volume is given as input. Finally, the runoff generated from these watershed is given as input along with precipitation in non-glacerised area (with the deduction of evapo-transpiration) to simulate the discharge of the whole basin. From the delineation of snow and glacier, it can be seen that Koshi basin (1.62%) is covered by larger areas of glacier than that of the Mahakali Basin (1.16%). Among different sub-basins Dudhkoshi basin has the highest area of glacier whereas Indrawati has a lowest cover of the glacier.

Snow melt and Ice ablation from the debris covered glacier area has a remarkable influence over the river discharge of Koshi and Mahakali basin. The analysis indicates that snow and glacier melt alone contribute about 7.87% of total annual flow in Koshi basin at Chatara. The contribution is more significant in dry period in which snow and glacier contribute about 12.02%. For smaller basin like Dudhkoshi River at Rabuwa bazar, the contribution of snow and glacier melt becomes 59.03% of total annual flow and 20.83% during dry period.

The study demonstrates that the impact of climate change (i.e. temperature) to stream flow is significant, which is in increasing trend resulting from snowmelt contribution. Analysis carried out using future climate scenario suggests that in most of the sub-basins of Nepal the annual flow volume as well as 40% reliable flow will decrease in 2030 but increase in 2050. The result thus indicates that the available flow for RoR (Run-of-River) and storage hydropower project will decrease in 2030 A.D. but the flow will increase in all the sub-basins in 2050 A.D. provided the area of snow and glacier cover remain unaltered. In most of the basins there will be shift in hydrograph indicating delay in monsoon. Also there will be irregular peak of flow caused by short burst of high intensity rainfall. For further study and refinement of this research, it is recommended that the meteorological station be established within the catchment itself, and more precise and more numbers of satellite images be used. The outputs of this study are important guidance for water resources managers to make and implement appropriate strategy for water resources planning and management.

The impact of climate change on water resources depends not only on changes in the volume, timing, quality of the stream flow and the recharge but also on system characteristics, changing pressures on the system, the manner in which the management of the system evolves and the adaptation measures implemented for climate change. The creation of mass awareness, formation of climate change sectorial policies on matters related to water resources, stringent judicial enforcement, reliable information of the Himalayan snow fields and glaciers, capable human resources development and the creation of opportunities for higher studies are some of the other necessities that are to be planned and carried out in a proper and effective manner.

Snow and glaciers of Nepal have substantial contribution in sustaining flow in lean period. The impact of global warming on the glaciers and ice reserves of Nepal will have serious implications for the freshwater reserve and consequently for low flows. This will adversely affect the dry season flow in future and will have serious adverse impacts on electricity generation from the hydropower plants, especially for Run of River type plant, water supply and irrigation planning and also on the biodiversity. Variation in seasonal flow significantly affects the electricity generation, even without the change in annual flow volume. In order to compensate the likely reduction in contribution of melt flow from snow and glacier to river system during lean flow season, appropriate basin scale intervention to enhance base flow of river system is imperative. Such conservation measures can only be possible once there are dedicated and precise policy and legislation related to Integrated Water Resources Management and River Basin Management. Similarly appropriate policy for the implementation of generation mix such as use of Run-of River (RoR), Peaking Run-of River (PRoR), storage project in conjunction with thermal plant is required in this regard.

Acknowledgements This work was supported by Nepal Academy of Science and Technology and Department of Civil Engineering, Institute of Engineering, Tribhuvan University, Nepal. The authors are grateful to DHM for providing data and information.

References

Chaulagain NP (2006) Impact of climate change on water resources of Nepal, The physical and socio-economic dimensions. University of Flensburg, Flensburg, Germany
Mool PK, Joshi SP, Bajracharya SR (2001) Inventory of Glaciers, Glacial Lakes and Glacial Lake Outburst Floods-Monitoring and Early Warning Systems in the Hindu Kush-Himalayan Region Nepal. ICIMOD, Kathmandu, Nepal
IPCC (2001) Climate Change 2001: The Scientific Basis. Cambridge University Press, Cambridge
Shrestha HM (1985) Water power potential. In: Majupuria TC (ed) Nepal: Nature's Paradise: Insight into Diverse Facets of Topography, Flora and Ecology. White Lotus Co. Ltd., Bangkok, pp 4–8
Shrestha M (2008) Runoff Modeling of Debris Covered Glacierized Watershed: A Case Study of Imja watershed. Water Resources Engineering Program IOE, Pulchowk
Sing P, Sing VP (2001) Snow and Glacier Hydrology. Springer, Dordrecht
WECS (2011) Water Resources of Nepal in the Context of Climate Change. WECS, Kathmandu
WWF (2005) An Overview of Glaciers, Glacier Retreat and its Subsequent Impacts in Nepal. WWF Nepal Program, Kathmandu, India and China, p 68

Chapter 17
Assessment of Variation of Streamflow Due to Projected Climate Change in a Water Security Context: A Study of the Chaliyar River Basin, India

S. Ansa Thasneem, Santosh G. Thampi, and N. R. Chithra

Abstract Results of many studies indicate that global water resources will be subjected to severe stress as a result of climate change; this is attributed to projected changes that would occur in the component processes of the hydrological cycle. The gap between water supply and demand for water is expected to widen. This will have serious implications on water supply for urban and agricultural uses. Analysing the interaction of climate and water resources helps to devise strategies for mitigating negative impacts by proper water management. In this paper, an assessment of likely changes in streamflow in the Chaliyar river in Kerala for the decade 2040–2050 due to projected climate change is performed. Soil and Water Assessment Tool (SWAT) is used for hydrological modeling of the river basin. The outputs of the regional climate model (RCM) Remo2009 (MPI) are corrected for biases and then input to the SWAT. The future scenarios considered are Representative Concentration Pathways, RCP4.5 and RCP8.5. Chaliyar river, the fourth longest river in Kerala, serves as the source of water predominantly for agricultural and domestic uses. Results obtained from the model can be used to assess the water supply scenario in the river basin in the future. The findings of this study would be very useful for devising strategies for the effective management of available water resources in the river basin.

Keywords Climate change · Streamflow · Water supply

17.1 Introduction

Climate change is one of the major threats faced by the world today and will affect all aspects of human life and natural systems. It has been projected that increased greenhouse gas emissions will lead to increase in temperature by the end of this century (IPCC 2007). In the fifth assessment report (IPCC 2013), it is stated that the increase in global mean surface temperature by the end of the twenty-first century (2081–2100) relative to 1986–2005 is likely to be 0.3 °C to 1.7 °C under RCP2.6, 1.1 °C to 2.6 °C

S. Ansa Thasneem (✉) · S. G. Thampi · N. R. Chithra
Department of Civil Engineering, NIT Calicut, Kozhikode, India
e-mail: ansathsnm@gmail.com

© Springer Nature Switzerland AG 2021
M. Babel et al. (eds.), *Water Security in Asia*, Springer Water,
https://doi.org/10.1007/978-3-319-54612-4_17

under RCP4.5, 1.4 °C to 3.1 °C under RCP6.0 and 2.6 °C to 4.8 °C under RCP8.5.This rise in temperature will affect the hydrology on a regional scale (Tavakoli and Smedt 2011). As a result of this, the hydrological processes such as streamflow, soil moisture, groundwater recharge etc. will be effected. All components in a watershed are interrelated so that change in any one of them will affect the other too. Adaptation strategies to climate change therefore demand an integrated watershed approach. Climate change impact assessment studies on a watershed scale enables sustainable decision making during the planning phase of watershed management.

Projected climate change and its impact on the hydrological cycle raise serious concerns about water security in the future. Water security of a watershed is dependent on the quantity and quality of water available in the river basin from various resources (Kibria 2016). Frequent occurrences of extreme events such as floods and droughts as a result of changing climate can adversely affect the water security of a region. Also, drastic fluctuations in precipitation and ice storage will cause variation in river flows and affect water supply and water security (Cisneros et al. 2014). Konapala et al. (2020) studied the combined effect of seasonal variation of annual mean precipitation and evaporation on water availability on a global scale by classifying regions into different hydro-climatic zones. They observed an increase in annual mean precipitation and evaporation over all the zones under projected climate change scenarios. Also, the seasonal variations in precipitation were found to be increasing and that of the evaporation were found to be decreasing implying that water availability would be seriously affected by seasonal variations in response to the changing climate. Stewart et al. (2020) analyzed the proportions of flow and storage in different parts of a watershed under various intensities of droughts during the period 2008–2018 and found that water storage and supplies to agricultural and urban users in some regions of the watershed were reduced. Gesualdo et al. (2019) investigated the influence of anticipated climate change on water security in the Jaguari Basin with the use projected streamflow under RCPs 4.5 and 8.5 and found that water scarcity is crucial in some months during the mid and far future periods.

Watershed is a bounded hydrologic system, within which all living things are inextricably linked by their common water course and where, as humans settled, simple logic demanded that they become part of a community (Powell 1890). Creation and implementation of plans, programs, and projects to sustain and enhance watershed functions within a watershed boundary are the main goals of watershed management. Distribution and availability of water plays a crucial role in the management of a watershed. Assessment of the water balance within a watershed boundary helps to identify the surpluses and deficits of water. It utilises hydrological and meteorological data to assess the past and forecast the future availability of water in a watershed. Challenges to water security can be tackled by an integrated watershed management program which could analyse the hydrological processes and its interaction with various components of a watershed (Mwangi et al. 2016). Tessema et al. (2015) used watershed modelling as a tool for sustainable water resources management. They studied the catchment characteristics and hydrological processes within a watershed using the Soil and Water Assessment Tool (SWAT) and utilized the results obtained in the process of decision making for the prevention of large floods in the watershed. Anitha et al. (2014) studied the spatial and temporal availability and demand of water in the Chaliyar river basin for the development of an integrated master plan and found that the river basin is facing deficit of water during the non-monsoon period. In the present study, the implication of climate change on future streamflows in the Chaliyar river basin is analysed and its consequences on future availability of surface water in the river basin is evaluated.

17.2 Materials and Methods

17.2.1 Study Area

The river Chaliyar is the fourth longest river in Kerala, India, with a length of about 169 km.The area of the river basin in Kerala is 2530 km^2, bounded by latitudes 11°06'07"N and 11° 33'35"N and longitudes 75°48'45"E and 76° 33'00". The basin comprises parts of four districts in two states viz. Kozhikode district over an area of 628 km^2 in the northwest, Wayanad district over an area of 114 km^2 in the north, Malappuram district over an area of 1788 k m^2 in the east and south and Nilgiris district of Tamil Nadu over an area of 378 km^2 in the northeast. It rises in the Elambaleri hills in the Wayanad plateau. Six major streams Chaliyarpuzha, Punnapuzha, Kanjirapuzha, Karimpuzha, Iruvahnipuzha and Cherupuzha constitute the Chaliyar river drainage system. The Chaliyar river basin can be physiographically divided into four well-defined units such as highland, midland, low land and coastal plains. The basin has a tropical humid climate with hot summer and high monsoon rainfall. Average annual rainfall of the river basin is about 3012 mm and streamflow is about 46,968 cumecs. Generally, March and April are the hottest and December and January are the coolest. The maximum temperature ranges from 34 °C to 42 °C and the minimum temperature ranges from 22 °C to 25 °C. The average annual maximum temperature is 34 °C and minimum is 23.7 °C. The temperature starts rising from January reaching its peak in April and it decreases during the monsoon months.

17.2.2 Details of Input Data and Their Sources

SWAT requires spatial data such as Digital Elevation Model (DEM), topographic map, land use/land cover map and soil map. It also requires daily weather data such as precipitation, temperature, solar radiation, wind speed, relative humidity and hydrologic data such as streamflow. Daily streamflow data from the Kuniyil gauging station of the Central Water Commission (CWC) for the period 1990–2003 is used in this study. Weather data for this period was collected from the Kottamparamba observatory of the Centre for Water Resources Development and Management (CWRDM), Kunnamangalam Kozhikode. Land use and soil maps were collected from the Kerala State Landuse Board. The scale of the land use map is 1:50,000 and that of the soil map is 1:25,000. Climate change impact on water resources differ by region based on regional geographical characteristics and climatic conditions. General Circulation Models (GCMs) are widely employed to obtain projections of future climate for use in impact studies including that on water resources (Fowler et al. 1999; Bergström et al. 2001; Graves and Chang 2007). Regional impacts could be effectively addressed only if the projections of global circulation models are brought down to a finer scale using downscaling techniques. For a sustainable outcome, management decisions should be based on regional analysis. In the present study, dynamically downscaled data from CORDEX (Coordinated Regional Climate Downscaling Experiment) is used (https://cccr.tropmet.res.in/home/index.jsp). This data is the downscaled output of RCM REMO2009 which is driven by the GCM MPI-ESM-LR. REMO has a resolution of $0.5° \times 0.5°$. Climate change scenarios are used as input to the climate model runs and as a basis for the assessment of possible climate change impacts, mitigation options, and associated costs. Representative Concentration Pathways (RCPs) are four greenhouse gas concentration trajectories adopted by the IPCC in its Fifth Assessment Report (AR5). The RCPs are consistent with a wide range of possible changes in future anthropogenic greenhouse gas (GHG) emissions (Moss et al. 2010). RCP8.5, characterised by a future of increased greenhouse gas concentrations, and RCP4.5, characterised by a future with relatively ambitious emission reductions, are used in this study.

17.2.3 *Methodology*

Predictions of future streamflow under a climate change scenario can be obtained using downscaled projections of climatic variables such as precipitation and temperature as input to a hydrologic model. Comparison of these predictions with observed streamflow values helps us to assess the impact of projected climate change. The overall methodology employed in this study is presented in Fig. 17.1.

17.2.4 *Description of the Hydrologic Model*

The hydrologic model SWAT (Soil and Water Assessment Tool) is used in the present study. It is a physically based, conceptual, semi-distributed model developed by the Agricultural Research Services of the United States of America and has been widely used (Gassman et al. 2007). Various processes including hydrology, transport of nutrients, erosion, vegetation, managerial methods, flow routing etc. are modeled in SWAT. SWAT was developed to predict the effects of different management practices on water quality, sediment yield and pollution loading in watersheds (Arnold et al. 1998). In SWAT, a watershed is divided into a number of sub-watersheds and these are again subdivided into hydrologic response units (HRUs). HRU is the smallest unit of the model which is characterized by similar land use, soil and slope characteristics

Fig. 17.1 Flow chart of the overall methodology

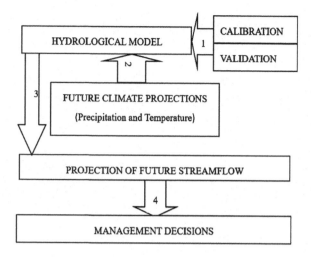

in a given sub-basin based on the thresholds provided by the user for each of these categories. In the present study, a 5% threshold on area was allowed for a particular landuse or soil in a sub-basin. The hydrologic cycle modelled by SWAT requires inputs, such as precipitation, maximum/ minimum air temperature, solar radiation, wind speed, and relative humidity. Based on the user's preference, SWAT will use observed weather data or it will simulate the weather data. Surface runoff, infiltration, evapotranspiration, lateral flow, return flow, canopy storage etc. are some of the hydrological processes modelled by SWAT. Hydrology in SWAT is governed by the water balance equation (Eq. 17.1).

$$SW_t = SW_0 + \sum_{i=1}^{t} \left(R_{day} - Q_{surf} - E_a - w_{sweep} - Q_{qw} \right) \qquad (17.1)$$

where SW_t is the final soil water content, SW_0 is the initial soil water content on day i (mm), R_{day} is the amount of precipitation on day i in mm, Q_{surf} is the amount of surface runoff on day i in mm, E_a is the amount of evapotranspiration on day i in mm, w_{sweep} is the amount of water entering the vadose zone from the soil profile on day i in mm, and Q_{gw} is the amount of return flow on day i in mm.

The different steps in hydrologic modeling with SWAT include watershed delineation, HRU definition, weather definition, input table writing and SWAT simulation. Boundaries of the watershed and sub-watersheds are delineated automatically using a DEM based approach. Figure 17.2 shows a delineated watershed with 31 sub-basins. Land use/ soil/ slope characterization of a watershed and creation of HRUs are performed under HRU analysis. The percentage of area of the watershed occupied by each landuse class is presented in Table 17.1. The predominant land uses in the watershed are agriculture and forests. Soil classification of the watershed obtained from SWAT shows that 38% of the area has loam soil, 29% of the area has clay loam,

Fig. 17.2 Delineated watershed

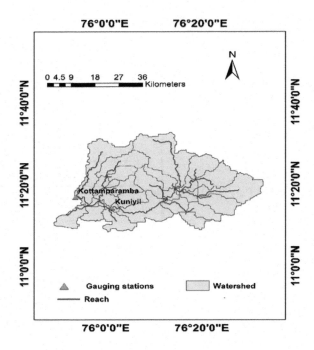

Table 17.1 SWAT Landuse classes

Landuse class	% Watershed area
Agricultural Land-Generic (AGRL)	34.52
Agricultural Land-Close-grown (AGRC)	5.13
Agricultural Land-Row Crops	7.6
Orchard (ORCD)	18.9
Forest-Deciduous (FRSD)	26.72
Forest-Mixed (FRST)	6.48
Residential-High Density (URHD)	0.65

27% of the area has clay and 5% of the area has sandy loam. Observed data of rainfall, temperature, solar radiation, and relative humidity are imported as the initial step for writing the input tables after distributing and defining the HRUs. Figures 17.3 and 17.4 present the landuse and soil classification of the Chaliyar river basin obtained from SWAT.

Fig. 17.3 Landuse classification of the Chaliyar river basin from SWAT

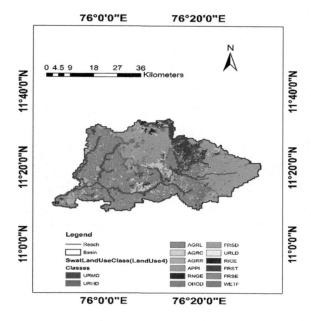

Fig. 17.4 Soil classification of the Chaliyar river basin from SWAT

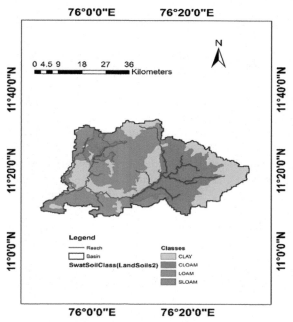

17.2.5 Calibration and Validation Using SWAT CUP

SWAT-CUP is a software that can be used for calibrating the SWAT model. SWAT-CUP accesses the SWAT input files and runs SWAT simulations by modifying the parameter sets. In this study, various SWAT parameters related to discharge were estimated using the SUFI-2 algorithm. It accounts for all sources of uncertainties such as uncertainty in driving variables, conceptual model, parameters and measured data. In SUFI-2, the model output uncertainty is quantified at the 95% prediction uncertainty (95PPU). The cumulative distribution of an output variable is obtained from Latin hypercube sampling. Initially, a large parameter range is assumed and based on that SUFI2 analyses whether the measured data falls within the 95% prediction uncertainty (95PPU) or not. The 95PPU can be estimated from the 2.5% and 97.5% levels of the cumulative distribution of an output variable obtained through Latin hypercube sampling. Goodness of fit measures such as coefficient of determination (R^2) and Nash Sutcliffe (NSE) coefficient can be evaluated for each iteration performed during the calibration (Abbaspour 2013).

17.2.6 Correction of Bias

Bias correction is the process of adjusting the climate model simulations to eliminate the systematic errors present in it. A number of methods are available for this purpose. In the present study, the delta change method is used to correct the biases in climate variables owing to its simplicity. In this method, the RCM predicted future change is obtained as a perturbation of observed data (Bosshard et al. 2011). The delta change uses observations as a basis and generates the future time series with the dynamics of the observed period. Generally, a multiplicative correction is used to adjust precipitation whereas an additive correction is used to adjust temperature (Teutschbein et al. 2012). In this study also, a multiplicative correction is applied for precipitation (Eq. 17.3), and an additive correction is applied for temperature (Eq. 17.5). As shown in the Eqs. 17.2 and 17.4, delta change approach assumes the value of the climate variable during the control period as the observed climate data series.

$$P^*_{cont}(d) = P_{obs}(d) \tag{17.2}$$

$$P^*_{scen}(d) = P_{obs}(d) \times \frac{\mu_m(P_{scen}(d))}{\mu_m(P_{cont}(d))} \tag{17.3}$$

$$T^*_{cont}(d) = T_{obs}(d) \tag{17.4}$$

$$T^*_{scen}(d) = T_{obs}(d) + \mu_m(T_{scen}(d) - T_{cont}(d)) \tag{17.5}$$

where μ_m is the mean within the monthly interval, $P^*_{cont}(d)$ is the final bias corrected precipitation in the control period (RCM simulated), $P_{cont}(d)$ is the daily precipitation in the control period, $P_{obs}(d)$ is the observed daily precipitation, $P_{scen}(d)$ is the daily precipitation while considering the scenario period of the climate model, $P^*_{scen}(d)$ is the final bias corrected daily precipitation during the scenario period, $T^*_{cont}(d)$ is the final bias corrected temperature, $T_{obs}(d)$ is the observed daily temperature, $T_{cont}(d)$ is the daily temperature in the control period, $T_{scen}(d)$ is the daily temperature while considering the scenario period of the climate model, and $T^*_{scen}(d)$ is the final bias corrected daily temperature during the scenario period.

17.2.7 Management Planning

Agriculture and orchards together occupy almost 66% area of the watershed. This river is a major source for meeting irrigation and domestic water demands. At present, there is deficit of water during the non-monsoon season and surplus during the monsoon season (Anitha et al. 2014). Although, many new schemes aimed at meeting irrigation and domestic water demands utilising this river have been proposed, these are still in the planning stage. Also, there are no major storages in the basin at present. Precipitation and temperature patterns in recent years seem to indicate considerable change from the pattern established over the years, creating an alarm in the community and among water managers. In this context, it is extremely important that a scientific study is carried out to assess the likely change in precipitation and streamflow, which would influence water security and availability, and would be helpful in informed decision making.

17.3 Results and Discussion

17.3.1 Calibration and Validation of SWAT

SWAT was calibrated using observed streamflow data for 10 years from 1990–1999. Calibration was performed using SWAT CUP. The parameters selected for calibration are CN2, ESCO, SOL_AWC, GW_REVAP, HRU_SLP, CH_K2, REVAPMN, SURLAG, OV_N, SLSUBBSN, CH_N2, ALPHA_BF, ALPHA_BNK, GWQMN, and SOL_BD. During calibration, the Nash–Sutcliffe efficiency, NSE is 0.86 and the coefficient of determination, R^2 is 0.92. The model was further validated for streamflow using data for four years from 2000–2003. The NSE and R^2 for the validation phase are 0.71 and 0.75 respectively. Overall, there is reasonably good agreement between the observed and predicted flow hydrographs (Fig. 17.5).

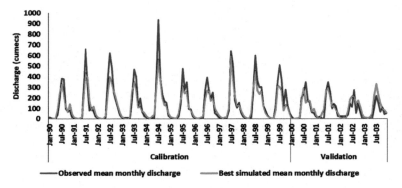

Fig. 17.5 Comparison of observed vs. predicted streamflow in calibration and validation

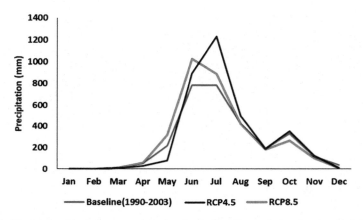

Fig. 17.6 Plot of monthly average precipitation for the baseline period and under the RCPs

17.3.2 Prediction of Future Streamflow

The average monthly values of observed and bias corrected precipitation are presented in Fig. 17.6. Precipitation shows an increasing trend with respect to baseline (1990–2003) during the monsoon season (June to November) under both the RCPs considered. For the non-monsoon season (December-May), decrease in precipitation is observed for RCP4.5 scenario whereas an increase in precipitation is observed for RCP8.5 scenario. The maximum temperature under both the future scenarios shows an increase in all the months. The minimum temperature also exhibits the same trend (Figs. 17.7 and 17.8).

Figure 17.9 shows the plot of monthly average streamflow during the base line period and under the RCPs4.5 and 8.5. The predicted change in streamflow during the monsoon and non-monsoon season by the decade 2040–2050 is presented in Table 17.2. From Table 17.2 it can be inferred that the future streamflow under both

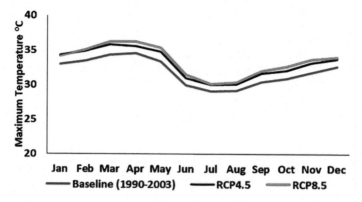

Fig. 17.7 Plot of monthly average maximum temperature for the baseline period and under the RCPs

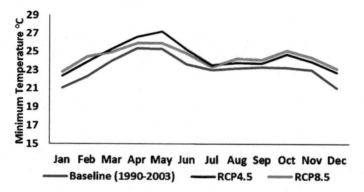

Fig. 17.8 Plot of monthly average minimum temperature for baseline period and under the RCPs

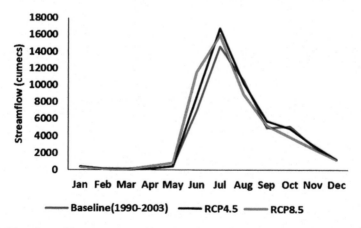

Fig. 17.9 Plot of monthly average streamflow for baseline period and under RCP scenarios

Table 17.2 Variation of streamflow under RCP 4.5 and RCP 8.5 with respect to baseline period

Variable	RCP4.5		RCP8.5	
	Monsoon (June-November)	Non-Monsoon (December-May)	Monsoon (June-November)	Non-Monsoon (December-May)
Precipitation	25%	−59%	9%	22%
Minimum Temperature	0.9 °C	1.5 °C	1.1 °C	1.4 °C
Maximum Temperature	1.1 °C	1.3 °C	1.5 °C	1.6 °C
Streamflow	9%	−5%	6%	15%

Table 17.3 Water availability under RCP scenarios

Season	Streamflow under RCP4.5 (MCM) (2040–2050)	Streamflow under RCP8.5 (MCM) (2040–2050)	Water Demand by 2040 (MCM)	Surplus/Deficit w.r.t water demand under the RCPs
Monsoon	4248	4131	87	Surplus
Non-Monsoon	216	261	413	Deficit

the RCPs would increase during the monsoon season, decrease under the RCP4.5 during the non-monsoon and increase under the RCP 8.5 in the non-monsoon period.

The estimates of surface water that will be available in the river basin during the period 2040–2050 is presented in Table 17.3. Anitha et al. (2014) projected the water demand of the Chaliyar river basin for the year 2040 and found that 87 MCM is required during monsoon season and 413 MCM is required during non-monsoon period. Estimation of groundwater potential is out of scope of this study and hence overall evaluation of water availability in the future is not attempted. Compared to the demand of water by the year 2040, the projected streamflow in the decade 2040–2050 under the RCPs 4.5 and 8.5 is estimated to be higher during the monsoon season (Table 17.3). But, in the non-monsoon period, the availability surface water is much less than the demand implying that the river basin will experience water scarcity. As mentioned previously, there are no major storage systems in the river basin and hence proper utilization of the increased streamflow during the wet months would also be difficult. The situation is likely to be more grave in view of the enhanced demand for water in future and the changes in landuse pattern in the last few years observed all over the State.

17.3.3 Suggestions for Effective Management of the Watershed

Results of this study clearly indicate that streamflow during the monsoon period is likely to increase consequent to increase in precipitation under both the RCPs considered in this study. However in the absence of major facilities for storage in the river basin, and in view of the topographical setting of the river basin, storing this water for use in the non-monsoon period will not be possible and most of it will flow into the Arabian Sea. Also, local incidents of flooding are likely during the monsoon season, causing damage to infrastructure, agricultural losses, impairment of water quality, landslides etc. Climate related risk is the product of climate hazard, exposure and vulnerability. Management of climate related risks depends upon the choices we make under conditions of uncertainty that would prevail in future. Adaptation to climate change involves adjustments by the community to reduce risk on the ecosystem. As an adaptation measure against flooding, improvement in drainage facilities and construction of river bank protection works should be initiated. Flood forecasting capabilities have to be enhanced. To improve water availability and enhance water security in the non-monsoon period, appropriate water harvesting structures to collect excess runoff and soil moisture conservation works to improve water availability for plant growth have to be urgently implemented. Failure to do so can have catastrophic effects. Changes in cropping pattern can also enhance overall water security in the river basin. Planting of appropriate varieties of trees helps to reduce the risk of soil erosion and to some extent, flooding during heavy rainfall events. It also helps to reduce the adverse effect of heat waves by keeping the environment cool. A limitation of this study is that this was performed with projections of only a single climate model. In order to quantify uncertainties in the predictions pertaining to water security, projections of multiple climate models have to be used. This study has been taken up and is in progress.

17.4 Conclusions

For effective watershed management, climate change impacts should also be taken into account. Measures to improve water availability and water security as well as adaptation measures could be identified and implemented only if the likely impacts are studied on a regional scale.In this study the variation in streamflow of Chaliyar river basin due to climate change is assessed. The output of the RCM REMO2009 was used in the hydrological model SWAT to predict streamflow in the Chaliyar river in the decade 2040–2050 under climate change. Results of the study indicates that water availability would not be a serious issue during the monsoon season since the predicted streamflow during the monsoon period under both the RCPs are surplus compared to the demand. But during the non-monsoon season the predicted streamflow could not meet the water demand for that period. Measures need to be adopted for the conservation of surplus water during the monsoon season so that the adverse impacts due to water shortage during non-monsoon season could be tackled efficiently.

References

Abbaspour KC (2013) Swat-cup 2012. SWAT Calibration and uncertainty program—a user manual
Anitha AB, Shahul Hameed A, Narasimha Prasad NB (2014) Integrated river basin master plan for Chaliyar. In: International symposium on integrated water resources management (IWRM-2014), CWRDM, Kozhikode, Kerala, India
Arnold JG, Srinivasan R, Muttiah RS, Williams JR (1998) Large area hydrologic modeling and assessment part I: Model development1
Bergström S, Carlsson B, Gardelin M, Lindström G, Pettersson A, Rummukainen M (2001) Climate change impacts on runoff in Sweden assessments by global climate models, dynamical downscaling and hydrological modelling. Climate Res 16(2):101–112
Bosshard T, Kotlarski S, Ewen T, Schär C (2011) (2011) Spectral representation of the annual cycle in the climate change signal. Hydrol Earth Syst Sci 15:2777–2788. https://doi.org/10.5194/hess-15-2777-2011
Cisneros, JT, Oki BE, Nigel W, Arnell, GBJ. Cogley G, Döll P, Jiang T, et al (2014) Freshwater resources
Fowler D, Cape JN, Coyle M, Flechard C, Kuylenstierna J, Hicks K, Stevenson D (1999) The global exposure of forests to air pollutants. In: Forest growth responses to the pollution climate of the 21st century, Springer Netherlands, pp 5–32
Gassman PW, Reyes MR, Green CH, Arnold JG (2007) The soil and water assessment tool: historical development, applications, and future research directions. Trans ASABE 50(4):1211–1250
Gesualdo, GC, Oliveira PT, Rodrigues DBB, Gupta HV (2019) Assessing water security in the São Paulo metropolitan region under projected climate change
Graves D, Chang H (2007) Hydrologic impacts of climate change in the Upper Clackamas River Basin, Oregon, USA. Climate Res 33(2):143–158
Kibria G (2016) Climate change & water security: impacts, future scenarios, adaptations & mitigations-a book summary. Research Gate online
Konapala G, Mishra AK, Wada Y, Mann ME (2020) Climate change will affect global water availability through compounding changes in seasonal precipitation and evaporation. Nat Commun 11(1):1–10
Moss RH, Edmonds JA, Hibbard KA, Manning MR, Rose SK, Van Vuuren DP, Meehl GA (2010) The next generation of scenarios for climate change research and assessment. Nature 463(7282):747–756
Mwangi HM, Julich S, Feger KH (2016) Introduction to watershed management. In: Pancel L, Köhl M (eds) Tropical forestry handbook. Springer, Heidelberg
Pachauri RK, Reisinger A (2007) (2007) IPCC fourth assessment report. IPCC, Geneva
Powell JW (1890) Institutions for the arid lands. Century 40:111–116
Stewart IT, Rogers J, Graham A (2020) Water security under severe drought and climate change: disparate impacts of the recent severe drought on environmental flows and water supplies in Central California. J Hydrol X: 100054
Stocker TF, Qin D, Plattner GK, Tignor M, Allen SK, Boschung J, Midgley BM (2013) IPCC, 2013: climate change 2013: the physical science basis. Contribution of working group I to the fifth assessment report of the intergovernmental panel on climate change
Tavakoli M, De Smedt F (2012) Impact of climate change on streamflow and soil moisture in the Vermilion Basin, Illinois. J Hydrologic Eng 1059–1070

Tessema SM, Setegn SG, Mörtberg U (2015) Watershed modeling as a tool for sustainable water resources management: SWAT model application in the Awash River basin, Ethiopia. In: Sustainability of integrated water resources management. Springer, Cham, pp 579–606
Teutschbein C, Seibert J (2012) Bias correction of regional climate model simulations for hydrological climate-change impact studies: review and evaluation of different methods. J Hydrol 456:12–29

Chapter 18
Addressing Water Resources Shortfalls Due to Climate Change in Penang, Malaysia

N. W. Chan, A. A. Ghani, Narimah Samat, R. Roy, M. L. Tan, and Haliza Abdul Rahman

Abstract Penang State is one of the most developed and urbanised states in Malaysia, but is a "water-stressed" state with limited water resources availability. In the light of global climate change and El Nino, prolonged droughts have resulted in water crises that are expected to be exacerbated. The objective of this paper is to examine various strategies that can be adopted to address shortfalls in water resources and cope with water crises. The methodology used is a mixture of questionnaire survey, qualitative interviews and secondary data. Results indicate that the Penang State Government should protect water catchments, restructure water tariffs, encourage water conservation, provide incentives to install water recycling plants and rainfall harvesting systems, improve non-revenue water performance, source alternative water resources, and run a campaign to raise awareness and educate the public. Likewise, nongovernmental organizations (NGOs) can work on intensifying awareness, education, conservation, protection and the creation of a "water saving society". NGOs also need to work on efforts to protect the Ulu Muda forest catchment which serves the northern region of Peninsular Malaysia, and push for the implementation of water saving devices (WSDs) for all new houses/buildings. The Federal Government could also have played a role in approving water tariff increases, drawing up regulations for water-saving fittings, and giving incentives for WSDs such as tax rebates. The paper concludes that climate change is unavoidable and that all stakeholders need to be committed in implementing coping strategies to ensure water sustainability in Penang to face the uncertainties of climate change.

Keywords Water resources management · Drought · Water tariff · Water saving devices · Adaptation

N. W. Chan (✉) · A. A. Ghani · N. Samat · M. L. Tan
Universiti Sains Malaysia, Penang, Malaysia
e-mail: nwchan@usm.my

R. Roy
Sher-e-Bangla Agricultural University (SAU), Dhaka, Bangladesh

H. A. Rahman
Universiti Putra Malaysia, Selangor, Malaysia

18.1 Introduction

Climate change is currently affecting many countries negatively. Some countries are afflicted by flooding while many are suffering water stress due to prolonged dry spells, El Niño effects and water crises (Chan 2016). Climate change exacerbates water problems, as more and more of the world's major rivers dry up, thereby degrading surrounding ecosystems, and threatening the health and livelihoods of people who depend upon these resources. Chan et al. (2016) emphasised that freshwater resources are an essential component of the Earth's hydrosphere and an indispensable part of all terrestrial ecosystems, intricately linking the hydrological cycle to all aspects of for the human population. However, there is widespread scarcity, gradual destruction and aggravated pollution of freshwater resources in many world regions, along with the progressive encroachment of incompatible activities.

In Malaysia, water is a crucial issue as rivers which supply 97% of the country's water needs are being severely degraded, polluted and exploited by over-abstraction (Phang and Chan 2013). Ironically, Malaysia is some country rich with water resources but seasonal variation, climate change and poor governance vis-à-vis rapid population growth, agricultural and industrial growth have increased water demands exponentially, thereby threatening the country's water security. Annual rainfall averaging 3,000 mm gives the country a richness in water resources estimated at 566 billion m^3 of water (runoff in river systems each year) (bin Abdullah 2002). However, despite plentiful water resources, the country is still besieged by countless water woes such as droughts, floods, pollution and water rationing (Nor Azazi et al. 2016).

Penang State is one of the most developed and urbanised states in Malaysia, but is a "water-stressed" state with limited water resources availability (Chan 2016). The state has an annual rainfall averaging 2,505 mm which is equivalent to 2.58 million m^3 of water resources annually. However, its evapo-transpiration is about 1.55 million m^3, reducing its effective rainfall to about 1.03 million m^3, which translates to a surface runoff of 1.03 million m^3. Groundwater availability is low. Penang State's total water demand is about 0.67 million m^3 or roughly 65.0% of its surface runoff.

Penang's major water source is the Muda River (a river originating from the Ulu Muda catchment in Kedah State) which supplies 80% of the state's water supply. This river is estimated to be able to supply enough water to both Kedah and Penang until the year 2020 (Chan 2016). During the so-called "Super El Niño in the beginning of 2016", it was reported that a contingency plan would be needed to avoid water crisis in the 4 states of Peninsular Malaysia (*The Sun Daily*, 18 April 2016). This El Niño is likely to have been exacerbated by global climate change. All dams in the state of Penang were down to their lowest levels during the dry spell. El Niño reduced the water levels of the Muda and Beris dams on the Muda River to 44.9% and 37.5% respectively.

In 1997/98, there was also a similar El Niño occurrence which impacted severely not only the 4 northern states but also the whole country. The effects of El Niño in 1997/98 and 2016 caused prolonged droughts that resulted in water crises. Many strategies were put in place by the Penang State Government to address these water crises. These include the protection of water catchments, raising water tariffs to encourage water conservation, incentives to install water recycling plants and rainfall harvesting systems, improving non-revenue water performance, sourcing alternative water resources, and a campaign to raise awareness and educate the public. On the part of nongovernmental organizations (NGOs) work on intensifying awareness, knowledge, conservation, protection and the creation of a "water saving society" was implemented. NGOs also worked on efforts to protect the Ulu Muda forest catchment which serves the northern region of Peninsular Malaysia, and pushed for the implementation of water saving devices (WSDs) for all new houses/buildings. The Federal Government can play a role in approving water tariff increases, drawing up regulations for water-saving fittings, and giving tax –free incentives for WSDs. This paper examines the effectiveness of these strategies to ensure water sustainability in Penang and what more needs to be done.

18.2   Material and Methods

The methodology used to derive the findings of this paper is based on both primary and secondary data sources. Primary data sources include a major quantitative questionnaire survey and qualitative in-depth interviews with key stakeholders, whereas secondary data sources are based on published data. The quantitative questionnaire survey was administered on 810 respondents in Penang State, who are all domestic water consumers stratified from different ethnic groups, age groups, levels of education and family sizes. Sampling was carried out via random and convenience sampling from February to December 2012. In addition, respondents were selected from urban areas, urban-fringes and rural areas. Respondents were asked to fill in the questionnaire with the interviewer on stand-by to answer questions or explain if the respondent did not understand a question. Respondents typically take between 10 to 15 min to complete a questionnaire which encompasses questions on personal information, water quality, type of water fittings installed, total usage of water volume, water tariffs, water conservation practices, willingness to pay, perception, water ethics, etc. The collected data was analysed with the SPSS statistical pacakage. Secondary data on water was collected from published journal papers, magazines, books, theses, newspapers, websites etc. related to water issues. In-depth qualitative interviews were conducted with relevant key stakeholders such as government officers, water service providers, officers of NGOs, etc.

18.3   Results and Discussion

The following sections examine the effectiveness of various strategies that can be employed to ensure water sustainability in Penang State and what more needs to be done to cope with the uncertainties of climate change which is expected to exacerbate water resource shortfalls. A combination of technical and human behaviour change strategies are examined in detail.

18.4 Water Resources Availability and Climate Change

Penang State is located in a country with copious rainfall averaging 3,000 mm annually, giving Malaysia a richness in water resources estimated at 566 billion m^3 of water (runoff in river systems each year) (Keizrul bin Abdullah 2002). Due to its exposure to monsoon winds, seasonal floods routinely affect many parts of the country, thereby giving people a false sense of security in terms of the richness of its water resources (Chan 2016). Unfortunately, Penang State is "poor" in terms of water resources. More alarmingly, Penang is highly dependent on its neighbouring state of Kedah for 80% of its water needs. Ironically, due to low water tariffs and apathy, Penangites use the most water in the country, averaging nearly 300 L per capita per day (l/c/d). Furthermore, Penang State is one of the most developed and urbanised states in Malaysia, being a leading state in manufacturing, tourism and business, all of which need water. The state has the lowest water tariffs in the country, but unfortunately, Penangites use the highest amount of water. Penang's domestic per capita daily water usage was 291 L in 2015. This is higher than the Malaysian national average of 211 L (2014), United Nation's recommended usage of 165 L, Singapore's 152 L and Tokyo's 230 L. This high water usage per capita in a "water-stressed" state means that Penang's water resources may not be sustainable in the long run.

Climate change and El Niño are expected to exacerbate Penang's water problems (Nambiar 2016). Hence, it is imperative that Penang reduces its per capita water use, and a reasonable proposed target is 230 L (based on an average family of 5 persons using 35 m^3 per month). To achieve this target by 2020, Penang must expedite three strategies to be implemented simultaneously. First, the authorities, private sector and NGOs must intensify their work on creating awareness, knowledge, conservation, protection and the creation of a "water saving society". NGOs can focus their efforts on water saving programmes/initiatives to create greater awareness and educate the public, and address the key challenges. Second, water tariffs must be restructured to ensure they encourage water savings. The current domestic tariffs do not encourage water savings as they are too low. A tariff structure that is based on cost recovery while maintaining the same charges for the first 20 m^3 per month so as not to burden the poor should be implemented. Third, authorities need to implement the mandatory use of water saving devices (WSDs) for all new houses/buildings. The Federal Government can play a role in giving incentives for WSDs such as tax rebates or a Goods and Services Tax (GST) waiver. To ensure families use below 35 m^3 of water per month, the Federal Government could incur the GST for usage above this amount. State Government can also adopt new regulations in the Building By Laws of the two local councils MPPP and MPSP to make WSDs mandatory for new houses/buildings. This paper examines these three proposed strategies and their impact on reducing the domestic per capita daily water use to 230 L in Penang by 2020.

Climate change is projected to impact severely on Malaysia, with wet areas becoming wetter (more floods) and dry areas becoming drier (more droughts). Based on climate change impact assessment studies by the National Hydraulic Research Institute of Malaysia (NAHRIM), it is projected that the north-western region of Peninsular Malaysia will experience more frequent and severe droughts. NAHRIM projected that the Muda Agriculture Development Authority (MADA) granary area will experience significant water deficits that would affect 10 out of 40 planting seasons. NAHRIM suggested that this prolonged deficit of irrigation water may warrant the cancellation of paddy planting in either some parts, or at worst, all of the MADA area. During other periods, excess water, which is expected for about 76 percent of the 240 months studied, could impact crop yields (Mohd Syazwan Faisal Bin Mohd 2014). In another study, NAHRIM (2006) indicates that based on the capacities of the Klang Gates Dam, Batu Dam, Sg. Selangor Dam and Tinggi Dam in Selangor, 28 (nearly 12%) out of the total 240 months are projected to face water supply deficits. These monthly water deficits range from 3 to 214 MCM. The highest surplus could be as high as 2,137 MCM. These estimates were based on demands from the projected population that is estimated to increase from nearly 5 million in 2010 to nearly 7 million in 2050. Penang State lies in the boundary/vicinity of the MADA region and is therefore expected to experience similar deficits in water resources.

Arguably, a concerted effort by all stakeholders need to be expedited to tackle climate change, including getting people prepared for water shortages, water cuts and droughts. Malaysia has had one bad experience where hundreds of thousands had to go without continuous water supply for months in 1998. This water crisis that came simultaneously with the El Niño dry spell cut water production by 50%, affecting more than 600,000 people in the Klang Valley (*The Sun*, 28 March 1998). If water consumers do not want to endure such a painful experience again, they need to change the way they look at water and the way they use water. Contrary to popular belief and misconception, conservation of water resources and water saving is not the sole responsibility of the government or the water authority. For water conservation to be totally effective, the public has an important role to play. This is where all stakeholders, including government, the private sector, industry, agriculture, businesses, NGOs and the public need to play their roles towards the creation of a water saving society.

18.5 Enhancing Awareness and Education towards a "Water Saving Society"

In general, Malaysians, especially Penangites, are insensitive to water issues and rather apathetic towards water conservation largely because of widespread misconception that country and state are rich in water resources due to the infrequent occurrence of droughts and lack of water crises in comparison to the frequent occurrence of floods, and pampered by low water tariffs that are heavily subsidized by government (Chan 2006). In a previous study, Phang and Chan (2013) analysed consumers' awareness, concern and willingness to pay for water services in Penang and found that generally Malaysians have low awareness and concern in water issues in the country, largely because water tariffs in the country are amongst the lowest in the world as they are heavily subsidized by government. Low water tariffs do not encourage water conservation but instead lead to water wastage, overuse and abuse, all of which undermine the sustainability of water resources and threaten water security in the country. However, Phang and Chan (2013) also found that Malaysians have a high willingness to pay for water services provided good water quality and services can be guaranteed to be provided. These findings indicate that water service providers who provide good services and good quality water can impose higher water tariffs, which in turn can be used to improve water quality and services further. However, it is not only domestic water tariffs that need to be restructured, but also commercial/industrial tariffs as currently hotels, factories, large businesses and other large water users do not find it economically feasible to install recycling plants, tap groundwater or use rain harvesting systems (Netto 2014).

The survey results found that the volume of water usage per month by most domestic water consumers in Penang State (m^3) varies between 10 m^3 to 30 m^3. Alarmingly, the vast majority of 719 respondents (88.8%) do not know how much water they use. It was also found that the majority of water consumers do not know the amount of money they pay for their water bills. This is a yet another reflection on their lack of concern regarding water issues and therefore apathy in terms of water issues. In general, Penangites were found to have low awareness in terms of water issues. The low domestic water tariff of MYR 0.31/1,000 L (for the first 20,000 L) in Penang was the main reason why Penangites are unconcerned about their water usage. The survey results also show the main types of major water problems in the households as reported by the respondents. An alarming fact is that 68.1% of respondents in Penang did not even want to make any comments about water problems. This can either mean that they are unaware, uninterested or unconcerned. Either of these reflects a poor awareness and concern about water resources. The major water problems highlighted were poor water quality (13.6%), leakages (4.9%) and low water pressures (4.4%).

Currently, Penang society can largely be described as a "Water Wasting Society" as the per capita daily water usage is very high. Penang State has traditionally recorded the highest water usage per capita in Malaysia, with an average increase in usage of about 1.7% per year between 2000 to 2010. Such high domestic per capita water usage can be exacerbated by future population increases resulting in insufficient water. Hence, it is imperative that not only domestic water consumers start reducing their water usage but also water consumers from agriculture, industry and all other businesses. Water demand management (WDM) needs to be inculcated amongst water all water consumers towards ensuring water security. WDM is needed when water resources availability is limited for a state such as Penang but popuplation, industry, tourism and other sectors are booming. Penang society is a "Water Wasting Society" because few people practice WDM or water conservation practices. Survey results show that only 1 in 5 respondents in Penang, whether male or female, practices water conservation. Overall, the vast majority of Penangites, more than 75%, do not carry out any water saving practices.

Research results indicate that 8 out of 10 Penangites have installed the traditional single flush toilet which uses 9 of water per flush. This old flush system is not water-friendly as it uses too much water, considering that modern water-friendly flush systems use only 4 to 5 L of water per flush. The other option is to install the dual-flush system whereby a single flush is between 3 L (small flush) to 5 L (big flush). Normally, 30% of total water use in a household in a day is from toilet flushing. Hence, installing a water-friendly toilet has great potentials for WDM. For example, if the traditional flush is replaced by the dual-flush system, one can save 4.5 L of water for every flush. If a consumer visits the toilet 6 times a day, he/she can save up to 27 L of water a day. This will translate to about 9855 L of water a year. If a family has 5 members, the amount of savings is 135 L/day or 6,652,125 L/year.

The above findings clearly indicate that Penangites are unconcerned about water and related issues and rather apathetic. Domestic consumers pay only a small fraction of their monthly income for their water bills, as the level of water charges as a proportion of household expenditure is only around 1.04% (Casey 2005). As water tariffs are heavily subsidised by the state government, it is imperative that such subsidies be removed to ensure water consumers feel the pinch and start water conservation. The wide consensus amongst government officials, NGOs and the private sector is that such heavily subsidised water tariffs have negative effects on water conservation. The fact that Penang has the country's lowest water tariffs is largely responsible for Penangites using the most water, i.e. 291 L per capita per day (l/c/d) in 2015, which is more 84 l/c/d (40.2%) more than Malaysia's 209 l/c/d national average. Compared to Kelantan State (146 l/c/d), Penang's water usage is about double. This is not sustainable for Penang as climate change and El Niño are expected to reduce water resources availability in the near future. Even if the reduction is about 10%, Penang would face disastrous water crises if water consumers do not do their part and reduce their water demands. According to Penang Water Supply Corporation CEO Ir. Jaseni Maidinsa, reducing Penang's per capita consumption to 260 l/c/d would help to ensure that Penang has sufficient water beyond 2020. The CEO cited five positive impacts if Penang could achieve the 260 l/c/d target: (i) Penang would be able to avoid water rationing even in extreme dry seasons caused by climate change; (ii) sufficient water supply to cater for future population growth; (iii) enough water for future socio-economic growth; (iv) prolonging the lifespan of its existing water supply infrastructure, delaying the need for additional multi-million ringgit infrastructure projects that will inevitably lead to higher water tariffs; and (v) "buying" Penang time to realise a second or alternative raw water resource for the long-term, such as the SPRWTS, rainwater harvesting, water recycling or desalination. The CEO went on to stress that 260 l/c/d was not a difficult or impossible target to achive as the Malaysian national average was 209 l/c/d in 2013. Singapore has achieved a 150 l/c/d in 2014. Also, it was pointed out that 14 years ago, Penang's per capita domestic consumption in Penang was only 262 l/c/d in 2001. The state government must intensify efforts to lower the per capita daily water consumption via awareness, education, restructuring water tariffs, and making water saving devices mandatory. A 260 l/c/d target is a pre-requisite towards becoming a "Water Saving Society".

18.6   Restructuring Water Tariffs

In general, water tariffs in Malaysia are amongst the lowest in the world. The national average is only MYR0.55 (1US$ = MYR4.11233 on 29 September 2016) per 1,000 L for the first 20,000 L, considered the basic water "lifeline". Penang State has the cheapest tariffs averaging MYR0.22 per 1,000 L for the 1st 20,000 L. In comparison, desalinated water (US$2.00/1,000 L) costs 20 times more than Malaysia's tariffs. Denmark's tariffs (US$9.21/1,000 L) are 97 times more expensive than Malaysia's, and Germany's water tariffs (US$7.35/1,000 L) are 74 times more expensive than Malaysia's. Comparing the water tariffs with Penang, the costs of water in Denmark, Germany and desalinated water are 172 times, 137 times and 37 times more expensive respectively. Despite such low water tariffs, in a press release recently, the Penang Water Supply Corporation (2015) claimed that Penang State's domestic tariffs are the most "people-friendly" in Malaysia for households who use less than 35,000 L per month. Penang's average tariff of MYR0.32 per 1,000 L is less than 50% of the national average of MYR0.66 per 1,000 L. Kedah State charges MYR0.67 per 1,000 L, Perak State charges MYR0.73 per 1,000 L and Johor State charges MYR1.05 per 1,000 L.

18.7   Other Strategies for Water Security under Climate Change

It is also imperative that Government impose the mandatory installation of water saving devices (WSDs) for all new buildings and all new water fittings. New houses that do not have WSDs should not be approved. Chan (2016) has documented in detail the various types of WSDs that can be installed in households and office buildings to reduce water consumption. Many projects in Penang have been successful in implementing WSDs in reducing water demands, including university, apartments and individual houses. There are many other strategies that the Penang State Government has implemented to cope with fluctuations in water resources due to climate change. The Penang Water Supply Corporation has recently launched its new RM11.9 million Bukit Dumbar Pumping Station 2, with a designed capacity of 270 million litres per day (MLD) to improve water supply services for the benefit of 315,000 water consumers on the island. The corporation has also implemented 3 water supply infrastructure projects: (i) a MYR54.9 million Package 12 of the Sungai Dua Water Treatment Plant, encompassing a new water treatment module, an environment-friendly sludge processing unit and a new water testing laboratory; (ii) two new reinforced concrete reservoirs costing MYR16.7 million in Jawi, Seberang Prai Selatan; and (iii) a 900 mm pipeline from Sungai Keluang, along Tun Dr. Awang, to the Penang International Airport roundabout costing MYR6.0 million. In addition, the Federal Government has approved funds to implement the Mengkuang Dam Expansion Project, or MDEP, with a Federal grant of RM1.2 billion. This will ensure more water available to Penangites during times of water stress caused by climate change. The MDEP costing US$200 million involves raising the central clay core dam by 11 m and extending it by 1.7 km. After completion, the reservoir capacity will be increased to 78 million m^3 from 23 million m^3. This will increase Penang's protection against climate change.

One future proposed plan is to look for another source of water, viz. the Sungai Perak Raw Water Transfer Scheme (SPRWTS), an inter-state water transfer project proposed by Penang since 1999 but have yet to be implemented (Penang Water Supply Corporation 2015). NGOs can play a vital role in Penang's efforts to achiecve water security. NGOs have a big role to play in IWRM, especially under dwindling water availability caused by climate change. NGOs are now increasingly offering the government to involve and use them. One of the major challenges of NGOs is to convince the water authorities to discard their age-old "top-down" water management approach that is almost single-handedly based on supply management. It is therefore imperative that users play an equally vital role in reducing water demand via conserving water. This is where NGOs can help. For example, a good example of smart-partnerships is between the National Water Services Commission and NGOs such as Federation of Malaysians Consumers' Associations (FOMCA), Malaysian Environmental NGOs (MENGOs), WWF Malaysia and Water Watch Penang (WWP) in and the private sector companies in coming out with regulations on WSDs and water-friendly fittings. WSDs proved to be valuable in the case of Penang which needed to reduce its domestic per capita daily water consumption. In Penang, WWP, the first "true" water NGO in Malaysia, has launched a State-wise campaign targeted at consumers to reduce water use since 1997. WWP started off with the schools, first running inter-school competition on water reduction methods and total water use reduction. Classes and field trips are organised for school children on the importance of caring for rivers, reducing water use in homes and schools. Essay writing and painting competitions (always with water as the theme) are organised year-round. These activities worked well because WWP worked in partnership with the Penang State Education Department and the private sector. Over the years, WWP has managed to reach thousands of students under their programme.

18.8   Conclusion

In the light of global climate change and El Niño, prolonged droughts are expected to hit Penang. Many strategies can be adopted to address the shortfalls in water resources and cope with climate change. The Penang State Government needs to protect water catchments, restructure water tariffs, encourage water conservation, provide incentives to install water recycling plants and rainfall harvesting systems, improve Non-Revenue Water performance, source for alternative water resources, and run a campaign to raise awareness and educate the public. Likewise, NGOs can work on intensifying awareness, education, conservation, protection and the creation of a "water saving society". NGOs also need to work on efforts to protect the Ulu Muda forest catchment which served the northern region of Peninsular Malaysia, and pushed for the implementation of WSDs for all new houses/buildings. The Federal Government could also have played a role in approving water tariff increase, drawing up regulations for water-saving fittings, and giving incentives for WSDs such as tax rebates.

Comprehensive coping strategies are needed to ensure water security in Penang and to face the uncertainties of climate change. In terms of increasing water availability, the following recommendations are proposed: (i) protection and gazettement of all water catchments (existing and potential); (ii) sourcing of new water sources, including from neighbouring states of Kedah and Perak, groundwater and rainwater, and desalination; (iii) greater monitoring and more effective enforcement of river and groundwater pollution control (including the employment of smart-partnerships between government and NGOs); (iv) fines for destruction of water catchments, water pollution and water-related crimes need to be increased substantially, including jail sentences for repeating offenders; and (v) industries, farms and other agricultural activities, aquaculture and squatters located near important water sources need to be relocated. Other coping strategies to climate change are to: (i) implement a comprehensive approach of water supply and water demand management, rather than the current lopsided water supply management; (ii) implement water recycling for industries, agriculture, domestic and other water users; (iii) carry out a water saving campaign (year round) by government, NGOs and industry; (iv) review water tariffs to recover costs, encourage conservation and penalize wasters; (v) reduce non-revenue water (NRW) to 10% from the current 20%; (vi) encourage the use of water-saving devices; (vii) look for alternative sources of water supply such as rainfall harvesting, wells, spring water, river water etc.; and (viii) Government to work with industry and NGOs as partners.

Acknowledgements The authors would like to acknowledge the Ministry of Higher Education Malaysia LRGS Research Grant 203/ PHUMANITI/ 67215003 for the data and information used in this paper.

References

Casey L (2005) Water tariff and development: the case of Malaysia. Institute of Southeast Asian Studies. https://www.researchgate.net/publication/23512205. Accessed 3 Nov 2017
Chan NW (2006) water tariffs and the poor. Water Malaysia (12), January 2006. ISSN 1675–2392, pp 22–23
Chan NW (2016) Raising awareness and it's impact on water usage in Penang, Malaysia. Paper presented at the Asia Water 2016 Conference, 6–8 April 2016, Kuala Lumpur

Chan NW, Mahamud KRK, Karim MZ, Lee LK, Bong CHJ (2016) Chapter 15: Hydrological cycle and cities. In: Chan NW, Imura H, Nakamura A, Ao M (eds) Sustainable urban development textbook. Water Watch Penang & global cooperation institute for sustainable cities, Penang, pp 95–104

bin Abdullah K (2002) Integrated river basin management. In: Chan NW (ed) Rivers: towards sustainable development, Penerbit Universiti Sains Malaysia, Penang, pp 3-14

Bin Mohd MSF (2014) Impact assessment studies & regional climate change scenarios data requirements in Malaysia. In: Paper presented at the 2nd Southeast Asia regional climate downscaling workshop, Bangkok, 9–10 June 2014, https://www.un.org/esa/sustdev/documents/agenda21/english/agenda21chapter18.htm. Accessed 7 May 2004

NAHRIM (2006) Study of the impact of climate change on the hydrologic regime and water resources of Peninsular Malaysia. National Hydraulic Research Institute of Malaysia, Kuala Lumpur

Nambiar P (2016) PBA: Protect water resources before next Super El Nino, 26 August 2016, https://www.freemalaysiatoday.com/category/nation/2016/08/26/pba-protect-water-resources-before-next-super-el-nino/. Accessed 28 Sept 2016

Netto A (2014) Raise commercial water tariffs in Penang as well. https://anilnetto.com/environmentclimate-change/sustainable-development-2/businesses-gulp-40-per-cent-total-penang-water-consumption/. Accessed 27 Sep 2016

Zakaria NA, Ghani AA, Sidek L, Abdul Talib S, Chan NW (2016) Urban water cycle processes, management & societal interactions: crossing from crisis to sustainability. River Engineering and Urban Drainage Research Centre Publication, Penang

Penang Water Supply Corporation (2015) Penang's Water Tariffs the most "People-Friendly" in Malaysia. Press Release on 26th April 2015 in Commemoration of World Water Day 2015 at the Air Itam Dam, Penang

Phang WL, Chan NW (2013) A study of consumers' awareness, concern and willingness to pay in water issues in Malaysia. In: Proceedings of persidangan kebangsaan geografi & Alam Sekitar Kali Ke 4 - Geografi & Alam Sekitar Dalam Pembangunan dan Transformasi Negara. Universiti Pendidikan Sultan Idris, pp 181–189

The Sun, 28 March 1998

The Sun Daily 18 April 2016

Chapter 19
Assessing the Physical Water Availability and Revenue Aspect of WUAs Under New Governance System: Case Studies of Pakistan

S. Ahmad, S. R. Perret, M. Imran, and S. A. Qaisrani

Abstract Irrigation policy makers and managers need information on the water productivity and irrigation performance for developing well-informed national water management strategies at various scales considering available water resources for food security of the rising population. The development of national policies on water is based on understanding the water footprint of a nation. New water governing arrangements were adopted to manage the scarce water resources more effectively and to improve production performance in the irrigation sector of Pakistan. Theoretically, the choice and implementation of the new water governance system should play a central role in achieving these objectives, both by improving water supplies and by recovering water costs. This study investigates the impact of the governance changes on the physical availability of water and revenue collection efficiency of newly established Water Users' Associations in Pakistan. A questionnaire survey was conducted to realize the farmers' opinions for self-governing irrigation scheme (farmers-managed) as sustainability of this scheme is under pressure in many countries, and results were compared with agency-managed irrigation scheme. Total 72 farmers were interviewed along with officials' discussions and field observations from two different governance situations. Major indicators such as water sufficiency and Irrigation Service Fee (ISF) were implied to evaluate the impact of new governance system. Statistical tools standard deviation, frequency and percentage were used to investigate the impressions of the different variables. The findings showed 22% improvement in water sufficiency and 16% decline in IF collection efficiency under new governance system. An effective enforcement system is needed to improve the monitoring and accountability of governing bodies and officials accountable.

Keywords Water governance · Institutional reforms · Irrigation service fee · Water sufficiency · Enforcement system

S. Ahmad (✉) · M. Imran · S. A. Qaisrani
COMSATS University Islamabad, Vehari Campus, Pakistan
e-mail: sajjad.ahmad@cuivehari.edu.pk

S. R. Perret
CIRAD-Agricultural Research for Development, Montpellier, France

© Springer Nature Switzerland AG 2021
M. Babel et al. (eds.), *Water Security in Asia*, Springer Water,
https://doi.org/10.1007/978-3-319-54612-4_19

19.1 Introduction

The limited freshwater resources across the globe are under high pressure in the form of consumptive water use and pollution. Pakistan is among the ten most affected countries by climate change in the world (Kreft et al. 2016). Water production and consumption are employed for water footprints of a nation (Hoekstra and Chapagain 2006). Considering the global population growth rate, the availability of freshwater resources is estimated as 4380 m^3 per capita by 2050. At a glance, it does not show any water shortage conditions but on the other hand, the 3 billion world's population would get water stressed by 2025 due to uneven spatial and temporal distribution of water resources (Johansson et al. 2002). Despite of water shortage, misuse of available water resources and deterioration of surface and groundwater resources is widespread.

The development of well-informed national policies on water is based on understanding the water footprint of a nation (Hoekstra and Mekonnen 2012). Pakistan has an area of 796,096 km^2 and its irrigation system is the largest integrated irrigation system in the world with approximately 88% water consumption in agriculture sector (PILDAT 2003). Pakistan has diverse agro-ecological settings and its irrigated agriculture largely depends on Indus River Basin for its surface water. The country is water scarce; on per capita basis water availability has been diminishing at a disturbing rate and has declined from 5,000 m^3 per capita in 1951 to about 1,038 m^3 per capita in 2015, verging on the international scarcity rate. It is estimated to decrease to 700 m^3 by 2025 (MoE 2008). Currently, irrigation system of Punjab province, Pakistan is facing ~32% water shortages to fulfil crop water requirements (Ahmad 2012). Efficiency of physical water supplies in the Indus irrigation system has an important concern. However, irrigation system of Punjab as being the part of IBIS has only 35–40% delivery efficiency from the canal head to the crop root zone due to its age and poor maintenance (Tarar 1995). This is advocated as the main reason for the low water supply at the end users of the canal system. Further, accurate and reliable information are not available on the distribution of irrigation water in the various parts of canal commands. It is considered as the major constraint for the efficient management of the scarce water resource (Ahmad et al. 2004). Almost 20–30% daily fluctuations have been reported in the discharge of irrigation canals within irrigation system of Punjab (Sarwar et al. 1997). Irrigation water is distributed among farmers through a rotational turn system, locally known as "*warabandi*", normally based on the fixed seven days. It means farmer is permitted to use the entire flow of water from the outlet once a week for the time allocated according to the size of his land holding by the irrigation department. It is very difficult to irrigate the entire land size due to the insufficient allocation of water (Qureshi et al. 2008). In this research study, physical water supply was assessed with a parameter of water sufficiency at various canal commands levels.

In most of the Asian countries, where agricultural production largely contributes in their national economy, irrigation management is an important concern. However, despite of efforts given to the irrigation development and management, the performance of government managed irrigation sector is not satisfactory (Barker and Molle 2005). It is remarkable that water is the most restrictive factor in the extensive irrigation system of Pakistan. Modernization of the existing irrigation system is more likely to enhance efficiency and to furnish new institutional structures (Bandaragoda 2000). Pakistan adopted new governance arrangements in its irrigation system in 1997 through institutional reforms in its irrigation sector focusing the global changes in water policy for sustainable water management. These institutional reforms in irrigation sector were practiced in Lower Chenab Canal (LCC-east) in February 2004 in order to improve governance, to equitably distribute available water resources, and to strengthen the local institutions financially. Because of the implementation of the institutional reforms in LCC, the governance structure of the irrigation system shifted from Agency Managed Irrigation System (AMIS) to Farmers Managed Irrigation System (FMIS). The newly established local institutions such as Farmers organizations (FOs) and Water Users Associations (WUAs) were responsible to ensure the equitable water distributions along the entire command area of canal. At the third level canals, WUAs were responsible to collect the Irrigation Service Fees (ISF) from all water users. Current information points towards the deprived performance of the Punjab's irrigation system. The main causes of the poor performance of the irrigation system of Punjab include an inadequate institutional capacity, insufficient database for planning of large irrigation projects, deteriorated century old canal infrastructure and intractability of institutional, technical and socioeconomic aspects (David 2004).

Recovery of Irrigation Service Fee (ISF) is very crucial to operate and to maintain the irrigation system for appropriate water delivery. According to a study (Latif and Pomee 2003), recovery of ISF was only 44% before governance change in the irrigation system of Punjab, Pakistan. The major reason for reduced ISF collection efficiency is decline in the provision of better water delivery services. In this concern, this study articulates the impact of institutional reforms on the physical availability of water at various canal reaches and efficiency of local institutions (WUAs and FOs) in ISF collection.

19.2 Material and Methods

A questionnaire was developed to conduct a field survey. Some variations were considered in the questionnaire according to different governance situations in the related case studies of irrigation systems. The respondents were selected using systematic random method from the commands of two different irrigation schemes under different governance situations, namely, Farooq Irrigation System (a third level canal) under FMIS and 2L Irrigation System (third level canal) but from a different governance situation under AMIS. Map of both study areas have been highlighted in Fig. 19.1.

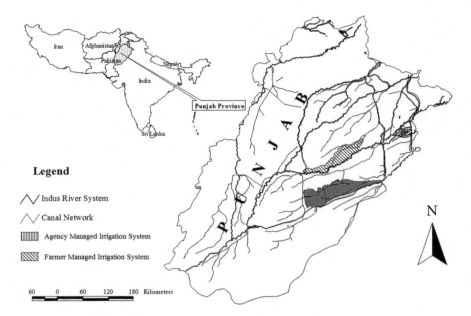

Figure 19.1 Map of the study area, Irrigation System of Punjab, Pakistan

Table 19.1 Irrigation scheme-wise distribution of respondents

Name	Irrigation schemes	Number of respondents	%
Farooq Irrigation System	FMIS	36	50
2L irrigation System	AMIS	36	50

The distribution of samples (respondents) and their proportion in both canal commands is shown in Table 19.1. Total number of respondents were 72 and those were equally distributed among the commands of both irrigation schemes.

19.2.1 Location-Wise Classification of Respondents

In order to make a comprehensive analysis, both irrigation systems were further categorized into three different parts namely as head, middle and tail as shown in Table 19.2.

This classification was processed based on the number of outlets due to various characteristics on different parts of canal systems. The total number of outlets of each irrigation canal was divided into three equal categories. The first category from

Table 19.2 Location-wise classification of respondents

Location	FMIS		AMIS	
	Frequency	%	Frequency	%
Head	12	33.33	12	33.33
Middle	12	33.33	12	33.33
Tail	12	33.33	12	33.33
Total	36	100	36	100

Table 19.3 General characteristics of the study areas

Parameters	FMIS	AMIS
Climate	Seasonal, Winter temperature range, 15–27 °C, Summer temperature range 32–46 °C	Same
Soil & topography	Medium to moderately textured, Rachna Doab, Surface drainage slow	Silty and clayey loams, Bari Doab, Surface drainage slow
Major crops	Wheat, Sugarcane	Wheat, Cotton, Maize, Vegetables
Nearer Market	Grain market	Grain and Vegetable market
Land holdings	3.74 (ha)	5.88 (ha)
Occupation		
Agriculture	63%	60%
Service	16%	9%
Business	18%	17%
Labor	3%	15%
Literacy rate	60%	62%
Age (year)	45 (±9.05*), 28–65 years	43 (±9.817*) 23–65
Ethnicity	Caste based	Caste based

* = Standard deviation

the start of canal is named as head, second as middle and third as tail at the end of canal. Further, equal number of respondents were selected from the all three different regions of the both irrigation systems. General characteristics of the both study areas are given in Table 19.3 to better understand the local socio-economic situation.

Primary data for this research study were collected through employing different techniques such as households' surveys, survey from key informants, observations, farmers group discussions, and some discussions with the officials from the both irrigation systems during the field survey.

The questionnaire contained both open ended and close-ended questions asked from the farmers, who either belonged either to the area, where the irrigated schemes were implemented, or to the areas without irrigated schemes for comparison. The questions were mostly related to farmers problems, institutional problems, benefits of the new system, and farmers participation.

Semi-structure interviews were conducted with an open framework from the key informants in the study area. The old farmers with knowledge of the old and the new system, members of WUA or FO organization were selected as key informants and questioned regarding the efficiency issue, local farmer participation, problems and benefits of the new system.

Focal group discussion was carried out with the group of farmers belonging to different regions of the same irrigation system. Discussion was mainly focused on the water availability in the system, problem faced by them, and their participation in the system. Observation technique was used to get an overview of the existing situation of the canals, head, tails, and other discrepancy in the system.

19.2.2 Data Processing and Analysis

All data from the questionnaires and other sources were compiled and then analyzed using Microsoft Excel. Following are the description of the methods that were used in data analysis.

Different types of descriptive statistical methods were used to formulate percentage, frequency, and average to analyze the statistical difference of social characteristics of household members on gender, education, and occupation. In addition, observations of the local situations, interviews and group discussions with water users were helpful to identifying the problems and issues related with irrigation system and their potential to pay the Irrigation Service Fee.

19.3 Results and Discussion

19.3.1 Water Sufficiency

In the FMIS and AMIS, water supply was not sufficient in order to meet the crop prerequisite. Water discharge variations were practiced in the period of crop cultivation. Particularly, in the FMIS, the allocation of water in the command area and other water management services were observed as successfully performed by the communities, but the key restrictions faced were physical and economic features. The stipulation of irrigation services was appraised in terms of sufficiency of water availability. It was observed that the organizational system and rules were practiced appropriately.

Questionnaire survey was implied to get farmers opinion about the irrigation water concerned matter. The farmers were inquired to rate their satisfaction level as virtually no water, very insufficient and sufficient. Farmers from both irrigation systems articulated their views about the water sufficiency. Figure 19.2 reveals that most of the farmers from head reach of the canal were receiving sufficient water whereas the majority of the tail end farmers were not found satisfied with water sufficiency. It had been found that the trend of water sufficiency was in declining order from head to tail ends of the canals.

In order to concern with the water sufficiency for irrigation in FMIS, it was found that 33% respondents were getting very sufficient supply of water at the overall command area, whereas only 11% farmers were getting very sufficient water supply in case of AMIS. So, 22% increase in the water sufficiency was seen in the FMIS. This physical water availability can be clearly linked with the determination of the water users towards newly defined rules and principles under governance system after institutional reforms.

If we move from the head reach of the canal towards middle and tail reaches, the situation of the water sufficiency looks entirely different. Very clear picture of the water availability at the different reaches can be seen from the Fig. 19.2. The number of the water users getting no water at all was higher in case of AMIS in comparison with FMIS, which shows better provision of water supplies at the tail end of the canal. Likewise, in the middle region of the canal under FMIS, situation of water sufficiency is better than the AMIS. These results show comparatively better outputs of governance structure in terms of water availability at farm level.

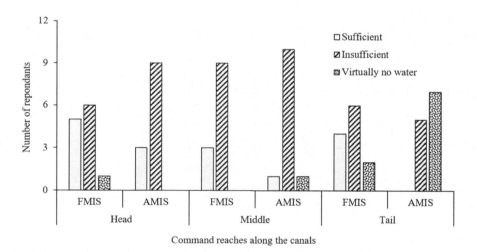

Fig. 19.2 Water sufficiency at different command reaches along the canals

19.3.2 Efficiency of Institutions for Collection of Water Charges

Collection of irrigation service fee (ISF) is a major component in the efficient and sustainable performance of any irrigation system. In the AMIS, 66% farmers responded that collection of ISF is efficient. So, most of the farmers were paying their ISF at proper time. On the other hand, 50% water users argued that the ISF collection from farmers under WUAs was efficient as shown in Fig. 19.3.

In this way, 16% increase can be seen in the efficiency of institutions for collection of water charges from farmers under AMIS. It was also noted that 49% farmers were not anyhow agreed with working efficiency of ISF collection under WUAs. However, majority of the water users believed that the system for the collection ISF under AMIS was working better and more efficient than under FMIS.

Irrigation department showed better enforcement system for the collection of water charges from farmers than the farmers-managed irrigation system. In farmers-managed irrigation system, chairmen of WUAs were noted to give relaxation in the collection of ISF in terms of extension in deadline to submit ISF. It was also noticed that system of ISF collection under WUAs is not transparent and efficient.

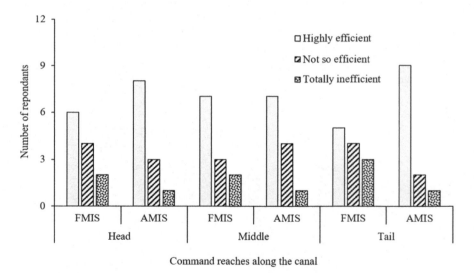

Fig. 19.3 Efficiency of institutions for revenue collection along the different canal commands

19.4 Conclusions

In this study, farmers managed and agency-managed irrigation systems (FMIS and AMIS) were compared to analyse the impact of institutional reforms on the overall performance of the Indus basin irrigation system (IBIS) in Punjab, Pakistan. In FMIS, 33% farmers are getting very sufficient volume of water in comparison with AMIS, where only 11% farmers are getting very sufficient amount of irrigation water. In AMIS, 63% farmers at middle and tail reaches of the canal are getting insufficient water as compared with FMIS (47%). Furthermore, FMIS is performing more efficiently and farmers are getting relatively better services (ownership, moral support, self-respect and conflict resolution) from WUAs and FO as compared with AMIS as also assessed by Mustafa et al. (2016) in Jordan. However, FMIS has been found less productive for ISF collection under WUAs than the AMIS due to mismanagement in collection of money. In both governance situations, the main reasons behind inefficient water delivery are unscheduled canal closure, water theft, and changes in the size of water outlet.

Meanwhile, FMIS is not providing efficient services regarding legal support and control over corruptions due to lack of strict policy framework. Farmers of the FMIS do not need to register their complaints for irrigation related issues direct to courts and police stations like in AMIS. In order to achieve water security in the area. Some amendments are needed in the system minimizing the political influence in the election of the chairperson of WUA and other members by the Area Water Board (AWB) at main canal level and Provincial Irrigation and Drainage Authority (PIDA). The monitoring cell of AWB and PIDA should recommend strict penalties against the actors involved in water theft. There must be transparency in ISF collection from the farmers, and WUAs and FOs should be held accountable by competent authority (AWB and PIDA). Additionally, the canals should be monitored frequently both in day and night times as most of the illegal activities were reported during night times.

A further research may be conducted on the actual measurement of water sufficiency, farm performance, magnitude of corruption in water delivery, and their associations with the actual performance of the water supply system under newly defined governing arrangements.

Acknowledgements The authors acknowledge the Higher Education Commission (HEC) Pakistan for providing financial support for the research study.

References

Ahmad S (2012) Investigating irrigation systems' performance under two different governance systems in Pakistan. Doctoral thesis, Asian Institute of Technology, Thailand

Ahmad MD, Stein A, Wim GMB (2004) Estimation of disaggregated canal water deliveries in Pakistan using geomatics. Int J Appl Earth Obs 6(1):63–75

Bandaragoda DJ (2000) A framework for institutional analysis for water resources management in a river basin context, vol 5. International Water Management Institute, pp 3–42

Barker R, Molle F (2005) Perspectives on Asian irrigation. In: Shivakoti GP et al (eds) Asian irrigation in transition: responding to the challenges. Sage Publications, India, pp 45–78

David EP (2004) Water resources and irrigation policy in Asia. Asian J Agric Dev 96(1):121–131

Hoekstra AY, Chapagain AK (2006) Water footprints of nations: water use by people as a function of their consumption pattern. Water Resour Manag 21(1):35–48

Hoekstra AY, Mekonnen MM (2012) The water footprint of humanity. PNAS 109(9):3232–3237

Johansson RC, Sur YT, Roe TL, Doukkali R, Dinar A (2002) Pricing irrigation water: theory and practice. Water Policy 4:173–199

Kreft S, Eckstein D, Dorsch L, Fischer L (2016) Global Climate Change Risk Index. German Watch, Briefing Paper, p 6

Latif M, Pomee MS (2003) Impacts of institutional reforms on irrigated agriculture in Pakistan. Research report. Centre of excellence in Water Resource Engineering, University of Engineering and Technology, Lahore, Pakistan

MoE. (2008) Pakistan country paper SACOSAN III. Ministry of Environment. Government of Pakistan

Mustafa D, Altz-Stamm A, Scot LM (2016) Water Users' Associations and the politics of water in Jordan. World Dev 79:164–176

PILDAT (2003) Issues of water resources in Pakistan, Briefing session for parliamentarians, Islamabad, Pakistan

Qureshi AS, McCornick PG, Qadir M, Aslam Z (2008) Managing salinity and water logging in the Indus Basin of Pakistan. Agric Water Manag 95(1):1–10

Sarwar S, Nafeez HM, Shafique MS (1997) Fluctuation in canal water supplies, Case study report, no. 27. IWMI. Lahore, Pakistan, p 72

Tarar RN (1995) Drainage system in Indus plains-an overview. In: Proceedings of the national workshop on drainage system performance in the Indus plains and future strategies, vol II, Tandojam, Pakistan, pp 1–45

Chapter 20
Simulation of Kathmandu Valley River Basin Hydrologic Process Using Coupled Ground and Surface Water Model

S. Basnet, N. M. Shakya, H. Ishidaira, and B. R. Thapa

Abstract Achieving water security is challenging for Kathmandu Valley, Nepal. Conjunctive use of ground and surface water is inevitable for the sustainable use of water within the valley. The focus of this study is to develop a robust groundwater model that can be used for better understanding surface and ground water interaction. For this purpose, a three-dimensional transient model is constructed using the U.S. Geological Survey (USGS) integrated model GSFLOW. This study tries to provide some insights on groundwater extraction volume especially that of Deep Aquifer system. A pumping sensitive analysis of groundwater system in Kathmandu valley shows that the area near Dharmasthali, Dhapasi, Maharajganj Sankhu and Gokarna has a decline of 0.02 to 0.12 m in head with per unit rise in pumping (m^3/s) whereas the area near Balaju, Samakhusi and Shywambu showed more decline of upto 0.12 m to 0.23 m. The proposed extraction rate map prepared through this analysis also indicates that the northern part of the ground water basin has more volume of water available per unit decline in head per year and the value of the extraction rate is decreasing as we move from northern part of groundwater basin to the southern part. Finally, a Village Development Committee (VDC) wise extraction rate map is prepared using the proposed extraction rate map which showed that Sangla, Baluwa and Danchi VDC have higher value of proposed extraction rate where, Danchi VDC is showing highest extraction rate of upto 6,273,967 m^3/yr.

Keywords Aquifer · Extraction rate · Groundwater model · GSFLOW

20.1 Introduction

Drinking water quality and quantity have been the biggest concerns in water sector in the Kathmandu Valley. The valley has more than 2.5 million population demanding

S. Basnet · N. M. Shakya (✉)
Institute of Engineering, Tribhuvan University, Lalitpur, Nepal
e-mail: nms@ioe.edu.np

H. Ishidaira · B. R. Thapa
University of Yamanashi, Kofu, Japan

© Springer Nature Switzerland AG 2021
M. Babel et al. (eds.), *Water Security in Asia*, Springer Water,
https://doi.org/10.1007/978-3-319-54612-4_20

370 million liters of water per day (mld) (KUKL 2015), which can only meet 19% and 31% of total demand during dry and wet season respectively (Thapa et al. 2016a). Valley's population is increasing at the rate of 4.7% along with which the water demand is also increasing (CBS 2011) which may increase up to 322 mld by year 2021 (Udmale et al. 2016). Kathmandu Upatyaka Khanepani Limited (KUKL) is the main responsible agency that supplies drinking water to the residents of Kathmandu Valley. KUKL is tapping water from hill of the Kathmandu Valley where the fresh water is available and suitable for drinking (Thapa et al. 2016b). The deficit is met through private groundwater pumping, wells, traditional stone spouts, water tanker, and private vendors (Thapa et al. 2016a) and the groundwater recharge is very less (Thapa et al. 2017). This has caused extensive use of groundwater resulting in depletion in both shallow and deep aquifers. Data from secondary sources indicate that the extraction rate has increased from 2.3 Million-liters-a-Day (MLD) in 1979, 29.20 mld in 1999, 59.10 mld in 1999 to 70.9 mld in 2009 (Pandey et al. 2010, Metcalf and Eddy 2000, Binni and Partner 1998, Dhakal 2010). It has been reported that the Water table of Kathmandu valley is continuously declining (Shrestha 2006) which ranges from 2.57 m to 21.58 m between 2003 and 2014 (Gautam and Prajapati 2014). Further Cresswell et al. 2011 by analyzing isotope tracer, suggested that groundwater recharge for deep aquifer is very small relative to that of withdrawal rate and is estimated to be at least 20 times less, implying that groundwater resource will be depleted below extraction level within 100 years. As the groundwater level declines, it will induce cascading effects through terrestrial and aquatic ecosystems (Palmer 2003) increasing the drilling cost, pumping cost and causing land subsidence. The groundwater level data are hard to monitor accurately and continuously but the ability to monitor resource use is often critical for the effective development and management of these resources (Pradhanang et al. 2012).

Many studies have been carried to understand the groundwater potential of Kathmandu valley. Pandey and Kazama (2012) in their study have found that for shallow aquifer the storage potential is 1.5 BCM (Billion cubic meter) and that in case of deep aquifer is 0.6 BCM. This study was done using Geographic information System (GIS) as a primary tool using readily available secondary data's and information and hence was unable to incorporate the dynamic nature of the groundwater system. Similarly, other studies have been made in understanding the groundwater system of Kathmandu valley. In one of those studies Cresswell et al. (2001) used radioisotope study for quantifying groundwater recharge rates and residence time. Also, JICA (1990) carried out modeling of deep aquifer system to estimate the sustainable extraction rate. However, these studies are unable to explicitly couple the groundwater system with surface water system. Usually Ground and surface water model can be loosely linked outside of the model (Hunt and Steuer 2000) but it becomes evidently difficult to simulate these models in smaller time scale where model can represent optimal system dynamics. Hence, the importance of considering integrated surface water and groundwater model for evaluating the groundwater resources was felt increasingly apparent.

This study is an attempt to utilize the numerical modelling approach to simulate the groundwater system of Kathmandu Valley through coupled surface and groundwater hydrological model. Here the two system are iteratively coupled at each time steps which accounts for feedback loop between evapotranspiration, surface runoff, and soil-zone flow using Groundwater and Surface-water FLOW (GSFLOW) in order to replicate the actual groundwater scenario and behaviour. The main objective of this study is to evaluate the spatial distribution of groundwater potential of deep aquifer system in Kathmandu Valley using integrated model GSFLOW based on aquifer response. Additionally, this model can also be used in addressing climate change issue as climate change impact surface water directly through changes in the major long-term climate variables such as air temperature, precipitation, and evapotranspiration. Also, groundwater resources are related to climate change through direct interaction with surface water resources, such as lakes and rivers, and indirectly through the recharge process. Therefore, the model may help in quantifying the impact of groundwater resource by not only reliably forecasting of changes in major climatic variable also thorough accurate estimation of groundwater recharge. Thus, this may prove as a strong tool for future management and development of groundwater resources.

20.2 Material and Methods

20.2.1 Study Area

The Kathmandu valley watershed area is located in central part of Nepal surrounded by the Mahabharat Hills. The Kathmandu Valley lies between 27°32'13"–27°49'10"N latitudes and 85°11'31"–85°31'38" E longitudes. The mean elevation of valley is about 1,300 m above sea level. The Kathmandu valley covers an area of 651.83 km^2 and is surrounded by four major hills: Shivapuri, Phulchoki, Nagarjun, and Chandragiri. Kathmandu Valley is part of three districts (Kathmandu, Lalitpur, and Bhaktapur). Hanumante,

Manohara, Bagmati, Dhobikhola and Bishnumati are the five largest rivers flowing from north to south of the valley catchment area and leaving the basin at Chobar gorge as shown is Fig. 20.1. The valley is a bowl shaped formation which center has relatively flatter lands in comparison to the rim of the valley and the city stands at an elevation of approximately 1,400 m above the sea level. Also, the lowest and highest elevation of Kathmandu valley is around 1212 and 2722 m respectively.

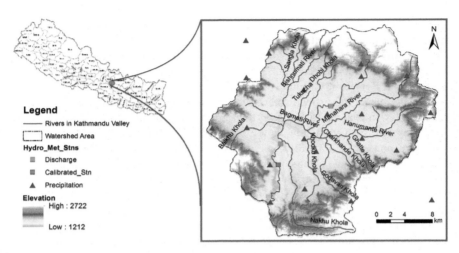

Fig. 20.1 Location map of Kathmandu valley River basin system

20.2.2 Hydrogeolocial Setting and Hydraulic Properties

The groundwater system of Kathmandu Valley is considered as closed and isolated system. The valley is physically separated by rock surface in all sides of the basin resulting in no groundwater flow across valley surfaces. The valley consists of the basement rocks on the bottom and surrounding the valley (Shrestha 2012). The aquifer system in the valley according to Gautam and Rao (1991) can be categorized into four zones as (a) Unconfined aquifer zone, (b) Two aquifer zone, (c) Confined aquifer zone, and (d) No groundwater zone. The aquitard layer sits between the shallow aquifer layer and deep aquifer layer preventing any direct recharge to the deep aquifer layer. The thickness of the aquitard layers is about 200 m in the central part of the valley and is in decreasing fashion as we move to the outer edge of the valley. Some areas in the north and southeastern part, the aquitard layer is so minimal that those areas are considered to be recharge areas for the deep aquifer. Also, shallow aquifer thickness is more toward the northern part of the groundwater basin whereas the thickness of deep aquifer is towards the southern part (Pandey and Kazama 2012).

The Hydraulic properties of an aquifer system are the key parameters that governs the aquifer response to different stress conditions and helps determine the flow of water in groundwater system. Hydraulic conductivity, storage coefficient, transmissivity, permeability in many cases are the critical parameters necessary for carrying out calibration of groundwater flow model. These parameters may vary because of geological heterogeneity. But with proper estimation of these parameters it will allow in quantitative prediction of hydraulic response of an aquifer system. The hydraulic properties for of the aquifers in Kathmandu valley were obtained from previous studies carried on in the valley. According to the study carried out by Pandey and

Kazama (2010) the hydraulic conductivity of shallow aquifer in Kathmandu valley ranges from 12.5 to 44.9 m/day. In case of deep aquifer, the values ranges from 0.3 to 8.8 m/day averaging around 4.55 m/day. Storage coefficient for shallow and deep aquifer were also estimated as estimated in their studies is around 0.2 for shallow aquifer and the value ranged between 0.00023 to 0.07 in case of deep aquifer. Similarly, transmissivity of shallow aquifer varies from 163.2 to 1056.6 m²/day with average value of 609.9 m²/day whereas that for deep aquifer varies from 22.6 to 737 m²/day with average value of 379.9 m²/day.

20.2.3   Model Description

GSFLOW is a coupled Groundwater and Surface-water FLOW model based on the integration of the USGS Precipitation-Runoff Modeling System (PRMS) and the USGS Modular Groundwater Flow Model (MODFLOW) (Markstrom et al. 2008) Where, PRMS is essentially a land-surface hydrologic model that simulates water cycling at land surface where precipitation, climate, land cover/use, and hydrologic processes occurs at the land surface. The model also provides reliable estimates of stream discharge, water balance relations and groundwater recharge rate. PRMS can be run using either distributed or lumped parameters (Leavesley et al. 1983). The basic element of PRMS is Hydrologic response unit (HRU). To study the area properly the model domain was discretized into numbers of HRUs. The HRUs of PRMS are homogenous in features related to landuse, distribution of rainfall, temperature, land type, and other factors that govern the hydrologic responses. However, groundwater is only included in a very rudimentary and empirical way. With GSFLOW, this empirical component is replaced with MODFLOW, which provides a powerful capability to simulate groundwater flow in a physically realistic manner, including flow processes.

MODFLOW is the USGS's three-dimensional (3D) finite-difference groundwater model. MODFLOW-2005 is extremely powerful and is capable of simulating the complex problems encountered in practice. It is able to represent complex heterogeneity in hydrogeological properties in three dimensions. The model can simulate inflows and outflows of water fully representing processes like pumping, stream recharge/discharge, drainage, and evapotranspiration. A user has tremendous flexibility in outputting a variety of different kinds of results including hydraulic head distributions, drawdowns and fluxes. The groundwater movement in MODFLOW is simulated using following groundwater flow equation.

$$\frac{\partial}{\partial x}\left[K_{xx}\frac{\partial h}{\partial x}\right] + \frac{\partial}{\partial y}\left[K_{yy}\frac{\partial h}{\partial y}\right] + \frac{\partial}{\partial z}\left[K_{zz}\frac{\partial h}{\partial z}\right] + W = S_s$$

where K_{xx}, K_{yy}, and K_{zz} are the values of hydraulic conductivity along the x, y, and z coordinate axes (L/T), h is hydraulic head (L), W is a volumetric flux per unit volume representing sources and/or sinks of water, where negative values are extractions, and positive values are injections. (T-1), SS is the specific storage of the porous material (L- 1), t is time (T) (Harbaugh 2005).

20.2.4 Methodology

This study aims to estimate the groundwater potential of Kathmandu valley using integrated surface and groundwater model GSFLOW. The interaction of surface and groundwater of system of the Kathmandu valley is simulated in daily time step using necessary data required for the model in same time step. Hydrological and Meteorically data such as temperature, precipitation, discharge, evaporation were collected for surface model PRMS. Similarly, data's like location and extraction rate of pumping well, groundwater level (GWL) data were collected from their respective sources. These initial data's were then analysed filtered and then modified to suite model input formats. Initially separate calibration strategy was adopted for surface model PRMS and ground water flow model MODFLOW. After achieving a good calibration and validation results, these two models were integrated using GSFLOW model. And with few parameter adjustments finally the GSFLOW model was also well calibrated and validated. With satisfactory output in respect to simulated and observed hydrographs, water table, the model was now capable of depicting existing groundwater situation and the trend that it was following. Further different pumping sensitive analysis was carried out in estimating the groundwater potential of the Deep aquifer system. Figure 20.2 represents flow chart of sequence of methodology.

20.2.5 Model Input Data

The GSFLOW model requires two sets of input data. The first sets of data are prepared in PRMS input format for surface water modeling. These data describe surface characteristics such as digital elevation data (DEM), Landuse/Landcover data and soil composition data. The second sets of data are prepared in MODFLOW input format using hydrogeological data which includes features of the geologic setting such as aquifer thickness data, hydraulic conductivity, aquifer storativity values and water levels. The initial input to define topographic surfaces was developed with the help of a 30 m resolution raster, Digital Elevation Model (DEM) developed by ASTER, a Japanese mission. Streams and HRU used in surface water model were generated with Arc Hydro tool. Sub basin parameters such as slope gradient, slope length of terrain and the stream network characteristics such as channel length, width and slope were also derived from DEM. The Land-use map of year 2010 was obtained from ICIMOD Regional Database System, Nepal. Similarly, soil map for the study area was obtained from soil and terrain map (SOTER) as well as from Food and Agricultural Organization of the United Nations (FAO).

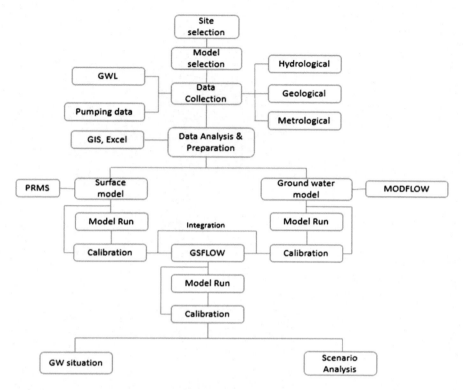

Fig. 20.2 Flow chart showing sequence of methodology

GSFLOW simulates on a daily time step, therefore, daily climate records form 20 precipitation station, and 7 temperature station was obtained for Department of Hydrology and Meteorology (DHM). Similarly, daily average stream flow data at Khokhana station was also obtained from DHM and was used for calibrating and validating the surface model. For calibration and validation of subsurface model groundwater level monitory data of 16 observation well along with the pumping well extraction rate of year 1999 was obtained from Groundwater Resource Development Board (GWRDB). Similarly, data for assigning stream reach properties in MODFLOW model were either assumed or generated using Arc Hydro tools.

20.2.6 Model Setup

The model domain is discretized into a network of land-surface called Hydrological Response Unit (HRU). Altogether of 44 sub-basins (refer Fig. 20.3) were generated using Arc hydro tool. Each HRU was then given a distinct hydrologic and physical

Fig. 20.3 HRU map for
PRMS

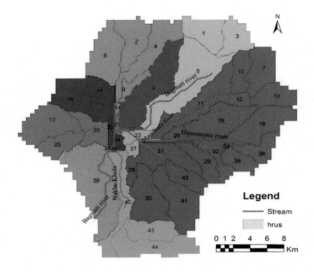

Fig. 20.4 Modflow Grid cell
Map

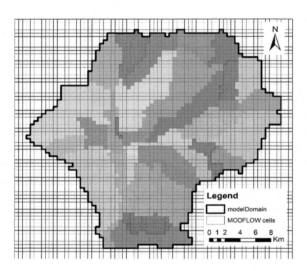

characteristic such as land-surface elevation, slope, aspect, vegetation cover, land
use, soil type. The integration of PRMS HRU with MODFLOW finite difference
cell required the spatial extent of active MODFLOW domain to be same with that of
PRMS HRU (refer Fig. 20.6). Hence the HRU map for PRMS model was clipped with
active model domain boundary of MODFLOW grid cell. As the connection between
each pair of HRUs or between HRUs and stream must be specified explicitly these
connections was manually identified. The resulting surface runoff routing contained
44 connections. These model related dimensions and parameters were organized in
a parameter file according to the Modular Modeling System Parameter File format.

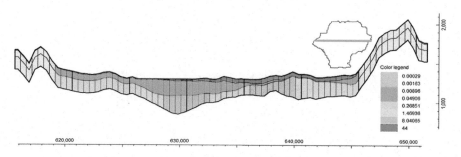

Fig. 20.5 Cross-sectional at row 30 showing different aquifer layers with hydraulic conductivity

The GSFLOW model uses climate data as input to the hydrologic system. A data file containing precipitation, temperature and discharge on daily time scale was prepared for model inputs. Finally, a control file for handling model related control parameters was also prepared.

Similarly, for subsurface model the entire model domain was discretized into grids which consist of 68 rows and 72 columns, with each grid cell being 500 m on a side (refer Fig. 20.4). The model was also vertically discretized into 3 layers, Shallow Aquifer, Clayey Aquitard and Deep Aquifer. The altitude of top layer is fixed by averaging the elevation value for each grid cell from DEM and the elevation of other layer was calculated based upon thickness of each aquifer layer obtained from 109 borehole logs from Pandey and Kazama (2010). Similarly, while assigning hydraulic conductivity different values were assigned, though well remaining within the hydraulic conductivity value range as provided in previous literature for different aquifer system. And after best matching the simulated and observed head during steady state calibration in modflow those values were fixed for different layers as shown in Fig. 20.5 for future simulation in GSFLOW. Finally, other model related parameters such as storage coefficient, specific yield, specific storage, pumping rates, infiltration rates and stream properties were also assigned by selecting their respective packages and programs.

20.2.7 HRU Attributes

HRU attributes generally refer to various input parameter sets that will uniquely define each HRU. Several of these attributes were determined using a DEM and other GIS coverages including coverages of vegetation and soil data. These coverages were projected into the same projection, same extent and same cell resolution as the local DEM used to delineate the basin. The "Environments" setting was set to be of the same extent and cell size used in the DEM for all coverage's generated. Finally, the list of attributes created was assigned to each HRU.

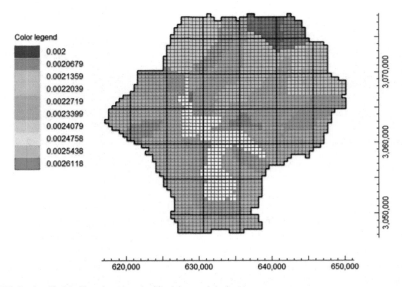

Fig. 20.6 Applied infiltration rate (m/day) in modeled area

20.2.8 Stream Flow Routing (SFR)

The SFR package is used to simulate streams in the MODFLOW. The SFR package accumulates surface and groundwater flows into the streams and route the water (Niswonger et al. 2006). For the MODFLOW simulation, there are 15 stream segments made up of 220 reaches. Various stream reach properties was assigned while using of SFR package such as reach length, stream bed top, stream slope, thickness.

20.2.9 Infiltration Rate

Recharge in groundwater is simulated with the use of Unsaturated Zone Flow (UZF) package. Percolation of precipitation through unsaturated zones is important for recharge. The package is capable of partitioning precipitation into runoff, infiltration, evapotranspiration unsaturated zone storage and recharge. UZF package is used in substitution of recharge package. Instead of applying recharge directly to groundwater it is rather applied to land surface as precipitation. Since evapotranspiration is not simulated using UZF package a scale variable is used to reduce the applied precipitation to account for surface and vadose zone Evapotranspiration (ET) and IRUNBND variable in UZF is used for routing rejected infiltration to stream during Steady state stress period. A good estimate of scale variable is calculated based upon long term average streamflow draining the basin (PPT/Q). And also, average low flow is calculated from available records which would represent recharge. The precipitation distribution is based upon HRU delineation and calculated using PRMS precipitation distribution module. The estimation of precipitation and low flow is based upon 17 years of data starting from year 1996 to 2013 for steady state simulation. Figure 20.6 shows the infiltration rate (m/day) used in steady state simulation.

20.2.10   Pupming and Observation Wells

Pumping rates and observation wells were assigned by selecting their respective packages. Pumping wells were simulated using well package and daily pumping rates of year 1999 were assigned for deep aquifers. The coordinates, pumping rates along with their screen location were analysed for 248 pumping wells whose location and pumping rates can be seen in Fig. 20.7. These pumping well location were converted to shape files and imported in modelmuse. Observation well is assigned in model using Head Observation Package. Altogether of 16 observation well is used for calibrating the groundwater flow model during steady state analysis as shown in Fig. 20.8. The head data at each observation well is obtained from GWRDP monitory wells.

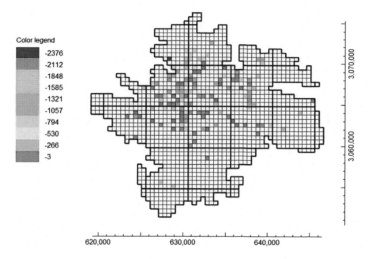

Fig. 20.7 Pumping rates and their locations

Fig. 20.8 Observation well
and their locations

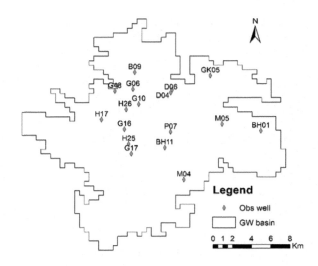

20.3 Results and Discussion

20.3.1 Calibration and Validation

The basic calibration strategy takes and advantage of individual calibration of PRMS only and MODLOW only uncoupled model available in GSFLOW. First the calibration of PRMS model was done manually by trial and error adjustment, where the sensitive parameters were identified with each iterative model runs and also from literature review. Second, calibration in MODFLOW was done through direct comparison of observed and simulated groundwater levels and stream flow values. With each adjustment of model parameters mostly hydraulic conductivity, the model was run at steady state conditions, while comparing simulated groundwater levels from MODFLOW and measured groundwater levels from monitory wells. And from the output of each run, different error parameter such as correlation coefficient, residual mean error was estimated to quantify overall error of the calibration refer (Table 20.1).

Finally, GSFLOW model was setup, calibrated and validated for Kathmandu Valley River Basin system. For the surface part the simulated discharge was calibrated using measured time series data at Khokhana station. The data selected for calibration was from year 2001 to 2005 and for validation, year 2010 to 2012 was selected. The groundwater part of the calibration was done using simulated and observed ground water table data from 16 ground water monitory stations. As expected, after initial calibration of PRMS and MODFLOW minor adjustment had to made for bringing the simulated and observed value of discharge and head in a good agreement. Figure 20.9 shows the measured and simulated hydrographs plotted during calibration year and validation year. For calibration period Nash Efficiency of the model run was 76.9%. As from visual comparison it can be seen that there is a good agreement between observed and simulated results. A simple plot of measured

Table 20.1 Measured and simulated hydraulic head

Well ID	Observed Head(RL)	Simulated Head(RL)	Difference (meter)
B09	1286.35	1290.82	−4.47
BH01	1318.57	1317.51	1.06
BH11	1290.88	1286.94	3.94
D04	1291.54	1295.46	−3.92
D06	1305.56	1295.1	10.47
G06	1294.56	1286.52	8.04
G10	1289.13	1285.96	3.17
G16	1281.03	1276.72	4.31
G17	1308.2	1300.78	7.42
G48	1294.16	1291.19	2.97
GK05	1320.4	1321.06	−0.66
H17	1291.37	1290.97	0.4
H25	1294.17	1291.62	2.55
H26	1288.12	1284.55	3.57
M05	1299.5	1306.31	−6.81
P07	1284.66	1285.49	−0.83

versus observed discharges provides an R^2 value of 0.812 with a volume deviation in calibration year of 12.9% but due to good R^2 value and good visual resemblance the calibration parameter adopted is acceptable. After calibration the predictive power of code GSFLOW was tested for the validation period where NSE is found to be 69.0% with volume deviation of 7.5% and R^2 value of 0.73. Table 20.2 shows the results of model performance criteria for PRMS and GSFLOW simulation.

Similarly Flow Duration Curve (FDC) for observed and simulated discharge was prepared as shown in Fig. 20.10. The striking resemblance of simulated and observed FDC also indicates that the model is quite properly calibrated. Also, the peizometric head generated by the GSFLOW model as shown in Table 20.1 is validated for the year 2001. The residual head was minimized for these wells (refer Fig. 20.11). From Fig. 20.12 it is observed that a good fit with R^2 value of 0.85 is also obtained from GSFLOW model.

20.3.2 Hydraulic Head

In Shallow aquifer layer, the hydraulic head is decreasing as we move from Northern to Southern part of groundwater basin. It is observed the head is quite constant at the central part of the valley and then gradually converges to the Bagmati River. Some of the cells in Shallow Aquifer are going dry, especially the cells on those layer that

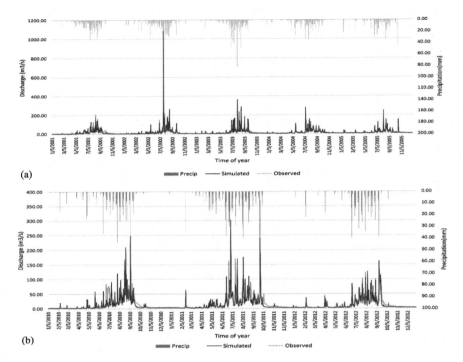

Fig. 20.9 Calibration (**a**) and Validation (**b**) of discharge at Khokhana station

Table 20.2 Model performance criteria

Model	Nash Efficiency		R²		Volume Balance	
	2001–2005	2010–2012	2001–2005	2010–2012	2001–2005	2010–2012
PRMS	79.7%	72.2%	80.9%	72.4%	6.7%	1.1%
GSFLOW	76.9%	69.0%	81.0%	72.0%	12.9%	7.5%

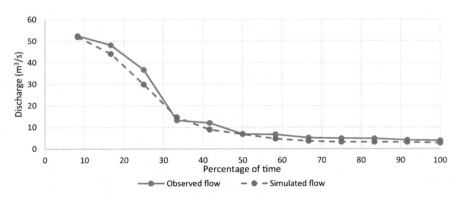

Fig. 20.10 FDC of simulated and observed flow at Khokhana station

Fig. 20.11 Plot of Residual head

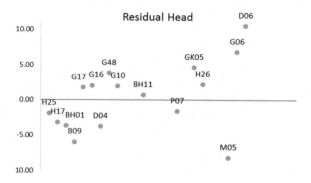

Fig. 20.12 Correlation plot between observed and simulated head

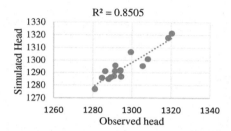

are relatively thinner with very low conductivity value and steeper slope. In case of deep Aquifer, low peizometric head seems to concentrate around the central part of the valley and it is mainly because most of the deep well extraction is occurring at those locations (Refer Fig. 20.13).

20.3.3 Groundwater Table Fluctuation

Groundwater table fluctuation map of Kathmandu Valley groundwater basin for 16 observation well is prepared. Figure 20.14 shows the change in peizometric head for four observation well over calibration and validation period. From the simulated results we can say that the peizometric head of deep aquifer is continuously decreasing even with the ground water pumping rate of year 1999. Although the decline in groundwater table of deep aquifer is quite small it can be said that this continuous decline in groundwater table is attributed to the over extraction rate from deep aquifer. We can see that in overall the decline pattern for all these observation wells is similar except for the well G48 which is an observation well located at shallow aquifer. The decline in this well is next to negligible as we see that the groundwater level is completely rejuvenated. This gives an indication that shallow aquifer is being recharged from seasonal rain whereas the deep aquifer is continually

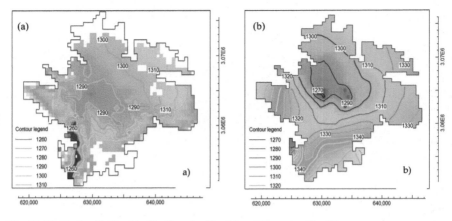

Fig. 20.13 Hydraulic Head in Shallow Aquifer (**a**) and Deep Aquifer (**b**)

overstressed resulting in its head decline. The water table have decreased from 1.57 to 4.18 m in these observation wells.

20.3.4 Assessement of Pumping Sensative Analysis on Groundwater System Based on Steady State Simulation

This analysis was carried out using steady state simulation in MODFLOW. The objective was simply to identify the areas which were more sensitive to pumping conditions. In this analysis entirety of model domain was subjected to total pumping rate of year 1999 that was equally divided among the cells. Then the pumping rate was increased gradually by 10% until it reached three times the actual pumping rate. Change in hydraulic head (ΔH) with respective change in pumping rate (ΔQ) is then obtained for each cell of entire model domain. With this it can be seen (refer Fig. 20.15) that Northern groundwater District (NGD) had more area that were less sensitive to groundwater pumping. The area near Dharmasthali, Hattigauda, Maharganj and also the area near Sankhu and Gokarna shows a decline of 0.02 to 0.12 m in head with per unit rise in pumping (m^3/s). The Area near Balaju, Samakhusi and Shywambu side showed more decline up to 0.12 m to 0.23 m with per unit rise in pumping. These areas with relatively low decline per unit pumping are specially the area where recharge is occurring showing that these places can be ideal for groundwater extraction.

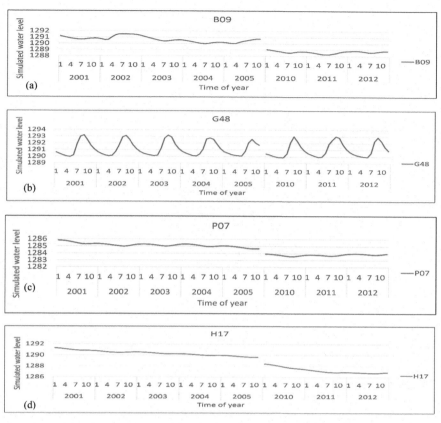

Fig. 20.14 Simulated peizometric head of observation well B09 (**a**), G48 (**b**), P07 (**c**) and H17 (**d**) over calibration and validation period

20.3.5 Assessement of Pumping Sensative Analysis on Groundwater System Based on Transient State Simulation

Pumping sensitive analysis in GSFLOW provided the capability to observe decline in water table at a given pumping condition with respect to time. Withdrawal rate is then estimated by fixing a critical drawdown level for each of these pumping wells. The GSFLOW model during these analyses was executed by subjecting deep aquifer system to same pumping conditions and by observing overall conjunctive response of the system. For this initially 35 pumping station spaced 3 km apart were positioned over the groundwater model domain (refer Fig. 20.16) then long term daily average climate data such as temperature and precipitation were inputted in the

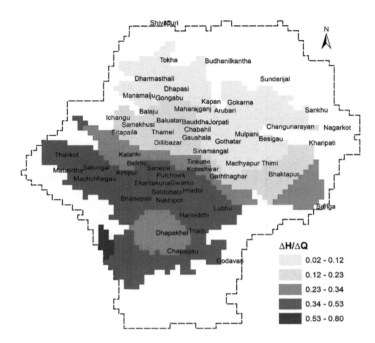

Fig. 20.15 Steady State pumping sensitive map

model. Finally, equal pumping rate is applied to all the virtual well and a drawdown over seven year of time period is noted with each increase in pumping rate.

Since all the virtual pumps are pumping at equal rates it was necessary to realize the possibility that withdrawal from the northern aquifer may affect the withdrawal rates in the southern aquifer. It was also necessary to estimate certain degree of interconnectivity of pumping condition. For this each of the pumping well stationed all over the model domain was also used as an observation well. These observation wells were then classified into three major groups Northern, Central and Southern groundwater observation well (refer Table 20.3).

Now each individual well was pumped one at a time where the pumping rate so applied was estimated by calculating existing pumping rate divided by provided number of virtual well. The observation of drawdown with respect to peizometric head at no pumping condition for all 35 virtual wells was then noted. Finally, the drawdown on northern, central and southern well due to pumping at any one of these regions was calculated. The value presented in Table 20.4 is the average drawdown for each grouped observation wells.

As seen in the Table 20.4 pumping in the northern region resulted in very lesser drawdown in southern region and pumping in southern region also did not much effect the drawdown in northern region. Here the average drawdown was seen upto 0.02 m. But the drawdown effect was more prominent in their particular region indicating that the drawdown effect in rather localized and there is low degree of

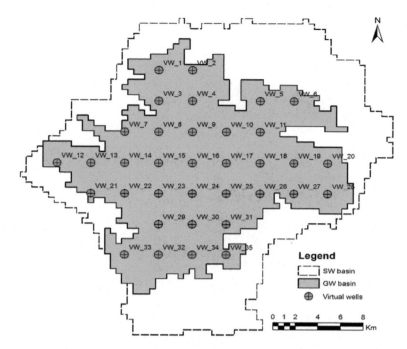

Fig. 20.16 Virtual well setup for sensitivity analysis

Table 20.3 Grouped observation wells

Observation well	Virtual well
Northern	VW_1 to VW_11
Central	VW_12 to VW_28
Southern	VW_29 to VW_35

interconnectivity of pumping dynamics between these two regions. The table also illustrates that the drawdown in northern region due to pumping of northern well is quite low in comparisons to drawdown in southern region due to pumping of southern wells. This points out to the fact that northern wells are more suitable for withdrawal then their southern counterpart. This also gives an insight into the fact that the lower possibility of withdrawal rate at southern region may not be due to extraction going on in the northern region but due to aquifer properties at southern region itself.

After ensuring certain degree of interdependency of pumping well the results of pumping sensitive analysis in terms of extraction rate of volume per year is computed as shown in Fig. 20.17. The map is prepared based upon extraction volume corresponding to a drawdown of 1 m/yr for each virtual well established in the model domain. It can be noted that the northern part of ground water basin is showing availability of more volume of water per unit decline in water table per year. The

Table 20.4 Drawdown effect of grouped region

		Pumping on		
		Northern	Central	Southern
Effect on	Northern	0.46	0.12	0.02
	Central	0.13	0.93	0.19
	Southern	0.02	0.16	2.30

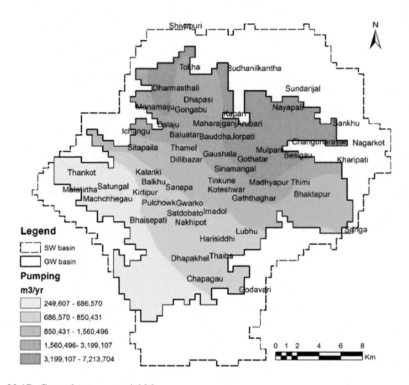

Fig. 20.17 Groundwater potential Map

map indicates that this proposed extraction rate value is decreasing as we move from northern part of groundwater basin to the southern part (refer Fig. 20.17). This result also complies with previous analysis done with individual pumping as shown in Table 20.4. The area near Dharmasthali, Budhanilkantha, Nayapati and Mulpani side are showing maximum proposed extraction rates ranging from 3,199,107 m^3/yr to 7,213,704 m^3/yr represented by dark patches of area. The light patches in the groundwater basin can be seen in southern ground water districts implying that those area are not suitable for groundwater extraction. The minimum extraction rate represented by light patches ranges from 249,607 m^3/yr to 686,570 m^3/yr.

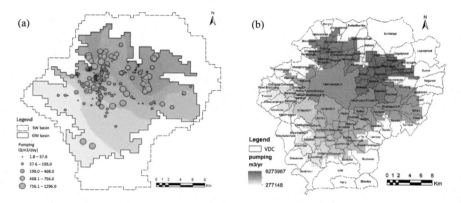

Fig. 20.18 Comparison of existing pumping scenario with Groundwater potential map (**a**), VDC wise groundwater potential map (**b**)

A pumping comparison with proposed extraction zone as shown in Fig. 20.18(a) indicates that most of the heavy pumping wells whose extraction ranges from 756 m³/day to 1296 m³/day is situated in Northern part of groundwater basin while the sheer number of pumping well is more at Thamel, Sitapaila and Dillibazar which lies at Central Groundwater District. Finally, Village Development Committee (VDC) wise extraction is shown in Fig. 20.18(b). The map is prepared by overlapping VDC area map with the proposed extraction rate map. It can be observed that the VDC lying in Northern groundwater district have higher value of proposed extraction rates but it is to be noted that some of these VDC like Sanga, Baluwa, Nayapati although having higher value of extraction rate the contributing area of groundwater extraction is low. Danchi VDC is showing high extraction rate of 6,273,967 m³/yr and Daxinkali VDC is showing the least extraction rate.

Finally, in order to identify effected region due to growing population and pumping condition. Pumping was concentrated in the area near where population is more as shown in Fig. 20.19. Initial total pumping rate of 2828 m³/day which is 5% of the actual pumping occurring in the entire basin was assigned. Then by gradually increasing the pumping rates and then observing the drawdown at different region refer Table 20.5 it is clear that pumping at this region will cause more drawdown in northern region than southern region which may be attributed to the fact the populated area is more near to the northern region.

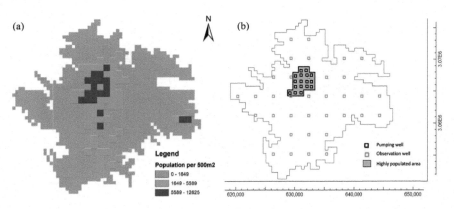

Fig. 20.19 Population map (2010) of valley within groundwater basin (**a**), pumping location at populated area (**b**)

Table 20.5 Drawdown table due to pumping at populated area

	Drawdown (m) with % increment		
	5	10	20
Northern	0.87	1.72	3.19
Central	0.53	1.06	2.38
Southern	0.07	0.14	0.32

20.3.6 Conclusions

The main purpose of this study was to develop a robust surface–groundwater model that can simulate the interaction between two different systems, i.e. surface water system and groundwater system. The objective here was to reliably quantify variables associated with each of these systems. An integrated approach of surface and groundwater flow model, GSLFOW was adopted to represent Kathmandu Valley river basin hydrologic process. For GSFLOW model, large amount of available geological, meteorological and hydrological data was integrated to construct a 3-D groundwater model in MODFLOW and a distributed surface water model in PRMS. These data were collected and modified so that they could be fit into their respective model. Preparation of data file and parameter file was carried out for PRMS model whereas a separate and functional groundwater flow model was prepared in MODFLOW with clear understanding of flow process and boundary conditions. These two separate and independent surface and groundwater flow model hence could be coupled with GSFLOW model. The execution of model and parameter control was possible by preparing a control file. The model was first separately calibrated and then again calibrated and validated in GSFLOW model. The calibration year taken was from 2001 to 2005 and for validation 2010 to 2012 was selected. The results of calibration and validation showed that even without adopting any auto calibration techniques such as LUCA for PRMS and PEST for MODFLOW the calibration of all three model have obtained good results. Overall the entire analysis points out to the two key lessons. First one being the understanding of groundwater hydrogeology of Kathmandu valley. The hydraulic head generated by the model showed that for shallow aquifer head is more in the northern part and is decreasing towards the southern part following the natural gradient of the topography. In case of deep aquifer, the head is significantly low near Shamakhusi area which can be attributed to heavy concentration of pumping well near those location. It can also be seen that the maximum proposed extraction rate from deep aquifer is highest at the Northern part of groundwater basin and the value of extraction rating is decreasing as we move from northern to southern part of groundwater basin. Also, a VDC wise extraction rate map showed that the highest allowed extraction rate was for Danchi VDC averaging the extraction rate to 6,273,967 m^3/yr.

The second lesson has a larger scope. It points out to the possibility of aiding in proper decision making for future expansion of groundwater extraction in Kathmandu valley. Over the year social, economic and demographic changes has undertaken within the valley. The result of which is that overall water demand pattern is also constantly changing. An appropriate response is necessary for better assessment of available resources, understanding the nature of vulnerability and future proofing. This can be only achieved through proper integration of research outcomes, policy making and management. As in our case if we plan to extend deep boring well fields in the southern region especially near Nakhipot, Bhaisipati, Dhapakhel and Chapagaun it has to be considered that these area are not suitable for groundwater extraction and licencing on these area should be strictly monitored whereas higher feasibility of groundwater extraction in northern region makes it ideal location for extensive pumping activities to be carried out. Also, with increasing population in the Kathmandu valley resulting in higher extraction rate, it is recommended to shift more pumping activities to northern part of the valley. Alternatively, in order to reduce stress in groundwater development of urban centre outside of the valley, macro level rain water harvesting structures and demand management can be implemented (Shrestha 2012). Hence, an integrated approach in understanding the system along with well formulated plan can be the key factor in ensuring water security for prosperity of Kathmandu valley.

Acknowledgements The authors are grateful to Dept. Of Civil Engineering, Institute of Engineering, Tribhuvan University Nepal for providing an opportunity to conduct this study. The research was partly supported by Science and Technology Research Partnership for Sustainable Development (SATREPS) as a Japan International Cooperation Agency (JICA) project in Nepal. Special thanks to Department of Hydrology and Meteorology (DHM), Kathmandu Khanepani Upatyaka Limited (KUKL), Groundwater Resource Development Board (GWRDB) for providing data and information.

References

Binne and Partners (1998) Water supply for Kathmandu-Lalitpur from outside the valley. Final report on Feasibility study, Appendix-L (Ground water resource within Kathmandu valley)

CBS (2011) Preliminary Result of National Population Census 2011. Central Bureau of Statistics, Kathmandu

Cresswell RG, Bauld J, Jacobson G, Khadka MS, Jha MG, Shrestha MP, Regmi S (2001) A first estimate of groundwater ages for the deep aquifer of the Kathmandu Basin, Nepal, using the radioisotope chlorine-36. Ground Water 39:449–457

Dhakal HP (2010) Groundwater depletion in Kathmandu valley (presentation in Water Environment partnership in Asia-Nepal Dialouge, Kathmandu) in 14th December 2010

Gautam R, Rao GK (1991) Groundwater resource evaluation of the Kathmandu Valley. J Nepal 7:39–48

Harbaugh AW (2005) MODFLOW-2005, the U.S. Geological Survey Modular Ground- Water Model - the Ground-Water Flow Process: U.S. Geological Survey Techniques and Methods 6-A16

Hunt RJ, Steuer JJ (2000) Simulation of the recharge area for Frederick Springs, Dane County, Wisconsin. U.S. Geological Survey Water-Resources Investigations report 00–4172, 33 p

JICA (1990) Groundwater Management project in Kathmandu valley. Final report Main report and supporting reports

KUKL (2015) Kathmandu Upatyaka Khanepani Limited Annual Report. Kathmandu Upatyka Khanepani Limited, Kathmandu

Leavesley GH, Lichty RW, Troutman BM, Saindon LG (1983) Precipitation- Runoff Modeling System: User's Manual: U.S. Geological Survey Water- Resources Investigations Report 83–4238, 207 p

Markstrom SL, Niswonger RG, Regan RS, Prudic DE, Barlow PM (2008) GSFLOW-coupled ground-water and surface-water flow model based on the integration of the Precipitation-Runoff Modeling System (PRMS) an the Modular. Ground-Water Flow Model (MODFLOW-2005): U.S. Geological Survey Techniques and Methods 6-D1

Metcalf and Eddy (2000) Urban water supply reform in the Kathmandu valley (ADB TA Number 2998-NEP). Completion reports, vols I, II, Executive summary, main report and Annex 1 to 7

Niswonger RG, Prudic DE, Regan RS (2006) Documentation of the Unsaturated-zone Flow package for modeling unsaturated flow between the land surface and the water table with MODFLOW-2005: U.S. Geological Techniques and Methods Book 6, Chapter A 19, 62 p

Palmer T (2003) Endangered rivers and the conservation movement. Rowman and Littlefield

Pandey VP, Chapagain SK, Kazama F (2010) Evaluation of groundwater environment of Kathmandu valley. Environ. Earth Sci. 60:1329–1342. https://doi.org/10.1007/s12665-009-0263-6

Pandey VP, Kazama F (2010) Hydrogeologic characteristics of groundwater aquifers in Kathmandu Valley, Nepal. Environ Earth Sci 62(8):1723–1732

Pandey VP, Kazama F (2012) Groundwater Storage potential in the Kathmandu Valley's shallow and deep aquifers. In: Shrestha S, Pradhananga D, Panday VP (eds) Kathmandu Valley groundwater outlook, pp 31–38

Pradhanang SM, Shrestha SD, Steenhuis TS (2012) Comprehensive review of ground water research in the Kathmandu Valley, Nepal. In: Shrestha S, Pradhananga D, Panday VP (eds) Kathmandu Valley groundwater outlook, pp 6–18

Shrestha MN (2006) Assessment of effect of landuse change in Groundwater. J Hydrol Meterol 3(1):1–13

Shrestha SD (2012) Geology and hydrogeology of groundwater aquifers in the Kathmandu valley. In: Shrestha S, Pradhananga D, Panday VP (eds) Kathmandu Valley groundwater outlook, pp 21–30

Stocklin J, Bhattarai KD (1977) Geology of Kathmandu area and central Mahabharat Range, Nepal Himalaya, HMG/UNDP Mineral Exploration Project, Kathmandu

Thapa BR, Ishidaira H, Bui TH, Shakya NM (2016b) Evaluation of water resources in mountainous region of Kathmandu Valley using high resolution satellite precipitation product. J JSCE, Ser G Environ Res 72:27–33

Thapa BR, Ishidaira H, Pandey VP, Shakya NM (2016) Impact assessment of Gorkha Earthquake 2015 on portable water supply in Kathmandu Valley: Preliminary Analysis. J Jpn Soc Civ Eng Ser 72:61–66

Thapa BR, Ishidaira H, Pandey VP, Shakya NM (2017) A multi-model approach for analyzing water balance dynamics in Kathmandu Valley Nepal. J Hydrol Reg Stud 9:149–162. https://doi.org/10.1016/j.ejrh.2016.12.080

Udmale P, Ishidaira H, Thapa B, Shakya N (2016) The Status of Domestic Water Demand: Supply Deficit in the Kathmandu Valley Nepal. Water 8:196. https://doi.org/10.3390/w8050196

Chapter 21
Estimation of Water Availability in Rivers of Stung Sreng Basin, Cambodia, Using HEC-HMS

P. Hok, C. Oeurng, and S. Heng

Abstract River system of the Tonle Sap Lake plays an important role in ensuring a long-term economic growth in Cambodia, meaning that water supply for domestic consumption, agriculture and environment, etc. must be sufficient. Stung Sreng is one among 11 main rivers of the lake system and it comprises of four ungauged tributaries: Stung Srang, Stung Sreng, Stung Tanat, and Stung Phlang. Stung Sreng has the biggest drainage area and lies in the world tourism site of Siem Reap province. A remarkable increase in number of tourists of over two million visited Siem Reap in 2015 and the government's target of exporting one million tons of rice per annum have posed a great concern on water security. Water from Stung Sreng might be used for tourism sector only. Therefore, analysis of water availability from each tributary is absolutely indispensable. HEC-HMS was applied to calibrate and to validate daily streamflow at a station where observed data are available. Indicated by three error indices, accuracy of the model prediction was concluded as satisfactory. The successfully calibrated HEC-HMS model was then employed to estimate water availability. The annual streamflow of Stung Srang, Stung Sreng and Stung Tanat River was estimated at 419, 686 and 230 MCM (Mm^3), respectively. The total yield from the whole basin is about 1572 MCM/year. The available water of 686 MCM/year from only Stung Sreng will be sufficient for tourism sector in Siem Reap, whereas the remaining amount of 886 MCM/year can be used for agriculture and other purposes.

Keywords Water availability · Stung sreng · HEC-HMS

21.1 Introduction

Water resources in Cambodia is heavily dependent on inflow from other countries, receiving 70% of the water this way (ADB 2014). Chanthy and Someth (2013) stated that the Tonle Sap Lake of Cambodia is the potential water resource for the

P. Hok (✉)
Asian Institute of Technology, Bangkok, Thailand

C. Oeurng · S. Heng
Institute of Technology of Cambodia, Phnom Penh, Cambodia

© Springer Nature Switzerland AG 2021
M. Babel et al. (eds.), *Water Security in Asia*, Springer Water,
https://doi.org/10.1007/978-3-319-54612-4_21

surrounded sub-basins, which means very significant to Cambodia. Most of these catchments are considered to have abundant water in rainy season. Unfortunately, this resource was reported to respond to new and serious challenges due to rice exports and knowledge gaps in all development sectors (MOWRAM 2015). This clearly demonstrated that water shortage will be occurring and facing owing to an increase in water demand for agriculture, domestic uses, commercial and industrial uses, hydropower production, the improper management of water uses, the effects of climate change, and so on. Such issues would lead to greater food insecurity for millions of people in the country.

A considerable attention should be given not only in the operation and management of reservoir and watersheds, but also in an accurate perception of assessing water availability both in upstream and downstream, especially the development of infrastructures within the system in order to deal with water related-problems. Yet, ADB (2014) reported that Cambodia's water resources assessment and monitoring are totally inadequate for management, particularly the model for river basin management is not being formalized well. ADB (2014) also detailed that the assessment of water resources in catchments around the Tonle Sap Basin has not well conducted yet, which has revealed a lack of comprehension on water resources systems in those river basins with their potential water availability to expand Cambodia's economic mainstay as an agrarian country. Likewise, because of a development of technology in computational capacity with supercomputer speeds, numerous researchers have applied mathematical equations into models to represent the real world phenomena, which utilized in hydrologic studies of rainfall-runoff simulation, flood forecasting, water quality, landuse change, climate change, etc., in order to simulate, to evaluate, and to estimate the effects of urbanization on hydrological response of the watersheds due to adverse impacts on the surface runoff quantities and quality through the reduction of land cover, loss of plant nutrients, deterioration of river water quality, and an increase of impervious surface area.

One of the most challenging areas in hydrology is denoted by the availability of data, so a comprehensive understanding and accurate prediction of runoff response to precipitation and its discharge to the outlet will be extremely difficult to achieve with data constrained area (Chu and Steinman 2009; Halwatura and Najim 2013; Todini 1996). Decision support tools can help in better developing an approach to this challenge, that is the use of suitable hydrologic models for the efficient management of watersheds and ecosystems (Yener et al. 2007). A continuous simulation of a hydrologic model is a suitable tool for studies of impacts of land use changes, river basin planning and management, and accounts for hydrologic process including evaporation, canopy interception, depression storage, percolation, and shallow subsurface flow, which are neglected in an event simulation model (McColl and Aggett 2007). It is generally used to predict the watershed's hydrological response to rainfall and to better estimate the watershed management practices, particularly to understand their impacts (Kadam 2011). In addition, a comprehensive literature review has shown that the studies on comparative assessment of watershed models for hydrologic simulations have not well yet studied in developing countries including India, particularly Cambodia (ADB 2014; Chanthy and Someth 2013; Kumar and Bhattacharjya

2011; Putty and Prasad 2000). Tackling study on hydrological simulation through development of a suitable model is truly desideratum.

The Hydrologic Engineering Centres - Hydrologic Modelling System (HEC-HMS), which is one among many watershed models supporting both lumped and distributed model used to simulate rainfall-runoff correlation (Madsen 2000), has become a popular and reliable hydrologic model due to its capability in short-time simulation, ease to use, and the use of common methods (Arekhi 2012), the less required input parameters, economics, the capacity in runoff simulation in ungauged catchment (Choudhari et al. 2014), and low flow prediction (De Silva et al. 2013). A number of studies have also reported a successful use of HEC-HMS model in generating runoff in different regions and climatic conditions around the globe (Anderson et al. 2002; Fleming and Neary 2004; Halwatura and Najim 2013; Majidi and Shahedi 2012; Yener et al. 2007; Yusop et al. 2007).

In order to address the current problems and to better start planning and managing the water resources within the river system of Tonle Sap, Cambodia the HEC-HMS model was used to estimate water availability in rivers of the Stung Sreng Basin. The study consists of two specific objectives: (1) calibration and validation of daily streamflow and (2) analysis of water availability without considering water withdrawal.

21.2   Materials and Methods

21.2.1   Study Area

Stung Sreng, the third largest sub-basins of the Tonle Sap Basin, is located in the northern part of Cambodia, and its catchment area is approximately 8,716 km^2. Its land elevation ranges from 3 m to 643 m above mean sea level in the northern region (Fig. 21.1). The land use is characterized by 26% of agriculture, 59% of forestry, and 15% of grassland, shrublands, water bodies, and build-up areas. The major soil types are cambisol and acrisol. Cambisol is the most fertile soil, while acrisol is the poorest quality soil. The study area is influenced by a tropical climate, a highly variable dry and wet seasons. The average annual rainfall across the basin varies from 1089 mm in the lower southern regions to 1322 mm in the northern Dangrek Mountain ranges. Average maximum temperature ranges from 31 °C in December to 37.8 °C in April; and the average minimum temperature ranges from 19.1 °C in December to 25.1 °C in July. The average monthly evapotranspiration varies from 4.3 mm/day during the wet season to 5.1 mm/day during the dry season.

The Stung Sreng River Basin (SSRB) is comprised of four main sub-river basins: Stung Srang, Stung Sreng, Stung Tanat and Stung Phlang. Ninety-three sub-basins were delineated by using the Soil and Water Assessment Tool (SWAT) in combination with GIS technology. Five sub-basins lie in Stung Phlang. Stung Sreng is one among the four containing majority of sub-basins (44). Stung Srang and Tanat share a similar

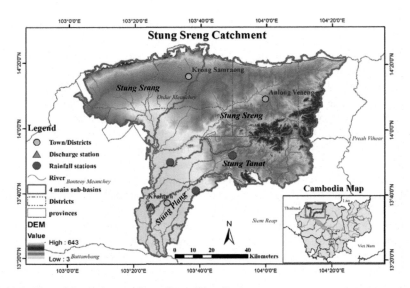

Fig. 21.1 Map of the study area, Stung Sreng River Basin

number of sub-basins, 23 and 21, respectively. The most northern boundary of the plan is located along the border of Thailand and its most southern point is within the Tonle Sap Lake. The territory of SSRB covers parts of four provinces of Cambodia: Otdar Meanchey 57%, Siem Reap 35%, Banteay Meanchey 7%, and Battambang less than 1%. The total population is 483,672, of which 50% are living in Siem Reap, 40% in Otdar Meanchey, and 10% in Banteay Meanchey. The majority of people work in the agricultural sector and approximately 30% are living in poverty.

21.2.2 Data

The data being used in this study was basically obtained from MOWRAM (Ministry of Water Resources and Meteorology) and only climate data was downloaded from global weather data (Table 21.1).

21.2.3 HEC-HMS Model

HEC-HMS is a product of the Hydrologic Engineering Centre, the U.S. Army Corps of Engineers, developed in 1992 as a replacement for HEC-1, which has long been considered a standard for hydrologic simulation. The Hydrologic Modelling System is designed to simulate the hydrologic processes of dendritic watershed systems (precipitation-runoff) (HEC 2015). It is widely used in a broad range of hydrologic problems varying from the analysis of large river basin water supply and

Table 21.1 Spatial and secondary data

Data type	Description	Source
Topography map	Digital Elevation Model (50 m × 50 m)	Ministry of Water Resources and Meteorology
Land use map	Land use classification (2003)	
Soil map	Soil types (2003)	
Hydrology data	Daily precipitation, Discharge, Water level (2000–2009)	
Geospatial data	Basin, sub-basins, river, province, 4 rainfall and discharge station(s)	
Climate data	Temperature, humidity, etc. (2000–2009)	Global weather data (NCEP 2015)

flood hydrology to the study of small urban or natural watershed runoff. It has been also used for studies of water availability, urban drainage, flow forecasting, future urbanization impact, reservoir spillway design, flood damage reduction, flood plain regulation, and systems operation.

The canopy and surface methods are intended to represent the presence of plants intercepted some precipitation in the landscape and the ground surface, where water may accumulate in the surface depressional storage, respectively (HEC 2015). The soil moisture accounting method (SMA) was used to represent the five-layer soil profiles for the continuous modelling of complex infiltration and evapotranspiration environments. The SMA model simulates both wet and dry weather behaviour, and is based on the Precipitation-Runoff Modelling System of Leavesley et al. (1983). In this model, the river basin is represented by a series of interconnected storage layers (Fig. 21.2).

Clark unit hydrograph is a synthetic unit hydrograph method requiring few parameters to compute direct runoff: time of concentration (T_c) and storage coefficient (R). Roseke (2013) stated that T_c is defined as the time duration for a drop of water falling in the most remote point of a drainage basin to travel to the outflow point. T_c and R are calculated as the Eq. (1) and (3), respectively:

$$T_c = L^{0.8} \frac{(S' + 25.4)^{0.7}}{4238 \times s^{0.5}} \tag{1}$$

$$S' = \frac{25400}{CN} - 254 \tag{2}$$

$$R = 142.2 \times (DA)^{0.081} \times S^{0.539} - T_c \tag{3}$$

$$CN_{composite} = \frac{\sum A_i CN_i}{\sum A_i} \tag{4}$$

Fig. 21.2 Structure of the
soil moisture accounting
model in HEC-HMS (HEC
2000)

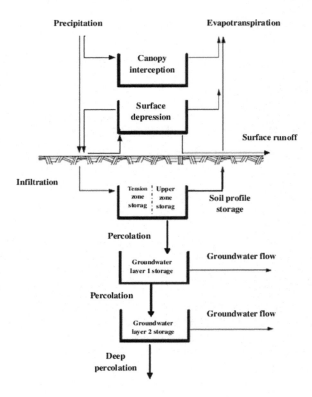

where S' the maximum retention, T_c the time of concentration (h), L the flow length (m), CN the curve number, s an average watershed land slope (%), DA a drainage area (km^2), S a channel slope (m/km), $CN_{composite}$ the composite CN used for runoff volume, i an index of watersheds subdivisions of uniform land use and soil type, CN_i the CN for subdivision I, and A_i the drainage area of subdivision i.

Among the alternatives offered in HEC-HMS, the linear reservoir baseflow model was employed in conjunction with the SMA model. It simulates the storage and movement of subsurface flow as storage and movement of water through reservoirs. The reservoirs are linear: the outflow at each time step of the simulation is a linear function of the average storage during the time step. Mathematically, this is identical to the manner, in which Clark's unit hydrograph model represents watershed runoff (HEC 2000). Indeed, the outflow from groundwater layer 1 of the SMA is inflow to one linear reservoir and outflow from groundwater layer 2 of the SMA is inflow to another. The outflow from the two linear reservoirs is combined to compute the total baseflow for the watershed.

Muskingum method for channel routing is chosen to look for X and K parameters. Theoretically, K parameter is time of passing of a wave in reach length and X parameter is a constant coefficient that its value varies between 0 and 0.5. Therefore, parameters can be estimated with the help of observed inflow and outflow hydrographs. Parameter K is estimated as the interval between similar points on the inflow and outflow hydrographs. Once K is estimated, X can be estimated by trial and error (HEC 2015).

21.2.4 Model Calibration and Validation

The model was calibrated to achieve good agreement between the simulated and observed data (Choudhari et al. 2014), so users have greater confidence in the reliability of the model (Muthukrishnan et al. 2006). For the present study, the Nash-Sutcliffe method was used in objective function with the Univariate Gradient method as a search method for optimization. After calibration, the capable model is demonstrated by model validation to produce acrate estimations for period outside the calibration period.

21.2.5 Model Evaluation

Three error indices are applied to evaluate the performance of HEC-HMS model are Nash-Sutcliff efficiency (NSE), RMSE-observation standard deviation ratio (RSR) and percent bias (PBIAS).

$$NSE = 1 - \frac{\sum_{i=1}^{n} \left(Q_i^o - Q_i^e \right)^2}{\sum_{i=1}^{n} \left(Q_i^o - Q_m^o \right)^2} \tag{5}$$

$$RSR = \frac{RMSE}{STDEV_{obs}} = \frac{\sqrt{\sum_{i=1}^{n} \left(Q_i^o - Q_i^e \right)^2}}{\sqrt{\sum_{i=1}^{n} \left(Q_i^o - Q_m^o \right)^2}} \tag{6}$$

$$PBIAS = \frac{\sum_{i=1}^{n} \left(Q_i^o - Q_i^e \right) \times 100}{\sum_{i=1}^{n} \left(Q_i^o \right)} \tag{7}$$

Table 21.2 Performance measures of the model in the calibration and validation period

No	Performance rating	NSE	RSR	PBIAS (%)
1	Very good	0.75–Unity	0–0.5	$<\pm10$
2	Good	0.65–0.75	0.5–0.6	±10–±15
3	Satisfactory	0.50–0.60	0.6–0.7	±15–±25
4	Unsatisfactory	<0.50	>0.7	$>\pm25$

where Q_i^e and Q_i^o the estimated and observed discharges (m³/s), respectively, Q_m^e and Q_m^o the mean of the estimated and observed discharges (m³/s), respectively, and n the number of data pairs.

The range adopted for performance rating using statistical criteria as listed in Table 21.2 was used to determine the performance evaluation of the HEC-HMS model in the study area (Moriasi et al. 2007).

21.2.6 Framework of the Study

Figure 21.3 illustrates the procedures to estimate water availability by generating runoff from rainfall in HEC-HMS.

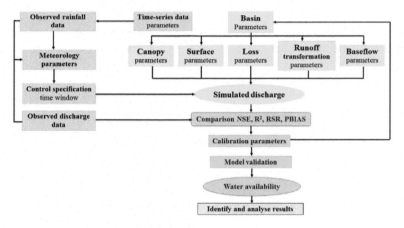

Fig. 21.3 Flowchart of runoff simulation in HEC-HMS

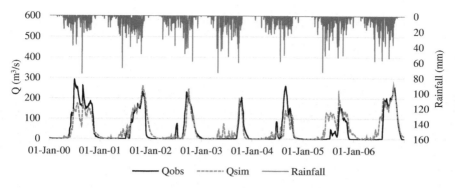

Fig. 21.4 Comparison between simulated and observed flow during calibration (2000–2006)

21.3 Results and Discussion

21.3.1 Model Calibration

The calibration of model parameters was conducted using observed flow data at Kralanh station between 2000 and 2006. A comparison of hydrographs representing simulated and observed streamflow during the calibration period is shown in Fig. 21.4. Figure 21.4 illustrates a close agreement between simulated and observed streamflow in 2001, 2002, 2003 and 2006, while other two years (2000 and 2004) indicated an under-prediction. These errors may result from the observed flow data, in which the baseflow contributed to the total flow, the external sources entering the system, or the inaccurate measurement of rainfall. Besides, an over-prediction was seen in 2005. This error may be from the observed flow data, which might be lost by infiltration, stored in natural or engineered ponds, or diverted out of the stream.

In addition, almost the whole calibration period has shown an overestimation of runoff from late dry season to early wet season. The simulated discharge did not well capture for low flow during the dry season in 2000, whereas the rest of the years were better in predicting low flow. Furthermore, the correlation between observed and simulated flow in Fig. 21.5 can assist in identifying model bias. Once the plotted points or ordinates were above the line 1:1, they represent an over-prediction, while below that represent an under-prediction, but when they fitted to line 1:1, the quality of simulated and observed discharge was represented.

21.3.2 Model Validation

Model validation involves running a model using the same input parameters as determined by the calibration process. With the optimized parameters used in calibration, the model was executed for the validation period between 2007 and 2009 in order to

check the ability of the model in predicting runoff for a period outside the calibration
stage at Kralanh station before it is recommended for use. Figure 21.6 illustrates a
comparison of hydrographs between simulated and observed flow, while Fig. 21.7
represents the correlation between them.

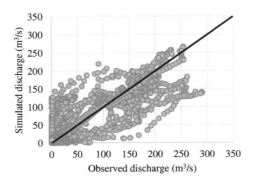

Fig. 21.5 Scatter plot between simulated and observed flow during calibration (2000–2006)

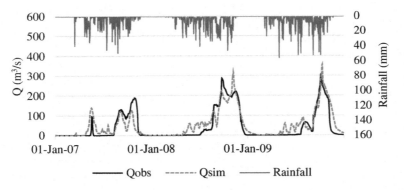

Fig. 21.6 Comparison between simulated and observed flow during validation (2007–2009)

Fig. 21.7 Scatter plot
between simulated and
observed flow during
validation (2007–2009)

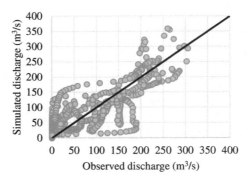

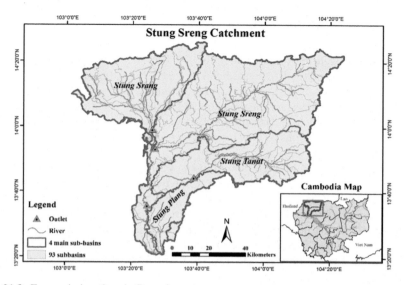

Fig. 21.8 Four main junctions in Stung Sreng

In the validation period, the simulated and observed discharges were not far different from each other during the rainy season, whereas they progressively differed from each other between the late dry season and the early rainy season. In fact, the simulated discharge was likely to overestimate in the early rainy season from May to June. In 2007, the simulated discharge was underestimated of the peak flow in the wet months, whereas the rest was likely to overestimate. The problems were coming from the observed flow data, which already stated in the calibration periods.

21.3.3 Model Evaluation

Evaluation of the model performance was undertaken on the calibration period as well as validation period in their entireties. Time-series of simulated and observed flows were taken from the results of simulation run of the HEC-HMS model of the Stung Sreng catchment, and were then analysed in Excel to evaluate the model performance. The evaluation was undertaken based on three statistical indicators which are NSE, RSR and PBIAS. The model was evaluated using the criteria given in Table 21.2. The assessment on model performance for the calibration and validation period was illustrated in Table 21.3. Overall, the performance of the HEC-HMS model was revealed as satisfactory.

21.3.4 Water Availability

Understanding water resources availability is absolutely vital, since it will help to properly manage and to plan the potential resources for a long term and in a sustainable manner. Figure 21.9 shows the average monthly flow volume at Stung Srang, Stung Sreng, Stung Tanat and catchment outlet, respectively, as depicted in Fig. 21.8. The results have indicated that the flow volume seems to be decreased from November to April, whereas it begins to pretty rise in May until it reaches the peak volume in October.

In Stung Tanat, the average monthly flow volume fell down from 36 MCM in November to 1 MCM in late dry season, while the early wet months grew between 8 MCM and 57 MCM in October. In Stung Srang, the average monthly flow volume started decreasing from 57 MCM in November to 0 MCM in February and a slight increase to 2 MCM in April. During the wet season, it differed from 15 MCM to 106 MCM. In Stung Sreng, it dropped from 103 MCM in November to 1 MCM in March, while 3 MCM in April, and it varied from 23 MCM to 190 MCM during wet months.

Table 21.3 Performance measures of the model in the calibration and validation period

No	Statistical evaluation	Results	
		Calibration	Validation
1	NSE	0.72	0.74
2	RSR	0.53	0.51
3	PBIAS	−6.69%	−18.84%

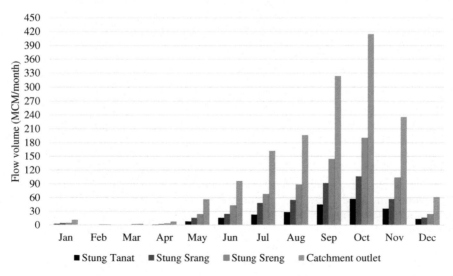

Fig. 21.9 Average monthly flow volume at Stung Tanat, Srang, Sreng and calibration outlet (2000–2009)

In catchment outlet at Kralanh station, it was conveyed from those tributaries and sub-basins along the main rivers. The average monthly flow volume decreased from 236 MCM to 2 MCM in February, and it again moderately rose between 3 MCM to 8 MCM in late dry months. It had shown an increase from 57 MCM to 414 MCM in wet season.

The result also demonstrated that the average annual flow volumes between 2000 and 2009 in Stung Srang, Stung Sreng and Stung Tanat were 419 MCM, 688 MCM and 230 MCM, respectively. The total yield from these three main tributaries and other sub-basins along the river will convey to the catchment outlet of 1572 MCM/year.

21.4 Conclusions

The HEC-HMS semi-distributed model has been successfully used for calibration and validation within the Stung Sreng catchment on simulating rainfall-runoff. The model efficiency given by Nash-Sutcliffe Efficiency criteria are 0.72 and 0.74 for calibration and validation period, respectively. Percent bias' (PBIAS) for the calibration and validation period were found to be -6.69% and -18.84%, respectively, which ranges from very good to satisfactory model fit. On the basis of RMSE-observation standard deviation ratio criteria, which gives a value 0.53 and 0.51 for the calibration and validation period, respectively, and the overall performance of the model it can be rated as satisfactory.

The outcome also found that the water availability in the rivers has been falling down from November to April, but it has started to slightly increase in the early May up to the maximum flow volume in October. The average annual flow volumes were at 419 MCM for Stung Srang, 686 MCM for Stung Sreng and 230 MCM for Stung Tanat, and the total amount of 1572 MCM/year was coming from the three main tributaries and the flow volume from sub-basins along the main river. The potential water availability from these three main upstream tributaries will significantly serve in various sectors, but mainly in agriculture, and may also attract the investment to develop the irrigation infrastructures and reservoirs to meet the sustainability of water uses inside the river basin for present and in the future. Based on overall evaluation, it could be concluded that the HEC-HMS model can be used to generate rainfall-runoff in Stung Sreng catchment, Cambodia, and this case study can be also used as a baseline study for future sustainable water resources planning and management strategy.

Acknowledgement The authors are extremely grateful to MOWRAM for great support on both hydrological and spatial data and supporting documents, to Mr. Tom Brauer for his helpful suggestions with HEC-HMS model, and to other supporters who have involved to make this research complete.

References

ADB (2014) Cambodia National Water Status Report 2014, Phnom Penh

Anderson M, Chen Z-Q, Kavvas M, Feldman A (2002) Coupling HEC-HMS with atmospheric models for prediction of watershed runoff. J Hydrol Eng 7(4):312–318

Arekhi S (2012) Runoff modeling by HEC-HMS Model (Case Study: Kan Watershed, Iran). Int J Agric Crop Sci 4(23):1807–1811

Chanthy S, Someth P (2013) Simulation of the rainfall-runoff process by using HEC-HMS Hydrological Model applied to the Stung Chrey Bak Catchment, Cambodia, Phnom Penh

Choudhari K, Panigrahi B, Paul JC (2014) Simulation of rainfall-runoff process using HEC-HMS model for Balijore Nala watershed, Odisha, India. Int J Geomatics Geosci 5(2):253–265

Chu X, Steinman A (2009) Event and continuous hydrologic modeling with HEC-HMS. J Irrig Drain Eng 135(1):119–124

De Silva M, Weerakoon S, Herath S (2013) Modeling of event and continuous flow hydrographs with HEC–HMS: case study in the Kelani River Basin, Sri Lanka. J Hydrol Eng 19(4):800–806

Fleming M, Neary V (2004) Continuous hydrologic modeling study with the hydrologic modeling system. J Hydrol Eng 9(3):175–183

Halwatura D, Najim M (2013) Application of the HEC-HMS model for runoff simulation in a tropical catchment. Environ Model Softw 46:155–162

HEC (2000) HEC-HMS Technical Reference Manual, California

HEC (2015) HEC-HMS User's Manual Version 4.1, California

Kadam AS (2011) Event based rainfall-runoff simulation using HEC-HMS model. (Unpublished P. G.)

Kumar D, Bhattacharjya RK (2011) Distributed rainfall runoff modeling. Int J Earth Sci Eng 4(6):270–275

Leavesley GH, Lichty RW, Thoutman BM, Saindon LG (1983) Precipitation-Runoff Modeling System: User's Manual, Colorado

Madsen H (2000) Automatic calibration of a conceptual rainfall–runoff model using multiple objectives. J Hydrol 235(3):276–288

Majidi A, Shahedi K (2012) Simulation of rainfall-runoff process using green - ampt method and HEC-HMS model (case study: Abnama Watershed, Iran). Int J Hydraul Eng 1(1):5–9

McColl C, Aggett G (2007) Land-use forecasting and hydrologic model integration for improved land-use decision support. J Environ Manage 84(4):494–512

Moriasi DN, Arnold JG, Van Liew MW, Bingner RL, Harmel RD, Veith TL (2007) Model evaluation guidelines for systematic quantification of accuracy in watershed simulations. Trans ASABE 50(3):885–900

MOWRAM (2015) Pilot Stung Sreng River Basin Plan, Phnom Penh

Muthukrishnan S, Harbor J, Lim KJ, Engel BA (2006) Calibration of a simple rainfall-runoff model for long-term hydrological impact evaluation. J Urban Reg Inf Syst Assoc 18(2):35–42

NCEP (2015) Global Weather Data for SWAT. http://globalweather.tamu.edu/. Accessed 30 May 2015

Putty M, Prasad R (2000) Understanding runoff processes using a watershed model—a case study in the Western Ghats in South India. J Hydrol 228(3):215–227

Roseke B (2013) How to Calculate Time of Concentration using the SCS Method. http://culvertde sign.com/how-to-calculate-time-of-concentration-using-the-scs-method/. Accessed 30 Aug 2015

Todini E (1996) The ARNO rainfall—runoff model. J Hydrol 175(1):339–382

Yener M, Sorman A, Sorman A, Sensoy A, Gezgin T (2007) Modeling studies with HEC-HMS and runoff scenarios in Yuvacik basin, Turkiye. Int River Basin Manag 4:621–634

Yusop Z, Chan C, Katimon A (2007) Runoff characteristics and application of HEC-HMS for modelling stormflow hydrograph in an oil palm catchment. Water Sci Technol 56(8):41–48

Part IV
Modelling Studies for Water Security Dimensions

Chapter 22
Predicting Low Flow Thresholds of Halda-Karnafuli Confluence in Bangladesh

A. Akter and A. H. Tanim

Abstract Eco-hydraulic modeling for flow assessment has increased in recent years due to complex hydraulic factors that control different life stages of ecological habitat. Both the Halda and Karnafuli Rivers play a vital role in the south-eastern part of Bangladesh. In almost every dry season they experience lower inflow. In this study, a 1D eco-hydraulic model, which is representing a Physical Habitat Simulation System (PHABSIM), and a 2D eco-hydraulic model for inflow regimes (CASiMiR), are applied to selected areas. To study low and minimum flow regimes two key factors, the Weighted Usable Area (WUA) and the Habitat Suitability Index (HSI) were applied. Based on the flow during flood tide and long term flow variability, a flow series was investigated to simulate suitable environmental flow regimes. Both models predicted similar trends in incremental discharge variation during minimum inflow and average minimum inflow operating in the in a range of 25–30.1 m^3/s. Although, difficulties arise while acquiring river bed topography data in the 2D eco-hydraulic model set up, reasonable prediction accuracy and geometry of regime could be obtained. However, insufficient bathymetric data necessitated the application of 1D eco-hydraulic simualtion which yielded reasonable performance while taking suitable eco-hydraulic factors into account.

Keywords CASiMiR2D · Low flow · Karnafuli river · PHABSIM · Physical habitat

22.1 Introduction

Low flow is the minimum flow requirement to sustain the ecological habitat of a riverine ecology. Since 1980, river managers have tried to correlate habitat change with flow using the inflow habitat model that can define a minimum flow while maintaining a healthy ecology. Usually, a habitat model that is based on predefined flow of a hydraulic model can predict water velocity and depth. In terms of velocity, size

A. Akter (✉) · A. H. Tanim
Chittagong University of Engineering and Technology (CUET), Chittagong, Bangladesh
URL: http://aakter.weebly.com

© Springer Nature Switzerland AG 2021
M. Babel et al. (eds.), *Water Security in Asia*, Springer Water,
https://doi.org/10.1007/978-3-319-54612-4_22

and depth of a certain range might be suitable for biota (Leclerc et al. 2003, Pasternack et al. 2004, Smakhtin 2001). A 2D or 3D hydraulic model can be sensitive to river geometry, if one considers depth, magnitude, and velocity distribution in X, Y (2D) or X, Y, Z (3D) direction. 1D models only predict mean velocities. To yield progress in modelling, the assessment of low flow regimes can be based on four major methods: a) hydrological methods (Mathews and Richter 2007), b) hydraulic methods (Lamouroux and Capra 2002), c) physical habitat methods (Muñoz-Mas et al. 2014), and d) holistic methods (McClain et al. 2014). All of those methods require hydrological, hydraulic, and biological data that maybe statistically or data driven. Physical habitat methods have a wider applicability due to consideration of alternative management methods, restoration actions, and climate change effects (Yi et al. 2014). This also allows the identification of habitats, which are exposed to flow changes. Main key assumptions include and refer to river velocity, river depth, river temperature, etc. They can become limiting factors and determine the distribution of habitats, which can be considered in the model in terms of Weighted Usable Area (WUA) (Milhouse and Waddle 2012). Based on those key assumption, model developments like PHABSIM (Nagaya et al. 2008), RHYHABSIM (Jowett, 1996), EVHA (Ginot 1995), and Mesohabitat (Parasiewicz and Dunbar 2001), or 2D model like Hydro2de, River 2D (Muñoz-Mas et al. 2014) were possible. However, few of these models hardly consider factors related to habitat availability and environmental factors like water quality and nutrient availability. In order to consider these factors, fuzzy logic based Computer Aided Simulation Models for inflow regimes (CASiMiR) (Yao et al. 2015) were developed to take 1D and 2D habitat modeling, environmental, and biotic and abiotic factors into account. In this study, preference curve methods used by PHABSIM and CASiMiR2D were used to assess low flow regimes and thresholds of selected areas of Karnafuli River.

22.2 Materials and Methods

22.2.1 Study Area

The study area is situated in the downstream part of Halda River where joining Karnafuli River (Fig. 22.1). This restoration site is located along Karnafuli River, 27 km away from the delta (Bay of Bengal). The restoration site is 3 km long. The study reach has 450~700 m width; the water depth varies from 3.28 to 7.5 m. Since 1980, researchers have started investigating ecological responses of Carp fish spawning ground at natural flow regimes, including the loss of suitable ecological factors, species and disruption of habitat for spawning fish (Tsai et al. 1981, Akter and Ali 2012). Fish monitoring revealed that three species Catla (*Catlacatla*), Mrigala (*Cirrhinusmrigala*), and Rohi (*Labeorohita*) dominate in this area. Low flow regimes are vital for a healthy and intact spawning environment.

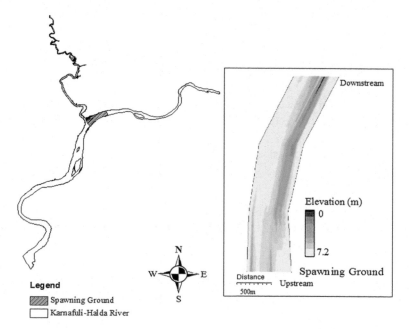

Fig. 22.1   Study Area

22.2.2   Data Collection

The Halda-Karnafuli confluence represents a natural spawning site. Spawning duration for three native fish species ranges from April to June (Akter and Ali 2012). During the spawning season, they migrate from upper to the lower parts of the Halda River due to suitable spawning environment. Thus, the carp fish was considered as critical ecological indicator, and establishment of minimum inflow regime was most important to sustain those ecological habitats. For the 1D PHABSIM model, a setup of total 4 cross-sections (i.e. transects obtained from Bangladesh Water Development Board in 2006 and Chittagong Port Authority in 2004 survey) were applied. Thus, acquired database was classified to establish symmetry of the CASiMiR 2D and PHABSIM model based on lateral cell boundaries. Totally, 231 surveyed points were used for the CASiMiR 2D model grid setup. Long-term flow characteristics (2010–2015) of the selected critical reach were taken from the flow pattern study of selected study sites conducted by Akter and Tanim (2017). The Habitat Suitability Curve (HSC) method was developed from comprehensive studies (Tsai et al. 1981). Microhabitat variables (depth, velocity, and substrate) were considered in this study as important factors of carp fish spawning. Different bed roughness coefficients were estimated in accordance with observations of bed material and bed form size. Substrate size was visually determined and a three-substrate type consisting of silt,

Table 22.1 Channel Index code (Schneider et al. 2010)

Code	Substrate type	Sizes (mm)	Code for model input	Substrate type	Sizes (mm)
0	Organic material, detritus	Visually identifiable	5	Large Gravel	20–60
1	Silt clay, loam	0.00024~0.062	6	Small stones	60–120
2	Sand	0.062~2	7	Large stones	120–200
3	Fine Gravel	2–6	8	Boulders	>200
4	Medium Gravel	6–20	9	Rock	Visually identifiable

sand, and vegetation type substrate was identified. The recommended channel index (Table 22.1) as per Schneider et al. (2010) is assigned in PHABSIM and CASiMiR 2D model.

22.2.3 Flow Characteristics of Restoration Site

Tides are the major driving force in Karnafuli River, along with freshwater outflow and saline water inland flow. Tides entering from upstream (where the river joins at mouth) are gradually distorted with distance, and increasingly extinct due to channel bottom friction (Devkota and Fang 2015). Thus, overall flow pattern prominently depends on tidal process of the river. Due to diurnal tide fluctuations, usually two tide cycles are observed in Karnafuli River and thus two directions of flow occur. During flood tide, saline water flow from the sea to the inland and direction changes in reverse order during ebb tide. Flow hydrograph and its nature changes with tide cycle and water level fluctuations. To determine low flow thresholds, the flood tide is considered and analysed as a unidirectional river. Based on the percentage of flow that exceeded the threshold during the flood tide, several incremental discharges were selected to cover the entire flow range (Table 22.2).

22.2.4 Habitat Simulation Method Based on Preference Curve

The habitat simulation method considers physical components of river hydraulics and predicts the optimum flow regime while analysing maximum suitability of factors like depth, velocity, temperature, and substrate size. From a global perspective, among 207 methodologies of inflow assessment in 44 countries the habitat simulation method is the second most frequently applied method (Tharme 2003). This method is employed in the habitat simulation model throughout several modules like

Table 22.2 Description of the selected discharge to assess habitat suitability

Discharge (m³/s)	Description
15.8	Minimum flow of flood tide in Karnafuli River during winter season(Akter and Tanim 2017)
25	10% of flow of annual runoff reported as per Tennant method (Tennant 1976)
30.1, 35.6	Minimum discharge of flood tide during monsoon season(Akter and Tanim 2017)
60.7	Q_{95} reported according to Q_{95} method obtained from flow duration curve (Arthington and Zalucki 1998)
75	30% of mean annual runoff as per Tennant method (Tennant 1976)
100, 153, 175	Discharges were selected in incremental order based on the change of stage
200, 225	Flooding discharge (Akter and Tanim 2017)

hydraulic module, hydrologic module, and habitat module. In the hydraulic module flow, components like geometric cross sections, depth, discharge, and velocity need to be introduced in the model interface through either manual input (PHABSIM) or developing algorithm (CASiMiR 2D). The hydrologic module consists of suitable ecological factors like substrate size and habitat preference curve. Finally, the habitat module takes decision of suitable minimum inflow based on univariate preference function in terms of WUA and Habitat suitability Index (HSI).

22.2.5 Habitat Suitability Curve

Due to lack of ecological data, the generation of proper Habitat Suitability Curves (HSC) is quite challenging while conducting eco-hydraulic modeling. During the study period (i.e., 2010–2015), an expert judgement was applied to work out proper HSC. In doing so, also data from literature are assessed (Akter and Ali 2012; Tsai et al. 1981). In the habitat simulation method, HSCs correlate the hydraulic module and the habitat module. However, HSC development usually needs comprehensive biological and environmental data collection. In Fig. 22.2, typical habitat suitability curves are shown for carp fish species.

22.2.6 PHABSIM Model Setup

To study low flow regimes in incremental discharge, the 1D eco-hydraulic model PHABSIM is the most widespread and preferred habitat model. It integrates a habitat simulation model with a biological model of habitat selection and relies on habitat suitability criteria like WUA using a discharge function (Ayllón et al. 2012). In this

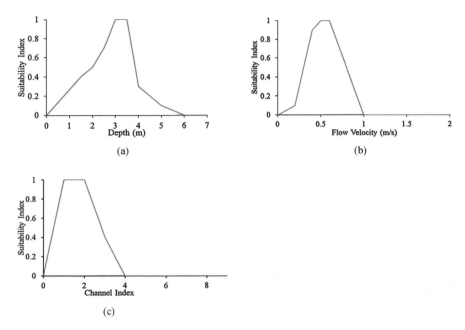

Fig. 22.2 HSC curves of major carp fishes based on (a) depth, (b) flow velocity, and (c) channel index (Schneider et al. 2010)

study, all of the 4 transects were placed along 1000 m intervals from the upstream consisting boundary conditions of defined flow and mean column velocity. Fish and other aquatic organisms can tolerate a certain range of stream velocity, water depth and bed substrate. An eco-hydraulic model like PHABSIM has key assumptions that reflect a habitat suitability of a regime based on velocity, depth, and substrate size, i.e., the Habitat Suitability Index (HSI). HSI has preference curves of targeted species on a scale of 0 to 1. The water level in the 1D eco-hydraulic model is usually predicted by three methods, whereas the Water Surface Profile (WSP) method is one of them. The WSP code uses a standard step-backwater method to determine the water surface elevations on a cross section and adjust manning's "n" (roughness coefficient for channels, closed conduits flowing partially full, and corrugated metal pipes) from the calculated discharge. Further, boundary conditions for the individual discharge can be provided. Based on predefined flow and discharge ranges, the WSP method - in this study – was modified for hydraulic modelling purposes.

The inflow is provided at upstream hydraulic section of restoration site. For 1D-HEC-RAS hydraulic simulations, the acquired water surface elevation was used as downstream boundary condition. During the HEC-RAS model setup, the average velocity ($\bar{V}$) was modified for application in each of the PHABSIM model cells. In order to minimize field survey for average velocity collection, Nikghalb et al. (2016) developed a relevant approach in the PHABSIM model. The cell velocity $(Vi)_{mod}$ obtained from the average velocity (Eq. 22.2) depends on the Manning's roughness coefficient (n), the hydraulic radius of each cell (HD), and the weighted discharge (Q') of each cell as shown in Eq. (22.3).

$$(V_i)_{mod} = V_i + \frac{\Delta Q}{nA_i} \,(1) \tag{22.1}$$

$$V_i = \overline{V} \times \left(\frac{\overline{HD}}{HD_i}\right)^{\frac{2}{3}} \tag{22.2}$$

Where $HD = R = \frac{A}{P} \approx \frac{A}{b} \approx \frac{A}{T}$
$\Delta Q = Q' \sim Q_{actual}$

$$Q' = \sum_{i=1}^{n} V_i \times A_i \tag{22.3}$$

22.2.7   CASiMiR2D Model Setup

CASiMiR (Schneider et al. 2010) is a fuzzy logic based eco-hydraulic model, developed by a group at the University of Stuttgart, Germany, to assess habitat suitability. The sub-model consisting of hydrodynamic parts can execute 1D, 2D and 3D hydraulic computation. It can calculate inflow values for the habitat using preference functions and fuzzy rules (Ahmadi-Nedushan et al. 2008). In this study, the CASiMiR 2D model was applied as 2D eco-hydraulic model using same univariate preference functions as for the PHABSIM model. The boundary conditions (discharge and water surface elevation) were established from HEC-RAS simulations. An algorithm in ASCII (American Standard Code for Information Interchange) languages was developed, which is based on Schneider et al. (2010). The algorithm consists of 231 numbers of X, Y, Z coordinates and channel indices to create a grid for hydraulic sub-model input. This was further interpolated in longitudinal scale (100 m) and vertical scale (0.3 m) to remove abrupt topography as well as to improve hydraulic performance of the channel. The WSP method was employed for the PHABSIM model, in which the boundary condition provides discharge throughout the water surface elevation at each sub-section. However, this boundary condition is limited between two adjacent transects. Another algorithm based on the WSP method (HEC-RAS 1D hydraulic simulation model) was developed to set the upstream and downstream boundary condition.

22.2.8   Low Flow Thresholds Establishment

In the Habitat Simulation Method (HSM), the Weighted Usable Area (WUA) and
the Habitat Suitability Index (HSI) can be used to describe the integrated habitat
suitability and low flow threshold of the whole river investigated. The WUA value
can be obtained by multiplying the area of each grid (CASiMiR) or cell (PHABSIM)
by its HSI value. Suitable environmental flow or low flow thresholds will be achieved
at maximum WUA. The WUA (m^2) function can be described as:

$$WUA = \sum_{i=1}^{i=n} A_i HS_i = f(Q), \tag{22.4}$$

where, i is the order number of cell; n is the number of all cells; A_i is the area of the
i^{th} cell; SI_i is HSI value of the ith cell; Q is discharge (m^3/s).

22.3   Results and Discussion

22.3.1   Habitat Suitability Assessment

Due to the topographic variation and sand bar formation, flow diversion occurred
within the first kilometer (Fig. 22.3). PHABSIM predicts this zone as most suitable
area of carp fish spawning, due to river confluence and less turbulence. The habitat
suitability criteria assumed that habitat suitability could be ascribed in a weightage
scale of 0 to 1. The effect of substrate size shows less prominent effect rather than flow
components (depth and velocity). The overall habitat diversity obtained was relatively
low in comparison to the total area. Spawning suitability of carp fish shows more
heterogeneity near the bank and around the place where flow contraction occurred.
Thus, the flow velocity is relatively low and makes the velocity most dominant.
The habitat simulation model was developed based on a calibrated hydraulic model

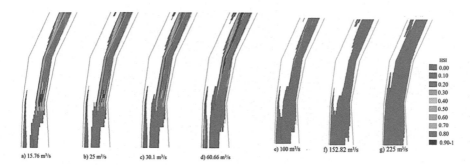

Fig. 22.3 Habitat Suitability Index obtained from CASiMiR 2D

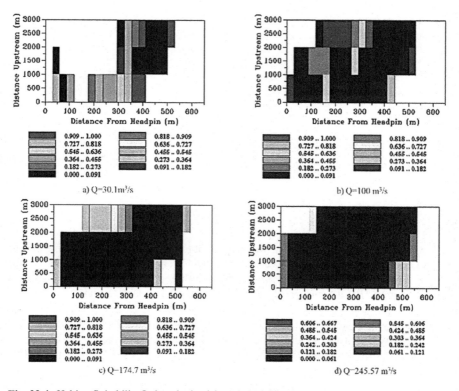

Fig. 22.4 Habitat Suitability Index obtained from PHABSIM

HEC-RAS, and the boundary condition was incorporated from the previously applied hydraulic model. The HSI variability makes the difference in WUA computation between 1D and 2D models that influence environmental flow prediction (Fig. 22.4). This result is expected because the hydraulic modules of both, the 1D and 2D eco-hydraulic model, have different approaches of considering transverse flow (Ayllón et al. 2012). However, the results might be influenced by optimum PHABSIM cross-section that depends on data availability. The PHABSIM hydraulic module is usually unstable in turbulent flow (Kondolf et al. 2000). However, the CASiMiR2D model was used to handle unstable turbulent flow. To maintain the required regime condition in a tidal river, the spatial flow variability is one of the common characteristics of tidal rivers.

22.3.2 Low Flow Thresholds

The WUA and flow relationships were obtained from model simulation (Fig. 22.5). Both models predict that 30.1 m³/s discharge can be established as low flow threshold.

Fig. 22.5 WUA vs. Discharge Relationship

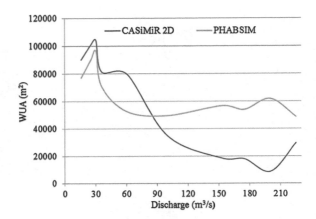

WUA in view of the flow relationship shows that low flow regime is sensitive to flow influencing parameters. The CASiMiR 2D model predicts that during flood tide the habitat suitability of carp fishes can maintain sufficient discharge between 15 m³/s and 60.7 m³/s. By increasing the discharge, the HSI reduces rapidly and reveals that spawning suitability is sensitive to flow parameters. In any hydraulic model, the accuracy of results relies on the topographic and hydrographical data that are influenced by river velocity and depth. Thus, related to the HIS, the variation of depth and velocity differed from 1D to 2D. This also resulted in variation of the WUA. Waddle et al. (2000) found that a 2D model can capture complex flow situations while significant transverse flow regimes are present. A significant variation of WUA in the 2D model was observed after 90 m³/s (Fig. 22.5). This indicates that the velocity and discharge prediction significantly vary when exceeding this threshold.

22.4 Conclusions

In this study, low flow threshold was established using the WUA concept, 1D, and 2D eco-hydraulic simulations for the Karnafuli-Halda river confluence. It was found that spawning conditions for carp fish relied on flow velocity and depth. Further, climate changes (storm surges or sea level rise) will influence the tidal flow and negatively affects the spawning ground. Migration and spawning of fish might occur in upper less turbulent parts of the Halda or Karnafuli River.

With respect to the spatial variability during the habitat and flow simulation, the CASiMiR 2D model is more coherent than the 1D PHABSIM model. The 2D eco-hydraulic model is suited for rapidly varying river topographies and tidal regimes with complex flow patterns. The improved interpolation techniques of the CASiMiR 2D model allows any 3D or 2D simulation at desired scales.

Acknowledgement This research was supported by funds provided by the Center for River, Harbor & Landslide Research (CRHLSR), Chittagong University of Engineering and Technology (CUET), Bangladesh. The authors gratefully acknowledge the logistic supports provided by Chittagong Port Authority and Bangladesh Water Development Board, Bangladesh.

References

Ahmadi-Nedushan B, St-Hilaire A, Bérubé M, Ouarda TBMJ, Robichaud É (2008) Instream flow determination using a multiple input fuzzy-based rule system: a case study. River Res Appl 24:279–292

Akter A, Ali MH (2012) Environmental flow requirements assessment in the Halda River Bangladesh. Hydrol Sci J 57:326–343

Akter A, Tanim AH (2017) Environmental flow assessment of diurnal tidal river using synthetic streamflow (submitted)

Arthington AH, Zalucki JM (1998) Comparative Evaluation of Environmental Flow Assessment Techniques: Review of Methods. Occasional Paper No. 27/98. Land and Water Resources Research and Development Corporation: Canberra

Ayllón D, Almodóvar A, Nicola GG, Elvira B (2012) The influence of variable habitat suitability criteria on PHABSIM habitat index results. River Res Appl 28:1179–1188

Devkota J, Fang X (2015) Numerical simulation of flow dynamics in a tidal river under various upstream hydrologic conditions. Hydrol Sci J 60:1666–1689

Ginot V (1995) EVHA, A Windows software for fish habitat assessment instreams. B. Fr, PechePiscic

Jowett IG (1996) RHYHABSIM. River Hydraulics and Habitat Simulation. ComputerManual, National Institute of Water and Atmospheric Research (NIWA) Report, Hamilton, p 50

Kondolf GM, Larsen EW, Williams JG (2000) Measuring and modeling the hydraulic environment for assessing instream flows. North Am J Fisheries Manage 20:1016–1028

Lamouroux N, Capra H (2002) Simple predictions of instream habitat model outputs for target fish populations. Freshwater Biol 47:1543–1556

Leclerc M, Saint-Hilaire A, Bechara J (2003) State-of-the-art and perspectives of habitat modelling for determining conservation flows. Canadian Water Res J Revue canadienne des resources hydriques 28:135–151

Mathews R, Richter BD (2007) Application of the indicators of hydrologic alteration software in environmental flow setting. JAWRA J Am Water Res Assoc 43:1400–1413

Mcclain ME, Subalusky AL, Anderson EP, Dessu SB, Melesse AM, Ndomba PM, Mtamba JOD, Tamatamah RA, Mligo C (2014) Comparing flow regime, channel hydraulics, and biological communities to infer flow–ecology relationships in the Mara River of Kenya and Tanzania. Hydrol Sci J 59:801–819

Milhouse RT, Waddle TJ (2012) Physical Habitat Simulation. PHABSIM) Softwarefor Windows Fort Collins, CO, FortCollins Science Centre, FortCollins

Muñoz-Mas R, Martínez-Capel F, Garófano-Gómez V, Mouton AM (2014) Application of Probabilistic Neural Networks to microhabitat suitability modelling for adult brown trout (Salmo trutta L.) in Iberian rivers. Environmental Modelling & Software, 59, 30–43 (2014)

Nagaya T, Shiraishi Y, Onitsuka K, Higashino M, Takami T, Otsuka N, Akiyama J, Ozeki H (2008) Evaluation of suitable hydraulic conditions for spawning of ayu with horizontal 2D numerical simulation and PHABSIM. Ecol Model 215:133–143

Nikghalb S, Shokoohi A, Singh VP, Yu R (2016) Ecological regime versus minimum environmental flow: comparison of results for a river in a semi mediterranean region. Water Resources Management, 1–16 (2016)

Parasiewicz P, Dunbar MJ (2001) Physical habitat modelling for fish: a developing approach. Arch Hydrobiol 135:1–30

Pasternack GB, Wang CL, Merz JE (2004) Application of a 2D hydrodynamic model to design of reach-scale spawning gravel replenishment on the Mokelumne River, California. River Res Appl 20:205–225

Schneider M, Noack M, Gebler T, Kopecki L (2010) Handbook for the Habitat Simulation Model CASiMiR. http://www.casimir-software.de/data/CASiMiR_Fish_Handb_EN_2010_10.pdf

Smakhtin VU (2001) Low flow hydrology: a review. J Hydrol 240:147–186

Tennant DL (1976) Instream flow regimens for fish, wildlife, recreation and related environmental resources. Fisheries 1:6–10

Tharme RE (2003) A global perspective on environmental flow assessment: emerging trends in the development and application of environmental flow methodologies for rivers. River Res Appl 19:397–441

Tsai C, Islam MN, Rahman KUMS (1981) Spawning of major carps in the lower Halda River, Bangladesh. Estuaries 4:127–138

Waddle TJ, Steffler P, Ghanem A, Katopodis C, Locke A (2000) Comparison of one and two-dimensional open channel flow models for a small habitat stream. Rivers 7:205–220

Yao W, Rutschmann P, Sudeep, S (2015) Three high flow experiment releases from Glen Canyon Dam on rainbow trout and flannelmouth sucker habitat in Colorado River. Ecol Eng **75**, 278–290 (2015)

Yi Y, Cheng X, Wieprecht S, Tang C (2014) Comparison of habitat suitability models using different habitat suitability evaluation methods. Ecol Eng 71:335–345

Chapter 23
Stream Flow Forecasting with One Day Lead Using υ-Support Vector Regression

Sameer Anipindiwar and Umamahesh V. Nanduri

Abstract Hydrological forecasting plays a significant role in effective operation of water resources planning and flood mitigation systems. In the present study υ-Support Vector Regression method is implemented to develop a model to forecast stream flow at Tekra site on River Pranahita with one day lead time. River Pranahita is a tributary of River Godavari, the largest river in South India. Coarse grid search method is applied to estimate the parameters of model, υ and γ. Study focuses on the development of four models. For Model A and Model B, the average weighted rainfall series and discharge series at Tekra site is considered as input while for the development of Model C only discharge series at upstream sites, Sirpur and Asthi, is taken into account as input. In addition to the inputs of Model C, discharge at Tekra itself is considered as input to Model D. Performance of the models is judged by evaluating mean square error (MSE), squared correlation coefficient (R^2) and Nash Sutcliffe (E) coefficient. Overall it is to be noted that discharge at Tekra site, when considered as input to model D, improved the performance of the model.

Keywords Flow forecasting · υ- support vector regression · River pranahita · Conference · Bangkok

23.1 Introduction

The rainfall runoff is a complex, dynamic and nonlinear process which is affected by many interrelated physical factors. The influence of these factors and many of their combinations in generating runoff is an extremely complex physical process. Being complex nature of hydrological processes, hydrological forecasting becomes substantial for effective operation of water resources systems. River flow forecasting is required to provide basic information on a wide range of problems related to the

S. Anipindiwar
Mott MacDonald Pvt Ltd., Mumbai, India

U. V. Nanduri (✉)
National Institute of Technology, Warangal, India
e-mail: mahesh@nitw.ac.in

© Springer Nature Switzerland AG 2021
M. Babel et al. (eds.), *Water Security in Asia*, Springer Water,
https://doi.org/10.1007/978-3-319-54612-4_23

design and operation of river systems. Hydrologic prediction helps improve planning and operation of hydrosystems, especially under flood and drought conditions (Parasuraman and Elshorbagy 2007). This may have significant economic value in decision control of reservoirs and hydropower stations. With reliable forecasting, water resources personnel can allocate water supplies optimally for water users. Therefore, the need for timely and accurate streamflow forecasting is widely recognized and emphasized by many in water resources fraternity.

Streamflow forecasting models may be broadly classified as physical models and data driven stochastic models. Physical models are good at providing insight into catchment processes but are difficult to implement (Luo et al. 2019). Stochastic models like autoregressive models (AR), autoregressive moving average models (ARMA), autoregressive moving average with exogenous inputs (ARMAX) have been widely used for flood forecasting and found satisfactory with reasonable accuracy. These stochastic models attempt to statistically estimate the underlying structure that generates the flow sequence.

Machine learning techniques which have the capability of modelling non-linear and complex processes are increasingly being used in hydrology. Recent studies have demonstrated the applicability of artificial intelligence techniques in hydrological forecasting. Govindaraju (2000) has given an extensive review of the literature on the applications of Artificial Neural Networks (ANN) in hydrology. ANN were extensively used for flow forecasting (Adamowski and Sun 2010; Dibike and Solomatine 2001; Nagesh Kumar et al. 2004; Parasuraman and Elshorbagy 2007; Sarkar and Kumar 2012). Adaptive Neuro-Fuzzy Inference System (ANFIS) is another technique which has been used widely for hydrological forecasting (He et al. 2014; Lohani et al. 2012; Nayak et al. 2004; Shiri and Kisi 2010). Adnan et al (2019) and Wang et al. (2009) reviewed the applications of soft computing techniques and various artificial intelligence techniques in streamflow forecasting and hydrological forecasting.

Support Vector Machine (SVM), a novel artificial intelligence method, based on the principle of structural risk minimization is another tool widely used in hydrology. This learning machine has been found to be a robust and competent algorithm for both classification and regression in many disciplines (Yu et al. 2006). SVM overcomes some of the limitations of ANN models. Sivapragasam et al. (2001) discussed the strengths of SVM over ANN. SVM has attracted the attention of hydrologists and has been used for several applications in hydrology over the last two decades. Sivapragasam et al. (2001) used singular spectrum analysis coupled with SVM for forecasting rainfall and runoff. Yu et al. (2006) demonstrated the application of Support Vector Regression (SVR) in flood stage forecasting with one to six hours lead. Han et al. (2007) compared the performance of SVM in flood forecasting with traditional flood forecasting models like trend model and linear transfer model and showed that SVM outperforms over these models. Maity et al. (2010) demonstrated the superiority of SVM over traditional ARIMA model in forecasting monthly flows. He et al. (2014) used ANN, ANFIS and SVM for forecasting river flow and proved that SVM performs better than the other two techniques. Granata et al. (2016) used SVR for rainfall-runoff modelling in an urban catchment and compared the performance with

Storm Water Management Model (SWMM), a physical model. The performance of SVR was comparable with that of SWMM. A hybrid model integrating SVR with generalized regression neural network model was developed for monthly streamflow forecasting by Luo et al. (2019) and they showed that the hybrid model performed better that ARIMA model for monthly streamflow forecasting. Raghavendra and Deka (2014) presented a detailed review of the applications of SVM in hydrology.

Daily flow in streams is considered as nonlinear and dynamic and daily streamflow forecasting is always a major challenge for hydrologists (Li et al. 2019). Moreover, daily streamflow process is influenced by several parameters and identifying the appropriate input variables for forecasting the daily streamflow is also a major challenge. In the present study υ-Support Vector Regression (a less frequently used variant of SVR in hydrology) is used to develop a one day lead time daily streamflow forecasting model for Tekra gauging site located on River Pranahita, a tributary of River Godavari in India. Four different models are developed considering different input variables and the performance of these models is compared to identify the best model.

The rest of the paper is organized as follows. A brief description of the Support Vector Machines is given in the next section, followed by the description of study area and model development. The results and discussion and the conclusions are presented in the subsequent sections.

23.2   Materials and Methods

23.2.1   Support Vector Machine (SVM)

The support vector machine (SVM) is based on the hypothesis of structural minimization, which minimizes both the empirical risk and the confidence interval and thus achieving a better generalization capability. The basic idea of SVM is to use linear model to implement nonlinear class boundaries through nonlinear mapping of the input vectors into a high dimensional feature space. The transformation is done through a kernel function Though SVM was initially developed for classification, it can be used for classification and regression. Support vector regression (SVR) is based on the computation of a linear regression function in a multidimensional feature space (Baydaroğlu and Koçak 2014).

Support vector regression maps the input data into a higher dimensional feature space. Linear regression is performed in this feature space. Given a training data $\{x_i, y_i\}$, the goal of SVR is to find a function $f(x)$ that has a deviation of less than ε from the targets y_i for all the training data and at the same time is as flat as possible. Let x_i be mapped into a feature space given by a nonlinear function $\phi(x)$. Let the regression function be expressed as

$$f(x) = w.\phi(x) + b \qquad (1)$$

where w and b are parameter vectors of the regression function.

The goal of SVR is to find $f(x)$ that has a deviation less than ε from the targets y_i for all the training data, while keeping the function as flat as possible. Flatness of $f(x)$ means w should be as small as possible, while the absolute difference between $f(x_i)$ and y_i has to be less than ε.

The function $f(x)$ can be found by minimizing the risk function given by

$$R = \frac{1}{2}w^2 + C\sum_{i=1}^{n} L_i \tag{2}$$

where C is a positive constant, n is the number of training samples and L_i is ε sensitive loss function defined by

$$L_i = \begin{cases} 0 & for |y_i - f(x_i)| \leq \varepsilon \\ |y_i - f(x_i)| - \varepsilon & otherwise \end{cases} \tag{3}$$

The parameter C indicates the degree of penalty imposed when the deviation of $f(x_i)$ by more than ε. The first term in Eq. (2) represents the confidence interval of the learning machine, while the second term represents the empirical risk. Thus SVR avoids under fitting and over fitting by minimizing both confidence interval and the empirical risk (Yu et al. 2006).

The regression problem can be stated as a convex optimization problem as given below

$$Min\frac{1}{2}w^2 + C\sum_{i=1}^{n} \left(\xi_i + \xi_i^*\right) \tag{4}$$

subjected to $y_i - (w.\phi(x_i) + b) \leq \varepsilon + \xi_i$

$$w.\phi(x_i) + b - y_i \leq \varepsilon + \xi_i^*$$

$$\xi_i, \xi_i^* \geq 0, i = 1, 2, \ldots, n$$

where ξ_i and ξ_i^* are slack variables.

The dual form of the optimization model given by Eq. (4) can be expressed as

$$Min\frac{1}{2}\sum_{i,j=1}^{n} (\alpha_i - \beta_i)(\alpha_j - \beta_j)\left(\phi(x_i).\phi(x_j)\right)$$
$$+ \varepsilon \sum_{i=1}^{n}(\alpha_i + \beta_i) - \sum_{i=1}^{n} y_i(\alpha_i - \beta_i) \tag{5}$$

subjected to

$$\sum_{i=1}^{n} (\alpha_i - \beta_i) = 0$$

$$0 \leq \alpha_i \leq C \quad i = 1, 2, \ldots, n$$

$$0 \leq \beta_i \leq C \quad i = 1, 2, \ldots, n$$

where α_i and β_i are dual variables.

Identifying an appropriate nonlinear function $\phi(x)$ may be difficult. However, it is not necessary to determine $\phi(x)$ explicitly. The kernel function $K(x_i, x_j) = \langle \phi(x_i).\phi(x_j) \rangle$ is sufficient to define the feature space. Common kernel functions used in SVR are linear, polynomial, radial basis and sigmoid functions. In the present study radial basis function is used as the kernel function and is given by

$$K(x_i, x_j) = exp\left(-\gamma |x_i - x_j|^2\right) \tag{6}$$

where γ is the parameter of the kernel function.

The dual optimization problem can be solved using kernel function and $f(x)$ is given as

$$f(x) = \sum_{i=1}^{n} (\alpha_i - \beta_i) K(x_i, x) + b \tag{7}$$

The three important parameters of SVR are the penalty cost C, error tolerance ε and the kernel parameter γ. The parameter C decides the smoothness/ flatness of the function. The parameter ε effects the number of support vectors. Small ε will cause more support vectors, while a large ε may cause loss in some important information hidden in data (Yu et al. 2006). Determining appropriate values of C and ε is often a heuristic trial and error process (Raghavendra and Deka 2014).

An alternate form of SVR is known as υ-SVR. The optimization model given by Eq. (4) can be modified to make ε as a variable in the optimization model.

$$Min \frac{1}{2}w^2 + C\left[v\varepsilon + \sum_{i=1}^{n} \left(\xi_i + \xi_i^*\right)\right] \tag{8}$$

subjected to

$$y_i - (w.\phi(x_i) + b) \leq \varepsilon + \xi_i$$

$$w.\phi(x_i) + b - y_i \leq \varepsilon + \xi_i^*$$

$$\xi_i, \xi_i^* \geq 0, i = 1, 2, \ldots, n$$

The dual of this model is similar to the one given by Eq. (5), except that an additional constraint given by Eq. (9) is added.

$$\sum_{i=1}^{n} (\alpha_i + \beta_i) \leq C\nu \qquad (9)$$

The optimization problem is similar to that of the original SVR, which is known as ε-SVR.

The ν-SVR improves upon ε-SVR by adjusting ε automatically based on data (Smola and Schölkopf 2004). The motivation of ν-SVR is that it may not be easy to decide the parameter ε (Chang and Lin 2002). In ν-SVR the parameter ε is replaced by ν. The parameter ν represents an upper bound on the fraction of training samples which are errors and a lower bound on the fraction of samples which are support vectors and thus has a more meaningful interpretation (Langhammer and Česák 2016).

A detailed description of Support Vector Regression and its variants can be found in Chang and Lin (2002); Raghavendra and Deka (2014); Smola and Schölkopf (2004); Yu et al. (2006).

In the present study ν-SVR is used for daily streamflow forecasting.

23.2.2 Study Area

The river Godavari, the largest of the peninsular rivers, and third largest in India, drains about 10% of India's total geographical area. Godavari basin comprises of seven sub-basins which includes upper Godavari, Godavari middle, Manjra, Wardha, Wenganga, Pranhita, Indravati and Godavari lower. Pranahita is the largest sub-basin of River Godavari and drains about 34% of the total catchment of River Godavari. The Pranhita River has a catchment area of 109,078 km^2. Pranahita begins at the confluence of River Wardha and River Wainganga. Present study aims to forecast discharge at Tekra site with one day lead time. Tekra is located just upstream of the confluence of River Pranahita with River Godavari. The coordinates of Tekra are latitude $18°58'42"$ North and longitude $79°56'49"$ East. Total catchment area is 108,780 km^2. Upstream sites namely Sirpur on Wardha River and Asthi on River Wenganga are considered for forecasting flow at Tekra. Figure 23.1 shows the location of these gauging stations. The daily flow data observed at these three sites are collected from Central Water Commission (CWC). The daily flow data observed at these three stations during 1998 to 2005 is considered for constructing the daily flow forecasting model.

23.2.3 Model Construction

Daily streamflow process is influenced by several parameters and identifying the appropriate input variables for forecasting the daily streamflow is always a major challenge. Systematically identifying both the lags of the input variables and the

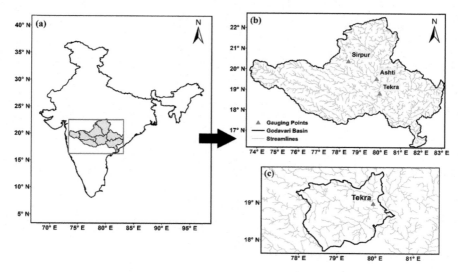

Fig. 23.1 Location of the study area

parameters of the model may be computationally expensive (Yu et al. 2006). Cross correlation among various variables with different lags is used to identify the significant variables influencing the flow at Tekra gauging site. Based on this analysis four model structures are considered for forecasting flow at Tekra. Model A and Model B consider both rainfall and flow at Tekra are considered as input variables. Model A considers the data over the entire year, while Model B considers only the monsoon season (June to October). Model C and Model D consider the flows observed at the upstream stations, namely, Asthi and Sirpur, as input variables. Model D considers the previous days flow at Tekra along with the flows at upstream stations as input variables, while Model C considers only the flows at upstream stations as input variables. The structure of the four models is given in Table 23.1.

The observed discharge and rainfall data consists of values varying over a significant range. To prevent the model from being dominated by the variables with large values, the variables are normalized to the interval [0, 1] using the following scheme.

Table 23.1 Details of identified input

Model	Model A	Model B	Model C	Model D
Considered period	Whole year	Monsoon season	Whole year	Whole year
Type of input data	Rainfall and discharge at time (t-1,t-2,t-3,t-4) at Tekra	Rainfall and discharge at (t-1,t-2,t-3,t-4) at Tekra	Discharge at (t-1,t-2) at two sites Sirpur & Asthi	Discharge at (t-1,t-2) at sites Sirpur & Asthi and at (t-1) at Tekra

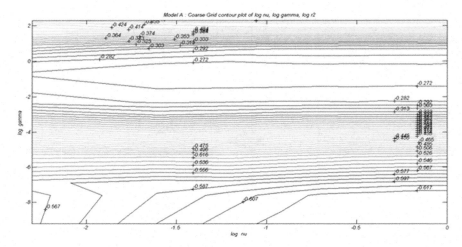

Fig. 23.2 Coarse grid Contour plot between log υ (X-axis), log γ (Y-axis) and log R^2 (each parameter set)

$$x_{scaled} = \frac{x}{x_{max} + 0.1} \tag{10}$$

The parameters of υ-SVR model are regularization parameter C, parameter υ, and Kernel function parameter γ. Radial basis function is used as Kernel function. Choy and Chan (2003) have shown that C does not significantly affect the optimization model. Hence, in this study C is fixed as 1 and the other two parameters, namely, ν and γ are determined using grid search technique. An open software package LIBSVM (Chang and Lin 2011) is used to develop the models.

As stated earlier the parameters υ and γ are estimated using grid search method. The goal is to identify good (υ, γ) so that the classifier can accurately forecast the data. The parameter ν ranges between [0.1 to 1.0], while parameter γ is assumed in the range of [0.0001 to 10.0]. Doing a complete grid range is time consuming (Hsu et al. 2010) and hence a course grid search is performed first. The better region of the grid is identified based on the course grid search and a fine grid search is performed over this region.

Three performance indicators, namely, Nash-Sutcliffe coefficient (E), Mean Square Error (MSE) and Coefficient of determination (R^2) between observed and forecasted flow, are used for evaluation of the four models. These parameters are given by

$$E = 1 - \frac{\sum_{i=1}^{n} \left(Q_i^o - Q_i^m\right)^2}{\sum_{i=1}^{n} \left(Q_i^o - \overline{Q_o}\right)^2} \tag{11}$$

$$MSE = \frac{1}{n} \sum_{i=1}^{n} \left(Q_i^o - Q_i^m\right)^2 \tag{12}$$

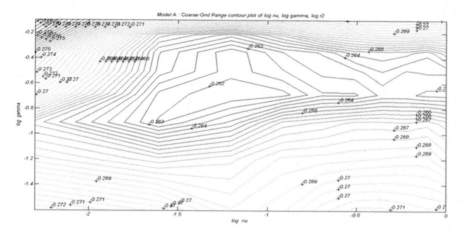

Fig. 23.3   Fine grid Contour plot between log υ (X-axis), log γ (Y-axis) and log R^2 (each parameter set)

$$R^2 = \left[\frac{\sum_{i=1}^{n} (Q_i^o - \bar{Q}_o)(Q_i^m - \bar{Q}_m)}{\sqrt{\sum_{i=1}^{n} (Q_i^o - \bar{Q}_o)^2} \sqrt{\sum_{i=1}^{n} (Q_i^m - \bar{Q}_m)^2}} \right]^2 \tag{13}$$

where, Q_i^o is ith observed discharge, Q_i^m is ith predicted discharge, $\bar{Q}_o$ is mean of observed discharge and $\bar{Q}_m$ is mean of predicted discharge.

23.3   Results and Discussions

23.3.1   Results

The coarse grid search method is applied to obtain the parameters for Models A, B, C, and D. Initial coarse range of γ is taken as [0.0001 to 10] and that for υ as [0.1 to 1.0]. Using LIBSVM, each model is trained for discrete parameter sets. For each parameter set, the squared correlation coefficient (R^2) and mean square error (MSE) are computed. Then a logarithmic contour plot between υ, γ and R^2 is obtained as shown in Fig. 23.2 (for Model A). Based on R^2 value, significant finer range of (γ, υ) is selected and new set of parameters is obtained. The model is trained using new set of parameters and coefficients such as R^2, MSE are computed. A logarithmic plot between υ, γ and R^2 is plotted as shown in Fig. 23.3 (for Model A) and a new finer range of parameters is obtained. The procedure is repeated till good regression coefficient and MSE is obtained and the resulting set of parameters is noted. The

Table 23.2 Model A-List of parameters and performance indicators

Parameters		Training			Validation		
υ	γ	MSE	R^2	E	MSE	R^2	E
0.233	0.52	0.0012	0.7704	0.7698	0.0007	0.7260	0.7215
0.198	0.41	0.0012	0.7697	0.7689	0.0007	0.7249	0.7207
0.304	0.49	0.0012	0.7698	0.7688	0.0007	0.7266	0.7211

model is validated and its performance is evaluated by the three indicators given by Eq. (11), Eq. (12) and Eq. (13).

During parameter estimation several effective sets of parameters are obtained. Parameters along with performance indicators during training and validation for Models A, B, C and D are listed in Tables 23.2, 23.3, 23.4 and 23.5 respectively. Comparative plots are obtained against observed discharge and forecasted discharge

Table 23.3 Model B-List of parameters and performance indicators

Parameters		Training			Validation		
υ	γ	MSE	R^2	E	MSE	R^2	E
0.642	0.27	0.0019	0.7496	0.7478	0.0011	0.7052	0.7002
0.496	0.27	0.0019	0.7494	0.7477	0.0011	0.7048	0.7002
0.715	0.27	0.0019	0.7495	0.7477	0.0011	0.7049	0.6995
0.788	0.26	0.0019	0.7494	0.7475	0.0011	0.7058	0.7004
0.300	0.25	0.0019	0.7482	0.747	0.0011	0.7042	0.7026

Table 23.4 Model C-List of parameters and performance indicators

Parameters		Training			Validation		
υ	γ	MSE	R^2	E	MSE	R^2	E
0.109	1.18	0.0015	0.7632	0.4206	0.0002	0.7843	0.5944
0.109	1.32	0.0015	0.765	0.4157	0.0003	0.7794	0.5744
0.109	1.34	0.0015	0.765	0.413	0.0003	0.7788	0.5693

Table 23.5 Model D-List of parameters and performance indicators

Parameters		Training			Validation		
υ	γ	MSE	R^2	E	MSE	R^2	E
0.691	0.84	0.0006	0.8144	0.7991	0.0003	0.9424	0.9412
0.719	0.85	0.0006	0.8144	0.799	0.0003	0.9423	0.9412
0.662	0.86	0.0006	0.8144	0.7989	0.0003	0.9422	0.941
0.577	0.88	0.0006	0.8144	0.7987	0.0004	0.9421	0.9409
0.100	0.85	0.0006	0.8149	0.7631	0.0004	0.9405	0.9365

Fig. 23.4 Model A-Comparative plot between observed flow and forecasted flow (during validation)

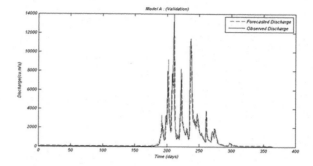

for each model during validation using effective set of parameters as shown in Figs. 23.4, 23.5, 23.6 and 23.7.

Fig. 23.5 Model B-Comparative plot between observed flow and forecasted flow (during validation)

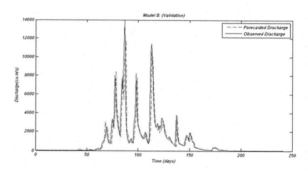

Fig. 23.6 Model C-Comparative plot between observed flow and forecasted flow (during validation)

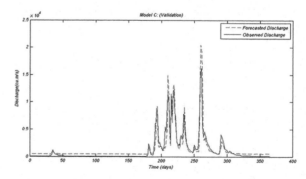

Fig. 23.7 Model D-
Comparative plot between
observed flow and forecasted
flow (during validation)

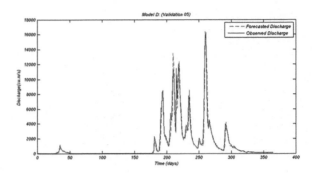

23.4 Discussions

From Tables 23.2, 23.3, 23.4 and 23.5, it is observed that MSE is small in all models.
Squared correlation coefficient is satisfactory in both training and validation. Nash
Sutcliffe coefficient is higher than 0.7 in all the models, except Model C (highest
in Model D). The whole year rainfall and discharge data were used as the input for
Model A, whereas only monsoon season data were taken as input for Model B. From
Fig. 23.4, it is observed that Model A performed well in forecasting high flows in
the stream but failed to estimate low flows. The values of performance indicators
are also found to be satisfactory with E and $R^2 = 0.72$. From Fig. 23.5, it is seen
that Model B forecasted low flows with good accuracy. Also it performed well in
forecasting high flows. In the development of Model C, only discharges at upstream
sites Sirpur and Asthi are considered for the input. From Fig. 23.6 and Table 23.4, it
is clearly seen that Model C failed to produce observed discharge series. For Model
D, along with inputs of Model C the discharge at Tekra is also taken into account. It
can be seen that Model D forecasted discharge close to the observed discharge with
higher value of Nash Sutcliffe coefficient.

It is to be noted that the Model A and Model B aimed to establish a relationship
between rainfall and runoff while Model C and Model D aimed to identify the runoff
pattern either by relating discharges at the same site or at upstream site. Models A, B
and C did not predict well the discharge with one day lead wheares Model D showed
the best predicted results.

23.5 Conclusion

The present study demonstrates the applicability of ν-SVR for one day ahead daily
flow forecasting at Tekra gauging site on River Pranahita in Godavari River Basin,
India. Four alternate models are developed for flow forecasting and their performance
is compared. The results show that SVR can be considered as a potential tool for
hydrological regression modeling.

Among the four models developed, Model D performed well with nearly matching flow pattern of observed target discharge series. Model C did not prove much efficient in forecasting the stream flow. It could be due to the type of inputs considered. Flow data at upstream sites Sirpur and Asthi alone does not prove significant for forecasting flow at Tekra.

Model A succeeded to some extent to establish the physical significance of rainfall runoff relationships. It was able to forecast high flows while it failed to forecast low flows. Input data for Model A consists of several zero values as rainfall series of whole year is taken into account. This can be the reason that Model A does not prove efficient.

Model B established relation of rainfall and discharge for isolated season. It is found that the model performed well in forecasting high flows and low flows but it could not achieve satisfactory accuracy. It can be due to the discontinuity in the input sequence considered.

References

Adamowski J, Sun K (2010) Development of a coupled wavelet transform and neural network method for flow forecasting of non-perennial rivers in semi-arid watersheds. J Hydrol 390:85–91. https://doi.org/10.1016/j.jhydrol.2010.06.033

Adnan RM, Yuan X, Kisi O, Yuan Y, Tayyab M, Lei X (2019) Application of soft computing models in streamflow forecasting. Proc. Inst. Civ. Eng. Water Manag. 172:123–134. https://doi.org/10.1680/jwama.16.00075

Baydaroğlu Ö, Koçak K (2014) SVR-based prediction of evaporation combined with chaotic approach. J Hydrol 508:356–363. https://doi.org/10.1016/j.jhydrol.2013.11.008

Chang C-C, Lin C-J (2011) LIBSVM. ACM Trans. Intell. Syst. Technol. 2:1–27. https://doi.org/10.1145/1961189.1961199

Chang C-C, Lin C-J (2002) Training v -Support Vector Regression: Theory and Algorithms. Neural Comput 14:1959–1977. https://doi.org/10.1162/089976602760128081

Choy KY, Chan CW (2003) Modelling of river discharges and rainfall using radial basis function networks based on support vector regression. Int J Syst Sci 34:763–773. https://doi.org/10.1080/00207720310001640241

Dibike YB, Solomatine DP (2001) River flow forecasting using artificial neural networks. Phys. Chem. Earth. Part B Hydrol. Ocean. Atmos. 26:1–7. https://doi.org/10.1016/S1464-1909(01)85005-X

Govindaraju RS (2000) Artificial neural networks in hydrology. II: Hydrologic applications. J Hydrol Eng 5:124–137. https://doi.org/10.1061/(ASCE)1084-0699(2000)5:2(124)

Granata F, Gargano R, de Marinis G (2016) Support vector regression for rainfall-runoffmodeling in urban drainage: a comparison with the EPA's storm water management model. Water (Switzerland) 8:1–13. https://doi.org/10.3390/w8030069

Han D, Chan L, Zhu N (2007) Flood forecasting using support vector machines. J. Hydroinformatics 9:267–276. https://doi.org/10.2166/hydro.2007.027

He Z, Wen X, Liu H, Du J (2014) A comparative study of artificial neural network, adaptive neuro fuzzy inference system and support vector machine for forecasting river flow in the semiarid mountain region. J Hydrol 509:379–386. https://doi.org/10.1016/j.jhydrol.2013.11.054

Hsu, C., Chang, C., Lin, C.: A practical guide to support vector classification (2010)

Langhammer J, Česák J (2016) Applicability of a Nu-Support vector regression model for the completion of missing data in hydrological time series. Water 8:560. https://doi.org/10.3390/w81 20560

Li X, Sha J, Wang Z-L (2019) Comparison of daily streamflow forecasts using extreme learning machines and the random forest method. Hydrol Sci J 64:1857–1866. https://doi.org/10.1080/02626667.2019.1680846

Lohani AK, Kumar R, Singh RD (2012) Hydrological time series modeling: A comparison between adaptive neuro-fuzzy, neural network and autoregressive techniques. J Hydrol 442–443:23–35. https://doi.org/10.1016/j.jhydrol.2012.03.031

Luo X, Yuan X, Zhu S, Xu Z, Meng L, Peng J (2019) A hybrid support vector regression framework for streamflow forecast. J Hydrol 568:184–193. https://doi.org/10.1016/j.jhydrol.2018.10.064

Maity R, Bhagwat PP, Bhatnagar A (2010) Potential of support vector regression for prediction of monthly streamflow using endogenous property. Hydrol Process 24:917–923. https://doi.org/10.1002/hyp.7535

Nagesh Kumar D, Srinivasa Raju K, Sathish T (2004) River flow forecasting using recurrent neural networks. Water Resour Manag 18:143–161. https://doi.org/10.1023/B:WARM.0000024727.947 01.12

Nayak P, Sudheer K, Rangan D, Ramasastri K (2004) A neuro-fuzzy computing technique for modeling hydrological time series. J Hydrol 291:52–66. https://doi.org/10.1016/j.jhydrol.2003.12.010

Parasuraman K, Elshorbagy A (2007) Cluster-based hydrologic prediction using genetic algorithm-trained neural networks. J Hydrol Eng 12:52–62. https://doi.org/10.1061/(ASCE)1084-0699(2007)12:1(52)

Raghavendra S, Deka PC (2014) Support vector machine applications in the field of hydrology: a review. Appl. Soft Comput. J. 19:372–386. https://doi.org/10.1016/j.asoc.2014.02.002

Sarkar A, Kumar R (2012) Artificial neural networks for event based rainfall-runoff modeling. J. Water Resour. Prot. 04:891–897. https://doi.org/10.4236/jwarp.2012.410105

Shiri J, Kisi O (2010) Short-term and long-term streamflow forecasting using a wavelet and neuro-fuzzy conjunction model. J Hydrol 394:486–493. https://doi.org/10.1016/j.jhydrol.2010.10.008

Sivapragasam C, Liong S-Y, Pasha MFK (2001) Rainfall and runoff forecasting with SSA–SVM approach. J. Hydroinformatics 3:141–152. https://doi.org/10.2166/hydro.2001.0014

Smola AJ, Schölkopf B (2004) A tutorial on support vector regression. Stat. Comput. 14:199–222. https://doi.org/10.1023/B:STCO.0000035301.49549.88

Wang W-C, Chau K-W, Cheng C-T, Qiu L (2009) A comparison of performance of several artificial intelligence methods for forecasting monthly discharge time series. J Hydrol 374:294–306. https://doi.org/10.1016/j.jhydrol.2009.06.019

Yu P-S, Chen S-T, Chang I-F (2006) Support vector regression for real-time flood stage forecasting. J Hydrol 328:704–716. https://doi.org/10.1016/j.jhydrol.2006.01.021

Chapter 24
Flood Discharge Estimation in Baddegama Using Pearson Type III and Gumbel Distributions

T. N. Wickramaarachchi

Abstract Gin catchment located in southern region of Sri Lanka is frequently subjected to flooding. Flood events have significantly disrupted livelihoods and severely damaged infrastructure in the area from the onset to the end of the southwest monsoon. Estimates of flood magnitude and frequency are of prime importance in flood plain water resource planning and management for the Gin catchment. This study presents results of flood frequency analysis carried out for Baddegama area based on thirty-three-year observed discharge data at Baddegama gauging station (6°11′23″ N, 80°11′53″ E) in the Gin River. Daily river discharge data were analysed to obtain flood discharges for different return periods of 2, 5, 10, 25, 50 and 100 years using Pearson Type III and Gumbel probability distribution functions. The Chi-square test was used to evaluate the correlation between the observed and estimated discharge data. It was found that both the Pearson Type III and Gumbel distributions are applicable for determining the flood discharges in the Gin River. The 100-year flood peaks estimated by the Pearson Type III and Gumbel distribution are 592 m³/s and 615 m³/s, respectively, which are about two times larger than the 2-year flood. This indicates that the absolute increased rate in the estimated flood discharge is about four times less than that of the flood recurrence interval. The findings of the present study support local authorities in the Gin catchment in developing their flood plain management plans.

Keywords Baddegama · Chi-square test · Flood discharge · Gumbel distribution · Pearson type III distribution · Recurrence interval

24.1 Introduction

Flooding is one of the severe natural disasters in Sri Lanka which causes loss of life and significant damage to infrastructure and property. During the southwest monsoon, floods are commonly triggered, across southern region of Sri Lanka. In this region,

T. N. Wickramaarachchi (✉)
Department of Civil and Environmental Engineering, Faculty of Engineering, University of Ruhuna, Hapugala, Galle, Sri Lanka
e-mail: thushara@cee.ruh.ac.lk

© Springer Nature Switzerland AG 2021
M. Babel et al. (eds.), *Water Security in Asia*, Springer Water,
https://doi.org/10.1007/978-3-319-54612-4_24

unplanned patterns of human settlement, development activities and land use change have resulted in severe encroachments into floodplains (DMC 2005). Located in the southern region, Gin catchment comprises of extensive floodplains, which are frequently affected by floods from the onset to the end of the monsoon. There is a significant increase in built-upland in the lower part of the Gin catchment (Wickramaarachchi et al. 2015); paddyfields which serve as flood detention areas in suburbs, are being converted to commercial and residential lands. Owing to consequences of continuous loss of pervious surfaces, the area is becoming more prone to flooding.

Knowledge of flood frequency analysis is essential for the estimates of flood magnitudes at various return periods and thereby for flood damages mitigation. The flood frequency analysis which relates the magnitude of extreme events to their frequency of occurrence, is commonly carried out by using the probability distributions (Chow et al. 1988). Pearson Type III and Gumbel extreme value probability distributions are the statistic tools that have been widely applied for flood estimations (Abdul Karim and Chowdhury 1995; Al-Mashidani et al. 1978; Canfield et al. 1980; Kamal et al. 2016; Millington et al. 2011).

This study aims to investigate the flood frequency for the Baddegama area in the lower Gin catchment by using the Pearson Type III and Gumbel probability distributions. The Chi-square test was performed to evaluate the accuracy of the proposed techniques. The study provides insights into the probability of flood occurrence in the study area that support local authorities in dealing with flood risk mitigation and management.

24.1.1   Study Area Description

Gin catchment located in the southern region of Sri Lanka covers about 932 km^2 area and entirely lies within the wet zone of the country. It is geographically located within longitude 80°08' E to 80°40' E and latitude 6°04' N to 6°30' N encompassing Galle, Matara, Kalutara and Rathnapura Districts. About 600,000 people are living in the catchment, and majority is concentrated in the middle and lower reaches of the catchment.

Gin River originates from the Gongala mountains in Deniyaya having an elevation of over 1300 m and flows to the Indian Ocean at Gintota in Galle district. The rainfall pattern in the catchment is of bimodal. The south west monsoon lasts from May to September, and the north east monsoon prevails from November to February. Inter monsoon rain occurs during the remaining months of the year. Rainfall varies with altitude with mean annual rainfall ranges from above 3500 mm in the upper reaches to less than 2500 mm in the lower reaches of the catchment.

The Gin catchment consists of two distinct parts, i.e., the mountainous areas located in upper reaches of the Gin River, which are mostly covered with natural rain forest, and the floodplain downstream. As shown in Fig. 24.1, Baddegama falls within the extensive floodplain of the Gin River. Land use in Baddegama area primarily consists of human settlements and agriculture. Here the terrain is relatively flat; the average slope is less than 2% (Fig. 24.2).

24.2 Materials and Methods

24.2.1 Daily Discharge Data

The daily river discharge data observed at the Baddegama gauging station (6°11'23" N, 80°11'53" E) during period of 1978–2010 were used to perform the flood frequency analysis. These flow data were provided by the Department of Irrigation, Sri Lanka. The annual maximum discharge data at the Baddegama varied from 158 m³/s to 596 m³/s. Baddegama gauging station is having an upstream catchment area of 780 km².

24.2.2 Probability Distributions

Probability distribution helps to relate the magnitude of extreme events with their number of occurrences, so that the probability of occurrence of the event can be predicted successfully (Patra 2002). The probability distributions used in this study

Fig. 24.1 Gin catchment and location of Baddegama discharge gauging station

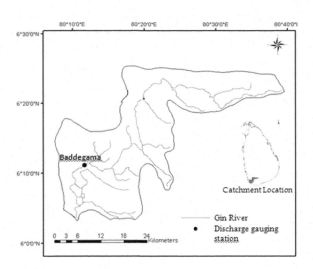

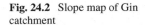 **Fig. 24.2** Slope map of Gin catchment

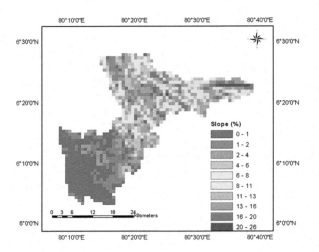

include Pearson Type III and Gumbel. Fitting method was based on parameter estimation technique of Method of Moments (MOM). Parameters of the distributions were estimated based on data given by the MOM.

24.2.3 Pearson Type III Distribution

Pearson Type III probability distribution is a three-parameter gamma distribution, which was first applied by Foster (1924) to describe the probability distribution of annual maximum flood peaks. In Pearson Type III, three major sample moments, i.e. the mean, the standard deviation, and the coefficient of skewness, are transformed into the three parameters λ, β and $\in$ of the probability distribution. According to Bobee and Robitaille (1977), this is a very flexible distribution assuming a number of different shapes as λ, β and $\in$ which are supposed to vary.

24.2.4 Gumbel Distribution

Gumbel (1941) developed the Extreme Value Type I distribution, which is also known as Gumbel distribution. Gumbel probability distribution is widely used for extreme value analysis of hydrologic and meteorological data like floods, maximum rainfalls and other events (Patra 2002). In the Gumbel probability distribution, x is unbounded and α and u are parameters to be determined.

Table 24.1 summarizes, the probability density function and the range of the variable for each of the probability distribution type.

Table 24.1 Probability distributions for fitting hydrologic data

Distribution	Probability density function and the range of the variable	Equations for parameters
Pearson Type III	$$f(x) = \frac{\lambda^{\beta}(x-\epsilon)^{\beta-1}(e)^{-\lambda(x-\epsilon)}}{\Gamma(\beta)}$$ $$x \geq \epsilon$$	$$\lambda = \frac{s_x}{\sqrt{\beta}}$$ $$\beta = \left[\frac{2}{C_s}\right]^2$$ $$\epsilon = \bar{x} - s_x\sqrt{\beta}$$
Gumbel	$$f(x) = \frac{1}{\alpha}\exp\left[-\frac{x-u}{\alpha} - exp\left(-\frac{x-u}{\alpha}\right)\right]$$ $$-\infty < x < \infty$$	$$\alpha = \frac{\sqrt{6}s_x}{\pi}$$ $$u = \bar{x} - 0.5772\alpha$$

Recurrence interval or return period or frequency (T) is defined as the average interval of time T within which a flood of given magnitude will be equaled or exceeded at least once (Patra 2002). Many probability distributions used in hydrologic frequency analysis use the frequency factor equation (Eq. 1) proposed by Chow et al. (1988).

$$x_T = \bar{x} + K_T s \tag{1}$$

where x_T is the value of the variate x of a random hydrologic series with a recurrence interval T, $\bar{x}$ is the mean of the variate, s is the standard deviation of the variate, and K_T is the frequency factor which depends upon the recurrence interval T and assumed frequency distribution.

24.2.5 Evaluation of Performance of the Fitted Probability Distributions

In this study, the Chi-square test was applied to evaluate the accuracy of the flow computation against observed datasets. According to Patra (2002), Chi-square test for goodness of fit includes a relation between the observed number of occurrences O_i and the expected number of occurrences P_{ei}:

$$\xi_v^2 = \sum_{i=1}^{v} \frac{(O_i - P_{ei})^2}{P_{ei}} \tag{2}$$

ξ_v^2 distribution has v degrees of freedom. The value v is equal to $(N - h - 1)$, where N is the total number of sample data, h is the number of parameters used in filling the proposed distribution.

24.3 Results and Discussion

This study established relationships between flood magnitude and recurrence interval using Pearson Type III and Gumbel probability distributions (Table 24.2). The thirty-three-year (1978–2010) observed discharge data from the Baddegama station were used to estimate flood discharges were estimated with different recurrence intervals of 2, 5, 10, 25, 50 and 100 years. The estimated flood discharge values are given in Table 24.3.

Flood discharges estimated by Pearson Type III and Gumbel probability distributions for the Baddegama station, as given in Table 24.3, were used to develop the flood frequency curves. Figure 24.3 shows the comparison between observed and estimated discharge data for the Baddegama station.

The statistical computation of Pearson Type III indicated that the flood discharge will increase from 300.8 m^3/s at recurrence interval of 2 years to 591.9 m^3/s at recurrence interval of 100 years. Gumbel distribution showed a 2-year and 100-year flood discharge of 297.8 m^3/s and 614.9 m^3/s, respectively. Comparison of two distributions showed that Pearson Type III distribution appears to under predict the estimated flood values for high recurrence intervals whereas Gumbel distribution appears to under predict the estimated flood values for low recurrence intervals.

Flood estimates by the two methods showed large differences particularly at the larger recurrence intervals. For both distributions, the 100-year flood is about two times large as the 2-year flood, highlighting that the increase in the estimated flood discharge is proportionally less than the increase in the recurrence interval.

In order to examine the computation accuracy, the Chi-square test was applied. The computed Chi–square statistics were found to be 2.96 and 1.45 for Pearson Type III and Gumbel distribution, respectively. These values were lower than the critical value of Chi–square statistics at 0.05 significance level (3.84 for Pearson Type III and 5.99 for Gumbel distribution). Therefore, both Pearson Type III and Gumbel

Table 24.2 Relationship between the magnitude of expected flood (y) and the recurrence interval (x) with coefficient of determination (R^2)

	Relationship between the magnitude of flood (y) and the recurrence interval (x)	Coefficient of determination (R^2)
Pearson Type III distribution	$y = 73.354ln(x) + 263.24$	$R^2 = 0.9924$
Gumbel distribution	$y = 80.308ln(x) + 249.09$	$R^2 = 0.9984$

Table 24.3 Estimated flood discharge values (m^3/s) at Baddegama for different recurrence intervals

	Recurrence interval (Years)					
	2	5	10	25	50	100
Pearson Type III distribution	300.8	388.5	442.0	505.3	549.7	591.9
Gumbel distribution	297.8	382.7	438.9	509.9	562.6	614.9

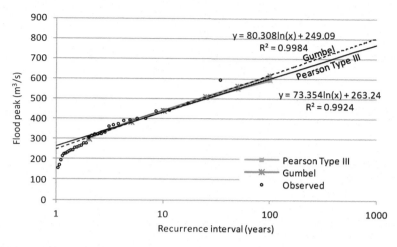

Fig. 24.3 Plot of observed and estimated flood discharges

distributions are acceptable for the estimation of flood discharge at the Baddegama site. Although the computed Chi-square value is appreciably below the critical region in Gumbel distribution, it is difficult to say that Gumbel distribution is superior to Pearson Type III distribution.

The consistency and accuracy of the frequency analysis depend upon the length of data. Generally, a minimum of thirty years of flood record is considered as essential (Subramanya 2000) to obtain satisfactory flood estimate which satisfied in the present study. In accordance with Davie (2008), present study made two assumptions in the flood frequency analysis: each flow event is independent from another flow event in the data set and the hydrological regime has remained static during the complete period of analysis.

According to the Interagency Advisory Committee on Water Data (1982), errors in flood flow measurements are usually random, and the variance introduced is usually small in comparison to the year-to-year variance in flood flows. Therefore, the effects of measurement errors could be neglected in flood frequency analysis.

24.4   Conclusions

Pearson Type III and Gumbel probability distributions were used to estimate flood discharges at the Baddegama station in the lower Gin catchment. The good fit between the observed data and the estimated discharge obtained using the Chi-square method indicated that both Pearson Type III and Gumbel distributions are suitable to estimate flood discharges in the Gin River.

Flood discharge values to be expected in various recurrence intervals presented in this study are important for local authorities in planning and designing of water related infrastructure, managing flood plains and managing water resources systems in the Gin catchment.

This study is an initial phase of a more comprehensive flood frequency analysis for the Gin catchment that is underway.

Acknowledgement Financial support provided by University of Ruhuna, Sri Lanka is gratefully acknowledged.

References

Abdul Karim MD, Chowdhury JU (1995) A comparison of four distributions used in flood frequency analysis in Bangladesh. Hydrol Sci J 40(1):55–66

Al-Mashidani G, Lal BB, Fattah Mujda M (1978) A simple version of Gumbel's method for flood estimation. Hydrol Sci Bull 23(3):373–380

Bobee BB, Robitaille R (1977) The use of Pearson Type 3 and log Pearson Type 3 distributions revisited. Water Resources Res. 13(2):427–443

Canfield, R.V., Olsen, D.R., Hawkins, R.H., Chen, T.L.: Use of extreme value theory in estimating flood peaks from mixed populations. Hydraulics and Hydrology Series, UWRL/H-80/01. College of Engineering, Utah State University (1980)

Chow, V.T., Maidment, D.R., Mays, L.W.: Applied Hydrology, p. 570. McGraw-Hill, New York (2008)

Davie, T.: Fundamentals of Hydrology, 2nd Edition, Routledge Fundamentals of Physical Geography-Taylor and Francis Group, London and New York (2008)

DMC: Towards a Safer Sri Lanka: A Road Map for Disaster Risk Management. Disaster Management Centre, Ministry of Disaster Management), Sri Lanka (Supported by United Nations Development Programme-UNDP) (2005)

Foster HA (1924) Theoritical frequency curves and their application to engineering problems. Trans. Am. Soc. Civil Eng. 87:142–173

Gumbel EJ (1941) The return period of flood flows. Ann. Math. Stat. 12(2):163–190

Interagency Advisory Committee on Water Data (1982) Guidelines for determining flood flow frequency. Hydrology Subcommittee Bulletin #17B, U.S. Department of the Interior Geological Survey, Reston, VA (1982)

Kamal V, Mukherjee S, Singh P, Sen R, Vishwakarma CA, Sajadi P, Asthana H, Rena V (2016) Flood frequency analysis of Ganga river at Haridwar and Garhmukteshwar. Appl. Water Sci. https://doi.org/10.1007/s13201-016-0378-3

Millington, N., Das, S., Simonovic, S.P.: The Comparison of GEV, Log-Pearson Type 3 and Gumbel Distributions in the Upper Thames River Watershed under Global Climate Models. Water Resources Research Report No. 77, Department of Civil and Environmental Engineering, The University of Western Ontario (2011)

Patra KC (2002) Hydrology and Water Resources Engineering, 1st edn. Narosa Publications, New Delhi

Subramanya K (2000) Engineering Hydrology, 2nd edn. Tata McGraw-Hill Publishing Company Limited, New Delhi

Wickramaarachchi TN, Ishidaira H, Magome J (2015) Spatial-statistical approach to evaluate land use change in galle DSD, Sri Lanka. Res. J. Appl. Sci. Eng. Technol. 11(10):1041–1047

Chapter 25
Impact of Climate Change in Rajshahi City Based on Marksim Weather Generator, Temperature Projections

M. Tauhid Ur Rahman, A. Habib, R. Tasnim, and M. Fida Khan

Abstract Climate change is an increasingly important issue affecting natural resources. Rising temperatures, reductions in snow cover, and variability in precipitation depths and intensities are altering the accepted normal approaches for predicting runoff, soil erosion, and chemical losses from upland areas and watersheds. It is, therefore, essential to comprehend the future possible scenario of climate change in terms of global warming. Previous standards for natural resources planning may no longer apply in the near future, if local weather varies greatly from long-term historical records. General Circulation Models (GCMs) operate at large scales to estimate changes in atmospheric conditions (temperatures, wind speeds, precipitation, etc.) due to projected increase in greenhouse gas concentration in the atmosphere. The project described in this paper utilized the MarkSim DSSAT Weather Generator, Google Earth and a Microsoft Excel Spreadsheet. In this study, future impact of climate change on temperature of Rajshahi city in the year of 2030, 2050, and 2080 have been assessed based on four RCP scenarios (2.6, 4.5, 6.0, and 8.5) and ensemble 17 GCM models of a new MarkSim web version for IPCC AR5 data (CMIP5). As a result of multi-model combinations (RCPs and GCM projections) of future average temperature changes in 2030, 2050, and 2080 with respect to present climate condition, it has been observed that the winter months in Rajshahi city might become warmer in future at a faster rate than other months, and the trend of temperature increase might continue on a monthly basis in future.

Keywords Temperature change · Climate model · MarkSim · RCP

M. Tauhid Ur Rahman (✉) · A. Habib
Climate Lab, Military Institute of Science and Technology, Dhaka, Bangladesh

R. Tasnim
The Hongkong University of Science and Technology, Clear Water Bay, Hongkong

M. Tauhid Ur Rahman
Shahjalal University of Science and Technology, Sylhet, Bangladesh

M. Fida Khan
CEGIS, Dhaka, Bangladesh

© Springer Nature Switzerland AG 2021
M. Babel et al. (eds.), *Water Security in Asia*, Springer Water,
https://doi.org/10.1007/978-3-319-54612-4_25

25.1 Introduction

Bangladesh is globally recognized as one of the most vulnerable countries to climate change impacts. It has a history of extreme climatic events claiming millions of lives and destroying past development interventions (DOE 2007). The exposures to different risks get aggravated because of varying high population density, poverty level and concentration of economic activities in different parts of Bangladesh (Ahmed 1999). Though climate change consequences cannot be predicted with complete certainty, but efforts are dedicated to proper risk assessment (intensity and magnitude of natural hazards/disasters) using multiple sophisticated tools/models for more logical understanding of the impact of climate change variables (Ahmed 1999; Islam 2009; CEGIS 2014). Warming of the climate system is now an established fact (IPCC 2007, 2013). It is now evident from observations that, among other things, global average air and ocean temperatures are rising resulting in widespread melting of snow and ice that eventually translates to global sea level rising (IPCC 2013). Assessment of average temperature over Bangladesh has been done using a new version of MarkSim DSSAT weather file generator (Jones and Thornton 2000, 2013; Jones et al. 2011; Moss et al. 2010; Rao et al. 2015) based on over a dozen recent GCMs. This paper discusses the impacts of climate change on temperature of Rajshahi city of Bangladesh using MarkSim DSSAT weather file generator, where the prediction of climate change is grounded on four scenarios - RCP2.6, RCP4.5, RCP6.0, and RCP8.5 (Thornton and Heinke 2009). The objective of the study is to predict the change of temperature of Rajshahi city for different future time slices compared to present conditions by applying MarkSim weather data predicting model so that the severity of the adverse effects of the future temperature change could be assessed.

25.2 Material and Methods

25.2.1 *Four Representative Concentration Pathways (RCPs)*

Four RCPs are produced from IAM scenarios: one high pathway, for which radiative forcing reaches >8.5 W/m^2 by 2100 and continues to rise for some amount of time; two intermediate "stabilization pathways", in which radiative forcing is stabilized at approximately 6 W/m^2 and 4.5 W/m^2 after 2100, and one pathway, where radiative forcing peaks at approximately 3 W/m^2 before 2100 and then declines (Table 25.1). These scenarios include time paths for emissions and concentrations of the full suite of GHGs, aerosols, and chemically active gases as well as land use/land cover.

Table 25.1 Median temperature anomaly over pre-industrial levels (source: adopted from Moss et al. 2010)

Name	Radiative forcing	CO_2 equiv. (ppm)	Temp. anomaly (°C)	Pathway
RCP 8.5	8.5 W/m² in 2100	1370	4.9	Rising
RCP 6.0	6 W/m² post 2100	850	3.0	Stabilization without overshoot
RCP 4.5	4.5 W/m² post 2100	650	2.4	Stabilization without overshoot
RCP 2.6 (RCP3PD)	3 W/m² before 2100, declining to 2.6 Wm² by 2100	490	1.5	Peak and decline

25.2.2 Emissions and Concentrations, Forcing and Temperature Anomalies

Each Representative Concentration Pathway (RCP) defines a specific emissions trajectory and subsequent radiative forcing (a radiative forcing is a measure of the influence that a factor has in altering the balance of incoming and outgoing energy in the earth-atmosphere system, measured in watts per square meter).

25.2.3 Approach and Methodology

General Circulation Models (GCMs) utilizes 3D climate models and it uses columns of atmosphere covering about 200 by 300 km at ground level (Jones and Thornton 2000, 2013; Jones et al. 2011). So they do not simulate the weather on the ground at a particular place very accurately. The mean deviation from the baseline for each atmospheric column (pixel) is re-evaluated to take into account the ground terrain and the characteristics of the expected weather. Under the present study to generate future temperature and precipitation, for different scenarios MarkSim DSSAT weather file generator was used. MarkSim is a third-order Markov rainfall generator that has been developed over 20 years. It downscales GCM weather data by a stochastic process.

MarkSim interface is integrated with Google Earth (Fig. 25.1). There, it is possible to select the area of interest of weather data. Moreover, it is also possible to put the value of latitude and longitude to navigate the location. Under the present study, Rajshahi city of North-west region of Bangladesh was selected. To generate different temperature file, GCM model, RCP scenario, year of simulation, and number of replications were selected in the MarkSim interface.

In the present study, future weather file was generated with RCP2.6, RCP4.5, RCP6.0, RCP8.5 scenario or base period, 2030, 2050, and 2080 using ensemble 17

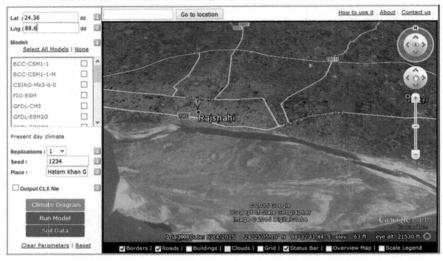

Fig. 25.1 Location of Rajshahi City in Google Earth

GCM models BCC-CSM 1.1, BCC-CSM 1.1, CSIRO-Mk3.6.0, FIO-ESM, GFDL-CM3, GFDL ESM2G, GFDL-ESM2M, GISS-E2-H, GISS-E2-R, IPSL-CM5A-LR, HadGEM2-ES, IPSL-CM5A-MR, MIROC-ESM, MIROC-ESMCHEM, MIROC5, MRI-CGCM3, NorESM1-M. Finally, the 1-D model was run with changing boundaries to see the future impact of climate change on temperature of Rajshahi City.

25.3 Results and Discussion

Under climate change scenario RCP2.6, from the projected graph (Fig. 25.2), it is evident that maximum temperature will rise above 38 °C in May. January was predicted to experience the maximum temperature rise among all months of almost 1.5 °C, and the temperature would rise up to 24 °C by 2080.

The analysis also predicted that in every month from January to December maximum temperature will increase at least 1 °C until 2080 under the emissions of 2.6 W/m^2.

Under climate change scenario RCP4.5, from the projected graph (Fig. 25.3), apparently the maximum temperature will rise above 39 °C in May in 2080. In March and April, the maximum temperature will increase almost 3 °C, which is more than all other months. Graph also shows that in every month from January to

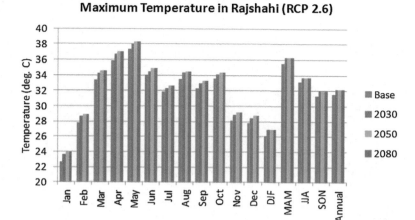

Fig. 25.2 Maximum temperature projections of present and future of Rajshahi under RCP2.6 scenario

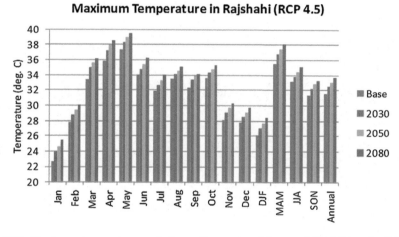

Fig. 25.3 Maximum temperature projections of present and future of Rajshahi under RCP4.5 scenario

December maximum temperature will increase at least 2 °C until 2080 under the emissions of 4.5 W/m^2.

Under climate change scenario RCP6.0, from the projected graph (Fig. 25.4), it is evident that the maximum temperature will rise above 39 °C in May in 2080. In January, the maximum temperature was again predicted to have a highest rise of almost 2.5 °C. Maximum January temperature can go up to 25 °C by 2080. The study also concluded that in every month from January to December maximum temperature will increase at least 2 °C until 2080.

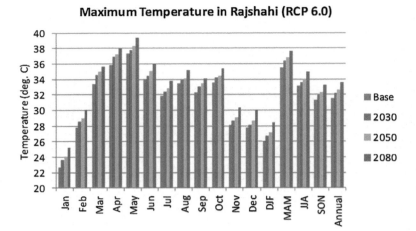

Fig. 25.4 Maximum temperature projections of present and future at the emissions of 6.0 W/m^2 of Rajshahi under RCP6.0 scenario

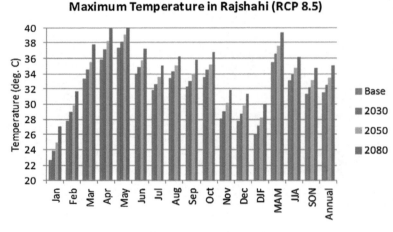

Fig. 25.5 Maximum temperature projections of present and future of Rajshahi under RCP8.5 scenario

Under climate change scenario RCP8.5, from the projected graph, it was evident that the maximum temperature will rise up to 40 °C in April and May in 2080. In January, March, and April a highest rise in maximum temperature was projected as 4 °C. The graph clearly shows that in every month from January to December maximum temperature should increase at least 2.5 °C until 2080 under the emissions of 8.5 W/m^2 (Fig. 25.5).

Under climate change scenario RCP8.5, from the projected graph (Fig. 25.6), it can be observed that the minimum temperature will rise up to 30 °C in August 2080.

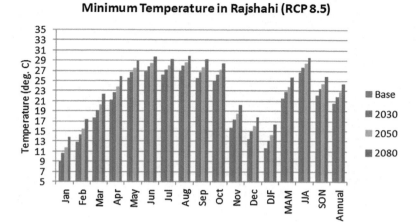

Fig. 25.6 Minimum temperature projections of present and future of Rajshahi under RCP8.5 scenario

In the graph (Fig. 25.6), it is also clearly shown that in every month from January to December minimum temperature will increase at least 2.5 °C until 2080 under the emissions of 8.5 W/m^2.

The results listed in the Tables 25.2 and 25.3 indicate that a significant change in temperature will likely be happened by the end of 2080. Specifically, the winter season will be warmer since the maximum temperature will increase in the range of 0.8 °C to 4 °C, and the minimum temperature will increase in the range of 1.45 °C to 4.5 °C.

25.4 Conclusions

In this study, MarkSim online software, Google Earth, and Microsoft Excel software were applied to generate and to predict the temperature data of Rajshahi division. The paper's outcome emphasizes climate change assessment that the temperature will gradually increase in the coming years significantly. It was predicted that in every month from January to December maximum temperature will increase at least 1 °C until 2080 under the emissions of 2.6 W/m^2. The month of January was predicted to experience the maximum temperature rise among all months of almost 1.5 °C, and the temperature would rise up to 24 °C by 2080. This trend poses a great threat in the form of already evident global warming. Temperature increase shown in the projections in different seasons especially in winter season is very alarming for the geographic, social and economic conditions for the human, animal, and agricultural productivity of the Rajshahi division and entire Bangladesh. Being a climate hit country, the North West part of Bangladesh will suffer much from the increased temperature.

Table 25.2 Projected changes in maximum temperature at 2030, 2050, and 2080 under different climate change scenarios

Increased maximum temperature

Scenarios	RCP2.6			RCP4.5			RCP6.0			RCP8.5		
*Seasons	2030	2050	2080	2030	2050	2080	2030	2050	2080	2030	2050	2080
DJF	0.847	0.847	0.847	0.986	1.656	2.353	0.706	1.135	2.359	1.13	2.13	3.94
MAM	0.769	0.769	0.769	1.294	1.975	2.549	0.888	1.319	2.15	1.08	2.06	3.82
JJA	0.549	0.549	0.549	0.7	1.259	1.959	0.441	0.889	1.865	0.76	1.64	3.06
SON	0.657	0.657	0.657	0.907	1.466	1.899	0.662	0.973	1.967	0.91	1.82	3.48
Annual	0.706	0.706	0.706	0.972	1.589	2.19	0.674	1.079	2.085	0.97	1.91	3.57

*[Seasons = DJF = December–January–February (Winter), MAM = March–April–May (Pre-monsoon), JJA = June–July–August (Monsoon), SON = September–October–November (Post-Monsoon)]

Table 25.3 Projected changes in minimum temperature at 2030, 2050, and 2080 under different climate change scenarios

Increased minimum temperature

Scenarios	RCP2.6			RCP4.5			RCP6.0			RCP8.5		
*Seasons	2030	2050	2080	2030	2050	2080	2030	2050	2080	2030	2050	2080
DJF	1.099	1.411	1.45	1.255	1.939	2.599	1.093	1.589	2.74	1.446	2.589	4.55
MAM	1.024	1.309	1.347	1.459	2.063	2.532	1.189	1.662	2.46	1.332	2.324	4.23
JJA	0.795	1.010	1.055	0.896	1.332	1.741	0.702	1.110	1.913	0.957	1.709	2.95
SON	1.033	1.265	1.296	1.098	1.696	2.224	0.91	1.348	2.368	1.283	2.238	3.77
Annual	0.988	1.249	1.28	1.177	1.758	2.274	0.97	1.427	2.371	1.255	2.215	3.87

*[Seasons = DJF = December–January–February (Winter), MAM = March–April–May (Pre-monsoon), JJA = June–July–August (Monsoon), SON = September–October–November (Post-Monsoon)]

To combat the future adverse environment, adequate adaptation measures should be adopted from now on. Building proper awareness in protecting the environment should also be started.

Acknowledgements Financial support received from HEQEP (CP No 3143), UGC and World Bank to carry out this research are sincerely acknowledged. Contribution of the Climate Change Division of Center for Environmental and Geographic Information Services (CEGIS) are gratefully acknowledged here for providing necessary data support regarding for this project.

References

Ahmed AU, Alam M (1999) Development of climate change scenarios with general circulation models. In: Huq S, Karim Z, Asaduzzaman M, Mahtab F (eds) Vulnerability and adaptation to climate change for Bangladesh Dordrecht. Kluwer Academic Publishers, The Netherlands, pp 13–20

CEGIS (2014) Development of national and sub-national climate change model for long term water resources assessment. Final Rep 1:1–12

DoE (2007) Climate change and Bangladesh. Climate Change Cell, Department of Environment, Government of the People's Republic of Bangladesh, Dhaka. Final Report

Islam MN (2009) Analysis of land surface temperature for temporal and spatial variability model for climate change: a case study of Bangladesh. In: Rahman MH, Badruzzaman ABM, Alam MJB, Rahman MM, Ali MA, Noor MA, Wadud Z (eds) Climate change impacts and adaptation strategies for Bangladesh. International Training Network (ITN) Centre, BUET, Dhaka

IPCC (2007) Climate change: impacts, adaptation and vulnerability. Summary for Policy Makers

IPCC (2013) The physical science basis. Summary for policy makers. Contribution of working group I to the fifth assessment. Report of the intergovernmental panel on Climate Change. IPCC Secretariat, WMO, Geneva, Switzerland, vol 3

Jones PG, Thornton PK (2000) MarkSim: software to generate daily weather data for Latin America and Africa. Agron. J. 93:445–453

Jones PG, Thornton PK, Giron E (2011) Web application. MarkSim GCM-A weather simulator. https://gismap.ciat.cgiar.org/MarkSimGCM

Jones PG, Thornton PK (2013) Generating downscaled weather data from a suite of climate models for agricultural modelling applications. Agric. Syst. 114:1–5

Moss RH et al (2010) The next generation of scenarios for climate change research and assessment. Nature. https://doi.org/10.1038/nature08823

Thornton PK, Heinke J (2009) Generating characteristic daily weather data using downscaled climate model data. IPCC fourth assessment

Rao MS et al (2015) Model and scenario variations in predicted number of generations of Spodoptera litura Fab. On peanut during future climate change scenario. PLoS One **10**(2). https://doi.org/10.1371/journal.pone.0116762

Part V
Urban Water Security

Chapter 26
Water Supply of Dhaka City: Present Context and Future Scenarios

Mohammed Abdul Baten, Kazi Sunzida Lisa,
and Ahmed Shahnewaz Chowdhury

Abstract Dhaka, the capital of Bangladesh, has been expanding both horizontally and vertically since last three decades with an estimated annual growth of six percent. Over the past several decades, concerns have been raised regarding the amount of water used in Dhaka as with higher rates of personal water use, which mostly collected from ground water resulting in declined aquifers across the area. This study investigated household consumption and management behaviour, more specifically household water management of different income groups of Dhaka city. Based on the present water demand and supply data of Dhaka city, this study attempted to build three scenarios for the year 2050. The study finds that despite Dhaka Water and Sewerage Authority's (DWASA) existing capacity of supplying required water to its population, quality-water scarcity has been increasing at household level mostly due to management inefficiency. Under a business-as-usual scenario, the situation signals for a worrying future where 12.37 million people may be deprived of basic water requirement by 2050. In addition to existing inequality, unplanned rapid urbanization and over extraction of groundwater make the inhabitants more vulnerable to access to safe water. This research predicts that unless substantive actions are taken to increase surface water supply as well as reducing Unaccounted for Water (UfW) in response to increased demand, a considerable number of Dhaka's population will be deprived of access to minimum water with increased inequality even under ambitious future roadmap .

Keywords Access to water · Dhaka city · DWASA · Groundwater depletion · Water demand and supply

M. A. Baten (✉) · K. S. Lisa · A. S. Chowdhury
Independent University (IUB), Dhaka, Bangladesh
e-mail: a.baten@iub.edu.bd

M. A. Baten
University of Maine, Orono, USA

© Springer Nature Switzerland AG 2021
M. Babel et al. (eds.), *Water Security in Asia*, Springer Water,
https://doi.org/10.1007/978-3-319-54612-4_26

26.1 Introduction

Urban public services provision has been struggling to keep pace with population growth in developing countries, where access to safe water is a debateable issue (Bakker 2010). An estimate suggests that globally 142 million urban dwellers are lacking improved drinking water sources (UN-Habitat 2008). However, the situation is worse in low-income urban areas due to their poor infrastructure and inefficient management. Kjellén and McGranahan (1997) predict that two-thirds of the world's population will experience water stress condition by 2025 and some countries may experience high water stress condition where water withdrawal against available resources will exceed 40%. The alarming news is that the withdrawal rate against available resources is 48% in South Asia (Ariyabandu 1999). Bangladesh, being a riverine country, confronts dual challenges from water: firstly, unlimited flood water during wet season, and secondly, increasing scarcity during dry season. Bangladesh also faces tremendous challenges in providing safe and adequate water to its large population, most importantly its' rapidly growing urban population.

Water is a cross cutting resource due to its social and ecological importance. But in many cases, water resource experiences mismatch in management, where the scale, at which decisions are made, does not match the scale, at which it functions (Cash et al. 2006). The urban water management involves complicated and systematic process that includes planning, research, design, engineering, regulation, and administration. Mismatches among these processes may give rise to inadequate water supply. Such situation creates spaces for private sector to intervene in the water supply, either in the form of large corporation or small vendor based on micro treatment plants, whose overwhelming focus on profit may exert additional price pressure to the consumers (Bakker 2010). Those, who cannot afford costly private water supply or out of public water supply network coverage, are often found collecting water through illegal tapping or compelled to use polluted water from urban lake or other common water bodies (Singh 2004).

Water right, inherently linked to other fundamental human rights, is not merely a right issue rather it is a highly political and policy issue (Watkins 2006). The constitution of Bangladesh reserves the rights of its citizen and acknowledges country's obligation to provide the basic necessities of life such as food, clothing, shelter, education and medical care [Constitution B (1972) Article 15, clause (a)]. It seems that the water right is considered under food, but no article or clause deduces the water right issue conspicuously.

In Bangladesh, the National Policy for Safe Water Supply and Sanitation (GoB 1998) declares state ownership to water. Even though the policy acknowledges that access to safe water is essential for socio-economic development of the country, but no special provision was made to ensure citizen's right to water. Rather, article 4.3 (paragraph f) addresses water as an 'economic good' by keeping provision of conferring water right to private or community bodies to provide secure, defensible and enforceable ownership/usufructuary rights for attracting private investment. Moreover, the Bangladesh Water Act (2013) in its preamble defines 'right to water' as acquired access and use rights. The act does not acknowledge 'citizen right to water' and state's obligation to supply safe water; rather it encourages privatization of water through a general authorization or license. These provisions defy the spirit of constitution's deceleration on rights to the basic necessities. On the other hand, the Government of Bangladesh (GoB 2011) kept provision of developing and to managing water resources efficiently in its Sixth Five Year Plan (2011–2015), but 'right to water' issue remained overlooked there as well.

Since Independence in 1971, Bangladesh has been maintaining an upward economic growth with non-linear GDP figures. Economic opportunities attract more people in rural areas to urban growth centres and, therefore, country's urban population is growing on an average six percent per annum (Islam 2015). Rasheed (2000) predicted that the urban population growth would continue to be higher than the rural population in this decade, although it might decrease somewhat from the present level. Based on current growth trend, Rasheed (2000) estimated that in the next decade the relative proportions of the urban population would be 47% by 2020, and 53% by 2025 in Bangladesh. Dhaka, capital of Bangladesh, has been experiencing a tremendous growth in last few decades both horizontally and vertically, but the earlier rate has slowed down in current decade, albeit maintaining an annual rate of 3.5% without following any systematic plan (Islam et al. 2010). The percentage of urban population in Dhaka district was 21.9% in 1981, which has reached to 27.8% in 2011 (BBS 2011).

Despite being recognized as a megacity based on population number, Dhaka's urban basic services provision is well behind the required standard (Moinuddin 2013). Specially, in the last five/six years the situation has reached an alarming state due to unregulated urban sprawl that dispenses immense pressure on the infrastructures of the city resulting in deprivation of basic amenities to a large population, where water supply has appeared as the most critical service. Nishat et al. (2008) estimate that 25% population of Dhaka city has no direct access to potable water.

Dhaka Metropolitan Area (DMA) constitutes 27% (360 km^2) of Dhaka Statistical Metropolitan Area (DSMA), which covers an area of 1353 km^2. Until 1989, Dhaka Water Supply and Sewerage Authority's (DWASA) operation was limited to DMA, but since 1990 it has extended operating area to adjacent Narayanganj metropolitan area (DWASA 2011). Even though this paper will focus on DMA, but it would consider demand and supply of DWASA as a whole. Based on the DWASA demand and supply data, this research has empirically examined Dhaka city's (Fig. 26.1) present water demand and supply context, built future scenarios both at spatial and temporal scale, and finally identified challenges that need to be addressed for an efficient water supply system.

26.1.1 Dhaka City's Water Supply System

History of piped water supply in Dhaka city is dated back in 1874, when a small water treatment plant was built on the bank of the Buriganga River to provide water to some rich households of old Dhaka through roadside hydrants (DWASA 1991).

Fig. 26.1 Map of Dhaka
Water and Sewerage
Authority (DWASA)
operation area and survey
area

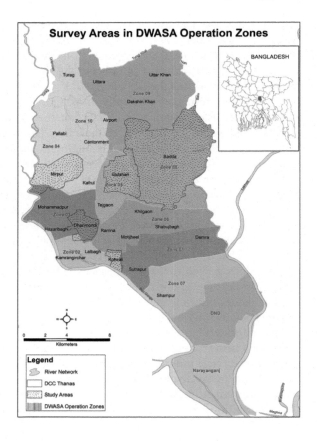

Before establishing DWASA in 1963, the responsibility for supplying water to Dhaka city residents was laid to Dhaka Municipal Corporation (DMC). Later DWASA was created in 1963 under the Ordinance No 19 to provide water and to manage sewerage of Dhaka city. On the other hand, DMC was renamed as Dhaka City Corporation (DCC) in 1990, which has been mandated to regulate private sources of water supply. However, the DCC became inactive in water, sewerage and drainage services after establishment of DWASA, although DCC is yet shown active performing these functions in the stature book (Siddiqui et al. 2000).

The DWASA ordinance was amended in 1996, and a vision plan was prepared to ensure 100% water supply coverage by 2005 and 80% sanitation coverage by 2020 (Haq 2006). At present, both Dhaka city and Narayanganj city are under the service coverage of the DWASA, and they have a plan to further extend services to adjacent suburb of Gazipur and Savar in the near future (Siddiqui et al. 2000). For management purposes, the DWASA service area is divided into 8 zones, of which 7 are located in Dhaka city and only one is in Narayanganj city (Siddiqui et al. 2000).

Since establishment, the DWASA has expanded both in coverage area and water supply capacity, yet its' standard of service is under critical scrutiny. World Bank (1999) conducted a survey and found that over the years, water supply situation in Dhaka did not improve much compared to the minimum requirement of the city dwellers. Although DWASA estimates water demand considering 150 L per Person per Day (lpd), but about one-third of the total city dwellers even do not get minimum water requirement of 100 lpd as set by WHO (2010). During dry season, when groundwater level goes down and water demand increases, water supply situation in some parts of the city becomes unbearable. Varis et al. (2006) estimated that DWASA could supply 0.51 km^3 of water per year against the demand of 0.73 km^3, serving around 72% of the city dwellers. The quality of the supplied water is very much in question. More than 1,300 private wells abstract another 0.40 km^3 of groundwater annually, mainly for industrial purposes (DWASA 2014).

Like many other urban areas of developing countries, getting water is becoming more difficult and often more expensive for the poorest people in Dhaka city. Women and children of poor households in urban settings spend hours - in extreme cases up to six to eight hours - each day to collect water from different sources (World Bank 2013). In most cases, the poor do not have piped water supply in their property. Instead, they have to buy or take water from other sources. People, who buy water from other sources, may have to pay three to ten times higher than the price of DWASA supplied piped water (World Bank 2013). In the slums of Dhaka city, the average user to water-point ratio is 1,000:1 and only 20% of the people have some form of sanitary latrine (Ahmed 2004). Lack of sanitation, long queuing times for water collection, and unhygienic surroundings are the appalling concerns in the slums of Dhaka city. Other than these squatters, permanent residents of Dhaka city have also been experiencing a year-round water scarcity, more acutely in dry season (Khan 2013). In response to water scarcity, many affluent residents and small water-vendors have installed private deep tube wells, and pump groundwater illegally that results in lowering the groundwater table further. Water quality is another major concern for dwellers of Dhaka city. Even though DWASA claims to maintain the quality of water according to WHO requirements, but the consumers are apprehensive regarding the quality. Consequently, the urban citizens seldom drink untreated piped water, rather boil the supplied water to kill dangerous bacteria making it potable.

Over the years, surface water sources have been considered as unreliable for drinking water that requires more capital investment to serve the citizens' demand. Accordingly, emphasis was given to groundwater extraction (Crow and Sultana 2002). Water supply in Bangladesh is overwhelmingly dependant on groundwater. In rural areas, more than 90% of the population extracts groundwater to meet drinking water demands (World Bank 2008), whereas 87.60% of the Dhaka city's water supply is coming from groundwater sources (DWASA 2014). Moreover, rapid urbanization and resultant increased built area has been contibuting to the increased surface run-off retarding infiltration. Thus, groundwater recharge rate is decreasing day by day. Due to unregulated abstraction and lower recharge of the aquifers, the water table in Dhaka has been lowering down at the rate of more than two meter per year (Akther et al. 2010; UNEP 2005).

26.2 Materials and Methods

This study has examined performance of the DWASA for the present and future. Data on water demand and supply were obtained from DWASA annual report 2012–13, 2013–14 and DWASA Monthly Information Report (MIR) June 2014. To gain understanding on the dynamics and management context of water supply at Dhaka city, and to build possible future scenarios, three experts from DWASA, Bangladesh University of Engineering and Technology (BUET) and Dhaka University were interviewed by both face-to-face and telephone calls.

To investigate household water consumption and management behaviour, 300 households were surveyed based on income groups from five different areas of Dhaka city (Fig. 26.1). Study areas were primarily selected using cluster sampling based on the Household Income and Expenditure Survey (HIES 2010) income range and RAJUK[1] (*Rajdhani Unnayan Kortipokkho*) recognized settlement areas based on income groups. For instance, *Gulshan* and *Dhanmondi* areas were selected for high-income residences, *Banasri* and *Wari* (Old City) were selected for middle-income, and *Mirpur* was selected for lower middle income and low-income groups. From each study area, 60 households were then surveyed using systematic random sampling.

Residential water use for the DWASA was calculated using the number of households on public water supply and the average per-household water use per day. Per-household water use is a function of demographic, socioeconomic, or climatic variables (Arbues et al. 2003). DWASA estimates its demand as 150 lpd (DWASA 2014).

Residential water use for the years 2020, 2030, 2040, and 2050 was projected by a linear-predictive model using an estimated model intercept (inelastic demand) and linear coefficients for demographic growth. For water demand projection, the average population growth of Dhaka city was estimated as 3.5% for the years 2010, 2020, 2030, 2040 and 2050.

Total residential water use for the years 2020, 2030, 2040, and 2050 is determined as follows:

$$Qy = (qy * hy)/10^6 \tag{26.1}$$

where,

Q_y is the total residential water use in year (y), in million liter per day;
q_y is the per-household water use in year (y), as determined; and
h_y is the number of persons served by public supply in year (y).

---

[1]RAJUK is a statuary body authorized for Dhaka city's settlement planning and implementation.

26.3   Results and Discussions

26.3.1   Current Water Demand and Supply in Dhaka City

Water supply in Dhaka city is predominantly groundwater based. Although four rivers surround Dhaka city (namely, Buriganga, Balu, Turag and Tongi *Khal*), only 12.40% of supplied water is coming from these rivers (ADB 2004). Surface water sources from surrounding rivers and lakes have already exceeded the standard limits of many water quality parameters because of the discharge of huge amount of untreated and municipal wastes. Treatment of this water has become so expensive that water supply authorities have to depend on groundwater aquifer for drinking water production (Biswas et al. 2010). Other than these four over-polluted rivers, the nearest water body is the Padma River and the Meghna River, which have acceptable water quality and ability to meet the demand; however, those rivers are located within a distance of 17 km and 50 km respectively from Dhaka.

Since 1990, DWASA had been projecting total water demand considering 160 lpd. But from the year 2010, it has started projecting demand as 150 lpd, and accordingly supplies water to the city dwellers. Total water demand in Dhaka city varies from 2,100 to 2,300 Million Liter per Day (MLD) with seasonal variation. Arguably, since 2014 DWASA total production capacity is 2,420 MLD (both groundwater and surface water) (DWASA 2014). Apparently, DWASA is able to fulfil current water demand through their capacity. However, DWASA has never been able to reach its production target. If we account 31.68% unaccounted for water (UfW)[2] or system loss between production and end-user level then end-user water availability would be 1,653.34 MLD. The statistics implies that approximately 25% of the population in Dhaka city are deprived of getting DWASA projected standard water requirement of 150 lpd. To supply water in Dhaka city, DWASA runs 702 Deep Tube Wells (DTWs) and four Surface Water Treatment Plants (SWTPs) (DWASA 2014) (Table 26.1).

---

[2]Unaccounted-for-water, expressed as a percentage, is calculated as the amount of water produced by the DWASA minus the metered customer use divided by the amount of water produced multiplied by 100. They collect water from the mainline without permission or in contract with corrupted person of the supply authority. During the financial year 2002–2003, almost 52% system loss was accounted that indicates very poor management scenario. The situation has been fluctuating but never came to an admissible limit. Present Government has taken a few initiatives to lessen system loss in water supply. With this and by the increasing effort of the DWASA officials, the system loss is gradually decreasing. During the FY 2009–2010, the system loss was accounted for 35.7% while in FY 2010–2011 the UfW goes down to 31.7%. It is an indication that if Government and authority show their willingness to improve management, the situation becomes well off. In addition, awareness campaign regarding misuse of water also contributes to the lowering down of the system loss (DWASA 2011).

Table 26.1 Current water supply information of DWASA (*Source* DWASA 2014)

	2010–2011	2011–2012	2012–2013	2013–2014	2014–2015
Deep tube well	586	615	644	672	702
Water treatment plant	4	4	4	4	4
Water production/day (MLD)	2,150	2,180	2,420	2,420	2,420 (+)
Water demand (MLD)	2,180	2,240	2,250	2,280	2,300
Ground water (MLD)					2,120
Surface water (MLD)					300
Water line (km)	2,800	3,040	3,040	3,040	3,461.56
Public hydrant (active)	38	38	38	38	38
Roadside tap	1,643	1,643	1,643	1,643	1,643
Connection to religious organizations	1,898	1,898	1,898	1,898	1,898
Residential consumer	–	–	–	–	89.59%
Total consumer	–	–	–	–	3,50,772 connections

26.3.2   Future Water Demand and Supply Scenarios of Dhaka City

Providing safe and adequate water to city dwellers are challenges to the DWASA. Considering population growth and expansion of the city areas, DAWSA plans to adjust its infrastructure and management practices accordingly. Rasheed (2000) projected that Dhaka city's water demand would reach to 3,200 MLD by the year 2025 (the author estimated water demand as 160 lpd). The author warned that unless undertaking more comprehensive and long-term schemes by the DWASA, the city would experience greater shortage of domestic water supply by 2025 compared to the present condition.

In this study, based on the existing water supply data and DWASA (2014) projected future road map, three scenarios on future water demand and supply status in Dhaka city are constructed for the years 2020, 2030, 2040 and 2050. For these projections, the average population growth of Dhaka city is estimated as 3.5% and water demand is calculated as 150 lpd. The first scenario is based on the current production status of DWASA (2014) under a business-as-usual situation; but second and third scenario considered UfW and Downtime Loss or Machine Loss (Cs) anchored on the ambitious future road map.

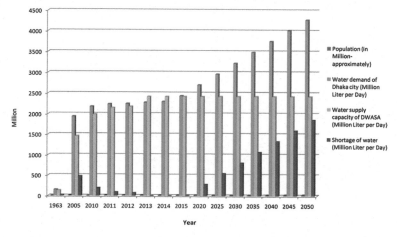

Fig. 26.2 Demand and supply of water supplied by DWASA in 2050 (*Source* DWASA 2014)

26.3.3   Scenario 1 (Obscure)

Based on demand and supply data as collected from the official document of DWASA (2014), considering 2014 as *status quo* in water production, a projection is constructed, which illustrates that by the year 2050 the demand and supply gap would be 1,855 MLD. The supply gap implies that 12.37 million people of Dhaka city may suffer from acute water crisis in 2050, unless the production increases comprehensively (Fig. 26.2).

26.3.4   Scenario 2 (Enigma)

DWASA has undertaken a number of initiatives to address increasing trend of present and future water demand. Many of those initiatives are still at the planning stage. The second phase of the Saidabad Surface Water Treatment Plant (SWTP) came under operation in 2012 with a supplying capacity of 225 MLD along with previous First Phase of 225 MLD. Moreover, several other projects are under consideration to lessen the dependency on groundwater and enhance the surface water extraction.

The current study has made another projection for 2050 based on DWASA's future roadmap. Taking into account the UfW or system loss as 30% from production to end-user level, the projection shows that despite huge investment plan in water production, DWASA will not be able to meet future demands without reducing UfW to 10% or less from the current percentage of 31.68% (Table 26.2).

Table 26.2 Water supply scenario based on future water production road map

Year	2012	2015	2020	2025	2030	2035	2040	2045	2050
Total user-end demand (MLD)	2,250	2,437.5	2,700	2,962.5	3,225	3,487.5	3,750	4,012.5	4,275
UfW (%)	30%	30%	30%	30%	30%	30%	30%	30%	30%
Ground water (MLD)	1,880	2,120	2,120	2,120	2,120	2,120	2,120	2,120	2,120
Sayedabad SWTP I (MLD)	225	225	225	225	225	225	225	225	225
Sayedabad SWTP II (MLD)	225	225	225	225	225	225	225	225	225
SWTP III (Khilkhet) (MLD)			500	500	500	500	500	500	500
SWTP IV (Padma) (MLD)			500	500	500	500	500	500	500
SWTP V (Sayedabad) (MLD)				500	500	500	500	500	500
SWTP VI (MLD)					500	500	500	500	500
SWTP VII (MLD)					500	500	500	500	500
Total production capacity (MLD)	2,330	2,570	3,570	4,070	5,070	5,070	5,070	5,070	5,070
Total user-end gap (30% UfW)	699	771	1,071	1,221	1,521	1,521	1,521	1,521	1,521
Shortage (Total production-(30% UfW + Total user-end demand) MLD	-619	-638.5	-201	-113.5	324	61.5	-201	-463.5	-726

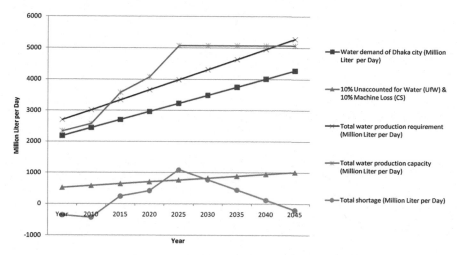

Fig. 26.3 Water demand and supply scenario considering 10% production loss (Cs) and 10% Unaccounted for Water (UfW)

26.3.5   Scenario 3 (Utopia)

The third scenario is a combination of ambition and reality. This scenario is based upon DWASA's future production roadmap and perceives that DWASA will have improved their management system so that they would have been able to reduce UfW up to 10% from current 31.68%. Nevertheless, the pump cannot operate to its full capacity due to power shortage, depleted ground water table, and pump-overhauling time. The scenario, therefore, considered 10% machine loss (Cs). The groundwater table is declining at a rate of 2.81 m/yr (UNEP 2005), the upper Dupi Tila aquifer has reached its limit, and it is likely that 223 numbers of DTW will need to be replaced as the existing ones would go out of service by 2013 because of depletion of groundwater table (DWASA 2011). Under this scenario, it is assumed that the DWASA would be able to fulfil Dhaka city's water demand by 2018. Moreover, DWASA supply would supersede the demand in 2020 through maintaining increased trend of production status and improved management system. However, unless new treatment plants are coming into operation in response to increased demand starting at 2045, DWASA will experience 209.63 MLD supply deficit by 2050 even in this highly ambitious scenario (Fig. 26.3).

26.3.6   Challenges of Water Supply in Dhaka City

Taking into account DWASA's incapacity of supplying required water, sewerage and drainage facilities to the growing population, Siddiqui et al. (2000) identified two incongruities. First, they showed their reservation in extending service area of DWASA without improving existing service quality. Second, they have noticed that DWASA has a strong preference to the higher income groups than lower income groups in case of service delivery. Their study found that about 98 to 100% of the households in *Banani, Baridhara, Gulshan* and *Dhanmondi* residential areas have water and sewerage connections, whereas only 50 to 75% households have these facilities in lower middle-income areas such as *Rayer Bazar, Bashabo, Jurain* and *Mirpur*. However, this current study has found a slightly different scenario for water connection. Almost 100% of the household in *Mirpur, Banashri* and Old Dhaka, considered as middle and low income residence, are connected to DWASA. But, in case of residence of *Gulshan* and *Dhanmondi*, the percentage is 67 and 87, respectively, though DAWSA has pipeline in those places (Table 26.3). This is because of DWASA's poor service quality that provokes high income residents of *Gulshan* and *Dhanmondi* areas to install private DTWs, and develop their own supply system. This study also finds that almost 90% of the people, who are dependent on DWASA as a primary source of household water, take any of the precautionary measures, either boiling or using filter or combination of both, to make the supplied water drinkable. It implies that most of the city dwellers do not rely on DAWSA water for drinking, but in *Gulshan* and *Dhanmondi* areas one third of the respondents were found drinking water directly collecting from their own source other than DWASA's supply system (see Table 26.3 for full results). World Bank (1996) found that only 42% have access to reasonably safe drinking water sources in urban areas of Bangladesh, and the rest others have to depend on contaminated sources.

Challenges regarding access to water vary from area to area and with the changing socio-economic status. Social wellbeing, economic development, and environmental quality are dependent on water resources, and a little change in water supply can largely affect the development process (Forslund et al. 2009). Siddiqui et al. (2000) surveyed over 3000 slums and squatter settlements, and found that access to the DWASA service was restricted to less than 2% of the population. The same study also interviewed 175 journalists on their experience of services provided by the DWASA including DMP, RAJUK, and DESA. The study found that almost 80% of the respondents were dissatisfied with the performance of the DWASA in terms of transparency, accountability, corruption, and people's participation. In an interview, DWASA's the then Managing Director also admitted that corruption was rampant in DAWSA, and sometime they fall short to reach the standard quality in water supply (The Daily Janaknatha July 7 1997).

Haq (2006), Khan and Siddique (2000), Rasheed (2000), and Siddiqui et al. (2000) have identified serious lack of coordination among DWASA, RAJUK, Titas Gas, DCC, and DESA, who are responsible for providing different services to Dhaka city dwellers. Such mismatches result in poor services and sufferings for the city dwellers. Haq (2006) and Siddiqui et al. (2000) also noticed DWASA's lack of comprehensive planning on tapping rainwater including enhancing ground water recharging. Both the authors identified system loss and faulty revenue collection as the most important governance challenges. Siddiqui et al. (2000) have noticed 74% revenue collection against the billed amount.

Table 26.3 Household water management behavior of Dhaka city residents of different income groups (expressed as percentage of population)

Study area		High income groups		Middle income groups		Lower middle and low income groups
		Dhanmondi	Gulshan	Rampura-Banashri	Old Dhaka (Wari)	Mirpur
Source of water at household	DWASA (%)	87	67	98	100	100
	Private well (%)	12	33	2	0	0
Primary source of drinking water	Piped water	98	60	95	87	95
	Bottled water	2	2	0	7	0
	Both piped and bottled water	0	12	2	0	2
Drinking water treatment	Filter attached to water tap	0	27	12	2	22
	Filter after collection	8	0	2	0	8
	Boiled water	5	0	15	42	15
	Use purification tablet	0	0	0	2	0
	Use carbon filter	20	0	10	2	0
	Boiled and Filter	32	53	52	47	38
	Direct use of tap water	35	20	10	7	17

The DWASA projects water demand taking into account the water consumption at a rate of 150 lpd, compared to the internationally accepted standard of 110 lpd (DWASA 2011). DWASA produced few scenarios to meet future demands considering water demand as 140 lpd for residential consumers in 2010, with a vision of reducing water demand to 110 lpd by 2025 through increasing water use efficiency (DWASA 2010). On the other hand, for slum dwellers, current water demand has been projected as 35 lpd (DWASA 2010) against basic water demand of 50 lpd[3] (Gleick 1996). However, under a long-term vision, DWASA plans to increase water supply to slum dwellers to 50 lpd by 2050 (DWASA 2011). The current scenario on water access implies that poor people are, mostly living in slum areas, neglected both at the demand and supply side, more deprived of having access to potable water, thus suffering from health hazards.

26.3.7 Groundwater Depletion of Dhaka City

With rapid urbanization, the paved area of Dhaka city has been increasing without following any regulated and structured trend (Aziz et al. 2012) that affects percolation of water. Moreover, paved area increases total surface run-off, which also affects the drainage system (UNEP 2005). In many parts of the city, the main aquifer has changed from a confined to an unconfined condition. Such changes in the hydrodynamic condition can make the aquifer vulnerable to possible groundwater contamination.

Dhaka city's water supply is overwhelmingly dependent on ground water that results in depletion of water table. Akther et al. (2010) estimated that DWASA withdrew 1.6 Million Cubic Meters per Day (MCM/d) of groundwater in 2005 against city's total water demand of 2.1 MCM/d. UNEP (2005) reported a 20 m decline of ground water table between 1997 and 2005 at a rate 2.81 m/yr. However, the depletion rate is uneven and the worst case recorded at the central part of the city due to high population pressure and increased commercial activities. Water table also depletes due to the overlying hard clay and increased urbanization that retards vertical recharge to aquifer. For instance, the groundwater table at *Tejgaon* area, located in the central part of Dhaka city, is about 65.7 m below the ground surface, at *Mirpur* about 55.4 m below the ground surface, but in *Mohammadpur, Dhanmandi* and *Sutrapur* areas, which are close to the river periphery, 20 to 34 m below the ground surface (Zahid et al. 2009).

In addition to the 702 DTWs (average capacity 3,000 L per month) operated by the DWASA, there are more than 1,300 wells of various capacities are under operation by the private Sector (DWASA 2014). Haq (2006), with reference to Sir McDonald and Partners (1991) and BUET (2000) studies, warned that such groundwater abstraction may result in land subsidence in Dhaka city, similar to the incidence that happened in Bangkok and Mexico City. Therefore, the study suggested reducing the contribution of groundwater to maximum 50% in the total water supply. However, the

---

[3] Water requirement (e.g. drinking, removing or diluting waste materials, producing manufactured goods, growing food, producing and using energy) varies with weather, lifestyle, culture, tradition, diet, technology, and wealth. The type of access to water is an important determinant in gross water use where use of water is the commutation of withdrawal (intake), recirculation, and reuse (Gleick 1996). Using a minimum level of 15 lpd for bathing and 10 lpd for cooking, the international organisations and water providers recommended adopting an overall Basic Water Requirement (BWR) of 50 lpd for meeting domestic basic needs, irrespective of climate, technology, and culture (Gleick 1996).

author was aware that building a treatment plant for surface water has a much longer gestation period and high initial cost, which may present challenge to water supply system. Moreover, lack of locally available manpower and equipment along with highly polluted surface water sources of the city (rivers and wetlands around Dhaka city) make it difficult to shift to surface water sources quickly from current higher dependency on ground water.

26.4   Conclusions

Access to safe water is critical to economics as well as to the ecosystems; and a scarcity of safe water can directly affect the long-term prospects for sustainable development (Forslund et al. 2009). Without an adequate water supply, living organisms may die, factories depending on water may have to close temporarily, crop yields may decline, workers may be unproductive, and fisheries may destroy. The present water supply of Dhaka city is heavily dependent on groundwater, which signals for a murky future with acute water crisis. Though the DWASA has already started to shift its present groundwater based production system to surface water production, but the shift demands huge investment and time. Moreover, the status of peripheral rivers of Dhaka city is so degraded and a major portion is under illegal encroachment that it is highly unlikely to fulfil the future demand just only relying on these sources. Considering the present crisis and future demand, it is high time to seek additional sources and improve present management system through reducing system loss and using water efficient appliances.

Acknowledgements This research received a financial grant from Christian Aid under "Addressing Inequality" project implemented by the Unnayan Onneshan and another grant from the Independent University Bangladesh (IUB) to conduct a survey on water management behaviour of residents of Dhaka city. This paper is a part of a larger report titled 'Water Supply of Dhaka City: Murky Future'. A part of the report titled "Diagnosing water supply of Dhaka city: present and future" paper was presented in the IWA World Congress on Water, Climate and Energy 2012, 13–18 May - Dublin, Ireland. However, this version of paper is a completely reworked and revised version. The authors would like to thank Mr. Rashed Al Mahmud Titumir, Chairperson of Unnayan Onneshan for his constant support, careful supervision, and continuous encouragement to carry out this research.

References

ADB (2004) Technical assistance consultant's report; Dhaka water supply project. Asian Development Bank

Ahmed M (2004) Development and management challenges of integrated planning for sustainable productivity of water resources. Proc Bangladesh J Polit Ecol 21(2):105–142

Akther H, Ahmed MS, Rasheed KBS (2010) Spatial and temporal analysis of groundwater level fluctuation in Dhaka City, Bangladesh. Asian J Earth Sci 3(4):222–230

Ariyabandu RDS (1999) Development of rainwater harvesting for domestic water use in Rural Sri Lanka. Asia-Pac J Rural Dev 9(1):1–14

Arbués F, García-Valiñas MÁ, Martínez-Espiñeira R (2003) Estimation of residential water demand: a state-of-the-art review. J Socio-Econ 32(1):81–102

Aziz SS, Ferdosh J, Khalid M (2012) Governance of urban services in Dhaka city. In: State of cities: urban governance in Dhaka. Institute of Governance Studies, Brac University

Bakker K (2010) Privatizing water: governance failure and the world's urban water crisis. Cornell University Press, Ithaca

BBS (2011) Population & Housing Census 2011. Bangladesh Bureau of Statistics, Dhaka

Biswas SK, Mahtab SB, Rahman MM (2010) Integrated water resources management options for Dhaka city. In: Proceedings of international conference on environmental aspects of Bangladesh (ICEAB10), Japan

Cash DW, Adger WN, Berkes F, Garden P, Lebel L, Olsson P, Young O (2006) Scale and cross-scale dynamics: governance and information in a multilevel world. Ecol Soc 11(2):8

Crow B, Sultana F (2002) Gender, class, and access to water: three cases in a poor and crowded delta. Soc Nat Resour 15(8):709–724

DWASA (1991) Annual Report 1990–1991. Dhaka Water and Sewage Authority, Dhaka

DWASA (2010) Annual Report 2009–2010. Dhaka Water and Sewage Authority, Dhaka

DWASA (2011) Existing situation & service delivery gap report. Dhaka Sewerage Master Plan Project (Package DS-1A), Dhaka Water and Sewage Authority, Dhaka

DWASA (2013) Annual Report 2012–2013. Dhaka Water and Sewage Authority, Dhaka

DWASA (2014) Annual Report 2013–2014. Dhaka Water and Sewage Authority, Dhaka

DWASA (2014) MIR June 2014 (management information report). Dhaka Water and Sewage Authority, Dhaka

Forslund A, Renöfält BM, Barchiesi S, Cross K, Davidson S, Farrell T, Smith M (2009) Securing water for ecosystems and human well-being: the importance of environmental flows. Swedish Water House Report, vol 24

Gleick PH (1996) Basic water requirements for human activities: meeting basic needs. Water Int 21(2):83–92

GoB (2011) Accelerating growth and reducing poverty. Sixth Five Year Plan FY2011-FY2015, General economics division, planning commission, Ministry of Planning, Government of Bangladesh, Dhaka

GoB (1998) National policy for safe water supply & sanitation. Ministry of local government, rural development and cooperatives, Government of the People's Republic of Bangladesh, Dhaka

Haq KA (2006) Water management in Dhaka. Int J Water Resour Dev 22(2):291–311

HIES (2011) Household income and expenditure survey 2010. Bangladesh Bureau of Statistics, Dhaka

Islam MS, Rahman MR, Shahabuddin AKM, Ahmed R (2010) Changes in wetlands in Dhaka city: trends and physico-environmental consequences. J Life Earth Sci 5:37–42

Islam N (2015) Urbanization in Bangladesh: challenges and opportunities. In: Conference towards sustained eradication of extreme poverty in Bangladesh, Dhaka

Khan A (2013) Conserve water, save life. The daily star. https://archive.thedailystar.net/beta2/tag/water-shortage/. Accessed 05 Oct 2013

Khan HR, Siddique QI (2000) Urban water management problems in developing countries with particular reference to Bangladesh. Int J Water Resour Dev 16(1):21–33

Kjellén, M, McGranahan G (1997) Urban Water-towards Health and Sustainability. Comprehensive Assessment of the Freshwater Resources of the World, SEI

Moinuddin G (2013) Metropolitan government and improvement potentials of urban basic services governance in Dhaka City, Bangladesh: rhetoric or reality? Theor Empir Res Urban Manag 8(4):85–106

Nishat B, Rahman SMM, Kamal MM (2008) Environmental assessment of the surface water sources of Dhaka City. In: Proceedings of international conference on sustainable urban environmental practices, Southeast Asia urban environment applications (SEA-UEMA), Chiang Mai, pp 13–37

Rasheed KBS (2000) Bangladesh towards 2025. In: Ahmad QK (ed) Bangladesh water vision 2025: towards a sustainable water world. Bangladesh Water Partnership, Dhaka

Siddiqui K, Ahmed J, Awal A, Ahmed M (2000) Overcoming the governance crisis in Dhaka City. University Press Limited, Dhaka

Singh R (2004) Water for all: is privatisation the only solution? Combat Law 3:2

The Daily Janaknatha (1999) Interview of DWASA managing director. Accessed 07 July 1997

UNEP (2015) Dhaka City - State of Environment (SOE). United Nations Environment Programme, Dhaka (2005). Accessed 10 Dec 2015

Un-Habitat (2008) State of the world's cities 2008–2009: Harmonious Cities, Earthscan

WHO (2010) The right to water, fact sheet no. 35. WHO, OHCHR, UN-Habitat, UN, Geneva

Varis O, Biswas AK, Tortajada C, Lundqvist J (2006) Megacities and water management. Int J Water Resour Dev 22(2):377–394

Watkins K (2006) Human development report 2006-beyond scarcity: power, poverty and the global water crisis. UNDP Human Development Reports, UNDP

World Bank (1996) Fourth Dhaka water supply project. Staff Appraisal Report, Washington DC

World Bank (1999) Toward an urban strategy for Bangladesh: infrastructure unit. South Asia Region, Washington DC

World Bank (2008) Bangladesh - water supply program project: restructuring. Washington, DC (2008). https://documents.worldbank.org/curated/en/2008/09/9880147/bangladesh-water-supply-program-project-restructuring. Accessed 22 Dec 2015

World Bank (2013) Access to safe water. Washington DC (2013). https://www.worldbank.org/dep web/english/modules/environm/water/. Accessed 09 Oct 2015

Zahid A, Hassan MQ, Karim MA, Islam MA (2009) Excessive withdrawal of groundwater for urban demand of Dhaka city: emergency measures needs to be implemented to protect the aquifer. In: Proceeding of the international symposium on efficient groundwater resources management, Bangkok, pp 219–222

Chapter 27
Quantification of Municipal Water Supply and Role of Metering in Large Indian Cities

P. Sampat and Y. Alagh

Abstract Water saving is both necessary and possible. State Governments have developed good infrastructure for water supply under various missions. Infrastructure facility can't generate new water but facilitates in water management. Most of the states have constructed a large number of storage and supply reservoirs, transmission network for many cities to improve the water supply services in the state; but haven't been able to achieve it successfully due to huge losses in supply system and poor water management.

Experts say that by 2030 more than 50% population of India shall be urban. It means water demand is going to increase continuously and resources are limited. Most of the urban water supply system is not performing well and facing water losses in the range of 30 to 50%; for example, UFW up to 42% in the city of Bangalore (case study). This is the state of affairs in spite of sophisticated metering systems and a considerably good database. Meanwhile, in Ahmedabad (case study), a city without metering and poor database leads to a failure in controlling high UFW 40%. This results in inefficient financial recovery and a low operating ratio. No system can sustain with high amount of water and monetary losses. Policies on water management must be based on the idea that accountability breeds responsibility. Therefore, the nature of 'Municipal Water Quantification' and role of metering in controlling unaccounted water (UFW) was explored in the research undertaken from 2012 to 2016.

Outcome of this research suggested that the role of metering is important but limited in terms of overall system reform, UFW reduction and overall efficient water management. Alternative methods of measurement and statistical techniques for approximations appeared useful too on an appropriate reliability scale for improving system's efficiency. This study inferred that a scientifically quantified database and various analytical reports are necessary for all the stakeholders to decide, maintain and enhance the system. Quantification can be practiced with or without metering

P. Sampat (✉)
CEPT University, Ahmedabad, India

Y. Alagh
Central University - Gujarat, Gandhinagar, India

© Springer Nature Switzerland AG 2021
M. Babel et al. (eds.), *Water Security in Asia*, Springer Water,
https://doi.org/10.1007/978-3-319-54612-4_27

by using tabulation or accounting methods, preferably accounting as it addresses accountability issues.

A sample water account was developed following United Nations Standard Economic-Environmental Accounting (SEEA) (United Nations 2014). Integrated Water Accounting Platform-IWAP was conceptualized as a tool.

Keywords Domestic water supply · Water management · Water quantification · ULB · Accounting

27.1   Setting the Context

Water is the essential life-giving natural resource. Its availability is a key component of socio-economic growth and poverty reduction. Today, global freshwater reserves (which account for 2.5% of the total global water supply of ~1400 MM litres) are rapidly depleting and are projected to have a major effect on many heavily populated areas of the world. Low to medium income developing regions as well as highly developed countries will face water stress in the future unless current water supplies are efficiently handled. While low and middle income developing countries currently have low per capita water consumption, rapid population growth and inefficient use of water across sectors are expected to lead to water shortages in the future. Traditionally, advanced countries have high per capita water consumption and need to emphasize on reducing their consumption and increasing system efficiency by minimizing water losses through enhanced water management using tools, techniques and practices in a proactive manner. It is in this environment that the world's water managers have to handle what is an increasingly scarce resource.

Cities are heavily dependent on regional water resources for the development and so any disruption on quantum or quality of water shall be detrimental. Owing to the current and futuristic adverse impacts of climate change along with the aging infrastructure, it becomes difficult to strike a balance between economic growth and water resource sustainability. It has therefore become important for water managers to incorporate modern principles for water usage and introduce efficient solutions using advanced technologies, resource management strategies such as tabulation or water accounting systems, etc. to better measure water supply and improve the water usage efficiency. These can provide a framework for the effective and dynamic management of the water conservation database.

27.2   Indian Urban Water Supply System Scenario

According to Indian Census 2011 data, almost 91.95% of urban households had access to drinking water (within premises-more than 71.2%, within 100 m of their premises-20.7% and over 8% beyond 100 m of their premises). Providing the access of urban drinking water supply to these households which is reliable, equitable and sustainable is currently an issue of concern for all the Urban Local Bodies (ULBs). Inadequate coverage, intermittent supplies, low pressure, and poor quality (of water and services) are some of the most prominent features of water supply in the cities of India. (High Powered Expert Committee (HPEC), March 2011) Additionally, high commercial and physical losses in the water distribution network which are as high as 30–40% (NRW) along with unwillingness to charge and collect user fees result in water utilities failing to improve service levels. Water utilities in India, typically recover only a third of their operations and maintenance (O&M) cost, which is lower than peer Asian city counterparts. (PEARL 2015).

Historically for domestic purposes, water is supplied by time measurement in Indian cities for 15 min, 30 min, 1 h, 8 h, etc. when water is in a liquid state. It is neither measured in terms of water flow in the city pipe distribution system nor in the quantity served to the customer by means of service connecting metres in a house or a commercial establishment. Generally, the amount of water served during any time span (day/month/year) is not reflected in the bill provided to the consumer. Leak identification and prevention surveys are not standard activity in most water departments (only limited to the visual leakages reported).This has resulted in lack of trust of people and engineers command over the water supply system. There is therefore a lack of availability of scientific knowledge and data, a lack of sense of quantitative aspects about the quantity of water being processed in the system as well as being used by the user, a lack of water accounts and a holistic audit of water with the Municipal authorities.

In a country like India which is densely and relatively uniformly populated, the growing water demand for the large metro cities are reached at alarming level. The resultant search for newer sources of water is bound to come face-to-face with ecological limits or deadlocks. Hence it has turned out to be significant for water managers to embrace and execute realistic measures such as asset administration systems like tabulation or accounting methods, and so on to enhance water use proficiency and protection as these can give a structure to proficient, dynamic and proactive water management in India.

It is in this context that the role of metering and its linkages with quantification aspects and conceptual framework required for water accounting to develop the quantification practice for water supply system of the city were explored in the research. Bengaluru-metered city and Ahmedabad-non metered city were taken up as case studies for the in depth analysis. On the basis of insights gained, the framework has been developed to do the accounting for the water in the system using the stock and flow theories. It has been tested with the hypothetical city named CEPTABAD (Sampat 2017).

27.3   The Selected Cities' Profile

BENGALURU: Widely known as "Silicon Valley of India", the capital city of Karnataka has a population more than 80 lakhs (Census 2011) and is located on the Deccan Plateau, at a height of over 900 m above sea level making it the highest among the major Indian cities. Bengaluru has very limited groundwater because of its hilly topography; more than 900 MLD raw water is procured from the river Cauvery, which is 97 km away. It is done by pumping against a head of 500 m, treated and supplied at a huge expenditure. The metering system has been developed over a period of time in a planned manner, with infrastructure changes at a huge cost in the entire city. The municipal water supply is operated and quantified for the billing and collection purpose by Bangalore Water Supply & Sewage Board (BWSSB) - an autonomous body formed by the State legislature.

AHMEDABAD: It is the seventh largest Indian metropolis had an urban population of 56 lakhs (Census 2011) situated on the bank of Sabarmati, a non-perineal river. It is a leading industrial city and the commercial capital of Gujarat state. Currently, major sources of water are rivers Narmada and Mahi. Ground water is used as a secondary source; seven French wells exist in the Sabarmati river. Several tube wells have also been developed inside the city area as a standby arrangement. More than 1000 MLD water is treated and supplied across the 5 distribution zones at a cost of ₹300.00 core on operation only by Ahmedabad municipal corporation (AMC). Ahmedabad is non-metered city and thus water supply is not quantified in totality.

27.4   A Comparison Between Bengaluru and Ahmedabad

- Bengaluru's water supply is in a more difficult situation compared to Ahmedabad; its source is 98 km away with a pressure head of 500 m, whereas Ahmedabad has the benefit of a closer source-canal Narmada's is only 5 km away. In Bengaluru, therefore the cost of energy and capital expenditure per connection is ₹19,166.00 as compared to ₹1899.00 for Ahmedabad. As a result, the tariff structure in Bengaluru is also found to be higher.
- BWSSB has precise measurements to determine the UFW and NRW to be 33% and 47% respectively. While, AMC takes a judgmental approach, results in inaccurate calculations.
- BWSSB has ample measurements necessary for quantification, which facilitates proper billing against the quantity served and thus allows the management to retain better financial control through proper billing and collection. Since AMC does not have the measurements, management is left to choose from other complicated billing options to retain financial power, e.g. by combining utility charges with property tax bills on the basis of the form of usage of water, the size of the connection and the area of the building/unit.

- NEWS Analysis showed that Bangaloreans have adopted metering and measurements for payment against the service, high public awareness-people accurately quote bulk water quantity while filing complaints. This is not the case observed among Amdavadis.
- The budget analysis showed that BWSSB has achieved 99.90% billing recovery, good financial control over O&M and capital expenditure. Whereas AMC has a poor billing recovery (8 months outstanding) and has hardly met the O&M expenditure, capital cost is subsidized.
- BWSSB practices water balancing on the basis of few assumptions for adjusting the losses, thus have yet to succeed in bringing UFW within national standard limit of 15%. Whereas AMC does not practice, thus resulting into undesirable situation.
- BWSSB has developed partial quantification practice with which it reviews water zones for any monetary or water losses on monthly basis, thereby strengthen its system and accountability towards public. Not following such a robust practice, just periodic quantification practice at City level was largely found unsatisfactory in the case of AMC.
- BWSSB's data reflects a reasonable level of equity in distribution in the city (94 to 138 LPCD with a mean of 125 LPCD) unlike AMC which has no specific data to discuss about equity which is one of the key social issues related to the equal distribution of water as natural resource.
- Bengaluru has a board level organization set up under the State named as BWSSB, which exclusively provides water related services to the city. Whereas, Ahmedabad has a department level set up under ULB named as AMC, which provides all the statutory services including water. This is a major level difference in governance, which is one of the major affecting factors for the difference in performance between two city water utilities.
- The BWSSB has an excellent decision support system with an IT centre, which keeps the database up to date and offers the requisite analytical reports for decision-making; while the AMC has yet to establish a full-fledged scientific support system therefore present working is more discretionary and subjective.
- A detail financial analysis of the latest five annual water budgets of both the organisations were done, which shows the deficit, increasing capital cost and revenue expenditure, lack of political will to raise tariff and correct the financial deficit. These are the matter of serious concern for both the organizations. They are dependent on government grants and loans. Ultimately, making difficult for the state to achieve sustainable development goals.
- Money can be saved by reducing system losses and increasing overall system operation efficiency by adopting proactive water management. This is one of the ways to improve the system and a major driver to practice quantification of the water in the system.

27.5   Results and Discussion

Comparative study of metered and non-metered city showed that overall metering provided required measurements and by keeping scientific records of the flow measurements in a data base form, various analytical reports required for the system monitoring could be generated. These reports make it easier for management to review the distribution of water in the city and to retain control over billing and collection. However, metering has a limited role to play in monitoring UFW primarily because of the improper design of the pipe network and a weak approach to leak detection and prevention. Therefore, new approach for quantification is required which address accountability and efficiency aspects mainly during water distribution.

Quantification is a concept-based solution which can meet the ULB requirements, but its implementation requires system reforms and scientific knowledge for the cities having metered or non-metered water supply arrangements. Quantification of domestic water supply can be done by 'Tabulation' or 'Water Accounting' methods. 'Tabulation' can give only a one-day account; linking of day-to-day accounts; while 'water accounting' (similar to financial accounting) can be done on an annual basis.

Measurements are a necessary quantification prerequisite. Where precise measurements are not available or are expensive, approximate measurements can be used for accounting with regard to the scale of reliability. It's meaningful and it serves the intent. Water meters also have limitations of errors and functional problems. Developing a fully metered system is expensive in developing countries like India where financial resources are major constrain; it takes a lot of time, political will and social acceptance. Currently only a handful of Indian cities have achieved a 100% metering facility. This in fact demands alternate methods or approaches other than metering for quantification and requires exploratory and a systematic study. Australia, Spain, China and Africa have adopted water accounting in line with financial accounting at basin level, linked to the national economy.

Water accounting is not in practice in India; therefore, a conceptual framework was developed based on Stock and Flow theories in line with financial accounting and methodology used for UN SEEA accounting structure to demonstrate how quantification can be done using the water accounting method. A sample water account was also developed for imaginary city named CEPTABAD to test the framework.

27.6   Inferences from Hypothetical Testing of Research
Concept

Water Accounting Framework. Tabulation method can only include a one-day account; this method is not suitable for connecting day-to-day accounts and producing reports. While 'water accounting' can be done on an annual basis and facilitates various features like; stocks and flows at any point of time, at any component of the water supply system. Financial accounting and stock and flow theory,

Double Entry system and Cash transaction method are found suitable. Practice needs the conceptual framework, the rules, and guidelines and setting up of a system which may be run by setting standards, regulations, and assurances.

CEPTABAD water account- Profit and Loss Account Fig. 27.1 shows that the details like source wise total water input in the system, section wise losses, supply and balance can be recorded meaningfully. Accounting structure, modular design facilitates implementation in parts and offers consistent data and useful analytical reports.

CEPTABAD water account showed that water accounting can be done for both—metered and non-metered water supply systems as the accounting requires numbers (value of the measurements) to account irrespective of which method of measurement

PROFIT AND LOSS ACCOUNT IN DETAIL					
CEPTABAD CITY			**CEPTABAD CITY**		
PARTICULARS	1-4-2012 to 10-4-2012		**PARTICULARS**	1-4-2012 to 10-4-2012	
WATER SOURCES			**WATER STORAGE**		
Sea Sources					
			WATER LOSSES		**2 440 0000**
Ground Sources		**200 0000**	Network Loss	1 200 0000	
Bor-1 source	100 0000		Transmission Loss	200 0000	
Bor-2 source	100 0000		Trunk Main Loss	960 0000	
Recycled Sources			**BALANCE**		**1 360 0000**
			DC ESR	8 0000	
Surfaced Source		**10 000 0000**	DC GSR	32 0000	
Canal			Network	320 0000	
Lake-1	5 000 0000		WTP	1 000 0000	
Lake-2	5 000 0000				
			DELIVERED		**6 400 0000**
			Residents	6 400 0000	
GROSS PROFIT c/o			**GROSS PROFIT b/f**		
		10 200 0000			**10 200 0000**
NET PROFIT					
TOTAL			**TOTAL**		

Fig. 27.1 CEPTABAD water account *Source* Author

or type of instruments used to measure the flow of water. Which shows that the details of the system inputs, various losses, water distribution and stock can be recorded meaningfully and various analytical reports like Balance Sheet, Ledger Accounts etc. can be generated which are useful for the water management. Accounting structure, modular design facilitates implementation in part/s and offers consistency as well ease in data recording, keeping and report generation.

27.7 Conclusion

The study has yielded quiet a lot of data, fair to draw general considerations and conclusions:

Metering-Pipe network design for the distribution in Bengaluru has been developed over a period of time in a planned manner and achieved 100% 'metering' for the flow measurement throughout the system. Despite of such expensive and extensive infrastructure changes, the authority has not been able to develop citywide functional district metering areas (DMAs). Therefore, proper quantification approach has not been established. Hence, it has been difficult to address accountability issues, reduce the water losses and restricting BWSSB to achieve desirable improvement in water management. Thus, only metering is not good enough to have optimum water use as a natural resource, proper water distribution through designed pipe network backed by water accounting, efficient O & M are equally important which can address the issue of accountability.

The role of Meter in quantification is to facilitate the measurements of the water flow in the pipe network and to indicate what is the stock in the storage system. Measurements are necessary to practice quantification, but any suitable method can be adopted. Would like to quote 'we cannot manage that which we cannot measure'. Water meter provides quick accurate flow measurement unlike other engineering methods for flow/quantity measurement; but all these can be used for quantification purposes with an appropriate reliability scale. Thus, quantification with approximate or accurate measurements can bring some control even in the non-metered system, which is far better than no control. This approach can facilitate ULB to improve water supply system cost effectively.

Authority is Responsible. The pie chart in Fig. 27.2 shows that the main cause of leakages even in the metered city Bengaluru are Main pipe-38.1%, Service Pipe-32.8% and Stand post-17.6% which is the responsibility of the ULB, as these losses occurs before the water connection, where the citizens/customers are not responsible! This is common observation among Indian cities.

The major challenges faced by authorities in making cities metered are funding and financing as citizens are also mainly concerned with the household meter cost and increase in amount of the water bill. This makes transformation of the water supply system from non-metered to metered very difficult and time-consuming lengthy process. Therefore, by the time authorities should proactively control the losses

Fig. 27.2 Types of leakages
in Bengaluru

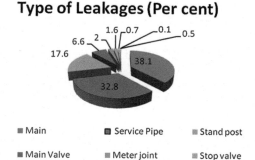

by taking appropriate actions like adopting real time leak detection and prevention techniques to improve the system efficiency.

Metering and Development Stages. Evidence proves that '100% Metering' is not the final solution, at least ULBs can start the quantification practice in stages. At least start with the 'tabulation' method using other modes of measurement, which helps reasonably to control and understand the system losses with minimum infrastructure changes at negligible expenditure. Later on, metering can be done in two stages-bulk and user to improve the measurement accuracy.

Implementation and Goals. Quantification practice can be done in stages and as per the requirements and scale of the ULB. Its implementation requires a multidisciplinary approach. Water audit provides required information to control subsidy and to keep equity in the distribution of the water as a natural resource. Subsidy should be issued to the poor only based on the urban poor-BPL guideline. Thus, quantification supports to achieve social and other organization goals.

Planning and Reforms. Planning issue analysis indicates that the central goal of water resource management is that the achievement of the optimal use of water keeping the public interest at center. The inherent simplicity of this aim is misleading, primarily due to concerns related to technology, administration, financing, planning and regulation. Possible Solutions propose requirements for changes in infrastructure and capacity building for management. These changes are important for improving the existing structure of domestic water supply. Ideally it should be carried out in such a manner that the transaction cost remains low and the system delivers high pay-offs. The system framework can be transformed by making new organization structure design, long-term planning, strengthened by good governance.

Ahmedabad case study shows that despite of political will and administrative decisions to develop a metered water supply system, it would be a Herculean task as the fact is that it is the city level reforms which may affect individual expenditure and the habit of water use. Present practice has failed in service delivery and financial cost recovery, and thus the system can be transformed by planning and public policy which can anchor public trust.

As per lump sum cost estimation, AMC requires a large investment of about ₹12,500.00 Cr. for infrastructure change to develop a fully metered water supply system.

The aim of this study is that the water supply system faces quantification challenges that cannot be addressed in the current practice of the ULB and the state policy framework. The reforms include not only the reorientation of current practices and policies, but also the restructuring of the system by making it a multi-disciplinary system with proper planning.

This thesis inferred that a statistically quantified database and numerous empirical reports was essential for the decision-making phase to sustain and improve the method. These are useful for all stakeholders. Quantification can be carried out with or without metering using tabulation or accounting methods, preferably accounting, as it addresses issues of accountability better. From another point of view, a substantial part of the concerned citizen is eager and looking forward to know how much water can be saved cost effectively by reducing UFW and how quickly utility can do it.

27.8 Recommendations

Quantification cannot be carried out in isolation; public awareness gaps, ULB financial-statistical power, consumer willingness and ability to pay by the citizen are interlinked. Its relationship with other stakeholders are mutually interactive. Following actions are suggested considering the possibility of making system efficient in a cost-effective manner in three different timeline:

27.8.1 For Immediate Actions

- Water accounting Principles based on Standards and Procedures should be framed by the national government.
- Declaring water quantification mandatory for the local bodies by the State governments (Water is a State Subject in India).
- Establishing control mechanism for who accounts to whom and for what.
- Promote correct terminology - Accounted for Water (AFW) replacing Unaccounted for Water (UFW).

27.8.2   Actions Within Next Two to Five Years

ULBs-quantify using bulk flow meter measurements, publishing of daily water balance, third party Water Audit for limited aspects. ULBs-quantify by water accounting, water audit for all the aspects and publishing audited reports.

27.9   Way Forward

The role of metering is important but limited in terms of system reform. A water account was developed on the idea that accountability breeds responsibility. Integrated Water Accounting Platform-IWAP was conceptualized as a tool. Developing an Integrated Water Accounting Platform (IWAP) where leak detection and prevention using GIS and GPS technology is integrated with water accounting. This shall be useful to optimize municipal water supply system by reducing water losses through proactive management approach and capacity building.

The project will involve use of the state-of-the-art Internet of Things (IoT) technology for gathering the data. The visualization of the integrated data with geo-informatics and water accounting will help in quick decision making, improve water supply system, urban governance and shall provide the large database required for the research work. This may ultimately lead the water managers in making water use sustainable for the Indian cities.

Reference

High Powered Expert Committee (HPEC) (March 2011) Report on Indian urban infrastructure and services, India

Ministry of Water Resources-GoI (2012) National water policy

Sampat P, Alagh Y (2017) Quantification of municipal water supply and role of metering in large Indian cities. World Water Congress, Cancun

PEARL (2015) Compendium of good practices-urban water supply and sanitation in Indian cities. NIUA

TERI India & NASSCOM (n.d.) Sustainable tomorrow: harnessing ICT potential

https://censusindia.gov.in/DigitalLibrary/MFTableSeries.aspx

Chapter 28
Towards Holistic and Multifunctional Design of Green and Blue Infrastructure for Climate Change Adaptation in Cultural Heritage Areas

Zoran Vojinovic, Weeraya Keerakamolchai, Arlex Sanchez Torres, Sutat Weesakul, Vorawit Meesuk, Alida Alves, and Mukand S. Babel

Abstract The traditional way of approaching flood risk mitigation is heavily based on the so-called grey infrastructure (i.e., hard core engineering measures such as pipes, channels, underground storages, etc.) with the primary aim of conveying flood waters from urban areas as soon as it is possible. The notion of holistic design of multifunctional green infrastructure, which is as a central concept in this paper, is to enforce the framework in which a climate adaptive design of urban drainage infrastructure for flood risk mitigation requires a combination of grey and green measures to meet the stakeholder preferences. The present paper deals with a systematic, holistic approach for multifunctional design of green infrastructure. The multifunctionality of green infrastructure is realised through a multipurpose design of a detention facility in Ayutthaya UNESCO heritage site (Thailand). This was done by combining a state-of-the-art landscape and hydraulic engineering practice through 1D–2D hydrodynamic model simulation. The following objectives were used in the design work: flood risk mitigation, aesthetics, recreational use, touristic activities and economic benefits.

Keywords Green and blue infrastructure · Multifunctional design · Climate change adaptation · Flood risk mitigation

28.1 Introduction

Over the past years, traditional engineering solutions for flood mitigation, also known as "hard" or "grey infrastructure", have shown some serious limitations in their effectiveness and ability to provide multiple benefits. Furthermore, their design solutions

Z. Vojinovic (✉) · A. S. Torres · V. Meesuk · A. Alves
IHE Delft, Westvest 7, 2611AX Delft, The Netherlands
e-mail: z.vojinovic@un-ihe.org

W. Keerakamolchai · S. Weesakul · M. S. Babel
Asian Institute of Technology, 58 Moo 9, Km. 42, Paholyothin Highway, Klong Luang,
Pathum Thani 12120, Thailand

© Springer Nature Switzerland AG 2021
M. Babel et al. (eds.), *Water Security in Asia*, Springer Water,
https://doi.org/10.1007/978-3-319-54612-4_28

lack flexibility and robustness to adapt to future climate change challenges and land development (Babovic 2014; Spiller et al. 2015; Kabisch et al. 2017). Therefore, the search for optimal solutions represents a great challenge for researchers and practitioners. As a result of that, green and blue infrastructure measures (a.k.a. Nature-Based Solutions) and their hybrid combinations with "grey infrastructure" are lately receiving significant attention, see for example Alves et al. (2016).

While traditional drainage solutions are focusing on efficient collection and fast conveyance of runoff through pipes and open channel networks, the nature inspired infrastructure measures, besides effective flood protection, also aim at providing multiple (or multifunctional) benefits to ecosystem services. "Multifunctionality" has also become a popular term in landscape design and spatial planning practices. In the landscape design and spatial planning context this means developing a landscape that can serve at least two or more purposes. From the land-use perspective, multifunctional land use is the implementation of more functions at a particular location over a certain period of time, see also Lagendijk and Wisserhof (1999). According to Lagendijk and Wisserhof (1999), multifunctional land use is achieved through the following four conditions: (1) increase in the efficiency of land use (intensification of land use); (2) interweaving of land use (which they define as the use of the same area for several functions); (3) use of the third dimension of the land (i.e., vertical space such as the below and/or above ground level along with the surface area); and (4) use of the fourth dimension of the land (i.e., over a certain time frame). Therefore, it is important to consider not only the number of functions (diversity) but also their use in space and time in the development of multifunctional landscapes. To put this into the context of flood risk mitigation, it would mean that we could design a site that is not only effective in flood protection but also in enhancing biodiversity and providing social benefits. The present paper uses stormwater retention and detention ponds as an example to discuss some of the key aspects concerning multifunctional design for flood risk mitigation in cultural heritage areas.

28.2 Materials and Methods

28.2.1 Case Study Area Description

The study area used in the present work is Koh Mueng in Phra Nakorn Si Ayutthaya province which is located about 80 km north of Bangkok (geographical location is $14°21'08''$N latitude and $100°33'38''$E longitude). This is an important region due to its cultural heritage character. It encompasses an area of approximately 7 km^2 which is surrounded by three rivers namely Chao Phraya River, Pasak River, and Lop Buri River. Besides the cultural heritage sites, the land use is mainly residential with some minor commercial activities. Since it is surrounded by three rivers, the same area is also referred to as Ayutthaya City Island and it is often subject to floods. There are also situations where the flooding is caused by heavy rainfall in urban

Fig. 28.1 Existing and potential pond locations

area, resulting in a long inundation (SPAN 1997). The current flood protection of the area is primarily relying on the dike that surrounds the area. Despite the dike, the Ayutthaya Island was greatly affected during the severe flood in 2011. The lowest area is situated on the southwest of the Island where the flood water accumulates during flood events. In a project carried out by the authors and funded by the Asian Development Bank, one of the solutions that showed great effectiveness in flood mitigation is enlargement of an existing pond and creation of new ones (see also Vojinovic et al. 2016). These locations are given in Fig. 28.1.

The following section describes the methodology that was used to design a multifunctional pond area within the Ayutthaya City Island.

28.2.2 Methodology

The first step in the methodology was associated with data collection and interviews with stakeholders. The data was used to set up the 1D/2D MIKE flood model, analyse various flood scenarios and derive the measures (see also Meesuk et al. 2017). Selection of measures was discussed with key stakeholders and it came up that there was a great interest for flood detention and retention ponds. Further to that, the analysis to achieve multifunctionality of the area was carried out. The final stage of the analysis involved a landscape design with a layout plan, cross-sections, and computer animations which were used in public communication and stakeholder consultations. The stakeholders consulted throughout this process were: Ministry of Culture Fine Arts Department, local communities and authority, the government, UNESCO Bangkok, and the Royal irrigation department.

28.3   Results

28.3.1   Analysis Concerning Pond Locations

The potential pond locations in the Ayutthaya City Island are given in Fig. 28.2. These locations are situated within the historical site and as such they are very attractive for tourists. Thus, the landscape design of the entire area, including the ponds, requires multifunctional features that can respond to different users throughout the year. This area also offers the possibilities for a recreational park with attractive water features, Art and Cultural village, rice farms, floating market and multipurpose open space for festivals and other events. In addition to the ponds, the same area also has the potential for restoration of ancient canals.

The entire province of Ayutthaya has an area of approximately 2,547 km^2. The average elevation of Ayutthaya is above mean sea level at 3.50 m. The topographical configuration in Ayutthaya is a flat region surounded with rivers and canals, where neither forests nor mountainous terrains can be found. The lowest area is situated on the southwest of the island which has an elevation at about 2.5 m.a.s.l. and it often gets flooded during floods (Fig. 28.3). The pond location was defined in accordance to the topography, land use and drainage system characteristics. The area with lowest elevations proved to be the best location for the multifunctional pond.

Fig. 28.2 Potential pond locations within the Ayutthaya City Island

Fig. 28.3 Terrain
topography in Ayutthaya

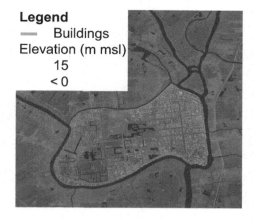

Fig. 28.4 Ayutthaya City
Land use map

28.3.2   Land Use Assessment

Most of the land area of Ayutthaya City is a low density residential zone as shown in
yellow colour in Fig. 28.4. However, about half of the Ayutthaya Island is conserved
as World Heritage Site and it is situated on the west side of the island. On the east side
of the island there are medium and high-density residential zones, while education
and government areas are sporadically located.

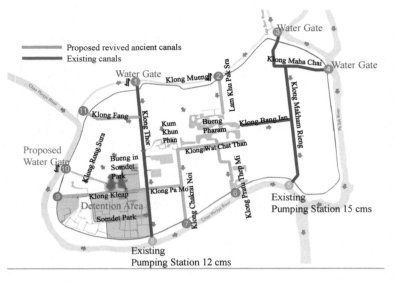

Fig. 28.5 An overview of modelled flood mitigation measures

28.3.3 Final Selection of Pond Areas and Multifunctional Design

From the numerical modelling work, using the 2011 flood event scenario, the effectiveness of various flood mitigation measures was evaluated and the most effective and preferred measures are illustrated in Fig. 28.5.

The measures evaluated include improvements of the local drainage system, enhancement of a detention pond facility (i.e., detention pond areas) as well as the restoration of ancient canals. Detention pond areas were designed in a manner that enhance the aesthetic and environmental quality of the area. Figure 28.6 depicts design features of the multifunctional detention facilities in the study area.

The proposed area would have different uses during dry and wet (i.e., flood) periods. During the dry periods, this area can be used as a recreational park, farm education center and art & cultural village, among other possible uses. During flood periods, possible uses are a pool for boating, Loykrathong festival and a living with water event, etc. Figures 28.7 and 28.8 provide illustrations of the pond area during dry and flood periods.

The restoration of ancient canals has the objective to increase the capacity of the drainage system. Besides, these canals would also enhance aesthetics of the area. Figures 28.9 and 28.10 provide impression of how the area would look like after restoration (or reviving) of ancient canals.

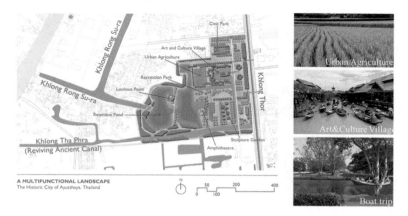

Fig. 28.6 Final layout of the multifunctional pond area

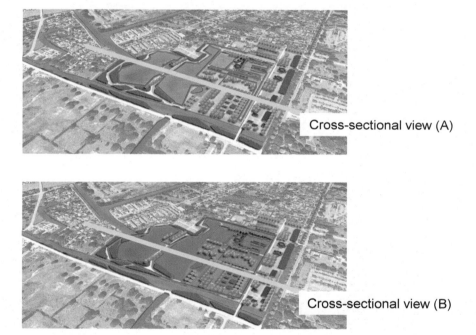

Fig. 28.7 Multifunctional detention facilities during dry and wet (i.e., flood) periods

The work performed demonstrates how the landscape analysis of engineering measures can enable achievement of multi-functionality in an area with a mixed landuse. The work also shows that collaboration between different disciplines is crucial for successful and holistic design of multifunctional flood protection measures.

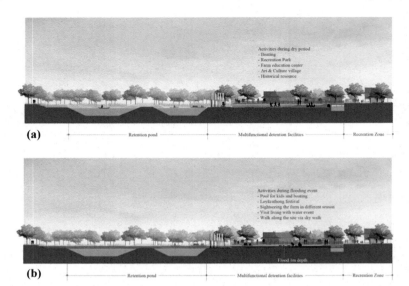

(a)

Fig. 28.8 (a) Cross-sectional view of the multifunctional detention facilities during dry periods (cross-sectional view A); (b) Cross-sectional view of the multifunctional detention facilities during flood events (cross-sectional view B)

Fig. 28.9 Perspective of the Arts and Craft Village

Fig. 28.10 Perspective of Ancient wall at the entrance of Khlong Klaep

28.4   Conclusions

The present work highlights the need for innovative and multifunctional design of flood protection measures that are necessary not only for flood control but also for other benefits and co-benefits within a mixed land use area. Although, there are several researches who demonstrated a design of detention facilities, none of them presented the design of a detention facility for a mixed land use area which also contains cultural heritage assets.

The present work followed several steps ranging from data collection and modelling, site selection and evaluation, to design of multifunctional detention facilities (i.e., detention pond areas) in the Ayutthaya City Island (Thailand). The preferred locations were determined through extensive stakeholder consultations by taking into consideration their preferences, local topography, land use planning and drainage system characteristics.

The work performed demonstrates that the collaboration between different disciplines is a key for successful design of multifunctional flood protection measures.

Acknowledgements This research was funded by the Asian Development Bank, under RETA 6498 Knowledge and Innovation Support for ADB's Water Financing Program (RETA 6498). This work was also partially funded by the European Union Seventh Framework Programme (FP7/2007–2013) under Grant agreement No. 603663 for the research project PEARL (Preparing for Extreme and Rare events in coastaL regions).

References

Alves A, Sanchez A, Vojinovic Z, Seyoum S, Babel M, Brdjanovic D (2016) Evolutionary and holistic assessment of green-grey infrastructure for CSO reduction. Water 8:402. https://doi.org/10.3390/w8090402

Babovic V (2014) Uncertainty, flexibility and design: real-options-based assessment of urban blue green infrastructure. In: 11th international conference on hydroinformatics, New York, US (August 2014)

Kabisch N, Korn H, Stadler J, Bonn A (2017) Nature-based solutions to climate change adaptation in urban areas. Springer. ISBN 978-3-319-56091-5

Lagendijk A, Wisserhof J (1999) Meer ruimte voor kennis. Verkenning van de kennisinfrastructuur voor Meervoudig Ruimtegebruik. (Give knowledge to space, give space to knowledge, part 1: exploration of knowledge infrastructure for multifunctional land use) Series, Part 1, RMNO Report 136. NLRO, Den Haag

Meesuk V, Vojinovic Z, Mynett AE (2017) Extracting inundation patterns from flood watermarks with remote sensing SfM technique to enhance urban flood simulation: the case of Ayutthaya, Thailand. Comput Environ Urban Syst 64:239–253. https://doi.org/10.1016/j.compenvurbsys.2017.03.004

Spiller M, Vreeburg JHG, Leusbrock I, Zeeman G (2015) Flexible design in water and wastewater engineering - definitions, literature and decision guide. J Environ Manag 149:271–328

SPAN Co., Ltd. & VEGA Engineering Consultants Co., Ltd. (1997) The study and detailed design of flood protection system of Phra Nakorn Si Ayutthaya urbanized area, Volume 1 & 2, Public Works Department, Ministry of Interior

Vojinovic Z, Hammond M, Golub D, Hirunsalee S, Weesakul S, Meesuk V, Medina NP, Abbott M, Sanchez A, Kumara S (2016) Holistic approach to flood risk assessment in urban areas with cultural heritage: a practical application in Ayutthaya, Thailand. Nat Hazards. https://doi.org/10.1007/s11069-015-2098-7

Chapter 29
Adaptation to Flood Risk in Areas with Cultural Heritage

Zoran Vojinovic, Daria Golub, Weeraya Keerakamolchai, Vorawit Meesuk, Arlex Sanchez Torres, Sutat Weesakul, Alida Alves, and Mukand S. Babel

Abstract The present paper discusses a novel approach for flood risk assessment and mitigation in areas with cultural heritage. The ambition of the present paper is to provide a 'road map' of the holistic way of working towards climate change adaptation which was introduced in some earlier publications of authors. It is designed to provide the reader with some basic ideas of the holistic view of flood risk, its practicalities and supporting frameworks for implementation. The work was undertaken in Ayutthaya heritage site in Thailand. The approach combined qualitative and quantitative data and methods. The qualitative part of analysis involved a more active role of stakeholders whereas the quantitative part was based on the use of numerical models and engineering principles. Based on the results obtained, this paper argues that perceptions of flood hazard and flood risk (i.e., qualitative part of analysis) yield a richer understanding of the problems and should be incorporated into the engineering analysis (i.e., quantitative part of analysis) to achieve more effective climate change adaptation and flood risk mitigation. Several benefits can be achieved applying the approach advocated in this paper. First, the combination of qualitative and quantitative data and methods opens up new views for risk analysis and selection of measures. Second, since it is based on a more active stakeholder participation the potential for success of this novel approach should be higher than any of the traditional approaches. Finally, design of measures can generate more favourable alternative as it employs a combination of measures that can deliver multiple benefits to stakeholders.

Keywords Adaptation · Holistic flood risk assessment · Qualitative data and methods · Quantitative data and methods

Z. Vojinovic (✉) · D. Golub · V. Meesuk · A. S. Torres · A. Alves
IHE Delft, Westvest 7, 2611AX Delft, The Netherlands
e-mail: z.vojinovic@un-ihe.org

W. Keerakamolchai · S. Weesakul · M. S. Babel
Asian Institute of Technology, 58 Moo 9, Km. 42, Paholyothin Highway, Klong Luang, Pathum Thani 12120, Thailand

© Springer Nature Switzerland AG 2021
M. Babel et al. (eds.), *Water Security in Asia*, Springer Water,
https://doi.org/10.1007/978-3-319-54612-4_29

29.1 Introduction

The world is experiencing a growing number of natural hazards-induced disasters where floods are most dominating the records. One way of responding to natural disasters, and particularly to flood related disasters is by collecting spatial and temporal data and developing numerical models (see for example, Mynett and Vojinovic 2009). With reliable data and models it is possible to analyse effects of various scenarios concerning urbanization, population growth and climate change and identify mitigation measures (Alves et al. 2016). However, as invaluable as they may be, a number of issues associated with data and models need to be carefully considered in order to gain full benefits, such as reliability of data, models calibration, computing time, etc., see for example Vojinovic and Tutulic (2009) and Vojinovic et al. (2011).

In 2011, following five consecutive storm events, Thailand experienced one of the worst floods in its history. The World Bank estimated 1,425 billion baht (US$ 45.7 billion) in economic damages and losses due to this flood event. As reported in Thailand Flood Executive Summary by Hydro and Agro Informatics Institute (HAII), 90 billion square kilometres of land was inundated by floodwaters, which is approximately more than two-thirds of the total country area, and 65 of Thailand's 77 provinces were announced as flood disaster zone. Six provinces were particularly affected, namely, Nakornsawan, Nakorn Sri Thammarat, Pra Nakorn Sri Ayutthaya, Suphan Buri, Pichit, and Phisanulok. Ayutthaya province, which is located about 80 kms north from Bangkok, has been submerged for approximately two months with flood depths ranging from 1 to 3 m. Floodwater entered the Ayutthaya island from the north, inundating the World Heritage Site and major industrial estates and forced evacuations. Figure 29.1 depicts aerial views of the Chao Phraya river basin before and during floods in 2011.

Most of the land area in Ayutthaya province is used for agricultural purposes. The historical park, which is part of UNESCO World Heritage sites, is located in the Ayutthaya Island. In the period 14th to 18th century, this area was a capital of Siam and grew into an important economic center. The 2011 flood event had caused significant damage to the historical sites and buildings (see for example ICOMOS 2014). Besides severe flooding, some areas also suffered from land subsidence. Since the historical sites and buildings contain precious cultural value for local population the damages incurred are immense and practically impossible to express in monetary terms.

Following the 2011 flood event, the government of Thailand collaborated with international and national institutions in the assessment of flood damages and identification of possible mitigation measures at both regional and local levels. However, most of the proposed measures can be regarded as "traditional engineering solutions" which are sometimes also referred to as "hard" engineering or "grey" infrastructure measures. Examples of such measures are various kinds of defences that provide barriers flood waters, construction and/or amplification of pipe networks and channels to increase conveyance and storage characteristics, etc. These measures are well established but lack in integration with surroundings and conservation of the heritage

Fig. 29.1 Aerial view of Chao Phraya river basin before (left) and during (right) floods in 2011 (*Source* Nasa Earth observatory_25102011)

character. To conserve the integrity of monuments and heritage compositions, finding the appropriate flood protection at a World Heritage Site requires more appealing structures that can better fit with the character of the area. In 2013, UNESCO lunched the project "Developing a Flood Risk Mitigation Plan for Ayutthaya World Heritage Site" which was funded by the Asian Development Bank (ADB) and aimed at developing a flood risk mitigation plan for Ayutthaya World Heritage Site. The authors of this paper had the privilege to be involved in that work.

29.2   Materials and Methods

Figure 29.2 depicts the framework which was used to address the main objective and aim of the ADB-funded project which is the Community Based Disaster Risk Management (CBDRM). There are six main components of this framework: stakeholder identification and mapping, risk communication study, participatory flood risk assessment, participatory disaster risk management plan, participatory monitoring and evaluation. This paper focuses on the two main aspects: stakeholder analysis and the use of numerical modelling for hazard delineation, vulnerability and risk assessment. An important step for the development of the numerical model is the acquisition of the necessary data through experts and stakeholders' interviews, field inspection and measurements. Data examples include: hydrological and hydraulic data (i.e. rainfall time series, catchment characteristics, drainage system, etc.), geographical data (i.e. DEM, surface features and land use), the knowledge of existing flood

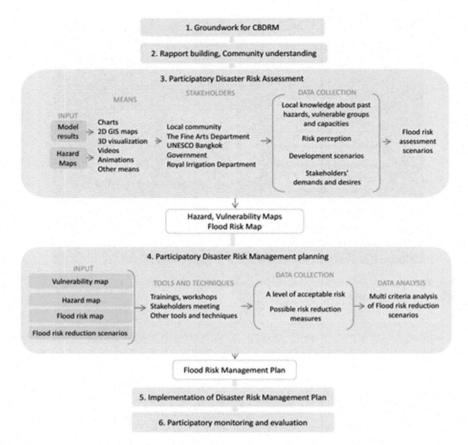

Fig. 29.2 Framework used in the present work

protection measures, and long-term cultural and heritage plans for the case study area. From these data sets, coupled 1D/2D models were instantiated which were subsequently used for the participatory flood risk assessment, Fig. 29.2. The actual risk assessment work combines hazards and vulnerabilities. The models were used to test various scenarios combining climate change, urbanisation and population growth projections. Moreover, the analysis also considered the terrain data for identifying the location of measures and further modification of the 2D model domain to introduce measures such as detention and retention ponds. The model results were presented to stakeholders and their feed-back was obtained. The feedback from stakeholders was necessary in order to come up with solutions that are more acceptable and appealing.

The key stakeholders for the Ayutthaya case study are the Fine Art department, local communities and authority, the government, UNESCO Bangkok, and the Royal irrigation department. The flood mitigation measures are classified into two groups; regional flood mitigation measures and local flood mitigation measures.

Fig. 29.3 Modelling
framework: 1D-2D model
where 1D is used for the
river network and 2D for the
terrain in the Ayutthaya
Island

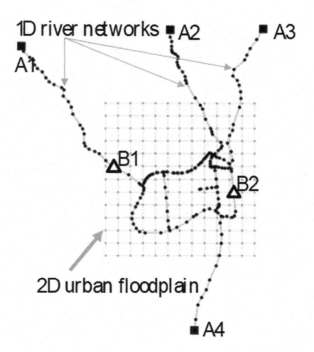

The hydrodynamic models used in the present work are MIKE 11 and MIKE 21 models and the flood event used for simulation is the event from year 2011 (Fig. 29.3).

29.3   Results and Discussion

The results from this study focus on two main aspects of the general framework presented previously. First aspect deals with identification of the key stakeholders and their feedback about the preferred measures. Second aspect deals with discussion of results obtained from the numerical model.

Stakeholder Analysis
Stakeholders' analysis included stakeholders' characterisation and production of a stakeholders' interactions/dependences map depicting their positions and interrelations. Stakeholders' analysis within the present work aimed at providing the insight of existing administrative system, legal action, and practical action systems with respect to flood risk management.

The following categorisation of stakeholders is based on the role and relevance of the stakeholders for flood risk management on Ayutthaya Island. Three groups were differentiated for stakeholders' categorisation:

- core stakeholders;
- secondary stakeholders;
- tertiary stakeholders;

Through this analysis the most important stakeholders were identified and these are: Fine Art department (Ministry of Culture), local communities and authority, the government, UNESCO Bangkok, and the Royal irrigation department.

Flood Modelling Work

As mentioned previously, 1D and 2D models were built in MIKE 11 and MIKE 21. The 1D model was calibrated with discharge and water level data collected at points A1, A2, A3 and A4 (Fig. 29.4). The calibration results are discussed in Meesuk et al. (2017). The 20 × 20 m DEM resolution was used as an input into the 2D model. The time step used for simulation was 5 s and the simulation period was between 1/07/2011—30/11/2011.

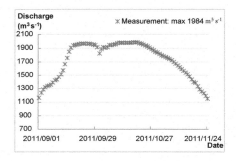 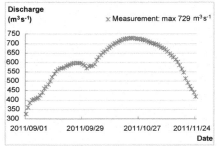

(a)

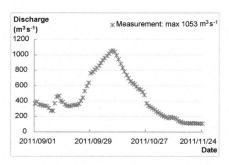 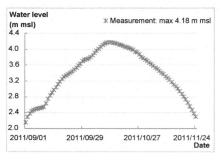

(b)

Fig. 29.4 Time series input at boundaries: (a) discharges of Chao Phraya River (location A1) and discharges of Lopburi River (A2), (b) discharges of Pasak River (location A3), and water levels of Chao Phraya River (location A4)

By merging computer vision techniques with advanced photogrammetry, high-resolution topographic maps were created to support this project. The so-called Structure from Motion (SfM) technique was applied and identified structures on the surface which were important for numerical modelling purpose.

Example of model results obtained is given in Fig. 29.5 (see also Vojinovic et al. 2016) (Fig. 29.5). The flood hazards were mapped on the basis of computed water depths and velocities (Fig. 29.6).

The overall simulation results have shown that the combination of measures such as detention pond, improvements of the local drainage and raising the UThong Road (which currently acts as a dyke) would substantially reduce the flood risk in the Ayutthaya area. The following Fig. 29.7 illustrates locations for some of the key drainage and flood protection measures in the Ayutthaya area. Special characteristics of the measures selected, which were of interest for this heritage area, were identified through discussions with local stakeholders. For instance, the relevance of reviving ancient canals, or the interest on retention and detention areas which could be designed to be used as multifunctional spaces.

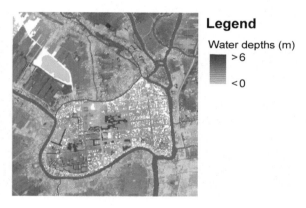

Fig. 29.5 Example of the 1D/2D model result

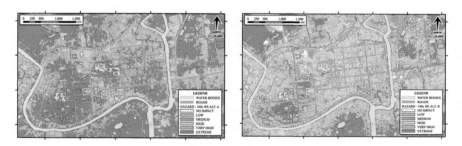

Fig. 29.6 Example of hazard assessment (existing situation—left, and incorporation of new flood protection measures—right); The hazards were substantially reduced with the proposed measures

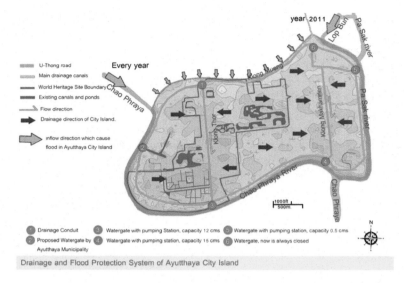

Fig. 29.7 Illustration of the key drainage and flood protection measures in the Ayutthaya heritage area

29.4   Conclusions

Destruction of heritage properties through disasters has a considerable effect on national and local communities, not only because of the cultural importance of heritage assets but also because of their socio-economic value. The present work addresses the issue of adaptation to flood risk in areas with cultural heritage. The work was undertaken in Ayutthaya, Thailand. The work performed highlights the importance of stakeholder participation, numerical modelling and innovative design in order to achieve adaptive capacity of flood protection measures while preserving the heritage character of the area.

The work undertaken fills the gap in the current methodologies by advancing the flood risk mitigation practice at heritage sites. It aims to provide engineers, local authorities and utility managers with an example of how to approach flood risk mitigation at such locations.

The framework developed and implemented was effective in the evaluation and selection of flood risk management measures and some of the proposed works are currently being implemented.

Acknowledgements This research was funded by the Asian Development Bank, under RETA 6498 Knowledge and Innovation Support for ADB's Water Financing Program (RETA 6498). This work was also partially funded by the European Union Seventh Framework Programme (FP7/2007–2013) under Grant agreement No. 603663 for the research project PEARL (Preparing for Extreme and Rare events in coastaL regions).

References

Alves A, Sanchez A, Vojinovic Z, Seyoum S, Babel M, Brdjanovic D (2016) Evolutionary and holistic assessment of green-grey infrastructure for CSO reduction. Water 8:402. https://doi.org/10.3390/w8090402

ICOMOS (2014) Report on the ICOMOS advisory mission to historic City of Ayutthaya, 28 April–2 May 2014. https://whc.unesco.org/en/documents/136457

Meesuk V, Vojinovic Z, Mynett AE (2017) Extracting inundation patterns from flood watermarks with remote sensing SfM technique to enhance urban flood simulation: the case of Ayutthaya, Thailand. Comput Environ Urban Syst 64:239–253. https://doi.org/10.1016/j.compenvurbsys.2017.03.004

Mynett A, Vojinovic Z (2009) Hydroinformatics in multi-colours – part red: urban flood and disaster management. J Hydroinf 11(3–4):166–180

Vojinovic Z, Tutulic D (2009) On the use of 1D and coupled 1D–2D approaches for assessment of flood damages in urban areas. Urban Water J 6(3):183–199

Vojinovic Z, Seyoum SD, Mwalwaka JM, Price RK (2011) Effects of model schematization, geometry and parameter values on urban flood modelling. Water Sci Technol 63(3):462–467

Vojinovic Z, Hammond M, Golub D, Hirunsalee S, Weesakul S, Meesuk V, Medina NP, Abbott M, Sanchez A, Kumara S (2016) Holistic approach to flood risk assessment in urban areas with cultural heritage: a practical application in Ayutthaya, Thailand. Nat Hazards. https://doi.org/10.1007/s11069-015-2098-7

Chapter 30
Rainwater for Domestic Use in Urban Area: A Simulation of Rainwater Harvesting System for Surabaya, Indonesia

C. Kusumastuti and H. P. Chandra

Abstract Many urban areas in Indonesia currently are taken for granted for the availability of clean water from Local State Drinking Water Enterprise (Perusahaan Air Minum Daerah or PDAM) of Indonesia. The number of costumers of PDAM, specifically in Surabaya City increases in the last decades. It raises an issue of maintaining the sustainability of good quality and enough quantity of water delivered to the costumers. On the other hand, flood has become annual events in Surabaya City (2015). Rainwater harvesting, which has been practised in several rural areas in Indonesia, is tried to be simulated for urban area, specifically in Surabaya City to overcome those problems. It is expected to replace some amount of tap water withdrawal as well as to reduce the surface runoff in urban area. An analysis of Analytic Hierarchy Process was done to observe the main aspect needs to be prioritized on installing a Rainwater Harvesting System (RWHS) in a house. A survey to 55 respondents in a real estate in West Surabaya results that the main aspect to be prioritized is environmental impact. This result indirectly is along with the purpose of designing RWHS to conserve water. Rainwater collected through the roof of 31.14 m^2 results in a maximum inflow of 432.9 L/day rainfall. 370 L/day of rainwater is used for gardening and toilet flushing. The complete simulation shows that RWHS installed in a house enables to reduce 21.5% of runoff due to the construction of the selected model of house.

Keywords Analytic Hierarchy Process · Rainwater harvesting · Surface runoff · Urban area

30.1 Introduction

Local State Drinking Water Enterprise (Perusahaan Air Minum Daerah or PDAM) of Indonesia has the responsibility to provide clean water to all resident in urban areas in Indonesia including Surabaya City. Data presented in Surabaya in Figures 2015

C. Kusumastuti (✉) · H. P. Chandra
Department of Civil Engineering, Petra Christian University, Surabaya, Indonesia
e-mail: cilcia.k@petra.ac.id

© Springer Nature Switzerland AG 2021
M. Babel et al. (eds.), *Water Security in Asia*, Springer Water,
https://doi.org/10.1007/978-3-319-54612-4_30

by Statistics of Surabaya City (Surabaya 2015) show the number of costumers of PDAM has increased by 56.6% within the last decade (2004–2014). Previously, it has significantly increased by 105.98% within five years (1999–2004). The number of customers of PDAM is mainly from household which take 91.9% from the total number of customers in 2014. The increasing of PDAM customers occurs along with the land use change in Surabaya, where most area has been developed into residential area. In 1990, an area of 36.4 km^2 in Surabaya was used for residential purpose. This number has become larger into 102.2 km^2 in 2009 (Pemerintah Kota Surabaya 2012).

To meet the water need in Surabaya, the increasing number of PDAM customers in Surabaya should be counterbalanced with the increasing volume of water produced by PDAM. PDAM of Surabaya has six sites of water treatment plants, i.e. Ngagel I, Ngagel II, Ngagel III, Karangpilang I, Karangpilang II, and Karangpilang III. Those stations produced clean water 7,731 L/s in 2004 and increased into 9,338 L/s in 2014.

Annual floods in the city is another problem which increase the burden of the city to deal with in addition to the increasing need of clean water and the land conversion to residential area. This situation has become a great concern which needs to be solved. Rainwater harvesting, a simple method and an alternative source of clean water, has been considered to be a solution for the problem faced by Surabaya City. Rainwater harvesting has been practised in some rural areas in Indonesia, such as in some villages in Kupang Regency, East Nusa Tenggara, and Gunung Kidul Regency, Yogyakarta Province. In those areas, access to clean water from springs or groundwater costs huge amount of money therefore, the people tend to utilize rainwater during the rainy season to fulfil their daily need of water. A concrete tank was built near the house to store the water, which is delivered through the gutter.

A study (Abdulla and Al-Shareef 2009) revealed that rainwater collected from roofs of residential buildings in Jordan is equal to 5.6% of total domestic water supply in 2005. In a newsletter and technical publication by United Nations Environment Programme (United Nations Environment Programme 2016), some countries have utilized rainwater through rainwater harvesting, i.e. Singapore; Tokyo, Japan; Berlin, Germany; Thailand; Capiz Province, The Philippines; Bangladesh; Gansu Province, China; Africa; Dar es Salaam, Tanzania; Botswana; Brazil; Bermuda; St. Thomas, US Virgin Islands; and Island of Hawaii, USA. Most of those countries use rainwater for toilet flushing. Many other technical reports or guidance on rainwater harvesting are also published, such as a manual of rainwater harvesting for domestic use by Texas Commission on Environmental Quality (White et al. 2007), a handbook on rainwater water harvesting in the Caribbean Islands by The Caribbean Environmental Health Institute (The Carribean Environmental Health Institute 2009), and a manual of rainwater harvesting for landscape use by (Waterfall 2006).

The simulation of rainwater harvesting system for Surabaya City presented in this paper aims to provide information of potential usage of rainwater for domestic use and its potential to reduce surface runoff as flood prevention.

30.2 Material and Methods

30.2.1 Study Area

Surabaya City is located in low laying area in East Java at 07°12′–07°21′ South latitude and 112°36′–112°54′ East longitude. Surabaya City is the second biggest city in Indonesia after Jakarta. It has a total land area of 326.38 km². 42% of the land area is utilized for residential area, 16.24% for rice fields, 15.2% for fishponds, 10.76% for business and commerce, 7.3% for industry, 5.5% bare land, and 2.8% for other purposes (Pemerintah Kota Surabaya 2012). Administratively, Surabaya City is divided into four, i.e. Central Surabaya consists of 4 districts, North Surabaya consists of 5 districts, East Surabaya consists of 7 districts, South Surabaya consists of 8 districts, and West Surabaya consists of 7 districts. Specifically, the study area of the research is West Surabaya.

30.2.2 Climate and Rainfall Characteristic

Surabaya as a part of Indonesia has a tropical climate with the temperature varies between 22.6 and 34.1 °C. It has two seasons, i.e. dry and rainy season. The average annual rainfall based on rainfall data from three meteorological stations at Perak I, Perak II, and Juanda in the last five years (2010–2014) was 1897.9 mm.

For data analysis purpose in the chosen study area (West Surabaya), data of daily rainfall during 2006–2015 have been collected from three rainfall stations in West Surabaya, i.e. Kandangan Station, Simo Station, and Gunung Sari Station. The average annual rainfall during 2006–2015 of 2131.8 mm is slightly higher than the average annual rainfall for the whole area of Surabaya City. An anomaly occurred in 2010, when the annual rainfall reached 3365.4 mm. That depth is more (150%) than the normal annual rainfall in the area. The complete average annual rainfall in West Surabaya is presented in Fig. 30.1.

For sustainability reasons, i.e. to provide clean water from rainwater during the whole year, monthly rainfall during 2006–2015 has been analyzed. Figure 30.2 shows the average monthly rainfall in West Surabaya. It shows that during the rainy season, West Surabaya receives relatively similar depth of rainfall from one month to another. At this state, the amount of rainwater is very potential to be used for replacing some water need in a household, where recently 100% is dependent from clean water delivered by PDAM Surabaya.

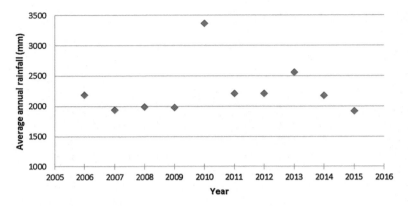

Fig. 30.1 Average annual rainfall in West Surabaya (2006–2015)

Fig. 30.2 Average monthly rainfall in West Surabaya (2006–2015)

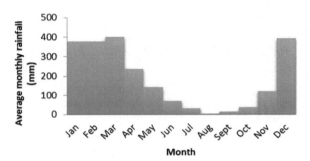

30.2.3   Preliminary Study

A Preliminary study has been conducted in a real estate in West Surabaya involving 55 respondents, who are the residents in the real estate. The survey was done to obtain the information of the most important aspect for them to accept the installation of Rainwater Harvesting System (RWHS) in their house. A single question has been asked to the respondent, i.e. "If a rainwater harvesting system will be installed in your house, what aspect is your main priority?".

In order to limit the answer of the respondents, five aspects have been given for consideration, i.e. costs, quality, maintenance, aesthetics, and environmental impacts. The form of the questionnaire was a matrix, which the respondents should choose in a scale from one to nine. Scale one indicates that both aspects compared is equally prioritized; scale three indicates that one aspect compared is slightly more prioritized than the other aspect; scale five indicates that one aspect compared is more prioritized than the other aspect; scale seven indicates that one aspect compared is more prioritized and more dominant than the other aspect; scale nine indicates that one aspect compared must be prioritized than the other aspect. The matrix of the questionnaire is presented in Table 30.1.

Table 30.1 Matrix of the questionnaire

Column 1																		Column 2
Cost	9	8	7	6	5	4	3	2	1	2	3	4	5	6	7	8	9	Quality
Cost	9	8	7	6	5	4	3	2	1	2	3	4	5	6	7	8	9	Maintenance
Cost	9	8	7	6	5	4	3	2	1	2	3	4	5	6	7	8	9	Aesthetics
Cost	9	8	7	6	5	4	3	2	1	2	3	4	5	6	7	8	9	Environmental Impact
]Quality	9	8	7	6	5	4	3	2	1	2	3	4	5	6	7	8	9	Maintenance
Quality	9	8	7	6	5	4	3	2	1	2	3	4	5	6	7	8	9	Aesthetics
Quality	9	8	7	6	5	4	3	2	1	2	3	4	5	6	7	8	9	Environmental Impact
Maintenance	9	8	7	6	5	4	3	2	1	2	3	4	5	6	7	8	9	Aesthetics
Maintenance	9	8	7	6	5	4	3	2	1	2	3	4	5	6	7	8	9	Environmental Impact
Aesthetics	9	8	7	6	5	4	3	2	1	2	3	4	5	6	7	8	9	Environmental Impact

Source: (Kurniawan and Kosasih, 2016)

30.3 Results and Discussion

The result of the questionnaire was analyzed in order to obtain the most prioritized aspects on designing RWHS by using Analytic Hierarchy Process (AHP) approach. The summary of the questionnaire presented in Table 30.2.

The value presented in Table 30.2 can be interpreted as when it is bigger than one, the design aspects in the column one is more prioritized than the design aspects in row one. The results show that quality and environmental impacts are five times more prioritized than aesthetics. Having the result, the next step of the analysis is normalizing the value in Table 30.2. The value of each aspect is normalized by subtracting each value in each cell with the total value in each column. The result of the normalization is presented in Table 30.3.

The final step of the analysis is to determine the rank of each design aspect by calculating the weight of it. The weight is obtained by adding up each value in every row of the design aspect and dividing it by the number of design aspects. The weight of each design aspect is presented in the last column of Table 30.4.

Based on the normalized value, the rank of each design aspect can be determined from the total weight of each design aspect. The result shows that environmental impact has the biggest weight; therefore, it is in the highest place of the order, followed by cost and quality at equal rank, maintenance, and aesthetics in the last rank.

Table 30.2 Result of questionnaire on prioritized design aspects of RWHS

Design aspects	Cost	Quality	Maintenance	Aesthetics	Environmental impacts
Cost	1.00	2.00	1.00	3.00	1.00
Quality	0.50	1.00	3.00	5.00	1.00
Maintenance	1.00	0.33	1.00	3.00	0.33
Aesthetics	0.33	0.20	0.33	1.00	0.20
Environmental impacts	1.00	1.00	3.00	5.00	1.00
Total	3.83	4.53	8.33	1.00	3.53

Table 30.3 Normalized value of each design aspect of RWHS

Design aspects	Cost	Quality	Maintenance	Aesthetics	Environmental impacts
Cost	0.26	0.44	0.12	0.18	0.28
Quality	0.13	0.22	0.36	0.29	0.28
Maintenance	0.26	0.07	0.12	0.18	0.09
Aesthetics	0.09	0.04	0.04	0.06	0.06
Environmental impact	0.26	0.22	0.36	0.29	0.28
Total	1.00	1.00	1.00	1.00	1.00

Table 30.4 Weight of each design aspect of RWHS

Design aspect	Cost	Quality	Maintenance	Aesthetics	Environmental impacts	Total weight
Cost	0.26	0.44	0.12	0.18	0.28	0.26
Quality	0.13	0.22	0.36	0.29	0.28	0.26
Maintenance	0.26	0.07	0.12	0.18	0.09	0.15
Aesthetics	0.09	0.04	0.04	0.06	0.06	0.06
Environmental impacts	0.26	0.22	0.36	0.29	0.28	0.28
Total	1.00	1.00	1.00	1.00	1.00	1.00

The result of the preliminary study of the research is as expected and matched with the purpose of installing RWHS for urban area, specifically Surabaya City. The following step in the research was designing the RWHS. By using the developed design of RWHS at a certain house in the selected study area (real estate), the amount of surface runoff is determined.

30.3.1 Simulation of Rainwater Harvesting System

The total land area used for a house in the selected real estate is 200 with 145 m^2 of it covered with roof from zincalume, tile, and concrete. Before determining the catchment area, a test of rainwater quality had been done. The result of the water quality test shows that the rainwater contains more than 1600 coliform and 13 *E. coli* in 100 ml of sample of rainwater. Based on this finding, the utilization of the rainwater without treatment is limited to fulfil certain needs of water which neglect the quality of water. It is then determined to be used for gardening and toilet flushing. 90 L per day of rainwater is used for gardening and 70 L for toilet flushing. To simulate the daily water usage in a household, the need of water of four people are estimated. A total flow of 370 L/day is used as the design outflow from the rainwater storage.

Due to the quality of rainwater, which limits the utilization of the water usage in the house, 31.4 m^2 of roof area, specifically on the tile part, is chosen to estimate the runoff as the inflow to the rainwater storage. The catchment area is presented in Fig. 30.3 and the detail of RWHS is presented in Fig. 30.4. Using the chosen catchment area, the amount of inflow in the RWHS is then calculated. The inflow is estimated from the depth of average daily rainfall per month during 2006–2015 in Surabaya City. In detail, the depth of average daily rainfall and the inflow are presented in Table 30.5.

Fig. 30.3 Catchment area of RWHS in mm (not to scale) *Source* Kurniawan and Kosasih (2016) with modification

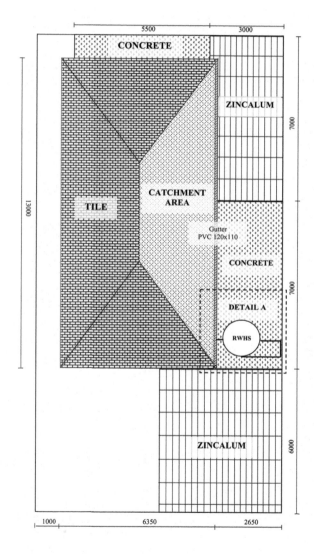

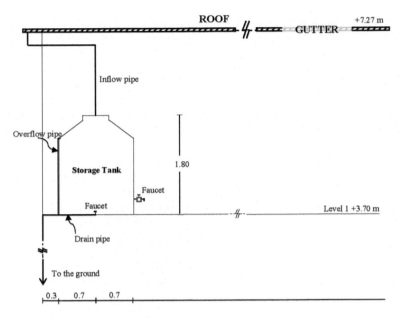

Fig. 30.4 Detail A of RWHS in meter (not to scale)

Month	Average daily rainfall (mm)	Catchment area (m²)	Inflow (litre/day)
January	11.97	31.14	372.78
February	13.90	31.14	432.91
March	12.64	31.14	393.54
April	8.73	31.14	271.98
May	5.38	31.14	167.42
June	2.13	31.14	66.42
July	0.48	31.14	15.03
August	0.03	31.14	0.82
September	0.06	31.14	1.77
October	0.98	31.14	30.47
November	4.10	31.14	127.72
December	12.31	31.14	383.40

Table 30.5 Average daily rainfall and inflow of RWHS per month

The inflow-outflow of rainwater is simulated based on the average daily rainfall and presented in Table 30.5. Since the depth of rainfall is varied from one month to another, as can be seen in Fig. 30.2, the simulation of inflow-outflow is done on monthly basis. The rainy season in West Surabaya in the last ten years, usually starts in November, therefore, November is used as the beginning month of the simulation.

Table 30.6 Simulation of inflow-outflow in RWHS

Month	Inflow		Outflow		Cumulative
	(Litre/day)	(Litre/month)	(Litre/day)	(Litre/month)	Δ storage per month (litre)
November	127.72	3831.64	90.00	2700.00	1131.64
December	383.40	11885.48	370.00	11470.00	415.48
January	372.78	11556.22	370.00	11470.00	86.22
February	432.91	12121.51	370.00	10360.00	1761.51
March	393.54	12199.74	370.00	11470.00	729.74
April	271.98	8159.33	280.00	8400.00	0.00
May	167.42	5190.10	90.00	2790.00	2400.10
June	66.42	1793.23	90.00	2430.00	0.00
July	15.03	120.24	90.00	720.00	0.00
August	0.82	0.82	0.00	0.00	0.82
September	1.77	3.54	0.00	0.00	3.54
October	30.47	731.35	0.00	0.00	731.35
Maximum Cumulative Δ storage (Litre/month)					2400.10

The monthly basis analysis, aside from determining how much rainwater can be captured and used, is also used to determine the size of water storage tank. The storage tank is determined from the maximum cumulative storage in the end of month during the year which is obtained by subtracting the inflow by outflow. The complete simulation of inflow-outflow as well as Δ storage is presented in Table 30.6.

During the rainy season, the outflow is optimum at 370 L/day and used for gardening and toilet flushing. It occurred during the month of December into March continuously. The inflow reduces in April and from August to October since the rain is less and it impacts to the amount of water which can be utilized.

The maximum cumulative Δ storage of 2400.10 L/month is then used to design the capacity of rainwater storage tanks. Considering the available water tank in the market in Indonesia, the price of the water tank as well as the process of installation, a plastic tank with a capacity of 2200 L is chosen. A dead storage of the tank is also determined as 50 L. The dead storage is given in order to maintain the storage tank in a good condition due to the exposure of sunlight in dry season. A new scenario of inflow-outflow is then simulated and the result is presented in Table 30.7.

The final simulation shows that the cumulative rainwater harvested in a year can be used for gardening for the whole month of November, May, June, and half month of July. During December to March, it can be used for gardening and toilet flushing. However, no rainwater remained to fulfil domestic needs in August to October.

Table 30.7 Final scenario of inflow-outflow in RWHS

Month	Inflow (Litre/day)	Outflow (Litre/day)	Cumulative Δ storage per month (Litre)	Overflow (Litre/day)
November	127.72	90 [c]	1131.60	0.00
December	383.40	370 [a]	1547.00	0.00
January	372.78	370 [a]	1633.18	0.00
February	432.91	370 [a]	2200.00	62.91
March	393.54	370 [a]	2200.00	23.54
April	271.98	280 [b]	1959.40	0.00
May	167.42	90 [c]	2200.00	77.42
June	66.42	90 [c]	1359.76	0.00
July	15.03	90 [c]	50.00	0.00
		15.03 [c,#]		
August	0.82	0.00	50.00	0.00
September	1.77	0.00	50.00	0.00
October	30.47	0.00	50.00	0.00

Note: [a](gardening and toilet flushing); [b](toilet flushing); [c](gardening); [#]the available water is less than water need

30.4 Conclusions

The main purpose of rainwater harvesting in West Surabaya is mainly to reduce the use of tap water delivered by PDAM Surabaya and to reduce the surface runoff which contributes to annual urban flooding in the city. The simulation of inflow-outflow in the proposed RWHS has shown that the scenario is feasible to be applied in Surabaya City. It reduces up 21.5% of runoff where 145 m^2 of building is constructed. The reducing of surface runoff could be higher when water treatment devices are installed in the system. The installation of water treatment devises enables the increasing of the inflow to the storage tank of RWHS which further can be used to meet all domestic needs in the household. In the proposed scenario, bigger capacity of the storage tank will be needed to store the bigger inflow and the outflow regulation should be adjusted accordingly.

Acknowledgements The authors would like to extend their gratitude to Kurniawan and Kosasih for the permission given to use some data from their thesis for data analysis in this paper.

References

Abdulla FA, Al-Shareef A (2009) Roof rainwater harvesting systems for household water supply in Jordan. Desalination 243:195–207

Kurniawan KH, Kosasih W (2016) Perencanaan sistem rainwater harvesting yang berbasis value engineering. Petra Christian University, Surabaya

Pemerintah Kota Surabaya DL (2012) Profil keanekaragaman hayati Kota Surabaya tahun 2012. Dinas Lingkungan Hidup Pemerintah Kota Surabaya, Surabaya

Surabaya BPSK (2015) Surabaya Dalam Angka 2015. Badan Pusat Statistik, Kota Surabaya

The Carribean Environmental Health Institute C (2009) Handbook For rainwater harvesting for the Caribbean, The Morne, Catries, St. Lucia, The Caribbean Environmental Health Institute

United Nations Environment Programme U (2016) Newsletter and technical publications: Rainwater harvesting and utilization

Waterfall P (2006) Harvesting rainwater for landscape use. The University Of Arizona, Arizona

White K, Soward L, Shankle G (2007) Harvesting, storing, and treating rainwater for domestic indoor use. Texas Commission on Environmental Quality, Austin

Chapter 31
Water Security and Climate Change: A Periurban Perspective

V. Narain

Abstract While most research on climate change and water security focuses on purely agrarian or urban contexts, periurban contexts need special attention in terms of research and capacity-building. This is because they suffer from the effects of both rural and urban stressors, that are aggravated by the effects of climate change. Further, they are rapidly growing in geographical spread and importance and sustain an urban metabolism in the context of growing cities of the global south. Providing platforms for dialogue between the state and water users, and building human capital to promote occupational diversification can build the coping capacity of periurban communities and reduce their vulnerability to water insecurity caused by the combined effects of urbanization and climate change. This paper examines the impacts of urbanization and climate change on the water security of periurban populations in Gurgaon, North-West India and their adaptive responses. Efforts to enhance their adaptive capacity by promoting platforms for stakeholder dialogue with the state agencies and supporting occupational diversification are described. The methodology, leaning on an interpretive and social constructivist paradigm, comprises ethnographic tools, semi-structured interviews and key informant interviews (KIIs). Research for this paper was carried out between the periods 2009 to 2015 and shows that the population of periurban Gurgaon lost private and common property land and water resources to support the expansion of the city. There is growing pressure on the water table from the expanding city. The effects of these trends have been aggravated by a change in the seasonal distribution of rainfall and the duration of seasons, rising temperatures and disappearance of the monsoon season. Periurban communities have responded by switching to sprinkler irrigation sets, changes in cropping choices and increased reliance on wastewater. The paper concludes by developing a typology of urbanization and climate change induced periurban water insecurity and by making a case for increased research and policy attention to periurban areas that suffer the combined effects of urbanization and climate change on their water sources.

Keywords Adaptation · India · Periurban · Urbanization

V. Narain (✉)
Management Development Institute, Gurgaon, India
e-mail: vishalnarain@mdi.ac.in

© Springer Nature Switzerland AG 2021
M. Babel et al. (eds.), *Water Security in Asia*, Springer Water,
https://doi.org/10.1007/978-3-319-54612-4_31

413

31.1 Introduction

There is a growing body of research focusing on vulnerability and adaptation to the effects of climate change in urban contexts, such as the research conducted under the aegis of the UCCRN (Urban Climate Change Research Network) and the ACCCRN (Asian Cities Climate Change Resilience Network). There is also a rapidly growing body of work on vulnerability studies in rural contexts; some examples of this are Jalan and Ravallion (1999), Leichenko and O'Brien (2002), Eakin (2005) and Saldana-Zorilla (2008). However, relatively little attention is devoted to the issues of vulnerability and adaptation in periurban contexts, especially in relation to water security. This chapter looks at the relationship between climate change and water security from a periurban perspective.

Urbanization processes result in the creation of periurban spaces, that combine features of both rural and urban areas. 'Periurban' represents a transition zone between the rural and the urban, that combines some features of both. Broadly, in the development literature the word has been used as a place, process or concept (Narain and Nischal 2007). More recent research on the periurban points to the periurban as a sociological phenomenon (Srestha 2019), while highlighting the limited utility of the term to denote a peripheral space around the city. In so far as it represents the co-existence of the rural and the urban this co-existence could even occur in the heart of the city (Singh and Narain 2020). Owing to increasing competition over such natural resources as groundwater, periurban spaces face specific needs of capacity building for resolving conflicts and engaging stakeholders (Gomes and Hermanns 2018).

Rather than a geographically demarcated area, it is better to see the word as representing a conceptual lens to look at rural-urban relationships and flows of goods and services (Narain 2014; Singh and Narain 2020). In this chapter, therefore, we do not use the word 'periurban' to represent an area at the periphery of a city; rather we use it to represent a process of transformation between the rural and the urban and the concomitant changing flows of goods, services and resources between the rural and urban areas. We examine how these changes coupled with climate change shape periurban water insecurity and how periurban communities adapt to these changes. Focusing on three villages in periurban Gurgaon, we seek to develop a typology of the ways in which climate change and urbanization interact to shape periurban water insecurity. We also describe the adaptation responses of the periurban residents and other efforts to build their adaptive capacity. The paper concludes with relevant messages for policy-makers, urban planners and rural development specialists in building the adaptive capacity of periurban inhabitants.

Water security is an emerging and debated concept and paradigm (for a review of different approaches to defining water security, see Cook and Bakker 2012). There is no consensus definition of the word and it is used differently at varying scale levels (regional, national, watershed or river basin, and community or village level). In fact, it has been considered to be a debatable and emerging paradigm (Cook and Bakker 2012). It has different meanings and connotations across disciplines and hydraulic levels. The word was used first in the 1990s in relation to subjects of military, food or environmental security. The Global Water Partnership in 2000 gave a first comprehensive definition of the concept to emphasise the integration of different aspects—both quality and quantity—of water use and access for human well-being.

The paper does not work with an objective measure or yard-stick of water security; given the diversity in periurban contexts, and the wide variations and diversity in access to water sources, both spatially and temporally, such a yardstick would be difficult to conceptualize. Instead the term water insecurity is used in the paper to denote the changing access to water in periurban contexts, as also the uncertainty attached to sources of water supply. Thus, the term is used to capture aspects both of insufficient water, as well as excess water, for instance, in times of high rainfall and rainfall flooding. It is used to capture the daily lived experience of periurban communities' changing access to water. Vulnerability to water insecurity caused by the effects of urbanization and climate change needs to be seen as a chronic phenomenon; this means that there is a need for sustained capacity building of periurban communities, rather than knee-jerk responses to occasional extreme events.

The research presented in this paper suggests that periurban communities are not passive recipients of the impacts of urbanization and climate change on their water security, but innovate both technologically and institutionally to adapt to these changes. The Inter-Governmental Panel on Climate Change (IPCC) (2007) defined climate change adaptation (CCA) as an adjustment in natural or human systems in response to actual or expected climate stimuli or their effects, which moderate harm or exploit benefit opportunities. CCA strategies aim to reduce vulnerability to expected impacts of climate change, even though this change may only be one of the factors shaping vulnerability (Mercer 2010). Adaptation analyses capture the role of human agency in responding to a negative external environment to reduce the degree of harm or to exploit benefit opportunities.

In recent discourses on periurban water insecurity, there has been a critique of the notion of "community resilience" in the context of the periurban. This is because given the wide social and economic differences within periurban populations, the extent of heterogeneity is high. This renders such expressions as 'community resilience' redundant. In this backdrop, the chapter further makes an effort to identify the differential vulnerabilities of periurban populations. The concept of vulnerability is used to describe people and organizations that are negatively affected, directly or indirectly, by a single process or event (O' Brien et al. 2009). In general, it is used to draw attention to the specific contextual factors that influence exposure and the capacity to respond to change in order to explain how and why some groups and individuals experience negative outbreaks from shocks and stressors (Leichenko and O'Brien 2002). So, it is to be seen as a concept that captures the changing nature of risks as well as the variable capacity to cope with both risk and change (Kirby 2006; Adger 2006).

While looking at the responses of the communities to water insecurity, the paper uses concepts of social capital and legal pluralism. The term social capital attempts to describe features of populations such as levels of civic participation, social networks and trust that shape the quality and quantity of social interactions and the social institutions underpinning society (Mackenzie and Harpham 2006). Legal pluralism denotes the co-existence of multiple regulatory and legal systems with regard to the same set of activities (von Benda Beckmann 1989). In this paper, we look at how in a context of urbanization and climate change induced water insecurity, water users mobilise different legal and normative systems to improve their access to water. Water rights may be granted by the state, that is, they may have a basis in statutory systems of resource allocation. However, they may be realized through normative systems outside state law. Social relationships and mutual norms of co-operation have an important role to play in this mediation of water insecurity.

Social capital - in particular, as measured by forms of civic engagement (Putnam 2000), can be weak in peri-urban contexts. This means that there is poor integration of periurban communities with state agencies. There is an absence of mechanisms for representing their interests. Building social capital - by providing platforms for dialogue between the state and water users can be an important way of building the coping capacity of periurban communities in dealing with the effects of urbanization and climate change on their water (in) security. This is to be distinguished from efforts to simply augment physical supply of water.

31.2   Material and Methods

31.2.1   Study Area and Methodology

The research is located in Gurgaon, a growing residential, outsourcing and recreation hub of North-West India. Gurgaon was once a sleepy town. It emerged after a real estate boom of the 1980s which led to massive acquisition of land by real estate. In many ways, Gurgaon could be considered to be an accidental city, in the sense that it grew autonomously through private enterprise. In fact, the city grew more or less autonomously till 2008 when it got its first municipal corporation. An important aspect of the growth of the city was that it was led mainly by private real estate developers (Narain and Singh 2019).

In the 1990s, in the post-liberal era, the state government of Haryana provided policy support for the promotion of special economic zones (SEZs). The growth of Haryana was promoted by three factors, namely, the proximity to New Delhi, the national capital, the proximity to the national airport and the adoption of a liberal policy environment by the state government of Haryana (Narain et al. 2016; Narain 2009). These factors led to a rapid transformation of the city's landscape. This process created a dual economy wherein shopping malls and residential areas are located interspersed with rural settlements whose land and water were acquired to support the expansion of the city. We see rural settlement areas alternating with urban gated communities. The gated communities have been built over the lands—private agricultural lands as well as common lands—acquired from the residents of the villages, while the rural settlement areas are still intact. This represents the co-existence of the rural and the urban that was referred to in earlier parts of this paper. In the South-East Asian literature, this is referred to as the 'Desakota (McGee 1991)'—the word 'desa' means village or country side, while 'kota' mean town. Thus the word "desakota" refers to a place where features of the town as well as the village co-exist.

The process of acquisition of land and water from the periurban villages had several social and ecological effects. There has been growing dissent against the land acquisition process; in particular, the low compensations received (Narain 2009). The loss of water bodies and increase in built-up area resulted in disturbance to the natural drainage of the city, often leading to rainfall flooding with a few hours of heavy rainfall. In fact, in the year 2018, the city of Gurgaon experienced what came to be known as "Gurujam", where immense waterlogging following a period of torrential rains caused the city's traffic to come to a standstill for several hours. The increase in the built-up area of the city as well as the loss of natural drainage routes are ascribed as important reasons for the repeated incidence of urban flooding during the monsoon season.

This research is located in three villages namely, Budhera, Sadhraana and Sultanpur. They are located about 15 kms from the city. They have witnessed several land use changes over recent years. Land has moved out from agriculture to other purposes, such as the construction of canals and a water treatment plant to provide water to the city, the construction of a national park to house the avian biodiversity and the building of farm-houses for recreation and weekend get-aways of the urban elite. Land use change in periurban spaces is a key driver of changes in water allocation and use. Following the changes in land use, changes have also occurred in the water use patterns. These include a fall in water tables, the loss of common property water sources and the loss of tubewells and water sources located on the lands acquired for urban expansion and infrastructure provision. These changes are the result of the expansion of the city. This is how urbanization has affected access to land and water in the villages.

The research draws on an interpretive paradigm. It takes a social constructivist approach to climate change and urbanization induced water insecurity, emphasizing the people's daily lived experience of the phenomenon. Prior to the presentation of this paper at the conference on Water Security and Climate Change, organized by AIT, Bangkok, in the year 2016, the author had been engaged with the research sites for a period of about six years. The research on which this paper is based was conducted over the period 2009 to 2015.

The methods of data collection used were mainly qualitative. Aspects of climate change were captured by listening to people's narratives. Listening to people's narratives of climate change is important in so far as that shapes their adaptation choices. Semi-structured interviews with the residents of these villages were the main sources of data collection. Direct observation was used as an important means of gaining insights into changing access to water and of immersing into the daily struggles of the respondents to access water. Key informant interviews were carried out along with some semi-structured interviews. The findings that relate to Sultanpur village draw on an Action Research Project whose approach and methodology are described in Narain et al. (2017).

31.3 Results and Discussion

31.3.1 Perceived Climate Change

On the one hand, these communities are impacted by changes brought on by urbanization as described above—losing land and water resources to support urban expansion; on the other hand, they are impacted by climate change and variability. The residents of these villages have experienced a decline in the frequency and intensity of rainfall and a change in its seasonal distribution, disappearance of the *'chaumasa'* (the four-month monsoon season), rising temperatures, shorter winters and prolonged summers. These narratives are supported by the analysis of hydro-meteorological data as well as through the use of PRA tools, that is described in Narain et al. (2016). The respondents reported rains to be 'good' till the year 1977–when they experienced high rainfall and rainfall flooding–and since then there has been a decrease in the intensity and frequency of rainfall. The year 2010 was considered an exception; it rained quite heavily then.

It is important to understand in the field the respondents' lived experience of climate change; how they interpret it and the meanings that they attach to it. These meanings get translated into narratives of climate change and influence the choice of adaptive responses and strategies. The residents of these villages perceive a changing climate in the form of increase or decrease in intensity as well as duration of winter and summer seasons. In general, respondents said that the winter season had reduced in terms of intensity and duration. Winter is now confined to only two months, December and January, unlike earlier when this season spanned four to five months. Besides, people also perceive a changing climate through changes in cropping patterns. Especially for farmers, their association with increase or decrease in the intensity of seasons is not in terms of temperature–as is the common yardstick for assessing climate change at the global level–but in terms of increase or decrease in the number of summer, winter, or rainy months and the corresponding impact on their cropping choices. Hydro-meteorologists and communities therefore have different framings of climate change, different ways in which this process is articulated.

In particular, the reduction in intensity of winter is perceived in a very interesting manner. Some respondents mentioned that there was a time when a pot of water kept out in the open would freeze overnight. This no longer happens. They also mark a specific year or a particular decade since when the climate parameters have not been stable. In the case of rainfall, people keep 1977 as the benchmark year. For seasons, the 1980s are considered the last period up to which seasons were timely (Narain et al. 2016). In fact, the winter season and fog were considered to be more intense and consistent until then.

The following sub-sections of this paper describe how urbanization and climate change have impacted the water security of the population living in these villages and how they have adapted to these changes.

31.3.2 *Budhera Village; Where the Urban Shadow Falls*

Budhera village lies about 15 kms from the city of Gurgaon. The village bore the ecological foot-print of the city by providing private and common lands for the expansion of the city's water infrastructure. The village's common property grazing lands were acquired to build a water treatment plant for meeting the drinking water requirements of the city. At the same time, two canals—the National Capital Region (NCR) Channel and the Gurgaon Water Supply Canal were dug to carry water from Sonepat district—about 80 kms away—to bring water to the Basai Water Treatment plant, that is the major source of supply for the city's water needs and for a new water treatment plant at Chandu-Budhera. When field work for this research was conducted, the construction of the new water treatment plant had just been initiated. The residents of the village lost land for the construction of these canals (Narain 2014). When the Gurgaon Water Supply Channel was dug, however, some farmers installed tubewells to benefit from the rise in the water table level. These tubewells were removed when the NCR channel was dug parallel to it. This demonstrates the aspect of periurban water insecurity, in other words, the uncertainty attached to the sources of water supply. Since the two canals were built parallel to each other, the same group of people were impacted. Thus the location of agricultural fields, in relation to sites of construction of urban infrastructure, is an important factor shaping the differential vulnerabilities of periurban populations.

The impacts of creation of urban infrastructure in the periurban spaces are gendered. Men and women's vulnerabilities are differential. Budhera is a livestock dependent village with a high preponderance of Balmeeks, - who rear livestock and do not have private agricultural land. These households were dependent on the common property grazing lands of the village. When the grazing lands were acquired, they had to cut back on the livestock population, or switch from grazing to stall feeding. While taking the cattle to graze is the domain of men in the household, collecting fodder is the responsibility of women. Thus, the acquisition of the grazing lands increased women's work loads around cattle rearing and fodder collection (Vij and Narain 2016). This shows how changing rural–urban interactions and the periurbanization process can change gender relations around natural resources. This is a black-box in the periurban literature, and further research should unravel the implications of the expanding ecological foor-print of cities on gender relations in periurban spaces.

There are multiple ways in which rural–urban water supplies are related, and a periurban conceptual lens helps us understand these relationships. As noted earlier in this paper, we do not use the word periurban in this paper simply to denote rural areas at the periphery of cities; rather we use it as a conceptual lens to study rural–urban links and flows of goods, services and resources. On the one hand, fresh water canals pass through the village, while on the other hand, a wastewater canal—called the Sewage Treatment Plant canal or Gurgaon Jhajjar canal - passes through the village as well, carrying the domestic sewerage of the city. This wastewater canal is a major source of irrigation for paddy and wheat in the village. Unlike the conventional treatment of 'rural' and 'urban' water supply as distinct categories, a periurban conceptual lens helps us see the relationships between them.

On the one hand, the village lost land and fresh water sources (like tubewells) located on those lands for the expansion of the city's infrastructure, which represents one aspect of water insecurity experienced in the village. On the other hand, there has been a long-term decline in rainfall over the years after the 1980s. 1977 was a year of heavy rainfall and the village experienced rainfall flooding. After that, locals report a decline in the rainfall, except for the year 2010, when high rainfall was experienced. At the same time, there was a change in the seasonal distribution of rainfall, and what the locals call the *chaumasa*–or four-month monsoon period—disappeared. The area is not served by an irrigation canal and the groundwater is saline. Under these circumstances, wastewater has emerged as a major source of irrigation. In a context wherein land and water sources are being acquired by the state for urban expansion, and there is a decline and change in the seasonal distribution of rainfall, the reliance on wastewater has increased.

Wastewater is used for irrigation, using a wide diversity of sources, such as pipe outlets, diesel and electric pump-sets and tractors. The choice of technology is shaped mainly by topographical factors (Narain and Singh 2017). When the fields are at the same level or below the bed level of the canal, pipe outlets are used. When the fields are at a higher level, wastewater is pumped from the canals using diesel or electricity pumps-sets. Farmers apply to the Irrigation Department for the installation of a pipe outlet. Once this application is sanctioned, they pay to the Irrigation Department a nominal fees annually for the use of the wastewater. However, wastewater is used by a large number of farmers below the outlet—as many as 20 farmers from each outlet based on mutual norms of cooperation, locally called *Bhaibandi* (meaning brotherhood) (Narain and Singh 2017).

This points to a difference in the concretisation of a water right and the materialization of a water right. Concretization of the water right refers to the manner in which the water right is defined. Materialisation of water rights refers to the manner in which water rights are realised. In other words, this points to a difference between the allocation of wastewater and its actual distribution.

When a farmer applies for a pipe outlet to use wastewater and the outlet gets sanctioned, it is a case of the concretisation of the water right. The water right has a sanction of the state authorities, that is, the Irrigation Department. However, its actual distribution below the outlet is shaped by informal norms of cooperation. Farmers below the outlet co—operate in terms of the collective building of the furrows and cleaning them once in two years. This points to the existence of legal pluralism in the use of wastewater; wastewater use is formally sanctioned by the state's Irrigation Department, but its use is shaped by other normative repertoires outside state sanction. Both non-statutory and statutory bases of legitimacy shape access to wastewater in the field.

In a context of urbanization and climate change, wastewater plays an important role in maintaining farmers' water (and livelihood) security. When a farmer was asked in the course of field interviews why he uses wastewater, he replied *"to aur kya karoon?* (what else should I do?)"*. However, there are also many aspects of water insecurity associated with the use of wastewater. First, the level of wastewater in the canal varies and there can be a mismatch between the supply in the wastewater canal and the farmers' requirements. Farmers also do not always have information regarding when the wastewater canal would be closed or shut. Finally, as the residential areas of the city expand and new settlements are constructed, there is an effort to cover the wastewater canal such that it flows under the ground and its sight and smell are not visible. This means that farmers are not sure about how long wastewater would be available to them as a resource. This represents another aspect of water insecurity.

The research in Budhera suggests that as rainfall becomes uncertain, the reliance on wastewater will increase. This is because farmers use wastewater as a substitute for rains. That is, they wait for the rains, and when rains fail, they use wastewater. In a context of urbanization and climate change, both of which make periurban residents water insecure, wastewater will play an important role in shaping water security in periurban agriculture. However, there are several aspects of insecurity attached to the use of wastewater itself, such as variations in the supply, uncertainty about its availability and the uncertainty about losing access to it as the city expands and the wastewater canal gets covered.

It is important to note however that none of the wastewater irrigated produce is consumed by the residents of the village themselves. It finds its way to the wholesale market. These villages are connected to two wholesale markets, namely, at Gurgaon and at Farukhnagar. This makes consumers of the produce vulnerable to negative health impacts. While the producers do not consume the produce, they are still vulnerable to the health impacts arising from exposure to their skin while standing in the wastewater, ankle and wrist deep. They safeguard themselves against the health impacts on their skin by applying mustard oil on their hands and legs.

Another adaptive response to water scarcity in this village has been institutional. The pundits of this village have organised themselves into collective tubewells. That is, they pool in resources to collectively dig and install tubewells. This represents an institutional innovation in the form of new collective institutions. This shows how communities mobilise their social capital to improve their access to water when urbanization processes and climate change weaken their water security. Details of this response are described in Narain et al. (2019).

As noted earlier in this paper, there is a need to unpack the notion of "community" in periurban contexts, given the huge diversities in the social and economic conditions of the inhabitants, that render connotations such as "community resilience" futile. In the context of Budhera, tenants and share-croppers are more vulnerable to the effects of high rainfall and flooding resulting in crop failure than are land—owners. Water insecurity is not only about deficit but also excess water. 2010 was a year of high rainfall. The agricultural fields of Budhera village are located in a low lying area, a depression, locally called *jheel*. In the event of heavy rainfall in 2010, the area experienced rainfall flooding. This caused damage to the paddy crop, which is the main crop of the *kharif* (monsoon) season. Besides, the rains and flooding were such that the flooding continued until the sowing time for wheat. The worst affected - and most vulnerable - to the impacts of flooding were tenants and sharecroppers who pay to the owner up-front an amount at the time the land is taken on contract for tilling. In the event of crop failure, thus, tenants and sharecroppers suffer the most. Besides, compensation is given by the state as a matter of policy only to land-owners, while tenants and sharecroppers receive no compensation. This is an interesting point in the Indian context where there has been much debate on the subject of land acquisition and fair compensation, focusing on the plight of land owners. However, it is the landless tenants and share-croppers who are more vulnerable as a result of the combined effects of urbanization and climate change.

There was another negative impact of the construction of the GWS canal and the NCR channel and the Chandu-Budhera water treatment plant on the population of Budhera. Seepage from the two canals and the water treatment plant rendered much of the fields alongside these canals unfit for cultivation. This was another way in which the negative impact of the growth of the city was borne by this village. This, too, represents the ecological foot-print of the city. Land owners have lately responded by crop or livelihood diversification. Owners of the lands adjacent to the water treatment plant have converted their lands to fish farms and given them out on contract. Some land-owners have switched to the farming of water chestnut.

31.3.3 Sadhraana: Chasing the Water Table

Sadhraana village lies adjacent to the well-known Sultanpur National Park, that came up in 1972 to protect the avifauna of the region. Sadhraana was a major supplier of land for the development of this National Park. Unlike Budhera, Sadhraana does not lie along the fresh water or wastewater canals. The major water insecurity issue here as a result of the combined effects of urbanization and climate change is a steady fall in the water table. The major stress on groundwater in this village comes from the large number of farm-houses of the urban elite, that are located in this village. These farm-houses are owned by the urban elite who use these as weekend get aways and for holding social get togethers and events. They have sprawling orchards that grow water guzzling fruits, flowers and vegetables. The effects of this increased stress on groundwater caused by the presence of the farm houses are aggravated by a change in the precipitation patterns. This is how urbanization and climate change interact to shape periurban water insecurity.

This village has two types of land, land overlying what locals call *meetha* (or sweet) water and lands overlying what locals call *khara* or saline water. Most farm-houses have been built on lands overlying sweet water. When farm-houses are built over saline groundwater, they have acquired small plots of land over the fresh water, installed submersible pump-sets there and transport water to their farm-houses, located as much as 4 kms away (Narain 2014). This is the result of a policy and institutional framework that ties access to groundwater to the ownership of land. It relates also to the common pool character of the resource, characterized by subtractability and non-excludability.

Local farmers find it hard to compete with the submersible pump-sets used by farm-houses and are thus left chasing the water table. Most farmers have switched from two to one crop a year, or cope by leaving their lands fallow. An important aspect of the adaptation strategies of the farmers in this village has been the use of sprinkler irrigation sets. This is because of a mix of agro-ecological and institutional factors. The terrain in the village is undulating, the soil is sandy and sprinkler irrigation is less labor intensive (Narain 2014). Thus it is very suited to periurban contexts where occupational diversification is on the rise and there is relatively little interest among youth to spend time on the agricultural fields.

The case of Sadhraana highlights a dimension of periurban water insecurity different from that of Budhera. Unlike Budhera, it shows that periurban water insecurity is represented not only by the physical flows of water from the country side to the city, but also local appropriation of groundwater by the urban elite, who move to the periphery and are able to dig deep into the aquifers. This places the resource out of the reach of small and marginal farmers who are unable to afford the high costs of extraction. These effects are aggravated by a decline in rainfall. This case also shows that periurban contexts, on account of the multiple stresses operating in them can be fertile grounds to study technical innovation. While farmers adopt new technologies - such as sprinklers in this case—it is necessary to study the social and agro-ecological conditions that facilitate the adoption of such technologies. In this case, the use of sprinklers is facilitated by the fact that occupational diversification is on the rise in periurban contexts, and that makes the adoption of sprinklers an attractive proposition. Further research on adaptation to water insecurity in periurban contexts should focus on the socio-technical mediation of water insecurity.

31.3.4 Sultanpur: Building Social and Human Capital

In Sultanpur, farmers have made important changes in the cropping pattern in response to changing climatic conditions. The main crops grown during the *Rabi*(winter) season are wheat and mustard, and during the *Kharif* (monsoon)season sorghum and pearl millet. But due to the use of saline ground water and reduced rainfall since 1977, farmers now grow only one crop yearly. Also, if they grow a *Kharif* crop, then the productivity of crops grown in the *Rabi* season of the same cropping year reduces.

About 10 or 12 years ago farmers used to grow vegetables and flowers; however, very few are able to do so now. Again, some 25 to 30 years ago farmers also used to grow pulses but they are not grown anymore. Reduction in rainfall was cited as the main reason behind these changes in the cropping patterns. The impact of decreased rainfall on crops was poignantly expressed by a village respondent. He said that due to less rainfall and a resulting dryness of land, termites move to top soil and harm the crops. That is, a decline in moisture increases the presence of termites in the field, which is detrimental to the crops.

People constantly referred to 1977 as the benchmark year for a comparison of the volume of rainfall across decades (Narain et al. 2016). With the reduction in rainfall, the production of pulses went down and eventually ceased. For some villagers, the high rainfall of 2010 compares to the kind of rainfall observed in 1977. Others believe the rains in 2010 were only relatively better than those in preceding years and pitched the rainfall of 2010 at about one-third the level of rains in 1977.

Sultanpur village saw some efforts at capacity-building by the project team under the IDRC supported project on water security in periurban South Asia (Narain et al. 2017). These were centered around building social and human capital. When the action research project was initiated, the team sensed a lack of accountability of the PHED, the Public Health Engineering Department to the residents of the village. In this context the project team organised a series of stakeholder dialogues and workshops between the representatives of the village and the PHED, the Public Health Engineering Department. This provided a forum for civic engagement and played an important role in reorienting the agencies of the state and making them more responsive to the water users. Thus, it sought to build social capital by providing avenues for civic engagement. The efforts were matched by capacity-building efforts through a series of workshops to build the capacity of state officials to be more responsive to the needs of water users.

Another aspect of building community resilience in peri-urban contexts in the face of urbanization and climate change is building their human capital. Peri-urban livelihoods face threats from land acquisition as well as climatic variability and change. Training villagers in alternative livelihood skills can help build their resilience. This was accomplished in the project by supporting the vocational training of village youth. Six of them were trained by the GMR Varalakshmi Centre for Livelihood and Empowerment. Five were trained in Refrigeration and Air-conditioning and one in an electrical engineering course. With support from the *panchayat* - the local level village governance body -the project team selected eight village youth six of whom were finally chosen by the Centre based on their educational and socio-economic background. At the end of the programme, one had dropped out for personal reasons. The remaining five received certificates and felt that this would enable them to secure their livelihoods.

31.4 Conclusion

This paper looks at the relationship between climate change and water security from a periurban perspective. Urbanization and climate change present multiple threats to water sources in periurban contexts. While most scholars focus on vulnerability in agrarian contexts or on building the resilience of cities, this paper focuses on peri-urban contexts that can be subject to both rural and urban stresses.

The study of three villages, namely, Sadhraana, Budhera and Sultanpur presents a typology of the different ways in which periurban water insecurity is experienced by periurban inhabitants as a result of the effects of urbanization and climate change.

This includes a fall in the water tables, caused by the movement of the urban elite into the periurban spaces; the construction of canals to transfer water from rural to urban areas, resulting in a loss of land and water resources, and a process of institutional neglect caused by state apathy. At the same time, periurban residents are not passive recipients of these changes, but as demonstrated in the case of Budhera and Sadhraana, innovate both technologically and institutionally to mediate water insecurity.

Most approaches to action research in periurban contexts focus on mobilising the community, for instance, through Participatory Action Planning. The state, however, remains a black box. Since periurban spaces suffer from institutional neglect, building social capital by strengthening civic engagement can be an important ingredient for improving the resilience of periurban communities. Since climate change and urbanization processes both place stress on periurban agriculture - through land acquisition and changes in precipitation patterns, building human capital by providing alternative livelihood and vocational skills can be other measures to improve the resilience of periurban communities.

Further research in periurban contexts should seek to identify the different ways in which urbanization and climate change intersect to shape water insecurity and the mix of technologies and institutions through which the mediation of water insecurity takes place. This approach can be used to develop archetypes of vulnerability and periurban water insecurity. Socio-technical perspectives, looking at the relationship between technology and institutions, as well as the social construction of technology can provide insights into the mix of technologies and institutions through which water insecurity is mediated. Perspectives of legal pluralism and social capital throw insights into how both statutory and non-statutory institutions are mobilized in the mediation of water insecurity. The concept of social capital can throw light on the mobilization of social relationships to improve water security, as well as the role of social relationships in shaping differential water (in) securities. Finally, listening to people's narratives of climate change, the way in which they perceive climate change and the meaning that they attach to it, is necessary to understand people's adaptive responses. These should not be simply dismissed as "perceptions" as their cognition of the experienced changes is necessary to understand the autonomous adaptation choices made by them. Building social and human capital can be important ways of building periurban resilience, allowing periurban residents to diversity occupationally and improve the representation of their interests in civic agencies.

Acknowledgements This paper is based on research carried out under two projects. The first is the project 'Water security in periurban South Asia: adapting to climate change and urbanization, supported by IDRC, International Development Research Center, Canada. The second is a project on periurban water security funded by NWO, the Dutch Scientific Organization under the CoCOON-CCMCC programme. Thanks are expressed to IDRC, Canada and the NWO. Thanks are also expressed to SaciWATERs, South Asian Consortium for Inter-Disciplinary Water Resources Studies for organizational support.

References

Adger NW (2006) Vulnerability. Glob Environ Chang: Hum Policy Dimens 16(3):268–281
Cook C, Bakker K (2012) Water security: debating an emerging paradigm. Glob Environ Chang 22:94–102
Eakin H (2005) Institutional change, climate risk and rural vulnerability: cases from Central Mexico. World Dev 33(11):1923–1938
Gomes SL, Hermans LM (2018) Institutional function and urbanization in Bangladesh: how peri-urban communities respond to changing environments. Land Use Policy 79:932–941
Intergovernmental Panel on Climate Change (IPCC) (2007) Climate change 2007: the physical science basis. Contribution of working group I to the fourth assessment report of the intergovernmental panel on climate change. Cambridge University Press, Cambridge
Jalan J, Ravallion M (1999) Are the poor less insured? Evidence on vulnerability to income risk in rural China. J Dev Econ 58(1):61–81
Kirby P (2006) Vulnerability and violence: the impact of globalization. Pluto Press, London
Leichenko R, O'Brien K (2002) The dynamics of rural vulnerability to global change: the case of Southern Africa. Mitig Adapt Strat Glob Chang 7(1):1–18
Mackenzie K, Harpham T (2006) Social capital and mental health. Jessica Kingsley Publishers, London
McGee TG (1991) The emergence of Desakota regions in Asia: expanding a hypothesis. The extended metropolis: settlement transition in Asia
Mercer J (2010) Disaster risk reduction or climate change adaptation: are we reinventing the wheel? J Int Dev 2:247–264
Narain V (2014) Whose land? Whose water? Water rights, equity and justice in periurban Gurgaon India. Local Environ: Int J Justice Sustain 19(9):974–989
Narain V, Nischal S (2007) The peri-urban interface in Shahpur Khurd and Karnera India. Environ Urban 19(1):261–273
Narain V, Singh AK (2017) Flowing against the current: the socio-technical mediation of water (in) security in periurban Gurgaon, India. Geoforum 81:66–75
Narain V, Ranjan P, Vij S, Dewan A (2017) Taking the road less taken: reorienting the state for periurban water security. Action Res 18:528–545. https://doi.org/10.1177/1476750317736370
Narain V, Ranjan P, Singh S, Dewan A (2016) Urbanization, climate change and water security in periurban Gurgaon India. In: Narain V, Prakash A (eds) Water security in periurban South Asia: adapting to climate change and urbanization. Oxford University Press, New Delhi
Narain V, Vij S, Dewan A (2019) Bonds, battles and social capital: power and the mediation of water insecurity in peri-urban Gurgaon India. Water 11(8):1607
Narain V, Singh AK (2019) Gurgaon. In: The wiley blackwell encyclopedia of urban and regional studies, pp 1–4
Narain (2009) Growing city, shrinking hinterland. Land Acquisition, transition and conflict in periurban Gurgaon. India. Environ Urban 21(2):501–512
O'Brien K, Quinlan T, Ziervogel G (2009) Vulnerability intervnetions in the context of multiple stressors: lessons from the Southern Africa vulnerability initiative. Environ Sci Policy 12(1):23–32
Putnam R (2000) Bowling alone: America's declining social capital. In: Culture and politics, pp 223–234
Saldana-Zorilla SO (2008) Stakeholders' views in reducing rural vulnerability to natural disasters in Southern Mexico: hazard, exposure, coping and adaptive capacity. Glob Environ Chang: Hum Policy Dimens 18(4):583–597
Shrestha A (2019) Urbanizing flows: growing water insecurity in peri-urban Kathmandu Valley, Nepal (Doctoral dissertation). Wageningen University, Wageningen

Singh AK, Narain V (2020) Lost in transition: perspectives, processes and transformations in Periurbanizing India. Cities 97:102494

Vij S, Narain V (2016) Land, water and power: the demise of common property resources in periurban Gurgaon, India. Land Use Policy 50:59–66

von Benda Beckmann (1989) Scape-goat and magic charm: law in development theory and practice. J Legal Pluralism Unoff Law 21(8):129–148

Chapter 32
Selecting Multi-Functional Green Infrastructure to Enhance Resilience Against Urban Floods

A. Alves, A. Sanchez, B. Gersonius, and Z. Vojinovic

Abstract Climate change and population growth are increasing pressure on urban drainage infrastructure, incrementing the level of flood risk in particular in urban areas. Traditional approaches to cope with urban floods offer low adaptation capacity to the uncertain future. As a response, the use of sustainable drainage measures, also called green infrastructure (GI), have been increasingly suggested in the last years. One important reason for their increasing popularity has been the multiple benefits that they offer to the environment. These benefits include environmental and socio-economic aspects, besides sustainable stormwater management. However, an important restriction for GI application in urban areas is the difficulty of decision making. These difficulties to select innovative drainage technologies are based mainly on lack of information and physical complexity of urban drainage systems. Moreover, there are many types of GI with particular characteristics, requirements and limitations. This work proposes a procedure for selection of promising sustainable measures to cope with urban floods, considering local constraints and environmental aspects through the evaluation and integration of multiple benefits. In order to facilitate the application of the proposed methodology, the process has been coded into a software program. The method is tested in two case study sites in Thailand to prove its effectiveness. The outcome of this work is seen as a useful approach for helping decision making processes with the aim of reducing urban flood risk in a sustainable way, allowing the improvement of other environmental aspects.

Keywords Green infrastructure · Urban drainage · Multiple benefits · Decision making

A. Alves (✉) · A. Sanchez · B. Gersonius · Z. Vojinovic
UNESCO-IHE, Institute for Water Education, Delft, The Netherlands
e-mail: a.alves@unesco-ihe.org

© Springer Nature Switzerland AG 2021
M. Babel et al. (eds.), *Water Security in Asia*, Springer Water,
https://doi.org/10.1007/978-3-319-54612-4_32

32.1 Introduction

32.1.1 Background

There are two main drivers, which affect the performance of urban drainage systems. On one hand, climate change is expected to have significant impacts on future rainfall characteristics (Vojinovic and van Teeffelen 2007; Singh et al. 2016). On the other hand, population growth is related with changes in land use and urbanisation impacting runoff characteristics (Kumar et al. 2013; Sanchez et al. 2014). Both, the combined effect of these two drivers as well as the tendencies followed for them suggest a probable increment of future flood risk (Yazdanfar & Sharma 2015; Vojinovic 2015).

However, even without taking into account future changes, nowadays flooding is a world-wide issue, which causes extensive destruction and economic damages, and this is among all natural disasters the one that occurs most frequently (Mynett and Vojinovic 2009). In particular, urban areas appear as the most impacted zones when results about recent flooding occurrence are considered (Jha et al. 2012). Even though it is well known that economic losses from weather and climate related disasters have increased in the last 30 years, estimations of losses are still undervalued. The reason of this lays in that many impacts, such as loss of human lives, cultural heritage (Vojinovic et al. 2015), and ecosystem services are difficult to value in monetary terms, hence they are poorly reflected in estimations of losses (IPCC 2012). The use of numerical models is invaluable for analysis of potential scenarios. However, a great care needs to be given in order to maximise the use of numerical models, and this relates to both data collection and processing activities as well as the model complexity and modelling approach (Vojinovic et al. 2006b, 2011; Abdullah et al. 2009,2011a,2011b; Seyoum et al. 2010,2012; Meesuk et al. 2015; Vojinovic and Tutulic 2009).

Cities are dynamic and self-organising systems, where the concept of sustainability goes beyond the idea of developing ecosystem services to the concept of building resilience capacity. Historically, urban planners have focused mainly in the spatial configuration of the urban form based on an equilibrium point of view. However, the sustainability of cities concept needs to involve a resilience approach; reaching this point the city can deliver ecosystem services over time (Ahern 2013). It is in this context where the concept of sustainable stormwater management measures becomes essential. These measures have the capacity to enhance benefits related with urban ecology, energy, landscape and socio-economic systems (Hoang and Fenner 2015).

The development of resilient urban spaces capable to cope with extreme climate is based on an integrated and interdisciplinary approach, which lacks sufficient knowledge. Whereas traditional pipe and storage based approaches count with enough technical support and tools for decision making, the sustainable approaches or non-traditional measures for stormwater management lacks sufficient supporting documentation and tools. In particular, this support is missing regarding the evaluation of additional benefits and long-term performances. Moreover, there is a deficiency of methods to identify how to combine different measures in order to achieve sustainable solutions (Voskamp and Van de Ven 2015). Besides, it is important to analyse and to understand the characteristics of each context in order to develop effective system rehabilitation and adaptation strategies for current and future scenarios (Vojinovic et al., 2006a, 2008, 2014; Barreto et al. 2006, 2008, 2010; Alves et al. 2016b). Regarding this, the measures should be evaluated under different conditions, analysing their applicability for different contexts, and their potentiality to achieve multiple benefits related with other ecosystem services and urban well-being rather than just the management of stormwater (Vojinovic 2015). Therefore, it is important to define methods capable of identifying measures that provide benefits now, but can also address projected trends of climate extremes. In order to achieve this, it is needed to recognise that many drainage strategies offer co-benefits, such as improvements in livelihoods, human well-being, and biodiversity conservation, and the design of solutions needs to incorporate these aspects (IPCC 2012).

32.1.2 Selection of sustainable measures

In the area of urban drainage management, similar concepts are named with different terms in different parts of the world. Terms such as BMPs (best management practices), LIDs (low impact development), WSUD (water sensitive urban design), SuDS and GI are broadly used. Taking this into account, a careful use of terminology should be achieved, to minimise the possibility of miscommunication (Fletcher et al. 2014). In this work, the names GI, non-traditional measures and sustainable solutions, are used to name all the different sustainable measures (BMPs, LIDs, SUDS, WSUD, GI, etc.), while traditional or grey infrastructure are used to refer to conventional solutions.

It is through the analysis of the multiple aspects of applying GI, that the multiple benefits from them obtainable in urban spaces can be visualised. These benefits are, for instance, stormwater detention, reduction of extreme heat events and energy consumption, improvement of water and air quality, reduction of potable water consumption, biodiversity enhancement, opportunities for education, and health benefits. Furthermore, the synergies between GI for stormwater management and other benefits have the capacity of saving costs. Considering this, GIs appear as cheaper in terms of investment and maintenance costs than strictly conventional strategies (Tzoulas et al. 2007; Ashley et al. 2011; USEPA 2013; European Commission 2012; CIRIA 2013).

Beyond the factors to be considered during the selection of measures, it is also important to evaluate the impact of different combinations of measures, and to develop criteria to define the best group of solutions for each case. The achievement of resilience is based on the improvement of four aspects oriented to reduce vulnerability: adaptive, threshold, coping and recovery capacity. Besides that, different GI have impact on different capacities. Consequently, it is combining measures that

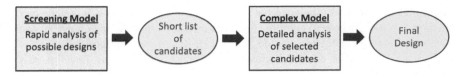

Fig. 32.1 Recommended steps for design processes (adapted from De Neufville and Scholtes 2011)

a complete climate vulnerability reduction and resilience capacity can be reached (Voskamp and Van de Ven 2015).

Decision-making can be defined as the process of making an informed choice to select among possible alternatives. Some different stages involved in this process are: the definition of objectives, collection of relevant information, identification of alternatives, and the establishment of criteria for the decision (Simonovic 2012). Decision support systems (DSS) are helpful tools to make safe and reliable decisions when trying to solve complex problems. These systems are a category within the group of information systems, and are composed of the mixture of databases and models in combination with a user-friendly interface (Price and Vojinovic 2011; Abbott et al. 2006; Abbott and Vojinovic 2009).

According to de Neufville and Scholtes (2011), it is a good practice to use screening and detailed models together for designing processes. In the first stages of a project, a screening model can be used to develop a rough understanding of the possible designs. Through this procedure, the range of design alternatives is narrowed down. Afterwards, complex models can be used to redefine, to validate and to modify the general results of the screening models. Figure 32.1 presents this process, in which screening models are more useful at the start, and detailed models are applicable as the design is in an advanced stage. The authors argue that screening model should have two principal characteristics: fast implementation, so that it can provide a quick answer for designs under different scenarios; and it should rank alternatives through comparison among them, so a short list with the best candidates can be extracted for detailed examination.

This work proposes a procedure for selection or screening of promising measures to cope with urban floods, considering local constraints and environmental aspects, through the evaluation and integration of multiple benefits. In order to facilitate the application of the proposed methodology, the process has been coded into a software program. The method is tested in two case study sites in Thailand in order to prove its effectiveness. The outcome of this work is seen as a useful approach for helping decision making processes with the aim of reducing urban flood risk in a sustainable way, allowing the improvement of other environmental aspects.

32.2 Materials and Methods

32.2.1 Methodology

This work develops and applies a framework, which includes the screening and ranking of options among all possible measures for stormwater management. The measures to be analysed are selected from an extensive Knowledge Base (KB) being developed for the PEARL Project (https://pearl-kb.hydro.ntua.gr, Karavokiros et al. 2016).

The methodology is divided into two steps. The first step consists of measures screening, the criteria considered here are the type of floods and local physical constraints. The first criterion for screening removes the options that do not fulfil the requirements according to the flood type affecting the area, while the second one considers local characteristics, which could prevent the use of some measures. For instance, if soil in the area is impervious, infiltration measures are most likely eliminated from this list.

The second step will consist of ranking the remaining measures according to two factors. The first ranking analyses local space characteristics to identify the most suitable measures according to the urban shape. The second ranking considers the co-benefits that the screened measures can provide, and which of them are identified as more relevant for the case according to local preferences.

Figure 32.2 shows the different steps to be followed in this methodology:

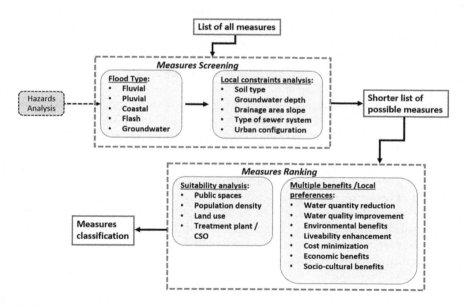

Fig. 32.2 Selection of measures considering local constraints and multiple benefits (adapted from Alves et al. 2016a)

- Measures Screening
- Measures Ranking

The complete analysis of measures is done from a qualitative point of view, based on desk study and using several available sources, which describe and analyse sustainable solutions for stormwater management (Alves 2014; Center for Neighborhood Technology 2009; CIRIA, 2007; UDFCD 2010; and fact sheets obtaiend from: https://nepis.epa.gov and https://www.stormwatercenter.net/).

With the objective of enabling the implementation of this methodology in an easy process, a tool was coded in Pascal using Embarcadero RAD studio X6. Through this tool the user can apply the method following the different steps and answering questions about local characteristics and benefits preferred. The achievement of good results depends on the availability of local experts in order to get proper answers about local characteristics. After following all required steps, a short and ordered list of measures is obtained.

32.2.2 Case Study Areas

With the objective of comparing the application of this methodology on different study areas, two cases from Thailand were selected. These cases have different characteristics, which are seen as a good condition for results comparison. The first study area (Study Area 1) is a high urbanised zone located in Eastern sub urban part of Bangkok which is one of the parts of Central Business District, called Sukumvit Area. As a result of climate change, it is expected that the urban drainage system in the area becomes inadequate in capacity in coming years, resulting in higher flood risk and more requirements for maintenance. Although pluvial flood in urban areas resulting from low capacity of drainage systems does not pose high hazard in terms of velocity and depth, the impacts imply an increment of flood volumes and affected area for the future. This type of floods in urban areas affect daily lives of people and socio-economic environments. As an example of negative impacts, contact with polluted water is commonly experienced by pedestrians during pluvial flooding, which results on high bacteriological risk. Mitigation and adaptation plans could prevent serious consequences in socio-economic environments (Shrestha 2013; Polania 2015). Figure 32.3 shows the flood map for the area corresponding to the rainfall of two years return period.

The second study area (Study Area 2) considered is Koh Mueng in Phra Nakorn Si, Ayutthaya province, which is about 80 km north of Bangkok. It has an area of 7 km^2 and is surrounded by three rivers: Chao Phraya River, Pasak River and Lop Buri River. As a consequence of this physical characteristic, Koh Mueng is so-called Ayutthaya City Island. From the study of flood problems in Ayutthaya urbanized area, it can be concluded that the most devastating flood events are a consequence of high water levels of these three rivers. However, the area also experiences pluvial floods. Almost half of the study area is covered by a World Heritage Site, while the remaining areas

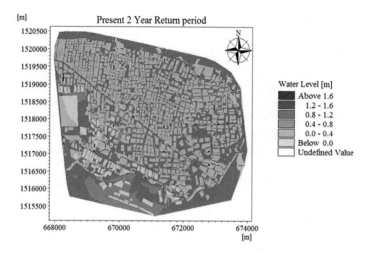

Fig. 32.3 Flood map for Sukumvit Area corresponding to 2 years return period event (Shrestha 2013)

Fig. 32.4 Flood map for Ayutthaya during 2011 flood events (Meesuk et al. 2015)

are used as residential zones, education purposes facilities, commercial installations, and public building (Keerakamolchai 2014). Figure 32.4 presents the flood map obtained for the large flooding experienced in the area during the 2011 events.

The study areas selected have significantly different characteristics; the one is a highly urbanised area, with mainly commercial land use and high percentage of impervious surface, while the other is a less urbanised area, with principally residential and touristic land use. Besides, the flood problems affecting both areas are different, while the first one suffers mainly pluvial flood problems, the second one faces fluvial floods as main problem.

32.3   Results and Discussion

The presented methodology was applied for Sukumvit Area, obtaining answers about local characteristics from a local technical stakeholder. Table 32.1 and Fig. 32.5 show the results obtained. In Fig. 32.5, the selected measures are presented (using reference numbers, which measures correspond is presented on Table 32.1) according to the two rankings considered, suitability and preferred benefits. Considering these results, a categorization of measures is done. Those measures with high score in both rankings included into category 1 (red zone in the graph); the measures performing good for one ranking but not good for the other one, are category 2 (yellow zone in the graph); and the measures that are performing bad in both cases are category 3 (green zone in the graph). Table 32.1 presents the names of selected measures, the score values and category in each case.

Analysing the selected measures, it can be observed that best cases selected (category 1) are decentralised measures (green roofs and walls), and green solutions (trees, parks and paved surfaces reduction). This result obtained in this case appears as proper, considering that the area has low space availability, where to locate large centralised solutions. Besides, all measures selected as preferred are green options, which can offer important co-benefits in a highly paved zone, such as heat stress reduction and public green spaces expansion.

From the application of the same methodology in Ayutthaya case, significantly different results are obtained, which are in agreement with the difference in flood type and local physical conditions. The results obtained in this case are showed in Fig. 32.6

Table 32.1 Selected measures for Sukumvit Area

	Measure	Score		Category
		Suitability	Preferences	
1	Wet Flood Proofing	14	11	2
2	Dry Flood Proofing	14	13	2
3	Infiltration Trenches (drain system)	13	15	2
4	Rainwater harvesting	13	15	2
5	Green walls	13	18	1
6	Green roofs	13	20	1
7	Pervios Pavements	13	15	2
8	Blue Roofs	13	14	2
9	Hollow roads	13	12	2
10	City trees / Parks	11	20	1
11	Paved surfaces reduction	11	20	1
12	Infiltration basin	9	19	2
13	Rainwater disconnection	9	17	2
14	Infiltration Trenches	7	15	3
15	Soakaways	7	15	3
16	Rain Gardens	7	21	2
17	Vegetated Filter Strips	5	21	2
18	Open gutters	11	12	2
19	Closed storages	9	10	3
20	Tunnels	9	11	3
21	Pipes	9	10	3

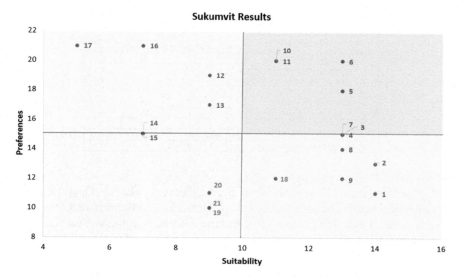

Fig. 32.5 Rankings scores for selected measures in Sukumvit Area case

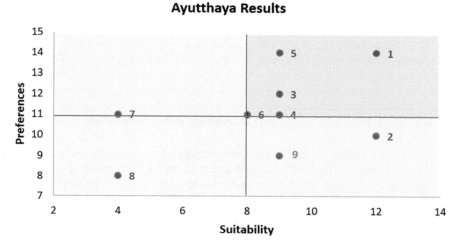

Fig. 32.6 Rankings scores for selected measures in Ayutthaya Area case

and Table 32.2. Again, the selected measures are divided into three different categories and presented with colours on the graph (red for the best measures, yellow for medium preferred measures, and green for the least preferred). Table 32.2 shows the correspondence between the reference numbers used in the graph and the measures names, the measures classification is also presented (categories 1, 2 and 3, for best, medium and least preferred measures, respectively).

Table 32.2 Selected measures for Ayutthaya Area

	Measure	Score		Category
		Suitability	Preferences	
1	Open Detention Basin	12	14	1
2	Floating/Amphibious Buildings	12	10	2
3	Temporary/Demountable Barriers	9	12	1
4	Wet Flood Proofing	9	11	2
5	Dry Flood Proofing	9	14	1
6	Non Return Valves	8	11	2
7	Polder	4	11	3
8	Pumping System	4	8	3
9	Raising existing dikes	9	9	2

In this case, there is a notoriously preferred measure, which is Open Detention Basin. Additionally, temporary solutions as Demountable Barriers and Dry Flood Proofing are chosen. In this case, the preferred solution to cope with pluvial floods is a centralised option. This measure requires availability of space, which in not a major constraint in the area.

32.4 Conclusions

This work presents a methodology for screening and ranking measures against urban floods. The selection process is based on the analysis of local flood problems and local conditions. This includes local physical characteristics as well as features of the urban space. Furthermore, the method provides rankings among the applicable measures, which consider not only the co-benefits achievable through the implementation of green infrastructure, but also local preferences for those benefits. In order to facilitate the implementation of the developed methodology, a decision support tool or desktop application was coded.

The developed tool was applied in two different study areas, obtaining different favourite measures, which are in accordance with the characteristics of each case. The outcome of this work is seen as a useful approach for helping decision making processes with the aim of reducing urban flood risk in a sustainable way, allowing the improvement of other environmental aspects.

Further work will be focused on the validation and improvement of the process through its application to other case studies. Moreover, future improvements will include the analysis of measures combination in order to develop long term sustainable strategies.

Acknowledgements The research leading to these results has received funding from the European Union Seventh Framework Programme (FP7/2007-2013) under Grant agreement n° 603663 for the research project PEARL (Preparing for Extreme and Rare events in coastaL regions). The study reflects only the authors' views, and the European Union is not liable for any use that may be made of the information contained herein

References

Abbott MB, Tumwesigye BM, Vojinovic Z (2006) The fifth generation of modelling in Hydroinformatics. In: Proceedings 7th International Conference on Hydroinformatics, 2091–2098, Nice, France

Abbott MB, Vojinovic Z (2009) Applications of numerical modelling in hydroinformatics. J Hydroinf 11(3–4):308–319

Abdullah A, Rahman A, Vojinovic Z (2009) LiDAR filtering algorithms for urban flood application: review on current algorithms and filters test . Int Arch Photogramm Remote Sens Spat Inf Sci 38:30–36

Abdullah AF, Vojinovic Z, Price RK (2011) A methodology for processing raw LIDAR data to support 1D/2D urban flood modelling framework. J Hydroinf 14(1):75–92

Abdullah AF, Vojinovic Z, Price RK (2011) Improved methodology for processing raw LIDAR data to support urban flood modelling - accounting for elevated roads and bridges. J Hydroinf 14(2):253–269

Ahern J (2013) Urban landscape sustainability and resilience: the promise and challenges of integrating ecology with urban planning and design. Landsc Ecol 28(6):1203–1212

Alves A (2014) Model based multi-objective evaluation of BMP system configurations for CSO reduction. Master of Science Thesis. UNESCO-IHE Institute for Water Education & Asian Institute of Technology, Delft

Alves A, Sanchez A, Gersonius B, Vojinovic Z (2016a) A model-based framework for selection and development of multi-functional and adaptive strategies to cope with urban floods. In: Proceedings 12th International Conference on Hydroinformatics, 877–884, Incheon, Korea

Alves A, Sanchez A, Vojinovic Z, Seyoum S, Babel M, Brdjanovic D (2016) Evolutionary and holistic assessment of green-grey infrastructure for CSO reduction. Water 8(9):402

Ashley RM, Nowell R, Gersonius B, Walker L (2011) Surface water management and urban green infrastructure: a review of potential benefits and UK and international practices. Foundation for Water Research, Marlow, Marlow

Barreto W, Vojinovic Z, Solomatine DP, Price RK (2008) Multi-tier modelling of urban drainage systems: muti-objective optimization and parallel computing. In: Proceedings 11th International Conference on Urban Drainage, Edinburgh

Barreto W, Vojinovic Z, Price RK, Solomatine DP (2010) A multi-objective evolutionary approach for rehabilitation of urban drainage systems. J Water Resour Plann Manage ASCE 136(5):547–554

Barreto W, Vojinovic Z, Price R, Solomatine D (2006) Approaches to multi-objective multi-tier optimisation in urban drainage planning. In: Proceedings 7th International Conference on Hydroinformatics, Nice

Center for Neighborhood Technology: Green Values: National Stormwater Management Calculator. https://greenvalues.cnt.org/national/calculator.php

CIRIA: The SUDS manual, CIRIA, London (2007)

CIRIA: Demonstrating the multiple benefits of SuDS - a business case, https://www.susdrain.org/resources/ciria-guidance.html

De Neufville R, Scholtes S (2011) Flexibility in Engineering Design. The MIT Press, Cambridge, Massachusetts/London, Engineering Systems

Commission E (2012) The Multifunctionality of Green Infrastructure. Science for Environment Policy, European Commission

Fletcher TD, Shuster W, Hunt WF, Ashley R, Butler D, Arthur S, Trowsdale S, Barraud S, Semadeni-Davies A, Bertrand-Krajewski J-L, Mikkelsen PS, Rivard G, Uhl M, Dagenais D, Viklander M (2014) SUDS, LID, BMPs, WSUD and more – the evolution and application of terminology surrounding urban drainage. Urban Water J 12(7):525–542

Hoang L, Fenner R (2015) System interactions of stormwater management using sustainable urban drainage systems and green infrastructure. Urban Water J 13(7):739–758

IPCC: managing the risks of extreme events and disasters to advance climate change adaptation. Cambridge University Press, Cambridge (2012)

Jha AK, Bloch R, Lamond J (2012) Cities and Flooding: A Guide to Integrated Urban Flood Risk Management for the 21st Century. The Word Bank, Washington

Karavokiros G, Lykou A, Koutiva I, Batica J, Kostaridis A, Alves A, Makropoulos C (2016) Providing evidence-based, intelligent support for flood resilient planning and policy: the PEARL knowledge base. Water 8(9):392

Keerakamolchai W (2014) Towards a framework for multifunctional flood detention facilities design in a mixed land use area the case of ayutthaya world heritage site. Thailand, Master of Science Thesis, Asian Institute of Technology and UNESCO-IHE, Thailand

Kumar DS, Arya DS, Vojinovic Z (2013) Modelling of urban growth dynamics and its impact on surface runoff characteristics. Comput Environ Urban Syst 41:124–135

Meesuk V, Vojinovic Z, Mynett A, Abdullah AF (2015) Urban flood modelling combining top-view LiDAR data with ground-view SfM observations. Adv Water Resour 75:105–117

Mynett A, Vojinovic Z (2009) Hydroinformatics in multi-colours – part red: urban flood and disaster management. J Hydroinf 11(3–4):166–180

Polania J (2015) A methodology for health impact assessment during pluvial flooding: a case study of the Sukhumvit area in Bangkok, Thailand. Master of Science Thesis, Asian Institute of Technology and UNESCO–IHE, Thailand

Price RK, Vojinovic Z (2011) Urban Hydroinformatics: Data, Models and Decision Support for Integrated Urban Water Management. IWA Publishing, London

Sanchez A, Medina N, Vojinovic Z, Price R (2014) An integrated cellular automata evolutionary-based approach for evaluating future scenarios and the expansion of urban drainage networks. J Hydroinf 16(2):319–340

Seyoum S, Vojinovic Z, Price RK (2010) Urban pluvial flood modeling: development and application. In: Proceedings 9th International Conference on Hydroinformatics, Tianjin

Seyoum SD, Vojinovic Z, Price RK, Weesakul S (2012) Coupled 1D and non-inertia 2D flood inundation model for simulation of urban pluvial flooding. J Hydraul Eng ASCE 138(1):23–34

Shrestha A (2013) Impact of climate change on urban flooding in sukhumvit area of Bangkok. Master of Science Thesis, Asian Institute of Technology and UNESCO – IHE, Thailand

Simonovic SP (2012) Floods in a Changing Climate: Risk Management. Cambridge University Press, Candbridge

Singh R, Arya DS, Taxak A, Vojinovic Z (2016) Impact of climate change on rainfall intensity-duration-frequency (IDF) curves in roorkee. India Water Resour Manage 30(13):4603–4616

Tzoulas K, Korpela K, Venn S, Yli-Pelkonen V, Kaźmierczak A, Niemela J, James P (2007) Promoting ecosystem and human health in urban areas using Green Infrastructure: a literature review. Landsc Urban Plan 81(3):167–178

UDFCD: Urban Storm Drainage Criteria Manual Volume 3, Stormwater Best Management Practice. Water Resources Publications, Denver (2010)

USEPA (2013) Case studies analyzing the economic benefits of low impact development and green infrastructure programs. Washington, DC

Vojinovic Z, Golub D, Hammond M, Hirunsalee S, Weesakul S, Meesuk V, Medina NP, Abbott M, Sanchez A, Kumara S (2015) Holistic approach to flood risk assessment in urban areas with cultural heritage: a practical application in Ayutthaya Thailand. Nat Hazards 81(1):589–616

Vojinovic Z, Sahlu S, Seyoum S, Sanchez A, Matungulu H, Kapelan Z, Savic D (2014) Multi-objective rehabilitation of urban drainage systems under uncertainties. J Hydroinf 16(5):1044–1061

Vojinovic Z, van Teeffelen J (2007) An integrated stormwater management approach for small islands in tropical climates. Urban Water J 4(3):211–231

Vojinovic Z, Solomatine DP, Price RK (2006) Dynamic least-cost optimisation of wastewater system remedial works requirements. Water Sci Technol 54(6–7):467–475

Vojinovic Z (2015) Floor Risk: The Holistic Perspective. IWA Publishing, London

Vojinovic Z, Tutulic D (2009) On the use of 1D and coupled 1D–2D approaches for assessment of flood damages in urban areas. Urban Water J 6(3):183–199

Vojinovic Z, Bonillo B, Chitranjan K, Price R (2006b) Modelling flow transitions at street junctions with 1D and 2D models. In: Proceedings 7th International Conference on Hydroinformatics, Acropolis - Nice

Vojinovic Z, Seyoum SD, Mwalwaka JM, R.K. (2011) Price effects of model schematization, geometry and parameter values on urban flood modelling. Water Sci Technol 63(3):462–467

Vojinovic Z, Sanchez A, Barreto W (2008) Optimising sewer system rehabilitation strategies between flooding, overflow emissions and investment costs. In: Proceedings 11th International Conference on Urban Drainage, Edinburgh

Vojinovic Z, Seyoum S, Salum MH, Price RK, Fikri AF, Abebe Y (2012) Modelling floods in urban areas and representation of buildings with a method based on adjusted conveyance and storage characteristics. J Hydroinf 15(4):1150–1168

Voskamp IM, Van de Ven FHM (2015) Planning support system for climate adaptation: composing effective sets of blue-green measures to reduce urban vulnerability to extreme weather events. Build. Environ 83:159–167

Yazdanfar Z, Sharma A (2015) Urban drainage system planning and design – challenges with climate change and urbanization: a review. Water Sci. Technol 72(2):165–179

Chapter 33
Climate Change and Sustainable Urbanisation: Building Urban Water Security in a Metro City of India

Shailendra K. Mandal and Gregg M. Garfin

Abstract India is experiencing rapid urbanisation and escalating urban water demand. The city of Patna, located in the central part of the Gangetic plains, is one of the fastest growing metro cities in India. Despite the fact that Patna is located on the banks of the river Ganga, residents are dependent on only groundwater aquifers for domestic water supply. Increasing pressure on groundwater supplies is exacerbated by the unregulated construction of deep wells, along with the development of apartment complexes to accommodate a mushrooming urban population. A comparison of data sets from 1960 to 2010 shows deep aquifer declines (Saha et al. 2013). This raises concerns about Patna's water supply security, its economic vitality and sustainability. Projected temperature increases will add to stress on water supplies, through increased water demand and evapotranspiration. The increasing exposure to climate change will be superimposed on existing vulnerabilities. In order to address these vulnerabilities and increase the Patna's water security, the research recommends building the capacity of city water managers to use climate information in urban planning and development. According to our case studies of United States urban water management practices, consideration of long-term climate variations are key to informing sustainable water planning. One simple adaptation measure is to foster the sharing of climate information and data with water managers. The study concludes that these rudimentary measures to address Patna's non-environmental water management challenges are a necessary stepping-stone to transformative pathways for dealing with climate change risks.

Keywords Urban experience · Water supply · Climate change · Groundwater resource · Water security measures

S. K. Mandal (✉)
National Institute of Technology Patna, Patna, India
e-mail: shailendra@nitp.ac.in

G. M. Garfin
Tucson, USA

© Springer Nature Switzerland AG 2021
M. Babel et al. (eds.), *Water Security in Asia*, Springer Water,
https://doi.org/10.1007/978-3-319-54612-4_33

33.1 Introduction

Access to safe water for all people is a United Nations Sustainable Development Goal (SDG) and is key to a successful development strategy. The most significant resource for reducing poverty and disease and improving the life of the poor is through adequate water access and increased food security (Reid & Vogel 2006; UNDP 2006). Many towns and cities in developing countries have unreliable piped water systems. Service delivery is deteriorating in these centers mainly because of rapid population growth and urbanisation.

As per ministry of housing and urban affairs, India is urbanising rapidly, out of the total population of 1210.2 million as on 1st March 2011, about 377.1 million are in urban areas. The percentage of urban population to the total population of the country stands at 31.6. There has been an increase 3.35-percentage point in the proportion of urban population in the country during 2001–2011. Further the number of million plus cities/urban agglomeration UA has increased from 35 in Census 2001 to 53 in Census 2011. Approximately 45% of the total population in India will be living in urban areas as per the report on Indian Infrastructure and Services. It also indicates that the unhappy state of urban service delivery for water and sanitation (Central Ground Water Board 2011).

Growth occurring in an unplanned manner, cities are stressed to keep up with the fast pace of urbanisation. Water distribution remains a tenacious issues in almost all the Indian cities. Some households receive water every other day for 30 minutes at a time (Institute for Social and Environmental Transition (ISET) 2011). Occurrence of contamination is very common in Indian cities due to leaks occurring where water supply pipes tend to run next to the sewage lines. Adding to the inefficiency and insufficiency issues in water systems is the combined impact of decrepit infrastructure and inadequate municipal finances; - for example, recovery of operation and maintenance costs is typically only about 30–35% for Indian cities.

Patna, the historical city of India (Fig. 33.1) is situated in the north-eastern part having the same story of many cities in the developing world. Due to the population growth in the cities and urbanisation around the world, institutions and infrastructure are experiencing the added impact of climate change (Christensen et al. 2013; Hijoka et al. 2014). In South Asia, observed urban climate change effects include statistically significant increases in average annual temperatures (Christensen et al. 2013; Hijoka et al. 2014) and increases in the numbers of warm days and nights (Christensen et al. 2013). The city is bordered by rivers on three sides nonetheless groundwater is Patna's chief source of potable water. Patna's shallow groundwater aquifer is polluted and the city relies on pumping from the deep aquifer, to provide potable water. Patna lacks groundwater regulation; thus, the city is at risk of overexploitation of this critical resource. Consequently, studies have demonstrated a decrease in deep aquifer levels. Closely monitoring of the piezometric level of the bore well shows decline in water table. In summary, there is high confidence in projected rise in temperature. There is medium confidence in summer monsoon precipitation increase in the future over South Asia (Sharma et al, 2015)

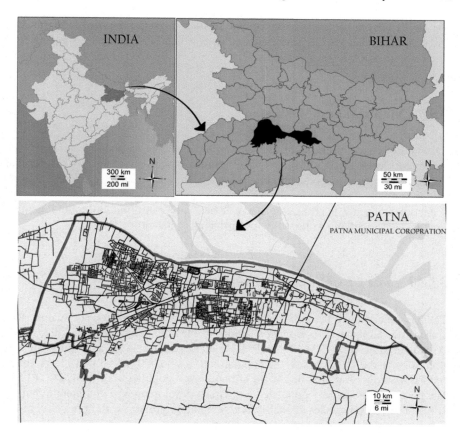

Fig. 33.1 Location of city of Patna, India

The Patna Municipal Corporation (PMC) is the government entity responsible for managing water supply and distribution for the city of Patna. Patna Jal Parishad is responsible for operations and maintenance and the Bihar Rajya Jal Parishad is responsible for capital works. 60% of city's population receive water through municipal piped water connection (City Development Plan). Inadequacies in water quality and delivery of water by local urban body leads to groundwater withdrawal by individuals, which leads to the declining in groundwater table. This creates challenges and inefficiencies within the water delivery system. It also creates issues like inequitable water distribution within the city. Disconnected decision-making structure ultimately leads to a more fragile and (often) biased system.

The objective of this study is to identify a set of strategies, to make Patna's water supply more sustainable and resilient in the face of urbanisation and climate change. Based on this study, some of the interventions have been identified like mapping the area most suitable for groundwater recharge with in the municipal boundary of the city. The vulnerability assessment and resilient strategies represent the first step towards developing the city's resilience strategy which can improve the sustainability and resilience of city's water supply.

33.2 Material and Methods

The methods used for the study are focus group discussions, interviews that include personnel working in the PMC and water utilities department of the city, residents of the wards and water experts working in the academic institutions and research institutions in Patna. These discussions and interviews helped in understanding the water scenario in the city and the needs that residents had for information, communication or connection with the water utility departments of the city.

33.2.1 The City of Patna – Urban Experiences

Patna is the capital of the Indian state of Bihar. It is the second largest city in eastern India, after Kolkata. The municipal limits of the PMC form part of Patna Urban Agglomeration Area (PUAA). The PMC boundaries cover an area of 100 sq km and according to the 2011 census, the population is 1,683,000 (16.83 lacs). In contrast, the PUAA covers an area of 146 sq km and has a population of 2,047,000 (20.47 lacs). The city is densely populated and is fast developing as a commercial hub of Bihar. The city comprises 72 wards, and population growth, with in the city, varies as a function of available amenities and infrastructure.

The most densely populated wards are located on the natural embankment (Fig. 33.2) of the river Ganga. Geographically the city has a width of 9.5 km. on the western side which gradually reduces to 2.5 km on eastern side. The city is situated on southern banks of river Ganga and extends linearly over a length of 25 km. The western periphery of the PUAA is bounded by the Sone River. The Punpun River runs parallel to the city, approximately 20 km to the south, and joins the Ganga, southeast of the city.

The city forms part of the Indo-Gangetic alluvial plains and has fertile soil. The region is flat, which permits wide spread of flood waters. Rain water percolates through the soil relatively rapidly. However, since the ground water table in the region varies from 2–5 m bgl (below ground level), following the monsoon seasons, and 5–10 m prior to the monsoon, absorption of water by the soil is reduced during the monsoon period.

As per a recent survey by the World Bank, Patna is one of the fastest growing cities in world in terms of Infrastructure development (Meihnardt Singapore Pte. Ltd. 2013).

Patna Population Density

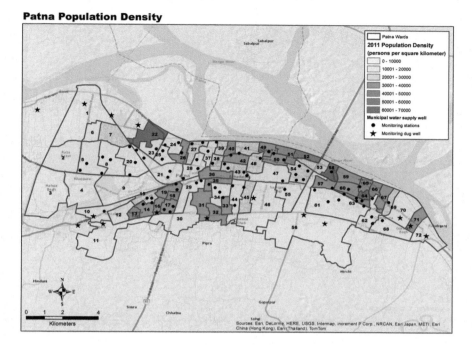

Fig. 33.2 Population Density of Patna, India (Census, 2011)

Today, all major industries include leather, handicrafts, and agro processing has a base in Patna, reflecting the growing importance of the city. There has been significant enhancement in the GDP of Bihar in the last decade. The growth of economy, along with urbanization and population trends, are indicators that city will continue to develop rapidly in next two to three decades. As per the Master Plan 2031, Population of Patna Planning Area is estimated to 60, 25, 232 of which urban population estimated to 48, 77, 129 for year 2031. It is also fast emerging as a hub of higher education with institutes like Indian Institute of Technology, All India Institute of Medical Science, National Institute of Fashion Technology of national repute being started in Patna.

33.2.1.1   Existing Land Use – Patna Municipal Corporation Area

Patna has emerged as a trade and business center, in last ten-fifteen years. As mentioned above, the city has experienced rapid population growth, primarily as a result of in-migration from surrounding rural areas throughout the state of Bihar. This has resulted in rapid urbanization of areas neighbouring the Patna municipal area (Fig. 33.3), which have become outgrowths of Patna City. In the absence of planning interventions since 1981, rapid growth has led to haphazard development, which

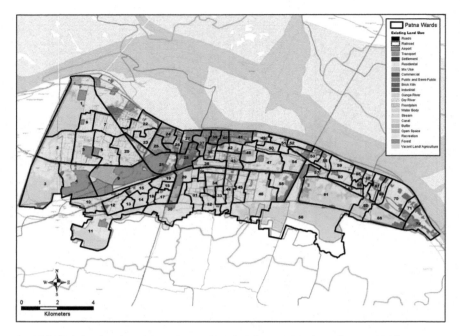

Fig. 33.3 Existing Land use Map – Patna Municipal Corporation Area (Source; PMC)

resulted in the deterioration of open space and forested area (only 2.34 sq.m. per capita), uncontrolled and unregulated construction activities and brick kilns in and along the Ganga floodplain, and the formation of slums and unregulated construction within the city core.

33.2.1.2   Flooding

Floods are a recurring problem in Patna, as a result of a combination of factors that include natural factors, such as high rates and amounts of precipitation during monsoon and tropical cyclone storms, along with bank erosion, and human-caused factors, such as decreases in vegetative cover due to deforestation, high rates of urbanization putting pressure on the drainage system, urbanization in low-lying areas, and land use changes along the river course (Sharma and Priya 2001). Two major floods recorded in the city were in 1984 & 1975 with the High Flood Level (HFL) recorded as 51.30 m to 51.51 m respectively (Patna Master Plan 2031). In many places, development has expanded the city's urban footprint to areas that are than the typical HFL of river Ganga; specifically, the majority of the city is located at heights between 48 and 51 m, whereas the HFL of the Ganga is 51.3 m. The linear bund, constructed on northern periphery of the city, prevents the entry of river Ganga water into the city and hence, this provides major flood protection. Similarly, the bund along Punpun River prevents the entry of floodwaters into the city from eastern side, particularly when the Ganga flows at high level and water backflows into Punpun's stream channel. The Patliputra neighbourhood in the north-western part of Patna, lies at 50–56 m above sea level, and hence, does experience major flooding. The city is further linearly bifurcated into northern and southern Patna, by the railway line through the middle of the city. This railway line is on a high embankment and acts as a flood protection measure. The new Patna by-pass, south of the railway line, which is under construction, provides additional flood protection measures.

33.2.1.3 Climate Conditions

Patna is located in the north-eastern part of India. It experiences transitional climate that fluctuates between tropical and humid subtropical, with hot summers to moderately cold winters. As per the data obtained from the Indian Meteorological Department data, the daytime maximum temperature generally ranges from 21.1 °C in January to 38.7 °C in May; night-time low temperatures range from 7.3 °C in December to 27.7 °C in June. The summer begins in April and peaks in June/July with the daytime extreme maximum temperatures soaring up to 43 °C, until the moisture laden monsoon winds bring precipitation. The summer rains typically continue through August and September, sometimes into early October.

The average of minimum and maximum temperature of the city during the last two decades is increasing (Figs. 33.4 and 33.5). According to Kothawale, D. R. and et al. (2005), on the recent changes in surface temperature trends over India based on the measurements of the India Meteorological Department (IMD) has shown that the all-India mean annual temperature has increased by 0.5 °C in the period 1901–2003. The warming trend has been associated with a rise in maximum temperature by 0.7 °C during the same period. However, during the last two decades the warming trend has been contributed by both maximum and minimum temperatures. The study has also shown that the rise in minimum temperature varies significantly between the winter and post-monsoon seasons, where it is 0.4 °C and 0.7 °C respectively.

Fig. 33.4 Average of Maximum Temperature during 1989–2009

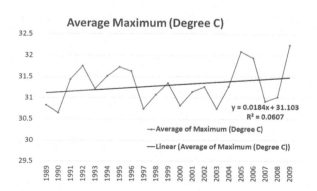

$y = 0.0184x + 31.103$
$R^2 = 0.0607$

Fig. 33.5 Average of
Minimum Temperature
during 1989–2009

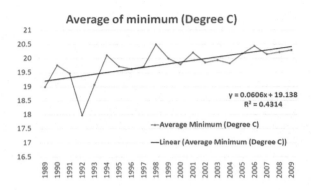

Average of minimum (Degree C)

$y = 0.0606x + 19.138$
$R^2 = 0.4314$

→ Average Minimum (Degree C)
— Linear (Average Minimum (Degree C))

Total Precipitation in seasons

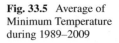

■ pre-monsoon   ■ monsoon   ■ Post-Monsoon   ■ Winter

Fig. 33.6 Total Precipitation in seasons

If we see the trend in precipitation in different season in the city, it clearly shows that the maximum precipitation received by the city is during the monsoon. As per the graph (Fig. 33.6), in year 1977, 1978, 1979 and 1985 receives the maximum rainfall during post monsoon season compare to other years. But if we see the different seasonality then we observe more clearly the different trend and climate variability in different seasons in the year between 1975 to 2003.

33.2.1.4 Groundwater

The ground water table is decreasing during the period from 1986 to 2012 (Fig. 33.7). It has very unusual trend during this period. From the year 1986 to 1995 it shows the decreasing trend and from 1996 to 2010 shows very unusual increasing trend. In year 2011 it shows clearly drop down of the water table approx. 2 m to approx. 6 m. As per the public health Engineer, the year 2011 the city experienced the lot of tube wells have been failed because of the suddenly drop in the ground water. The ground

Fig. 33.7 Ground Water
Table in pre-monsoon

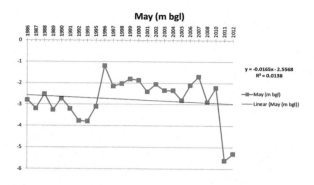

Fig. 33.8 Ground Water
Table in post-monsoon

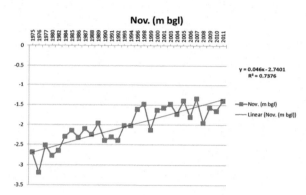

water table is increasing during the period from 1975 to 2011 (Fig. 33.8). It shows very strong trend during the post monsoon period. Pre-monsoon and post-monsoon depth of piezometric level of year 2010 can be referred in Figs. 33.9 and 33.10.

33.2.1.5 Climate Change Projections for Patna Region

General Climate Models and Regional Climate Models have a difficult time in replicating key features of historically-observed monsoons due to this the scientific community's ability to project how the monsoon system might evolve under various climate scenario may be weak. Despite these challenges, projections from multiple models are starting to converge and agree upon changes in trends of temperature and rainfall for various regions of Asia (Christensen, et al. 2007).

It is always advisable while using the climate projection data for infrastructure planning cautiously without taking into account model uncertainties.

Regional annual average monthly maximum temperatures are projected to increase 2.5 °C by 2049 (Fig. 33.11), based on the ensemble average of 41 CMIP5 models and assumptions of moderate future increases in greenhouse gas emissions (i.e., RCP 4.5; data source: https://climexp.knmi.nl).

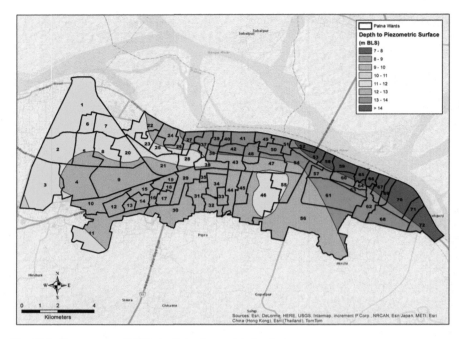

Fig. 33.9 Pre-monsoon (2010) depth to piezometric level

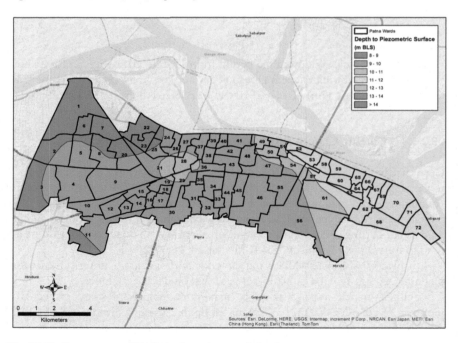

Fig. 33.10 Post-monsoon (2010) depth to piezometric level

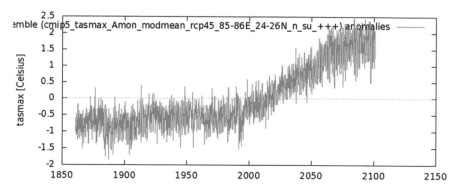

Fig. 33.11 Future Maximum Temperature projection for the city of Patna (*Source:* https://climexp. knmi.nl)

33.2.1.6 Urban Water Supply System of Patna City

According to data available, at present, Patna's urban area has 85 deep tube wells (Fig. 33.2); in addition, there are individual tube wells (privately owned), in the outskirts of the city, and, tube wells in water scarcity areas, that have been developed by some affluent people, to meet their own water demand.

In general, the PMC area has been divided broadly into 5 water districts (zones), referred to as the eastern, western, central, southern and Guljarbag zones. The water distribution network has been framed and executed by grouping a number of tube wells as and when required on piece meal basis. As a result, there is unequal distribution of water, with an excess water supply in some areas, and inadequate, water supply in other parts of Patna. To remedy this situation and assure equal-distribution of water in all, water districts, it is essential to the areas, it is essential to analyze the pipeline networks in all five zones wise and to develop an optimal network, by route rationalization. There is a very strong possibility for developing of negative pressure during non-supply hours and this may cause help in entry of polluted and/or contaminated water through leaky joints in waterlines.

Further, PMC area has been divided into 25 water districts with in five zones of the city, having individual water source. The distribution lines are interconnected with all 25 water districts, so that water can be diverted from one district to another district with in city if it is required to fulfil the water demand during the summer period when the water table goes down and which lead to water problem in some of the district with in the city.

There are 23 overhead reservoirs (OHR) in existence with in the city, but all of them are non-functional due to non-maintenance and repaired by the city authority. Out of these 23 OHRs, only 15 can be reused immediately and cracks have already been developed in other OHRs. At present, pumping is done from zonal deep tube from deep tube wells twice a day i.e., the supply is intermittent. Pumping Hours are between 5 am to 1 pm (8 h) and 3 pm to 11 pm (8 h). Replenishment occurs during night time by infiltration from the River Ganges and Son. Turbine pumps ranging from 35 to 120 hp are in use. Water us available at a depth between 30 and 300 m.

About 400 km. of pipelines varying from 350 to 50 mm diameter are in existence in Patna Municipal corporation (PMC) area. In water scarcity areas and in slums about 20 mld of water is supplied through 1500 stand posts connected to street water main in PMC area. A few slums have shallow tube wells of about 30 m depth.

33.2.1.7 Water Supply Source

The Patna town is divided in two parts by Delhi-Howrah railway line. The part of the town north of the Delhi-Howrah railway line will be supplied with surface water from river Ganga, where as part of town south of railway line shall be continued to supply ground water from Tube wells. (Source: December 2011, Bid Document, Design, Build, Operate, Manage and Maintain Water Supply System in Patna). The proposal has not been taken by the PMC yet.

All the towns have reasonably assured sources, as Patna has excellent ground water resources, and a large number of private tube wells exist for domestic as well as non-domestic use. The private tube wells are used to supplement or even as an alternative to the public water supply. The ground water table is overexploited in the city region. River Ganga in north and Sone in south are surface water sources, which are not utilized for the drinking water purpose due high investment cost.

33.2.1.8 Key Issues

After reviewing available literature and interviewing the city officials and interacting with the residents of the different wards of the city, the main problems with the existing system are non-uniform supply across the city, and contamination due to various leakages. Some of the key understandings brought by this study of water scenario in Patna city are:

33.2.1.9 Lack of Regulation and Monitoring of Water Supply and Quality

There is a lack of awareness related to quality of water among the residents of the different wards of the PMC. The genuine efforts to create awareness around water quality among the residents of the city is also lacking from the city administration side. The outbreaks of waterborne disease during the summer and early rainy season have been widely witnessed among the city residents especially among the urban poor. It remains challenge for the PMC to improve the quality of water supplied and sustaining it.

Table 33.1 Vulnerabilities and water security measures for the city of Patna (Adapted from ISET)

Core vulnerabilities	Interventions to improve resilience
Dependence on the ground water source only	1. Diversifying water supply sources 2. Managing groundwater resources efficiently and effectively 3. Addition of water conservation and efficiency measures in city byelaws.
Poor management of water utility	1. Infrastructure improvements and leak detection programs 2. Metering 3. Improved complaint redress system
Lack of groundwater management	1. Proper monitoring of groundwater table in all wards with in the city and around the city 2. Increase groundwater recharge through rainwater harvesting, dug wells and artificial recharge methods. 3. Mandatory provision of rainwater harvesting in all building of the city.
Lack of water quality monitoring and regulation	1. Water quality monitoring at household, community and ward level
Lack of information and understanding of climate change impacts	1. Capacity building programs for the personnel working in the area of water sector. 2. Sharing the climate information and data with water managers

33.2.1.10 Declining Water Levels

Increased water demand due to population increase and groundwater extraction by the city residents have contributed in the declining water lever in the city. The attention to groundwater decline is also lacked in the city development plan (2010-2030) and Patna Master Plan 2031 documents.

33.2.1.11 Increasing Energy Demand

The energy demand increases as the more households depend on the individual extraction of groundwater to meet the water demand and that leads to increase in energy demand. Increase in energy demand contributes to the increase in greenhouse gas emissions (ISET 2011).

33.2.1.12 Absence of Community Organisation

Due to absence of community organisation, the issues related to quality of potable water and monitoring and regulation of groundwater table are not addressed properly. As per the document master plan for Patna - 2031, the unaccounted for water loss

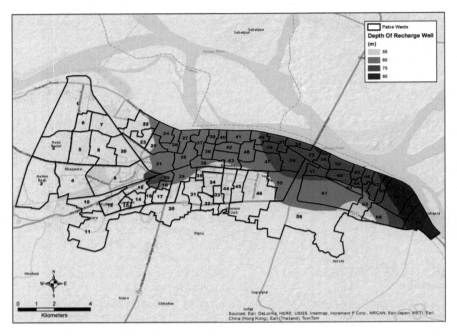

Fig. 33.12 Area suitable for recharge

is above 40% due to poor and old supply of network. Problem of arsenic content in water in surrounding areas second layer of geological strata. The pipes are in the center of the road due to road widening and facing heavy traffic, resulting in loss of carrying capacity, contamination of water, problem in repair and maintenance. In many colonies the drinking water and sewerage pipelines are interceding each other, with sewer line on top of water line increasing the possibility of contamination. Multilateral institutional arrangement for management of water supply system. Highly subsidized water supply and high operation and maintenance cost.

33.3 Results and Discussion

33.3.1 Water Security Measures

Improving the water supply in a way which will contributes towards sustainability and resiliency of water infrastructure will required a strong political will, social and financial investments and time for the PMC. This study focuses on the pragmatic solutions, which can easily be implemented by the city administrations. Many residents especially from slum areas and low-income groups struggle to meet their water demand and their water needs are immediate. "Waiting for the overall structure of water governance in Indian cities to change may take far longer than these residents have to address the issues of water governance" (ISET, 2011). Based on the study, some of the key water security measures which are the initial steps towards water solution that would contribute towards water resiliency in Patna shown in Table 33.1. After reviewing the literatures on various sources, the study proposes the interventions to improve resilience.

33.3.2 Rainwater Harvesting

Rainwater harvesting is suitable for Patna's urban area. The area suitable for recharge with minimum depth of the recharge well, is indicated in Fig. 33.12. As the shallow aquifers are barely used and the deeper aquifers are extensively exploited, recharge through wells may be a feasible option.

33.4 Conclusions

Impact of climate change has major significances for water resources. It also complicates long and short term planning for water supply. Cities of developing countries are more prone to this, which have already experiences the fast urbanisation. The city of Patna has already started experiencing it in water delivery. The city administration needs to enact a more resilient water system to counter the issues faced by its residents. Water utility department of the city of Patna does not meet the water needs of all residents that increases the use of private bore well to meet the demands. Slum dwellers are particularly more vulnerable as they have much lower capacity to adapt to changes. Local body of the city needs to adopt a resilience approach to address these issues.

Each city around the world will have its own set of resilience strategies based on their local needs, what is most needed is that to bring all stakeholders to discuss the issues and agree on the set of resilience strategies to address it. This study represents a process, which can be applied to arrive at the set of interventions to address it. It also highlights the departments' work as silos and it leads to very confusing and conflicting situations sometimes. As such, measures for resilience building need to be initiated at all levels. The interventions developed in this study could assist in the resilience intervention process with in the city.

Climate change and fast growing population in urban areas posing a very difficult situation to the city authority to deliver the water needs of the population. This poses both a threat and an opportunity in the city of Patna. Resilience strategies suggested in this study for water utilities department could be helpful for other cities around the world to cope with water insecurity as a result of climate change. This study also highlights the residents of slums are more vulnerable to climate change impacts on water sector. Effective tools and policies are needed to improve water quality, encourage connections among all stakeholders and reduce energy dependence. Climate change adaptation portrays another chance to implement a shared vision for a more sustainable and equitable water system.

Some key lessons emerging from this study, which are applicable in other cities of the worlds, are firstly, water management are not solely responsibility of the city administration but also the residents of the city. Secondly, effective tools and policies are needed to improve water quality and equitable distributions of water. Thirdly, it is significant to draw notice to the significant role of groundwater and the need for better management of it.

In the city of Patna, local government needs to create better systems of monitoring and conserving groundwater resources, including rainwater harvesting, and improving conservation and efficiency. Finally, water utilities department in the cities need ways to communicate and share information including climate change related data in the face of water availability.

Acknowledgements The author expresses sincere thanks to Prof. Gregg M. Garfin, The University of Arizona, Tucson, USA for reviewing earlier versions of the book chapter manuscript.

References

Central Ground Water Board (2011) Ground water scenario in major cities of India. Government of India

CEPT (2014) Master Plan for Patna - 2031. Patna

Christensen JH, Hewitson B, Busuioc A, et al (2007) Global climate projections. In: Climate Change 2007: The Physical Science Basis. Contribution of Working Group I to the Fourth Assessment Report of the Intergovernmental Panel on Climate Change. Cambridge

Christensen JH, Kanikicharla KK, Aldrian E, et al (2013) Climate phenomena and their relevance for future regional climate change. In: Climate Change 2013: The Physical Sci- ence Basis. Contribution of Working Group I to the Fifth Assessment Report of the Intergovernmental Panel on Climate Change Stocker, T.F. Cambridge

De US, Rao GSP (2004) Urban climate trends - the Indian scenario. J Indian Geophys Union 8:199–203

Hijioka Y, Lin E, Pereira JJ (2014) Climate change 2014: impacts, adaptation, and vulnerability. In: Part B: Regional Aspects. Contribution of Working Group II to the Fifth Assessment Report of the Intergovernmental Panel on Climate Change Barros, V.R., C.B. Field, D.J. Dokken, M.D. Mastrandre. Cambridge

Hunt J (2004) How can cities mitigate and adapt to climate change? Build Res Inf 32:55–57. https://doi.org/10.1080/0961321032000150449

Institute for Social and Environmental Transition (2011) Climate Change and Urbanisation: Building Resilience in the Urban Water Sector - A Case Study of Indore, India

Intercontinental Consultants & Technocrats private limited (2010) City Development Plan (2010–30)

Kothawale DR, Rupa Kumar K (2005) On the recent changes in surface temperature trends over India. Geophys Res Lett 32:1–4. https://doi.org/10.1029/2005GL023528

Meihnardt Singapore Pte. Ltd. (2013) Draft social management plan for pahari sewerage projects under NGRBA programme. Patna

Ramesh R, Yadava MG (2005) Climate and water resources of India. Curr Sci 89:818–824

Saha D, Dwivedi SN, Singh RK (2013) Aquifer system response to intensive pumping in urban areas of the Gangetic plains, India: the case study of Patna. Environ Earth Sci 71:1721–1735. https://doi.org/10.1007/s12665-013-2577-7

Sharma A (2015) Sustainable and socially inclusive development of urban water provisioning: a case of Patna. Environ Urban ASIA 6:28–40. https://doi.org/10.1177/0975425315583757

Sharma B, Jangle N, Bhatt N, Dror DM (2015) Can climate change cause groundwater scarcity? An estimate for Bihar. Int J Climatol 35:4066–4078. https://doi.org/10.1002/joc.426

Shukla P, S.K S, N.H R, et al (2003) Climate change and India: vulnerability assessment and adaptation

Sharma VK, Priya T (2001) Development strategies for flood prone areas, case study: Patna. India. Disaster Prev Manag 10(2):101–110. https://doi.org/10.1108/09653560110388852

Reid P, Vogel C (2006) Living and responding to multiple stressors in South Africa—Glimpses from KwaZulu-Natal. Glob Environ Change 16(2):195–206

UNDP (2006) Human Development Report 2006. Beyond Scarcity: Power, Poverty and the Global Water Crisis, 440 pp. United Nations Development Programme, New York

Chapter 34
An Analysis on Relationship Between Municipal Water Saving and Economic Development Based on Water Pricing Schemes

W. C. Huang, B. Wu, X. Wang, and H. T. Wang

Abstract In recent years, China has been facing water scarcity extensively, thus water saving has become an increasingly popular and crucial topic. In this study, it is analyzed how economic development has effect on municipal water saving in terms of water price changes. The proportion of water prices in residents' incomes and expenses was calculated, and then their relationship was investigated across the nation. Afterwards a phase-division model between water consumption and water price was established to have a closer analysis on the residents' attitude towards water price variation and their corresponding action for water saving. It is also indicated that an Increasing Block Tariff (IBT), with distinction between intensive and non-intensive consumers in the exponential phase, can be a better way compared with simply technical efforts. With these findings water communities may have a better idea on how user consumption reacts with the change in water price, along with how economical and reasonable water consumption could be achieved in a water-scarce environment through price adjustment.

Keywords Economic development · Elasticity theory · Increasing Block Tariff (IBT) · Phase-division model · Water price

W. C. Huang · B. Wu · H. T. Wang (✉)
College of Environmental Science and Engineering, Tongji University, Shanghai, China
e-mail: hongtao@tongji.edu.cn

X. Wang
UNEP-Tongji Institute of Environment for Sustainable Development, Sino-U.S. Eco Urban Lab, Tongji University, Shanghai, China

H. T. Wang
Key Laboratory of Yangtze River Water Environment, Ministry of Education, Shanghai, China

B. Wu · H. T. Wang
State Key Laboratory of Pollution Control and Resource Reuse, Shanghai, China

© Springer Nature Switzerland AG 2021
M. Babel et al. (eds.), *Water Security in Asia*, Springer Water,
https://doi.org/10.1007/978-3-319-54612-4_34

461

34.1 Introduction

In the past decades, China has been witnessing water scarcity, to which the extent can no longer be ignored. In the year of 2014, income flow of major rivers in this country was reported to be less than normal years. This problem was especially serious in north China, where the Yellow River collected more than 30 percent less water than normal while in its middle reaches precipitation even increased by 30% to 60% (MoWR 2015). Direct economic loss of equivalently 0.29% GDP was resulted from drought during the past nine years since 2006, on average (MoWR 2015).

Multiple examinations have been assessed for the imbalance between supply and demand in this country. Among these discussions listed the agricultural water dissipation due to the extensive employment of paddy fields for rice as a major grain in this nation of cuisine (Chapagain and Hoekstra, 2010; OH et al. 2017). More concerns were raised in the low water-use efficiency within Chinese agricultural departments for lack of effective control in groundwater extraction and economical solution to lower water footprint for chiefly small-scale tillage in China, where Mao and Zhong (2002) estimated that inefficient use of cultivation water could be up to 60%. What stayed also among the criticisms was the insufficient water resources in this nation, resulting from both climate and water quality attributions. From this aspect water quality issues, along with their influence on efficient water consumption and water price tariff, are of intensive concerns and arguments, especially in recent years and in better developed urban areas.

Urban areas have led the country's economic development for almost forty years. Despite the fact that municipal water supply was merely the least component in this industry, occupying 12.6% (MoWR 2015) of the total use of water resources, tap water price has always been raising attention. As a kind of public service that monopolies the need as a whole, the price of tap water holds an absolute domination over how much water the users would consume, by contrast the industrial need can be met through water right trades, and agricultural need is satisfied by individual water intake from groundwater to a large extent (Herzfeld et al. 2017).

Various potable water tariffs have been studied, in which Single-Price Tariff (SPT), Two-Part Tariff (TPT) and Increasing Block Tariff (IBT) are the most introduced applications (Chan 2015). Nevertheless, common principles as meeting the expectation of household food, drinking and hygiene needs apply to all tariffs. Detailed price composition was divided into three parts: resource value, engineering value, and public value. With the resource value as the base value of water extraction, this part of water price was generally regarded as non-adjustable by market. On the other hand, the engineering value from extraction itself and the public value introduced as a common welfare which stands for sewage treatment costs, were more or less fluctuating with market demand (Wu 2001).

The water tariff in China has been developing ever since its industrialization. The first fifteen years since 1949 during the people's republic witnessed a government-paid water tariff, where water consumption was free for residents, as a public service. Explicit charging scheme was not in consideration until after the opening-up and economic revolution in 1978 of this country, where regulations were launched to bring resource and engineering value into water prices (Mao et al. 2002). With increasing social attention to environmental protection, the government issued laws by the end of last century on water resource protection, adding public value into water price. It was only in the latest decade that cities in China started to try on per-capita IBT schemes, which largely depends on strictly metered water consumption (Binet et al. 2014).

It has been generally perceived that the demand of tap water is in close relationship to water price as well as personal income (Renzetti 2002). It was also concluded by Liao et al (2016) that IBT could eliminate more notably on water dissipation, in view of the discrete strategy of household consumption (Lopez-Mayan 2013). With an integrated pricing framework where the governmental authority (Dong 2002; Wang 2007) and market efforts both work on a rising supply side, it still remains a question how water supply should be valued and priced for a more reasonable demand (Wang et al. 2011).

As an essential part of climate dynamics, water cycle at both global and watershed scale has been found to have a feedback mechanism to climate changes, where massive water consumption could have negative impact on a sustainable watershed-level climate (Mo et al. 2009), and deteriorating climate may in turn decrease water resource efficiency and availability (Haddeland et al. 2013; Hanjra and Qureshi 2010). The aim of this study is to analyze the relationship between municipal water saving and regional economic development. Also it is included in this paper how policy affects the relationship, and a modified model of water price-user response is suggested. With this water-price-economy framework, the article is expected to provide a theoretical approach for making water price policy, for an objective of sustainable feedback on watershed water-climate relationship.

34.2 Review on Existing Literatures

As a country with massive agricultural and industrial production, Chinese literatures have been focusing on water pricing for these two categories. Mao et al. (2002) has concluded that with an alternative pricing scheme, the peasants could make up the increased water cost through the reduction of water consumption. It is also analyzed by Jia et al. (2000) that agricultural water consumption is more elastic than industrial and municipal uses, with the increase in water price. On the other hand, industrial water cost can be negligible based on different industries since the average cost of water in budget could be as low as 1%. However, China still shows some distinguishing characteristics that municipal water demands illustrate fewer relationships with climate than reported by Rinaudo (2012), probably due to rare garden uses of water in this country than that in European countries and the U.S.

The relationship among population, economy and water resources was found to have spatial imbalance characteristic, that the spatial distributions of population and economy were inverted with that of water resources, through an analysis by Zhang et al (2015) in Northeast China, while Chen et al. (2006) established a model of an estimated water use under Increasing Block Tariff (IBT) for water. Lei et al (2002) provided a model between water use and water price as:

$$Q_1 = Q_0 \cdot \left(\frac{P_1}{P_0}\right)^{\varepsilon} \tag{34.1}$$

Where Q_0, P_0 stands for original water consumption and price respectively, and Q_1, P_1 stands for adjusted water consumption and price respectively. Another model based on elasticity theory was provided by Ma et al (2013) as follow:

$$Y = aX_1{}^{E_1}X_2{}^{E_2} \tag{34.2}$$

Where Y is the monthly individual water need; X_1 stands for actual water price, X_2 stands for monthly expenses on water services; E_1 and E_2 represents the elasticity of price and personal income respectively; and a works as a constant. As Rivers and Groves (2013) calculated, the general elasticity of water demand was around -0.4, giving a moderate regulation on water-price nexus. Moreover, Luckmann et al. (2015) found that the elasticity of potable water could be, on a proportion, affected by the corresponding marginal cost and income from water consumptions. It is determined by Chen et al. (2016) that with a higher marginal income, the elasticity of water price could decrease for lack of interest.

As we can see, a basic model is needed to analyze how water demand is related to its price. Nevertheless there is hardly a theory to figure out the correlation between economic development and water price within a region.

34.3 Methodology and Data Analysis

34.3.1 Economic Development in China

To reveal the relationship, an indicator must be chosen among the economic indexes to serve as a reference. In economics GDP is widely used to evaluate the standard of development within a certain region. It provides a comprehensive view of the regional development resulting from agricultural, industrial and servicing production. To be specific on individual water need, GDP per capita is required.

Note that data of Hongkong, Macau and Taiwan are not included in Fig. 34.1 and figures hereinafter.

In this figure, we can see a great difference among different regions in China. In north and east China, there was a relatively higher GDP per capita over 60,000 Yuan on average, while Southwest China had an average number even less than half of that in metropolises as Beijing, Tianjin and Shanghai. This phenomenon may cause conspicuous divergence of the attitude to water prices.

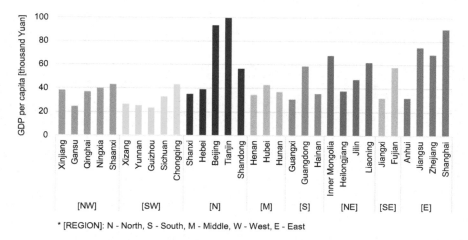

Fig. 34.1 GDP per capita (thousand RMB Yuan) by provinces in China in 2013 (CESDSD 2014)

Although GDP per capita gives a general view on regional economic development and its data is easy to access, this indicator does not suit well as a reference. It is much too broad and meanwhile its calculation covering every production unit within statistics rather than every municipal water consumer makes it incapable of the comparison with water prices. On this account, the indexes to assess the income of municipal residents come forth. In the database two indicators turned out to fit this assessment: Resident Disposable Income per capita in Cities and Towns (RDI_{CT}^0) and Resident Disposable Income per capita in Cities and Towns by family ($RDI_{CT,F}^0$). Between them the difference is mainly on that RDI_{CT}^0 contains only labour as a base number while $RDI_{CT,F}^0$ includes all residents from the infant to the senior. Since water is a kind of vital resource of which the absence is prohibited, to every single person alive, $RDI_{CT,F}^0$ is obviously better for analysis (Fig. 34.2).

We can find that the difference of regions has been reduced in comparison with GDP, since GDP reflects more on manufacturing and RDI is much closer to the residential possession, which is our expected base number. Only developed areas such as Beijing, Tianjin, Shanghai, Jiangsu, Zhejiang, Fujian and Guangdong stood out in this figure.

We can also assess the water prices on base of economic consumption. When we take Resident Average Consumption in Cities and Towns (RAC_{CT}^0) as the indicator, we can see a similar picture as RDI, except for some minor difference in ranking. Seeing from Fig. 34.3, the provincial difference based on RAC is relatively smaller than RDI since the deposit and investment mainly covers their difference.

There are also other indicators other than these three ones, like Consumer Price Index (CPI) and Resident Disposable Income of different income levels. Nevertheless, they all have problems as missing data or being too broad. Thus RDI and RAC was chosen in this paper to serve as the reference number of water prices.

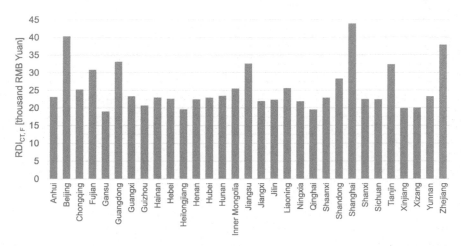

Fig. 34.2 Resident Disposable Income per capita in Cities and Towns by family [$RDI^0_{CT,F}$] (RMB Yuan) by provinces in China in 2013 (CESDSD 2014)

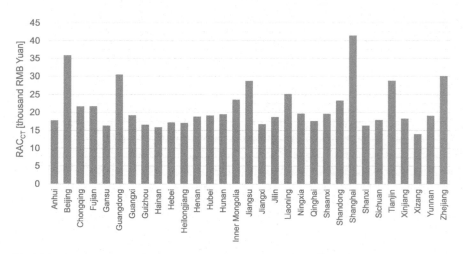

Fig. 34.3 Resident Average Consumption in Cities and Towns [RAC^0_{CT}] (RMB Yuan) by provinces in China in 2013 (CESDSD 2014)

34.3.2 Water Supply and Pricing in China

In China, the industry of water supplies spans both public service and manufacture market (Dong 1989). Under this circumstance water price can have a significant influence on water consumption, as illustrated in Fig. 34.4. Judging from the figure we can see the Daily Household Water Consumption per capita (DWC_0) drops with

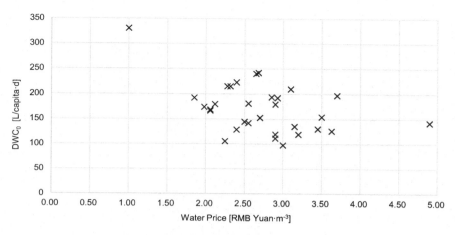

Fig. 34.4 Water price-consumption relationship in China in 2013 (CWN 2014)

the increase of Municipal Water Price (WP). However, their relationship remains unclear since it is non-linear and may have a complex mechanism.

Where DWC_0 stand for Daily Household Water Consumption per capita.

Now that water price rises do indeed help decrease water consumption, question comes out as to what extent can we and should we raise the price of water, to take water scarcity in control and maintain a similar living standard. To answer this question, we must figure out the relationship between economic development and water price.

First, the Monthly Water Cost (MWC) is calculated. Here we assume that all citizens have the same attitude to water prices, and take 30 days for one month. Thus MWC is taken as the money one single resident in cities or towns should pay for water using within a month (RMB Yuan):

$$MWC = DWC_0/1000 \times WP \times 30 \tag{34.3}$$

Afterwards a comparison can be made through water cost (MWC) and the economic indicators (RDI & RAC):

(i)   Take the proportion of Monthly Water Cost in Resident Disposable Income as the Index of Water Cost based on Resident Disposable Income,

$$IWC_{RDI} = MWC/RDI^0_{CT,F} \tag{34.4}$$

where higher IWC_{RDI} indicates a larger part of one's income goes to payment of water using (Fig. 34.5).

Where IWC_{RDI} = Monthly Water Cost /Resident Disposable Income.

While a composition of IWC_{RDI} as high as more than 0.08% can illustrate a lower development in this region or water price above average, a much smaller

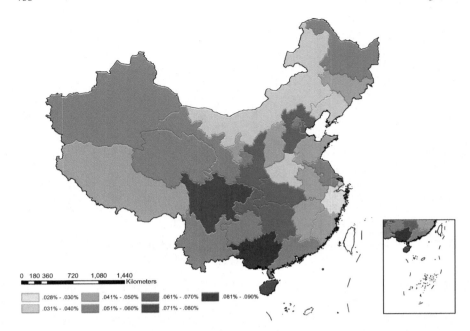

Fig. 34.5 IWC_{RDI} Distribution of Provinces in China in 2013

IWC_{RDI} can also state that the water in such region may be undervalued. Seeing from this figure, the highest IWC_{RDI} shows up in Guangxi and Sichuan, whose water price were still moderate (ranked 26th and 14th, respectively) in China, indicating a lower development of economy in these regions. After them comes the provinces in middle reaches of the Yellow River and the Yangtze River, additionally, Hebei and Tianjin. It appears that Shaanxi, Hubei and Hunan are in similar situations as Sichuan and Guangxi, where a low average income leads to relatively higher water prices; while in Hebei, Tianjin and Chongqing, things are different. Three positions in the highest four water prices were taken by these provinces, especially in Tianjin and Chongqing, where both price of water resources and of wastewater treatment are among the highest throughout the country (Table 34.1).

Of all provinces listed, Zhejiang is the lowest in IWC_{RDI}. This is owing to its elevated economic and living standard and plentiful of water resources such as Xin'anjiang Reservoir and the famous Qiandaohu. However, water resources in this

Table 34.1 Ranks of different parts in water price in 3 Provinces in China

Provinces/Cities	Ranked by price of water resources	Ranked by price of wastewater treatment	Ranked by Total Price
Hebei	2nd	30th	3rd
Tianjin	1st	6th	1st
Chongqing	4th	4th	4th

Table 34.2 Ranks of different parts in water price in Zhejiang Province

Provinces/Cities	Ranked by price of water resources	Ranked by price of wastewater treatment	Ranked by Total Price
Zhejiang	28th	27th	30th

province are absolutely not satisfying as there are still issues of poor water quality, north Zhejiang in especial. Water in Zhejiang is still much too cheap to sustain a clearer waterbody (Table 34.2).

(ii) Take the proportion of Monthly Water Cost in Resident Average Consumption as the Index of Water Cost based on Resident Average Consumption,

$$IWC_{RAC} = MWC/RAC_{CT}^0 \tag{34.5}$$

where higher IWC_{RAC} indicates that payment of water using shares a larger part of one's daily expenses (Fig. 34.6).

Where IWC_{RAC} = Monthly Water Cost / Resident Average Consumption.

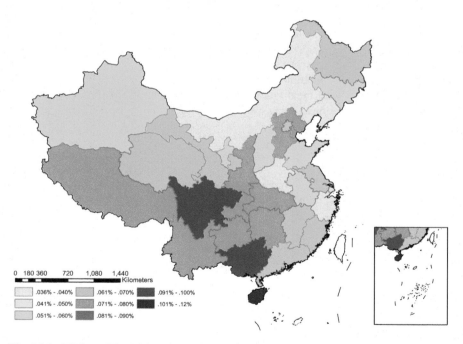

Fig. 34.6 IWC_{RAC} Distribution of Provinces in China in 2013

IWC_{RAC} shares a similar theory with IWC_{RDI} that if this figure goes too small, it implies that too cheap the water is; when it rises high, it can result from an insufficient economic development or a high cost in obtaining fresh water and processing wastewater.

When we use the residents' total cost in daily life (IWC_{RAC}) as a reference instead of their average income (IWC_{RDI}), the figure becomes more distinguishable as it wipes out factors other than merely expenditure. It can be noticed that in lower developed regions like Tibet, legends go darker more quickly than highly developed regions, indicating the cost of water prices occupies more proportion in expense than in earning.

34.3.3   Water Price and Water Consumption

As is mentioned in Review of existing analyses, the models to explain how water consumption reacts with its price are generally illustrated as an exponential function like in Fig. 34.7.

$$Consumption = k \cdot Price^{-\varepsilon} \qquad (34.6)$$

where ε stands for the elasticity of water price, and k serves as a constant.

In view of the characteristic of water as both public service and commodity, a phase-division model is proposed (Fig. 34.8). With water price per unit rises, residential water users can think differently on price increase during different phases. In the first phase called Linear Phase, people are merely aware of a rise in price and start to launch a reduction in using water parallel to the percentage of price elevation.

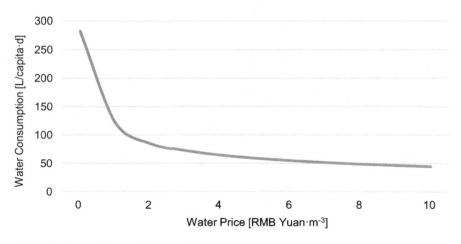

Fig. 34.7 Exponential correlation model

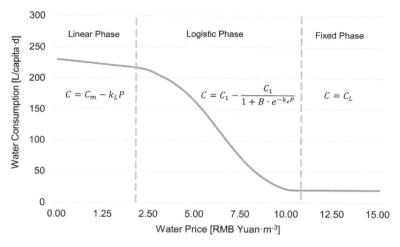

Fig. 34.8 Phase-division model

This phase may last long or short, depending on IWC_{RDI}. If water costs take quite an amount of income, an individual can be more sensitive and easier to enter the next phase. On the other hand, when IWC_{RDI} becomes too low, people may be ignorant of the price rises and hardly enter the next phase. The second phase is the Exponential Phase. Dramatic decline in water consumption occurs in this phase, as a similar way as an exponential function, on account of the water price beyond expectation. Finally it will come to the third phase named Fixed Phase, where water consumption sustains a fixed fundamental need as the least water income for an individual, as in Fig. 34.8.

Where C: Water Consumption; C_m: Maximum Water Consumption; C_L: Minimum Water Requirement; C_0: First (Linear) Diversion Point; k_L: Consumption Reduction Rate in Linear Phase; k_e: Consumption Reduction Rate in Exponential Phase; B: Destination Coefficient in Exponential Phase.

34.3.4 Measures on Water Price and the effect

Generally, there are two types of methods to control water consumption. One is technical, like Water Reuse, and the other one is economical, like Water Price Leverage. In this section, these two kinds of methods will be synergized with the phase-division model to analyze their effect on water prices and water saving. For a certain Phase-division Model where the function (indicating resident's need for water), if water is supplied with a certain price, the area below Resident Need and right of Price of Supply will be Consumer Surplus as illustrated in Fig. 34.9, which means the difference between the actual price and the highest price consumer (here the residents) would like to pay.

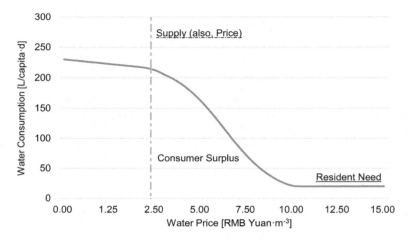

Fig. 34.9 Water consumption variation with a certain price

At the same time, the Producer Surplus, which means the difference between the actual price and the minimum profit they need, is zero in Fig. 34.9 since it should have been between the lines of Supply and the Price (Fig. 34.10).

(i).   When technical methods are adopted, residents require less water from the taps, illustrated as:

We can find that while the Producer Surplus remains zero, the Consumer Surplus goes down. It appears to be 'a loss' of the residents and that is part of the reason why citizens are sometimes reluctant to adopt technical methods, not to speak of time and sometimes even money to spend in these constructions. Nonetheless, the residents

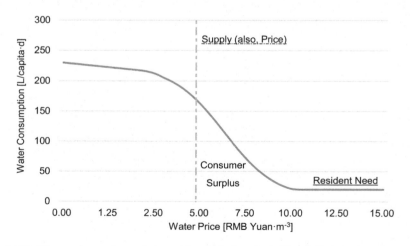

Fig. 34.10 Water consumption variation with technical methods (Supply Raise) adopted

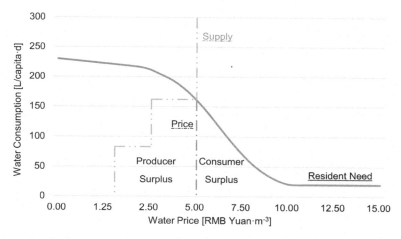

Fig. 34.11 Water consumption variation with increasing block tariff (IBT) pricing adopted

actually lose nothing (if technical methods are paid by the government) since the reduction of Consumer Surplus is just because of the reduction of their own need for water. In this situation, technical methods can be a long-lasting solution, but it is scarcely an easy one.

(ii). When economical methods are adopted, such as Increasing Block Tariff (IBT), where water price rises as the consumption increases (Fig. 34.11):

Things become different when IBT Pricing Scheme is adopted. According to economics, water will be produced where the Need and the Supply crosses, and on its left lies the Producer Surplus, on the right the Consumer's. Obviously IBT Pricing provides more motivation for tap water suppliers.

34.3.5 Policy Implications

There are three 'blocks' representing certain prices in this example for IBT Pricing Scheme, shown as the double-dot dashed lines in Fig. 34.12. We can deduct from this figure that:

1) If the first or second block moves right, that is to say, the producer now raise the price in a larger scale each time. The residents may come to the second block from the third, but they have to pay more for water instead of paying less.
2) If the first or second block extends in height, that is to say, the residents now consume more water to meet the next block. This actually encourages water use since when people use more water, they may even pay less.

The key to a suitable water price is actually a balance between Producer Surplus and Consumer Surplus, which is the key factor in commodity, and the control of total

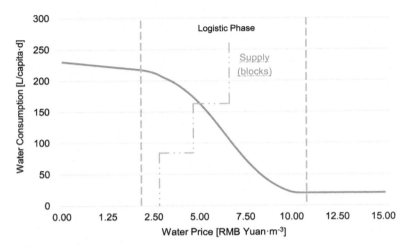

Fig. 34.12 Blocks put inside exponential phase

water consumption, which is the key factor in public service. Despite the general practice which places the first block at median water consumption (Rinaudo et al. 2012), three essential aspects are still open for discussion as how many blocks to set, where to set the blocks and what should the unit cost within each block (Liao et al. 2016). As we may suggest for the second question, the blocks in IBT Pricing Scheme should be put inside the exponential phase, where the effect of blocks can be sufficiently utilized.

34.4 Conclusions

Based on the analysis on the economic indicators and water prices of provinces in China, conclusions could be drawn that with distinct conditions of development and water scarcity levels, the water price in each region with respect to residential incomes and expenses could vary enormously, as a consequence of resource limitation or social-economical limitation, or sometimes both; yet a general trend stated that with increasing water price, residential water consumption will decrease, but within a limited elasticity alterability. Further assessment into pricing extended from a simple exponential relationship to a Phase-division model between water consumption and water price, proposed to be much closer to reality. With an Increasing Block Tariff (IBT) scheme embedded into this model, it was suggested from figures that not only IBT scheme could make social welfare in water economy on reducing overall water consumption, but also the blocks of IBT scheme should be set inside the exponential phase to make better benefits.

With these findings water communities and policy makers are supposed to get a better picture on the localization of water-pricing scheme, where highly developed economy in a certain region can prevent water saving from further progress, but insufficiently developed areas may also meet problems as difficulty in tap water withdrawal and insufficient water processing. It is also revealed from illustrations of both technical and economical methods that under this phase division model, economic method obviously gives more chance to call on residents to save water. Suggestion should be of policy-makers' interests comes that identical elasticity model put forward by existing literatures may not necessarily depict the actual pricing-consumption interaction, especially with limiting resources or mismatched schemes. Specific proposal of putting the 'blocks' inside the exponential phase is put forward, for reference of policy making within an Increasing Block Tariff pricing framework. We suggest that through the complementary schemes of IBT and phase division model, local communities would be able to have extensive control on reach- or region-based water balance for municipal water supply. We are also looking forward to a country-based coordination on watershed-level water balance in a water-economy nexus based on our findings in province distinctions of water scarcity, water pricing, and the relationship behind different schemes and their effects. We hope these mechanisms would benefit the sustainability on a larger extent in both spatial and temporal aspects, with positive feedback on water-climate relationship.

Nevertheless we still realized that the methods to calculate the margin of three phases and the deduction of residents' consumption reduction rate are not yet under sufficient research. These parameters are not easy to find out and requires vast labor in data collecting and analyzing, thus more is to be done to make this model better. Also there comes suggestions that a simple water nexus pricing may lead to increasing energy consumption (Wu et al. 2003), calling on a closer analysis on the relationship between pricing and consumption under an integrated water-energy nexus.

Acknowledgements The authors would like to thank financial support of DAAD through EXCEED/SWINDON project. This work was also partially supported by the Royal Academy of Engineering under the UK-China Industry Academia Partnership Programme Scheme "Global water scarcity: a case study on urban water crisis and its relation to businesses in China and UK".

References

Binet ME, Carlevaro F, Paul M (2014) Estimation of residential water demand with imperfect price perception. Environ Resour Econ 59(4):561–581

Chan NWW (2015) Integrating social aspects into urban water pricing: Australian and international perspectives. In: Understanding and Managing Urban Water in Transition 311–336. Springer Netherlands

Chapagain AK, Hoekstra AY (2011) The blue, green and grey water footprint of rice from production and consumption perspectives. Ecol Econ 70(4):749–758

Chen YY, Li TL, Bao CC, Li HF, Jiang JD (2016) Measurement of price elasticity on China's industrial water: based on marginal productivity model. J Zhejiang Inst Sci. Technol 36(3):232–237

Chen H, Yang ZF (2006) Scalar urban water pricing model based on utility function. Resour Sci 28(1):109–111

China Water Network (2010) Water Prices online database, Retrieved from https://price.h2o-china. com/

CNKI. (2015) China Economic and Social Development Statistics Database, Retrieved from www. cnki.net

Dong WH (2002) Discussion on formation mechanisms of water price. Water Resour Dev Res 2(2):1–5

Haddeland I, Heinke J, Biemans H, Eisner S, Flörke M, Hanasaki N, Stacke T et al (2014) Global water resources affected by human interventions and climate change. Proc Natl Acad Sci 111(9):3251–3256

Hanjra MA, Qureshi ME (2010) Global water crisis and future food security in an era of climate change. Food Policy 35(5):365–377

Lei SP, Wang N, Xie JC (2002) Discussion to the water price and its function in water resources management. J Lanzhou Railway Univ (Nat Sci) 21(4):132–135

Leibniz Institute of Agricultural Development in Transition Economies (IAMO) (2017) WATER AND AGRICULTURE IN CHINA: Status, Challenges and Options for Action. Herzfeld et al, Halle, Germany

Liao XC, Xia EL, Wang ZF (2016) The impact of increasing block water tariffs on residential water usage and the welfare of low income families in Chinese cities. Resour Sci 38(10):1935–1947

Lopez-Mayan C (2014) Microeconometric analysis of residential water demand. Environ Resource Econ 59(1):137–166

Luckmann J, Flaig D, Grethe H, Siddig K (2016) Modelling sectorally differentiated water prices-water preservation and welfare gains through price reform? Water Resour Manage 30(7):2327–2342

Jia SF, Kang DY (2000) Influence of water price rising on water demand in North China. Adv Water Sci 11(1):49–53

Ma T, Zhang X, Fan Y, Chen MQ (2013) Decision-making model of impact of water price for urban water-consumption. J Northeast Agric Univ 44(2):82–87

Mao XQ, Zhong Y (2002) Market oriented sustainable water resources management. China Popul Resour Environ 12(2):48–52

Ministry of Water Resources, P. R. China (MoWR) (2015) China Water Resources Communique 2014. Ministry of Water Resources Information Centre, Beijing, CHINA

Ministry of Water Resources, P. R. China (MoWR) (2015) Floods and Droughts in China, Communique 2014. Ministry of Water Resources Information Centre, Beijing, CHINA

Ministry of Water Resources, P. R. China (MoWR) (2015) Hydrology Situation Annual Report 2014. Ministry of Water Resources Information Centre, Beijing, CHINA

Mo X, Liu S, Lin Z, Guo R (2009) Regional crop yield, water consumption and water use efficiency and their responses to climate change in the North China Plain. Agr Ecosyst Environ 134(1):67–78

National Bureau of Statistics, Ministry of Environmental Protection of China (MoEP) (2015) Urban Water Supply and Use by Region, CHINA STATISTICAL YEARBOOK ON ENVIRONMENT 2014. Ministry of Environmental Protection Information Centre, Beijing, CHINA

Oh B-Y, Lee S-H, Choi J-Y (2017) Analysis of paddy rice water footprint under climate change using aquaCrop. J Korean Soc Agric Eng 59:45–55. https://doi.org/10.5389/KSAE.2017.59.1.045

Renzetti S (2002) Residential water demands. In: The Economics of Water Demands 17–34. Springer US

Rinaudo JD, Neverre N, Montginoul M (2012) Simulating the impact of pricing policies on residential water demand: a Southern France case study. Water Resour Manage 26(7):2057–2068

Rivers N, Groves S (2013) The welfare impact of self-supplied water pricing in Canada: a computable general equilibrium assessment. Environ Resource Econ 55(3):419–445

Wang F, Wang JH (2011) Empirical study on the performance evaluation of the municipal water industry's privatization in China. Collected Essays Financ Econ 5:9–18

Wang XY, Tan XX, Chen Y (2011) Research on composing an entire cost-based water price model. Water Resources Power 29(5)

Wang YH (2007) An Evaluation on the institutional reforms of water pricing, water right and water market in China. China Popul Resour Environ 17(5):031

Wu JS (2001) A tentative discussion on forming a proper water price system. China Water Resources 3:17–19

Wu PT, Feng H, Niu WQ, Gao JE, Jiang DS, Wang YK, Qi P et al (2003) Analysis of developmental tendency of water distribution and water-saving strategies. Trans Chinese Soc Agric Eng 19(1):1–6

Zhang C, Liu Y, Qiao H (2015) An empirical study on the spatial distribution of the population, economy and water resources in Northeast China. Phys Chem Earth, Parts a/B/C 79:93–99

Part VI
Water Governance and Management

Chapter 35
Improving Water Security to Mediate Impacts of Climate Change in the Ganges Basin

B. Sharma, P. Pavelic, and U. Amarasinghe

Abstract In spite of being water surplus, the 600[+] million population of the large Ganges basin spread over 1.09 m km^2 in South Asia is water insecure, poor, and highly exposed to water-induced stresses of floods and droughts. The contribution from the glaciers to the streamflow is ~70% in the Himalayan catchments though spatially distributed quantification is unavailable. An application of the *Water Evaluation and Planning (WEAP) model* with a sub-routine for snow and glaciers melt processes in the basin was set up. The model also examined the possible impacts of an increase in temperature of +1, +2 or +3°C over 20 yrs of the simulation period. The impact on stream flows was high in the upstream (+8 to +26% at Tehri Dam) and moderate in downstream (+1 to +4% at Farakka). These increases shall create flood events more frequently or of higher magnitude in the mountains and Upper Ganga flood plains. To moderate the climate-change induced impacts of floods and improve water security during the non-monsoon season the novel concepts of Underground Taming of Floods for Irrigation (UTFI) and Cranking up the Ganges Water Machine for Ecosystem Services (GAMES) were developed, and pilot tested in the Ramganga sub-basin. Analysis showed that there is an assured possibility of reducing the floods and enhancing sub-surface storage in the identified basins to the level of 45 Bm$^{3.}$ The demonstrated managed aquifer recharge interventions are technically feasible, operationally acceptable and economically viable.

Keywords The Ganges basin · Climate change · Water evaluation and planning model · Underground taming of floods for irrigation · Ganges water machine for ecosystem services

B. Sharma (✉)
International Water Management Institute, India Office, NASC Complex, New Delhi, India
e-mail: b.sharma@cgiar.org

P. Pavelic
International Water Management Institute, South-East Asia Office, Vientiane, Laos

U. Amarasinghe
International Water Management Institute, Colombo, Sri Lanka

© Springer Nature Switzerland AG 2021
M. Babel et al. (eds.), *Water Security in Asia*, Springer Water,
https://doi.org/10.1007/978-3-319-54612-4_35

35.1 Introduction

The Ganges basin is one of the largest and most populous river basins in Asia with a population of more than 655 million. It is distributed between India (79%), Nepal (13%), Bangladesh (4%), and China (4%). It covers a total area of 1,086,000 km^2 (Jain et al. 2007). The total length of the river Ganga from the source at Gangotri to its entry into the Bay of Bengal is 2,525 km. The Ganga River drains a basin of extraordinary variation in altitude, climate, land use and cropping patterns. Basin elevations range from 8848 m to sea level. Topographic contrasts in the basin are the largest compared to any other basin on earth and can be appreciated by the fact that within a distance of nearly 200 km, the elevation plunges from the highest mountain in the world at 8848 m (Mt. Everest peak in Nepal) to about 100 m (the elevation of the flat Ganga plains in India) (Fig. 35.1, Jain et al. 2016). The water availability in the Ganges basin is high, as on an average hydrological year around 1200 billion m^3 (BCM) of precipitation falls in the basin. Of this, around 600 BCM becomes streamflow with the rest directly recharging groundwater or returned to the atmosphere through evapotranspiration (World Bank, 2013). There is however a tremendous variability in the spatial and temporal distribution of water supplies. The South Asian monsoon largely defines the climate and hydrology of the Ganga system. Snow and ice melt together have been estimated to contribute about 9% (almost 2/3 from snow) of total annual flows in the Ganga (Immerzeel et al. 2010), though they contribute much more in some

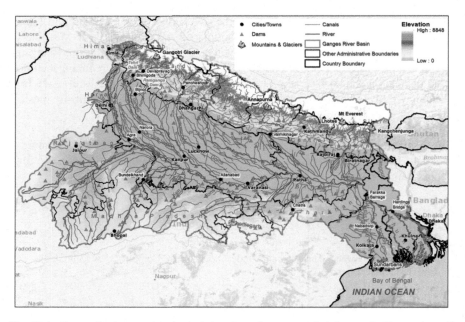

Fig. 35.1 Ganges Basin map showing existing dams, barrages and major canals (Source https://gisserver.civil.iitd.ac.in/grbmp/downloaddataset.aspx)

tributaries (e.g., 30% of annual flow in the Buri Gandak basin), and in small upstream catchments. Snow and ice comprise 2.4, 0.5 and 0.04% of the Ganges source (2004 km^2), Ghaghara (570 km^2) and Upper Yamuna (11 km^2) catchments, respectively. The Ganges basin has a total of 1,020 large glaciers. Mean annual precipitation varies significantly throughout the Basin: rainfall is highest in the eastern Himalayan belt and in the delta areas (>2000 mm), and lowest in the desert of Rajasthan in the west (<250 mm).

The predicted rise in temperature across the basin will alter the timing and magnitude of contributions from snow and glacier melt to the Ganges. In the medium term, melting in the Himalayas will begin earlier in the year and persist longer and thus increase the runoff (Rikiishi and Nakasato 2006). In the long term, the consistent contributions from meltwater will decrease as storage gradually declines (Alford and Armstrong 2010). Seasonal storage in the form of snow will decrease, and more variable rainfall and the resultant floods will then become increasingly dominant (Miller et al. 2012). There is an urgent need to understand these potential impacts of climate change on variability of the water resources in the basin and devise suitable strategies and interventions to mediate these impacts. By employing specific routines of the Water Evaluation and Planning (WEAP) model, this study simulates the impact of rise in temperature of 1, 2 and 3°C over a period of 20 yrs on the snow and glacier melt and stream flow in different parts of the basin. It then presents some novel innovations like Underground Taming of Floods for Irrigation (UTFI) and Cranking up the Ganges Water Machine for Ecosystem Services (GAMES) for making productive uses of the additional flood water available in the basin to improve the water security and the livelihoods.

35.2 Study Design

Water Evaluation and Planning (WEAP; https://www.weap21.org/) model, which contains an experimental glacier module that accounts for snow and glacier processes in the Ganges basin, i.e. seasonal mass variations and contributions to streamflow was used to have an estimate of the impact of temperature rise on snow and glacier melt. WEAP employs a unique approach where a database maintains water demand and supply information to drive a mass balance model on a link-node architecture. As climate change becomes an increasingly important challenge facing water managers, WEAP has expanded to include integrated hydrologic simulation functionality. With this enhanced version, dynamically integrated rainfall/runoff routines translate information on climate and catchment conditions into hydrologic fluxes that drive the existing water system simulation routines (Yates et al. 2005). The WEAP model helped to: (i) simulate the surface water resources in the Ganges basin with special focus on the contribution from snow and glacier melting, as well as the anthropogenic utilisation of the resources; and (ii) enable an effortless development of prospective scenarios to climate change (increase in temperature). The Ganges basin was discretised with respect to elevation to account for the variation with altitude of the glacier coverage and climate. In the second step, WEAP was calibrated and partly validated on observed streamflows. In a third step, WEAP analysed the current context of the surface water resources in the Ganges basin, in particular the contribution from the melting of glaciers. We also examined possible impacts of an increase in temperature of +1, +2, or +3°C over 20 years. We proceeded by river system (e.g., Ganges, Yamuna, Koshi), from upstream to downstream, with a monthly calculation time step. Whenever large time series of observed data were available, i.e., more than 10 recent years, part of this time series was kept aside for validation. We also evaluated the quality of the WEAP-Ganges and found that it was satisfactory (as it was possible to simulate average monthly and annual flow trends) for the Nepalese sub-basins, where observed time series were available and where most of the glaciers are present. As such, it was possible to consider monthly average trends for glacier-related analyses. In several other sub-basins, the quality was variable and as such was good for providing average trends. Other limitations of the current version of WEAP-Ganges include the following and as such the results may be considered preliminary (Sharma et. al. 2013):

i.   WEAP-Ganges does not cover the entire Ganges basin but just the part up to the Farakka barrage in India. Below that point the impact of glaciers on streamflows was rather small;

ii. no time series for glacier coverage was available; hence the calibration of the WEAP glacier module aimed at reproducing streamflows, while an additional calibration target could be the variation of the glaciers' area;

iii. the description of glacier behavior is based on a simple conceptual model, which may not capture all glacier processes.

As floods and droughts, along with over-exploitation of groundwater, were the major concerns in the Ganges basin, an approach referred to as the "Underground Taming of Floods for Irrigation (UTFI)" was developed and tested in the Ramganga sub-basin. This approach involves interventions at the river basin scale to strategically recharge aquifers upstream during periods of high flow, thereby preventing local and downstream flooding and simultaneously providing additional groundwater for irrigation during the dry season (Pavelic et al. 2015). The intervention mediates the flood impacts felt across one part of the basin to create groundwater recharge opportunities in another part of the basin.

A fresh look and analysis of the concept the Ganges Water Machine (GWM), a concept proposed 40 yrs ago, to meet the growing water demand through groundwater and to mitigate the climate- change induced impacts of floods and droughts in the basin was carried. The GWM provides additional sub-surface storage through accelerated use of groundwater prior to the onset of the monsoon season, and subsequent recharging of this sub-surface storage through monsoon surface runoff (Amarasinghe et al. 2016).

35.3   Results and Conclusions

The contribution from snowmelt and glaciers is important (60 to 75%) in Upper Ganga and in the Nepalese sub-division of Ghagra, Gandak and Koshi rivers (40 to 55%). The contribution however reduces significantly further downstream as flows from glaciated areas are diluted by stream flows generated by rainfall/ runoff processes. The average seasonal contribution is much contrasted, the flows from glaciated areas occur predominantly in the months of June to September, being almost nil in other months. Another interesting result is that glaciers are apparently buffers against inter-annual variability of rainfall. As evident from data under Table 35.1, annual streamflow contributions from glaciers have a smaller inter-annual coefficient of variation than annual flows generated by rainfall/ runoff and thus mediate the inter-annual variability of total annual streamflow.

This buffer characteristic is visible in sub-basins where the model setting is 'very good to good' during years with low rainfall (Table 35.2). In these cases, the proportion of streamflows which is generated from melting of snow and ice in glaciated areas is greater in dry years than during wet years, hence contribution from glaciated areas is important during years of weak monsoon.

Table 35.1 Partitioning of average annual streamflow generated from rainfall/runoff and glacier processes at Haridwar and Farakka (simulated for period 1982 to 2002)

Parameter	Haridwar sub-basin (upstream)			Farakka sub-basin (downstream)		
	Total	Rainfall/ runoff	Glaciers	Total	Rainfall/ runoff	Glaciers
Average flow, km^3/yr	30.4	12.0	18.4	335.2	275.0	60.2
Coeff. of variation, %	11	20	11	21	24	8

Table 35.2 Average contribution from melting of snow and ice in glaciated areas simulated for some sub-basins where WEAP-Ganga's setting is 'very good to good'

N°	Sub-basin	Average proportion of annual stream flows from glaciated areas		
		For years 1982 to 2002	For 3 wet years	For 3 dry years
44	Busti	44%	40%	46%
46	Rabuwa Bazar	45%	40%	50%
51	Chatara Kothu	24%	22%	26%

35.4   Impact of Temperature Rise on the Melting of Glaciers

IPCC report (2007 and later) indicate that in Tibetan range, the temperature rise at the end of the century could be +3.8 °C. Studies by ICIMOD (2009) and other agencies provide information of the same magnitude. The study considered the following three scenarios:

i.   an increase of 1 °C after 20-yrs, i.e., a rate of +0.05 °C/year (optimistic),
ii.   an increase of 2 °C after 20-yrs, i.e., a rate of +0.10 °C/yr (business-as-usual),
iii.   an increase of 3 °C after 20-yrs, i.e., a rate of +0.15 °C/yr (extreme).

The last scenario should be considered as the extreme scenario. As calibration of the glaciers parameters was based solely on observed streamflow data, we only analyzed simulated streamflows and not, for instance, on variations in glaciers' area. In each scenario, the temperature was raised gradually every year. Rise in temperature increases the quantity of snow and ice melt in the glaciated areas and thus augments the streamflows. However, the impact decreases from upstream to downstream, as (i) enhanced contribution from rainfall dilutes flows from glaciated areas, and (ii) increased temperature also leads to greater evapotranspiration in the plains and thus smaller streamflows. As contribution from glaciated areas is mainly during the high flow season, the increase in streamflow occurs predominantly during the high flow season of June- September. Although, there is little modification during the lean flow season, melting of snow and ice starts earlier (in April) and ends later (in November). Such a phenomenon is likely to create more flood events and of higher magnitude, whether in the Upper Ganga or in the mountainous sub-basins of Devghat and Tehri dam (Table 35.3). Such a perceived change is of great consequence for the water infrastructure and the water security in these regions.

However, on a long-term basis the combination of glacial retreat, decreasing ice mass, early snowmelt and increased winter streamflow suggest that climate change is already affecting the Himalayan cryosphere (Kulkarni et al. 2007). Reduced surface runoff will reduce groundwater recharge and affect groundwater dynamics in the

Table 35.3 Simulated average change in annual streamflow at selected locations of the Ganga basin when temperature increases gradually over 20 yrs (as compared to the reference scenario)

Scenarios	Change in annual streamflow, km³/yr					
	Tehri dam (Sub-basin n°1)	Haridwar (Sub-basin n°2)	Narora (Sub-basin n°3)	Tajewala (Sub-basin n°8)	Delhi (Sub-basin n°9)	Farakka (Sub-basin n°53)
+1 °C over 20 yrs	+0.6 (+8%)*	+1.9 (+6%)	+1.8 (+8%)	+0.2 (+2%)	Negligible	+4.2 (+1%)
+2 °C over 20 yrs	+1.2 (+17%)	+3.9 (+13%)	+3.6 (+15%)	+0.3 (+4%)	Negligible	+8.8 (+3%)
+3 °C over 20 yrs	+1.9 (+26%)	+6.0 (+20%)	+5.4 (+23%)	+0.4 (+5%)	Negligible	+13.4 (+4%)

region, which will be critical in the western region, where most irrigation is from groundwater.

The extra water from the glaciated areas in the short to medium term presents a set of potential opportunities and some serious threats in the form of floods and water congestion especially in the eastern part of the basin. These potential flooding threats can be converted into improved water security through basin wide interventions like UTFI and the GAMES.

35.5   Underground Taming of Floods for Irrigation (UTFI)

The analysis of flood duration suggests that the majority of floods in the Ganges basin are relatively short with an average duration of between 3 and 5 days. Flood durations of 20–60 days have occurred 20 times and there were 7 floods of over 60 days during the period of 1985–2011. The 2007–2008 floods in India and Bangladesh had the longest duration which is explained by abnormal heavy rains during the monsoon season. These floods cause heavy damages to crops, assets, livestock and human lives and need to be mediated for water security and livelihood resilience.

UTFI is a specific and unique application of managed aquifer recharge (MAR). It adds new value to often ad-hoc MAR efforts and puts it into larger scale perspective that offers a wider range of benefits to both upstream and downstream areas. Central to UTFI is distributing recharge-enhancing interventions across strategic parts of the basin to provide supplies to meet additional demand during the dry season, and for this water to be recovered via agricultural wells and tubewells rather than allowing surface water to concentrate and be problematic in the lower floodplain areas of the basin.

The Ganges River basin where problems related to seasonal flooding, groundwater depletion and food and water security are particularly serious and likely to exacerbate due to climate-change induced impacts was selected for development of the opportunity assessment. Factors influencing the occurrence and impacts of floods in the Ganges were identified based on flood-risk mapping, groundwater development and mapping of groundwater potential zones. Across the Ganges basin, the maximum inundation was highly variable with an area varying from 6,000 to 11,000 km^2 Results show that 24% of the inner basin had very high suitability for UTFI (Fig. 35.2, Brindha and Pavelic 2016). The Suitability Index (SI) distribution reveals differentiation in suitability across the landscape- not all areas have the same potential to support UTFI implementation. There was a tendency for the watersheds at the periphery to have a low SI and these upland areas contribute runoff that can lead to flood inundation downstream.

Ramganga sub-basin is one of the area with generally high potential for UTFI implementation. In a drainage area of 18,665 km^2, there are about 20 million people living in the sub-basin. As in other parts of the basin, the most challenging water resources management issue in this sub-basin is the difference between water demand and seasonal availability. During monsoon months, there is widespread flooding and

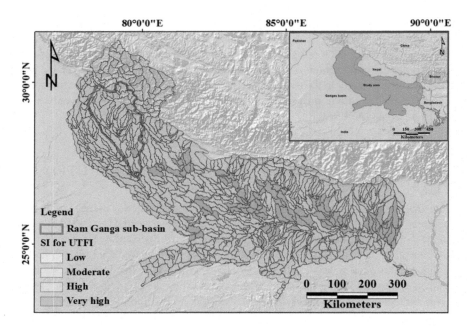

Fig. 35.2 Suitability Index (SI) rankings for the UTFI across the inner Ganges River Basin determined at the watershed level. The study site was located in the Ramganga sub-basin

during rest of the year, surface water flows are limited and groundwater levels are dropping across the basin. Flow-flood relationships showed that a 50% reduction would reduce flow of the highest magnitude with a recurrence interval of 16 yrs down to 2 yrs. For the 50% target, across the entire sub-basin, about 1,741 m³/ha would need to be intercepted, stored and recharged. For this to be achieved, and assuming a minimum infiltration rate of 10 m per wet season, less than 2% of the land area would need to be dedicated to recharge interventions across the entire Ramganga sub-basin.

A pilot trial of this concept was established in the Jiwai Jadid village of Moradabad district in the sub-basin. As the area is intensively cultivated and land is costly and scarce, the ponds within the village situated in local depressions, where there was scope to propose conversion of land for floodwater recharge became the obvious choice. Deep infiltration methods, which bypass the upper layers that are resistive to flow were used to transfer water directly under gravity to deeper more permeable layers. A set of 10 gravity-fed recharge wells with six-inch inner diameter PVC pipes slotted at the base were sunk into the bottom of the pond. The wells were coupled to a recharge filter consisting of gravel in a small brick masonry chamber as shown in Fig. 35.3 (Pavelic et al. 2015). Based on the initial testing of the performance of the recharge well, the anticipated recharge rates could be as high as 432 m³/day. Apart from improved livelihoods from enhanced irrigation, UTFI could be an effective means of disaster risk reduction when implemented at a larger scale. Though it is

Fig. 35.3 UTFI pilot site at Jiwai Jadid village, Ramganga sub-basin. India

too early to conclude from limited observations from the 2015 recharge season, the preliminary results from the pilot site shows an impressive improvement in the local groundwater table. Additionally, opportunities and risks of UTFI are being studied to provide the necessary degree of confidence for scaling up. It appears that there is enormous potential to apply the approach to help decision makers when planning investments in climate change adaptation and flood risk reduction to improve the water security in the Ganges and similar other basins.

35.6   Cranking up the Ganges Water Machine

Though the Ganges River Basin has abundant water resources, seasonal monsoon causes a mismatch in supply and demand which is likely to widen due to climate change impacts. Recurring floods and droughts associated with climate variability are common in the land and waterscape of the basin. The floods in the monsoon season (June to October), and of prolonged dry periods in the non-monsoon season (November to May) are serious concerns for the basin water security. More than 75% of the process depletion (evapotranspiration) from the irrigation, domestic and industrial sectors, in the basin at present is from groundwater withdrawals. The reliance on groundwater in the basin will increase further due to limited prospects for development of additional surface water storages.

The "Ganges Water Machine" (GWM) - a concept introduced in the early 1970's by Revelle and Lakshminarayana (1975) to enhance water storage-is a departure from traditional methods of storing water in surface reservoirs and tanks. In summary it entails:

- pumping water before the monsoon for depleting consumptively, in the irrigation and other sectors, to create an additional sub-surface storage in the Basin, and
- using carefully planned recharge structures to recharge the sub-surface from the monsoon runoff. The cycle of pump-recharge-pump increases storage for run-off water, which mitigates the impacts of floods and droughts.

Initially, the GWM envisaged capturing about 115 Bm^3/yr of monsoon runoff in sub-surface storage and irrigating about 38 Mha of cropland. However, over the last 40 yrs, that estimate of gross irrigated area has already been realized (Amarasinghe et al. 2007). As a result, some areas are experiencing falling groundwater tables (Chinnasamy 2017). In the Ganges river basin, demand for irrigation is low during the Kharif season, but high in the Rabi and hot-weather seasons. Rainfall meets only 30% of the total monthly consumptive water use between November and March. The crop evapotranspiration requirements are very high in April and May. Because of the gap between water supply and crop water requirement, the actual irrigated area at present is substantially lower than the potential. A recent analysis by Amarasinghe et al. 2016 showed that the application of GWM concept is possible partially in some sub-basins (above Ramganga confluence, upper Chambal, Kali Sind, Upstream of Gomti, Ghaghara confluence to Gomti confluence, Tons, lower Yamuna), and fully in others (Bhagirathi, Damodar, Gandak, Ghaghara, Gomti, Kosi, and Son). Due to the patterns of irrigation requirement, the additional irrigation in the Rabi and hot weather seasons has the highest potential for depleting groundwater resources and creating subsurface storage. The potential unmet demand under the two scenarios ranges from 59 to 119 Bm^3. However, due to supply constraints in some basins, the available water resources can meet only about 45 to 84 Bm^3 which is quite significant. A preliminary ex-ante benefit and cost analysis, which captures the potential gains and losses after implementing a project, show that the GWM is a financially viable intervention with a benefit to cost ratio of over 2.3. However, the actual realization of its potential depends on several other hydrological and socio-economic factors. Using solar energy for groundwater pumping, which is financially more viable than using diesel for groundwater pumping as practiced in many areas now shall help in better realisation of the twin targets of reducing floods and enhancing water security and the livelihoods.

35.7 Conclusions

i.   The Ganga basin is one of the largest and most populous river basins in Asia with a population of more than 655 million. Topographic contrasts in the basin are the largest compared to any other basin which cause serious flooding in the downstream sub-basins. The Ganges basin has a total of 1,020 large glaciers and mean annual precipitation varies significantly throughout the basin. Rainfall pattern and glacier melt will be further impacted by the impending climate change.

ii. The experimental glacier module of the WEAP model as applied to the Ganges basin showed that the contribution of glacier and snow melt decreases from upstream mountainous sub-basins to downstream flat sub-basins, with a magnitude variation of 75 to 3%. The glaciated areas contribute to streamflow predominantly during the wet period of June to September, i.e., when flows are actually already high, and glaciers are apparently a buffer against inter-annual variability of rainfall. The impact of climate change through temperature rise of 1 to 3 °C over a period of 20-yrs on the streamflow across the river course would be significant upstream causing serious flooding impacts but also presenting an opportunity for resource augmentation.

iii. 'Underground Taming of Floods for Irrigation' is a useful approach which involves interventions at the river basin scale to strategically recharge aquifers upstream during periods of high flow, thereby preventing local and downstream flooding and simultaneously providing additional groundwater for irrigation during the dry season. The pilot trial of UTFI was encouraging and indicated an enormous potential to help decision makers when planning investments in climate change adaptation and disaster risk reduction in the Ganges basin.

iv. The Ganges River Basin has abundant water resources, but seasonal monsoon causes a mismatch in supply and demand which shall be further accentuated due to climate change. Cranking up of the 'Ganges Water Machine' shall provide additional sub-surface storage through accelerated use of groundwater prior to the onset of the monsoon season, and subsequent recharging of this sub-surface storage through monsoon surface runoff. There is a potential unmet water demand of 59–125 Bm^3/yr under different scenarios of irrigation water use and GWM can satisfy this demand up to 45–84 Bm^3 in the selected sub-basins, and thus in the Ganges basin.

References

Alford D, Armstrong R (2010) The role of glaciers in stream flow from the Nepal Himalaya. Cryosphere Discuss 4:469–494

Amarasinghe UA, Shah T, Turral H, Anand BK (2007) India's water future to 2025–2050: business-as-usual scenario and deviations. In: IWMI Research Report 123. Colombo, Sri Lanka: International Water Management Institute (IWMI)

Amarasinghe UA, Muthuwatta L, Smakhtin V, Surinaidu L, Natarajan R, Chinnasamy P, Kakumanu KR, Prathapar SA, Jain SK, Ghosh NC, Singh S, Sharma A, Jain SK, Kumar S, Goel MK (2016) Reviving the Ganges water machine: potential and challenges to meet increasing water demand in the Ganges River Basin. In: Colombo, Sri Lanka: International Water Management Institute (IWMI) IWMI Research Report Vol. 167, p. 42

Brindha K, Pavelic P (2016) Identifying priority watersheds to mitigate flood and drought impacts by novel conjunctive water use management strategies. Env. Earth Sc. 75(5):1–17

Chinnasamy P (2017) Depleting groundwater-an opportunity for flood storage? A case study from part of the Ganges river basin. India. Hydrology Research 48(2):431–441

ICIMOD (2009) Climate Change in the Himalayas. Monthly Discharge Data for World Rivers, International Centre for Integrated Mountain Development, Kathmandu, Nepal, p 4

Immerzeel WW, Van Beek LPH, Bierkens MFP (2010) Climate change will affect the Asian water towers. Science 328(5984):1382

IPCC 2007 Climate change 2007: the physical science basis. Contribution of Working Group I to the Fourth Assessment report of the Intergovernmental Panel on Climate Change. (Eds.) Solomon S, Qin D, Manning M, Chen Z, Marquis M, Averyt KB,Tignor M, Miller HL Cambridge: Cambridge University Press, U.K

Jain SK, Agarwal PK, Singh VP (2007) Ganga Basin. Hydrology and Water Resources of India, Water Science and Technology Library, Springer, Heidelberg, Germany, pp 333–418

Kulkarni AV, Bahuguna IM, Rathore BP, Singh SK, Randhawa SS, Sood RK, Dhar S (2007) Glacial retreat in Himalaya using Indian remote sensing satellite data. Curr Sci 92(1):69–74

Miller JD, Immerzeel WW, Rees G (2012) Climate change impacts on glacier hydrology and river discharge in the Hindu Kush-Himalayas: a synthesis of the scientific basis. Mt Res Dev 32(4):461–467

Revelle R, Lakshminarayana V (1975) The Ganges water machine. Science 188:611–616

Pavelic P, Brindha K, Amarnath G, Eriyagama N, Muthuwatta L, Smakhtin V, Gangopadhyay PK, Malik RPS, Mishra A, Sharma BR, Hanjra MA, Reddy RV, Mishra VK, Verma CL, Kant L (2015) Controlling floods and droughts through underground storage: from concept to pilot implementation in the Ganges River Basin. In: Colombo, Sri Lanka: International Water Management Institute (IWMI) IWMI Research Report Vol. 165, p. 33

Rikiishi K, Nakasato H (2006) Height dependence of the tendency for reduction in seasonal snow cover in the Himalaya and the Tibetan Plateau region, 1966–2001. Ann Glaciol 43(1):369–377

Sharma BR, de Devraj C (2013) Opportunities for harnessing the increased contribution of glacier and snowmelt flows in the Ganges basin. Water Policy 15:9–25

World Bank (2013) Ganges strategic basin assessment: a regional perspective on opportunities and risks. D.C., The World Bank, Washington

Yates D, Sieber J, Purkey D, Huber-Lee A (2005) WEAP21 – A demand-, priority-and preference-driven water planning model. Part 1: model characteristics. Water International 30:487–500

Chapter 36
Equitable Distrubution of Water in Upper Godavari Sub Basin: A Case Study from Maharashtra

S. A. Kulkarni

Abstract Maharashtra state is located in the semi-arid climatic zone of India, where assured irrigation supply is essential for optimal crop growth. Since the last four decades, the state government has been giving emphasis on water resources development by constructing many large and small dams. The state now has the highest number of dams in the country, creating a storage capacity of about 40 billion cubic meters (BCM). At the state level, on an average about 70% of surface water storages are diverted for irrigation. However, in absence of integrated approach at basin and sub-basin level, there has been lopsided development and use of water resources in the state. Godavari basin, the second largest river basin in India, covers about half of Maharashtra's geographical area. The Upper Godavari Sub-basin is one of the most developed basins in terms of agricultural, urban, and industrial growth in the state. There are 31 large and 558 small irrigation projects in the sub-basin with a design live storage capacity of 5.5 billion cubic meters catering the growing water demand for irrigation, domestic, and industrial uses. As the water demand has continually been rising, the sub-basin has been experiencing water scarcity situation frequently leading to conflicts of water sharing between upstream and downstream stakeholders as also among different categories of uses. Maharashtra is the first state in India to enact legislation and to establish an independent water regulatory authority in 2005 to resolve water related conflicts among others. The case study presented in this paper describes the genesis of the conflict, methodology adopted in framing guidelines towards sharing the sub-basin's water, role played by the regulatory authority, and judicial institutions in addressing and resolving the dispute of water sharing.

Keywords Upper Godavari sub-basin · Equitable distribution · Water Resources Regulatory Authority · Jayakwadi project · Maharashtra

S. A. Kulkarni (✉)
Maharashtra Water Resources Regulatory Authority, Mumbai, India

© Springer Nature Switzerland AG 2021
M. Babel et al. (eds.), *Water Security in Asia*, Springer Water,
https://doi.org/10.1007/978-3-319-54612-4_36

36.1 Introduction

Many river basins in India are experiencing water scarcity due to its rapidly increasing demand from urban dwellers, farmers, private sector companies, industries, thermal and hydropower plants, recreational and cultural uses. When any two or more competing users share the same water resources, the chances of finding mutually acceptable solutions become a challenging task. In the absence of effective institutions conducive to conflict management, this becomes a flashpoint, heightening tension and regional instability that may require years or even decades, resolving it. The latest instance of violent protests by the states of Karnataka and Tamil Nadu over sharing of Cauvery river water has demonstrated that successive agreements, boards, and tribunals found to be ineffective in resolving the century-long dispute (Ganesan and Venkatesh 2016). In India, water is a state subject and the states are obliged to protect water resources as a trustee for the benefit of all its people. As the water scarcity is being increasingly experienced by many states of the country as also the inter-state and intra-state conflicts are rising, the Union Ministry of Water Resources has brought out a Draft National Water Framework Bill (MoWR 2016). As per the Section (30) of the Bill, all efforts shall be made through appropriate institutional arrangements at all levels to prevent a water-related dispute or conflict from arising among different water-uses, or different groups or classes of users, or different areas; and when a dispute or conflict does arise, to settle it through negotiations, conciliation or mediation, or other such means before the dispute or conflict becomes acute so as to avoid recourse to litigation as far as possible.

This paper presents a case study of equitable water distribution in the Upper Godavari sub-basin of the state of Maharashtra. The state has about 10% of India's population but has only 6% of the country's renewable water resources. Geographical constraints further impose the practical limit on impounding/harnessing of water within the state boundaries. Ever growing demand for irrigating parched lands coupled with rapidly increasing urbanization and industrialization has further exacerbated the pressure on the limited freshwater resources of the state. In Maharashtra, the government policy of water allocation prioritizes the domestic use first, followed by the agricultural and industrial uses (GoM 2019). The powers of sectoral allocation of water are vested with the state government. As per the Integrated State Water Plan (GoM 2018) the total surface and groundwater use for irrigation, domestic, and industrial purposes in 2016 was 63 billion cubic meter (BCM), 5 BCM and 1.7 BCM, respectively. The water demand is estimated to rise to 71 BCM, 8 BCM, and 3.7 BCM, respectively by 2030. Since the last decade, the state has been witnessing many conflicts among irrigation vs. non-irrigation users, upstream vs. downstream users, gravity vs. lift irrigation schemes, food crops vs. cash crops, and big projects vs. small schemes. This paper provides a brief genesis of the water distribution conflict, methodology adopted in framing guidelines towards sharing the sub-basin's water, role played by the state's water regulatory authority, and judicial institutions in addressing and resolving the dispute of water sharing.

36.2 Water-Sharing Dispute in Upper Godavari Sub-basin

The Godavari basin is the second largest river basin of India (30.2 million ha) and encompasses parts of six states viz., Maharashtra, Andhra Pradesh, Chhattisgarh, Madhya Pradesh, Odisha, and Karnataka. About 49% of Maharashtra's geographical area (15.26 million ha) is covered by the Godavari basin. In Maharashtra, the Godavari basin is sub-divided into 30 sub-basins. The Upper Godavari sub-basin has a geographical area of 2.18 million ha, and about 8.6 million population lives in its 45 towns and 1,883 villages. The sub-basin comprises the entire catchment of the Godavari River from its source to Jayakwadi dam including the catchment areas of the rivers Mula, Pravara and all other tributaries, which join the Godavari River in this reach. There are 17 major, 14 medium, and 558 small irrigation projects in the sub-basin with a design live storage capacity of 5.48 billion cubic meter (BCM). There are also many *bandharas* (weirs) constructed across the main river and its tributaries for storing water for irrigation during Rabi (winter) season. In Maharashtra, irrigation projects are classified into three broad categories, namely major, medium, and minor. The minor project has cultural command area less than 2,000 ha, the medium has between 2,000 and 10,000 ha and the major project has more than 20,000 ha. The location of Godavari basin and the Upper Godavari sub-basin are shown in Fig. 36.1.

The present case study deals with the equitable sharing of water between 23 major and medium dams located in the upstream part of the upper Godavari sub-basin, and the Jayakwadi dam located exactly at the downstream border of the sub-basin. The upstream projects are grouped into five complexes: (1) Mula, (2) Pravara, (3) Godavari-Darana, (4) Gangapur, and (5) Palkhed. The locations of these 23 dams/reservoirs with their groupings (complexes) in the sub-basin are schematically shown in Fig. 36.2.

The Jayakwadi irrigation project was planned in 1964 and commissioned in 1975. The designed live storage capacity of the reservoir created by the dam (called as Jayakwadi dam or Paithan dam) is 2.17 BCM and was planned to irrigate about 0.27 million ha, predominantly of Aurangabad/Marathwada region. However, since 1976 the Paithan dam has received the designed yield or more in 19 years out of 42 years. The state Water Resources Department (WRD) in 2004 carried out a study to assess the sub-basin yield based on the latest hydro-meteorological data. Table 36.1 shows the comparison of the virgin yield and utilization in the upstream dams as per the initial planning and the revised study.

It can be seen from the Table 36.1 that as per the revised study, the 75% annual dependable virgin yield at the Jayakwadi reservoir site was estimated at 4.45 BCM, which is 17% less than the project design. This may be due to the use of *Strange's Table* for run-off estimation in the earlier study made in 1964, when the river gauge data were not available. During the British rule, Engineer Strange has carried out an investigation in the then Bombay Province and evolved an empirical method for converting rainfall into run-off. Later in 2004, study carried out by the WRD was based on the historic rainfall data of 75 years along with tank gauge and river gauge data, and the yield worked out was considered as more realistic than that arrived at by

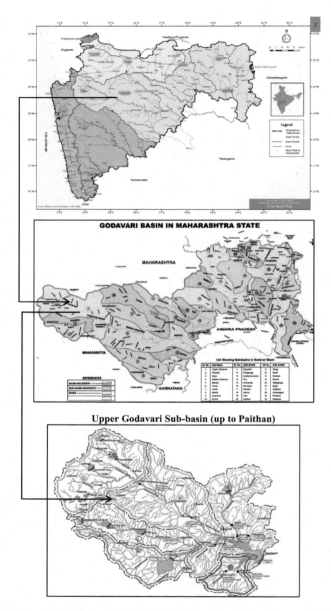

Fig. 36.1 Location of the Upper Godavari sub-basin in Maharashtra. *Source* (GoM 2018), (GoI 2014)

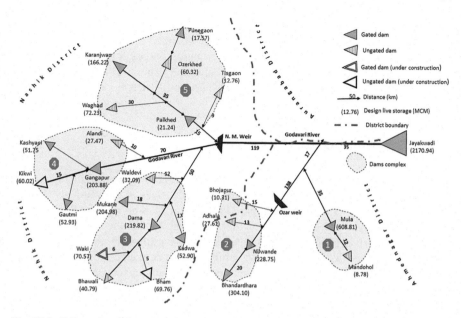

Fig. 36.2 Schematic of the reservoir complexes in the Upper Godavari Sub-basin considered in the case study

Table 36.1 Comparison of the virgin yield and upstream utilization in the Upper Godavari sub-basin (as planned and as per the revised study)

Estimates of yields and utilization	As per project Report, 1964 (BCM)	As per revised study by the WRD, 2004 (BCM)
75% dependable annual virgin yield at the Jayakwadi dam	5.366	4.450
Utilization at the upstream of Jayakwadi dam	3.270	4.073
Net yield available at Jayakwadi dam (including estimated regeneration)	2.296	0.671

the Strange method. Besides decrease in the virgin yield, the upstream utilization has increased almost by 25% compared to the project design. Further, as per the Godavari Study Group Report (GoM 2013) the non-irrigation use of water from upstream projects has increased from 247 million cubic meters (MCM) at the time of planning of the Jayakwadi project to 976 MCM in the year 2013. This has adversely impacted the yield at Jayakwadi dam resulting into significant reduction in irrigated area as envisaged in the project design. Incidentally, the upstream and downstream sub-basin areas including the command area of Jayakwadi project belong to different administrative regions and thus the conflict seemingly assumed inter-regional dimension. As the water demand for irrigation and non-irrigation purposes from both upstream and downstream of the sub-basin has continually been rising over the years, conflicts between the regions for water sharing are also becoming fierce.

36.3 Policy and Legal Provisions

Maharashtra state has always been the trailblazer in the country in introducing various policy measures, acts, institutional reforms, and initiatives in the water sector. The 'State Water Policy' formulated by the Government of Maharashtra (GoM 2019) envisages that the water resources of the state shall be planned, developed, and managed at the river basin/sub basin level adopting a multi-sectoral approach. The policy states that the distress in water availability during deficit periods shall be shared equitably amongst different water use sectors and also amongst upstream and downstream users. As per the Maharashtra State Water Policy, the priority of water usage is: (1) domestic uses like drinking, cooking, bathing, sanitation, personal hygiene and for domestic livestock, (2) agricultural (irrigation) and agro-based industries, (3) industry, thermal and hydropower generation, (4) ecosystem, (5) cultural and religious ceremonies, recreation, amusement, sports etc.

In the year 2005, the Maharashtra Water Resources Regulatory Authority (MWRRA) Act (GoM 2005) was enacted by the state Government. Subsequently, MWRRA was established with a mandate to regulate water resources within the state, facilitate and ensure judicious, equitable and sustainable management, allocation and utilisation of water resources. The key objectives of the regulatory authority are to promote efficient use of water by minimizing its wastage, enhancement and preservation of water quality and to promote integrated and multi-sectoral approach in water sector planning, development and management at the river-basin or sub-basin level.

Due to scanty rainfall during 2012–2015, there was a drought situation in most part of Maharashtra state. The live storage in the Jayakwadi reservoir in 2012 was only 3.3% of the design and there was an acute shortage of water even for domestic and industrial uses. The storage position in the upstream dams was higher than the Jayakwadi. So, the downstream stakeholders filed a Public Interest Litigation in the High Court praying to direct release of water from upstream reservoirs to Jayakwadi (Paithan) dam. In view of the severity of the drought in 2012, the High Court directed to release water from the upstream dams. Accordingly, the WRD released 298 MCM from upstream reservoirs in two spells, of which 62% had reached the Paithan dam. The aforesaid quantity of water was released on ad-hoc basis as there were no guidelines available as how much and from which upstream dams' water to be released. For the subsequent years (2014 and 2015) too, when the Jayakwadi dam was filled about 49% and 14%, respectively, the downstream stakeholders filed petitions with the water regulatory authority (MWRRA) requesting equitable distribution of water in the sub-basin as provided in the MWRRA Act 2005. In the meanwhile, and in view of the intricacies involved in deciding quantities of water to be released from the upstream reservoirs to Jayakwadi, the Government of Maharashtra constituted in January 2013 a "Godavari Study Group" for formulation of guiding principles towards integrated operation of the reservoirs in Upper Godavari sub-basin up to Jayakwadi dam (GoM 2013). The Group submitted its report to the government in August 2013.

There are two pertinent clauses in the MWRRA Act 2005, which empower the Authority for taking decision on the equitable distribution of water within a sub-basin viz., Clause (c) of Section 11: *to determine the priority of equitable distribution of water available at the water resources project, sub-basin and river basin levels during periods of scarcity;* and Clause (c) of Section 12(6): *in order to share the distress in the river-basin or sub-basin equitably, the water stored in the reservoir, in the basin or sub-basin, as the case may be, shall be controlled by the end of October every year in such a way that, the percentage of utilizable water, including Kharif use, shall, for all reservoirs approximately be the same. However, in order to apply the Clause (c), the conditions indicated in the interrelated Clause (a) and (b) of the Section (6) viz., the quota to be fixed at the basin-level, sub-basin level or project level needs to be fulfilled.* On the other hand, Section 11(c) is an independent provision for determining the priority of equitable distribution of water during periods of scarcity, and is separate from the function of fixation of the quota under Section 12 (6)(c). Similarly, neither the command under the Jayakwadi Project has yet been delineated, nor have any Water User Associations been formed under MMISF Act (GoM 2005), and there are no "Entitlement holders" for whom quota can be fixed. Obliviously, Section 12(6) (c) could not be applied for the present case.

As regards, the Section 11(c), earlier threshold of 33% of the designed live storage capacity considered for deciding the 'scarcity' in a reservoir has been negated and the Rules for MWRRA Act framed in 2013 were repealed in the state assembly due to non-unanimity on the threshold limit. In absence of the Rules to define the 'scarcity', the Authority relied on the term 'hydrological drought' to relate it with the 'water scarcity' as per the Manual for Drought Management (GoI 2009). The 'hydrological drought' is defined as a deficiency in surface and sub-surface water supply leading to a lack of water for normal and specific needs (minimum drinking, food crop requirement, and minimum industrial use, which generates employment). Such condition arises even in times of average precipitation, when increased usage of water diminishes the reserves. Since the Jayakwadi dam had shortage of water during 2014 and 2015, the Regulatory Authority considered it as the 'hydrological drought' situation, which is in consonance with the concept of 'scarcity' and exercised its powers of proposing equitable distribution of water within the Upper Godavari sub-basin. Based on the various petitions filed both by the upstream and downstream stakeholders, the matter was heard and the MWRRA issued the Order of 19 September 2014 directing the WRD to release 355 MCM water to Jayakwadi reservoir (MWRRA 2014).

36.4   Rationale Adopted in Equitable Distribution of Water

The MWRRA in its order adopted the guiding principles of the reservoir operation and equitable distribution of water among the upstream and the downstream reservoirs as proposed in the Godavari Study Group Report. The operating strategy for reservoir operation is based on quantity of water to be stored and released in such times depending on the water availability and water demands in the complex during that period. While computing the quantities of water that can be released from each of the upstream reservoirs, it is necessary to ensure that adequate water is available in each of the reservoirs to fulfill the basic and minimum needs of the people dependent on the upstream dams. A reservoir operating strategy is then decided based on the dependable flow (yield) at Jayakwadi reservoir (Table 36.2). In case of 100% dependable year to average yield year, there will be overall 20% reduction in the sanctioned demands for domestic and industrial water as also for irrigation in Kharif (monsoon) season. Similarly, there will be reduction in Rabi (winter) irrigation water supply from 68% (90% dependability year) to 20% (average yield year). No water will be supplied during Hot Weather (summer) irrigation for Strategy I to V. Thus, as the dependability of yield decreases, more water is expected to be available in the reservoirs and there will be lesser cuts to the water demands by all the use sectors. During drought years, available water will have to be used efficiently by all categories of uses. While planning a water resources project in Maharashtra, water withdrawal

Table 36.2 Operating strategies for reduction in non-irrigation and irrigation supplies based on the actual observed net inflow at Jayakwadi reservoir during different dependable yield situations

Operating Strategy	Observed net inflow at Jayakwadi dam in MCM, (% dependable year)	Reduction in the designed water supply (%)				
		Non-irrigation		Irrigation		
		Drinking	Industrial	Kharif (Monsoon)	Rabi (Winter)	Hot Weather (Summer)
Strategy I	1,178.7 (100%) (Worst year)	20	20	20	100	100
Strategy II	1,554.6 (90%)	20	20	20	68	100
Strategy III	1,790.4 (75%)	20	20	20	48	100
Strategy IV	2,027.1 (50%)	20	20	20	28	100
Strategy V	2,119.9 (Average yield)	20	20	20	20	100
Strategy VI	2,618.6 (Good year)	0	0	0	0	0

from dead (inactive) storage for irrigation purpose is not envisioned. Nevertheless, during drought/extreme water shortage the dead storage can be used for drinking purpose.

The Study Group has proposed operating strategies for the reservoir operation under six different storage scenarios to help achieve the approximate equitable and judicious distribution of water in the upstream and the downstream reservoirs for irrigation and non-irrigation uses. This is to be achieved by following a step-by-step synchronization of storages in upstream reservoirs with that of Jayakwadi reservoir under different annual rainfall situations viz. from bad or low rainfall year (Strategy I) to good rainfall year (Strategy VI) as shown in Table 36.3. In the Table 36.3, utilizable water of a given reservoir complex is the sum of water used for drinking, industrial, irrigation, and lost by evaporation from the reservoir during the Kharif (monsoon) season. Besides the balance live storage and exclusion of carryover, if provided, it could thus be more than the designed live storage. The figures in the parenthesis are the percentage of the actual live storage to that of design live storage for all the reservoirs in each of the complex. While determining the equitable distribution of water resources, the Regulatory Authority observed that there must be sufficient water available in upstream dams before it is released to downstream dam. Since water is released through a long river carrier system (about 150 km), there will be significant conveyance and evaporation losses on the way and need to be considered while computing the quantum of water to be released from the upstream dams.

If any reservoir on the upstream side is short of water to meet its own minimum needs (drinking and two irrigation rotations) governed by the respective strategy to be adopted for reservoir operation, no release of water from that reservoir will be allowed. Also, in case the natural storage position of the Jayakwadi reservoir in the first fortnight of October is above or equal to 65% of the live storage (Strategy III) then water from the upstream reservoirs is not required to be released. It means that, if the Jayakwadi reservoir is filled equal to or above 65% of its designed live storage, the situation is not considered as "water scarcity". Based on these guiding principles, MWRRA directed WRD to release 202 MCM in 2014 and 294 MCM in 2015 water from the upstream dams to Jayakwadi reservoir. During the water release period, the power supply was required to be stopped all along the river banks to prevent any unauthorized lifting of water. Strict policing was also done to avoid any physical conflicts or protest by the people. The volume of water actually reached in Jayakwadi reservoir in 2014 and 2015 was 140 MCM and 188 MCM, respectively. It shows that about 30–35% of water quantum was lost during conveyance through the natural river course. The water released from upstream reservoirs into the Jayakwadi reservoir helped alleviate the drinking, industrial, and irrigation water shortages to the extent possible of the downstream stakeholders. The Regulatory Authority has directed the WRD that the exercise of equitable distribution of water has to be carried out by the end of October every year so as to have minimum conveyance losses. Moreover, it coincides with the commencement of the Rabi season, which is the main irrigation season in the state.

Despite the MWRRA's endeavor to resolve the disputes of equitable distribution of water within a sub-basin in a rational manner, its order was challenged both by upstream and downstream stakeholders in the High Court of Judicature at Bombay.

Table 36.3 Strategies for upstream reservoir operation with respect to Jaykwadi reservoir storage for different dependability of flows during their filling (monsoon) period

Reservoir complexes	Jayak-wadi	Mula	Pravara	Godavari - Darna	Gangapur	Palkhed
Dams included in the complex	Jaykwadi (Paithan)	Mandhol, Mula	Bhandardara, Nilwande, Adhala, Bhojapur	Alandi, Kadwa, Bham, Waki, Bhawali, Darna, Mukane, Waldevi	Gangapur, Kashyapi, Gautami	Karanjwan, Waghad, Punegaon, Ojharkhed, Palkhed, Tisgaon
Design live storage (MCM)	2,170.9	617.6	570.8	718.4	308.6	350.3
Carry over (MCM)	381.7	28.3	0.0	0.0	11.6	0.0
Design water use (MCM)	2,618.6	717.8	835.8	1220.0	324.8	456.5
Operating Strategy	Utilizable water including monsoon use and excluding carry over (Mm^3) (figures in parenthesis show the % of the design live storage)					
Strategy I	797 (37%)	303 (49%)	320 (56%)	461 (64%)	187 (61%)	254 (73%)
Strategy II	1173 (54%)	402 (65%)	425 (74%)	604 (84%)	227 (74%)	254 (73%)
Strategy III	1,409 (65%)	489 (79%)	500 (88%)	736 (102%)	252 (82%)	287 (82%)
Strategy IV	1,645 (76%)	576 (93%)	575 (101%)	870 (121%)	277 (90%)	345 (99%)
Strategy V	1,738 (80%)	611 (99%)	605 (106%)	918 (128%)	287 (93%)	369 (105%)
Strategy VI	2,237 (103%)	689 (112%)	836 (170%)	1,220 (170%)	313 (101%)	457 (130%)

The High Court dealt with legal battles among the groups of people belonging to different regions on sharing of water. In the present case, as many as 33 public interest litigations and writ petitions were filed in the High Court. There were petitions challenging the constitutional validity of Sections 11 and 12 of the MWRRA Act 2005, under which the MWRRA had issued its 19 September 2014 Order. The Bombay High Court gave its final judgment (High Court 2016) by upholding the 19 September 2014 Order of the MWRRA. The judgment underscored that the water flowing through the rivers and stored in various reservoirs is the property of the state. The doctrine of public trust will apply to these water resources and the state is the trustee thereof and the public is beneficiary of water. Therefore, no citizen or entity is entitled to claim any preferential right to get supply of water in a particular manner or of a particular quantity except in accordance with the provision of the law. The entire proceedings of resolving the dispute of equitable distribution of water in the Upper Godavari sub-basin by the Regulatory Authority within a legal framework has become a landmark in the history of water governance in Maharashtra.

36.5 Conclusions and Way Forward

As the demand for water is continually increasing due to rapid population growth, rising urbanization and industrialization in India, competition among different categories of uses is on rise. As more and more water is being diverted to various use sectors, some sub-basins are getting closed, and as a result, conflicts among upstream and downstream users are cropping up. Water allocations planned in the most dams constructed in the past have significantly altered due to modification in the sub-basin/basin hydrology, unforeseen and increased water demand for irrigation and non-irrigation uses. Climate change is likely to exacerbate the water scarcity in the river basins further.

The case study of Upper Godavari sub-basin from the state of Maharashtra presented here is a classic example of the resolution of conflicts regarding sharing of water resources within a sub-basin. The guiding principles (operating rules) as specifically evolved for the Upper Godavari Sub-basin help achieve an approximate equitable and judicious distribution of available water among the different categories of uses as also among the users in upper and lower reaches. However, it is to be noted that the guidelines presented here need to be refined further on the basis of data validation and experience gained over the years. Water deficit within a sub-basin/basin needs to be shared equitably in proportion of the demands. Water regulatory institutions like Maharashtra Water Resources Regulatory Authority can play a crucial role in addressing and resolving such conflicts in the best possible manner. Various policies, acts, rules, regulations, study group reports, technical manuals relating to water resources development and management provide added strength to the regulatory authority in taking appropriate decisions. The acts and rules should be unambiguous to avoid legal complications. As the disputes like equitable allocation and distribution of water involves conducting scientific/hydrological/ technical studies, judicial institutions like High Court generally rely on the decisions of the technical expert body like that of regulatory authority by ensuring that the legal procedure and constitutional aspects are not compromised. The High Court Orders provided an added strength to the regulatory authority in enforcing its directions. As the complexity of water resources availability and its equitable allocation and distribution within a sub-basin or basin will increase with space and time in view of the climate change, advanced tools and techniques like modelling, sophisticated real time reservoir operation procedure will be required to address the issues. It is hoped that the case study presented here will be useful as a guide for resolving disputes related to equitable distribution of water resources under similar situation elsewhere. Nevertheless, key to sustainable management of water resources lies in how best communities, voluntary and professional organizations, policy makers, and politicians come together to resolve the conflicts amicably.

Reference

Ganesan R, Venkatesh S (2016) The 140-year-old conflict, Down to Earth magazine, New Delhi. https://www.downtoearth.org.in/feature/war-zone-cauvery-55848

Government of Maharashtra (2019) Maharashtra State Water Policy, Water Resources Department, Maharashtra. https://mwrra.org/wp-content/uploads/2019/09/SWP-2019.pdf

Government of Maharashtra (2018) Integrated state water plan for Maharashtra, Water Resources Department, volume I, Main Report. https://wrd.maharashtra.gov.in/Site/1324/Integrated-State-Water-Plan

Government of Maharashtra (2013) Godavari study group report on formulation of guiding principles on integrated operation of reservoirs for conservation uses in Upper Godavari (up to Paithan) Sub basin, Water Resources Department, Government of Maharashtra. https://mwrra.org/wp-content/uploads/2018/09/Godavari-Study-Group-Report.pdf

Government of Maharashtra (2005) Maharashtra Water Resources Regulatory Authority Act, 2005. https://mwrra.org/wp-content/uploads/2018/07/MWRRA-Act-Regulatory-English.pdf

Government of Maharashtra (2005) Management of Irrigation Systems by Farmers Act. https://mwrra.org/wp-content/uploads/2018/09/4-Maharashtra-Management-of-Irrigation-System-by-Farmers-Act-2005.pdf

Government of India (2014) Godavari Basin, Ministry of Water Resources, Central water Commission, New Delhi. https://indiawris.gov.in/downloads/Godavari%20Basin.pdf

Government of India (2009) Manual for drought management, Department of Agriculture and Cooperation, Ministry of Agriculture, New Delhi. https://nidm.gov.in/PDF/manuals/Drought_Manual.pdf

High Court of Judicature at Bombay (2016) Jayakwadi- PIL Order in PIL 173 of 2013. https://bombayhighcourt.nic.in/generatenewauth.php?bhcpar=cGF0aD0uL3dyaXRlcmVhZGRhdGE vZGF0YS9qdWRnZW1lbnRzLzIwMTYvJmZuYW1lPUNYRkVSODAxMDAxMi5wZGY mc21mbGFnPU4mcmp1ZGRhdGU9JnVwbG9hZGR0PTIzLzA5LzIwMTYmc3Bhc3NwaaH Jhc2U9MjIwOTIwMTE1NTQ2

Maharashtra Water Resources Regulatory Authority (2014) Order of Case No. 1 of 2014 In the matter of the Release of Water into the Jayakwadi Reservoir from Upstream Reservoirs for the Equitable Distribution of Water in the Godavari sub-basin, MWRRA, Mumbai. https://mwrra.org/wp-content/uploads/2018/11/Case-No-1-of-2014-Dt-19-September-2014.pdf

Ministry of Water Resources (2016) Draft National Water Framework Bill, Govt. of India, New Delhi. https://jalshakti-dowr.gov.in/sites/default/files/Water_Framework_18July_2016%281%29.pdf

Chapter 37
Watershed Conservation for Ecosystem Services and its Implication for Green Growth Policies in the Context of Global Environmental Change: A Case of Bhutan

Om Katel, Dhan B. Gurung, Kazuhiro Harada, and Dietrich Schmidt-Vogt

Abstract Globally, green growth policies, such as focusing on the development of hydropower are emerging rapidly. This represents a significant opportunity for countries to incorporate ecosystem services into land management, which can enable them to integrate environmental and developmental goals. Such integration would play a role in achieving sustainable economic growth which is so crucial for developing economies. This paper presents the case of Bhutan. Its objective is to (a) analyse water resources and forest cover characteristics, (b) assess potential of hydropower to enhance green growth, and (c) discuss the challenges of Bhutan striving for sustainable development in the face of global environmental change. For this purpose, the paper analyses the influence of watershed conservation on socio-economic development through hydropower development. The results reveal that the Bhutan Himalaya is rich in water resources and can provide significant ecosystem services for the socio-economic development in the region. Past experience in Bhutan also shows significant economic returns from ecosystem services through forest conservation. This paper is intended to document the land cover status and its contribution to economic development through the use of ecosystem services.

Keywords Ecosystem services · Forest cover · Watershed management · Water resources

O. Katel (✉) · D. B. Gurung
College of Natural Resources, Royal University of Bhutan, Lobesa, Punakha, Bhutan

K. Harada
Gradute School of Bio-Agricultural Sciences, Nagoya University, Nagoya, Japan

D. Schmidt-Vogt
Faculty of Environment and Natural Resources, Freiburg University, Freiburg, Germany

© Springer Nature Switzerland AG 2021
M. Babel et al. (eds.), *Water Security in Asia*, Springer Water,
https://doi.org/10.1007/978-3-319-54612-4_37

505

37.1 Introduction

Green growth policies, such as focusing on the development of hydropower are emerging rapidly all over the globe. Green growth can be defined as "economic progress that fosters environmentally sustainable, low carbon and socially inclusive development" (UNESCAP 2012). Pursuing green growth is expected to use a low level of resources and to reduce emissions while meeting the demands for food production, transport, construction and housing, and others. Green growth is characterized by substantially increased investments in economic activities that enhance the earth's natural capital such as forest management (Bartelmus 2013).

Forests provide several benefits as Ecosystem Services (ESs). These cannot be measured directly as in conventional markets and substituted with the perceived costs (Ninan and Inoue 2013; Vogl et al. 2016). However, ESs play an important role in enhancing human well-being and sustainable development (MEA 2005). At the global scale, Costanza et al. (1997) estimated the total annual value of the world's ESs amounting to an average of US$ 33 trillion, and that of global forests at US$ 969 per ha. This indicates that protecting forests can help protect ESs and ultimately contribute to human well-being.

ESs can be defined as the benefit people obtain from nature in various forms. ES can be categorized broadly into four categories: (i) provisioning; (ii) supporting; (iii) regulating, and (iv) cultural services (MEA 2003). Water as ESs is of prime importance to human well-being and water is therefore considered as the core of sustainable development (UN-Water 2014). Land cover like forests can have significant impact on the water cycle as forests are also the main interacting interface in the ecosystems (Salemi et al. 2012). However, views on the relationship between water flows and forests have been controversial and inconsistent. Many studies (Mark and Dickinson 2008; Mark and Dili et al. 2017; Madani et al. 2018) show that increasing the cover of low vegetation such as grasslands can increase annual water yield although water yield also depends on the type of species and the soil. Nevertheless, it can be argued that there is a close link between forest cover and water yield (Calder 2002; Bruijnzeel 2004).

In the case of Bhutan, the total value of ES is estimated at US$ 15.5 billion per year (Kubiszewski et al. 2013). The Royal Government of Bhutan has adopted various strategies to enhance forest protection. To support green growth is closely tied to the goal stated in the constitution of Bhutan to maintain a minimum of 60% forest cover in all times to come (MOF 2013; GNHC 2013; RGOB 2008). Valuation of ESs and understanding the link between forest cover and water yield is, therefore, of special importance in Bhutan.

This paper presents the overall status of forest conservation in the watershed areas of Bhutan, links forest cover with water resource status, and shows the contribution of water resources as ESs to economic development. This paper documents forest conservation in watershed areas also as adaptation and mitigation strategies to address the impacts of climate change. It is, therefore, divided into three sections: (i) status of forest conservation in Bhutan watersheds, (ii) water resources characteristics, and (iii) forest conservation as a strategy to address impacts of climate change. This paper is intended to make a contribution towards understanding how investing in forest cover may translate into investments in ESs as a means to contribute to sustainable economic development.

37.2 Geography of Bhutan

Bhutan is a landlocked country with a land area of 38,394 km^2 located in the Eastern Himalayas. It is surrounded by the Tibetan region of China in the north and by the Indian states of Arunachal Pradesh in the east, Assam and West Bengal in the south, and Sikkim in the west (Fig. 37.1) (WWF 2016). The country extends over about 200 km from east to west and about 150 km from north to south and has an elevation range from about 100 m in the south to over 7000 m in the north. The country can be divided into three broad physiographic zones: the southern belt is made up of the Himalayan foothills along the Indian border (about 100 m to over 2000 m); the inner Himalayas comprising the main river valleys and steep mountains (about 2000 to 4000 m); and the great Himalayas in the north encompassing snow clad peaks and alpine meadows (above 4000 m) (Ohsawa 1987).

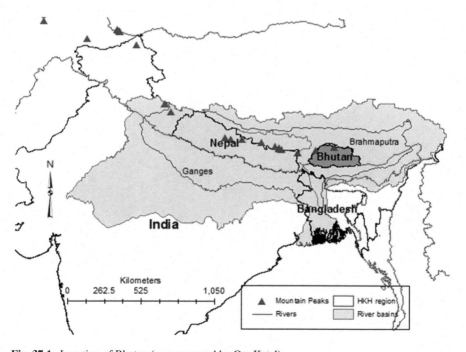

Fig. 37.1 Location of Bhutan (map prepared by Om Katel)

37.2.1 Forest Conservation and Management in Watersheds of Bhutan

Bhutan is especially famous for its extensive forests, which cover more than 70.46% of the total land area. It is also a predominantly agricultural country where traditional mixed farming is practiced, and where agriculture is the main economic activity. Agriculture contributes 18.5% to the government revenue (RGOB 2012) and more than 60% of the total population are engaged in this sector. That agriculture depends on forests as primary sources of organic manure adds another dimension to the environmental goal of protecting and maintaining an extensive forest cover in Bhutan. Recently, the government of Bhutan has declared that by the year 2025, farming in Bhutan would be 100% organic. To achieve this goal, the government has enacted and implemented conservation policies. One of the main conservation strategies is the designation of Protected Areas (PAs) that now cover more than half (51.4%) of the country's land area (Fig. 37.2) (RGOB 2010). PAs are considered as one of the best strategies for the conservation of biodiversity, and ultimately for enhancing ESs (IUCN 2003; Chape et al. 2005). Bhutan has four main river basins encompassing four major watersheds and 187 micro watersheds. Most of these watersheds are located inside the designated protected areas and the overall objective is to protect

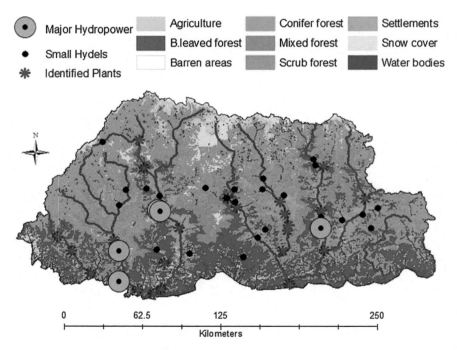

Fig. 37.2 Land use, completed hydropower plants and identified hydropower plants to be built in future (map prepared by Om Katel)

forest in those watershed areas to enhance ESs, especially water as an ES which contributes significantly to the country's economy. Currently, there are 16 projects that can produce the energy upto 13,632 MW capacity and in addition to these projects there are many other small hydels (Katel et al. 2015) accounting the total number of hydro power plants to more than 50 across the country (Fig. 37.2).

37.2.2   Status of Water Resources and Their Utilization for Economic Development

Bhutan has abundant water resources, which can be considered as an engine for economic growth not only in Bhutan but also in the Assam and West Bengal states of India and in Bangladesh affecting more than 35 million people living in this region of Bangladesh and India which is also known as Ganga-Brahmaputra-Meghna river basin. All the rivers coming from Bhutan drain into this basin. Till date, the total water resources are estimated to amount to 78,000 m^3 per year, with the per capita total actual water resource amounting to 109.244 m^3 per year (WWF 2016).

Bhutan has four main river basins (Amochu Basin; Wangchu Basin; Punatsangchu Basin; Manas/Drangmechu Basin) (Fig. 37.3). All rivers originating and passing through Bhutan drain into the Brahmaputra river and finally into the Bay of Bengal. Bhutan's water resources can be categorized as rich with the total annual internal renewable surface water of approximately 78,000 million m^3 (The surface water

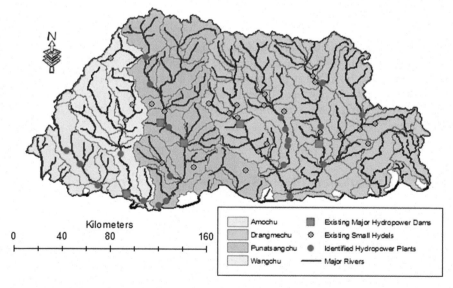

Fig. 37.3 Watershed boundaries and river basins in Bhutan (map prepared by Om Katel)

Table 37.1 Summary of glaciers and glacial lakes in Bhutan

Sl no	Sub-basins of rivers	Glaciers			Glacial lakes	
		Numbers	Area (km^2)	Ice reserves (km^3)	Numbers	Area (km^2)
1	Amo chu	–	–	–	71	1.83
2	Ha chu	–	–	–	53	1.83
3	Pa chu	21	40.51	3.22	94	1.82
4	Thimphu (Wang) chu	15	8.41	0.33	74	2.82
5	Mo chu	118	169.55	11.34	380	9.78
6	Pho chu	154	333.56	31.87	549	23.49
7	Dang chu	–	–	–	51	1.81
8	Mangdechu	140	146.69	11.92	521	17.59
9	Chamkhar chu	94	104.1	8.11	557	21.03
10	Kuri chu	51	87.62	6.48	179	11.07
11	Drangme chu	25	38.54	2.26	126	5.82
12	Nyera ama chu	–	–	–	9	0.08
13	Northern basin	59	387.73	51.72	10	7.81
Total		**677**	**1,316.71**	**127.25**	**2674**	**106.78**

Source Chhopel et al. (2011) (Climate Summit for living Himalaya document). Average annual flow in Bhutan is 73,000 million m^3 per year at per capita mean annual flow availability of 109,000 m^3 (Table 37.2) (Chhopel et al. 2011). The available water can be used for hydropower generation and for farming. However, water may not be accessible to all farmers as most of the settlements are located on the valley slopes and hill tops far away from the flowing rivers through the deep valley gorges

leaving to India is estimated at 78,000 million m^3 per year). These water resources are available in the form of rivers, streams, lakes, and other water bodies that are continuously fed by a rich high altitude wetland system in addition to over 2,674 glacial lakes and 677 glaciers (Table 37.1). However, ground water resources are assumed to be limited due to the topography, and are mainly drained by the surface water network.

37.2.3 Utilization of Water Resources for Hydropower

As water is one of the major sources of revenue for Bhutan, dams have been constructed or are under construction, and potential sites are identified for future construction of additional hydropower plants. The completed hydropower plants include Chhukha dam on the Wangchu river in Chhukha dzongkhag (district) in the southwest, which is 40 m high and the oldest dam in the country. Tala-Wankha dam, further downstream on the Wangchu river, is 91 m high. The Kurichhu dam in eastern Bhutan is 33 m high. Other and smaller hydropower projects include the Basochu

Table 37.2 Estimated water resources in Bhutan

Sl no.	Characteristics	Values
1	Country land area	38,394 km^2
2	Total population	634,982 as of 2015
3	Long term mean annual flow for entire country	2,325 m^3/s $=$ 73,000 million m^3/year
4	Per capita mean flow availability	109,000 m^3
5	Minimum 7 days flow of 10 year period	427 m^3/s $=$ 13,500 million m^3/year
6	Per capita minimum flow availability	21,207 m^3
7	Precipitation (long term average)	2,200 mm/year

Source Chhopel et al. (2011) (Climate Summit for living Himalaya document)

dam near Wangduephodrang town and the Dagachu dam in Dagana dzongkhag. With 141 m, Punatsangchu dam in Wangdue dzongkhag is the highest dam in Bhutan. The Mangdechu hydropower project is the latest project and has been commissioned only recently. Currently, total hydropower production in the country amounts to 1,500 MW. Hydropower energy contributes 96% of the total energy generated in the country (Fig. 37.3). The country's first hydropower project was installed in 1986. The second one was the Tala Hydro Power Project in 2007 which subsequently paved the way for other hydropower projects in the country. With the commissioning of the first hydropower project in 1986, and the second project in 2007, hydropower alone accounted for about 45% of the country's revenue. While Bhutan was one of the poorest nations in south Asia before 1980s, the installation of hydropower projects has contributed substantially to increasing its national revenue. Bhutan's GDP of Nu. 1,204.8 million in 1980 has increased significantly over a period of just five years (1980–1985) and reachedt Nu. 1,674.5 million (1 US$ $=$ 12.61 Nu) in 1985 with an annual growth rate of 6.8% (RGOB 1987). Total hydropower generation currently accounts for 1,500 MW of the feasible potential of 27,000 MW (BEA 2010; Biswas 2011; Bisht 2012; Dhakal and Jenkins 2013). Hydropower contributes as much as 45% to the country's revenue and is thus the major source of revenue for the Bhutanese economy (Singh 2013). Sixteeen additional projects –including many micro hydels –are planned to be on the grid until 2021 with a capacity of 13,632 MW (Biswas 2011; Bisht 2012). The hydropower generated will be exported to India. It is projected that Bhutan can earn revenue of more than US$ 100 million when all the planned hydropower projects are completed (Bisht 2012). Given the relatively small population and the low population growth rate of about two percent, the income earned from hydropower can support basic infrastructure and social service developments. As of 2019, about 90% of the rural households are connected to power supply.

37.2.4   Forest Conservation and its Role in Enhancing Ecosystem Services

Net forest cover has increased from 1990 to 2010 (Bruggeman et al. 2016). The current forest cover is 70.5% of the country's total land area (NSSC and PPD 2011; RGOB 2013) (Fig. 37.4). Many degraded forest areas were improved by implementing forestation and forest management programs such as private forestry development, community forestry development, and rehabilitation of degraded land. The most prominent aim of the national forestry policy is to protect water catchment and watershed areas in Bhutan in order to sustain hydropower generation for the country's economic development. Forest cover increase translates into a decline in unit cost of hydropower generation because improved forests reduce soil erosion and as a consequence, reduce the clogging up of hydropower turbines. Forest cover and hydropower generation is interlinked as erosion can affect the life of the hydropower turbine, and erosion is one of the main factors affecting water quality in the Himalayan region (Ives and Messerli 1989). Soil conservation and watershed management practices are extremely important not only for hydropower generation and sustenance, but also for adaptation to and mitigation of negative impacts of climate change and climate variability.

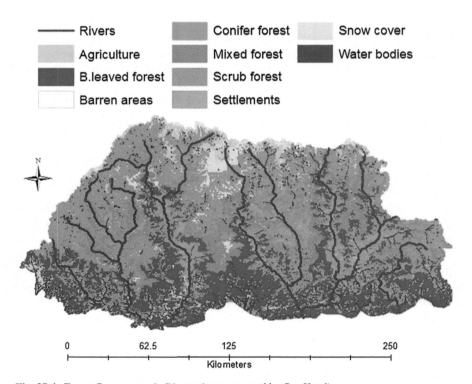

Fig. 37.4 Forest Cover status in Bhutan (map prepared by Om Katel)

37.2.5 Forest Conservation as a Strategy to Climate Change Adaptation and Mitigation

Forest conservation and management can be used as adaptation as well as mitigation strategies. Adaptation to climate change includes the use of forest resources by local people taking into account resistance options, resilience options, and response options (Miller et al. 2007). The resistance options include complete eradication of invasive species and management of forests for the reduction of fire as well as of insect and diseases infestation. The resilience options include increasing seed banks in forests and accelerating re-vegetation which would in turn provide protection against soil erosion and loss of soil moisture. The response options include facilitation of seeds dispersal, migration, and colonization. Such strategies are widely applied in Bhutan. Mitigation strategies, on the other hand, include sequestering carbon and reducing overall greenhouse gas emissions. Bhutan is a carbon neutral country today because it sequesters more carbon than what it emits.

37.3 Constraints to Economic Development

Climate change has been a major constraint on economic development with significant impacts on the agriculture and forestry sectors. It also poses a threat to the hydropower sector (Katel et al. 2015). A study conducted by Hoy et al. (2015) has shown an "increase of the annual precipitation rate by about 500 mm for the last 90 year i.e. an amplification by about 20–25% of the current country average". In addition, the precipitation pattern was highly variable in recent years with a tendency towards extremes. Rains were comparatively heavier during the monsoon months and at a minimum during the cold and dry winter. Not only has precipitation become more variable at higher elevations, temperatures are expected to increase by a rate that is three times higher than the global average (IPCC 2007; Xu et al. 2009). Changes of this magnitude can be expected to disrupt ecosystem services affecting livelihoods of people particularly those living in the Himalayan region. This is especially true for ecosystem services affecting the contribution of hydropower development to the economy of Bhutan.

While hydropower development is promising, there is, however, a lack of quality data and a need to study river sources, quality and dynamics in the context of global environmental change. Global environmental change affects Bhutan in two different ways: (a) it changes ecosystem structures and functions, affecting agriculture and depleting natural resources, and (b) it affects direct ES benefits such as drinking water provision, irrigation and hydropower. Generating data on river water dynamics and stability of river basins is crucial especially because hydropower plants are constructed over the rivers that are fed by glaciers that are more than 677 covering about 3.4% of the country's total surface area (Chophel et al. 2011). An average retreat rate of 8.1% was found between 1963 and 1993 in a study of 66 Bhutanese glaciers conducted by Karma et al. (2003), corresponding to the temperature increase during that time. In addition, Bhutan experiences a stronger monsoon in summer than the western Himalaya (Ageta et al. 2000). Melting of glaciers is associated with Glacial Lake Outburst Floods (GLOFs). GLOFs pose major hazards in the Himalaya and especially in Bhutan, where they are far more common than in other regions (Richardson and Reynolds 2000; UNDP 2012). These evidences show that people living in Bhutan are increasingly experiencing the impacts of climate change through ecosystem and environmental change which also mean that climate change can also pose a significant threat to the Bhutanese economy.

37.4   Conclusions and Recommendations

Bhutan has abundant water and forest resources. Both are interlinked in the sense that Bhutan's extensive forest cover protects watersheds and supports the provision of ESs by water resources. Provision of ESs by water resources is further enhanced by policies relating to watershed protection and by sustainable management. Bhutan's water resources, have contributed to the country's economic success mainly through hydropower development since the 1980s, and that its continued success will rest on enabling green growth policies. This success is, on the other hand, jeopardized and constrained by environmental change and especially by climate change. As suggested by Vogl et al. (2016) and Brogna et al. (2017), forest conservation can enhance adaptation as well as mitigation strategies in the face of climate change in addition to the enhancing ESs. The focus of this paper is on the current status of water and forest resources and on the role of Bhutan's enabling policies in enhancing the provision of ESs by these resources, especially by water resources in the face of climate change. We also recommend further research on the relation of water discharge and the forest change relating to soil erosion in the foothills, middle hills and high Himalaya, which would be useful to understanding whether or not forest conservation contributes to the reduction of soil erosion which would also have a positive effect on the operation of completed hydropower plants. Another area in which more research is needed is the link between seasonal water discharges, dynamics of forest cover change, and human land cover interactions. Assessing the characteristics of watersheds using hydrological models is also recommended in order to better understand the role of minimum environmental flow in the dynamics of water resources in the Bhutan Himalaya and its implications for economic development. Particular attention should be given to studies of river basins management at the trans-boundary level so that a governance system at the international level can be strengthened. Since climate change is a global phenomenon, trans-boundary collaboration is of utmost importance.

Acknowledgement Authors would like to thank the College of Natural Resources, Royal University of Bhutan for facilitating in writing this manuscript.

References

Ageta Y, Iwata S, Yabuki H, Naito N, Sakai A, Narama C, Karma: Expansion of glacier lakes in recent decades in the Bhutan Himalayas. Debris-covered Glaciers, IAHS Publications no. 264, pp. 165–175 (2000)

Bartelmus P (2013) The future we want: green growth or sustainable development? Environ Dev 7:165–170

BEA [Bhutan Electricity Authority] (2010) Druk Green Power Corporation Limited. Tariff Review Report, Thimphu Bhutan

Bisht M (2012) Bhutan-India power cooperation: benefits beyond bilateralism. Strateg Anal 36(5):787–803

Biswas AK (2011) Cooperation or conflict in trans-boundary water management: case study of South Asia. Hydrol Sci J 56(4):662–670

Brogna D, Vincke C, Brostaux Y, Soyeurt H, Dufrene M, Dendoncker N (2017) How does forest cover impact water flows and ecosystem service? Insights from real life catchments in Wallonia (Belgium). Ecol Indicators 72:675–685

Bruggeman D, Meyfroidt P, Lambin EF (2016) Forest cover changes in Bhutan: revisiting the forest transition. Appl Geography 67(2016):49–66

Bruijnzeel LA (2004) Hydrological functions of tropical forests: not seeing the soil for the trees? Agric Ecosyst Environ 104:185–228

Calder IR (2002) Forests and hydrological services: reconciling public and science perceptions. Land Use Water Resour Res 2:1–12

Chape S, Harrison J, Spalding M, Lysenko I (2005) Measuring the extent and effectiveness of protected areas as an indicator for meeting global biodiversity targets. Phil Trans R Soc B 360:443–455

Chophel GK, Dulal IR, Chhophel K, Wangchuk C, Rinzin U, Dupchu K, Kunzang, Tse-ring K, Tshethar K, Dorji U, Wangchuk T, Lhamo T, Nidup J (2011) Securing the Natural Freshwater Systems of the Bhutan Himalayas: climate Change and Adaptation measures on Water Resources in Bhutan. A climate summit for a living Himalaya. http://sa.indiaenvironmentportal.org.in/files/file/Water_Paper_Bhutan.pdf

Costanza R, d'Arge R, De Groot R, Farber S, Grasso M, Hannon B, Van Den Belt M (1997) The value of the world's ecosystem services and natural capital. Nature 387(6630):253–260

Dhakal DNS, Jenkins GP (2013) Risk sharing in hydropower development: case study of the Chukha Hydel Project, Bhutan

GNHC [Gross National Happiness Commission] (2013) Eleventh Five Year Plan–2013–2018. Vol. I. Thimphu, Bhutan

Hoy A, Katel O, Thapa P, Dendup N, Matschullat J (2015) Climatic changes and their impact on socio-economic sectors in the Bhutan Himalayas–an implementation strategy. Reg Environ Change. https://doi.org/10.1007/s10113-015-0868-0

IPCC (2007) Climate change 2007: impacts, adaptation and vulnerability. Contribution of working group II to the fourth assessment report of the intergovernmental panel on climate, Cambridge University Press, Cambridge, p 976

IUCN (2003) World Conservation Union (IUCN). World Parks Congress

Ives JD, Messerli B (1989) The Himalayan Dilemma: reconciling development and conservation. United Nations University. ISBN 0-203-26465-7

Karma T, Ageta Y, Naito N, Iwata S, Yabuki H (2003) Glacier distribution in the Himalayas and glacier shrinkage from + 30-to +33-in the Bhutan Himalayas. Bull Glaciol Res 20:29–40

Katel ON, Schmidt-Vogt D, Dendup N (2015) Transboundary water resources management in the context of Global Environmental Change: the Case of Bhutan Himalaya in (Eds) Managing Water resources under climate uncertainty. Springer

Kubiszewski I, Costanza R, Dorji L, Thoennes P, Tshering K (2013) An initial estimate of the value of ecosystem services in Bhutan. Ecosyst Serv 3:e11–e21

Li J, Liu D, Wang T, Li Y, Wang S, Yang Y, Wang X, Guo H, Peng S, Ding J, Shen M, Wang, L (2017) Grassland restoration reduces water yield in the headstream region of Yangtze River. Scientific Reports, 7: 2162. https://doi.org/10.1038/s41598-017-02413-9

Madani EM, Jansson PE, Babelon I (2018) Differences in water balance between grassland and forest watersheds using long-term data, derived using the CoupModel. Hydrolgy Research, 49:1. IWA publishing. Hydrology Research

Mark AF, Dickinson KJ (2008) Maximizing water yield with indigenous non-forest vegetation: a New Zealand perspective. Front Ecol Environ 6(1):25–34

MEA (2003) Ecosystems and Human Well-being: A Framework for Assessment. Island Press, Washington, DC

MEA (2005) Ecosystems and Human Well-being: Current States and Trends. Island Press, Washington, DC

Miller CI, Stephenson NL, Stephens SL (2007) Climate change and forests of the future: managing in the face of uncertainty. Ecol Appl 17(8):2145–2151

MOF (2013) Bhutan Trade Statistics 2012. Royal Government of Bhutan, Thimphu

Ninan KN, Inoue M (2013) Valuing forest ecosystem services: what we know and what we don't. Ecol Econ 93:137–149

NSSC and PPD Bhutan land cover assessment 2010. Thimphu: National Soil and Services Centre and Policy and Planning Division, Ministry of Agriculture and Forests, Royal Government of Bhutan. Technical report (2011)

Ohsawa M (1987) Life zone ecology of the Bhutan Himalaya. Chiba University, Laboratory of Ecology, Chiba

RGOB [Royal Government of Bhutan] (1987) Fifth Five-Year Plan Document. Gross National Happiness Commission, Thimphu

RGOB [Royal Government of Bhutan] (2003) Biodiversity Act of Bhutan. Ministry of Agriculture, Royal Government of Bhutan

RGOB (2004) Bhutan Biological Conservation Complex. Department of Forest and Nature Conservation Division, Thimphu Bhutan

RGOB (2008) The Constitution of the Kingdom of Bhutan. ISBN 99936-754-0-7

RGOB [Royal Government of Bhutan] (2010) Statistical Year Book of Bhutan. National Statistics Bureau. Royal Government of Bhutan, Thimphu. ISBN 978-99936-28-06-4

RGOB [Royal Government of Bhutan] (2013) RNR statistical coordination section. Policy and Planning Division, Ministry of Agriculture and Forests, Royal Government of Bhutan, Bhutan

Richardson SD, Reynolds JM (2000) An overview of glacial hazards in the Himalayas. Quatern Int 65–66:31–47

Salemi LF, Groppo JD, Trevisan R, Moraes JM, de Lima WP, Martinelli LA (2012) Riparian vegetation and water yield: a synthesis. J Hydrol (Amsterdam) 454–455:195–202

Singh BK (2013) South Asia energy security: challenges and opportunities. Energy Policy 63:458–468

UNDP (2012) Glacial Lake Outburst Flood, Reducing Risks and Ensuring Preparedness. The Report on the International Conference, Thimphu

UNESCAP (2012) Green Growth, Resources and Resilience. Environmental Sustainability in Asia and the Pacific. www.unescap.org/esd/environment/flagpubs/GGRAP

UN-Water (2014) A Post-2015 Global Goal for Water: Synthesis of key findings and recommendations from UN-Water. UN Water, pp. 2014

Vogl AL, Dennedy-frank PJ, Wolny S, Johnson JA, Hamel P, Narain U, Vaidya A (2016) Managing forest ecosystem services for hydropower production. Environ Sci Policy 61:221–229

WWF, Bhutan (2016) "What We Do. WWF Bhutan". http://www.wwfbhutan.org.bt/_what_we_do/. Accessed 20 Aug 2016

Xu J, Grumbine RE, Shrestha A, Eriksson M, Yang X, Wang Y, Wilkes A (2009) The melting Himalayas: cascading effects of climate change on water, biodiversity and livelihoods. Conserv Biol 23(3):520–530

Chapter 38
Testing Framing Effects on Subjective Wellbeing in Lao PDR

J. Ward and A. Smajgl

Abstract There is an increasing international reliance on measures of subjective well-being to guide policy formulation by identifying well-being dimensions prioritised by affected households. The cognitive dimension of subjective wellbeing represents how one's life measures up to expectations and resembles an individual's envisioned ideal life. Tversky and Kahneman (1981) established that survey questions framed as either a loss or a gain significantly bias responses. The effects of question framing have not been previously tested in subjective wellbeing research. We test the effect of framing difference (satisfied vs. dissatisfied) based on a survey of randomly selected households in two adjacent Lao PDR river basins: the Nam Ngum (n = 1000; well-being posed as "how satisfied") and the Nam Xong (n = 1000; well-being posed as "how dissatisfied"), using a multi-dimensional and location specific list of 38 well-being factors. Subjective wellbeing factors were quantified and prioritised for each household by calculating a widely tested index of dissatisfaction (IDS). The results indicate no significant differences (p < 0.05) in the pattern of ranked IDS factors across the sample cohorts and inconsistent valency in mean IDS values. Roads, electricity, food security, and domestic water supply had the highest ranked IDS scores for both basins and represent interventions points where well-being gains are most likely. Policies singularly focussed on increased household income represent suboptimal solutions to improve household wellbeing in the Nam Xong and Nam Ngum River basins.

Keywords Subjective well-being · Framing effects · Lao PDR

38.1 Introduction

Recent multi-governmental initiatives such as the Sustainable Development Goals have reinforced consideration of human wellbeing as a central priority of many national agendas. The outcomes of development projects are increasingly being

J. Ward (✉) · A. Smajgl
Mekong Region Futures Institute, Bangkok, Thailand
e-mail: john.ward@merfi.org

© Springer Nature Switzerland AG 2021
M. Babel et al. (eds.), *Water Security in Asia*, Springer Water,
https://doi.org/10.1007/978-3-319-54612-4_38

517

measured not just in financial terms but in terms of improvement to both livelihoods and human wellbeing (Scoones 1998; DFID 1999; Bamberger et al. 2009).

Improved wellbeing is a priority objective of Lao PDR policy formulations, including policy objectives for the Nam Xong and Nam Ngum river basins. Improved wellbeing is articulated as either explicit statements or implied through improvements to either health, child nutrition, education, personal safety or resource security, endowments and entitlements.

A common proposition articulated in an extensive and diverse corpus of wellbeing scholarship is the necessity of jointly meeting the imperatives of improved wellbeing and sustainable use of natural resources, including water (Stiglitz et al. 2009; Anand and Sen 2000; Gasper 2004a,b, 2007; White and Ellison 2007). Stiglitz et al. (2009 p. 12) argue for the development of a 'statistical system that complements measures of market activity by measures centred on people's wellbeing and by measures that capture sustainability'. Such a system must, of necessity, be multi-dimensional, focus on the household and capture individual life evaluations and priorities, propositions consistent with *inter alia* Campbell (1981), Doyle and Gough (1991), Gasper (2007), McGregor et al. (2007), Nussbaum (2000) and White and Ellison (2007). As a corollary, there is an increasing international reliance on measures of subjective well-being to guide policy formulation by identifying the well-being dimensions prioritised by affected households. Numerous instrument variants have been developed to elicit subjective wellbeing (Kahneman et al. 1999, Gasper 2007, White and Ellison 2007) and at the time of writing, wellbeing dimensions and questionnaires remain non-standardised and generally operationalised in the context of specific communities (Nussbaum 2000).

Tversky and Kahneman (1981) established that survey questions framed as either a loss or a gain significantly bias responses. That is people respond differently to decision making, preference ranking and choices depending on alternative, but objectively equivalent, descriptions of the problem. Tversky and Kahneman (1981, 1987, 1991) and Camerer (1995) argue that framing effects are an example of cognitive bias, whereby people respond differently to decision making and choices depending on alternative, but objectively equivalent, descriptions of the problem. Tversky and Kahneman (1981, 1987) focussed on loss aversion. When decision options appear framed as a likely gain, risk-averse choices predominate. A shift toward risk-seeking behaviour occurs when the equivalent decision is framed in negative terms. Levin et al. (1998) developed a framing effect typology to distinguish between three different kinds of valence framing effects. First the standard risky choice framing effect introduced by Tversky and Kahneman (1981) to illustrate how valence affects willingness to take a risk. Second, attribute framing, which affects the evaluation of an object or event characteristics, and third, goal framing, which affects the persuasiveness of a communication. The dissatisfaction/satisfaction framing of wellbeing corresponds mostly with the attribute class of framing effects. In the case of the Nam Ngum and Nam Xong, the level of individual (dis)satisfaction of wellbeing factors was the preference being ranked, framed as either satisfaction or dissatisfaction.

Development initiatives and interventions can be comparatively evaluated through time and space, or across case studies as an alternative to static analyses. With the increasing reliance on measures of subjective wellbeing coupled with non-static comparisons, the deployment of non-standardised instruments to elicit wellbeing introduces an increased risk of biased evaluations by failing to account for omitted variables that can influence individual questionnaire responses. The seminal loss aversion research of Tversky and Kahneman (1981, 1987, 1991) suggests framing wellbeing as either satisfaction or dissatisfaction is a likely candidate variable capable of biasing responses.

The effects of question framing have not been previously tested in subjective wellbeing research. We applied four analytical approaches to test the hypothesis H_1: significant differences will be detected in the numerical ranking of wellbeing factors when framed as either "how satisfied are you" compared to "how dissatisfied are you"? First a numerical wellbeing value was calculated as an Index of dissatisfaction (IDS) and wellbeing factors ranked from highest to lowest. Second the ranking pattern of the two sample cohorts was compared using the Wilcoxin signed rank test. Third the difference in mean IDS values for individual wellbeing factor was compared using Student's t-test. Fourth IDS scores were aggregated into the three wellbeing dimensions: environmental, social and economic and mean values compared. Finally latent constructs were identified by cluster analysis of the three wellbeing dimensions and differences in cluster membership and mean IDS values compared across the '*satisfied*' and '*dissatisfied*' cohorts.

38.1.1 Developing Instruments to Measure Subjective Wellbeing

Gasper (2007 Table 2.3), in an extensive review of the diverse conceptualisations and ontology of wellbeing and needs, defines objective wellbeing as the 'externally approved, normatively endorsed, non-feeling features of a person's life'; and subjective wellbeing as 'feelings of the person whose wellbeing is being estimated'. Diener et al. (1999 p. 277) regard subjective wellbeing as a general domain of scientific interest, comprised of a "broad category of phenomena that includes people's emotional responses, domain satisfaction and global judgements of life satisfaction". Frey and Stutzer (2002), Diener et al. (1999) and Kahneman and Kreuger (2006) distinguish three conceptual dimensions requiring independent assessment: positive and negative affects; life satisfaction (the cognitive element) and *eudaimonia* (the most fundamental concept going back to the Greek philosophy and referring to a good or virtuous life as whole).

The affective element represents a hedonic evaluation guided by emotions and feelings while the cognitive element is an information based appraisal of how one's life measures up to expectations and resembles an individual's envisioned "ideal life" (Rishi and Mudaliar 2014). Quantitative research on subjective wellbeing focuses on life satisfaction as an immediate concept that is a more than a transitory emotional reaction (Kahneman and Kreuger 2006, Diener et al. 1999), but refrains from claiming that it has any normative content. This is the concept mainly discussed in this paper.

Parfit (1984) proposed three concepts of wellbeing: hedonism, or wellbeing as pleasure; desire theories or wellbeing as the fulfilment of preferences/desires; and objective list theories or wellbeing as the attainment of the elements in a list of what makes a life well-lived. Gasper (2004a) argues substantive is a more appropriate term than objective and cites Nussbaum's list (Nussbaum 2000) derived through consultation and debate within a particular political community. "Nussbaum's list has aspects of all these types: it derives from the use of formal criteria combined with ethical intuitions, and is to be elaborated and operationalized in each political context" (Gasper 2004a, p. 8).

Designing multi-dimensional "lists" that correspond to political context is a common thread in the subjective wellbeing and development literature (*inter alia* Biswas-Diener and Diener 2001; Gough and McGregor 2007; McGregor et al. 2007; White and Ellison 2007; Camfield et al. 2009). For example, McGregor et al. (2007) distinguished wellbeing outcomes (happiness, life satisfaction, welfare) and processes (freedoms, rights, capabilities) when developing a wellbeing questionnaire trialed in Bangladesh and Peru that investigates the relationships between resources that individuals command, the needs they are able to satisfy and quality of life they are able to attain. Resource dimensions were defined as Social, Cultural, Material, Natural Resources and Human, combined with variables eliciting intermediate needs not met (for example food, housing, health, education, family relations).

Larson (2010) and Larson et al. (2013) extend the Australian Unity Wellbeing Index, developed by Cummins et al. (2003) as a three dimensional wellbeing index (economic and services, environment and social), aligned to local wellbeing perceptions through consultation and applied to natural resource management. The instrument deployed to elicit subjective wellbeing priorities of irrigators in the Nam Ngum and Nam Xong river basins followed Larson's approach, corresponds with Nussbaum's notion of a contextualised, substantive wellbeing list, and focussed on the sustainable management of water basins as the natural resource system.

The relationship between subjective wellbeing and water scarcity and security, either in the case of poor water quality, the lack of water access or both, has received limited attention in reviewed studies. Exceptions are Bookwalter and Dalenberg (2004), who used ordered logit techniques in a survey of South Africa respondents to demonstrate that water plays an important role in the formation of subjective wellbeing, at least in the rich quantiles. Guardiola et al. (2013) employed a two-staged method to relate water access variables to subjective wellbeing of residents of selected villages in Yucatan, Mexico.

38.2 Methods

38.2.1 The Study Sites

The subjective wellbeing research was conducted in the Nam Ngum and Nam Xong River basins, located in northern Laos. The Nam Ngum River Basin (16,700 km^2) is one of the most important in Lao PDR, in terms of annual flow (21 km^3, or 14% of Mekong River annual flow), population (9% of country's population, including parts of the Vientiane capital and Vientiane province) and food production. Elevation in the Nam Ngum Basin ranges from 155 m above sea level at the confluence of the Nam Ngum River with the Mekong River to 2,820 m in Phou Bia Mountain. The basin is mostly hilly and mountainous and includes two main flat areas: (i) the Vientiane Plain in the lower part of the basin where the Nam Ngum River forms numerous meanders before flowing into the Mekong and (ii) the Xiengkhouang Plateau, including the Plain of Jars located in the upper part of the catchment. One third of the basin is covered by bamboo, shrub land and re-growing young forest (Lacombe et al. 2014). Cropped areas cover about 10% of the catchment's surface area. Hydropower and mining are major contributors to the national economy. The Nam Xong is a sub-basin of the Nam Ngum Basin, located towards the west of the watershed (Fig. 38.1).

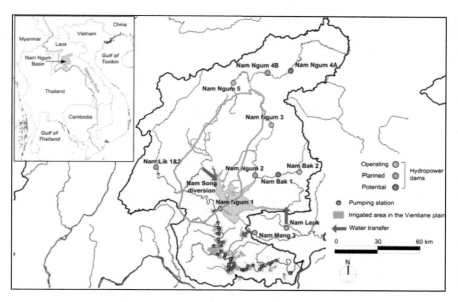

Fig. 38.1 Location of hydropower dams, water transfers and pumping stations in the Nam Ngum Basin. ——represents the approximate Nam Xong basin boundary (source Lacombe et al. 2014)

38.2.2 Household Sampling

A sample (n) of 1,011 households for the Nam Xong was randomly selected from a geographically referenced sample frame of 157 villages located in Kasi, Vangvieng and Hin Heup in 2015. Villages with less than 30 households (established from the 2015 Laos census) were omitted from the sample frame. A sample of n = 1,001 randomly selected households was drawn from a sample frame of 595 villages for the Nam Ngum survey in 2012.

20 households were randomly selected by lottery for interview from each of 50 randomly selected villages in each of the two sample cohorts. Interviewees were either household heads or members who have made decisions on behalf of their household members. Data were collected through voluntary, face to face interviews of 1 to 1.5 h duration, using the structured and pre-tested questionnaire. The Nam Ngum interviewers were conducted using paper questionnaires and data entered at a later date. The Nam Xong interviews were administered via digital tablets, where data was automatically uploaded to a central database. The Ministry of Planning and Investment approved the questionnaire and ethics approval for the survey was previously granted by CSIRO (see Ward and Poutsma 2013). Details of the questionnaire and results can be obtained from the corresponding author and Ward et al. (2016; www.merfi.org). All field enumerators completed a two day training course, including participation in a pilot survey of 60 households, conducted with an aim to ascertain interviewer and interviewee understanding, the clarity of questionnaire wording, consistent interviewing techniques and revision of the questionnaire before conducting field interviews.

38.2.3 Numerical Estimation of Subjective Wellbeing

Previous research in the area of regional subjective wellbeing indicates that wellbeing contributors are likely to be a mix of generic, common factors (such as health or safety) and region specific factors, unlikely to be known to researchers or policy makers without input from the community. Collection of data on perceived contributors to personal and family wellbeing in the region allows better understanding of factors of importance to regional populations. This information would not only enhance relevance of the sustainable development models/scenarios developed, but would also provide valuable insights for both decision makers and researchers concerned with participatory processes to manage water resources (see Ward et al. 2016).

Participatory processes were central to the development of the vector of subjective wellbeing factors posed to the respondents (Gasper 2007; Camfield et al. 2009; Larson 2010) and the focus of eventual research application. The methodological approach follows Larson's (2010) conceptual model of well-being that integrates factors of well-being from the natural environment, social, and economic dimensions, expressed as an Index of Dissatisfaction (IDS). The approach allows for subjective, self-assessed quantification of the factors of well-being most important to the respondent. Larson (2010) deployed a two-step process to select the final set of well-being factors. First, a generic list of factors related to the social, economic and services, and natural environment dimensions was compiled from the literature, influenced mainly by frameworks based on integration of the natural environment with human well-being (for example Alkire 2002; Hassan et al. 2005; McGregor et al. 2007; Nussbaum 2000; Clarke 2005).

Second, the original set of 27 factors established by Larson (2010) was revised by local experts and communities to generate a set of 38 wellbeing factors specific to both the Nam Xong and Nam Ngum River Basins. The process of stakeholder input renders the array of well-being factors included in the final questionnaire as idiosyncratic and salient to the region under study. Therefore, the participatory approach does not define a universally valid set of well-being factors. In accord with Larson (2010) and Nussbaum (2000) the process of developing a reflexive set of case study specific factors suggests that local communities determine their own sets of well-being factors as a first step in the process.

Thirty eight well-being factors included in the final questionnaire were grouped into three domains:

(1) *Family and community domain:* family relations; community relations; personal/family safety; cultural identity; personal/family health; civil and political rights; personal/family education levels; community relations; and sports, travel, and entertainment.

(2) *Natural environment domain:* air quality; water quality; soil quality; access to natural areas; biodiversity; recreational activities; fishing, hunting, and collecting produce; beauty of the landscape and beaches; and condition of the landscape.

(3) *Economy and services domain:* work; income; housing; health services; recreational facilities; condition of the roads; public infrastructure and transport; training and education services; and support services.

Respondents selected eight of the 38 randomly introduced wellbeing factors, scoring the factor importance and the level of factor satisfaction (Nam Ngum) or dissatisfaction (Nam Xong) according to 1–10 and 0–10 numerical scales respectively. Finally respondents were asked to score their overall life satisfaction according to a 1–10 numerical rating scale. Focussing illusions, that is anchoring effects or answer bias due to the sequence of questions (Kahneman and Tversky 1992, Kahneman et al. 2006), were minimised by placing the question "what is your overall assessment of satisfaction (Nam Ngum) or dissatisfaction (Nam Xong) with your life?" after respondents had scored their selected wellbeing factors. The calculation of IDS follows the approach developed by Larson (2010).

We hypothesise that significant differences will be detected in the IDS scores of ranked subjective wellbeing factors, when the wellbeing factors and overall wellbeing are framed either positively or negatively: that is as either "how satisfied are you with wellbeing factor k" posed to Nam Ngum respondents compared to "how dissatisfied are you with wellbeing factor k" posed to Nam Xong respondents.

38.2.4 Statistical Grouping of Individual IDS Scores

Hierarchical cluster analysis was used to segment respondents into relatively homogeneous groups of households/farmers based on their aggregated IDS scores of the economic, social and environmental wellbeing dimensions. Ward's minimum variance method (Ward, 1963; Kaufman and Rousseeuw 1990) represents an example of agglomerative hierarchical cluster analysis, which treats each case as a singleton cluster at the outset and then successively merges cluster pairs by minimising variance (the minimum increase in the error or sum of squared deviations) until all clusters have been merged into a single cluster that contains all cases (Everitt et al. 2011; Romesburg 2004).

Cluster analysis is sensitive to missing or dropped cases, is algorithm dependent and non-unique solutions are possible (Everitt et al. 2011). As each respondent selected eight of 38 possible factors, IDS scores were therefore compiled into three variables representing the summed family/social, economic and environmental dimensions of subjective wellbeing. The three components were standardised as z scores, cluster distances estimated by squared Euclidean distance and the final number of clusters derived by frequency analysis of membership and dendogram analysis. Centroid clustering produced similar cluster membership results and Wilk's Lambda canonical discriminant analysis was used as a further validation test of the clustering solution (Stevens 2002).

38.3 Results and Discussion

38.3.1 Ranked IDS Scores

The relative ranking and mean IDS scores of all wellbeing factors for the Nam Ngum and Nam Xong sample cohorts are detailed in Fig. 38.2.

Wellbeing factors associated with a high weighted importance score indicates selection by a large proportion of the sample and scored as important. Higher dissatisfaction scores are associated with low satisfaction levels for the respective wellbeing factor. A high IDS score is therefore associated with high levels of importance, a high proportion of selection and high levels of dissatisfaction.

At a significance level of 95%, the Wilcoxin signed rank test indicates there is no significant difference ($p = 0.098$) in the ranking of the mean IDS scores of the wellbeing factors between the Nam Ngum and Nam Xong sample cohorts. Therefore we accept the null hypothesis H_0 that there is no significant difference between framing

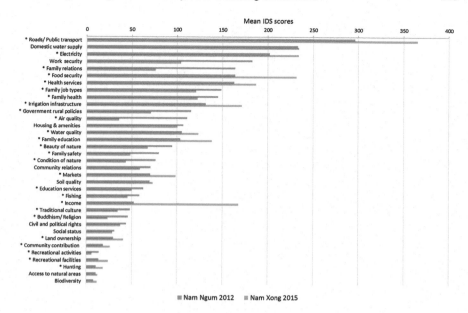

Fig. 38.2 IDS scores of 38 wellbeing factors for the Nam Ngum and Nam Xong river basins. *indicates significant difference in mean values (p < 0.05); Students t-test based on two sided tests assuming equal variance; adjusted for all pairwise comparisons using Bonferroni correction

wellbeing as either "how satisfied" compared to "how dissatisfied". In contrast, the null hypothesis would be rejected at a significance level of p < 0.10 and we accept H₁; that is question framing does significantly influence wellbeing responses at a 90% confidence level.

38.3.2 Differences in Mean IDS Values

T-test results indicate significant differences (p < 0.05) between the Nam Ngum and Nam Xong mean IDS scores for all wellbeing factors except: domestic water supply, work security, housing, community relations, soil quality, civil and political rights, social status, access to natural areas and biodiversity. The valency of IDS difference is approximately split between the two sample cohorts. The Nam Ngum IDS values of 20 of the wellbeing factors are greater than the Nam Xong, 15 are less than the Nam Xong and the valency differences are spread throughout the three wellbeing dimensions. Levin et al. (1998) and Tversky and Kahneman (1981, 1987) posit that when decision options are framed as a likely gain (satisfied), risk-averse choices predominate. A shift toward risk-seeking behaviour occurs when the equivalent decision is framed in negative terms (dissatisfied). Evidence to support H₁ suggests Nam Ngum mean values (satisfied) would be consistently lower than Nam Xong values (dissatisfied). The relatively even spread of both negative and positive values implies rejection of H_1.

Household income is a common metric to assess wellbeing (Easterlin 2003). Household income was derived from questionnaire responses and disaggregated into aggregate farm income, off-farm income and monetised subsistence income. Subsistence income was estimated as the monetised value of farm production used by the household for home consumption, calculated from the value of produce sold at market, or if no produce was sold by the respondent, from the mean of produce sold by adjacent households. Although income IDS was ranked 7th (Nam Xong) and 24th (Nam Ngum), improved household income has the potential to yield wellbeing gains for households. The correlation between reported household income and IDS scores was not significant for all income classes ($r < 0.1$, $p \geq 0.05$) for both sample cohorts. The non-significant correlations between household income and IDS income scores is an important finding consistent with global studies of subjective wellbeing and income correlations in developed economies reported by Easterlin 2003; Diener et al. (2010) and Kahneman et al. (2006). Linssen et al. 2011, Lissner et al. (2014) and Ward et al. (2016) report similar low correlations in India.

There was no significant difference ($t = 0.077$, $p < 0.05$) in responses to the question "how (dis)satisfied with your overall life?" (Table 38.1). The IDS scores of individual wellbeing factors were aggregated into the environment, social and economic wellbeing dimensions to estimate differences in composite values in contrast to individual wellbeing factors. The aggregate mean Nam Xong IDS scores for the economic dimension were significantly higher ($t = -23.65$, $p < 0.05$) than those observed for the Nam Ngum; the Nam Ngum social IDS dimension was significantly higher than the Nam Xong responses. There was no significant difference between the two cohorts for the environment wellbeing dimension (Table 38.1).

Table 38.1 T-test for equality of mean IDS scores of overall life dissatisfaction, and aggregated environmental, social and economic wellbeing dimensions

	Sample	N	Mean	Std. Deviation	Std. Error Mean	t^a	p
Overall dissatisfaction	Nam Ngum	1000	3.16	1.498	.047		
	Nam Xong	1011	3.15	2.767	.087	0.77	0.939
IDS environment	Nam Ngum	1000	201.83	174.642	5.523		
	Nam Xong	967	202.49	187.886	6.042	−.080	0.936
IDS social	Nam Ngum	1000	86.98	102.497	3.241		
	Nam Xong	967	68.86	100.715	3.239	3.954	0.000
IDS economic	Nam Ngum	1000	404.06	302.305	9.560		
	Nam Xong	967	981.72	708.509	22.784	−23.652	0.000

[a]Equal variance not assumed (Levine's test $p < 0.05$)

38.3.3 *Cluster Analysis*

Ward's hierarchical cluster analysis (Ward 1963) was used to identify latent constructs that may exist between the respondent IDS scores across all three wellbeing dimensions and reveal statistical differences not evident from an analysis of individual factors and aggregate wellbeing dimensions. Respondents were classified into four relatively homogenous groups who share common social, economic and environmental IDS scores based on the statistical associations between the three wellbeing dimensions. The number of respondents assigned membership to the four statistically distinct wellbeing clusters are reported in Table 38.2.

Nominal descriptive terms for the four clusters were: 1. Environmentally dissatisfied (n = 431); 2. Economically dissatisfied (n = 214); 3. Socially dissatisfied (n = 417) and 4. Environment and socially satisfied (n = 905).

The cluster analysis was validated using an alternate canonical discriminant analysis algorithm (Wilk's Lambda). 88% of cluster membership was correctly classified (Wilk's lambda: functions 1 through 3 = 0.119, 0.305 and 0.566; λ^2 significant at p < 0.001), supporting the cluster analysis classification. The variables age, household income, education, gender, farm size and household debts were entered using stepwise analysis and subsequently eliminated as significant variables predicting wellbeing cluster membership.

The proportions of cluster membership for the two river basins are reported in Table 38.2. There are significant differences in the economically dissatisfied and socially dissatisfied cluster constructs ($\lambda^2 = 293$, p < 0.001). The majority (97.7%) of the members of the economically dissatisfied cluster were located in the Nam Xong: conversely 73.9% of the socially dissatisfied cluster were located in the Nam Ngum (Table 38.2).

The equivalence of mean IDS values, aggregated for the three wellbeing dimensions and compared across clusters are reported in Table 38.3. For example, the environment and socially satisfied cluster are less dissatisfied with their overall life, the environment and social dimensions.

Table 38.2 Wellbeing cluster membership by River Basin

Wellbeing Cluster	Nam Ngum		Nam Xong	
	Column N %	Row N %	Column N %	Row N %
1. Environmentally dissatisfied	23.8%	55.2%	20.0%	44.8%
2. Economically dissatisfied	0.5%	2.3%	21.6%	97.7%
3. Socially dissatisfied	30.8%	73.9%	11.3%	26.1%
4. Environment and socially satisfied	44.9%	49.6%	47.2%	50.4%

Table 38.3 Mean values by wellbeing cluster

Wellbeing dimension	Wellbeing Cluster							
	Environmentally dissatisfied (a)		Economically dissatisfied (b)		Socially dissatisfied (c)		Environment -socially satisfied (d)	
	Mean	S.D	Mean	S.D	Mean	S.D	Mean	S.D
Overall dissatisfaction	3.36 d	2.18 d	3.81 d	2.70	3.54 d	1.90	2.75	2.12
IDS environment	**411 bcd**	177	239 cd	227	192 d	125	**99**	73
IDS social	34	46	125 ad	168	**204 abd**	74	**30**	37
IDS economic	496	369	**2015 acd**	631	468	320	567 ac	376

Students t-test based on two sided tests assuming equal variance with sig level (p < 0.05); adjusted for all pairwise comparisons using Bonferroni correction. For each significant pair the key of the smaller mean appears under the class with larger mean

38.4   Conclusions

Development initiatives and interventions comparatively evaluated through time and space, or across case studies, can reveal observable changes to wellbeing and livelihoods compared to static analysis. With an increasing reliance on subjective wellbeing as a primary development metric coupled with non-static assessments, the deployment of non-standardised instruments to elicit wellbeing introduces an increased risk of biased evaluations by failing to account for omitted variables that can influence individual responses (Tversky and Kahneman 1981, 1991).

We have argued that the effects of question framing have not been previously tested in subjective wellbeing research. We applied four analytical approaches to test the hypothesis H_1: significant differences will be detected in the numerical ranking of wellbeing factors when framed as either "how satisfied are you" compared to "how dissatisfied are you"? First a numerical wellbeing value was calculated as an Index of dissatisfaction (IDS), wellbeing factors ranked from highest to lowest for the Nam Ngum and Nam Xong sample cohorts and compared with the Wilcoxin signed rank test. The results indicate the ranking pattern of the two sample cohorts was not significantly different at a confidence level of 95%, but was significant at 90%.

Second the difference in mean IDS values for each pair of individual wellbeing factor was compared using Student's t-test. Levin et al. (1998) and Tversky and Kahneman (1981) posit that when decision options are framed as a likely gain (satisfied), risk-averse choices predominate and mean values are predicted to be reduced. A shift toward risk-seeking behaviour occurs when the equivalent decision is framed in negative terms (dissatisfied) and mean values predicted to increase. Significant differences in mean values were detected between the Nam Ngum and Nam Xong across the majority of wellbeing factors, although valency was not consistent. Twenty of the Nam Ngum mean values were greater than the Nam Xong values, 15 were less. Evidence to support H_1 suggests that Nam Ngum mean values (satisfied) would be consistently lower than Nam Xong values (dissatisfied). The relatively even spread of both negative and positive values implies rejection of H_1.

Third. IDS scores were aggregated into the three wellbeing dimensions: environmental, social and economic and mean values compared. Significant differences between the two River basins ($p < 0.05$) were detected for the Social and Economic wellbeing dimensions, but not for overall life dissatisfaction and the environmental dimension. Note that a mean life dissatisfaction value of 3.16 is equivalent to life satisfaction value of 6.64, consistent with previous studies of Larson (2010, 2013) and Ward et al (2016).

Finally latent constructs were identified by cluster analysis of the three wellbeing dimensions and differences in cluster membership and mean IDS values compared across the '*satisfied*' and '*dissatisfied*' cohorts. The Nam Xong respondents had a relatively higher membership in the socially dissatisfied cluster (30.8% compared to 11.3% for the Nam Xong), whilst 21.3% of the Nam Xong respondents were assigned membership to the economically dissatisfied cluster compared to 0.5% of Nam Ngum respondents. Significant differences in the wellbeing dimensions were detected across the clusters, however there was no clear framing effect.

The combination of both significant and non-significant results was not sufficient to accept the working hypothesis H_1 that framing wellbeing as either satisfaction or dissatisfaction will significantly influence responses. We therefore reject H_1 and accept the null, H_0. However the mixed results warrant further investigation of framing effects in subjective wellbeing research and instrument design; it may be that the ambiguity is specific to the two Lao PDR catchments.

The non-significant correlations between reported household income and IDS scores ($r < 0.1$, $p >= 0.05$) was an important finding not directly concerned with testing the research hypothesis. The results are consistent with global studies of subjective wellbeing and income correlations in developed economies reported by Easterlin (2003), Diener et al. (2010) and Kahneman et al. (1999, 2006). Linssen et al. 2011, Lissner et al. (2014) and Ward et al. (2016) report similar low correlations in India. The results for the two watersheds contrast with claims that improved household incomes equates with improved subjective wellbeing and are consistent with the arguments of Diener and Biswas-Diener (2002), and Kahneman et al. (2006) that a singular reliance on household income as an indicator of individually perceived wellbeing is not supported by existing evidence.

Acknowledgements The Nam Xong and Nam Ngum research was funded by The CGIAR Water Land and Ecosystems (Greater Mekong) and the CSIRO AusAid Alliance respectively. The field surveys were supported and approved by The National Economic Research Institute (Ministry of Planning and Investment) and the Department of Water Resources (Ministry of Natural Resources and Environment).

References

Alkire S (2002) Dimensions of human development. World Develop 30:191–205

Anand S, Sen A (2000) Human development and economic sustainability. World Develop 28(12):2029–2049

Bamberger M, Rao V, Woolcock M (2009) Using mixed methods in monitoring and evaluation: experiences from international development. BWPI working paper 107. Brooks World Poverty Institute. University of Manchester, ISBN 978-1-907247-06-04

Biswas-Diener R, Diener E (2001) Making the best of a bad situation: satisfaction in the slums of Calcutta. Soc Indicators Res 55:329–352

Bookwalter JT, Dalenberg D (2004) Subjective well-being and household factors in South Africa. Soc Indicators Res 65(3):333–353

Camerer C (1995) Individual decision making. In: Kagel JH, Roth AE (eds) The handbook of experimental economics. Princeton University Press, Princeton, pp 587–703

Camfield L, Crivello G, Woodhead M (2009) Wellbeing research in developing countries: reviewing the role of qualitative methods. Soc Indicators Res 90:5–31

Campbell A (1981) The sense of well-being in America: recent patterns and trends. McGraw Hill, New York

Clarke M (2005) Assessing well-being using hierarchical needs, Research Paper, UNU-WIDER, United Nations University (UNU), No 2005/22, ISBN 9291907014

Cummins RA, Eckersley R, Pallant J, van Vugt J, Misajon R (2003) Developing a national index of subjective wellbeing: the Australian unity wellbeing index. Soc Indicators Res 64:159–190

DFID (1999) Sustainable livelihoods guidance sheets. The Department for International Development, London. Accessed 20 Dec 2012. http://www.eldis.org/vfile/upload/1/document/0901/section1.pdf

Diener E, Ng W, Harter J, Arora R (2010) Wealth and happiness across the world: material prosperity predicts life evaluation, whereas psychosocial prosperity predicts positive feeling. J Pers Soc Psychol 99:52–61. https://doi.org/10.1037/a0018066

Diener E, Eunook M, Lucas RE, Smith HL (1999) Subjective wellbeing: three decades of progress. Psychol Bull 125(2):276–302

Diener E, Biswas-Diener R (2002) Will money increase subjective well-being? A literature review and guide to needed research. Soc Indicators Res 57:119–169

Doyal L, Gough I (1991) A theory of human need. Macmillan, London, p 365p

Easterlin RA (2003) Explaining happiness. Proc Nat Acad Sci 100(19):11176–11183

Everitt B, Landau S, Lease M, Stahl D (2011) Cluster analysis, 5th edn. Wiley and Sons, Sussex

Frey B, Stutzer A (2002) What can economists learn from happiness research? J Econ Lit XL, 402–435

Gasper D (2004a) Human well-being: concepts and conceptualisations. Discussion Paper 2004/06. United Nations University; WIDER. Helsinki

Gasper D (2004b) Subjective and objective well-being in relation to economic inputs: puzzles and responses. Wellbeing in Developing Countries ESRC Research Group Working Paper 09, University of Bath, U.K

Gasper D (2007) Conceptualising human needs and wellbeing. In: Gough I, McGregor JA (eds) Wellbeing in developing countries: from theory to research. Cambridge University Press, Cambridge

Gough I, McGregor JA (2007) Wellbeing in developing countries: from theory to research. Cambridge University Press, Cambridge

Guardiola J, Gonzalez-Gomez F, Grajales AL (2013) The influence of water access in subjective well-being: some evidence in Yucatan. Mexico. Soc. Indic. Res. 110(1):207–218. https://doi.org/10.1007/s11205-011-9925-3

Hassan R, Scholes R, Ash N (2005) Ecosystems and human wellbeing: current state and trends. Millennium ecosystem Assessment, vol 1. Island Press Washington D.C

Kahneman D, Kreuger AB (2006) Developments in the measurement of subjective well-being. J Econ Pers 20(1):3–24

Kahneman D, Tversky A (1992) Advances in prospect theory: cumulative representation of uncertainty. J Risk Uncertain 5(4):297–323

Kahneman D, Diener E, Schwarz N (1999) Wellbeing: the foundations of hedonic psychology. Russell Sage, New York

Kahneman D, Krueger AB, Schkade D, Schwarz N, Stone AA (2006) Would you be happier if you were richer? A focussing illusion. Science 312(6):1908–1910

Kaufman L, Rousseeuw PJ (1990) Finding Groups in Data. Wiley, New York

Lacombe G, Douangsavanh S, Baker J, Hoanh CT, Bartlett R, Jeuland M, Phongpachith C (2014) Are hydropower and irrigation development complements or substitutes? The example of the Nam Ngum River in the Mekong Basin. Water Int 39:649–670

Larson S, Stoeckl N, Neil B, Welters R (2013) Using resident perceptions of values associated with the Australian Tropical Rivers to identify policy and management priorities. Ecol Econ 94:9–18

Larson S (2010) Regional well-being in tropical Queensland, Australia: developing a dissatisfaction index to inform government policy. Environ Plann A 42:2972–2989

Levin IP, Schneider SL, Gaeth GJ (1998) All frames are not created equal: a typology and critical analysis of framing effects. Organ Behav Human Decis Process 76(2):149–188

Linssen R, van Kempen L, Kraaykamp G (2011) Subjective well-being in rural India: the curse of conspicuous consumption. Soc Indic Res 101(1):57–72. https://doi.org/10.1007/s11205-010-9635-2

Lissner TK, Reusser DE, Lakes T, Kropp U (2014) A systematic approach to assess human wellbeing demonstrated for impacts of climate change. Change Adap Soc Ecol Syst 1:98–110

McGregor JA, McKay A, Velazco J (2007) Needs and resources in the investigation of well-being in developing countries: illustrative evidence from Bangladesh and Peru. J Econ Methodol 14(1):107–131

Nussbaum M (2000) Women and human development. Cambridge University Press, Cambridge

Parfit D (1984) Reasons and persons. Clarendon Press, Oxford. ISBN 0-19-824615-3

Rishi P, Mudaliar R (2014) Climate stress, behavioral adaptation and subjective well being in coastal cities of India. Am J Appl Psychol 2(1):13–21

Romesburg HC (2004) Cluster analysis for researchers, pp 130–135. Lulu Press, Morrisville

Scoones I (1998) Sustainable rural livelihoods: a framework for analysis. IDS Working Paper No. 72. Brighton: Institute for Development Studies, University of Sussex, p 22

Stevens JP (2002) Applied Multivariate Statistics for the Social Sciences, 4th edn. Lawrence Erlbaum Associates, Inc, Mahwah

Stiglitz JE, Sen A, Fitoussi J-P (2009) Report by the Commission on the Measurement of Economic Performance and Social Progress. Paris. Accessed September 2015. http://www.insee.fr/fr/pub lications-et-services/dossiers_web/stiglitz/doc-commission/RAPPORT_anglais.pdf

Tversky A, Kahneman D (1981) The framing of decisions and the psychology of choice. Science 211:453–458

Tversky A, Kahneman D (1987) Rational choice and the framing of decisions. In: Hogarth RM, Reder MW (eds) Rational choice: the contrast between economics and psychology. University of Chicago Press, Chicago

Tversky A, Kahneman D (1991) Loss aversion in riskless choice: a reference-dependent model. Q J Econ 107:1039–1061

Ward JH Jr (1963) Hierarchical grouping to optimize an objective function. J Am Stat Assoc 58:236–244

Ward J, Poutsma H (2013) The compilation and summary analysis of Tonle Sap Household livelihoods: the exploring Tonle Sap futures project. CSIRO Climate Adaptation Flagship, Canberra Australia

Ward J, Varua ME, Maheshwari B, Oza S, Purohit R, Hakimuddin, Dave S (2016) Exploring the relationship between subjective wellbeing and groundwater attitudes and practices of farmers in rural India. J Hydrol 540:1–16

White SC, Ellison M (2007) Wellbeing, livelihoods and resources in social practice. In Gough IR, McGregor JA (eds) Wellbeing in developing countries: from theory to research, pp 157–175. Cambridge University Press, Cambridge

Chapter 39
Development Trade-Offs in the Mekong: Simulation-Based Assessment of Ecosystem Services and Livelihoods

A. Smajgl, J. Ward, and T. Nuangnong

Abstract This paper presents results from an analysis of development related trade-offs in the Nam Xong catchment, one of the Mekong tributaries in Lao PDR. The study presents results from an agent-based simulation model, which genuinely links bio-physical and socio-economic dynamics, and considers spatial impacts, such as deforestation, hydrological flow and human migration. The results indicate that mining expansions in the upper catchment are likely to trigger larger local income losses in existing livelihoods than income gains from the new employment created. Downstream dynamics are likely to increase flood peaks and lead to increased outmigration. Two system links emerge as key mechanisms for understanding development trade-offs, the link between livelihoods and ecosystem services, and migration, which emphasises the need to apply methodologies that effectively represent these system connections.

Keywords Mekong · Agent-based modelling · Trade-off analysis

39.1 Introduction

Economic development involves risks of trade-offs as activities by one sector are likely to impose constraints for other sectors. In the context of the Nam Xong—a Lao tributary to the Mekong—local stakeholders observed such trade-offs between upstream mining and rubber and banana plantations, mid-stream tourism and agriculture, and downstream fishing and agriculture. The substantial income generated by tourism activities in and around Vang Vieng were perceived to be endangered due to changes in water levels and water quality. Thus, stakeholders invited and engaged in a participatory research process to understand some of these development trade-offs.

A. Smajgl (✉) · T. Nuangnong
Mekong Region Futures Institute, Bangkok, Thailand
e-mail: alex.smajgl@merfi.org

J. Ward
Mekong Region Futures Institute, Vientiane, Lao PDR

© Springer Nature Switzerland AG 2021
M. Babel et al. (eds.), *Water Security in Asia*, Springer Water,
https://doi.org/10.1007/978-3-319-54612-4_39

The research design involves a mixed method approach that was implemented during a participatory process, the Challenge and Reconstruct Learning approach (Smajgl and Ward 2013; Smajgl et al. 2015). The methods included a household survey, a hydrological model (Kallio 2014), and the agent-based model MerSim (Smajgl, Toan et al. 2015; Smajgl, Ward et al. 2015; Smajgl, Xu et al. 2015). This paper summarises results from the agent-based approach, which genuinely links hydrological dynamics, agricultural growth, various ecosystem services, and livelihood related behaviours.

The results emphasise that taking a narrow sector perspective leads to wrong recommendations as income in the upstream district seemingly increase and poverty slightly declines. However, if considering land-use change and the resulting consequences through losses in ecosystem services other livelihoods face losses that substantially outweigh the gains from mining. Importantly, most mining employment is likely to be taken up by persons from outside the district, which were beforehand not under the poverty line. Widening the spatial area emphasises further consequences that add to the local trade-offs. Two types of downstream effects prove critical for the province and national perspective. First, changes in water levels, in particular an increase in flood peaks, trigger agricultural income losses in downstream districts that drive families to migrate out of the Province. The level of poor households in the remaining populations increases as most out-migrating families were not under the poverty line.

From a methodological perspective it becomes evident that the understanding of development outcomes and related trade-offs require a sophisticated approach that captures the socio-economic and bio-physical connections. Two aspects seem particularly important, first the link between livelihoods and ecosystem services, and second migration, which stresses the importance of spatial and temporal dynamics of poverty (Bohensky et al. 2013; Smajgl and Bohensky 2013).

39.2 Material and Methods

39.2.1 Development Context of the Nam Xong

The Nam Xong (also Nam Song) flows into the Nam Ngum, Laos' largest contributor to the Mekong (Lacombe et al. 2014). This sub-catchment is the home of about 200,000 people from various ethnic groups, including Thai Lao, Thai Neua, Thai Dai, Meo Khao (White Hmong) (Schliesinger 2003). Three development strategies are being unfolded by government strategies and foreign direct investments, tourism, mining, plantations, and agricultural intensification.

Tourism is largely centred around the town of Vang Vieng, mid-stream of the Nam Xong sub-catchment. Tourist numbers in this area have more than tripled between 2006 and 2013 (Tourism Development Department 2013) providing many new livelihoods to local households but also putting pressures on the environment as sewage volume increases and urban sprawl replaces natural vegetation, especially along river banks.

Mining investments in the upstream area of the Nam Xong target mainly deposits of gold, copper, and limestone (for downstream cement production) (Kallio 2014). The enormous increase of rubber prices between December 2008 and February 2011 triggered substantial Chinese investments into the northern provinces of Lao PDR. While prices have dropped since then there has been a resurge since early 2016. More recently, banana has emerged as a second cash crop covering wide areas in the Nam Xong and other parts of northern Laos.

The fourth development strategy resulted from the Government's declared goal to lift agricultural productivity (due to its under proportional contribution to GDP growth). Such agricultural intensification would include more suitable crops, irrigation, and improved management (incl. application of fertilizer, herbicides and pesticides).

In combination, these four development strategies harbour substantial potential for trade-offs, in particular when considering constraints related to ecosystem services. For instance, upstream development in mining, plantations, and irrigated agriculture reduce the water levels of the Nam Xong in the tourism areas. Given that many tourists come to Vang Vieng for water sport related activities, such declining flow can suddenly prevent tourists from rafting, canoeing and tubing. The economic gain of upstream expansions would thereby reduce tourism related income in and around Vang Vieng. Similarly, increased tourism can further deteriorate water quality for downstream households, their fishing, and for state of the Nam Ngum 1 reservoir. This pressure is accelerated by upstream development causing deforestation and erosion. Economically, costs can emerge and potentially outpace the gains of upstream gains. These trade-offs follow non-linear and rather complex relationships and understanding thresholds seems essential to manage these trade-offs wisely and to achieve sustainable development.

39.2.2   Participatory Process: Implementing the ChaRL Framework

The development challenges in the Nam Xong result from an array of independent interests and values. The provision of scientific evidence to such a contested decision making space requires an effective process design for engaging with the various stakeholders and interest groups. The Challenge and Reconstruct Learning (ChaRL) framework was chosen to effectively bridge science and policy. ChaRL combines in a five-step process visioning and evidence-based deliberations. The key difference to other processes is that after developing a shared vision assessment results are being presented to elicit causal beliefs, which will then be tested with the various data and methods developed so far. The step of challenging these underpinning causal beliefs is a critical psychological moment that creates the opportunity to redraw the

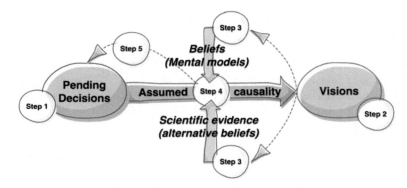

Fig. 39.1 The ChaRL (challenge and reconstruct learning) framework (based on Smajgl and Ward (2013) and Smajgl, Ward et al. (2015))

positions in the cross-sector negotiation and redesign development strategies that minimize trade-offs (Fig. 39.1).

The ChaRL process in the Nam Xong brought together various central ministries (e.g. Department of Water Resources from the Ministry of Natural Resources and Environment, and the National Economic Research Institute form the Ministry of Planning and Investment), various province and district level agencies (e.g. agricultural, environment, planning), and the tourism association. A total of five workshops were conducted between March 2015 and November 2016. The visioning (Foran et al. 2013) involved all participants to identify key drivers, their expected trends, and to develop two most desirable futures (=visions), one most likely future (no major shocks), and one most undesirable future (for the design of risk strategies) of the Nam Xong. The scientific evidence presented during this participatory process was based on a hydrological model, results from a household survey, and an agent-based social simulation model. The following explains the design of the agent-based model MerSim, which is the focus of this publication.

39.2.3   The MerSim Model

The description of the agent-based Mekong region simulation (Mersim) model follows the ODD (Overview—Design concepts—Details) protocol (Grimm et al. 2006; Grimm et al. 2010; Müller et al. 2014) and model details including Java code can be found in Smajgl et al. (2013).

39.2.4 Purpose of the Model

The model design was embedded in a participatory process, which follows the Challenge and Reconstruct Learning approach as outlined by Smajgl and Ward (2013) and Smajgl and Ward (2015). This participatory process helped eliciting the policy indicators and policy scenarios.

- Climate change and increase in flash floods
- Continued deforestation
- Rubber expansion and rubber price increase
- Mining expansion
- Tourism drop
- Tourism expansion.

The results aim to inform the basin development plan and specific sector strategies for mining, forest management and agriculture.

39.2.5 State Variables

The participatory process placed the stakeholder priorities at the core of the model design and determined the state variables as: poverty, forest cover, rubber production, water flow, water quality (dissolved oxygen), rice production, migration, land use, household livelihoods and fish catch.

39.2.6 Emergence

Corresponding with stakeholder-defined modelling goals, emergent phenomena include the temporal and spatial poverty patterns, the spatial extend of forest cover and rubber plantations, and water quantity and quality changes in response to the expansion of mining and rubber plantations upstream. One core policy focus is the trade-offs between upstream investments and mid-stream tourism income, which is emerging from modelled interactions.

39.2.7 Household Data for Parameterisation

The parameterisation process is described based on the framework provided in Smajgl et al. (2011).

Figure 39.2 shows the principle parameterisation steps required in an empirical model (boxes) and which particular options were implemented for this study (arrows).

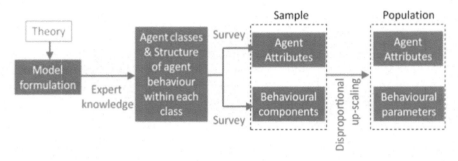

Fig. 39.2 Parameterisation sequence for the Mersim model, adapted from Smajgl et al. (2011)

The Mersim model formulation is based on theory articulated by Castellani and Hafferty (2009) that conceptualises social-ecological complexity, in particular the focus on a disaggregated systems approach that allows non-linear system components to interact and, thereby allowing for emerging phenomena, i.e. Funtowicz and Ravetz (1994) and Sawyer (2005).

Experts helped to identify principle agent classes, such as household agents, government agents and spatial agents. This expert-based process also identified principle agent behaviour such as the harvesting of tea and the tapping of rubber. These livelihood-related activities were put into annual calendars and linked to associated regions and altitudes where necessary.

The next step involved the specification of household attributes and household behaviours. A random sample of 1,000 households (20 randomly selected households from 50 randomly selected villages) across the Nam Xong sub-basin were surveyed to elicit their key characteristics (i.e. location, household size, livelihoods, production, and income), their self-selected attributes of subjective wellbeing, the principle human values that guide their lives, and their adaptation intentions. Intentions represent responses to questions that frame a specified hypothetical change. In this case the change households were asked to imagine:

- Flash floods started to occur frequently
- Deforestation continued
- Rubber prices would increase
- More mining jobs would be available
- Tourist numbers would drop
- More jobs in the tourism sectors would be available.

Households had four principle response options: either.

- To maintain their livelihood activity in their current household location;
- To change their livelihood in their current household location;
- To migrate out but maintain their current livelihood; or
- To migrate out and change their livelihood activity.

In each of these categories, responses to follow-on question informed estimates of the magnitude or type of livelihood response, the impediments to adaptation and/or the location for migration. The intentional data and behavioural changes elicited from household survey responses delimited the cognitive complexity of household agents to a more parsimonious depiction of largely reactive agents in the model.

The sample data for attributes and behavioural rules was mapped into the total agent population by disproportional up-scaling. Proportional up-scaling refers to a technique in which the proportions of responses in the sample is maintained to parameterise the whole population by simply replicating (or cloning) each response by sample size divided by population size (in this case multiplied by about 200). Disproportional up-scaling on the other hand changes proportions as some responses are used more often than others due to some scaling factors. In this case the proportions were amended to match the actual land use, in particular rubber plantations, rice paddy and the urban population. Otherwise a random approach would map intentional data from a tourism-dependent household into a non-tourism area and responses from a rubber farming household into an urban area. This GIS-based adjustment was intended to represent a more realistic spatial distribution of simulated household behaviour.

39.2.8   Adaptation and Objective

Given the way we reduced agent cognition, agents step through a simple adaptation process, which allowed a reduction in the run time of the model so that live runs were able to be performed during the participatory modelling process (Smajgl, Ward et al. 2015). Figure 39.3 depicts the steps for household agents.

Household agents respond to income levels derived from paid labour or agricultural activities. Households' objectives are implicit to their behavioural response functions (or rules). Modelled agents respond to livelihood related changes based on intentional data derived from the household survey responses. No additional optimisation or satisficing assumption is implemented. As a corollary, household expectations and learning are not explicitly represented but implicitly captured by the empirically derived response strategies.

39.2.9   Adaptation and Objective

Most parameters are assumed to be stochastic to resemble more realistic model assumptions, including crop prices, productivity, and wages. The ranges were developed by experts in conjunction with time series data.

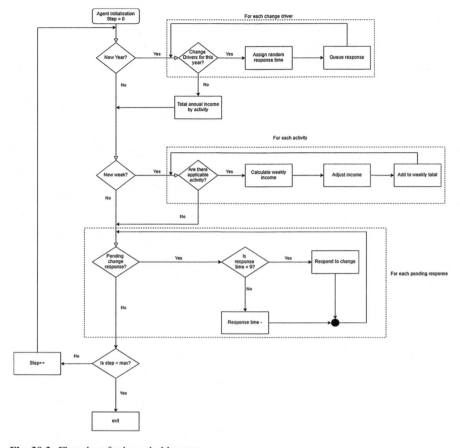

Fig. 39.3 Flowchart for household agents

39.2.10 Initialisation and Submodels

The Mersim model utilises five sets of GIS data: (1) administrative boundaries down
to administrative villages, (2) soil data, (3) land cover data, (4) rainfall projections,
and (5) a digital elevation model. These datasets were used to specify the artificial
landscape while the household survey provided the necessary data on household
attributes and behavioural responses.

Five essential submodels were integrated to deliver the processes stakeholders
had requested: hydrology, crop growth, water quality, livelihood, and income. The
hydrology module calculates in daily time steps the run off for each spatial polygon
based on rainfall, slope, inlet and outlet node, land cover and soil type. Based on this
method flood and drought risk for the tourist town of Vang Vieng can be estimated,
which triggers tourism numbers to divert from a projected trend. Based on rainfall,

soil type and land cover livelihood-relevant crops (e.g. rice, rubber, trees, grass) grow following established growth algorithms. The combination of water flow and particular land cover provides the necessary information for calculating dissolved oxygen, which is calculated for Vang Vieng town and for Hinheup where the Nam Xong is partly diverted into the Nam Ngum 1 dam and partly continues its flow into the Mekong. The livelihood module follows crop and job specific calendars, which involves, for instance planting and harvesting. Household livelihoods only change based on intentional data elicited through the household survey. The income module calculates the weekly income for all household members and assigns how many are below the poverty line. This calculation includes the monetisation of subsistence production to avoid a misleading, under-estimated quantification of poverty.

39.3   Results and Discussion

For this paper we selected as the scenario the expansion of mining activities in the Nam Xong based on pending investment proposals. This would involve a doubling of mined areas, mostly located in the district of Kasy. Figure 39.4 shows the projected poverty rate for 2015–2029 for the baseline and for the selected mining expansion scenario. From a narrow sector perspective mining income is likely to reduce local poverty in Kasy only marginally. However, as mining involves the replacement of forest cover and, thereby the loss of forest-based livelihoods there are also losses that impact on the poverty rate. In total, poverty is likely to increase slightly by about 1.3% across this timeframe.

Figure 39.4 also shows the impact of the mining expansion on poverty rate further downstream in the district Vang Vieng. Here the poverty rate is likely to increase by

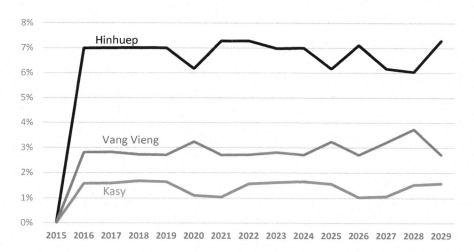

Fig. 39.4 Impact of Mining expansion on poverty in Kasy, Vang Vieng and Hinheup, 2015–2029

Households

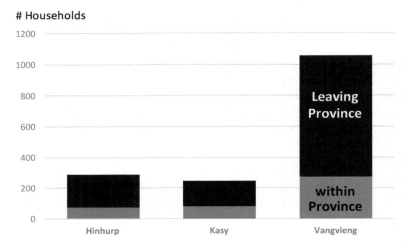

Fig. 39.5 Impact of Mining expansion on poverty in Vang Vieng

2.7% in average over the considered time period. In Hinheup, at the end of the Nam Xong sub-basin, poverty rates are likely to increase in average by 6.4%.

Figure 39.5 shows migration movements during this 15-year period. In total nearly 14% of households are likely to migrate due to the mining expansion strategy and its biophysical and socio-economic ripple effects. Some groups move between districts, which involves largely households seeking employment in the growing mining sector in Kasy. Additionally, it involves households that lose their previous livelihoods or land. A larger portion of migrating households decide to leave the Province, mostly heading for urban and industrial areas, including the capital Vientiane.

39.4   Conclusions

The modelling results emphasise (1) the relevance of social factors of poverty changes, and (2) the need to understand spatial and temporal dynamics of poverty. The expansion of mining activities is often promoted as a poverty alleviating strategy. This study shows that most households under the poverty line would either not apply for a mining job or would not be employed for educational reasons. The majority of households newly employed in the mining sector replace other livelihoods and were initially not under the poverty line. The cumulative impacts are that poverty is increasing due to the loss of existing forest-based livelihoods. This social constraint (mainly attitudes and skills) limits the ability of this mining-focused strategy to alleviate poverty in this province.

The dynamic perspective reveals that without an explicit inclusion of migration any analysis of poverty impacts is likely to deliver misleading results. In this case most households that seek mining employment in Kasy are non-poor households from Vang Vieng and Hinheup. This migration results already in increasing poverty rates in downstream communities (as proportionately more poor households are left behind). Concurrently, poor people lose their livelihoods and leave this area, which translates into rural outmigration. This process is likely to shift poverty into urban and peri-urban areas of Lao PDR if skill-sets of migrating households mismatch employment opportunities in urban areas. This study did not include any additional costs caused by these spatial changes in poverty over time, which can be substantial as many urbanisation experiences in southeast Asia have shown.

Acknowledgements The authors are grateful to the funding received from CGIAR's Research Program on Water, Land, and Ecosystems—Greater Mekong program

References

Bohensky E, Smajgl A, Brewer T (2013) Patterns in household-level engagement with climate change in Indonesia. Nat Clim Chang 3:348–351. https://doi.org/10.1038/NCLIMATE1762

Castellani B, Hafferty F (2009) Sociology and complexity science: a new field of inquiry. Springer, Heidelberg

Foran T, Ward J, Kemp-Benedict E, Smajgl A (2013) Developing detailed foresight narratives: a participatory technique from the Mekong region. Ecol Soc 18(4):6

Funtowicz SO, Ravetz JR (1994) Emergent complex systems. Futures 26(6):568–582

Grimm V, Berger U, DeAngelis DL, Polhill JG, Giske J, Railsback SF (2010) The ODD protocol: a review and first update. Ecol Model 221(23):2760–2768

Kallio M (2014) Effects of mining and hydropower on metals in surface waters: case: Nam Ngum. (Bachelor of Science), Helsinki Metropolia University of Applied Sciences, Helsinki

Lacombe G, Douangsavanh S, Baker J, Hoanh CT, Bartlett R, Jeuland M, Phongpachith C (2014) Are hydropower and irrigation development complements or substitutes? The example of the Nam Ngum River in the Mekong Basin. Water Int 39(5):649–670

Müller B, Balbi S, Buchmann CM, de Sousa L, Dressler G, Groeneveld J, Klassert CJ et al (2014) Standardised and transparent model descriptions for agent-based models: current status and prospects. Environ Model Softw 55:156–163. https://doi.org/10.1016/j.envsoft.2014.01.029

Sawyer RK (2005) Social emergence: societies as complex systems. Cambridge University Press, New York

Schliesinger J (2003) Ethnic groups of Laos Vol 1: introduction and overview. White Lotus, Bangkok

Smajgl A, Bohensky E (2013) Behaviour and space in agent-based modelling: poverty patterns in East Kalimantan, Indonesia. Environ Model Softw 45:8–14. https://doi.org/10.1016/j.envsoft.2011.10.014

Smajgl A, Brown DG, Valbuena D, Huigen MGA (2011) Empirical characterisation of agent behaviours in socio-ecological systems. Environ Model Softw 26(7):837–844

Smajgl A, Toan TQ, Nhan DK, Ward J, Trung NH, Tri LQ, Vu PT (2015) Responding to rising sea-levels in Vietnam's Mekong Delta. Nat Clim Chang 5:167–174. https://doi.org/10.1038/nclimate2469

Smajgl A, Ward J (2013) A framework for bridging science and decision making. Futures 52(8):52–58. https://doi.org/10.1016/j.futures.2013.07.002

Smajgl A, Ward J (2015) A design protocol for research impact evaluation: development investments of the Mekong region. J Environ Manag 157:311–319

Smajgl A, Ward J, Foran T, Dore J, Larson S (2015) Visions, beliefs and transformation: exploring cross-sector and trans-boundary dynamics in the wider Mekong region. Ecol Soc 20(2):15. https://doi.org/10.5751/ES-07421-200215

Smajgl A, Xu J, Egan S, Yi Z-F, Ward J, Su Y (2015) Assessing the effectiveness of payments for ecosystem services for diversifying rubber in Yunnan, China. Environ Model Softw 69:187–195

Tourism Development Department (2013) 2013 Statstical Report on Tourism. Retrieved from Vientiane

Chapter 40
Groundwater Institutions and Governance in North India: Cases of Hoshiarpur and Jammu Districts

Ishita Singh

Abstract Intensive use of groundwater, lack of adequate planning, legal framework and governance has posed a great threat to sustainable use of groundwater resources in Punjab. More or less similar has been the situation in Jammu District of Jammu and Kashmir. Keeping the above scenarios in view, a study of institutions for groundwater governance became essential in both these North Indian States. The primary data and information have been collected from 180 water users each in agriculture from Jammu district of Jammu and Kashmir State and Hoshiarpur district of Punjab State. The study reveals that energy subsidies to agriculture have resulted in over pumping of groundwater, declining water table levels, failure of tubewells, and increasing diesel run pumping costs, with serious environmental consequences. Minimum support price for paddy, wheat and sugar had greater influence on groundwater use. Uncertainty of monsoon and existence of groundwater markets add further stress to groundwater resources. Therefore, there became a need for a gradual phasing out of cross-energy subsidies. Conjunctive use of groundwater and surface water has been suggested for achieving much greater water-supply security, larger net water-supply yield and better timing of irrigation-water delivery.

Keywords Groundwater institutions · Governance · North Indian States

40.1 Introduction

The knowledge on groundwater use and the institutions and policies governing the use of these resources is limited (Mukherji and Shah 2005) in the North Indian States of Punjab and Jammu & Kashmir. No single paradigm, model, instrument or policy has been successful in effective and sustainable governance of groundwater. Groundwater is a local issue and need local solutions. A study has been conducted in Hoshiarpur District of Punjab and Jammu District of Jammu & Kashmir. Improving water governance is widely regarded as the key to solving water insecurity problems in developing countries (OECD 2011). The legal frameworks are essential for

I. Singh (✉)
Post Graduate Govt. College, Sector-11, Chandigarh, India

© Springer Nature Switzerland AG 2021
M. Babel et al. (eds.), *Water Security in Asia*, Springer Water,
https://doi.org/10.1007/978-3-319-54612-4_40

effective groundwater governance (Kerstin 2012). The reforms in the water sector have often been heavily influenced by non-binding water policy rather than law. Numerous water management institutions have been promoted by policy makers (Meinzen-Dick 2007). Local situations often influence groundwater users' level of compliance with applicable regulations. No one size fits all applies to groundwater governance. Groundwater overexploitation is due to uncountable individual decisions at the micro-level (World Bank 2010). Water markets helps to allocate water sustainably and economically (Zetland 2011). The trading of capped groundwater allocations can reduce the cost of limiting water use. The efficacy of community-based management of natural resources is widely recognized. In addition to it the demand and supply imbalance needs to be addressed at the local level, in these Districts as manifested by declining water tables.

40.2   Material and Methods

The primary data have been collected using well-structured and pre-tested question-naires, participatory rural appraisal (PRA) techniques and group meetings. Whenever possible and required, secondary data sources have also been utilized. In selecting the villages and water users, in these two Districts, random sampling technique has been used. Conclusions have been drawn by descriptive statistics. The sample size has been restricted to a total of 360 water users in agriculture by selecting 180 from Jammu district of the Jammu and Kashmir State and 180 from Hoshiarpur district of Punjab State. From the selected district, two blocks each viz. Ranbir Singh Pura and Bisnah from Jammu district and Dasuya and Mukerian from Hoshiarpur district have been selected. These blocks have been selected due to predominance of ground-water use along with surface water use in agriculture in Jammu district. However, in selected blocks of Dasuya and Mukerian from Hoshiarpur district, only groundwater is being used and there is no use of surface water and conjunctive water. In the selected blocks of Jammu district, three villages each have been selected with sample size of 30 each of groundwater users, surface water users and conjunctive water users and in the selected blocks of Hoshiarpur district, three villages each have been selected with sample size of 30 of groundwater users only as the surface water (e.g. canal water, river water) and conjunctive water (use of groundwater and surface water in combination) users' categories of households are non-existent due to non-use of canal water. The study has been done in comparative perspective for institutional-economic analysis of groundwater management and governance in agriculture.

40.3   Results and Discussion

The study conducted in Districts of Punjab and Jammu & Kashmir State is based on the tested questionnaire. The questionnaire has focused mainly on information like income status, economic capacity, water usage, source of water, cropping pattern and water management adaptation from the households' sample. The rapid increase in population has been putting pressure on the groundwater and surface water for the use in the domestic, agriculture and industrial sectors. The increasing pressure on water resources has led to tensions, conflicts among users, water pollution and water scarcity. Water scarcity has been linked with water shortage, especially in these regions. The indicators of water scarcity included severe environmental degradation and water pollution, declining groundwater levels and increasing problems of water allocation where some groups win at the expense of others. Agriculture water use have played significant role in poverty reduction, food security, and employment opportunities. However, rural poor population has suffered the most due to inadequate water supplies. The greater competition for water in agriculture posed a major threat to future poverty alleviation in these areas. It has been observed that high proportion of the rural poor competes for water entitlement and its access to grow food, and livestock as well as domestic purposes. There is high demand for water for irrigated agriculture. Based on the questionnaire collected through random sampling technique following results have been observed:

40.3.1 Groundwater Resource Conditions-1

Majority of sample households using conjunctive water in blocks of R. S. Pura and Bisnah reported that the surface water, especially rivers and canals reflect higher fluctuations than groundwater (70% and 86.6% respectively), and experiences reduced river and canal flows and dam storage resulted in severe droughts (76.6% and 90% respectively). In the case of surface water users' category, 63.3% and 56.7% of sample households respectively experiences reduced river and canal flows and dam storage resulted in severe droughts. None of the sample households reported that the surface water, especially rivers and canals reflect higher fluctuations than groundwater in groundwater users and surface water users' categories even in these two blocks of Jammu district. In the blocks of Mukerian and Dasuya in district Hoshiarpur responses to such resource conditions are virtually non-existing. Majority of the sample households in all the categories of water users across the selected blocks of districts of Jammu and Hoshiarpur have reported that high temperatures increase evaporative losses and both surface and groundwater resources shows diminishing trends with a bleak future for agricultural development. In order to ascertain the responses of the respondents, they were asked the question "Do you think that GW is most precious natural resource?" Majority of the sample households perceive that groundwater is the most precious natural resource for them. This is true across the selected blocks of both the districts. Even the category of sample households depending exclusively on surface water for irrigation and using both the surface water and ground water as conjunctive resource perceives groundwater as the most precious resource, 80% and 63.3% respectively and 76.7% and 90% respectively in the blocks of R. S Pura and Bisnah respectively. The sample households depending exclusively on groundwater for irrigation reported value of groundwater as the most precious natural resource in larger proportion across the selected blocks of both the districts. The respondents were asked to reveal that "why reliable groundwater supplies are necessary?" In response to this question, it is significant to note that almost all of them reported groundwater is highly essential for meeting drinking water needs and irrigation in categories of groundwater users across the selected blocks of R. S. Pura and Bisnah as well as the blocks of Mukerian and Dasuya. In the category of households depending on surface water and conjunctive water in blocks of R. S. Pura and Bisnah, the proportion of households depending on groundwater for meeting drinking water needs stood at 63.3% and 56.7% respectively and 76.7% and 90% respectively. Besides above, the groundwater is predominantly used for sanitation purpose across the sample households in all the blocks under study. It is significant to note that groundwater is used for sanitation purpose even in categories of households depending on surface water in both the blocks of R. S. Pura and Bisnah in Jammu district to the tune of 70% and 53.3% respectively. The respondents were also asked to reveal their perceptions why do they think that development of GW essential? For them, development of groundwater is essential for increased access to safe drinking water, meeting water use in agriculture, and to reduce poverty in all categories of water users' households across the selected blocks of both the districts.

40.3.2 Effective Management of Groundwater Using Economic Instruments-2

In order to know more about the perceptions of the sample households regarding value of groundwater and its management, they were asked "Are effective economic instruments to manage groundwater difficult?" This question was posed to all categories of water users viz. groundwater users, surface water users as well as conjunctive water users. It is significant to note that as high as 86.7% and 76.7% of conjunctive water users' households in blocks of R. S. Pura and Bisnah respectively perceives that it is very difficult to manage effectively the groundwater resources using only economic instruments. In case of surface water users' households comparatively low proportion of them (36.7% and 40% respectively in R. S. Pura and Bisnah) perceives that it is very difficult to manage effectively the groundwater resources using only economic instruments. In the category of households depending exclusively on groundwater resources, a significant proportion of them reveal that it is very difficult to manage effectively the groundwater resources using only economic instruments, and this is true across the selected blocks of both the districts of Jammu and Hoshiarpur.

The respondents who have perceived that it is very difficult to manage effectively the groundwater resources using only economic instruments, they were asked to state the reasons for the same. The main reasons are high cost and complexity of assessing groundwater resources, abstract nature of true supply, and near irreversibility of most contamination in varied proportion across the different categories of water users in selected blocks of both the districts. It is significant to note that groundwater pollution is very high in selected blocks of Mukerian and Dasuya in district Hoshiarpur. Thus, it is very difficult to manage groundwater pollution using economic instruments, rather it calls for greater awareness among all stakeholders including the farmers and industrial sector to not to pollute the groundwater resources and engage in big way for scientific management of the groundwater resources to make them sustainable.

Are cultivated lands under irrigation fed through groundwater sources? All the conjunctive water users' sample households in blocks of R. S. Pura and Bisnah use groundwater to irrigate cultivated land. Similar is the situation in case of all the ground water sample households in blocks of Mukerian and Dasuya. In surface water users' households, none uses groundwater to irrigate cultivated land. In rest of the groundwater users' households, significantly very high proportion of them uses groundwater to irrigate cultivated land. Thus, groundwater is intensively used in the villages where alternative sources of irrigation are not available, which results in more stress on the groundwater resources and declining groundwater tables, more so in the blocks of Mukerian and Hoshiarpur compared to blocks of R. S. Pura and Bisnah. This is also attributed to growing sugarcane in comparatively large proportion of cultivated land in district Hoshiarpur along with paddy cultivation.

The respondents, who have reported use of the groundwater resources to irrigate cultivated land, were asked to answer, "Is the use and distribution of groundwater highly uneven?" In the category of groundwater users' households, majority of them (70% and 53.3% respectively in R. S. Pura and Bisnah) perceive that groundwater use and its distribution is highly uneven, whereas in the category of conjunctive water users' household, comparatively a small proportion of them (43.3% and 46.7% respectively in R. S. Pura and Bisnah) perceive that groundwater use and its distribution is uneven. A more or less similar picture can be noticed in Mukerian and Dasuya, where comparatively less proportion of the groundwater users' households perceived that groundwater use and its distribution is uneven.

The respondents were asked to mention the types of agricultural groundwater schemes in their area. They mentioned two types of groundwater schemes, first, the shallow, small-scale systems that supply water for both crops and livestock, and the second, the deep and large-scale irrigation systems. In the category of conjunctive water users' households, all of them reported agricultural groundwater scheme of 'the shallow, small-scale systems that supply water for both crops and livestock' in both the blocks of R. S. Pura and Bisnah in district Jammu. In the category of groundwater users' households, majority of them reported agricultural groundwater scheme of 'the shallow, small-scale systems that supply water for both crops and livestock' in both the blocks of R. S. Pura (86.7%) and Bisnah (83.3%) in district Jammu. In blocks of Mukerian and Dasuya of district Hoshiarpur, the second type of scheme i.e., the deep and large-scale irrigation systems, is more prevalent compared to the blocks of R. S. Pura and Bisnah in district Jammu.

40.3.3   Groundwater Exploitation-3

In order to know about the level of groundwater exploitation in the study area, the sample households were asked to answer, "Is groundwater exploitation high in your area?" In both the categories of groundwater users and conjunctive water users in blocks of R. S. Pura and Bisnah, the level of groundwater exploitation is very low compared to the level of groundwater exploitation in blocks of Mukerian and Dasuya. For instance, it is as high as 86.7% and 76.7% respectively in Mukerian and Dasuya. The sample households who reported high groundwater exploitation were asked "Is vulnerability of groundwater to agricultural pollution high in your area? In the category of conjunctive water users, the vulnerability of groundwater to agricultural pollution is virtually nil and very low in the category of groundwater users in the blocks of R. S. Pura and Bisnah. Comparatively the vulnerability of groundwater to agricultural pollution is very high in blocks of Mukerian and Dasuya in all sample villages. Thus, it is evident that there is high vulnerability of groundwater to agricultural pollution in district Hoshiarpur than district Jammu.

In order to know about the prevalence of intensive groundwater use in the study area, the sample households were asked to answer, "Is there intensive groundwater use prevalent in your area?" In the category of conjunctive water users' households, there is no intensive groundwater use at all in blocks of R. S. Pura and Bisnah. However, 23.3% and 13.3% of the sample households in groundwater users' category reveals that there is intensive groundwater use in their respective villages in blocks of R. S. Pura and Bisnah, respectively. Comparatively there is more intensive use of groundwater in blocks of Mukerian and Dasuya, for instance as high as 70% of households in Dasuya and 63.3% of households in Mukerian. Thus, there is more intensive groundwater use in district Hoshiarpur than district Jammu. The sample households who reported prevalence of intensive groundwater use were asked "Does your area rely almost entirely on groundwater for irrigation?" Without exception, all the households across water users' categories and in the blocks of R. S. Pura and Bisnah in district Jammu and in the blocks of Mukerian and Dasuya in all sample villages in district Hoshiarpur have entire reliance on groundwater for irrigation though in varied proportion. Overall, it is evident that there is greater reliance on groundwater for irrigation in district Hoshiarpur than district Jammu.

The sample households who reported entire reliance on groundwater for irrigation were asked "Does irrigation of vast areas of arable land with groundwater from wells and deep boreholes widespread?" Without exception, all the households across water users' categories and in the blocks of R. S. Pura and Bisnah in district Jammu and in the blocks of Mukerian and Dasuya in all sample villages in district Hoshiarpur have widespread irrigation of arable land with groundwater using wells and deep boreholes though in varied proportion. Overall, it is evident that there is widespread irrigation of arable land with groundwater using wells and deep boreholes in district Hoshiarpur than district Jammu. Do you think that groundwater accounts for major freshwater withdrawals in your area? In blocks of R. S. Pura and Bisnah, fresh groundwater withdrawals are significant in the category of groundwater users' households compared to the category of conjunctive water users' category, whereas in the blocks of Mukerian and Dasuya, there has been very high proportion of households in all sample villages in district Hoshiarpur engaged in freshwater groundwater withdrawals. Thus, comparatively there is more freshwater groundwater withdrawals in district Hoshiarpur than district Jammu.

The respondents were asked "Do you think that groundwater a significant source of water for irrigated agriculture?" Nearly 86% and 70% of the groundwater users' households respectively reported that the groundwater is a significant source for irrigated agriculture in blocks of R. Pura and Bisnah. Comparatively less proportion of sample households in the categories of conjunctive water users and surface water users in these blocks have reported the groundwater as a significant source for irrigated agriculture, whereas in blocks of Mukerian and Dasuya comparatively very large proportion of the households in all sample villages in district Hoshiarpur perceive the groundwater as a significant source for irrigated agriculture. Thus, comparatively there is more use of groundwater for irrigated agriculture in district Hoshiarpur than district Jammu. Is information about groundwater scanty? In blocks of R. S. Pura and Bisnah as significant proportion of groundwater users' household perceive that information about groundwater is scanty compared to surface water users' households and conjunctive water users' households. In blocks of Mukerian and Dasuya, comparatively a very high proportion of groundwater users' households (56.7% and 76.7% respectively) reported scanty information about ground water. Thus, comparatively there is less information available about groundwater in district Hoshiarpur than district Jammu. This calls for provisioning of more sound and better data on groundwater availability in the selected districts for more scientific management and use of groundwater resources for agricultural development.

40.3.4 Groundwater and Rural Livelihoods-4

The respondents were asked "Is groundwater playing a limited role in sustaining rural livelihoods?" Nearly 76% and 70% of the surface users' households respectively reported that the groundwater plays a limited role in sustaining rural livelihoods. This is not at all surprising as all these households depend purely on surface water to meet their irrigation water requirements. Comparatively less proportion of sample households in the categories of conjunctive water users and surface water users in these blocks have reported the limited role of groundwater for sustaining rural livelihoods, as they considered groundwater as a significant source for irrigated agriculture and sustaining rural livelihoods, whereas in blocks of Mukerian and Dasuya comparatively very large proportion of the households in all sample villages in district Hoshiarpur perceive the groundwater plays a significant role in sustaining rural livelihoods. Thus, comparatively there is more role of groundwater in sustaining rural livelihoods in district Hoshiarpur than district Jammu. The sample households who reported significant role of groundwater in sustaining rural livelihoods were asked "Is there a need to expand groundwater and surface water systems to manage water resources?" Without exception, all the households in conjunctive water users' category in the blocks of R. S. Pura and Bisnah in district Jammu reported that there is cent per cent need to expand groundwater and surface water systems to manage water resources. At the same time, a very high proportion of the groundwater users' households and surface water users' households in blocks of R. S. Pura and Bisnah reported the need to expand groundwater and surface water systems to manage water resources. In the block Dasuya in all sample villages, the cent per cent of sample households have expressed the need to expand groundwater and surface water systems to manage water resources. In Mukerian, majority of sample households have expressed the need to expand groundwater and surface water systems to manage water resources. Overall, it is evident that there is widespread need in all sample villages to expand groundwater and surface water systems to manage water resources.

The sample households who reported the need to expand groundwater and surface water systems to manage water resources were asked to give reasons to expand groundwater and surface water systems to manage water resources. The main reasons to expand groundwater and surface water system to manage water resources include conjunctive use of groundwater and surface water necessary to mitigate impacts of droughts and to reduce the risk of over exploiting limited resources, groundwater resource assessment and management essential to address immediate societal needs, and groundwater closely linked to water resources management, especially surface water and non-conventional water resources. The conjunctive use of groundwater and surface water is considered necessary by majority of the sample households across the categories of water users in all the selected blocks of both the districts. Similarly, groundwater resource assessment and management is reported essential to address immediate societal needs by the significant proportion of the sample households. A significant proportion of the sample households reported that groundwater is closely linked to water resources management, especially surface water and non-conventional water resources.

40.3.5   On-Farm Water Management-5

Is on-farm water management (OFWM) essential? Majority of sample households in all the categories of water users in blocks of R. S. Pura and Bisnah reported that on-farm water management is essential for scientific water management. Similar is the situation in case of all the groundwater users' sample households in blocks of Mukerian and Dasuya with minor variations. However, comparatively large proportion of sample households in blocks of R. S. Pura and Bisnah reported importance of on-farm water management than blocks of Mukerian and Dasuya. The respondents, who have reported that on-farm water management is essential, were asked to state, "how on-farm water management (OFWM) essential to optimize soil water-plant outcomes and achieve high crop yields". A varied proportion of them reported that OFWM is essential to optimize soil water-plant outcomes and achieve high crop yields in selected blocks of the districts of Jammu and Hoshiarpur. They are of the opinion that OFWM helps in minimizing inputs and maximizing outputs of agricultural operations. It is significant to note that the OFWM covers water resources, irrigation facilities, by-laws and procedures, farmers' institutions, and soil and cropping systems. Besides, the OFWM requires measure to increase the water-use efficiency and water productivity. OFWM is essential for increasing water productivity compared to increasing water efficiency. This is reported by majority of sample households across the categories of water users in selected blocks of both the districts. In this context, it is necessary to mention that water-use efficiency commonly used to evaluate performance of an irrigation system and water productivity improved by increasing yield per unit of land area.

The respondents were asked to reveal their opinion about importance of smallholder irrigation development. Smallholder's irrigation development is essential for increasing food production, reducing poverty, mitigation against climate change, and providing sustainable rural development. It is significant to note that poverty reduction can be possible through smallholder irrigation development. This is reported by comparatively greater proportion of the sample households across the categories of water users in selected blocks of both the districts of Jammu and Hoshiarpur than increasing food production, mitigation against climate change, and providing sustainable rural development. The respondents were asked to reveal their opinion regarding inadequate irrigation development. They have given varied reasons, which hampered irrigation development and include high investment costs, poor agricultural policies, and inappropriate institutional framework. This is true across the sample households in all the categories of water users across the selected blocks of both the districts. However, majority of the sample households have cited high investment costs of irrigation development as a potential reason of inadequate irrigation development. The sample households were asked to reveal the benefits of appropriate management of irrigation water. A varied proportion of the sample households across the categories of water users in selected blocks of the districts under study reported that appropriate management of irrigation water leads to conservation of water supplies, reduction in negative water quality impacts, and improvement of producer net returns. Comparatively a high proportion of the sample households across the selected blocks reported that better irrigation water management lead to improvement of producer net returns than conservation of water supplies and reduction in negative water quality impacts.

Water savings through improved management of irrigation supplies are essential for meeting future water needs, and this has been reported by significantly very high proportion of sample households in category of conjunctive water users in blocks of R. S. Pura and Bisnah (86.7% and 70% respectively), surface water users (63.3% and 73.3% respectively), and ground water users (70% and 76.7% respectively). Whereas in blocks of Mukerian and Dasuya, comparatively a less proportion of the sample households in all the villages of groundwater users reported that improved management of irrigation supplies are essential for meeting future water needs. Irrigated agriculture affects water quality in several ways, which include higher chemical use rates, increased soil salinity and erosion, accelerated pollutant load with drainage flows, leaching of plant nutrients, and greater in-stream pollutant concentrations due to reduced flows. This has been reported varied proportion of the sample households across the categories of water users in the selected blocks of both the districts. Farmers may reduce water use by applying less than full crop-consumption requirements (deficit irrigation), shifting to alternative crops or varieties of same crop that use less water, and adopting more efficient irrigation technologies. Shifting to alternative crops or varieties of same crop that use less water is reportedly the more potent ways of reducing water use by the farmers for irrigating crops compared to applying less than full crop-consumption requirements (deficit irrigation) and adopting more efficient irrigation technologies. Even a significant proportion of the sample households across the sample households reported adopting more efficient irrigation technologies is more effective way of using less irrigation water compared to using deficit irrigation. This is true across the selected blocks of both the districts with minor modifications.

40.4 Conclusions

Sustainable development of ground water is essential to tackle the pressure on ground water resources and attending problems of its depletion. The governance and management challenges include these measures to be taken on priority like, formulation of appropriate institutional arrangements to identify and coordinate actions between relevant agencies, metering to promote efficient water use and energy demand, adopting block water rates with increasing rates for high-intensity users due to an escalating price structure for high water use, promotion of water-efficient technologies, reduction of water loss i.e. reducing the leakages through reducing pressure in systems, and drought risk management i.e. development of drought management plans. Besides strengthening water governance, there is a need to improve the knowledge and information system, collecting data, monitoring and evaluation, enhancing human and institutional capacity and developing integrated water resources management (IWRM) systems are suggested. Successful policy measures to address water scarcity include introduction of water pricing (full cost recovery), water pricing through installation of water meters, and relaxing land rent control. Full cost recovery followed by water pricing is the most successful policy measures to address water scarcity across the categories of water users in selected blocks of both the districts. The challenges to enhance adaptation capacity includes economic resources (poverty and economic status), technology development and dissemination, information and skills, infrastructure and governance structure, gender and equity, environmental and health issues, extension services and incentives, and conflicts among different interest groups. Farmer's participation in water governance in appropriate institutional framework including raising awareness among them, making investment, and use of appropriate technology are reportedly perceived to be more potent compared to cooperation and capacity building, though in varied proportion across the categories of water users in selected blocks of both the districts. The formulation and implementation of integrated water management strategies are essential for the poverty alleviation, economic growth, increasing agricultural productivity, ensuring food security, and robust environmental management in varied proportions by the sample households across the categories of water users in selected blocks of both the districts under study.

References

Kerstin M (2012) Thematic paper 6: legal and institutional frameworks. Groundwater governance, a framework for country action, GEF ID 3726

Meinzen-Dick R (2007) Beyond panaceas in water institutions. Proc Natl Acad Sci USA 104(39):15200–15205

Mukherji A, Shah T (2005) Groundwater socio-ecology and governance, a review of institutions and policies in selected countries. Hydrogeol J 13(1):328–345

OECD (2011) Water governance in OECD countries, a multi-level approach, 236 pp Paris. https://www.oecd.org/document/13/0,3746,en_2649_37465_48896205_1_1_1_37465,00.html.

World Bank (2010) Deep wells and prudence towards pragmatic action for addressing groundwater overexploitation in India. World Bank, Washington, DC

Zetland D (2011) The end of abundance; economic solutions to water scarcity, digital. Mission Viejo, Aguanom, Amsterdam

Chapter 41
The Effect of Cotton Management Practices on Water Use Efficiency and Water Security Challenges in Pakistan

F. Zulfiqar and G. B. Thapa

Abstract This research was carried out in cotton producing areas of Punjab province to investigate water security challenges faced by farmers. The investigation focused on how various cotton production management practices influence irrigation application and subsequent water use efficiency. Although cotton was making significant contribution to rural economies, the inefficient cotton production management practices added to water losses. Cotton farming was characterized by excessive use of irrigation water. It was found that there is much scope for improving financial return by improving water use efficiency in Pakistan. Particularly in Punjab, it is feasible to improve the irrigation water use efficiency by at least 14%. A disproportionate ground water extraction for irrigation has also caused perilously deeper water table. Moreover, inefficient management practices in cotton also result in water losses which leads to substantial degradation of biodiversity, freshwater irrigation resources, and other ecosystem services. The farmers in Pakistan generally apply irrigation through flood irrigation to uneven bunded cropland which result in longer time for irrigating a field, over-irrigation, and lower water productivity. Policy recommendations are made to improve the water use efficiency of cotton through improvements in several management practices. These are applicable to Pakistan and other south Asian countries producing cotton under similar climatic conditions.

Keywords Water use efficiency · Water security · Water resource management · Sustainable cotton production

41.1 Introduction

Cotton (*Gossypium hirsutum*) is mainly produced for its fibre. It also provides raw materials for oil and feed industries as its seed is rich in oil (18–25%) and protein (20–25%) (Saxena et al. 2011). Apart from meeting global fibre requirement, it fulfils

F. Zulfiqar (✉)
Asian Institute of Technology, Pathum Thani, Thailand
e-mail: farhad@ait.ac.th

G. B. Thapa
Kathmandu Forestry College, Kathmandu, Nepal

© Springer Nature Switzerland AG 2021
M. Babel et al. (eds.), *Water Security in Asia*, Springer Water,
https://doi.org/10.1007/978-3-319-54612-4_41

557

the demand for around a fifth of global vegetable oil (Gupta et al. 2012). Cotton is produced in more than 100 countries and traded by over 150 countries (Edwards et al. 2007). However, the top four cotton growing countries (China, United States, India and Pakistan) account for two-thirds of the global cotton production (Nazli 2010). It is grown as a cash crop by over 20 million farmers of Asia and Africa (Saxena et al. 2011).

Cotton accounts for 1.5% of GDP and 7.1% of the total value of agricultural production in Pakistan (GOP 2015). The country is world's fourth largest producer of cotton, which is also an important source of scarce foreign currency and raw material for the national textile factories (GOP 2014; Nadeem et al. 2014; Naheed and Rasul 2010). Cotton seed accounts for 63% of domestic edible oil (Amjad 2014). Cotton farmers constitute 26% of total farmers, and over 15% of total cultivated area is allocated to cotton of which 80% is in Punjab province and 20% in Sindh province (Nazli 2010).

Cotton production is an input intensive process requiring a delicate balance of inputs. Despite the fact that excessive irrigation is as harmful to the crop as is under-irrigation, cotton farming in Pakistan is characterized by excessive use of irrigation water (Watto and Mugera 2014, 2015). As a result, financial return from cotton irrigation in Pakistan is much lower than in other leading cotton producing countries (Shabbir et al. 2012). Maintenance of appropriate soil moisture level is also necessary for optimum returns from pesticide use in cotton (Tariq et al. 2006).

The findings of scientific studies have revealed that crop yields and profits can be maintained by reducing the amounts of inputs used (Abraham et al. 2014; Coulter et al. 2011; Sharif 2011; Zulfiqar and Thapa 2016). There is much scope for improving financial return by improving water use efficiency in Pakistan (Zulfiqar et al. 2017). Particularly in Punjab, it is feasible to improve the irrigation water use efficiency by 46–54% (Watto and Mugera 2015). By improving irrigation system, cotton producers in Punjab province can increase crop productivity by 19–28% without any increase in inputs use (Watto and Mugera 2014, 2015).

Furthermore, conventional cotton is characterized by excessive pesticide (Banuri 1998; Hasnain 1999) and fertilizer use (Makhdum et al. 2011), which needs to be optimized. It was in this context that the "better cotton" production system was introduced to enhance farmer's wellbeing as well as environmental protection. "Better cotton" refers to cotton production system which is socioeconomically and environmentally better for farmers and society (Zulfiqar and Thapa 2016). The first ever "better cotton" was cultivated in Punjab province of Pakistan in 2009 (BCI 2010). "Better cotton" farmers in China, India and Pakistan had experienced a marked reduction in pesticide, synthetic fertilizer and irrigation use during 2013. Despite reduced application of inorganic inputs and irrigation water, the yield of "better cotton" was found to be 11, 18 and 15% higher than the yield of conventional cotton in China, India, and Pakistan, respectively (BCI 2013). It is also important to note that cotton production in Pakistan is plagued by lower resource use efficiencies (Shafiq and Rehman 2000; Watto and Mugera 2014, 2015). Therefore, the purpose of this article is to investigate the efficiency in the water use of cotton production in Pakistan.

41.2   Material and Methods

41.2.1   Study Area and Sampling

This research was based on primary data collected from cotton farmers in Punjab province of Pakistan. A multistage stratified random sampling was used. At first stage, Punjab province was selected because of its biggest share in cotton production in Pakistan (Nazli et al. 2012). At second stage, Bahawalpur district was selected as it constituted the largest cotton area and production among all districts of Punjab (GOP 2012). At third stage, Bahawalpur, Yazman, and Ahmad Pur sub-districts were selected because of housing largest farming populations. At last stage, farmers were randomly selected among two stratum—"better cotton" and conventional cotton. Yamane formula (Yamane 1967) was used to estimate the number of farmers to be interviewed. This formula gave us 161 "better cotton" and 141 conventional cotton farmers.

41.2.2   Profile of the Study Area

Pakistan is predominantly an agricultural economy with agriculture striving to meet her food and fibre demands as well as supply necessary raw material to related agro-industries. Cotton, wheat, rice and sugarcane are major crops of Pakistan. Cotton crop contributes to Pakistan's economy through provision of raw material to the textile industry and vegetable oil production. A great majority of cotton is produced in Punjab province (80%) and the rest (20%) is produced in Sindh (GOP 2013; Nazli et al. 2012). The cotton cultivation in Punjab is carried out in cotton-wheat belt in the cotton-wheat and mixed cropping agro-ecological zones (FAO 2004; Khan 2004).

Within Punjab province, Bahawalpur, Bahawalnagar and Rahimyar Khan are three largest districts in terms of cotton area and production (GOP 2012). Bahawalpur district of Punjab was selected as this was the first district where BCI project was implemented in 2009 (BCI 2010) and ranked first in the province in terms of both area and production of cotton (GOP 2012). Cotton cultivation in Bahawalpur accounts for 11% of total national cotton area and 12% of overall cotton production (GOP 2012). The Pakistan rural household survey found that 53.4% of households in Bahawalpur were cotton growers (Nazli 2010). Bahawalpur district has low average rainfall, and a hot and dry climate. It lies in the cotton-wheat agro-ecological zone and has mostly sandy soil. Canal water is the leading source of irrigation, and desert covers almost two-third of the district (Kouser et al. 2015; Nazli et al. 2012). Bahawalpur district comprises of five sub-districts (tehsils) namely Bahawalpur, Ahmadpur, Yazman, Hasilpur, and Khairpur Tamewali. Out of these five sub-districts, Bahawalpur, Ahmadpur and Yazman sub-districts were selected because these were the largest sub-districts in terms of the number of farm households.

41.2.3   Methods and Techniques

Before going into details about the efficiency in water use by cotton producers, it is important to understand what kind of practices both conventional and "better cotton" farmers were using. The farmers in the study area were divided into three groups by farm size: 1) the farmers operating less than 2 ha were considered small farmers, 2) farmers operating between 2 to 5 ha were designated medium farmers, and 3) the farmers operating more than 5 ha were considered large farmers. The farmers were divided into different groups following the previous research done in the same study area (Nazli 2010; Nazli et al. 2012).

After looking at the farming practices of cotton producers, it is important to ask the question about the efficiency of resource use—whether the practices being used ensure efficient allocation and use of resources. In the wake of strong water demand in agriculture and growing scarcity of water in Pakistan, it is important to determine whether irrigation water is being used judiciously. As cotton is a water intensive crop, any crop practices are the future of cotton production which can reduce the irrigation use or increase production per unit of irrigation.

Therefore, efficiency in water use in cotton production was assessed. Any practices are desirable which enhance production per unit of irrigation or reduce irrigation use which maintaining production. Thus, efficiency in water use was assessed by using the following formula to determine which practices are more effective in water resource use between conventional cotton and "better cotton".

$$E_w = P_c / I_n$$

Where, E_w = Efficiency in water use (Kg of cotton produced by one irrigation).
P_c = Productivity of cotton (Kg per hectare).
I_n = Number of irrigations per cropping season (Number of irrigations per hectare).

The efficiency of water use was measured separately for both farming practice groups and a higher cotton production per unit of irrigation was an indicator of more efficient use of scare water resource. To determine whether these efficiency scores were statistically different from each other, two-group mean comparison T-test was employed following the example from other research (Zulfiqar and Thapa 2016).

41.3   Results and Discussion

This section presents the practices used by cotton producers in the study area and then explains the efficiency of water use based on different practice groups.

41.3.1 The Farming Practices of Cotton Producers

Cotton farmers in the study area, both conventional and "better cotton", use various customary cotton production practices (Table 41.1). The cotton is sown either manually, by drill or by planter. The manual sowing is done either on furrows or bed and furrows. The drill sowing is done on the level field and sometimes it is converted to ridges at later stages to save irrigation water (Makhdum et al. 2011). The irrigation is through canal, tube well or mixed. The cotton picking is entirely done by female cotton pickers. The cotton is sold to nearby ginning factories directly or through middleman. There is variation in the farming practices adopted by both farming systems as well as by farm size (Table 41.1).

The farmers in the study area prefer to cultivate cotton through manual sowing which is visible from on an average 60% farmers using this practice in conventional and "better cotton" groups. However, upon close examination it came to the fore that large farmers were less inclined to sow cotton manually. This is understandable as with larger area of operation it becomes more attractive to use machinery for sowing. This was true in the study area as large farmers' use of drill sowing and planter sowing

Table 41.1 Distribution of farmers' farming practices by farmer groups

Farming practices	Better cotton			Conventional cotton		
	Small	Medium	Large	Small	Medium	Large
Manual sowing	16 (53.3)	47 (65.3)	32 (54.2)	41 (71.9)	30 (62.5)	12 (33.3)
Drill sowing	13 (43.3)	23 (31.9)	33 (55.9)	13 (22.8)	14 (29.2)	15 (41.7)
If drill, convert to ridges	8 (26.7)	16 (22.2)	31 (52.5)	8 (14.0)	4 (8.3)	12 (33.3)
Planter sowing	3 (10.0)	14 (19.4)	16 (27.1)	6 (10.5)	7 (14.6)	19 (52.8)
Sowing on bed and furrows	2 (6.7)	13 (18.1)	17 (28.8)	6 (10.5)	10 (20.8)	19 (52.8)
Sowing on furrows	24 (80.0)	54 (75.0)	40 (67.8)	46 (80.7)	31 (64.6)	15 (41.7)
Irrigation sources: canal	12 (40.0)	23 (31.9)	30 (50.8)	5 (8.8)	9 (18.8)	20 (55.6)
Irrigation sources: tube well	0 (0.0)	1 (1.4)	2 (3.4)	0 (0.0)	1 (2.1)	0 (0.0)
Irrigation sources: mixed	18 (60.0)	48 (66.7)	27 (45.8)	52 (91.2)	38 (79.2)	16 (44.4)
Total farms	30 (18.6)	72 (44.7)	59 (36.6)	57 (40.4)	48 (34.0)	36 (25.5)

Source Derived from Survey Data 2014. The values in parenthesis are percentages

was higher than manual sowing in both farming systems. The second and third most preferred way of sowing was drill and planter sowing respectively for both farming systems.

Also, a higher proportion of "better cotton" farmers (twice) converted flat beds to ridges as compared to conventional cotton farmers. This has important implications for irrigation water saving as conversion to flatbed-furrows helps in saving 27–33% irrigation water (Makhdum et al. 2011). The conversion to ridges requires machinery and here again the large farmers' proportion of using it was more than double of small and medium farmers. A higher proportion of large farmers (45.3%) converted flat beds to ridges as compared to small (18.4%) and medium (16.7%) farmers. Thus, it can be deduced that the large farmers in the study area used more machinery for sowing purposes as compared to small and medium farmers.

The farmers in the study area either used normal furrows or flatbed-furrows for cotton cultivation. Both techniques help in water saving. However, flatbed-furrow is more labor intensive or requires machinery (tractor) use but results in more water saving than normal furrows (Makhdum et al. 2011). The highest proportion of farmers using bed and furrow technique was large farmers in both conventional and "better cotton" farming systems (Table 41.1). The small farmers made up the largest group using normal furrow sowing in both farming systems.

The sources of irrigation in the study area were canal and tube well. However, a majority of farmers used a combination of these. The canal water was a source of irrigation for a much higher proportion (40.4%) of "better cotton" farmers as compared to conventional farmers (24.1%). The reason behind such a variation was that "better cotton" farmers used water more efficiently because of training provided by BCI. They were trained to carry out water scouting (the estimation of water requirements of the crop by observing several water stress indicators (Makhdum et al. 2011)) before any irrigation application thereby reducing the required frequency of irrigation. Add with this the situation in the study area that canal water is available only once weekly due to which the "better cotton" farmers used a higher proportion of canal irrigations in their cotton production. Contrarily, the conventional cotton farmers had to rely more on mixing canal and tube well irrigations due to more frequent irrigations.

A closer look at the irrigation application from farm size perspective revealed that large farmers, in both farming systems, used a higher canal irrigation proportion as compared to small and medium farmers. In fact, more than half large farmers in conventional and "better cotton" cultivation only used canal water for irrigating cotton (Table 41.1). This can be explained from the irrigation distribution in Pakistan. The irrigation water is distributed once a week in proportion to total farm area of the farmer (Shaikh 2003). Thus, large farmers have higher canal water access and can irrigate more cotton through canal water if other crops at that time do not require irrigation. A majority of small and medium farmers, on the other hand, used mixed irrigation due to insufficient canal water availability.

Table 41.2 Distribution of farmers' farming practices by farmer groups

	Mean cotton yield	Mean no. of irrigations	Efficiency in water use[†]
Better cotton	2263.20	12.16	219.84**
Conventional cotton	2236.66	13.10	190.60**

[†]Efficiency in water use = Kg of cotton produced by one irrigation. The significance values are for two-group mean comparison T-test
**Significant at the level 0.05

41.3.2   The Efficiency in Water Use

The analysis of farming practices adopted by conventional and "better cotton" farmers revealed that the latter group used more water conserving practices than the former. Therefore, it was expected that they were producing more cotton per unit of irrigation. The results of the analyses revealed that the "better cotton" farmers' efficiency of water use was substantially higher than conventional cotton growers. The former group was producing 219.84 kg of cotton per irrigation while the latter group was only producing 190.60 kg per irrigation (Table 41.2). Thus, "better cotton" farmers' efficiency in water use was more than 14% higher than conventional cotton. The mean comparison tests also validated that these differences were statistically significant.

Thus, it can be deduced from Table 41.2 that the adoption of "better cotton" practices can enhance efficiency in water use by a little more than 14%. This was due to superior management practices which lead to water saving, moisture conservation and eventually less number of required irrigations by "better cotton" (Fig. 41.1).

The reasons for reduced number of irrigation applications and eventual higher efficiency in water use by "better cotton" were superior management practices. The main practices included cultivation of cotton on bed and furrows, furrow irrigation, and water scouting. Water scouting is a practice to estimate the need for irrigation through observation of sub-soil firmness, red stripe appearing on plant stem, stiffness of upper plant portion, internodes' length, presence of wrinkled leaves early morning, and incidence of white flowers on plant top (Makhdum et al. 2011; Zulfiqar and Thapa 2016). The farmers growing "better cotton" are trained to follow the above indicators before applying any irrigation (Zulfiqar and Thapa 2016).

Thus, it is evident from the result and discussion of the analyses that the management practices of "better cotton" farmers ensure higher efficiency of water use in cotton production. As Pakistan is facing numerous water scarcity challenges and water security is a burning issue not only for agriculture but also for other sectors of the economy. Agriculture is a major consumer of freshwater and surface water scarcity has forced farmers to exploit excessive groundwater which is not sustainable (Watto and Mugera 2013).

Therefore, the largest consumer of water, agriculture, cannot afford to continue using unsustainable irrigation practices in the face of growing demand to increase

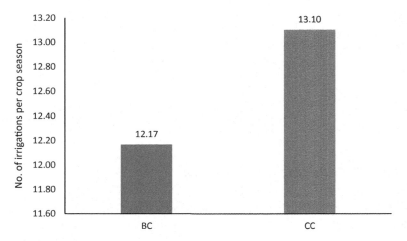

Fig. 41.1 Irrigation application (number of irrigations) by farmer groups

agricultural productivity. The common man need water for daily household needs and they are looking towards agriculture sector to get their due share to ensure their water security. Amid threats to water security in the face of water pollution, excessive use, and climate change, the efficient use of available water should be the highest priority.

41.4 Conclusions

Cotton is the most important cash crop of Pakistan along with meeting nearly 70% of edible oil requirements. It is also one of the most intensive water use crops. Thus, the efficient utilization of this scarce resource in cotton production can ensure sustainable exports and domestic edible oil supply. The introduction of "better cotton" in Pakistan since 2009 has provided us a huge potential for maintaining our productivity while reducing the use of irrigation water. Also, it is possible to produce more cotton by using less water under the current technological scenario. The analyses of results of this article revealed that "better cotton" farmers produce on an average 219.84 kg of cotton per irrigation while conventional cotton growers only produce 190.60 kg of cotton per irrigation.

This efficient use of water by cotton opens up more possibilities including producing more with same water input or saving irrigation water while maintaining current productivity level. If these management principle of "better cotton" are combined with other technologies such as precision agriculture, the water scarcity and security challenges can be partly addressed from the cotton production perspective. Thus, it is recommended that these improved management practices be promoted to conventional cotton farmers in the study area as well as in other cotton production areas of Pakistan. Also, as these are management practices which can also be adapted to other crops, dealing with water scarcity and security challenges entail extending these practices to other major crops in Pakistan such as wheat, rice, sugarcane and maize. Countries with similar biophysical and environmental conditions, such as India and China, can also look to adapt these practices for a collective response to global water security challenges. The climate change is expected to reduce water availability for agriculture in many countries. Thus, this study findings provide policy input for many of these regions to adapt to impending climate-induced water scarcity.

Acknowledgements The first author is extremely thankful to Higher Education Commission of Pakistan and Asian Institute of Technology, Thailand for supporting the Ph.D. research on which this paper is based.

References

Abraham B, Araya H, Berhe T, Edwards S, Gujja B, Khadka R, Koma YS, Sen D, Sharif A, Styger E, Uphoff N, Verma A (2014) The system of crop intensification: reports from the field on improving agricultural production, food security, and resilience to climate change for multiple crops. Agric Food Secur 3:4. https://doi.org/10.1186/2048-7010-3-4

Amjad M (2014) Oilseed crops of Pakistan: status paper. Plant Sciences Division, Pakistan Agricultural Research Council, Islamabad

Banuri T (1998) Pakistan: environmental impact of cotton production and trade. Paper prepared for UNEP project on Trade Environment. International institute of sustainable development. winnipeg, Manitoba, Canada

BCI (2013) Harvest report 2013. Better Cotton Initiative, Switzerland

BCI (2010) Harvest report 2010. Better Cotton Initiative, Switzerland

Coulter JA, Sheaffer CC, Wyse DL, Haar MJ, Porter PM, Quiring SR, Klossner LD (2011) Agronomic performance of cropping systems with contrasting crop rotations and external inputs. Agron J 103:182–192. https://doi.org/10.2134/agronj2010.0211

Edwards M, Osakwe C, Townsend T (2007) Cotton exporter's guide. International trade centre, Geneva

FAO (2004) Fertilizer use by crop in Pakistan. Rome, Italy

GOP (2015) Economic survey of Pakistan 2014–15. Islamabad

GOP (2014) Economic survey of Pakistan 2013–14. Islamabad

GOP (2013) Agricultural statistics of Pakistan 2012–13. Islamabad

GOP (2012) Crops area and production by districts 2011–12. Islamabad

Gupta SK, Pratap A, Gupta DS, Rajan N, Ganeshan VD (2012) Hybrid technology: chapter 1. In: Gupta SK (ed) Technological innovations in major world oil crops, vol 2: Perspective. Springer, New York, pp 1–21. https://doi.org/10.1007/978-1-4614-0827-7

Hasnain T (1999) Pesticide use and its impact on crop ecologies: issues and options. SDPI Working Paper Series No. 42. Islamabad

Khan AG (2004) Technical report on: the characterization of the agro ecological context in which fangr (farm animal genetic resource) are found. Development and application of decision support tools to conserve and sustainably use genetic diversity in indigenous livesto. Islamabad, Pakistan

Kouser S, Mahmood K, Anwar F (2015) Variations in physicochemical attributes of seed oil among different varieties of cotton (Gossypium Hirsutum L.). Pak J Bot 47:723–729

Makhdum AH, Khan HN, Ahmad S (2011) Reducing cotton footprints through implementation of better management practices in cotton production; a step towards Better Cotton Initiative. In: Fifth meeting of the asian cotton research and development network. Lahore, Pakistan

Nadeem AH, Nazim M, Hashim M, Javed MK (2014) Factors which affect the sustainable production of cotton in Pakistan: a detailed case study from Bahawalpur district. In: Xu J, Fry JA, Lev B, Hajiyev A (eds) Proceedings of the seventh international conference on management science and engineering management. Lecture Notes in Electrical Engineering. Springer, Heidelberg, pp 745–753. https://doi.org/10.1007/978-3-642-40078-0

Naheed G, Rasul G (2010) Recent water requirement of cotton crop in Pakistan. Pakistan J Meteorol 6:75–84

Nazli H (2010) Impact of Bt cotton adoption on farmers' wellbeing in Pakistan. Doctoral dissertation, The University of Guelph

Nazli H, Orden D, Sarker R, Meilke K (2012) Bt cotton adoption and wellbeing of farmers in Pakistan. Selected paper prepared for presentation at the International Association of Agricultural Economists (IAAE) Triennial Conference, Foz do Iguaçu, Brazil, 18–24 August 2012

Saxena D, Sharma S, Sambi S (2011) Comparative extraction of cottonseed oil. ARPN J Eng Appl Sci 6:84–89

Shabbir A, Arshad M, Bakhsh A, Usman M, Shakoor A, Ahmad I, Ahmad A (2012) Apparent and real water productivity for cotton-wheat zone of Punjab, Pakistan. Pakistan J Agric Sci 49:357–363

Shafiq M, Rehman T (2000) The extent of resource use inefficiencies in cotton production in Pakistan's Punjab: an application of Data Envelopment Analysis. Agric Econ 22:321–330. https://doi.org/10.1016/S0169-5150(00)00045-1

Shaikh IB (2003) Efficient and sustainable irrigation-management in Pakistan. In: Qurashi MM (ed) Water resources in the south: present scenario and future prospects. Commission on Science and Technology for Sustainable Development in the South, Islamabad, Pakistan, pp 41–50

Sharif A (2011) Technical adaptations for mechanized SRI production to achieve water saving and increased profitability in Punjab, Pakistan. Paddy Water Environ 9:111–119. https://doi.org/10.1007/s10333-010-0223-5

Tariq MI, Afzal S, Hussain I (2006) Degradation and persistence of cotton pesticides in sandy loam soils from Punjab, Pakistan. Environ Res 100:184–196. https://doi.org/10.1016/j.envres.2005.05.002

Watto MA, Mugera A (2014) Measuring efficiency of cotton cultivation in Pakistan: a restricted production frontier study. J Sci Food Agric 94:3038–3045. https://doi.org/10.1002/jsfa.6652

Watto MA, Mugera AW (2015) Econometric estimation of groundwater irrigation efficiency of cotton cultivation farms in Pakistan. J Hydrol Reg Stud 4:193–211. https://doi.org/10.1016/j.ejrh.2014.11.001

Watto M, Mugera A (2013) Measuring Groundwater Irrigation Efficiency in Pakistan: A DEA Approach Using the Sub-vector and Slack-based Models. Working Paper No. 1302, School of Agricultural and Resource Economics, The University of Western Australia

Yamane T (1967) Statistics an introductory analysis, 2nd edn. Harper and Row, New York

Zulfiqar F, Datta A, Thapa GB (2017) Determinants and resource use efficiency of "better cotton": an innovative cleaner production alternative. J Clean Prod 166:1372–1380. https://doi.org/10.1016/j.jclepro.2017.08.155

Zulfiqar F, Thapa GB (2016) Is "Better cotton" better than conventional cotton in terms of input use efficiency and financial performance? Land Use Policy 52:136–143. https://doi.org/10.1016/j.landusepol.2015.12.013

Chapter 42
Multi-stakeholder Negotiating Platforms for Effective Water Governance: A Case of Mashi Basin, Rajasthan, India

M. S. Rathore

Abstract Freshwater management (surface and groundwater) is both a global and local concern, and involves a plethora of public, private and non-profit stakeholders in the decision-making, policy and project cycles. Most countries are currently moving away from conventional forms of water governance. The trend now is for distributed water governance systems to supplement formal authority by an increasing reliance on informal authority, through genuine public—private coordination and co-operation. It seems that the policy makers, planners and executers find difficult to internalizing the concept of IRBM and implementing the concept at different levels. The paper discusses the new approach/model of community management of water resources in a river basin in Rajasthan, India. Mashi River Parliament as an institution and governance model is expected to emerge as a unique model of distributed governance in Rajasthan and can also be replicated in other parts of India.

Keywords River Basin Parliament · Distributed governance · User groups

42.1 Introduction

Rapid economic development and societal change are putting increasing pressure on water ecosystems and other natural resources. There is worldwide demand for changes that leads to more effective, more efficient and more sustainable water resource management. Efforts are being made to rethink water planning and management. Water management today poses multi-dimensional challenges, with complex geographical, ecological, social, political and economic factors. Also, water stress and water scarcity are challenges with far-reaching economic and social implications. Growth in population, increased economic activity and improved standard of living lead to increased competition for and conflicts over limited fresh water resources.

M. S. Rathore (✉)
Centre for Environment and Development Studies, Jaipur, India
e-mail: msr@cedsj.org

© Springer Nature Switzerland AG 2021
M. Babel et al. (eds.), *Water Security in Asia*, Springer Water,
https://doi.org/10.1007/978-3-319-54612-4_42

567

The deep appreciation to the complex issues surrounding water resource development has led to new approaches that seek to meet the ecological, social, political and economic challenges posed by the prevalent practices.

Water management has moved from the sectoral approach to an integrated approach. All water management techniques have complex and multidimensional implications, related to the existing geographical, ecological, socio-political and economic situations. However, these techniques need to be modified, updated and adapted in response to changes in existing order.

Globally water shortages, quality deterioration and flood impacts are among the problems which require greater attention and action at all levels. Accessible and high quality freshwater is a limited and highly variable resource. OECD projections show that 40% of the world's population currently lives in water-stressed river basins, and that water demand will rise by 55% by 2050 (OECD 2012a). In 2050, 240 million people are expected to remain without access to clean water, and 1.4 billion without access to basic sanitation. Freshwater management (surface and groundwater) is both a global and local concern, and involves a plethora of public, private and non-profit stakeholders in the decision-making, policy and project cycles.

In most countries water sector reforms are being carried out, governance and management needs more attention than water augmentation and access. In India, World bank and other donor agencies supported water sector reforms and under that water governance issues are addressed, but the outcome is less than the expectation mainly because of lack of political will and wider understanding on governance issues.

42.2   Water Governance

An important distinction should be made between governance and management, with management being a sub-category of governance. Effective management, although an important part of governance of natural resources, is not sufficient to secure good governance (Oviedo et al. (IUCN). The literature is rich with diverse and varied definitions of water governance (GWP 2002; UNDP 2001; Moench et al. 2003; Rogers and Hall 2003).There is no single definition of governance and therefore, different approaches need to be followed. Some may see governance as questions of financial accountability and administrative efficiency while other may focus on broader political concerns related to democracy and human rights and participatory process. However, good governance is the demand of the recent and future times to address the vital natural resource.

Most countries are currently moving away from conventional forms of water governance, which usually had *a top-down supply-driven approaches*, towards bottom-up demand-driven approaches, which combine the experience, knowledge and understanding of various local groups and people (UNDP 2007a). Governments are also moving towards better policy alignment in recognition of the fact that many policies outside the water sector can have a major bearing on levels and patterns of water demand and use (e.g. agricultural, trade and energy policies). These changes require improvements to water governance systems that include: more effective stakeholder dialogue, better vertical and horizontal sharing of information amongst stakeholders, conflict resolution at a range of different scales and planning procedures that are based on a vision that is common to relevant stakeholders (Batchelor, C.).

There is a growing perception that the governance of water resources and water services functions more effectively within an open social structure which enables broader participation by civil society, private enterprises and the media, all networking to support and influence government (Batchelor, C.). The ideology of a *command and control* or a *hierarchical* central State system caring for its citizens has to be replaced by *market-led* water governance models. The trend now is for distributed water governance systems to supplement formal authority by an increasing reliance on informal authority, through genuine public—private coordination and co-operation.

Achieving good water governance cannot be undertaken hastily using blueprints from outside any given county or region. Good governance needs to be developed to suit local conditions. Incremental improvement and flexibility are key (Batchelor, C.). Rogers and Hall (2003) argues that there is no single model of effective water governance; indeed, to be effective governance systems must fit the social, economic and cultural particularities of each country. Nevertheless, there are some basic principles or attributes that are considered essential for effective/good governance, such as, in Approach: Open and Transparent, Inclusive and Communicative, Coherent and Integrative, and Equitable and Ethical, while in Performance and Operation: Accountable, Efficient, Responsive, predictable, participative and Sustainable.

Governance can take many different forms depending on the economic, cultural and traditional political norms of a country and the behavior of the legislature and legislators. We want to have a governance system in Mashi River basin in Rajasthan, India where there will be a balance among politicians, people and government, cooperating within the given legal framework in the larger interest of the society on long term basis aiming at sustainable development and management of natural resources in the Mashi River Basin. This will require the politicians to move away from the mentality of severing the constituency to ensure reelection and look for long term development of the people by sustaining the health of natural resources. Our objective is to create a framework (institutional and administrative) within which people with different interests can peacefully discuss/debate and agree to co-operate and coordinate their actions to sustainably manage the natural resources of the River basin.

Generally, much doubt is expressed about the capacity of people to come together and act like community in larger interest of protecting environment in the present competitive world and ask question such as, can society coordinate and manage itself? In the River Parliament Model, distributed governance is visualized, where in, coordinated interaction of people, civil society organizations, government, and technocrats with a defined role and responsibility will participate in the larger interest to sustainably manage the resources with minimum conflicts. Though there will be well defined role for each of the stakeholders but more role for people so that it is a people centric governance system contrast to what have been presently practices as *State Driven Government Centric Decentralized Model*, which in most cases failed to deliver. The state will provide all legal, financial and other support with minimum interference in the working of River Parliament. Government should also play major role in financing the planned infrastructure and other activities by convergence of the line departments activities and allotted budgets and also create enabling environment for better democratic functioning of the Parliament. While the team of technocrats is assigned the role of providing all kinds of technical advice on different aspect of natural resources rejuvenation, conservation, protection and sustainable management. They will also build the capacity of stakeholders to facilitate equal and better participation in governance system and also facilitate selection of better livelihood options to reduce pressure on natural resources.

In India at national and state level there is greater attention on water management aspects both at the level of policy and practice and governance issues are sidelined or given low priority. The result is that there are number of line departments dealing with water resource management yet not getting the desired results. In the following section water resource management is discussed to identify gaps and lessons for our new approach.

42.3   Water Resource Management

India's water demand has increased many folds because of economic development and fast increasing urban and rural population. Agriculture sector continues to be the main user of water (85%), largely groundwater. Lack of access to safe water is an important impediment to the progress of public health, education and poverty reduction in the country.

Water resources management is big challenge because of multiple sources of supply, multiple stakeholders with competing demands, unclear ownership of water resources and unequal access to varied level of supply, and complex social and political context. The conventional model of water resources management is based on four practical elements: policies, laws and plans; an institutional framework of management and technical instrument; and investment in water infrastructure. In order to address the existing and emerging challenges change is necessary in the four basic strategies of water resources management.

The Integrated River Basin Management (IRBM) approach, globally accepted as solution to the overwhelming need for sustainable and equitable development and management of water resources has been accepted formally by the Indian national government in its National Water policy and gradually the states are also adapting it by enacting River Basin Management Acts.

It seems that the policy makers, planners and executers find difficult to internalizing the concept of IRBM and implementing the concept at different levels. The paper will discuss the new approach/model of community management of water resources in a river basin in Rajasthan, India.

The demand, supply, availability and access of water resources do not always match and more particularly in the Arid and Semi-Arid water scarce areas. As the population increases the demand for fresh water and irrigation water to grow more food goes up. Changes in the land use patterns, climate variability and diminishing efficiency to use of resources only strain the available reserves further. Ultimately the new water scarcity will shape how we live, how we work, how we relax. It will reshape how we value water, and how we understand it. There is mark difference in the perception and value system of traditional societies and modern world and that is reflected in emerging water problems.

User groups as institutions in parallel with and in preference to PRIs were promoted by Donor agencies as part of their approach to decentralized planning in India (EUSPP 2012). Since the institutional decentralization was not fully followed rather unrelated with decentralization of powers the system did not work efficiently and consequently could not deliver the desired results. This applies to all sectoral developmental programs attempted in India. Coordination and integration could not be practiced as these words remained on paper or policy documents. Despite the new understanding that community participation is central in improving effectiveness of government schemes in general, currently the level of community participation and thus local-level transparency, accountability and ownership of planning within the water sector vary considerably (EUSPP 2012).

42.4   Review of Water Resource Management Approaches

Water resources management has been a major challenge of all the societies in the past. Countries have adopted different approaches of management based on their resource availability and usage pattern. However, globally Integrated Water Resource Management (IWRM) approach has been recommended as universal solution to water related problems. In this section few well known approaches to water resources management are discussed to look for a alternative new approach based on Rajasthan socio-economic and political environment. The Few approaches are as follows:

(a)   **The Techno-economic Approach** to water resource management has been the conventional or mainstream approach throughout the latter half of the twentieth century. This approach has solved some of the short term crises of availability that plagued the countries of the Third World during the mid-twentieth century. Food production, availability of power, and access to water has increased for significant number of people. However, the long term adverse effects of such large scale interventions on the natural environment and on human communities raised doubts about such projects and this in turn led to a new way of looking at water management.

(b)   **The Integrated River Basin Management (IRBM) approach** is a concept that aims to conserve and utilize the natural resources within a river basin sustainably, through integrating the needs and skills of various stakeholders like farmers, industries, government departments, academics, NGOs and people and their representatives. IRBM has been accepted formally by the national government in the National Water policy 2012 but water management continues to be a centralized top-down approach causes more problems than solutions.

It seems that the policy makers, planners and executers find difficult to internalize the concept of IWRM and working out practical implications of implementing the concept at different levels. The resultant outcome is that even the integration of identified line departments to be involved in water resource management at state level has become difficult proposition. To address the emerging issues a new approach namely Negotiated Approach to IRBM is tried. The negotiated approach is a variant of conventional IRBM. It is aimed at creating space for negotiation, including with local stakeholders, on river basin management options. The negotiated approach calls for the reverse, allowing local actors to develop basin management plan and strategies specific to their local context, which are then incorporated in the larger basin management plan. This allows their knowledge to influence regional and national decisions and feel sense of ownership, responsibility and accountability towards the change in the management and implementation system. This ultimately results in a truly participatory bottom-up process of policy development and management.

(c)   **The IWRM Approach**

Integrated Water Resource Management (IWRM) approach has emerged at Global level from the United Nations Water Conference in 1977, with most governments later committing in 2002, to application of IWRM by developing IWRM and water efficiency plans. By 2012, more than 80% of countries had made progress towards meeting the target.

IWRM is a process which can assist country and within country different States in their endeavor to deal with water issues in a cost effective and sustainable way (GWP 2000). IWRM as an approach to manage water resources has been Globally accepted by most countries including India as it finds place in National Water Policy 2012 and in Rajasthan State Water Policy 2010. The recent announcement and promulgation of The Rajasthan River Basin and Water Resources Planning Act 2015 on April 24, 2015 is a big step in this direction.

General principles, approaches and guidelines relevant to IWRM are numerous and each has their areas of appropriate application. The Dublin principles are particularly useful set of such principles and out of the four principles particularly the principle; "Water development and management should be based on a participatory approach, involving users, planners and policy makers at all levels", is critical for sustainable management of water resources particularly at local level, i.e. basin or aquifer level. The requirements for IWRM is to have holistic approach to management, recognizing all the characteristics of the hydrological cycle and its interaction with other natural resources i.e. land, biomass and ecosystem. The effects of human activities lead to the need for recognition of linkages between upstream and downstream users of water. Upstream users must recognize the legitimate demands of downstream users to share the available water resources and sustain usability. This clearly implies that dialogue or conflict resolution mechanisms are needed in order to reconcile the needs of upstream and downstream users. Groundwater management also need understanding among users on regulating it to ensure sustainable present and future use.

The other important component of IWRM is participatory approach. Real participation only takes place when stakeholders are part of the decision making process. Participation occurs only when participants have the capacity to participate. It implies that, first, build the capacity of the stakeholders, particularly women and marginalized social groups. This may not only involve awareness building, confidence building and education, but also the provision of the economic resources needed to facilitate participation and the establishment of good and transparent source of information.

The main practical elements of IWRM can be listed as follows;

(i)   A strong enabling environment—policies, laws and plans that put in place "rule of the game" for water management that use IWRM

(ii)   A clear, robust and comprehensive institutional framework for managing water using the River Basin as the basic unit for management while decentralizing decision making.

(iii) Effective use of available management and technical instruments—use of assessments, data and instruments for water allocation and pollution control to help decision makers make better choices.

(iv)   Sound investments in water infrastructure with adequate financing available— to deliver progress in meeting water demand and needs for flood management, drought resilience irrigation, energy security and eco-system services.

The ultimate objective of IWRM is to make changes in water management approach in the complex social and political context. Besides water management and governance issues there are other issues and the most pertaining is the food or agricultural issues, as water use in irrigation is the highest among all other usages.

IWRM Approach in Rajasthan[1] at Gram Panchayat level was initiated by the Water Resource Department (WRD) of the Government of Rajasthan in 2007 with the following broad objectives under water sector reforms undertaking activities such as, human resource development, institutional reforms (Government and Panchayat Raj), legal reforms by undertaking review of existing laws and preparing comprehensive water laws, financial reforms by adopting Medium Term Expenditure Framework (MTEF), creating data bank, organising mass awareness campaigns, take initiatives to seek NGO participation, etc.

The Government of Rajasthan (GoR) developed a New Approach to multi-level Integrated Water Resource Management (IWRM). The approach was ensuring the emphasis of the GP plan by way of:

---

[1] For details see A J James, M S Rathore et al. (2015) Monitoring and Evaluation of EC-assisted State Partnership Programme (Rajasthan), Submitted to ICF International, UK, Institute of Development Studies, Jaipur.

– sustainable management of water resources;
– regulating major users and uses;
– allocation of water to primary needs;
– equitable and secure access of the poor and marginalised to water services;
– to ensure that issues of inter-sectoral, inter-village and inter-block equity and
  sustainability are picked up.

The review of the GP-level planning process indicated many fundamental Gaps.[2]
The main learning from the Pilot IWRM in Rajasthan were; (i) The GP-IWRM
Planning process was being ineffective as a result of the bypassing of the regular
structure of planning and monitoring of government activities that were operational
at district and sub-district levels. (ii) villagers' key water problems could not be
identified for lack of their active participation in the planning process.

42.5 Our Model of Mashi River Basin Parliament

42.5.1 General Features of Mashi River Basin

Mashi River Basin is part of the larger River Basin called Banas River Basin, which is
located in the middle of the Rajasthan. There are 11 sub basins in Banas River Basin
namely; Banas (1,174,039 ha), Dain (306,138.4 ha), Gudia (92,038.56 ha), Kalisil
(62,308.94 ha), Khari (639,052.9 ha), Kothari (229,852.1 ha), Mashi (647,615.8 ha),
Morel (572,250.7 ha), Sodra (151,942.2 ha) and Berach (830,788.6 ha). The catchment area including all upstream Major/Medium projects is 5,872.0 Km^2 where
as the differential catchment area (area excluding upstream catchment areas of
Major/Medium projects) is 3,641.4 Km^2 and falls in Tonk District.

The Mashi River Basin area falls in three districts namely Jaipur, Ajmer and Tonk
Districts. The two main tributaries of Mashi River are, namely Bandi and Mashi,
which originates from the hills of Samod and Ajmer district respectively. Mashi River
originates from the Silora hills about 6 km south of Kishangarh Town in Ajmer district
and passing through Phulera tehsil in Jaipur district. It flows initially in an eastward
direction and then towards south for about 96 km in partly hilly and partly plain
areas along the borders of Jaipur and Tonk districts between the tehsil of Malpura
and Phagi until it turns south to join the Banas River at Galod village near Tonk.
The catchment of the Mashi River is located between latitudes 26°11 and 26°16′ and
longitudes 74°48′ and 75°54′. It has got one tributary called Bandi. Bandi River the
tributary of Mashi River originates from hills located in the North-West of Jaipur and
passes through Kalwar town near Jobner and meets Mashi near Madhorajpura. These
tributaries are fed by large number of small rivulets originating from the plains of
tehsil Sanganer, Dudu, Chaksu, Malpura, etc. All of them are non-perennial rivulets.

---

[2]For details see A J James, M S Rathore et al. (2015), pp.6–8.

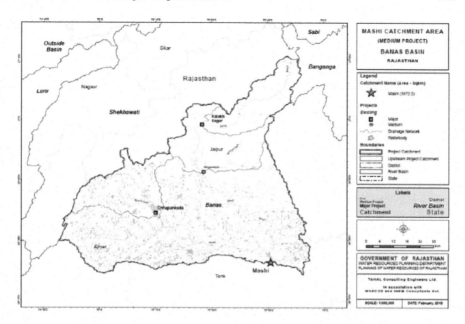

Fig. 42.1 Automatically delineated catchment of Mashi Sub Basin

The Fig. 42.1 shows the automatically delineated catchment of Mashi Sub Basin is shown in Fig. 42.1.

Water Management Issues in the Basin

- The soils of the region suffer variously in the different soil regions from excessive drainage, low water retentive capacity, moderate erosion by wind, and low fertility mostly in the upper northern part of the basin. Salinity, alkalinity, poor drainage accompanied moderate to severe erosion are the problem of the soils.
- Five hydrogeological formations viz; *Younger Alluvium, Older Alluvium, Phyllite& Schist, Quartzite and BGC (Banded Gneissic Complex)* are the main water bearing formation (aquifer) in the Basin.
- **Land Use**: The Cultivated area (including current and permanent fallow lands) accounts for 78.2% of total geographical area of the Basin. The forest area is around 3.6% and Barren/ un-culturable/ Wastelands 14.5%. Other categories are covering less than 5% area.
- **Surface Water**: The number of Water Harvesting Structures (WHS) constructed in with differential catchment is 3,087 with total water holding capacity is 112.23 Mm^3. Actual mean annual water yield to the Mashi sub basin is computed to be 203.95 Mm^3 (with all interventions).
- Rainfall occurs mainly during the monsoon season in Mashi Project catchment therefore, major portion of stream flow occurs only during these months. The

annual dependable water yield at 50% is 59.6 Mm^3, while water yield at 75% dependability is 6.4 Mm^3 (13.3% of gross storage capacity).

- Major and Medium Projects: There are 3 upstream projects in Mashi sub basin catchment. The live storage capacity of these three existing upstream projects in the Mashi Sub Basin catchment is 81.36 Mm^3.
- Minor Projects: There are 97 Minor projects in the catchment area of Mashi Sub Basin with total live storage capacity of 90.64 Mm^3. There are large numbers of minor projects constructed in the catchment of Mashi Dam capacity of which exceeds its design yield which may have substantial impact on inflow to project.
- **Groundwater:** Groundwater availability for long-term exploitation, clear of any current state of overdraft is the basic element. Since it is a derivative of rainfall, the dependability level of such rechargeable 'dynamic' groundwater availability relies on the statistic occurrence of precipitation. The total net annually assessed groundwater resource in the Mashi Basin is 2586.29 Mm^3 and groundwater draft 3497.64 Mm^3. The stage of groundwater development in the basin is 135.24% and the basin is categorized as overexploited basin.
- Groundwater Quality: The groundwater quality in the Mashi Sub Basin has been reported with reference to selected parameters, namely, concentration of Chloride, Fluoride, Nitrate and EC value.
- The average chlorides concentration was relatively stable and ranges from a minimum of 175 mg/l to a maximum of 474 mg/l during the period of 1984–2010.
- Fluoride concentrations are above the upper permissible limit for drinking water in most of the basin's area. The 100% non-potable water area belongs to quartzite aquifer unit in Mashi sub basin.
- The average nitrates concentration ranges from a minimum of ~25 mg/l to a maximum of ~267 mg/l during the aforementioned time period 1984–2010. The concentration rose from a value of ~41 mg/l during 1984 to a value of ~100 mg/l during 2010, a total rise of ~144% within 27 years. The average nitrate concentration is between the desirable and maximum allowed concentrations for drinking water (45 mg/l and 100 mg/l, respectively); nevertheless, the last average value (2010) is very close to the maximum allowed limit. Most of the area in the basin is affected by nitrates ion concentrations above permissible concentrations.

There are numerous problems of water resources i.e. of availability, distribution, equity in access, quality, competition in usage, water pollution, encroachment on water bodies and catchment areas, ownership and right issues, etc. It is for this reason that this basin was selected to attempt a new model of water resource management. Also, the State Government has enacted a River Basin Act without much understanding the implication of it in terms of governance of water. The proposed River Basin Parliament may help in understanding and addressing the future water governance and management needs of the State.

42.6 Key Governance Features of the Parliament

Based on the learning's from the above review of experiences in Rajasthan the CEDSJ is trying to adopt a new approach of Participatory Community Management of River Basin. It is planned to have two associated partner NGOs namely Gramin Navyuvak Mandal Laporia (GVNML) and Gramodaya Samajika Sansthan (GSS) as field level implementation partners to facilitate, i.e., community mobilization, formation of River Basin organisations in the Mashi sub-Basin starting from Gram Panchayat, micro watershed and sub-Basin level, Undertake capacity building activities jointly with CEDSJ in order to prepare IWRM plans. The information generated in the CEDSJ (2015) study with the input from Hydrologist or Hydro-geologist and Remote sensing expert etc. to map the land, water (water bodies and drainage system) and other natural resources in the Sub-Basin based on IWRM approach and document changes/obstructions caused in the sub-basin hydrological system by people, development agents and development activities by the State, will form the basis for capacity building training modules. Trainings will be imparted to the three group members, i.e., Stakeholder Group, Technical support Group and Public Representatives Group as shown in the organogram (Fig. 42.2). The details about the groups are as follows:

Stakeholder Groups: The Stakeholder group comprise of three sub groups; (i) Farmer and Non-farm sector members, (ii) Industrialist Group, and (iii) Unorganized sector members. The First sub group will be at the watershed level and as there are six watersheds in the Mashi Basin in total there will be six groups of 5 members each adding to a total of 30 members. The Second Sub group will be of Industrialist and there are two major industrial area, namely SEZ—Mahindra City and RIICO area. This group will have 5 representatives. The third sub group is of unorganized sector members representing business groups on road side and will have two representatives.

Technical Support Group: The group will comprise of representative of line departments at block level, subject matter specialist, such as Geologist, Geohydrologist, Agronomist, Watershed Specialist, Economist, Institutional expert, NGO representatives, CEDSJ representative, etc. In total this group will have 14 members. Technical support group will act like an advisory group to facilitate smooth working of parliament and help preparing IWRM plans and implementation of the plan.

Public Representatives: It has been observed that in most of the development groups formed by NGOs or State Government the public representative is either missing or are considered as passive members. Since each public representative has been allocated fund for development works in his/her constituency we thought of involving them in the River Basin activities and also take the issue at the state level. Their participation will ensure political support to the River Basin Parliament. Hence MLAs, Pradhans and Sarpanch's will be the member of this group and in total there will be 29 members.

Mashi River Basin Parliament: Mashi River basin Parliament will be constituted after discussion with the three group members. The process of formation will start

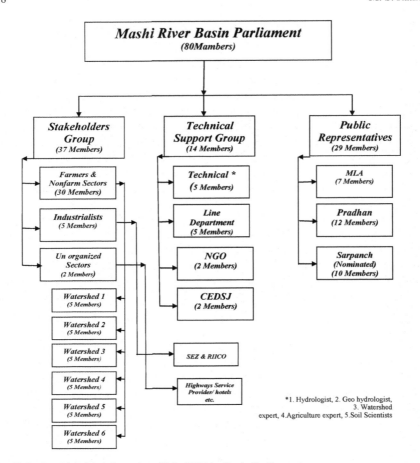

Fig. 42.2 Organizational Structure of Mashi River Basin Parliament

from below with capacity building trainings of all the stakeholders. The constitution and working procedures of the Parliament will be formulated in the stakeholder meetings though the draft document prepared by CEDSJ based on the review of community based organisations working in different parts of India. The experience of working with the Tarun Bharat Sangh in formation of Arvari River Parliament which is functioning for last 20 years will also be applied at each stage of its formation. In total there will be 80 members in the parliament. The River Parliament organisation is shown above (Fig. 42.2):

Acknowledgements Thanks are due to Mr. Shashai Kant Chopde for his valuable suggestion on the first draft of the paper.

References

Batchelor C (n.d.) Water governance literature assessment report, IIED

CEDSJ (2015) Report entitled climate resilient development- a case study of Mashi Sub Basin in Rajasthan. Centre for Environment and Development Studies, Jaipur

Fishman C (2012) **The Big Thirst** – The secret Life and Turbulent Future of Water. Free Press, New York

Global Water Partnership (2000) Towards water security: a framework for action, GWP

Global Water Partnership (2001) Tool Box for IWRM, GWP

Global Water Partnership (2002) Introducing effective water governance, Mimeo

GWP (2000) Integrated water resources management, TAC Background paper No 4. GWP Technical Advisory Committee

Government of Rajasthan (2012) Detail Project Report- IWMP IV, Rural Development & Panchayati Raj Department, Watershed Development & Soil Conservation Department, Rajasthan, Jaipur

Involving Communities (2011) A guide to the negotiated approach in integrated water resources management, ENDS and Gomukh environmental trust for sustainable development

James AJ, Rathore MS et al (2015) Monitoring and evaluation of EC-assisted state partnership programme (Rajasthan). Report 5, Submitted to ICF International, UK, Institute of Development Studies, Jaipur

Kooiman J (ed) (1993) Modern governance: new government-society interactions. Sage Publications, Thousand Oaks

Meinzen-Dick R, Chaturvedi R, Domènech L, Ghate R, Janssen MA, Rollins ND, Sandeep K (2016) Games for groundwater governance: field experiments in Andhra Pradesh, India. Ecol Soc 21(3):38

Moench M et al (2003) The fluid mosaic-water governance in the context of variability, uncertainty and change, NWCF Kathmandu and ISET, Boulder, USA

Mundle S, Chowdhury S, Sikdar S (2016) Governance performance of Indian States -changes between 2001–02 and 2011–12. Economic & Political Weekly, vol II, No 36

OECD (2015) Report on the OECD principles on water governance. Directorate for public governance and territorial development

OECD (2012a) OECD Environmental Outlook to 2050. OECD Publishing. https://doi.org/10.1787/9789264122246-en

Oviedo G, Mansourian S, Surkin J (n.d.) IUCN Contribution to "Environmental Justice and Global Citizenship" Governance of Natural Resources: Reconciling Local and National Levels, IUCN, UK

Rogers P, Hall AW (2003) Effective Water Governance, TEC Background Papers No. 2. Global Water Partnership Technical Committee

Tupepepa E (n.d.) Public-private and community participation in water resource management: the missing dimension - the power of three. Ritsumeikan Asia Pacific University, Japan

World Bank (2010) Deep wells and prudence: towards pragmatic action for addressing groundwater overexploitation in India. World Bank, Washington, D.C., USA

Chapter 43
Economic Benefits of EbA Measures to Assure Water Security: A Case Study on EbA for Sediment Trap Versus Dredging

R. Treitler, P. Kongapai, J. Ngamsing, and K. Sansud

Abstract Ecosystem based adaptation (EbA) has emerged as a key instrument widely applicable to help people to adapt to the adverse effects of climate change. EbA, similar to green infrastructure, uses services that river ecosystems are already providing, addressing the crucial link between sustainable resource management for the benefit of human populations and biodiversity conservation. EbA relevant case studies and literature address a broad range of water security topics, however, they lack knowledge on technical–economic comparison between EbA and alternative measures. This presentation sheds light on the major threat of water scarcity in the Huai Sai Bat River Basin in Khon Kaen, Thailand, comparing Sediment Pool Installation (EbA measure) with Dredging (Engineering measure), both designed to counteract this major threat by increasing water storage capacity. The comparative analysis, based on current land use in the catchment, demonstrates that the EbA measure has an economic benefit over the engineering measure from the first year onwards, increasing over the years. The overall costs for water storage can be reduced by up to 65% by applying the EbA measure. In addition, the presentation also addresses the subject of integrating land use management and water management.

Keywords Ecosystem-based adaptation · Cost effectiveness · Economic evaluation · Integrated water management

43.1 Introduction

Ecosystem-based adaptation (EBA) is defined by the Convention on Biological Diversity (CBD) as an increase in the population's adaptation capacities through the sustainable use and protection of ecosystem.

Non-sustainable land use (deforestation, overgrazing, pollutant input, etc.), lack of land planning (e.g. uncontrolled settlement and encroachment of floodplains, or the

R. Treitler (✉) · P. Kongapai · J. Ngamsing · K. Sansud
GIZ German International Cooperation, Bangkok, Thailand
e-mail: roland.treitler@giz.de

© Springer Nature Switzerland AG 2021
M. Babel et al. (eds.), *Water Security in Asia*, Springer Water,
https://doi.org/10.1007/978-3-319-54612-4_43

drying out of wetlands for golf courses) and water resources lead to many ecosystems already being damaged, restricting their services. Providing, regulating and purifying functions of ecosystems are crucial to protect the population from the effects of extreme events such as floods and droughts, which, as a result of negative consequences of climate change, will occur more frequently (compare Buyck et al. 2017; Muang et al. 2013; Perez et al. 2010).

Advantages of ecosystem-based adaptation measures in flood and drought prevention in comparison to traditional grey measures are usually lower costs and lower maintenance requirements. They can be implemented with locally available materials. Positive side effects are the protection of biodiversity. Although these functions require intact ecosystems, ecosystem-based adaptation does not concentrate on protecting the environment, but to use ecosystem functions as a service to increase the adaptation capacity of population to a changing climate. Therefore the EbA concept is improving the climate resilience of people.

When sensitive ecosystems, such as mangrove forests, mountain forests and swamps are protected, both, their resilience against weather extremes, as well as the resilience of the local population can be strengthened, sustaining living conditions and economic development possibilities.

Within ecosystems, the water cycle is connected with other system components such as soils and vegetation and is influenced by these in its functions. Both the quantity available and the quality of the water resources are determined by ecosystem services.

43.2   Case Study in Huai Sai Bat River Basin, Thailand

43.2.1   Huai Sai Bat River Basin, Thailand

The Huai Sai Bat River Basin comprises approximately 678 km^2. The highest point upstream has an elevation of about 550 masl. The elevation drops to about 150 masl within a distance of about 60 km. Only the headwaters area is covered with forest, where the watercourses stretch over 10 km with steep slopes generating high energy potential of water. The precipitation sums up to 1000 mm/yr inducing an average discharge of 200 mm/yr. Approximately 65,000 people live in the river basin which results in 95 inhabitants per km^2.

A prominent feature of the Huai Sai Bat river basin are the structures for water management including two reservoirs, weirs along Huai Sai Bat, some tributaries and countless small ponds scattered throughout the river basin and mainly used for livestock and irrigation.

Nong Yai Reservoir in the north is an important water resource for the local population and the agricultural sector. Sedimentation is a prominent issue in this area and leads to a considerable decrease of water storage volume as well as costs for extracting the sediment out of reservoirs. Furthermore, the sediment is contaminated by agricultural runoff, which leads to further costs for purifying. The picture below shows the Nong Yai Reservoir right in the middle of intense agricultural land use (Figs. 43.1 and 43.2).

The catchment area of the Nong Yai Reservoir comprises 66 km², the surface of the Nong Yai Reservoir is about 600.000 m², its average depth is about 5 m, which corresponds with a water storage volume of 3 mm³. The reservoir is surrounded by agricultural land. The water is used for irrigating about 4.700 rai (=752 ha).

Fig. 43.1 Nong Yai Reservoir after rainy season

Fig. 43.2 Nong Yai Reservoir before rainy season; *Source* (Lohr 2016)

Table 43.1 Source: FAO data Thailand 2014

Rice farmer				
Yield per ha	3.01	t/ha	481.6	kg/rai
Producer Price per ton	8,280	THB/t	240	US$/t
Total Irrigated Area	4,700	rai	752	ha
Area affected	58.75	rai	9.4	ha

43.3 Nong Yai Reservoir in Huai Sai Bat River Basin: EbA Measure Sediment Trap Versus Business as Usual Dredging

Currently, sediment is reducing the storage by 1.25% from its original capacity or minimum 38.000 m³ yearly. (Lohr 2016) (Meier et al. 2015) The business as usual scenario is dredging every 20 years, which means that the storage capacity is reduced by 1.25% every year over a time span of 20 years. The yearly loss of water storage leads to a reduced irrigation area of about 60 rai or 9.4 ha per year. Not having enough water reduces the yield of the farmers and therewith their income. Reducing the storage capacity by 38.000 m³ every year means that 9.4 ha each year are lost for irrigation. Consequently, the yield will be lower in this area. The yield reduction assumptions are based on interviews with the farmers in the region. The feedback varied between 10 and 50%.[1] What is the economic effect of the reduced water storage capacity?

For calculating the losses, the team used rice as an example and based the calculation on the following prices and yields (Table 43.1):

Assuming that the yield will be reduced by 10% the rice farmers will lose THB 23.430 (EUR 609) in the first year. An average famer household with 22.7 rai losses THB 9.080 (EUR 236). After 5 years the losses will be already THB 117.000 (EUR 3.039) per year (average famer THB 45.400 = EUR 1.179). Aggregated losses from year 1 to year 5 will be about THB 351.410 (EUR 9.128) and from year 1 to year 20 THB 4.9 m (EUR 127.273). The losses for an average farmer will be from year 1 to 5 about THB 136.000 (EUR 3.532) and from year 1 to year 20 THB 1.9 m (EUR 49.351).

An average farmer household with 22.7 rai would lose THB 113 (EUR 3) in year 1 and THB 570 (EUR 15) in year 5. However, if the average farmer is unlucky and all his land is in the affected area, a farmer can lose up to THB 9.080 (EUR 236) in year 1 and THB 45.400 (EUR 1.179) in year 5.

Simulating a reduced rice yield of 30% will triple the losses. Therefore, the farmers would lose THB 70.280 (EUR 1.825) in the first year and after 5 years THB 351.410 (EUR 9.128) per year. An average farmer losses THB 340 (EUR 9) per year and after 5 years THB 1.700 (EUR 44). These yearly losses would lead to an aggregated loss

---

[1] The author decided to calculate two scenarios: one with the minimum reduction and another one with the average of the feedback.

of THB 1.05 Mio (EUR 27.273) within 5 years and THB 14.8 Mio (EUR 384.416) after 20 years. A single household would lose THB 5.100 (EUR 132) within 5 years and THB 71.500 (EUR 1.857) after 20 years.

The Department for Water Resources (DWR) is managing the reservoir and addresses the sedimentation issues by dredging the reservoir every 20 years. This pattern is considered business as usual. After 20 years the storage capacity is reduced by 25% and the reservoir needs to be dredged. The current costs for dredging are THB 75 Mio (EUR 1.95 Mio).

DWR spends THB 75 Mio every 20 years for dredging. During the time from one to the next dredging intervention the farmers lose between THB 4.9 Mio and THB 14.8 Mio by the reduced water storage capacity caused by sedimentation. Hence, the private as well as the public sector are losing money without sustainably improving the situation, which could be achieved by an EbA sediment trap. Overall, both sectors lose between THB 79.9 Mio (EUR 2.1 Mio) and THB 89.8 Mio (EUR 2.3 Mio) every 20 years.

As an alternative to the dredging, a wetland restoration as a sediment trap at a confluence upstream of the Nong Yai Reservoir is proposed as an EbA measure to reduce the sediment in the reservoir (Lohr 2016). With such a measure the sediment can be reduced by 75%. Assuming the 75%, the yearly loss would be 9.500 m^3 (compared to 38.000 m^3). The proposed measure would cost THB 750.000 for construction with running costs for maintenance of THB 15.000 per year. (Meier et al. 2015) (Fig. 43.3).

The construction of the sediment pool will lead to slightly higher costs at the beginning. Ideally the trap will be constructed when the reservoir is dredged. Consequently, the overall costs will be THB 75.75 Mio (EUR 1.97 Mio) (Dredging: THB 75 Mio, Sediment pool: THB 750.000). The yearly maintenance costs for the trap are THB 15.000 (EUR 390). Until the next dredging in 20 years the overall costs for the sediment trap will be THB 1.34 Mio (EUR 34.680). Therefore, the overall costs for the sediment trap (dredging plus sediment poll plus maintenance) are THB 76.04 Mio, which is slightly higher than the dredging costs (THB 75 Mio). However, in 20 years the costs for the sediment trap will increase by the yearly maintenance rate of THB 15.000 (EUR 390), while without the trap further dredging is needed. The dredging costs again THB 75 Mio (EUR 1.95 Mio) neglecting inflation rate and price increases. After 21 years the overall costs of the sediment trap is significantly lower than the dredging costs—THB 76.05 Mio (EUR 1.98 Mio) vs. THB 150 Mio (EUR 3.9 Mio). The graph below shows the investments over 100 years. After 80 years the reservoir must be dredged even with the sediment trap measure being implemented, because the reservoir is filled with sediments. But until then, without the implementation the reservoir would have been needed to be dredged 4 times already.

As mentioned above, the reduction of water storage capacity can be lowered by the installation of a sediment trap, because less sediment is flowing into the reservoir. The sediment trap can reduce the sediment inflow by 75% (Lohr 2016). Hence, the storage capacity remains relatively high compared to the business as usual scenario (no sediment pool, but dredging every 20 years). While the water storage capacity is

EbA	Wetland restoration, sediment trap, nutrient trap
Location	Immediate upstream of Nong Yai Reservoir
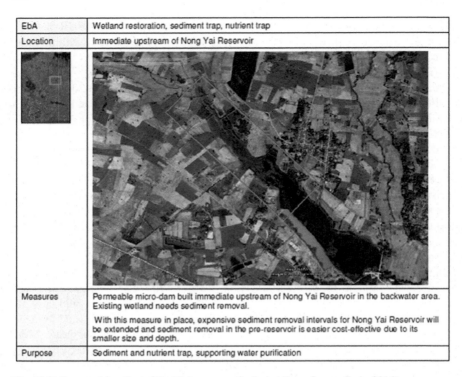	
Measures	Permeable micro-dam built immediate upstream of Nong Yai Reservoir in the backwater area. Existing wetland needs sediment removal. With this measure in place, expensive sediment removal intervals for Nong Yai Reservoir will be extended and sediment removal in the pre-reservoir is easier cost-effective due to its smaller size and depth.
Purpose	Sediment and nutrient trap, supporting water purification

Fig. 43.3 Proposed location of the EbA measure Sediment Trap; *Source* (Lohr 2016)

reduced by 25% within 20 years in the business as usual scenario, it needs 80 years to lose the same volume, when a sediment trap is installed (see graph below).

The absolute numbers (graph above) are indicating a higher storage volume when a sediment trap is installed. The average water storage volume per year should confirm this observation. For that the yearly water volume is calculated, added up and the aggregated numbers are divided by the years. With such a calculation the average water storage capacity of the business as usual scenario and the sediment trap installation can be compared (graph below).

The graph clearly shows that at any time the yearly average water storage capacity is higher when a sediment trap is installed. That means that more water is available, which further leads to higher economic output.

Calculating the impacts of the sediment trap installation on reduced water storage capacity and further on the lower yields of the farming as done above with the business as usual scenario (dredging every 20 years) demonstrates a remarkable advantage of the sediment trap.

Using the same input data and assumption as above (business as usual scenario), the overall economic losses are far lower despite a higher initial investment for dredging and installation of the sediment trap. While in the business as usual scenario 9.4 ha per year are affected by the lower water storage, the affected area can be reduced to 2.4 ha per year (−75%). Only 2.4 ha per year cannot be irrigated, which leads to much lower negative economic impacts. Assuming that the yield will be minus 10% the rice farmers lose THB 5.860 (EUR 152; 0.03% of the values produced in the whole area) per year, in 5 years THB 29.280 (EUR 760; 0.16%) and in 20 years THB 31.950 (EUR 3.040; 0.62%). On aggregate the losses will be THB 87.850 (EUR 2.280; 0.09%) after 5 years and THB 1.23 Mio (EUR 31.950; 1.31%) after 20 years. Again, assuming the yield will fall by 30% the negative economic impact will triple.

The table below compares the losses occurred by business as usual scenario (dredging) and by EbA scenario (sediment trap) (Table 43.2).

The benefits of the EbA measure compared to dredging increases over time, because the costs for dredging occur regularly and remain relatively high, while the overall costs for the sediment trap are rather low. The graph below compares the costs for 1 m³ of water between the EbA scenario (sediment trap) and the business as usual scenario (dredging) and shows the cost advantage per 1 m³ of water storage in %. In this case study the economic advantage of EbA measures increases in time and peaks at 75% in year 80, just before another dredging of the reservoir is needed despite the sediment trap. In other words, the costs per m³ of water can be reduced by up to 75%, if a sediment trap as an EbA measure is installed.

Over a time span of 100 years dredging costs THB 375 Mio. On average the water storage capacity is about 2.65 Mio m³ per year or 265 Mio m³ for the whole period. Hence, 1 m³ of water costs THB 1.42. With a sediment trap the overall costs can be reduced to THB 152 Mio, while the average water storage can be increased to 2.71 Mio m³. This corresponds with costs of THB 0.56 per m³ (Fig. 43.7).

Ecological Sustainability

The economic analysis demonstrated a clear advantage of the EbA measurement over the dredging. However, the economic analysis does not indicate, if the economically advantageous scenario (EbA measurement) is ecologically more sustainable than dredging. For the ecological analysis we introduce the

Innovative Evaluation Method
Intrinsic Ecological Value

The method is based on the photosynthesis, the oldest natural terrestrial process and the base for biodiversity. "The primary function of photosynthesis is to convert solar energy into chemical energy and then store that chemical energy for future use. For the most part, the planet's living systems are powered by this process." (https://www.canr.msu.edu/news/the_important_role_of_photosynthesis). Photosynthesis is not only one of the core natural processes, it reflects sustainability as the oldest terrestrial process.

Photosynthesis produces oxygen out of on three input parameters: sunlight, carbon, and water. Sunlight and the product oxygen are priceless, while carbon and water are priced based on various sources. For example, the price of carbon is based on the Emission Trading System and the costs of reducing emissions; water prices are informed by externalities and water tariffs.

Table 43.2 Direct economic impacts of water losses on rice farmers

Agriculture Rice, Thailand 2014 (FAO)		
Yield per ha	3.01 t/ha	481.6 kg/rai
Producer Price per ton	8,280 THB/t	240 US$/t
Total Irrigated Area	4,700 rai	752 ha

	Dredging	Sediment Pool
Water storage loss	38,069 m3	9,517 m3
Effected Area	58.75 rai	14.69 rai
	9.4 ha	2.4 ha
Yield reduction	10 %	10 %
Yearly loss	-23,427 THB	-5,857 THB
	-609 EUR	-152.13 EUR
in 5 years	-117,137 THB	-29,284 THB
	-3,043 EUR	-761 EUR
in 20 years	-468,549 THB	-117,137 THB
	-12,170 EUR	-3,043 EUR
Yield reduction	30 %	30 %
Yearly loss	-70,282 THB	-17,571 THB
	-1,826 EUR	-456.38 EUR
in 5 years	-351,411 THB	-87,853 THB
	-9,128 EUR	-2,282 EUR
in 20 years	-1,405,646 THB	-351,411 THB
	-36,510 EUR	-9,128 EUR
Aggregated numbers		
Yield reduction	10 %	10 %
Loss after 1 year	-23,427 THB	-5,857 THB
	-609 EUR	-152.13 EUR
after 5 years	-351,411 THB	-87,853 THB
	-9,128 EUR	-2,282 EUR
after 20 years	-4,919,761 THB	-1,229,940 THB
	-127,786 EUR	-31,946 EUR
Yield reduction	30 %	30 %
Loss after 1 year	-70,282 THB	-17,571 THB
	-1,826 EUR	-456.38 EUR
after 5 years	-1,054,234 THB	-263,559 THB
	-27,383 EUR	-6,846 EUR
after 20 years	-14,759,282 THB	-3,689,821 THB
	-383,358 EUR	-95,839 EUR

The core activities of nature used for the proposed approach are photosynthesis and water storage. Ecosystems store water and provide photosynthesis by absorbing CO_2. Carbon has a market value and price and water is considered as an economic good (Dublin principles 1994, https://www.gwp.org/contentassets/051 90d0c938f47d1b254d6606ec6bb04/dublin-rio-principles.pdf). The quantification of both carbon sequestration and water storage is possible by means of standard and well-established biological and hydrological methods and tools. The other two parameters of photosynthesis—sunlight as an input and oxygen as an output—can be neglected economically because they are provided (sunlight) or produced (oxygen) for free. However, the net volumes of carbon (absorption minus emission) and water (provision minus use) can be calculated. Carbon absorption and water provision are assets, while carbon emission and water use are liabilities. Such core services are provided by all ecosystems, irrespective of their kind or location.

Since carbon has a market price, a value can be allocated to its absorption. Further, water is withdrawn from the soil during the photosynthesis. Photosynthesis is a natural process and the economic value of photosynthesis should be zero. Hence, the positive value of the carbon absorption should offset by the negative value of the water use, which further leads to a monetary value of water. Therefore, water obtains an economic value as well. It is obvious that ecosystems are exposed to the climate and an IEV is directly connected to natural processes and phenomena like precipitation and droughts as well.

This method which is based on the oldest and most sustainable terrestrial process describes sustainability and the impacts of any process on nature. Hence, it can be used for measuring the sustainability of any process, which will be shown below. In addition, the IEV tool was applied to analyzing changes of the dredging periods.

The IEV is applied on both scenarios. The BAU scenario foresees dredging every 20 years, while the EbA measurement extends the dredging period by four times (every 80 years). In addition to the sediment pond the banks of the reservoir will be covered with vetiver grass with a width of 2 m. The same water volumes as shown in Fig 43.5 and 43.6, and the same investments costs as presented in Fig. 43.4 are used. In the EbA scenario the carbon sequestration benefits of the vetiver grass are added. We have chosen 1.45 kgC m^{-2} y^{-1} carbon sequestration benefit, which is the average of the range from 1.37 to 1.53 kgC m^{-2} y^{-1} published by Nopmalai et al. 2015.

In a first step, IEV of each scenario was calculated with the average water availability shown in Fig. 43.6 and a water price based on externalities of EUR 0.01334 per m^3. The carbon benefits are quantified according to Nopmalai et al. 2015 (1.45 kgC m^{-2} y^{-1}) and monetarized with the carbon price of US\$ 25, which was traded at the future exchange on October 1st, 2019 (https://www.investing.com/commodities/carbon-emissions-historical-data). We used a EUR/US\$ exchange rate of 1.09.

Formula: (C abs—C emi) * CP + (W volume—W loss) * WP.

Whereas

C abs = Carbon absorption in t
C emi = Carbon emission in t
CP = Carbon price in EUR per t (US\$ 25 = EUR 22.94)
W volume = Water storage capacity of the reservoir
W loss = Water losses by sediments
WP = Water Price (EUR 0.01334 per m^3)

The IEV, which describes the sustainability of the process, is decreasing over years in both cases because of the water storage losses caused by the sediment. However,

Investment in THB

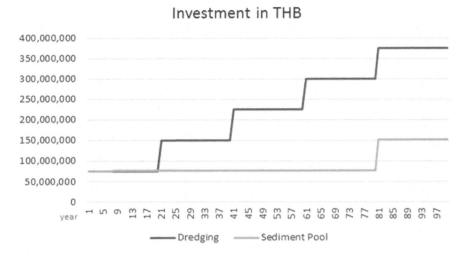

Fig. 43.4 Investments in THB (y-axis) over a period of 100 years (x-axis) for dredging and sediment pool

Waterstorage in m3 per year - absolute number

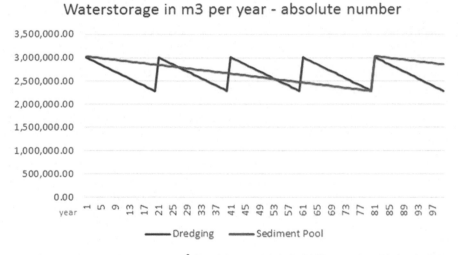

Fig. 43.5 Water storage capacity in m³ (y-axis) over a period of 100 years (x-axis) for dredging and sediment trap

the decrease in the BAU scenario (dredging) is bigger (~2.5% per year) than in the EbA scenario (~0.4% per year).

The ecological value based on the IEV describes the sustainability. A positive number reflects an ecologically sustainable situation, whereas the higher the number the more sustainable, while a negative number stands for an unsustainable situation.

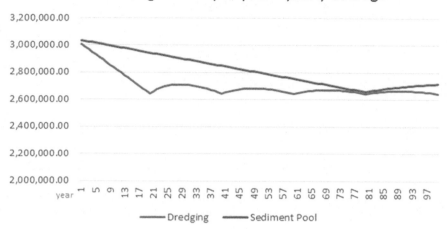

Fig. 43.6 Average water storage capacity per year in m³ (y-axis) over a period of 100 years (x-axis) for dredging and sediment trap

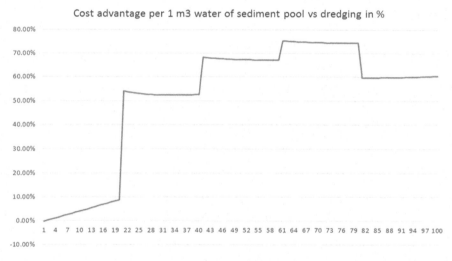

Fig. 43.7 Cost advantage for 1 m³ of stored water compared between sediment trap and dredging in % (y-axis) over a period of 100 years (x-axis)

In summary, both scenarios (BAU, EbA) are sustainable. However, the EbA approach has still an advantage over the dredging, because the IEV values are higher. Over the time, the advantage of the EbA approach increases (see Fig. 43.8). The ecological sustainability of the EbA scenario is about 3% higher in year one and increase to 77% in year 19 compared to the dredging.

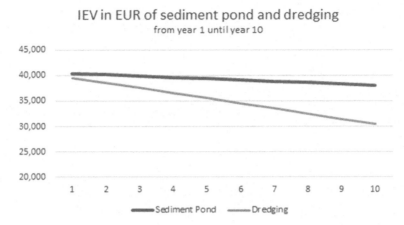

Fig. 43.8 Intrinsic Ecological Value in EUR of Sediment Pond and Dredging from year 1—year 10

As mentioned above the IEV reflects the ecological sustainability of a project or scenario. The ecological sustainability can be measured in absolute numbers by the IEV, which further can be analyzed and compared to other projects, measurements and scenario. This comparison offers powerful opportunities because different, diverse projects, measurements and scenarios can be compared with each other. This comparative approach offers additional opportunities. In our case, we used the IEV to optimize the dredging periods. We wanted to find the ecological most sustainable dredging period. It can be said that the shorter the dredging periods the more ecological sustainable the measurements become. The effects are bigger and immediate in the BAU or dredging scenario. The improvements in the EbA scenario are gradually and occur in the mid and long term.

The ecological sustainability can be analyzed economically, too. The IEV produces ecological values for each scenario, the costs for the scenarios are known as well. Hence, we can calculate how much the ecological sustainability costs: costs for dredging divided by IEV (dredging) or costs for EbA divided by IEV (EbA).

From year 1 until year 10, the accumulated costs in the dredging scenario are THB 75 m (~EUR 1.95 m). The accumulated ecological values for the dredging scenario are THB 13.5 m (EUR 0.35 m) for the period year 1 to year 10. By dividing the accumulated costs by the accumulated IEVs, we get the costs per one IEV. One THB of IEV costs 5.56 THB in the BAU scenario. In the EbA scenario one THB IEV costs 5.09 or 8.4% less than in the dredging scenario. This costs per IEV decreases in both scenarios over time. While the minimum costs in the dredging scenario are THB 3.25 per IEV, the costs in the EbA scenario falls to THB 1.11 per IEV. The costs advantage of the EbA compared to the dredging scenario are up to almost 70%.

43.4   Conclusion

The case study demonstrated that an EbA measure (wetland restoration as sediment trap) has a clear economic advantage to the grey measurement (dredging) even from year 1 on, although the initial investment is slightly higher. The higher installation costs are compensated by higher storage capacity, which leads to higher economic output of the farmers. Therefore, the EbA measure has an immediate positive and economically direct benefit, which increases over years.

Beside the economic benefit it became clear that such a measure has also a favorable impact on drought and flood management. Obviously, more water storage capacity can reduce the negative impacts of drought. With the higher storage capacity, the overflow of the reservoir can be delayed during heavy rain, which gives more time for preventive action downstream.

The Intrinsic Ecological Value is a simplified evaluation method based on few data that generates values for ecological sustainability. It was demonstrated that the EbA measurement has an ecological advantage over the dredging. (This ecological benefit can be achieved at much lower costs per IEV (up to 69%).

References

Buyck C et al (2017) Ecosystems protecting infrastructure and communities: lessons learned and guidelines for implementation, Gland IUCN

Meier G et al (2015) Economic evaluation of proposed ecosystem-based adaptation measures in tha Di and Chi river basins. Bangkok, Cologne

Lohr H (2016). Vulnerability analysis for the river basins of Huai Sai Bat, Tha Di and Trang. Bangkok, Weinheim

Munang R et al (2013) Climate change and ecosystem-based Adaptation: a new pragmatic approach to buffering climate change impacts. Curr Opin Environ Sustain 5:67–71

Pérez A, Muñoz MMM, Páez KS, Triana JV (2010) The ecosystem approach and climate change adaptation: lessons from the Chingaza Massif, Colombia

Nopmalai P et al (2015) Study on carbon storage and carbon balance in vetiver grass cultivation areas in Northern Thailand, Land Development Department, Chatuchak, Bangkok, Thailand

Part VII
Wastewater Engineering and Management

Chapter 44
Coping with Salinity Intrusion in South-Western Bangladesh: The Continuing Struggle

Mokbul Morshed Ahmad and Muhammad Yaseen

Abstract Bangladesh is one of the major countries vulnerable to climate change. The country is particularly at risk because of its geographical location, climatic variability, global warming and sea level rise. Sea level rise, frequent storm surge and recent cyclones such as A1LA and SIDR are intensifying salinity affecting the livelihoods of coastal region people especially in the south western part of Bangladesh. Along with salinity intrusion, corruption by local leaders and government officials, exploitation by big shrimp farmers, absence of drinking/sweet water have adversely affected local people. This paper explores the alternative mechanisms adopted by coastal communities to cope with increased salinity.

Keywords Salinity · Climate variability · Climate change · Adaptation

44.1 Introduction

This paper sets out to explore the alternative mechanisms adopted by coastal communities to cope with increased salinity. Bangladesh is recorded as one of the major countries vulnerable to climate change and also climate hazards, alongside the coastal area in the exposure frontline (Huq and Ayers 2008). The area has more than 47,000 km^2 or close to a third of the land area of the nation, which accommodate more than 35 million individuals within 6.85 million households or 28% of the entire population (BBS 2012). Moreover, the coastal region consists of many mangrove ecosystems, particularly the Sundarbans mangrove forest, including transitional zones between the marine water and fresh water (World Bank 2000). The ecosystems, water, and land of coastal regions are critically influenced by tropical cyclones, sea-level rise, salinity,

M. M. Ahmad (✉)
Department of Development and Sustainability, Asian Institute of Technology (AIT), Klong Luang 12120, Pathum Thani, Thailand
e-mail: morshed@ait.ac.th

M. Yaseen
School of Economics and Management, Neijiang Normal University, Neijiang, Sichuan, China

© Springer Nature Switzerland AG 2021
M. Babel et al. (eds.), *Water Security in Asia*, Springer Water,
https://doi.org/10.1007/978-3-319-54612-4_44

and storm surges. The south-west coastal area of Bangladesh is a special environment whereby a massive floodplain is supported to facilitate a profitable, but dangerous agro-ecosystem. The joint dependence on scarce natural resources, particularly water and land, including shared exposure to most natural hazards, particularly cyclones, are major attributes of this area. There are also two main trends, which have impacted and will always affect this region's natural resources. First, the water resources and coastal land resources, which have gone through some physical changes because of the key development interventions, particularly shrimp farming and pulverization, are developing both environmental hazards and economic opportunities. Second, climate change is presently impacting rural livelihoods as well as has huge effects in many decades to come. Salinity intrusion has turned out as an acute issue for the coastal Bangladeshi people, which will be aggravated by the rise in sea level and climatic variability. Also, salinity intrusion with respect to the decrease in freshwater that flow from the upstream, soil salinity fluctuation and salinity of groundwater are the main issues of concern for the South Western Bangladeshi people. Many studies have proven that Bangladesh is presently encountering the worst effects of saline intrusion because of the rise in sea level. The variability in the level of soil salinity has affected around 830 million hectares of productive land. The Bay of Bengal produces salt water, which is reported to enter 100 km or additional inland across the tributary channels amid the dry season as well (IPCC 2007). This will have some effects on agriculture (livestock, crops, and fisheries), water (drinking water availability, water logging), energy, livelihood pattern, human settlement, health of the communities in the coastal regions of Bangladesh.

Along with salinity intrusion, corruption by local leaders and government officials, exploitation by big shrimp farmers, absence of drinking/sweet water etc. have adversely affected the local people (Nasreen 2009). Climatic variability and frequent natural disasters are not only restraining and limiting the adaptability and normal livelihoods of the coastal people but also deteriorating the situation in the coastal regions. Salinity intrusion creates vulnerability among different groups of people in the society. The rural poor and disadvantaged people become day laborers who collect wild shrimp larvae (or fry) from coastal rivers and marshes. As salinity increases in the river, fewer fries are available in the open water bodies. The increasing salinity thus decreases the productivity of land and pushes the community members to unemployment and poverty. In the context of above background, the study was initiated to focus on the adaptation mechanisms and capacity of the people in mitigating the risk of salinity induced disasters.

44.2 Rationale of the Study

A detailed research conducted by Rawlani and Sovacool (2011) regarding the Bangladesh's vulnerability to climate change, recognized five major sectors that are critically susceptible to climate change. Coastal zones and even the water resources have been discovered as the major critical areas, since these are affected by changes in coastal morphology, saltwater intrusion, damage from natural disasters, and drainage congestion. Human settlement and infrastructure are prone to be affected and the development of about 25 million climate refugees who will come from the coastal districts have been projected. The Bangladeshi government records that the production of wheat and rice may be reduced by 2050 by 32% and 8%, respectively. Moreover, the Sundarbans, mangrove forest, can probably be critically affected; the health of humans in this region will also suffer (Rawlani and Sovacool 2011). Climate Change projects and programmes have been initiated in the recent years as funds have been earmarked for adaptation and mitigation projects and programmes. Some of the projects focus of infrastructural aspects such as building dykes and embankments for example, a total of 5,695 km of embankments, including 3,433 km in the coastal areas, 1695 flood control/regulating structures, and 4,310 km of drainage canals have been constructed by the Bangladesh Water Development Board (BWDB) during the last several decades (Shaw et al. 2013). Other programmes have developed cyclone shelters, disaster management mechanism at local level as well as disaster education programmes. Thus, endeavors have been undertaken to support disaster management, adaptation and now, to some extent, also mitigation programmes.

There has been some progression in the practical execution, when the required measures taking care of climate change risks are included within the current decision structures that relate to livelihood enhancements, land use planning, risk management, water management, including other development initiatives, resource management systems, that bring about "mainstreaming" adaptation. The economic resources can possibly support the execution of a new technology and guarantee the access to training opportunities and can possibly bring about higher political impact (Smit and Wandel 2006). A major case from Bangladesh exhibits whereby a community-based adaptation project has helped in raising and reinforcing homesteads, making them highly resilient against cyclonic activity and flooding activity, decreasing the need for individuals to leave their homes amid critical weather conditions and decreasing losses as well (Ayers and Huq 2009).The focus of the study is to explore the vulnerabilities of coastal communities to salinity related disasters and adaptation strategies to cope with it. Due to the geo-socio-economic context of coastal zone, the recurrent climatic variability and natural disasters put coastal communities entirely at risk. The majority of households that were part of the sample villages rely upon climate-sensitive fisheries and agriculture for their daily living. To a reasonable amount, their environment was defined based on saltwater intrusion, erratic rainfall, freshwater scarcity, waterlogging, critical climate hazards, and riverine erosions—these entirely have some impacts on the livelihoods of households. The sufferings induced by salinity intrusion further get worsened due to some manmade activities. It is evident from several studies that the socio-economic consequences of salinity intrusion in water are responsible for creating different types of vulnerabilities such as food crisis, drinking water crisis, diseases etc. Though Bangladesh has achieved a remarkable success in managing disaster risk reduction, unfortunately the social protection policy is not adequate to mitigate these vulnerabilities (Nasreen 2010). The community based salinity risk reduction strategies and dynamics of household and individual adaptation strategies with the pre-existing socioeconomic and geographical setting of the coastal communities have always been overlooked. Besides, it is also noticeable that very few studies have emphasized to focus on the dynamics of the adaptation strategies to salinity intrusion. The present study thus, has been designed to find out the alternative livelihood that can minimize individual's vulnerability to salinity intrusion. In this context, this study is a modest endeavor to explore the vulnerabilities the community faces during salinity intrusion and adaptation strategies.

44.3 Methodology

In order to get an insight on the nature of vulnerability and disaster experiences, the study followed a combination of quantitative and qualitative methods. A questionnaire was used to measure variables related to socio-economic condition to assess mitigation efforts and adaptation strategies taken. We have conducted some case studies as part of qualitative method to understand the experiences of the affected people. To adapt with salinity intrusion the coastal people follow some traditional coping strategies which cannot be measured by only quantitative approach. For example, some vulnerability of women in the locality generally remains unexplored. In this case, the qualitative approach is suitable to explore the real situation and vulnerabilities of women and their role and adaptation strategies.

The method of sampling for the research was not random rather selective based on purpose of the research. The target respondents for this study comprised of both women and men from nine villages of two different sub-districts namely Paikgacha and Kaligonj from Khulna and Shatkhira district as a coastal area in the southern region of Bangladesh. A total sample of 190 respondents was interviewed from 190 households from the study population and 5 case studies were conducted.

44.4 Findings and Discussion

44.4.1 Salinity Intrusion and Shrimp Cultivation

The government and many large entrepreneurs, back in the '70s considered shrimp a crop that is commercially valuable, and during the '80s and '90s, there was a boom in this industry. There were two major factors that contributed to this boom. First, during the '70s, there was a huge demand from the international market, while there was also a rapid increase in the costs of shrimp and other sea products. Second, during the '80s, the World Bank and the International Monetary Fund urged Bangladesh to take on the export-oriented agricultural policies with respect to the successive structural adjustment programs at the time (Paprocki and Cons 2014). The entire '80s experienced an era of commercial production of shrimp, which was facilitated and funded by key development agencies and global banks, alongside loans of around $30 Million USD, to fortify the supply chains directly to the entire global markets in particular (Adnan 2013). There were some steps identified by the national government towards the expansion of shrimp cultivation as seen in the Second Five Year Plan between 1980 and 1985 so as to generate employment, food, income, as well as related benefits to people living within the rural communities including the national economy as well. Meanwhile, the coastal embankments were set up to support the increased productivity in agriculture and were changed into a mechanism to ease the production of shrimp at the detriment of agriculture. For more than these 20 years to year 2008, the dominant land-use was the result of the brackish-water shrimp culture within the coastal zone, while there was a rapid rise in the frozen shrimp export industry of Bangladesh (Paul and Vogl 2011). Bangladesh's shrimp farming is very productive as opposed to rice farming; which was the traditional economic activity. According to an estimation by Miah and Bari (2002), the net returns provided by the enhanced types of shrimp cultivation within the Southwest coastal area amount to between BDT 90,000 and 111,000 per ha, just as the wet-season rice alongside modern varieties provided net returns between BDT 8,500 and 10,000/ha. A key challenge impacting on the small and medium farmers was based on the fact that the massive shrimp farmers were unable to run out the brackish water when necessary for the cultivation of rice during the wet season. Nevertheless, the appropriateness of shrimp aquaculture has been queried because of the accompanying ecological results (disease outbreaks, mangrove destruction, and saltwater intrusion), poor economic effects, including impacts on social relations. Most especially, the cultivation of shrimp in Bangladesh has resulted in critical conflicts existing between and within the entire local communities, agribusiness investors, state actors, and absentee landowners over the access and control on land. Salinity intrusion due to shrimp cultivation has had an adverse impact on the society and it affects the people in different ways. Hossain and Nasreen (2012) identified that coastal people of the southwestern Bangladesh are facing several challenges including social network, women's income and employment, migration to elsewhere, gender insecurity, spread of vector and water borne diseases etc. due to salinity intrusion see also Islam and Shamsuddoha (2017). Cropping in any of the seasons became impossible because of the decline that occurred to rice yield as well as other crop yield. Since the livelihood alternatives for landless workers and smallholders living in the village diminished, many relocated to the urban areas. However, we found that most people were unable to cope with the changing situation of salinity intrusion. The coastal communities are fighting to keep pace with the adverse situation posed by salinity. In order to get rid of it, the coastal people are now trying to cultivate portable water fish (8.5%) and they are also building high embankment to protect themselves from saline water. Shrimp cultivation has also created potable water crisis. The affected people are now installing deep tube-well for sweet water, using pond water and also drinking water from pond water filter. Occupational problem due to shrimp cultivation and salinity in the locality has become a major concern. The coastal people are now changing their occupation and migrating to other places especially to Dhaka and joining the Garment factories.

Since salinity levels are likely to change due to sea level rise, the construction of polders started in the 1960s under the Coastal Embankment Project to protect the area from flood and to prevent saline water intrusion (MOWR 1999). Vulnerability in the coastal zone would be acute due to coastal embankments along with the combined effects of climate change, sea level rise, changes of upstream river discharge and cyclone (World Bank 2000). we found that though the coastal embankments were set up to prevent the devastating impact of frequent disasters and climate change, it is being misused by local elites, political leaders and local representatives for shrimp cultivation where it has created several problems for the coastal people. It was found that most people are yet to take any measure to remove the problem induced by embankments. However, the different types of strategies were adopted by the affected people to mitigate the impacts of embankments. The survey shows that 6% of the respondents were using chemicals to reduce the impact of salinity whereas only 3% collect water from the tube well. Besides, it was also observed that 4% of the respondents organized social movements against shrimp cultivation.

44.4.2   Agriculture and Risk Reduction Process

The increasing importance of agricultural productivity led to the establishment of the coastal embankments, which were changed into a mechanism that will generate shrimp at the detriment of agriculture. For more than 20 years, the culture of blackish-water shrimp was prevalent in the coastal zone and this had a great impact on the export industry of Bangladesh's frozen shrimp. Traditional agriculture were no longer valued as shrimp farming dominated the area and led to environmental degradation, most especially through freshwater scarcity and increased salinization, it also affected native vegetation, livestock, fish, and crop yield. A research was carried out by Islam et al. 1999 on the salinity level of shrimp; he compared the salinity levels of both shrimp and non-shrimp areas and the result showed that shrimp farming could increase the salinity level of the soil by almost 500% with great constrain on the production of agriculture. According to Haq 2000, shrimp farming also affected groundwater. The socio-economic groups in the coastal area were affected by shrimp farming practice and it also led to considerable numbers of conflicts as the farmers and wage labourers were getting poorer and just a few business owners and influential landowners were getting richer. In the late 2000s, there was a political shift that offered poor farmers the opportunity to recover their land from large shrimp farms, these farmers could either revert into cropping during dry season or shrimp farming on a very small scale. However, small scale farmers sustained high vulnerability concerning the legacy of the shrimp boom. Salinity intrusion is causing reduction of agricultural production due to unavailability of fresh water, soil degradation, terminative energy and germination rate of some plants (Rasid and Haider 2003). In this study majority of the respondents (89%) argued that agricultural crops are at high risk due to salinity in the coastal regions. The different strategies and initiatives were taken by the affected people to reduce the economic and other problems. It was found that around 12% of the informants shifted their harvesting time so that the crops could grow properly while 11% of the respondents tried to cultivate crops within short time, and more than 3% cultivated maize to mitigate their economic losses. The salinity affected farmers were also cultivating hybrid saline tolerant crops provided by the NGOs (11.6%), indigenous saline tolerant crops (2.8%) and homestead vegetables (6.1%). The salinity affected people also used pond and tube-well water for production of crops. As one woman mentioned:

> "Minimum steps taken by government and NGO cannot reduce vulnerability due to nepotism and localization by the elite class who mainly control the politics and have good communication with the representatives. So, I sometimes adopt self-motivated method to cope with salinity in water. I cultivate vegetables to generate income during dry season and this brings empowerment." Fatima (28).

44.4.3   Coping Mechanisms to Overcome Food Insecurity

Salinity intrusion has an adverse impact on livelihood (Dalby 2002). The shrimp culture greatly affected the fisheries and livestock. Grazing land were reduced by shrimp farming, including fresh water usable by livestock during the dry season was reduced as well. Nearly every form of natural vegetation were ruined because of this salinity, hence, scarcity of fodder for the cattle, which made the farmers left with no other option than to lease or sell them. Moreover, the owners of the shrimp farm additionally disallowed the rearing of ducks since they were also consuming the shrimp. Concurrently, the high destruction and salinity of the fish fry amid the gathering of shrimp fry decreased the natural fish species availability. Those relying on fishing or livestock rearing for their income and/or consumption lost their essential livelihood sources. Ahmed et al. (2007) also illustrated that the food security of the salinity affected people is found to be completely shattered because of non-availability of land-based productive system and lack of sustained flow of income. The local economy is so much stressed and therefore, no appreciable employment is generated in the area. People, man and women of virtually all ages, religion and sect use handhold push-nets to catch small fish, mostly shrimp fries.

In order to overcome the food insecurity, the coastal people are taking several strategies to survive. We found that more than 21% of the household borrowed their necessary food from relatives and neighbors and nearly 14% households cultivated vegetables in the yard. On the other hand, about 9% of the households tried to introduce poultry farming. But, it should be mentionable that a significant number of households could (45.4%) hardly manage their food to survive and the managed amount of food was meager compared to their necessity.

The households in the studied villages were additionally getting their livelihoods diversified. Bigger farmers began to invest more in other activities not relating to farming, such as large-scale fish culture as well as other businesses. The small and medium scale farmers began small-scale fish culture, poultry, tree plantations, petty businesses, and homestead gardening. A few of the middle farmers were already investing their money to purchase an "easy bike" and motorcycle in order for a household member to get some income to provide transportation to other people within the village. Many landless and small farmers had their participation in daily wage labor, rickshaw or van pulling. Poor household women began to work as wage laborers, while the women from the small and medium farm households began to make handicrafts for sale, like local mats.

44.4.4 Strategies to Meet the Sweet Water Crisis

People were unable to consume ground and surface water, which was specifically meant for drinking, in many areas in Southwestern Bangladesh, ever since the proposed salinity level went beyond the 960 micromhos/cm level for potable water in the community (Roy 2004). Numerous households ensured that ponds are within their homesteads, which were utilized for domestic purposes, particularly for drinking during the dry season. Nevertheless, after the shrimp cultivation was introduced, the pond water turned out extremely salty to utilized, even to bath. Because of the shrimp cultivation, people had to move to another village to get fresh drinking water during the dry season. It was found that the coastal people encountered severe sweet water crisis. They solved the water problem through adopting different strategies. More than 29% of the respondents preserved rainwater while more than 18% respondents used pond snow filter and 1.2% preserved water in pitcher. It was also found that nearly 28% of the respondents were collecting pure drinking water from a distance and only 4.1% of the respondents used deep tube-well (Table 44.1). Besides these strategies, the coastal people used water filter to remove salt from the water, used water purifying tablet and made water tank. Even, some had to dig a ditch and preserved water there and also kept the ditch clean. A housewife told:

> "Before salinity intrusion, we could grow vegetables and earned extra money. But now it is impossible to grow vegetables due to saline water. I collect drinking water from the pond of my neighbor. The shrimp cultivators hatch snail as shrimp food in the pond. So, the water of

Table 44.1 Strategies taken to solve Sweet Water Crisis

Strategies	Frequency	Percent
Preserving rainwater	50	29.4
Using PSF water	31	18.2
Preserving water in pitcher	2	1.2
Collecting pure water from distance	47	27.6
Using water from deep tube-well	7	4.1
Using tube well through filtering	18	10.6
Construction of water tank	7	4.1
Using alum for purifying water	4	2.4
Renovated and reserved pond	4	2.4
Total	170	100.00

Source Field work

that pond is no longer fit for drinking. Now, I have to fetch water from far away. It is a great problem for us." Rizia (48).

44.4.5 Actions to Face Salinity Induced Diseases

The climatic change had a direct and indirect effect on humans. The major direct effect is through extreme events. Public health relies upon getting adequate food, safe drinking water, and other environmental conditions—these conditions can be possibly affected by a change in climate. The health impacts of a quickly changing climate are probably going to be shockingly negative, especially in many of the poorest areas. The increased salinity distribution and density had made different germs to get comfortably habitat and are able to spread in this coastal region. According to Chatterjee and Huq (2002), the majority of the main epidemics that have taken place over the past 50 years emanated from the coastal areas. We found that due to saline water more than 71% of the respondents encountered several types of diseases. The salinity affected people suffered from diarrhea (73.3%), cholera (5.2%), jaundice (20%), malnutrition (25.9%), skin diseases (27.4%), gastric (14%) and fever (1.5%) etc. So, it could be concluded that saline environment has close association with different diseases.

The salinity affected people took several mitigation strategies to cope with water borne diseases. More than 11% of the respondents used herbal medicine while a significant number of respondents (54.1%) took treatment from local village doctor (Table 44.2). Besides, nearly 2% of the respondents took saline whereas around 9% used soda to get cured from diarrhea. It should be mentionable that twenty percent of the respondents remained without any treatment due to poverty. On the other hand, more than 4% were unable to get proper treatment because of financial constraints. So, it can be argued that the salinity has created an adverse condition among the coastal people.

Table 44.2 The effects of Salinity on Health

Action	Frequency	Percent
Herbal treatment	15	11.1
No treatment	27	20.00
Unable to get treatment due to financial restraints	6	4.4
Local quack doctors	73	54.1
Taking soda	12	8.9
Taking drinking saline	2	1.5
Total	135	100.00

Source Field work

Table 44.3 Types of strategies taken by local representatives

Types of strategies	Frequency	Percent
Campaigning to raise awareness	19	46.3
Implementation of helpful measures	5	12.2
Teaching mechanisms to cope with existing situation	7	17.1
Applying local experiences	10	24.4
Total	41	100.00

Source Field work

44.4.6 Salinity and the Role of Local Representatives

Along with the community members, the local representatives are taking initiatives though their roles are not significant to reduce the risk and vulnerability to salinity intrusion. Only 22% of the respondents agreed that the local representatives are taking adaptation options to save the locality from the salinity intrusion whereas a significant number of the respondents (78.4%) told that the local representative did not take any measure against salinity intrusion and shrimp cultivation. However, it was stated by more than 46% respondents that the local authority was campaigning to raise awareness against salinity and shrimp cultivation whereas more than 17% opined that the local representative was teaching them how to cope with the existing devastating situation. On the other hand, more than 24% of the respondents stated that they were applying the indigenous experiences to prevent the salinity intrusion (Table 44.3). We found that initiatives taken either by local representatives or community members to cope with salinity at community level (such as community-based committee) are not organized at all to make the initiatives more effective. One woman told:

> "Shrimp cultivation has created anxiety as well as problems. There is no immediate step from the government or Non-government level organization to reduce the vulnerability. Though there are some steps taken to abate the vulnerability of people, those are not specific for women. Sometimes NGOs extend their hand by providing credit, but interest rate is not flexible. As a result, this credit cannot bring any fruitful outcome" Fatema (28), Khulna.

44.4.7 Salinity and Role of Non-Governmental Organizations (NGOs)

Along with government organizations, Non-governmental Organizations (NGOs) such as UTTARAN, BRAC, Nijera Kari, Ahsania Mission, Susilon, Red Crescent etc. were helping the coastal people to cope with salinity by constructing house, embankments, providing financial support, giving food, setting and reconstructing shelter center and generating employment opportunities. Besides, the non- government organizations also provide saline tolerant seeds and crops, water purification tablets, fresh drinking water and training on agricultural production etc. It is clearly seen that the non-government organizations are playing a role in developing local mechanisms of the people in case of protecting themselves from the salinity intrusion.

44.5　Concluding Remarks

This study focused on the mechanisms and alternative means for the people in the coastal areas to survive against salinity intrusion. However, the socio-economic condition, geographical location, climate variability and incessant disasters push the people of the coastal region of Bangladesh affected by salinity intrusion into further vulnerability and risk. From this study, it can be summed up that with the increase in number of disasters and climatic variability, all the alternative initiatives taken to reduce and get rid of vulnerability are very limited and quite insufficient compared to the necessity and level of vulnerability. The dimension and level of vulnerability of the coastal people to their livelihood and health hazards due to natural disasters often get worsened. The government of Bangladesh has already identified the coastal zones as 'vulnerable to adverse ecological process' and as one of the 'neglected regions' (Hossain and Nasreen 2012). But surprisingly no alternative mechanisms such as income generating activities, or any other initiatives have been taken by the government. The study reveals that very few and limited scale of programmers have been initiated and introduced by Non-Government Organizations and local voluntary organizations compared to the local necessity and those alternative means of livelihood is yet to be comprehensive in nature with wider coverage.

References

Adnan S (2013) Land grabs and primitive accumulation in deltaic Bangladesh: interactions between neoliberal globalization, state interventions, power relations and peasant resistance. J Peasant Stud 40(1):87–128

Ahmed AU, Neelormi S, Adri N (2007) Entrapped in a water world: impacts of and adaptation to climate change induced water logging for women in Bangladesh. Centre for Global Change (CGC), Dhaka

Ayers JM, Huq S (2009) Supporting adaptation to climate change: what role for official development assistance? Dev Policy Rev 27(6):675–692

BBS (2012) Statistical Yearbook of Bangladesh-2011, Dhaka: Bangladesh Bureau of Statistic, Statistics and Informatics Division, Ministry of Planning, Government of the People's Republic of Bangladesh

Brunori P, O'Reilly M (2010) Social protection for development: a review of definitions, European Report on Development

Chatterjee R, Huq S (2002) A report on the inter-regional conference on adaptation to climate change, mitigation and adaptation strategies for global change 7(4):403–406

Dalby S (2002) Environmental change and human security. Isuma Can J Policy Res 3(2):71–79

Haq RAHM (2000) Integrated wetland system for mitigation of the salinization of southwest region of Bangladesh', the paper presented at Eco Summit-2000. Halifax, Nova Scotia, Canada, pp 18–22

Hossain KM, Nasreen M (2012) Alternative livelihoods options in southwest region considering salinity intrusion. Paper presented on a seminar, May 2012 at CIRDAP Auditorium organized Voluntary Services Overseas Bangladesh (VSOB)

Huq S, Ayers J (2008) Climate change impacts and responses in Bangladesh: Note', Brussels, Belgium: European Parliament. DG Internal Policies, Policy Department Economy and Science

IPCC (2007) Climate change 2007: mitigation of climate change. Working group iii, fourth assessment report. Cambridge University Press: UK and New York, USA

Islam MA, Sattar MA, Alam MS (1999) Scientific report on impact of shrimp farming on soil and water quality of some selected areas in the greater Khulna District. Research and Development Collective, Dhaka, Bangladesh

Islam MR, Shamsuddoha M (2017) Socioeconomic consequences of climate induced human displacement and migration in Bangladesh. Int Sociol 32(3):277–298

Miah MG, Bari MN (2002) Ecology and management of the Sundarbans Mangrove ecosystem, Bangladesh. In: Sudhan M (ed) Managing transboundary nature reserves: case studies on Sundarbans Mangroves ecosystems, pp 1–44. UNESCO, New Delhi

MOWR (1999) National Water Policy. Ministry of Water Resources (MOWR), Government of the People's Republic of Bangladesh, Dhaka. https://www.warpo.org

Nasreen M (2009) Violence against women: during flood and post-flood situations in Bangladesh. Women's rights & Gender Equality Sector, Action Aid Bangladesh

Nasreen M (2010) Rethinking disaster management: violence against women during floods in Bangladesh. In: Dasgupta, S, Sinner I, Sarathi, P De Women's Encounter with Disaster. FrontPage Publications Limited, London

Paprocki K, Cons J (2014) Life in a shrimp zone: aqua- and other cultures of Bangladesh's coastal landscape. J Peasant Stud 41(6):1109–1130

Paul BG, Vogl CR (2011) Impacts of shrimp farming in Bangladesh: challenges and alternatives. Ocean Coast Manag 54(3):201–211

Rasid H, Haider W (2003) Floodplain residents' preferences for water level management options in flood control projects in Bangladesh. In: Mirza MMQ, Dixit A, Nishat A (eds) Flood problem and management in South Asia. Springer, Dordrecht. https://doi.org/10.1007/978-94-017-0137-2_5

Rawlani AK, Sovacool BK (2011) Building responsiveness to climate change through community based adaptation in Bangladesh. Mitig Adapt Strat Glob Change 16(8):845–863

Roy K (2004) Water logging in the south west coast. Reducing vulnerability to climate change project. CIDA-CARE-CDP, Climate Information Cell, Khulna, Bangladesh, p 68

Shaw R, Mallick F, Islam A (eds) (2013) Climate change adaptation actions in Bangladesh. Springer, Tokyo. https://link.springer.com/book/10.1007/978-4-431-54249-0

Smit B, Wandel J (2006) Adaptation, adaptive capacity and vulnerability. Glob Environ Chang 16(3):282–292

World Bank (2000) Bangladesh: climate change and sustainable development. World Bank, Washington, DC

Chapter 45
Reclaimed Wastewater Reuse for Irrigation in Turkey

M. E. Aydin, S. Aydin, and F. Beduk

Abstract Arid and semiarid regions have low precipitation levels and, therefore, in these areas water sources are limited. However, the water demand is very high because of high temperature and population. Due to limited or uneven precipitation and high temperature, over 80% of water demand is for agricultural use in Asia and Africa. Reclaimed water reuse options are important for both reducing freshwater demand and protecting water bodies from wastewater discharge. By employing wastewater in agricultural irrigation freshwater sources could be reserved for other uses, fertilizer requirement could also be reduced. Furthermore, discharge of reclaimed wastewater to receiving water bodies could be prevented. The reuse of wastewater for irrigation is considered as a solution to reduce the pressure on freshwater. In this work, general information about water distribution and wastewater treatment, discharge and reuse in Turkey is presented. Water usage in Konya region and the wastewater treatment and reuse practices are given. Along with brief results of scientific projects carried out on wastewater pollution monitoring and investigation of pollution caused by long term irrigation with untreated wastewater in the agricultural area of Konya are discussed.

Keywords Water scarcity · Wastewater reuse · Irrigation · Soil pollution · Crop contamination

45.1 Introduction

The reuse of treated wastewater for irrigation is increasingly being considered as a solution in water poor countries. Turkey is one of the water poor countries in the Middle East Region. Available annual water amount per capita was 1,600 m^3 in 2005, while it is estimated to decrease to 1,300 m^3 in 2023 (Karaaslan 2013). Agricultural irrigation accounts for about 70% of the freshwater use in Turkey, and high consumption of water sources for irrigation purposes impose a stress on water

M. E. Aydin (✉) · S. Aydin · F. Beduk
Necmettin Erbakan University, Konya, Turkey
e-mail: meaydin@erbakan.edu.tr

© Springer Nature Switzerland AG 2021
M. Babel et al. (eds.), *Water Security in Asia*, Springer Water,
https://doi.org/10.1007/978-3-319-54612-4_45

bodies. Treated wastewater in Turkey is discharged to surface water sources, and in some cases, these sources are used illegally by farmers for irrigation purposes for mitigation the water shortage.

Notification of technical rules for Wastewater Treatment Plants (WWTPs) regulates wastewater reuse criteria for irrigation purpose (NTR 2010). Reuse of treated wastewater is permitted for both non-processed and processed food in Turkish regulations. Categorization of treated wastewater for irrigation is given in Table 45.1. Treatment requirements, necessary parameters and monitoring periods are given according to the irrigation purpose. While secondary treatment, filtration and disinfection are necessary processes for reuse of wastewater for non-processed foods consumed directly and for recreational areas, secondary treatment and disinfection are necessary processes for reuse of wastewater for processed food, non-public green areas such as grass production, and grassland.

Sodium Adsorption Ratio (SAR) is given as critical parameter for treated wastewater irrigation. High SAR causes decrease of soil porosity, resulting in clogging of soil, and preventing the entrance of air and water into the soil. Plants have different tolerances to SAR, chloride and boron that are critical parameters for wastewater

Table 45.1 Categorization of treated wastewater for irrigation (NTR 2010)

Irrigation type	Treatment type	Quality of treated wastewater	Monitoring period	Application distance
Class A a: Agricultural irrigation for non-processed food; b: Recreational irrigation				
a) non-processed food irrigated by surface irrigation and sprinkler irrigation methods b) Recreational areas such as parks, golf areas etc	Secondary treatment Filtration Disinfection	pH 6–9 BOD < 20 mg/L Turbidity < 2 NTU Fecal Coliforms: 0/100 mL Specific viruses, protozoa and helmint in some cases Residual chlorine > 1 mg/L	pH: weekly BOD: weekly Turbidity: continuously Fecal Coliforms: daily Residual chlorine: continuously	Minimum distance to drinking water source wells
Class B a: Agricultural irrigation for processed food; b: Non-public green areas; c: Agricultural irrigation for non-edible foods				
a) fruit trees, grape yards irrigated by flood irrigation method b) grass production, monoculture agriculture, c) grassland	Secondary treatment Disinfection	pH 6–9 BOD < 30 mg/L SS < 30 mg/L Fecal Coliforms < 200/100 mL Specific viruses, protozoa and helmint in some cases Residual chlorine > 1 mg/L	pH: weekly BOD: weekly SS: daily Fecal Coliforms: daily Residual chlorine: continuously	90 m distance to drinking water source wells 30 m to public areas in case of rain type irrigation

Table 45.2 Parameters for chemical quality of irrigation water (NTR 2010)

Parameter	Units	Level of adverse effects		
		No effect	Small-medium	Dangerous
Salinity				
Conductivity	μS/cm	<700	700–3000	>3000
Total dissolved solids	mg/L	<500	500–2000	>2000
Permeability				
SAR	0–3	EC ≥ 0.7	0.7–0.2	<0.2
	3–6	≥1.2	1.2–0.3	<0.3
	6–12	≥1.9	1.9–0.5	<0.5
	12–20	≥2.9	2.9–1.3	<1.3
	20–40	≥5.0	5.0–2.9	<2.9
Specific Ion Toxicity				
Sodium (Na)				
Surface Irrigation	mg/L	<3	3–9	>9
Drip Irrigation	mg/L	<70	>70	
Chloride (Cl)				
Surface Irrigation	mg/L	<140	140–350	>350
Drip Irrigation	mg/L	<100	>100	
Boron (B)	mg/L	<0.7	0.7–3.0	>3.0

irrigation. Adverse effects of these specific parameters are given in Table 45.2. In case of surface irrigation, sodium has no effect below 3 mg/L and dangerous effects are expected above 9 mg/L. In case of drip irrigation, sodium is expected to have no effect below 70 mg/L. Chloride concentration above 350 mg/L and boron concentration above 3.0 mg/L are expected to have dangerous toxic effects on plant, while there is no effect below 700 μS/cm electrical conductivity, and dangerous effects are expected above 3000 μS/cm.

Heavy metals are one of the most important pollutants found in wastewater. Avci and Deveci (2013) reported about heavy metal contamination of soil and cultivated products irrigated with wastewater in Turkey. 200 mg/kg Nickel (Ni) was determined in soil samples. Some trace elements in crop samples exceeded threshold values given by Turkish regulation. Maximum admissible concentrations for heavy metals and toxic elements in irrigation water are given in Turkish regulations (Table 45.3).

Persistent Organic Pollutants (POPs) are another main concern for wastewater irrigation. The use wastewaters contain POPs for irrigation, results in depositions of these pollutants in soil. Hence, this affects the flora and fauna of soil, consequently, the human health through food-chain. Plant uptake of these contaminants differs according to the type of the plant. Nasir and Batarseh (2008) investigated the residues of poly aromatic hydrocarbons (PAHs), polychlorinated biphenyls (PCBs), chlorinated benzenes (CBs) and phenols in soil, wastewater, groundwater and plants. It was reported that environmentally relevant concentrations of targeted compounds

Table 45.3 Maximum admissible concentrations (MACs) for heavy metal and toxic elements in irrigation water (NTR 2010)

Element	Total MACs for unit area (kg/ha)	MACs	
		For every type of soil, mg/L	pH 6.0–8.5 clayey type of soil irrigated for less than 24 years, mg/L
Aluminum (Al)	4600	5.0	20
Arsenic (As)	90	0.1	2.0
Beryllium (Be)	90	0.1	0.5
Boron (B)	680	–	2.0
Cadmium (Cd)	9.0	0.01	0.05
Chromium (Cr)	90	0.1	1.0
Cobalt (Co)	45	0.05	5.0
Copper (Cu)	190	0.2	5.0
Fluoride (F)	920	1.0	15
Ferrous (Fe)	4600	5.0	20
Lead (Pb)	4600	5.0	10

were detected for wastewater much higher than for groundwater. The overall distribution profiles of PAHs and PCBs appeared similar for groundwater and wastewater indicating common potential pollution sources. As a consequence, there were some difficulties in evaluating the translocation of PAHs, CBs, PCBs and phenols from soil-roots-plant system. The uptake concentrations of various compounds from soil, in which plants are grown, were dependent on plant variety and plant part, and the compounds taken up by the plants changes. Among the different plant parts, roots were found to be the most and fruits the least contaminated parts.

Reuse of treated wastewater is not widely acceptable in the public. There is a resistance not only for anxiety about efficacy of wastewater treatment plants (WWTPs), but also for cultural and traditional reasons. However, soil fertility and crop productivity encourage farmers for prolonged use of wastewater for irrigation. In some cases, there is no other alternative. Farmers generally do not record that they use treated wastewater so as not to face an objection.

Turkey is a good example for constructing high number of WWTPs. By the year 2012, 72% of municipal population in Turkey took up wastewater treatment services. Giving this service to 85% of municipal population is planned until the end of 2017. There were 595 WWTPs in the year 2014 that treat a total amount of 3.5 billion m^3 wastewater. 41.6% advanced, 33.2% biological, 25% physical, and 0.2% natural treatment methods were used. While 50.5% of treated wastewater was discharged to the sea, 40.5% was discharged to streams. Strategic measures are taken for sustainable management and usage of water resources in the country. There are action plans for restoration of various basins (TUIK 2015).

Even there is a water stress in some basins in Turkey, wastewater reuse is still in quite low level. There are only few examples in the country. Refuges in Konya,

Table 45.4 Wastewater reuse potential of some countries (Cakmakci 2016)

Country	Amount of reused reclaimed wastewater/Amount consumed fresh water, %	Amount of treated wastewater, m^3/day
Kuwait	35.2	424,657
Israel	18.1	1,014,000
Singapore	14.4	75,000
USA	0.6	7,600,000
Turkey	0.1	136,966

golf courses in Antalya are irrigated with treated wastewater. There is a project for agricultural wastewater irrigation in Afyon with the support of Dutch government. Treated wastewaters are used for garden irrigation and car wash in Muğla. Ergene Basin is being evaluated for possible application of wastewater irrigation. A comparison of wastewater reuse percentages in some countries are given in Table 45.4. Israel is reusing wastewater considerably. Nearly 70% of treated wastewater is reused in Israel, mostly for agricultural irrigation (Cakmakci 2016).

45.2 Material and Methods

45.2.1 Study Area

Konya Closed Basin is one of the important basins in Turkey. There is an intense agricultural activity with a high water demand. The basin is the largest gross producer of wheat in the country. Domestic and industrial wastewaters of the Konya have long been collected by a combined sewage system and conveyed through the Main Drainage Channel (MDC) to the Salt Lake. Until 2010, this was practiced without any treatment. During the dry seasons, the wastewater in the channel had been used for land irrigation by the farmers. Besides, high number of illegally drilled wells resulted in lowering of the groundwater level in the basin.

45.2.2 Wastewater Management in Konya Region.

Konya's WWTP was designed for 1 million person equivalent and 200,000 m^3/d and constructed in 2010. It was designed for organic carbon and partial nitrogen (N) removal, including activated sludge basins working by the Bardenpho process. Inlet parameters for Biological Oxygen Demand (BOD), Chemical Oxygen Demand (COD), Suspended Solids (SS), and Total Nitrogen (TN) are in average 436 mg/L, 904 mg/L, 467 mg/L and 93 mg/L, respectively, while they decrease in the outlet to

38 mg/L for BOD and SS, 100 mg/L for COD, and 63 mg/L for TN. These parameters do not meet the Turkish quality standards for the reuse of treated wastewater in urban areas for irrigation of parks, landscaping areas, refuge, etc., as well as for vegetables eaten raw. Therefore, a wastewater reuse system was constructed including tertiary treatment process (filtration, UV treatment, and chlorination) for making use of the treated wastewater in Konya urban landscapes. Reuse of wastewater within the city has been developed for a capacity of 3,600 m^3/d. "Treated Wastewater Irrigation Network" of Konya is the first application in the country.

45.2.3 Effects of Long Term Irrigation with Untreated Wastewater in Konya

Long term irrigation with untreated wastewater can result in accumulation of pollutants in soil and agricultural products. Transport of heavy metals to the soil and wheat samples cultivated in the Konya city was examined by Aydin et al. (2015) in a joint research study of Turkish and German scientists. According to the obtained results, high alkaline properties and clay structure of Konya soil reduce the mobility of contaminants and has a potential to cause accumulation in top layer of soil. While non-irrigated soil samples were analyzed to define geologic background, well-water irrigated soil samples were analyzed as control samples. The highest concentrations of Pb, Cr, Cu, Cd, Zn, Ni, and Hg in wastewater irrigated soil were 5.32, 37.1, 31.5, 11.4, 91.5, 134, 0.34 mg/kg, respectively. According to the results, there is Cd and Ni pollution both in well-water irrigated soils and wastewater irrigated soils when compared with geologic background.

When geo-accumulation index (Igeo) was calculated by adopting from Zhiyuan et al. (2011), wastewater irrigated soils were strongly polluted by means of Cd (8.23–11.6 mg/kg) and moderately to strongly polluted by means of Ni (47.7–134 mg/kg), exceeding Maximum Admissible Concentrations for Trace Elements in Agricultural Soils and Sewage Sludge Regulation limit values of Turkey. For the application of sewage sludge, Cd and Ni parameters should be determined both in sewage and in soil. Cd pollution both in wastewater and well-water irrigated soils can be explained by polluted phosphate fertilizers. No heavy metal accumulation effect of wastewater irrigation is probably a result of good drainage in the region. On the other hand, intense effect of wastewater irrigation on soil Electrical Conductivity (EC) (130–865 μS/cm) was determined.

Maximum concentrations found for Pb, Cr, Cu, Cd, Zn and Ni in wastewater irrigated wheat grain were 8.44, 1.30, 9.10, n.d, 29.3, and 0.94 mg/kg, respectively. Hg was not detected in any samples of wheat grain. Pb contamination in wheat samples grown in the sampling site was compared with the limit values given in Turkish Food Codex.

In this work, persistent organic pollutants, i.e., polychlorinated biphenyls (PCB 28, 52, 101, 138, 153, 180) and polycyclic aromatic hydrocarbons (naphthalene, acenaphthalene, acenaphthene, fluorene, phenan-threne, anthracene, fluoranthene, pyrene, benzo[a]anthracene, chrysene, benzo[b]fluoranthene, benzo[k]-fluoranthene, benzo[a]pyrene, indeno[1,2,3-cd]pyrene, dibenzo[a,h]anthracene, benzo[g,h,i]perylene) were also determined in wastewater irrigated agricultural soil samples and the wheat samples cultivated in the region. High alkaline properties and the clay structure of Konya soil result in the accumulation of contaminants in top soil layer used for agricultural production. On the other hand, PCBs and PAHs compounds were determined in comparable concentrations in well water irri-gated reference soils as in wastewater irrigated soils. Naphthalene, Acenaphthylene, Acenapthene, Fluorene, Phenanthrene, Anthracene were determined in all separated parts of corn samples. Lower molecular weight PCB 28 and PCB 52 congeners were in higher concentrations in corn samples when compared with the other PCB congeners.

Vertical profile of soil did not suggest interpretable data, since the investigated area was periodically ploughed for agricultural activities.

The results of this study reveal that preventive measures should be taken for wastewater irrigation and the application of sewage sludge in soil. Pb values of wheat samples were above the limits given by Turkish food codex. Even no contamination of Cd and low levels of Ni and Cr were detected in wheat samples. High accumulation rates of these metals in soil reveals the need for continuous monitoring.

45.2.4 Effects of Long Term Irrigation with Untreated Wastewater in Braunschweig

Wastewater irrigation has been applied over 50 years in Braunschweig, Germany. In the city, untreated wastewaters were used between the years 1900–1960. After-wards, treated wastewater has been used for agricultural irrigation. Additionally, the sludge from wastewater treatment has been used in this area. In winter time only the effluent of the sewage treatment plant of Braunschweig was used for irrigation, while during summer digested sludge was mixed with the effluent. The irrigation has been performed during the vegetation period only. Main grown crops are cereals, sugar beet and corn.

In parallel with Konya, Braunschweig soils were analyzed for heavy metals, PAHs and PCBs. Heavy metal concentration of Cd, Cr, Pb and especially Zn were deter-mined above background levels in the upper layers 0–50 cm. However, soil concen-trations were still below the limit values for soil from the sewage sludge regulation. PAHs and PCBs were determined above the background levels of urban areas in the upper layers of soil (0–50 cm). The higher concentrations of PAHs and PCBs in comparison to farmland could be explained with the practice that the plants were not harvested but worked into the soil by rotary tillage. It was determined that PAHs and PCBs concentration of soil correlated with the Total Organic Carbon (TOC) ingredient of soil.

In a previous study, Ternes et al. (2007) indicated that most of the selected pharmaceuticals and personal care products were never detected in any of the lysimeter or groundwater samples, although they were present in the treated wastewater that is used for irrigation of the fields because of the sorption of compounds by soils.

45.2.5 *Important Aspects for Wastewater Irrigation*

The reuse of treated wastewater must be considered along with its environmental effects. There is a high probability for accumulation of domestic and industrial pollutants in soil and agricultural products. Wastewaters include a variety of organic and inorganic compounds, including heavy metals, surfactants and Persistent Organic Pollutants (POPs). Sufficient treatment and continuous monitoring is necessary for sustainable applications. Source control of industrial pollutants is important to avoid pollution load to sewer system. Non-essential heavy metals are often found in urban wastewater contaminated with industrial discharges (Bahadir et al. 2016).

One of the most important subjects for wastewater irrigation is the risk of infection caused by pathogens. The risk is high especially for non-processed and raw eaten foods. Neglecting microbial quality of wastewater may threat human health. Microbiological standards are given in Table 45.5. Turkish standards are comparable with those of US EPA.

Insufficient wastewater treatment may result in soil deterioration. Soil salinization, acidification, or alkalizations affect soil productivity. Soil geochemical properties, such as carbonatic, organic forms, sand, silt, and clay fractions are important factors for accumulation and mobility of pollutants in soil (Klay et al. 2010). Organic matter ingredients of wastewater irrigated sites significantly affect the mobility and bioavailability of environmental pollutants. The main challenge for wastewater irrigation is the data about soil deterioration and agricultural product contamination.

Continuous and excessive irrigation with wastewater result in contamination of underground and surface water sources. Acceptable quality of water and selecting proper irrigation method are important factors for sustainable application. Irrigation

Table 45.5 Coliform standards for wastewater irrigation (NTR 2010)

WHO	California (Title 22)	US-EPA	Turkey
<1000 FK/100 mL	2.2 TK/100 mL	0 FK/100 mL[*]	0 FK/100 mL[*]
–	–	200 FK/100 mL[**]	200 FK/100 mL[**]

[*]Non-processed food, [**]Processed food, FK: Fecal coliform, TK: Total coliform

methods such as flooding, sprinkler and drip irrigations, affect the plant contamination level. Distance to public areas is important for sprinkler irrigation. SS is a considered parameter for sprinkler and drip irrigation because of clogging problems.

Most of the water sources are used for agricultural irrigation. Withdrawal of fresh water must be reduced for agricultural sustainability. Operational changes are necessary for a more efficient irrigation. Either in case of wastewater or freshwater irrigation, it is important to decrease the amount of water used. Excessive use of wastewater results in the contamination of underground water sources. Proper drainage is necessary to avoid accumulation of contaminants in soil.

Even heavy metal accumulation effect of wastewater irrigation was evidenced in many studies. There is a lack of data on contamination of POPs in soils, ground water, and surface water systems. Neglecting accumulation effect of POPs may arise a serious risk for human health and sustainability. It is important to make a source control to avoid their adverse effects.

Advantages of wastewater irrigation are nutrient ingredient of wastewater that improves agricultural productivity and decrease the need of fertilizers. Consumption of high quality water sources can be lowered by using treated wastewater. Discharge to receiving water bodies is avoided. The soil itself is a treatment media for wastewater.

45.3 Results and Discussion

Despite numerous advantages for soil fertility and crop productivity, wastewater irrigation has several adverse effects on environment. Treated wastewater irrigation leads to changes in physicochemical properties of soil. These effects are highly correlated with the quality of treated wastewater, duration of application, method of irrigation, and soil physicochemical properties. To ensure safe and long term application of wastewater irrigation, periodic monitoring of soil fertility and water quality parameters are required.

There are regional water shortage problems in Turkey. While there is enough water in some areas, there is a serious water stress in some others. In arid regions, wastewater irrigation is a solution for sustainable agricultural activities. This option will probably be more common in the future as a result of increasing water stress. Besides, farmers are generally interested in using wastewater for its nutrient ingredients that decreases the need of chemical fertilizers. Konya is an example for wastewater irrigation in Turkey. Intense impact of wastewater irrigation on soil EC was determined. Cd and Ni contamination of soil, and Pb contamination of wheat samples were reported. PCBs and PAHs sources other than wastewater irrigation were evident for the studied field.

45.4 Conclusions

Inevitably, farmers will depend in future more on wastewater irrigation as a result of water shortage. Achievement of sufficient quality is the key factor for wastewater irrigation. Establishment of proper standards is necessary. It is essential to consider the soil structure, the tolerance of plants, and the irrigation methods in the decision making process.

References

Avci H, Deveci T (2013) Assessment of trace element concentrations in soil and plants from cropland irrigated with wastewater. Ecotox Environ Safe 98:283–291

Aydin ME, Aydin S, Beduk F, Tor A, Tekinay A, Kolb M, Bahadir M (2015) Effects of long term irrigation with untreated municipal wastewater on soil properties and crop quality. Environ Sci Pollut Res 22:19203–19212

Bahadir M, Aydin ME, Aydin S, Beduk F, Batarseh M (2016) Wastewater reuse in middle east countries – a review of prospects and challenges. Fresen Env Bull 25(5):1284–1304

Cakmakci: Wastewater Reuse. www.rewistanbul.com/files/AAEP

Karaaslan Y (2013) Water resource management in Turkey. In: DAAD-EXCEED regional workshop on wastewater treatment and reuse, 03–06 June 2013, Konya, Turkey. https://www.exceed.tu-bra unschweig.de/exceed/events-and-dates

Klay S, Charef A, Ayed L, Houman B, Rezgui F (2010) Effect of irrigation with treated wastewater on geochemical properties (saltiness, C, N and heavy metals) of isohumic soils (Zaouit Sousse perimeter, Oriental Tunisia). Desalination 253:180–187

Nasir FA, Batarseh MI (2008) Agricultural reuse of reclaimed water and uptake of organic compounds: pilot study at Mutah University Wastewater Treatment Plant, Jordan. Chemosphere 72:1203–1214

Notification of technical rules for WWTPs, Turkey (2010)

NTR (2010) Turkish Notification of Technical Rules for Wastewater Treatment Plants, Official Newspaper No: 27527

Ternes TA, Bonerz M, Herrmann N, Teiser B, Andersen HR (2007) Irrigation of treated wastewater in Braunschweig, Germany: an option to remove pharmaceuticals and musk fragrances. Chemosphere 66:894–904

TUIK (2015) Turkish Statistical Agency. https://www.tuik.gov.tr/PreHaberBultenleri.do?id=18778

Zhiyuan W, Dengfeng W, Huiping Z, Zhiping Q (2011) Assessment of soil heavy metal pollution with principal component analysis and geo-accumulation index. Procedia Environ Sci 10:1946–1952

Chapter 46
Domestic Wastewater for Climate Mitigation

K. Unwerawattana, T. Reinhardt, A. Michels, and C. Wongburana

Abstract Thailand intends to reduce greenhouse gas (GHG) emissions by 20–25% in the year 2030. The wastewater sector can make a significant contribution to climate change mitigation: better wastewater management can lead to reduction of GHG emissions produced from its treatment process, i.e., carbon dioxide, methane, and nitrogen oxide in addition to applying energy efficiency and/or energy saving equipment to the treatment system. On behalf of the German Federal Ministry for the Environment, Nature Conservation, Building and Nuclear Safety (BMUB), the Deutsche Gesellschaft für Internationale Zusammenarbeit (GIZ) and the International Water Association (IWA) are working together on the project "Water and Wastewater Companies for Climate Mitigation (WaCCliM)" as part of the International Climate Initiative (IKI). In Thailand, the WaCCliM project is implemented in partnership with the Wastewater Management Authority (WMA). The Chiang Mai wastewater treatment plant has been selected as the first pilot plant to assess the GHG reduction potential based on the GHG accounting tool that allows carbon accounting in the urban water cycle. Additional wastewater utilities are expected to join and to move towards low carbon utilities in 2016. This can be a stepping-stone for other water and wastewater utilities to adopt climate change mitigation measures in order to reduce their emissions and to contribute to nation's GHG reduction goal as a whole.

Keywords GHG reduction · Wastewater · GIZ WaCCliM · Climate change mitigation

K. Unwerawattana · A. Michels (✉)
Deutsche Gesellschaft für Internationale Zusammenarbeit (GIZ) GmbH, Bangkok, Thailand
e-mail: astrid.michels@giz.de

T. Reinhardt · A. Michels
Deutsche Gesellschaft für Internationale Zusammenarbeit (GIZ) GmbH, Eschborn, Germany

A. Michels · C. Wongburana
Ministry of Natural Resources and Environment, Wastewater Management Authority, Bangkok, Thailand

© Springer Nature Switzerland AG 2021
M. Babel et al. (eds.), *Water Security in Asia*, Springer Water,
https://doi.org/10.1007/978-3-319-54612-4_46

46.1   Introduction

Climate change manifests itself in heavy precipitation events, altered runoffs, and increased soil erosion, and is expected to exacerbate current stresses on water resources (Bates et al. 2008). Climate change affects human water security because it reduces the capacity to access adequate and acceptable water quality for sustaining well-being (Bigas 2013). With climate change making precipitation more variable, water scarcity is likely to become an even greater problem. Changing weather patterns increase the frequency and severity of extreme events, especially of floods and droughts (Gitay et al. 2001). During the past decades, weather pattern in Thailand have fluctuated from severe droughts to severe floods, which affected both residential areas and agriculture areas (Kisner 2008). The negative impacts of such disasters can be seen clearly from the great flood in 2011 caused by unprecedented heavy rainfall during the dry season and lasting over 158 days, affecting agricultural production, the industrial sector, and life in general (Promchote et al. 2016). During this great flood, maintaining the necessary water quality was a severe problem (Wongpat and Passananont 2012).

Water and wastewater utilities (WWUs) are among the largest consumers of energy in developing and newly industrialising countries (IEA 2016; Rothausen and Conway 2011). This is partly due to high water (50–60%) and energy losses (40%). Water managers are challenged to meet an increased demand for water and wastewater services of a growing population coupled with reduced water availability and the rising pressure to reduce costs, energy consumption and greenhouse gas (GHG) emissions. Therefore, water management and efficiency strategies that optimize the use of a scarce resource are a key requirement to ensure Thailand's water and energy security in future years.

Limiting climate change requires substantial and sustained reductions in greenhouse gas emissions in all sectors. To mitigate climate change, the United Nations Framework Convention on Climate Change (UNFCCC) was established in 1992 and the Conference of Parties (COP) has been held every year since 1995. The 2015 Paris Agreement charts a new course in the global effort to combat climate change (UNFCCC 2014). As a developing country highly vulnerable to the impacts of climate change, Thailand intends to reduce greenhouse gas (GHG) emissions by 20–25% compared to the business-as-usual level in year 2030. Within Thailand's National Determined Contributions, both water and wastewater sector are included and comprise energy efficiency and improved treatment process. According to the Thailand's Second National Communication, the treatment of wastewater accounts for 47.5% of energy consumption in the waste sector (4.1% of all energy consumption) (Office of Natural Resources and Environmental Policy and Planning 2000). Sources for GHG emissions include the use of outdated and energy-intensive treatment technologies and pumps, CH_4 and N_2O emissions during wastewater treatment process, and leave opportunities for recovering energy (biogas) and nutrients from wastewater unexploited. Energy efficient pumps, the production of biogas from wastewater, the reuse of treated wastewater and the nutrient recycling can enable WWUs to save GHGs.

On behalf of the German Federal Ministry for the Environment, Nature Conservation, Building and Nuclear Safety (BMUB), the Deutsche Gesellschaft für Internationale Zusammenarbeit (GIZ) and the International Water Association (IWA) are working together on the project "Water and Wastewater Companies for Climate Mitigation (WaCCliM)" as part of the International Climate Initiative (IKI). With the objective to develop a climate solution for water and wastewater utilities (WWUs), the WaCCliM project introduces GHG reduction technologies or processes to reduce GHG emissions and improve the carbon balance of water and wastewater utilities in Mexico, Peru, Jordan and Thailand while maintaining or improving their service levels. The WaCCliM project has been launched in 2014 and lasts till 2018. For the WaCCliM component in Thailand, domestic wastewater has been selected as a main focal point for mitigation measures working closely with the political partner, the Wastewater Management Authority (WMA).

46.2 Material and Methods

To reduce GHG emissions and to achieve carbon neutrality, WaCCliM project is addressing both water and wastewater sector and holistic water cycle, which is a main concept behind WaCCliM project (Fig. 46.1) because water are all interconnected.

By addressing the whole picture of water cycle, WaCCliM project developed an approach to guide utility staff to reduce GHG emissions and to become a climate smart utility. The WaCCliM approach can be explained in four major steps as can be seen in Fig. 46.2.

Additionally, WaCCliM developed a carbon accounting tool for utilities to address their combined water and climate issues: the Energy assessment and Carbon emissions Monitoring and Assessment Tool or 'ECAM Tool'. The ECAM Tool (https://wacclim.org/ecam/) is a free web-based tool that assesses GHGs and energy intensity of the urban water cycle and prepare these utilities for future reporting needs on climate mitigation. ECAM was developed by IWA, Institut Català de Recerca de l'Aigua (ICRA) and GIZ and is consistent with the Intergovernmental Panel on Climate Change (IPCC) Guidelines for National Greenhouse Gas Inventories (Doorn et al. 2006) and peer-reviewed literature. It offers a transparent and sound approach for emission calculation within the water sector. ECAM helps link Monitoring, Reporting and Verification of mitigation actions in the water sector to the national level (Fig. 46.3).

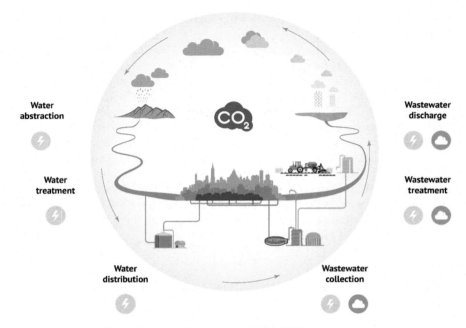

Fig. 46.1 Urban holistic water cycle (focusing area of WaCCliM project)

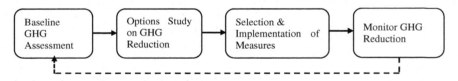

Fig. 46.2 Simplified diagram of WaCCliM approach

46.2.1 WaCCliM Approach

1. Baseline Assessment
   One of the very first things to reduce GHG emissions and to achieve climate neutrality in utilities is to understand the current situation and to establish the baseline data of current GHG emissions. Therefore, WWUs staff defines an assessment period and collects utility specific data using ECAM tool for calculate current GHG emissions. The results from this first application of the ECAM tool do not only show the current GHG emissions, but also show the highest areas of GHG emissions for further focusing, which can be the starting point for WWUs to understand their GHG emissions source.
2. Options Study

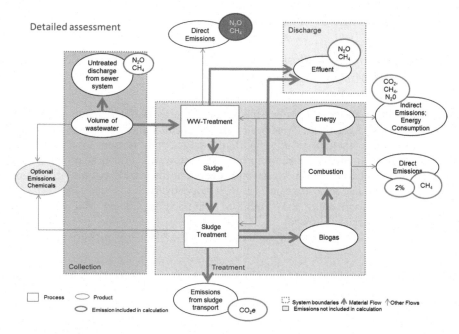

Fig. 46.3 System boundaries of ECAM Tool

After the Baseline data is available, WWUs can select the target area of improvement measures based on the results and conduct a detailed study or 'Options Study' to find possible improvement measures. Possible measures can range from operational measures, which can improve energy efficiency or efficiency of treatment process or GHG reduction measures, which can improve service quality or reduce GHG emissions from the process.

3. Selection and Implementation of Measures
   The detailed assessment on options for GHG reduction measures is followed by implementation. By developing scenarios for each identified measure, WWUs staff can select and implement measures based on a set of predefined criteria.

4. Monitoring the results
   After implementation, ECAM Tool can be used as a monitoring tool to track the progress of reduction, which can be done by entering a new assessment period data after the implementation and by comparing the results with the Baseline or scenario development. Since there is not only one opportunity to reduce GHG emissions, WWUs can continue to start WaCCliM approach. By continuously following the WaCCliM approach, WWUs will achieve climate neutrality and become a climate smart utility.

46.3   Results and Discussion

46.3.1   Example Thailand, Chiang Mai Case Study and Mitigation Options

In Thailand, the WaCCliM project has selected the wastewater treatment plant in Chiang Mai, managed by the Wastewater Management Authority. Chiang Mai is one of the most important touristic cities in Thailand. One of the reasons to select Chiang Mai was the high energy consumption of the wastewater collection system. Some characteristics of Chiang Mai wastewater treatment plant and layout of collection system is shown in Table 46.1 and Fig. 46.4, respectively.

The results from using the WaCCliM approach in Chiang Mai wastewater treatment plant are as followed:

The Chiang Mai wastewater system consists of the following infrastructure (Silva 2015):

- Houses and buildings send the blackwater to an on-site septic tank, which is installed and maintained by the house owner, due to the building control act B.E.2522 (1979). The municipality is the enforcement entity for new installations.
- The effluent from septic tanks overflows to the sewer together with the greywater from houses and buildings.
- The combined wastewater system (conveying storm water, septic tank overflows and greywater), only on West side of Ping river, is managed by the municipality until it reaches one of the nine pumping stations, which are under WMA responsibility as on Fig. 46.4.
- The combined sewer system is pumped to a lagoon for treatment via the last pumping station (P10).

Due to installation of septic tank and combined wastewater system, Chiang Mai wastewater treatment plant receives wastewater, which has low organic content of only around 20–30 mg/L during the Baseline period. Based on hydraulic and organic loads in the wastewater, it was estimated that currently approximately 90% of the

Table 46.1 Characteristics of Chiang Mai wastewater treatment plant

Treatment type	Aerated Lagoon
Number of pumping stations (collection system)	9
Treatment capacity	55,000 m^3/day
Averaged wastewater influent	10,000 m^3/day*
Chiang Mai inhabitants	150,000 people
Population in the service area	75,000 people
Current serviced population	8,250 people*

*Operational data from WMA (November 2014–May 2016)

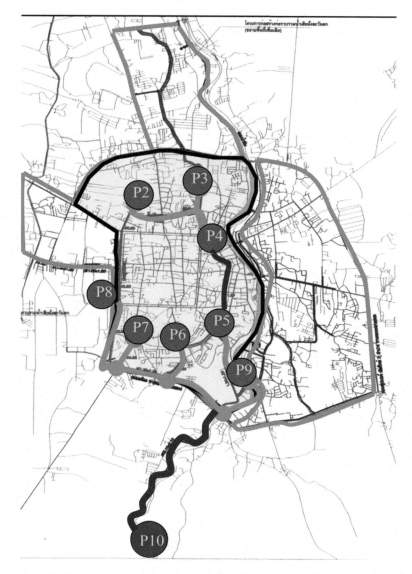

Fig. 46.4 Collection sewer network in Chiang Mai City (Wastewater Management Authority 2014)

domestic wastewater is bypassing the sewer network or diluting by other water sources, such as water from canal or rainwater, before reaching the wastewater treatment plant. Data from November 2014–May 2015 show that the main source for GHGs in Chiang Mai is untreated wastewater (Fig. 46.5). Untreated wastewater also adversely affects public and environmental health and the tourism industry of the city.

GHG emissions (Total ≈ 1,400,000 kgCO₂e)

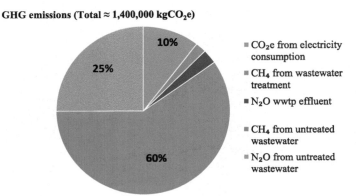

Fig. 46.5 Initial baseline GHG detailed assessment result from Chiang Mai wastewater utilities

Fig. 46.6 Initial baseline energy performance assessment result from Chiang Mai wastewater utilities. Sub-stages correspond to pumping stations (PS) and treatment plant

Energy consumption substages (Total ≈ 200,000 kWh)

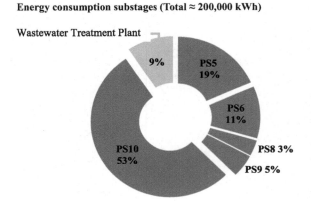

Figure 46.5 shows also that the majority of the GHG emissions from untreated wastewater can be attributed to methane and nitrous oxide emissions. The remaining 35% can be attributed to GHG emissions from treated wastewater, from which 10% can be attributed to electricity consumption.

For energy performance assessment in Baseline, as can be seen in Fig. 46.6, around 90% of energy consumption is related to energy use in the wastewater collection stage.

The options study for the Chiang Mai Wastewater Treatment Plant identified two main challenges of the plant: 1) Outdated equipment, which is energy intensive and results in high operation costs, and 2) Fractured pipes in the wastewater collection system. These two constraints are the major causes for the large amount of untreated wastewater flowing directly into the public canal of Chiang Mai City.

Proposed measures for the Chiang Mai wastewater treatment plant to reduce GHGs were identified into two categories;

a)   Operational and Maintenance measures, and

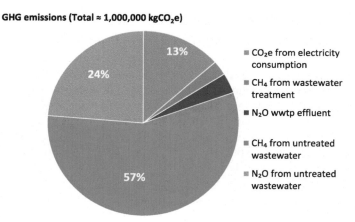

Fig. 46.7 Expected GHG detailed assessment result after implementations (Unwerawattana et al. 2016)

b)   GHG reduction measures.

Operational measures include reducing groundwater infiltration, improving electrical equipment, improving pump efficiency, and repairing the wastewater collection system. While GHG reduction measure is increase wastewater influent to treatment plant to reduce untreated wastewater.

Five scenarios were evaluated based on their GHG reduction potential, investment costs, payback period and energy intensity (Unwerawattana et al. 2016). GHG emissions could be reduced by up to $\approx 80\%$ if untreated wastewater is collected and treated in the Chiang Mai Wastewater Treatment Plant (Fig. 46.7).

Figure 46.7 shows the potential for GHG reduction through a 30% expansion of service area with increasing pumping efficiency. GHG emissions from Chiang Mai wastewater treatment plant could reduce 400,000 kg CO_2e in the same period.

The interesting lesson learned with the Chiang Mai pilot project is that there is a need to improve the operation of the system before energy optimization measures can be implemented and provide a return on investment. In addition, the reduction of GHG emissions is associated with two types of measures: 1) Energy efficiency measures and 2) Service level improvement, which increase the amount of sewage treated. The second set of measures has very high GHG reduction potential, but no financial savings. These measures correspond to public investment in sanitation infrastructure. This is a strong message for other utilities considering this path.

46.3.2 Scaling Up for Climate Smart Utilities; Co-benefits Challenge and Opportunities to Overcome Barriers

The findings from Chiang Mai show that significant GHG emissions are related to untreated wastewater in Chiang Mai. However, another source of GHG emission, not yet quantified, is faecal sludge management. According to the SFD promotion initiative, only 30% of faecal sludge in Chiang Mai is being managed properly (Michels 2016). As a result, it can be assumed that 70% of organic matter in wastewater has been handled unsafely providing a huge unexploited opportunity for domestic wastewater management to become a climate smart utility and city in the future. However, one challenge for improving Chiang Mai domestic wastewater management is the split responsibilities for domestic wastewater (WMA) and faecal sludge management (Chiang Mai Municipality), which belong to different entities. Therefore, the WaCCliM project intends to support this communication between WMA and Chiang Mai municipality to overcome these challenges since better domestic wastewater management can result in a variety of benefits, such as improved quality of services, reduced stress on water resources, and contributing to the country's climate targets.

46.4 Conclusions

Domestic wastewater sector can contribute to climate mitigation by applying the WaCCliM approach. The baseline assessment in Chiang Mai indicates that untreated wastewater has the biggest proportion of GHG emissions. In addition, faecal sludge management also produces significant amounts of emission. In order to improve domestic wastewater, both issues need to be taken into account. Chiang Mai represents a case study for developing smart climate utilities for other cities in Thailand or even in other countries that have similar wastewater systems as in Thailand. Reducing GHG emissions in pilot utility has multiple benefits such as improved human and environmental health and contributions to the country's climate goal. WaCCliM project will support further studies and training to decarbonize the water sector and to enable pilot utilities on the path to become a leading of climate smart utilities.

Acknowledgements WaCCliM project in Thailand has received well support from WMA and Chiang Mai municipality for providing an insight information and operational data during both baseline and options studies. The founding of current domestic wastewater management in Thailand, such as case study in Chiang Mai provide valuable information for project to support Chiang Mai wastewater treatment plant and achieve project objectives. Lastly, project will not succeed without contributions and efforts from group of experts from IWA who provide valuable knowledge and efforts for conducting study and research for WaCCliM project.

References

Bates B, Kundzewicz Z, Wu S, Palutikof J (2008) Climate change and water. IPCC Working Group 2. IPCC, Geneva

Bigas H (2013) Water Security & the Global Water Agenda: a UN-water analytical brief, 48

Doorn M, Towprayoon S, Manso Vieira S, Irving W, Palmer C, Pipatti R, Wang C (2006) Wastewater Treatment and Discharge. IPCC Guidelines for National Greenhouse Gas Inventories

Gitay H, Brown S, Easterling W, Jallow B (2001) Ecosystems and their goods and services. In: McCarthy JJ, Canziani OF, Leary NA, Dokken DJ, White KS (eds) Climate change 2001: impacts, adaptation, and vulnerability. Cambridge University Press, New York, pp 287–291

International Energy Agency (2016) Water Energy Nexus, 63

Kisner C Climate change in Thailand: impacts and adaptation strategies. https://climate.org/archive/topics/international-action/thailand.html.

Michels A Improving urban water services in Chiang Mai, Thailand using ECAM and SFDs. https://www.susana.org/_resources/documents/default/3-2635-7-1473414899.pdf.

Office of Natural Resources and Environmental Policy and Planning: Thailand's Second National Communication under the United Nations Framework Convention on Climate Change. https://unfccc.int/files/national_reports/non-annex_i_natcom/submitted_natcom/application/pdf/snc_thailand.pdf.

Promchote P, Wang SYS, Johnson PG (2016) The 2011 great flood in Thailand: climate diagnostics and implications from climate change. J Clim 29(1):367–379

Rothausen SG, Conway D (2011) Greenhouse-gas emissions from energy use in the water sector. Nat Clim Chang 1:210–219

Silva C (2015) Initial baseline assessment of energy performance and GHG emissions WMA Chiang Mai wastewater system, water and wastewater companies for climate mitigation (WaCCliM)

UNFCCC: Background on the UNFCCC: the international response to climate change. https://unfccc.int/essential_background/items/6031.php.

Unwerawattana K, Reinhardt T, Michels A, Trommsdorff C, Promes E (2016) Option study for reduction of carbon footprint and energy optimization Chiang Mai Wastewater Treatment. Water and Wastewater Companies for Climate Mitigation (WaCCliM), Thailand

Wastewater Management Authority (2014) Chiang Mai wastewater treatment plant presentation

Wongpat N, Passananon S Water quality management in Bangkhen water treatment plant for the flood crisis 2011. https://tappingtheturn.org/wp-content/uploads/2012/05/Water-Quality-Management-Bangkok-2011-Flood-Crisis.pdf.

Chapter 47
Making the Case for Wastewater Irrigation in Bangladesh

S. T. Mahmood

Abstract Agriculture contributing nearly 12% of GDP and providing approximately 47% of country's employment remains a major sector in Bangladesh's economy. Bangladesh in the recent years has seen an economic growth rate of over 6.5% and intends to become a middle income country by 2021. Ensuring growth in the agriculture sector therefore, is a precondition to ensure sustainable economic growth of Bangladesh. However, the country's agriculture is already affected by climate change which will likely become more severe in the coming years. Rainfall variability affects farmers heavily as many of them depend on rainwater for irrigation purposes. One way of adapting to climate change induced rainfall variability could be the use of wastewater for irrigation purposes. This research highlights the successful cases of wastewater irrigation in other countries in order to provide researchers and practitioners with a better understanding of the importance, applicability and feasibility of using wastewater irrigation in Bangladesh.

Keywords Wastewater · Irrigation · Agriculture · Climate-change

47.1 Introduction

Climate Change poses a great threat to social and economic development of developing countries (Misra 2014) as it significantly brings huge threat to agricultural sector (Zou et al. 2013). This very sector is quite dependable but also a climate sensitive sector (Misra 2014). Bangladesh being a developing country is highly dependent on Agricultural sector since the contribution of this sector in GDP is nearly 12%, a provider of around 47% of country's employment (GED 2015). Besides, Bangladesh itself a very vulnerable to climate change (IUCN, n.d.). According to (Kreft et al. 2014) Bangladesh stands 6th amongst 10 other most climate vulnerable countries. Climate-change impact cross-cuts vastly; impacts such as high changes in rainfall pattern, sea level rise, frequent flooding, cyclone, drought etc. (IUCN n.d.). Amongst

S. T. Mahmood (✉)
International Centre for Climate Change and Development (ICCCAD), Dhaka, Bangladesh
e-mail: Mahmood.tashfiq@gmail.com

© Springer Nature Switzerland AG 2021
M. Babel et al. (eds.), *Water Security in Asia*, Springer Water,
https://doi.org/10.1007/978-3-319-54612-4_47

all of the impacts, water is a major element which is distinguishable (IUCN n.d.). There is a high concern about future agricultural water requirements (Fischer et al. 2007) as agricultural water resource is highly influenced by Climate change as well as other environmental aspects (Nam et al. 2015). An attention has been given internationally on wastewater reuse as an alternative source of water resource (Jeong et al. 2016). Wastewater is being used in around 4 of every 5 cities of developing nations that makes this practice a very common reality for developing nations. China, India, Mexico, Pakistan, Ghana are the top most countries to utilize wastewater for irrigation (SAI Platform 2010).

In Bangladesh, agricultural sector uses the highest amount of water. In the changing climate, assessment of total demand of agricultural sector is necessary for a long term planning as there will be a massive change in the water supply for irrigation purpose in Bangladesh (Shahid 2011).

However, this study by providing with cross country evidences on wastewater irrigation might be a helpful knowledge for practitioners and researcher to u understand feasibility, challenges, necessity and applicability of waste water into agricultural sector. Socioeconomic aspects of Bangladesh rely on agriculture, conducting more research and development project on wastewater irrigation might be a catalyst for the further growth and economic contribution of agricultural sector in Bangladesh.

47.2 Methods and Background

The paper is absolutely secondary data based and more of descriptive nature. All the data presented here are gather by conducting distinctive internet research. Google Scholar was used as the main tool to find out relevant information. Different scientific literatures including journal articles and sector specific official reports are reviewed scrupulously in preparing the paper. This research highlights the successful cases of wastewater irrigation in other countries in order to provide researchers and practitioners with a better understanding of the importance, applicability and feasibility of using wastewater irrigation in Bangladesh.

47.3 Results and Discussion

47.3.1 Current Practices and Feasibility

Considerably the most recognized and traditional application in agriculture is the use of wastewater (Scheierling et al. 2010).Worldwide, 10 percent of total irrigated surface is covered by wastewater (Jimenez 2006). Across the world, farmers reflect that wastewater provides a trustworthy source of water supply considering the quality of water (Amerasinghe et al. 2013. For example, In Pakistan, 26% of country's domestic vegetable is produced by wastewater. Which is the result of utilizing country's 36% of total wastewater discharges. For irrigation purpose in Pakistan, around 240,000 m^3/day of untreated wastewater has been used directly (Weckenbrock 2010). Researchers also found that, variety of major crops has been produced in India by irrigating lands with wastewater; such as, Cereals, vegetables (e.g.: Cucurbits, eggplant, okra, Spinach, mustard, cauliflower etc.), fodder crops and many more (Kaur et al. 2012).

In developing countries, increased use of wastewater in agricultural sector is being driven by few principles; one of them is evolving shortage and stress on water resources (World Health Organization 2006).

According to the usage of wastewater in Agriculture, wastewater reuse can be classified into direct and indirect wastewater reuses (Jeong et al. 2016); though there are other several characteristics defined by professionals according to the use of wastewater irrigation by developing countries or elsewhere, such as, Direct or treated or reclaimed wastewater, direct use of treated wastewater, indirect use of untreated wastewater, planned or controlled use of wastewater etc. (Scheierling et al. 2010). In many countries of world are practicing any of these methods, for example, In Mexico about 260,000 ha are irrigated with untreated wastewater in Mexico (Mousavi et al. 2015). Peri-urban agriculture had been practiced in in Kumasi Ghana by applying untreated wastewater in around 11,900 ha of area (Rutkowski et al. 2007). The interest in using wastewater for irrigation grew rapidly within a country like Turkey (Kiziloglu et al. 2008). Most important finding of the review is that, soil and plant growth can have a positive effect from wastewater due to being rich of organic matters and nutrients such as nitrogen, potassium and phosphorus (Mousavi et al. 2015). It is also found in a research that using municipal wastewater in irrigation may help to minimize the risk of ecosystem from the direct disposal of contaminated wastewater into surface or ground water; this persona of wastewater irrigation makes this an environmentally sound practice (Kiziloglu et al. 2008).

47.3.2 Socio Economic Aspects: Opportunity and Challenges

Performance of national economy of many developing regions depend on agriculture, which makes the sector very important (Stevanovic et al. 2016). Being economically dependable yet climate sensitive sector (Misra 2014), agriculture sector poses a great threat from climate change, as climate change may cause irreversible damage to arable lands and water resources (Fischer et al. 2005). Agreeing to (IPCC 2008), water demand for irrigation would come under great pressure due to higher temperatures and increased unpredictability of rainfall.

While freshwater sources become scarcer, wastewater use in agricultural sector becomes an attractive option (Mokhtari et al. 2012), use of wastewater is also considered and well believed as a reliable resource for farmer's living when there is water shortage (WHO 2006).

According to Macmillan dictionary of Modern Economics 'Any good that is scarce and is something which one would choose more of if one could is an Economic good'; in that case, for developing countries, wastewater is an economic good too (Devi et al. 2007).

Irrigation through wastewater brings a great economic deal for the families in Nairobi who have mostly used wastewater for irrigation. They have generated an average annual income of nearly USD 280, even the production through irrigation continues through the year. A similar case was recorded in Ghana where farmers came out of poverty line by generating incomes which ranged approximately from USD 500 to 8,000 by using wastewater for agricultural irrigation. (Raschid-sally et al. 2005). Other studies have found that, in Haroonabad, Pakistan, gross margin was significantly high approximately USD 150/ha in vegetable production with untreated wastewater as expenditure on chemical fertilizer was less even achieved a great yield (Keraita et al. 2008).

Wastewater is also very useful in cities for agricultural practices (Jimenez 2006). For example, Vegetables grown using untreated wastewater for irrigation meets almost 60% requirements of perishable food in Pikine, one of the farming cities in Dakar (Keraita et al. 2008). On the other hand, it has been estimated that, in India, irrigation through wastewater covers around 73,000 ha of peri-urban agriculture (Kaur et al. 2012).

This 'Urban Agriculture' can be executed in arid or even wet countries depending on the accessibility of wastewater, local perishable food demand and people with no income or job (Jimenez 2006). For example, in some water scarce part of India, Wastewater irrigated fields produced crops, vegetables flowers, fodders which helped to generate a great employment and income prospect for female and male agricultural labourers (Kaur et al. 2012). Essentially, wastewater is a reliable source of water for irrigation as this influences farming practice and contributes towards urban food resource as well as livelihoods to many agricultural labours (Abaidoo et al. 2010).

Wastewater as a potential resource is often misjudged regardless of being highly valued. Numerous benefits to society, economy, environment, ensuring social equity and enhancing food security can be aided through wastewater if it is well-managed (The World Bank 2010) (Sagasta et al. 2013) as mentioned earlier. Often the management for wastewater remains undone due to ignorance about the financial benefits of wastewater treatment, as it was thought to be an expensive process (Mara 2003). Opportunities are there to reduce risk and for a better management of wastewater, such as, improved policies, collaboration amongst relevant public agencies, institutional dialogues and financial mechanisms (Qadir et al. 2010).

For a better planning and management of wastewater use in agriculture, availability of data on projected usage of wastewater in agriculture is needed (Qadir et al. 2010). There are several studies which have highlighted the significance of wastewater use in agriculture, but updated information on the amount of wastewater is generated nationally is missing (Sato et al. 2013). This kind of data is well required for a proper management and productive use wastewater in agriculture, as the information may help researchers, practitioners and public institution to prepare legitimate policy and national plan; however, wastewater may appear as a challenge but this brings opportunity as well (Sagasta et al. 2013).

47.3.3 Coping with Climate Change

Two perspectives can be drawn in term of the relationship in between wastewater and climate change. Number one is the water availability, quality and volume in a changing climatic condition in a certain time and space which and other one is the adaptation to climate change in terms of how wastewater is managed (Corcoran et al. 2010). For example, in Moshi, Tanzania, to cope with water unavailability in climatic changes, a recycling facility for wastewater has been provided to create and alternative source of water. Which has doubled the crop consequently, reduced poverty and eradicated conflict due to water limitation.

(UNFCCC 2011). In another research it's mentioned that, fodder crop can be grown by reusing wastewater that could take a part to improve resilience to climate changes and food security for small cities of developing country (Jiménez et al. 2010).

The Middle East and North Africa region has created an opportunity in terms of managed beneficial use of wastewater to cope with the climatic events that has substantially deteriorated water quality and quantity though the rate is still low in many MENA countries (Qadir et al. 2010).

Adaptation to climate change may be taken forward by reusing wastewater by recycling it. Hence, Adaptation to climate change will be constrained by many political, institutional, economic, social reasons in developing and Asian nations. Wastewater reuse is a potential and sustainable adaptive method for developing countries to climate change and can be very cost effective in the long run (IPCC 2008).

47.3.4 Bangladesh Perspective

One of the most potential impacts on Bangladesh due to climate change is the changes in hydrology (Shahid 2011).This is a crucial measure for Bangladesh because of its strong economic reliance on natural resources and rain fed agricultural practices (Anik and Khan 2012). It has been predicted by Global Circulation Model (GCM) that Bangladesh will face an average temperature increase climate change which is 1.0 °C by 2030 and 1.4 °C by 2050. Stressed water condition and declined agricultural production is likely to be noticed due to high temperature along with higher rates of evapotranspiration (Anik and Khan 2012).

As an alternative source to backup water scarcity and stress, wastewater use in agriculture is recognized as a reliable source of water globally and the use is well spreading. Wastewater for irrigation purpose is also very useful for containing necessary nutrients that enhances plant growth (Rizwan et al. 2009). Major changes in water supply because of climate change could be addressed through the use of wastewater as it presents a good package of components (WHO 2006).

In this regard, Bangladesh has a high potential as the annual production of wastewater is near about 725 Mm^3. This has been reported by The Economic and Social Commission for Asia and the Pacific that the source is only from urban areas of Bangladesh. It is also mentioned in studies that well planned and managed wastewater could be a great resource for more irrigation coverage, a great catalyst for socioeconomic elevation of farmers and a reliable method of food production (Mojid et al. 2010).

There have been several examples of wastewater irrigation in agriculture in some sub/districts of Bangladesh. Table 1 is presenting the recorded summary of wastewater irrigation;

Upraising awareness amongst farmers of some places in Bangladesh about persistent water crisis is driving their interest in utilizing wastewater for irrigation. Year around availability of wastewater helped nearly 37% of farmers to have high yield, three cropping in one year by some was reported too. Researchers have also found that in Bangladesh, wastewater is a reliable source with adequate nutrient at almost no cost (Mojid et al. 2010).

Table 47.1 Summary of wastewater irrigation and crops irrigated by wastewater;

Location	Crops irrigated with waste water	Wastewater irrigated area
Bogra	Rice, potato, potato, ginger and local vegetables	47–67
Chittagong	Rice, cauliflower, cabbage, tomato and spinach	Unidentified
Comilla	Rice	20–27
Dhamrai	Rice	67–80
Gazipur	Rice	27–33
Khulna	Rice	Unidentified
Mymensingh	Rice	Unidentified
Natore (sugar mill)	Sugarcane, rice	Unidentified
Pabna (sugar mill)	Sugarcane, rice, wheat, radish, carrot, cauliflower, cabbage and tomato	Unidentified
Rajshahi	Rice, wheat potato, maize, cabbage, cauliflower, papaya, spinach and local vegetables	27–33
Sherpur	Rice	27–33
Sylhet	Rice	267–334

Source (Mojid et al. 2010)

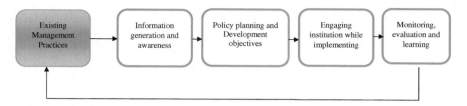

Fig. 47.1 Stages of implementing adaptation technologies (*Source* UNFCCC 2014)

47.4   Recommendation

Changing the behavioural pattern for adapting to climate change is at the paramount level. It often offers a 'hard' form of adaptation as well as 'soft' form of technologies. Any level of technologies and its application requires repetitive process rather than a one-off activity (UNFCCC 2006). This process illustrates in the Fig. 47.1 of this article.

The illustration shows that a planned adaptation as an idealized four stage sequence. Firstly, it needs generating information and awareness building to those who are the most responsible in the process. It may include stakeholders range from policy makers to end user. Secondly, policy planning design should incorporate country's development objectives as well as international goals of climate action. The policies should be articulated in a way which will make the technology cost effective, environmentally sustainable, culturally compatible and socially acceptable. In the third stage, while implementing a technology must be endorsed by national, formal, informal and community based institutions. The fourth stage may include a comprehensive monitoring process that will allow new consideration in the management system, course of corrections and modification, further innovation and learning.

Nonetheless, globally, the data on use of waste water in agriculture is very incomplete and uncertain. The practice has been going on in planed or unplanned, direct or indirect and treated or untreated basis (Mateo-Sagasta and Salian 2012). A framework should be developed in order to generate national level data on municipal wastewater production, collection, treatment, discharge or direct use in agriculture is very much of timely approach to be taken. This will potentially engage all the those are responsible, namely, policy maker, development practitioners, academic researchers working in national or international organizations, and particularly those dealing with water resources management in agricultural section in urban and peri-urban areas.

47.5   Conclusions

Global climate change likely to make the water security a harder, complex and a costly process to accomplish and sustain (Grey and Sadoff 2007). Alarming effects of climate change will struck the surface and ground water resources severely (Arnell 1999). As a consequence, country's' economy which is dependent on surface water for agricultural irrigation will be significantly impacted (Biswas 2013). For example, countries like Bangladesh, an agricultural economy based country which is highly influenced by seasonal characteristics and climatic variable would be highly impacted. Overall economy would be confounded by the cross cutting effects of climate change on agricultural production of Bangladesh (Sikder and Xiaoying 2014).

There will be a requirement for different strategies to be introduced over the coming decades to deal with the pressure on water, wastewater for irrigation could certainly be a way forward. The drivers of using wastewater irrigation can be expressed widely into social well-being and conservational security (Hamilton et al. 2007). Wastewater provides a reliable and possibly less costly water supply to agricultural sector. Wastewater in agricultural use is well fitted in urban agriculture offering a secure urban agriculture along with other benefits, such as greater employment for women or men, increased income and food security. This way of irrigation practices often help to get more yields than using freshwater for irrigation because of its nutrient contents. Using wastewater for agricultural sector is capable to serve for water resource management too (Scheierling et al. 2011).

Though the views among stakeholder's of wastewater management influenced highly on its risk and benefits as a resource (Evans et al. 2010). Actual opportunities and threats of wastewater irrigation can be justified through public investment on research & development and increasing the effort of public entities and researchers by providing them with better data on extent of wastewater use for irrigation (Qadir et al. 2010).

Involvement of private sector is also a very vital approach to make for upscaling irrigation technologies (e.g.: wastewater) and increasing irrigation investment strategies. A national guidance and regulation should be introduced and implemented for managing and reuse of wastewater (Kulkarni 2011). Availability on financial resources should be around to implement appropriate protection measure to mitigate some disadvantages faced by environment, user and consumer by wastewater irrigation (Buechler et al. 2006).

Acknowledgements Firstly, I would like to thank all of the respectedglobal researchers for their contribution in the following topic I worked on. Without their thought and contribution, the paper would not have been produced smoothly. Secondly I would like to thank my senior colleague Mr. Feisal Rahman, PhD for providing me with the opportunity to take an approach and to prepare a research paper. Last but not the least, my family which includes my mother, father, wife and sister, without their enormous support nothing would have been so meaningful right now. A special thanks to Mr. Todd H. McKay for his great support.

References

Abaidoo RC, Keraita B, Drechsel P, Dissanayake P, Maxwell AS (2010) Soil and Crop Contamination Through Wastewater Irrigation and Options for Risk Reduction in Developing Countries. In Dion P (ed.) Soil biology and agriculture in the tropics. Springer, pp 275–297

Amerasinghe P, Bhardwaj RM, Scott S, Jella K, Marshall F (2013) Urban Wastewater and Agricultural Resuse Challenges in India. International Water Management Institute (IWMI),Colombo

Anik SI, Khan MA (2012) Climate change adaptation through local knowledge in the north eastern region of Bangladesh. Mitig Adapt Strat Glob Change 17(8):879–896

Arnell NW (1999) Climate change and global water resources. Glob Environ Chang 9:S31–S49

Biswas M (2013) Climate Change and its Impacts on Bangladesh. Planned Decentralization: Aspired Development

Buechler S, Mekala GD, Keraita B Wastewater Use for Urban and Peri-urban Agriculture. Cities farming for the future: urban agriculture for green and productive cities, pp 243–273

Corcoran E (2010) Sick water?: the central role of wastewater management in sustainable development: a rapid response assessment. UNEP

Devi MG, Davidson B, Boland AM (2007) Economics of Wastewater Treatment and Recycling: An investigation of conceptual issues A paper presented at the 51st Australian Agricultural and Resource Economics Society on 13–16 February 2007, pp 13–16

Evans AE, Raschid-Sally L, Cofie OO (2010) Multi-stakeholder processes for managing wastewater use in agriculture. Wastewater Irrigation p 355

Fischer G, Shah M, Tubiello FN, Velhuizen HV (2005) Socio-economic and climate change impacts on agriculture: an integrated assessment, 1990–2080, Philosophical Transactions of Royal Society, 24

Fischer G, Tubiello FN, Velthuizen HV, Wiberg DA (2007) Climate change impacts on irrigation water requirements: Effects of mitigation, 1990–2080. Technol Forecast Soc Change 74(7):1083–1107

GED (2015) Seventh Five Year Plan FY 2016-2020. Government of the People's Republic of Bangladesh, Dhaka

Grey D, Sadoff CW (2007) Sink or Swim? Water security for growth and development. Water Policy 9(6):545–571

Hamilton AJ, Stagnitti F, Xiong X, Kreidl SL, Benke KK, Maher P (2007) Wastewater irrigation: the state of play. Vadose Zone J 6(4):823–840

IPCC (2008) Climate change and water. Intergovernmental Panel on Climate Change, Geneva

IUCN (n.d.) Climate change and agriculture: Information brief, Government of the People's Republic of Bangladesh, Dhaka

Jeong H, Kim H, Jang T (2016) IrrigationWater quality standards for indirect wastewater reuse in agriculture: a contribution toward sustainable wastewater reuse in South Korea. Water 8(4):169

Jimenez B (2006) Irrigation in developing countries using wastewater. Int Rev Environ Strateg 6(2):229–250

Jiménez B, Drechsel P, Koné D, Bahri A, Raschid-Sally L, Qadir M (2010) Wastewater, sludge and excreta use in developing countries: an overview. Wastewater Irrig 1:25

Kaur R, Wani S, Singh A, Lal K (2012) Wastewater production, treatment and use in India. In National Report presented at the 2nd regional workshop on Safe Use of Wastewater in Agriculture, pp 1–13

Keraita B, Jimenez B, Drechsel P (2008) Extent and implications of agricultural reuse of untreated, partly treated and diluted wastewater in developing countries. CAB Rev Perspect Agric Vet Sci Nutr Nat Res 3(58):1–15

Kiziloglu FM, Turan M, Sahin U, Kuslu Y, Dursun A (2008) Effects of untreated and treated wastewater irrigation on some chemical properties of cauliflower (Brassica olerecea L. var. botrytis) and red cabbage (Brassica olerecea L. var. rubra) grown on calcareous soil in Turkey. Agric Water Manag 95(6):716–724

Kreft S, Eckstein D, Doresch L, Fischer L (2014) Global climate risk index 2016, Who suffers most from extreme weather events, pp 1–31

Kulkarni S (2011) Innovative technologies for water saving in irrigated agriculture. Int J Water Resour Arid Environ 1(3):226–231

Mara D (2003) Domestic Wastewater Treatment in Developing Countries, Earthscan

Mateo-Sagasta J, Salian P (2012). Global database on municipal wastewater production, collection, treatment, discharge and direct use in agriculture. In Report on the Methodologies of the Food and Agriculture Organization's. (FAO) Aquastat

Misra AK (2014) Climate change and challenges of water and food security. Int J Sustain Built Environ 3(1):153–165

Mojid M, Wyseure G, Biswas S, Hossain A (2010) Farmers' perceptions and knowledge in using wastewater for irrigation at twelve peri-urban areas and two sugar mill areas in Bangladesh. Agric Water Manag 98(1):79–86

Mokhtari T, Bagheri A, Alipour MJ (2012) Benefits and risks of wastewater use in agriculture. In The 11th International and The 4th National Congress on Recycling of Organic Waste in Agriculture 26–27 April 2012, pp 1–12

Mousavi SR, Tavakoli MT, Dadgar M, Chenari AI, Moridiyan A, Shahsavari M (2015) Reuse of treated wastewater for agricultural irrigation with its quality approach. Biological Forum 7(1):9

Nam W-H, Choi J-Y, Hong E-M (2015) Irrigation vulnerability assessment on agricultural water supply risk for adaptive management of climate change in South Korea. Agric Water Manag 152:173–187

Qadir M, Bahri A, Sato T, Al-Karadsheh E (2010) Wastewater production, treatment, and irrigation in Middle East and North Africa. Irrig Drain Syst 24(1–2):37–51

Qadir M, Wichelns D, Raschid-Sally L, McCornick PG, Drechsel P, Bahri A, Minhas PS (2010) The challenges of wastewater irrigation in developing countries. Agric Water Manag 97(4):561–568

Raschid-Sally L, Carr R, Buechler S (2005) Managing wastewater agriculture to improve livelihoods and environmental quality in poor countries. Irrig Drain 54:1–12

Rizwan A, Robinson C, Clemett A (2009) Management and Treatment of Urban Wastewater for Irrigation in Rajshahi. Bangladesh- WASPA Asia Project, NGO Forum for Drinking Water Supply and Sanitation, Dhaka

Rutkowski T, Raschid-Sally L, Buechler S (2007) Wastewater irrigation in the developing world—two case studies from the Kathmandu Valley in Nepal. Agric Water Manag 1(91):83–91

Sagasta J, Medlicott K, Qadir M, Raschid-Sally L, Drechsel P, Liebe J (2013) Proceedings of the UN-Water Project on the Safe Use of Wastewater in Agriculture, UN-Water Decade Programme on Capacity Development (UNW-DPC)

SAI Platform (2010) Wastewater use in agriculture. Water Conserv. 7:1–30

Sato T, Qadir M, Yamamoto S, Endo T, Zahoor A (2013) Global, regional, and country level need for data on wastewater generation, treatment, and use. Agric Water Manag 130:1–13

Scheierling MS, Bartone C, Mara DD, Drechsel P (2010) Improving Wastewater Use in Agriculture: An Emerging Priority. World Bank Policy Research Working Paper Series, pp 111–107

Scheierling SM, Bartone CR, Mara DD, Drechsel P (2011) Towards an agenda for improving wastewater use in agriculture. Water Int 36(4):420–440

Scott CA, Faruqi NI, Sally LR (2004) Wastewater Use in Irrigated Agriculture: Management Challenges in Developing Countries. Wastewater Use in Irrigated Agriculture: Confronting the Livelihood and Environmental Realities, pp 1–10

Shahid S (2011) Impact of climate change on irrigation water demand of dry season Boro rice in northwest Bangladesh. Clim Change 105(3–4):433–453

Sikder R, Xiaoying J (2014) Climate change impact and agriculture of bangladesh. J Environ Earth Sci 4(1):6

Stevanovic M, Popp A, Lotze-Campen H, Dietrich JP, Muller C, Bonsch M, Weindl I (2016) The impact of high-end climate change on agricultural welfare. Sci Adv 2(8):10

The World Bank (2010) Improving Wastewater Use in Agriculture: An Emerging Priority. The World Bank

UNESCO. (2003) Water for people Water for life. UNESCO

UNFCCC (2006) Technologies for adaptation to climate change. UNFCCC, Bonn

UNFCCC. (2011) Climate Change And Freshwater Resources. United Nations Framework Convention on Climate Change

UNFCCC (2014) Background Paper on Technologies for Adaptation, Bonn.

UNW-DPC. (2013) Safe Use of Wastewater in Agriculture. UNW-DPC, Bonn

Weckenbrock P (2010) Making a virtue of necessity –wastewater irrigation in a periurban area near Faisalabad, Pakistan. Freiburg im Breisgau

WHO (2006) Volume 1: Policy and regulatory aspects. In Guidelines for the safe use of wastewater, excreta and greywater. World Health Organization, Geneva, p 114

WHO (2006) Wastewater use in agriculture. in guidelines for the safe use of wastewater, excreta and greywater, World World Organization, Geneva, p 222

World Health Organization (2006) Guidelines for the safe use of wastewater, excreta and greywater

Zou X, Li Y, Cremades R, Gao Q, Wan Y, Xiabo Q (2013) Cost-effectiveness analysis of water-saving irrigation technologies based on climate change response: a case study of China. Agric Water Manag 129:9–20

Chapter 48
Monitoring of Performance of Deammonification Process in Treating Wastewater in a Pilot Study, KTH, Sweden

M. T. Ur Rahman, U. R. Siddiqi, Md. A. Habib, and Md Rasheduzzaman

Abstract Due to the escalating demand of the sustainable biological nitrogen removal processes nowadays, deammonification has been investigated by several research groups in many developed countries. Since 2001, deammonification involving less energy and chemical demand has been studied in a laboratory scale pilot plant consisting of two reactors filled with Kaldnes rings as the carrier for biofilm development at the Royal Institute of Technology, Stockholm, Sweden. Based on the models developed for the collected data of this pilot plant, this article is concerned with the evaluation of the deammonification process using Multivariate Data Analysis (MVDA) software package SIMCA-P. Mainly four years' operational data including all physical, analytical, and derived parameters have been evaluated with MVDA in developing six trial models with PCA (principle component analysis) and then with PLS (partial least squares with latent structures). Interpretation of the output of these models would suggest the researchers to consider influencing operational parameters, which have the strongest positive or inverse relationships with other variables. However, overall efficiency of the pilot plant has been proved as the important assessing parameter while evaluating the performance of the deammonification process. Conductivity measured in reactor two, on the other hand, has been identified as the most important monitoring parameter during operational time period.

Keywords Efficiency · Multivariate Data Analysis (MVDA) · PCA · PLS

M. T. Ur Rahman · Md. A. Habib (✉) · M. Rasheduzzaman (✉)
Climate Lab, Military Institute of Science and Technology, Dhaka, Bangladesh

M. T. Ur Rahman
Shahjalal University of Science and Technology, Sylhet, Bangladesh

U. R. Siddiqi
Disease Control Unit, Directorate General of Health Services, Dhaka, Bangladesh

© Springer Nature Switzerland AG 2021
M. Babel et al. (eds.), *Water Security in Asia*, Springer Water,
https://doi.org/10.1007/978-3-319-54612-4_48

48.1 Introduction

Nitrogen removal is the main feature of wastewater management often carried out by microbial progression including nitrification and denitrification (Jetten et al. 1997; Rahman 2006). In recent times, a novel microbial procedure for nitrogen elimination was studied in a fluidized bed reactor in Delft (The Netherlands) (Mulder 2003). The treatment of nitrogen-rich water can be taken place fruitfully by an amalgamation of the Sharon and Anammox processes (Tsushima et al. 2007). In this joint process, more than 50% of the aeration energy is saved, no COD is needed, and the least amount of sludge is produced. This newer development could lessen the release of the greenhouse gas CO_2 during wastewater treatment by 88% while lowering the running cost of current treatment systems by 90% (Plaza 2003; Hulle 2005) compared to conventional methods.

In the Land and Water Resources Eng. Department of Royal Institute of Technology (KTH), pilot plant-scale experiments have been carried out to study the Anammox for nitrogen removal from wastewater since 2001 (Plaza 2003; Gut et al. 2007; Rahman 2006). Collected data of different physical and chemical operating parameters for the successive years have been utilized as the input in the Multivariate Data Analysis software with a view to obtaining a model to evaluate the performance of this Anammox conducting at the laboratory-scale pilot plant (Hulle 2005).

Amaralab et al. (2005) conducted a research related with activated sludge monitoring of a wastewater treatment plant using image analysis and partial least squares regression. They utilized PLS (partial least squares) multivariate statistical technique to treat the morphological data of the operating parameters (ex. TSS and SVI) of the biomass attending in the wastewater treatment plant.

Recently, the research group (Gut et al. 2007) of Water management of Land and Water Resources Dept. at KTH, Stockholm, Sweden, has applied the multivariate data analysis to make an assessment of a two-step partial nitration/anammox system. Both PCA and PLS have been utilized to obtain relationships between different controlling variables. Their findings show that nitrite to ammonium ratio (NAR) appears to be the key factor in the process control and monitoring. They have concluded that multivariate data analysis provides a powerful tool in assessing the partial nitration/anammox system.

The key target of this study is to evaluate the nitrogen removal system with partial nitration/anammox process investigated at the laboratory-scale pilot plant with application of Multivariate Data Analysis.

48.2   Material and Methods

48.2.1   Pilot Plant Description

In 2001, researchers of water engineering and resource management group of Land and Water Resources Engineering department at Royal Institute of Technology, Stockholm, Sweden set up a laboratory-scale pilot plant (Fig. 48.1) in order to investigate the nitrogen removal from wastewaters containing high nitrogen load and low biodegradable organic matter. The target of this experimental study is to investigate the nitrogen conversion pathways (deammonification method) and to assess the controlling factors having a role on the newly invented microorganisms' growth rate. Another important goal of this pilot plant is to apply the upshot of these experimental reactions for the treatment of the extremely ammonium-rich wastewaters such as leachate and supernatants from dewatered sludge in a larger scale. The pilot plant has two reactors filled with Kaldnes rings (Fig. 48.2) using as a carrier component for the biomass development in a fixed film. Table 48.1 present the characteristics of the pilot plant.

Mechanical stirrers are furnished in the reactors to provide proper oxygen supply and to discourage sedimentation process. Two heaters with thermostats in two reactors are installed to maintain the temperature at a constant level considering the seasonal variation and the indoor heating system of the plant room. The inlet tank of the pilot plant is continuously filled with diluted and dewatered supernatant collected from the industrial scale waste water treatment plant (WWTP) situated at Bromma.

Fig. 48.1 Laboratory-scale pilot plant (*Source* Plaza et al. 2003)

Fig. 48.2 Kaldnes ring used
as a carrier

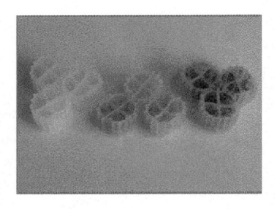

Table 48.1 Design
characteristics for the pilot
Plant at KTH (after: Plaza
et al. 2003)

Item	Magnitude	
	R-1	R-2
Volume (dm^3)	9.81	7.06
Kaldness filling (%)	40	40
Active Volume (m^3)	3.92	2.82
Average Flow (liter per day)	4	4

For supplying sewage from the inlet tank to the first reactor, a peristaltic pump is
employed. Gravitational method is used for supplying liquid from the first to the
second reactor and then from the second reactor to the outlet tank (Plaza et al. 2003).
To adjust the pH of the solution, reactor one has another peristaltic pump for applying
Na$_2$CO$_3$ (0.5 M) dose time to time. Dewatered supernatant produced in the centrifuge
sludge digesting system at Bromma WWTP is supplied to the laboratory-scale pilot
plant situated at KTH. The supernatant contains 706.6 mg/L of NH$_4$-N and 321.7 mg
O$_2$/L of COD on an average.

48.2.2 Measurement of Selected Parameters

Since July, 2001, researchers responsible for the laboratory-scale pilot plant have
been conducting the measurement of selected parameters such as pH, temperature
and dissolved oxygen. For measuring conductivity and nitrogen compounds (NH$_4$-N,
NO$_3$-N, and NO$_2$-N) samples have been collected once per week. ORION (model
210A) is used for pH measurement, while RUSSLEL electrode (model RL425) is
employed to take reading of DO. On the other hand, TEACTOR-AQUATEC 5400,
ANALYZER is used for measuring nitrogen forms in the specimens.

48.2.3 Collected Data of the Parameters

The database of the laboratory-scale pilot plant consists of almost four years' data starting from July´2001 to July´2005. In this study, 235 days (04-07-2001 to 20-07-2005) out of 4 years are considered for analysis excluding the missing measurements to make a homogeneous data set for all of the parameters. With some missing values, the database is sufficient enough to provide the necessary information for analysis in the multivariate data analysis package software. Using Microsoft Office Excel program and applying the regression equations as algorithms, abundant data set for the parameters like NH_4-N, NO_2-N, NO_2-N/NH_4-N are computed.

48.2.4 Modelling Software

Umetrics AB- Software like MODDE suitable for design of experimental and optimisation purpose and SIMCA for multivariate data analysis are the cutting edge products of Umetrics AB. MODDE and SIMCA are the windows based software and easy to operate and build models (Pell and Ljunggren 1996).

SIMCAP, the sophisticated "point and click" software for multivariate modeling and analysis, is capable of squeezing the huge data set into a few concrete graphical presentations. This software supports file formats like EXCEL, MATLAB 4.0, LOTUS JCAMP-DX, Image analysis and direct import from databases. Statistical parameters like sampling numbers, variables, minimum, maximum, mean, median, standard deviation, skewness, frequency histogram are used as built in functions in this software for data viewing purposes. Moreover, data sets having missing values up to 50% can be possible to handle with this tool. It is possible to generate new variables with this program. Transformation of variables and quick information about variables are possible to perform in the spreadsheet of the work set. There is an option to include or exclude the variables and observations, when it is required to do that operation. In model developing, PCA (Principal Component Analysis), PLS (Partial Least Squares), PLSDA (Partial Least Squares Discriminant Analysis), automatic fit of class model, Scores, Loadings, Coefficient plots and VIP are considered for analysis. In model validation, cross validation and permutation test are particularly performed.

48.3 Results and Discussion

48.3.1 Trial with PCA

Taking into consideration all available homogeneous data (04-07-2001 to 20-07-2005), trial was designed to get an overview of the whole observation data set by

PCA and then attempts were made to develop further model with PLS to obtain the interconnection between different physical parameters and derived variables (factors and responses). After accomplishing the data pre-treatment including both scaling and transforming, data are handled with PCA to develop PC model. Data are fitted with four principal components (A = 4). Model explained 81.4% (R^2X) and predicted 53.6% (Q^2). Two separated clusters of the observation data (total observation data = 232) are presented clearly in the Fig. 48.3, which need to classify the whole data in two classes such as class-one and class-two, and this two classes are plotted successfully with the help of Cooman's plot (Fig. 48.4). Strong outliers are eliminated from the data set to increase the predictive capability of the model. Distribution of observations of the influent ammonium nitrogen concentration (NH4-Nin-R1) is plotted as the score scatter plot (Fig. 48.5). Most of the concentrated data are skewed in the North-West direction. Then, score scatter plot showing distribution of observations of the overall efficiency (OA Effic.) of the pilot plant were plotted to get an idea of the distribution of the efficiency data for the selected experimental period. Temperature and oxygen in reactor two were discarded from the final model, as they were identified as insignificant. The score scatter plot (Fig. 48.6) of overall efficiency shows that most of the data are in the higher efficiency range (60–90%). It means that the efficiency of the pilot plant in removing the nitrogen is satisfactory enough.

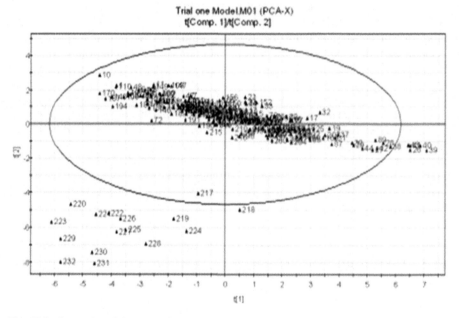

Fig. 48.3 Score plot of the data set by PCA

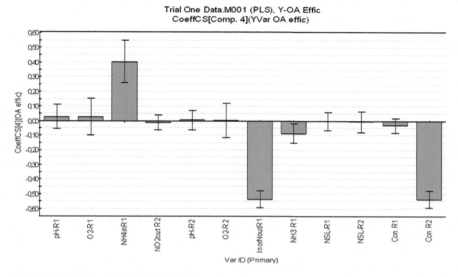

Fig. 48.4 Coefficient plot for OA efficiency

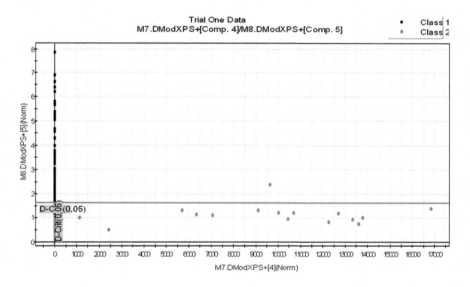

Fig. 48.5 Score scatter plot showing distribution of observations of the Over All Efficiency (OA Effic.) of the pilot plant

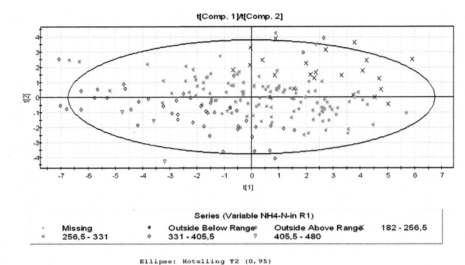

Fig. 48.6 Score scatter plot showing distribution of observations of the influent ammonium nitrogen concentration (NH4-Nin R1)

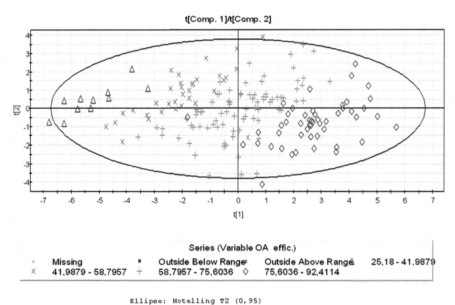

Fig. 48.7 Loading Scatter plot of the selected variables for the Trial one

Loading plot (Fig. 48.7) presents that conductivity has the strongest relationship with influent nitrogen concentration at R1 and effluent nitrite concentration (NO_2-N-R2), which shows that conductivity could be the important parameter to monitor the process performance. Oxygen in reactor one has a positive influence on both of the influent ammonium (R1) and effluent nitrite concentration (NO_2-N-R2). On the other hand, pH in reactor two might be another important parameter while measuring the performance of the Anammox process. The interesting finding of this figure is that overall efficiency has a strong adverse relationship with pH (both R1 and R2), which means increasing pH (R1 and R2) will ensure decreasing overall efficiency. And so, pH would be an important parameter while measuring the overall efficiency of the plant.

48.4 Trial one with PLS (Y = OA Efficiency)

Taking OA Efficiency as a response variable in the same trial for four PLS components, explained and predicted variables with the model such as R2X-77.5.7%, R2Y-95.8% & Q2-96.3% were found. Loading plot (Fig. 48.8) shows that efficiency of the pilot plant is strong negatively interrelated with inorganic nitrogen from reactor one and conductivity of the reactor two. Moreover, free ammonia in reactor two influences positively the overall efficiency. DO in reactor two also has negative influence on the efficiency. NO_2-N out from R2 has a strong negative influence on efficiency. Nitrite in effluent (R2) is found as the most important indicator showing inverse relationship in getting an idea about efficiency. Conductivity in the effluent (R2) is found as the most valuable monitoring parameter for the computation of the process

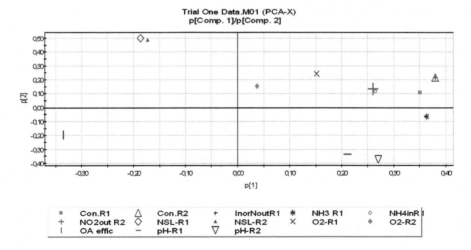

Fig. 48.8 Loading scatter plot for OA efficiency

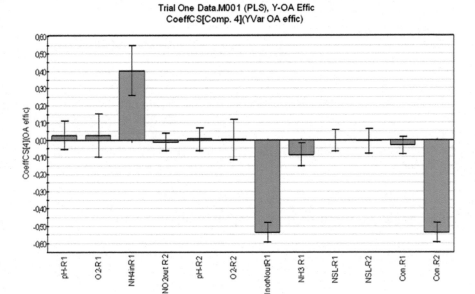

Fig. 48.9 Coefficient plot for OA efficiency

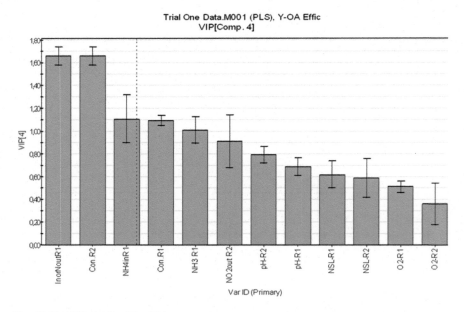

Fig. 48.10 VIP plot for OA efficiency

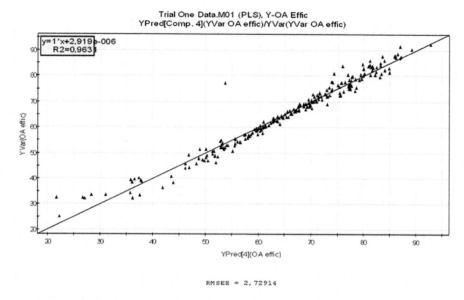

Fig. 48.11 Observed vs Predicted plot for OA efficiency

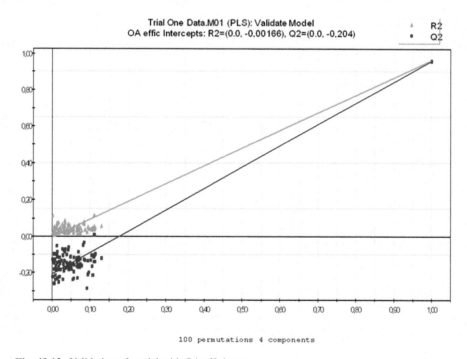

Fig. 48.12 Validation of model with OA efficiency

efficiency. Coefficient plot (Fig. 48.9) and VIP plot (Fig. 48.10) present the influencing variables on efficiency. The regression coefficient (R2) for the observed vs. predicted graph is found as 96%, which shows excellent fitting of the data considered (Fig. 48.11). Figure 48.12 presents that the model is excellently valid according to the model validation perspective.

48.5 Conclusions

Finally, it can be concluded that these six trials were devoted to explain the process performance of the partial nitration/Anammox process of the pilot plant. After critically interpreting the outputs of the six trail models, it can be stated that during the initial stage of the selected time period, partial nitration process was run successfully in the first reactor. Then, partial nitration (in R1) and Anammox process (in R2) have been running separately for the time being. After that, both processes have been occurring simultaneously (in the CANON form) in both of the reactors having a stable condition with a moderately nitrogen removal efficiency. Outcome of all trials harmonized in one idea that overall efficiency was the best parameter considering it as the response variables while assessing the process performance.

Acknowledgements The author would like to express his gratefulness to Swedish International Development Cooperation Agency for awarding research scholarship while staying in Sweden.

References

Amaralab AL, Ferreiraa EC (2005) Activated sludge monitoring of a wastewater treatment plant using image analysis and partial least squares regression. Anal Chim Acta 544(2005):246–253

Gut L, Plaza E, Hultman B (2007) Assessment of a two-step partial nitration/Anammox system with implementation of multivariate data analysis. Chemom Intell Lab Syst 86(1):26–34

Hulle VS (2005) Modelling Simulation and optimization of autotrophic nitrogen removal process. Doctoral Thesis in Environmental Technology. Applied Biological Sciences. University of Gent, Belgium

Jetten MSM, Horn SJ, van Loosdrecht MCM (1997) Towards a more sustainable municipal wastewater treatment system. Water Sci Technol 35(9):171–180

Mulder A (2003) The quest for sustainable nitrogen removal technologies. Water Sci Technol 48(1):67–75

Plaza E, Trela J, Gut L, Lowen M, Szatkowska B (2003) Deammonification process for treatment of ammonium rich wastewater. In Proceedings International optimisation of urban sanitation systems, Stockholm, pp 77–87

Pell M, Ljunggren H (1996) Coiviposition of the bacterial population in sand-filter columns receiving artificial wastewater, evaluated by soft independent modelling of class analogy (SIMCA). J. Adv Water Resour 30(1996):41–48

Rahman MT (2006) Application of Multivariate Data Analysis for Assessment of Partial Nitration/Anammox Process. MSc Thesis, KTH, Stockholm, Sweden

Tsushima I, Ogasawara Y, Kindaichi T, Hisashi S (2007) Development of high-rate anaerobic ammonium-oxidizing (Anammox) biofilm reactors. Water Res 41(8):1623–1634

Part VIII
Disaster Risk Assessment and Management

Chapter 49
Assessment of Urban Flood Resilience for Water, Sanitation and Storm Water Drainage Sectors in Two Cities of India

S. Thakur and U. Bhonde

Abstract The study appraised infrastructure, institutional, and financial gaps in two cities in different geo-climatic zones for water, sanitation, and storm water drainage (WSS) sectors. Both study cities have similar topographical conditions like bowl shape surrounded by hills. Storm water runoff with high velocity from steep hills results in urban flooding and impose threats to infrastructure like intake points, treatment plants, and distribution systems. There is no single institution responsible to respond to urban flooding in terms of planning and designing resilient infrastructure, to co-ordinate between agencies during urban floods, to assess damage to critical infrastructure, and importantly to document it. Risk profiling and vulnerability assessment of critical infrastructure needs to be undertaken and suitable adaptation measures are required, which are cost effective, hybrid and redundant in nature, and are essential parts of planning and designing new infrastructure and retrofitting existing.

There are multiple attempts to assess vulnerability and resilience of Indian cities; however, none of these studies looked into the cost of building resilient cities. This research study was designed to estimate approximate financial cost of urban resilience for WSS sectors and their resilience towards urban floods. The study indicates that cities resort to costly engineering solutions to reduce water logging and lack integrated planning approach. It emphasise that unless infrastructure, institutions, and finances are strengthened, Indian cities would not become flood resilient. Adapting to green infrastructure, doing conservation of wetlands and local water bodies, capacity building of city managers, and improving own financial resources could make WSS sectors flood resilient.

Keywords Urban floods · Resilience · Green infrastructure

S. Thakur (✉) · U. Bhonde
National Institute of Urban Affairs, New Delhi, India
e-mail: sthakur@niua.org

© Springer Nature Switzerland AG 2021
M. Babel et al. (eds.), *Water Security in Asia*, Springer Water,
https://doi.org/10.1007/978-3-319-54612-4_49

49.1 Introduction

Floods are high frequency high impact natural hazards in India, experienced every monsoon season in one or other parts of country as well as during cyclonic conditions. Majority of rivers in India are non-perennial in nature and carries enormous amount of storm water and sediments (silt load) from catchment areas with high rainfall. When the carrying capacities of rivers are lesser than surface runoff from catchment areas, flooding conditions occur in rivers/tributaries. Every five years, a major flood is witnessed in India (NDMA 2008) and results in life and property losses. Since past few years, occurrences of flash floods have also been reported in India associated with cloud bursts. This has resulted in life losses and damage to public and private infrastructure especially in the cities located in hilly terrains.

However, it was Mumbai and Surat cities floods in 2005 and 2006 respectively that highlighted urban flooding in country. These two important cities located in western part of India having high economic significance faced decelerated economic growth due to urban flooding. It was after these urban floods more attention has been given to this phenomenon in India. Recently (2015), Chennai city floods in south of India, which resulted in very high economic losses, clearly revealed that even metropolitan cities of India are not fully prepared against floods in all aspects, i.e. infrastructure security, institutions' capacities, and economic reliability to meet expenditures. Urban areas (cities) play a vital role in economic growth of the country as their share in Gross Domestic Product (GDP) is increasing. High Powered Expert Committee (HPEC) indicated that it is expected to be 75% by 2030 (HPEC 2011) in India. Thus, any adverse impact on urban economy due to natural or manmade disasters directly affects the economy of India.

Since floods are recurrent phenomena, this article has focused on financial aspects of urban flooding. In India, urban areas are flooded by the river floods due to high rainfall (mostly), flash floods in mountainous areas, coastal floods, and release of excess water from reservoirs, or reduced river bed/dam capacity to accommodate water due to high silt load from deforested catchments. Heavy rainfall is root cause for urban floods which is increasing due to changing climatic conditions. Due to intense and periodic rains, huge quantity of water flows beyond carrying capacity of drainage systems of the cities. Drainage systems remain blocked due to silting, dumping of solid waste material at the inlets of drainages, and encroachment on natural drainage and water bodies. Impacts of urban floods has been observed in 2016 monsoon in the Gurugram, a satellite town of Delhi, India. Due to short duration heavy rainfall, water logging affected traffic. Travellers including children in school buses were stuck for hours (6 h in some cases). Unanimously, poor urban planning was held responsible for this.

Urban floods increase flood peaks from 1.8 to 8 times and flood volumes up to 6 times (NDMA 2010). This is due to developed catchments, wherein significant land use land cover changes have occurred over the time due to poor urban planning. Floods with high velocity damage critical infrastructure like Water supply, Sanitation, and Storm water drainage systems (WSS) in urban areas. Damages to public properties due to drainage congestions are very common in many cities across India. Urban local bodies (government) have to face high economic losses. Despite of this, systematic documentation of damages to public properties are not maintained by authorities.

Govt. of India [ministry of environment, forest and climate change (MoEF)] had already communicated to UNFCCC that due to climatic changes frequency of urban floods is expected to increase in India and adversely affect water balance of the country and urban areas (NATCOM 2004). Researchers have also indicated increase in floods from 10 to over 30% from their existing magnitude across all river basins in India (Gosain 2011). Significant care for infrastructure is recommended for such increase especially for storm water drains, roads, dams, bridges, etc., on which modern cities are highly depended.

Flood management in India is not new as it has started around 1960s. Considerable progress has been done by authorities to minimize impacts of floods on communities, livestock, and infrastructure but with emphasis on structural measures (embankments, levees, dams etc.). However, since urban flooding is comparatively recently experienced and understood in India, planning to mitigate it is yet to be formalized starting with documentation of damages occurred to urban service sectors.

Globally, the concept of "flood resilience", which is an advanced curative approach to mitigate urban floods, is gaining momentum. In India presently the approach is more reactive in nature that is to respond after floods occur. It is more on managing the floods and not preventing them. Following section provides brief on differences between flood management and flood resilience from Indian perspective to form setting for this article.

49.2 Flood Management vs. Flood Resilience: Indian Perspective

In India, after the Disaster Management Act enacted in year 2005, National Disaster Management Authority (NDMA) was established. Under its guidance, all State Disaster Management Authorities (SDMAs) function. The Act made it statutory requirements for National Disaster Response Force (NDRF) to respond when natural disasters like floods occur in the country. Correspondingly, the National Disaster Management Institute (NDMI) was also formed to conduct research and capacity building programs for institutions.

As a part of flood management strategy, both structural and afterwards non-structural measures have been adopted in India. Construction of embankments, dykes, and dams are being done in many flood prone states in India like Assam, Bihar, Uttar Pradesh, and Odisha in order to control floods. This had increased after flood commission's report in year 1980. Other forms of structural measures adopted are flood levees, ocean wave barriers, flood retention walls, sand bags, evacuation centers/shelters, etc. However, non-structural measures, which do not involve any physical construction but uses knowledge, e.g., flood zoning maps, Early Warning Systems (EWS), planning and coordination, agreements, treaties to reduce flood risks, policies, acts, public awareness, and training and capacity building are yet in preliminary stage in India.

Term "resilience" is now widely referred particularly in natural hazard risk reduction and climate change adaptations programs. Web research indicates that use of this term has become prominent post year 2005 (Serre and Barroca 2013). Resilience is an interdependent parameter and is, therefore, complex in nature as well as difficult to practically apply. There are different definitions proposed for resilience, considering systems including regions, communities, households, economy, ecology, etc. Nevertheless, three basic properties of resilience, i.e. speedy recovery, tolerance limit before changing structural property, and capacity to learn from the failure and transform, are commonly observed most (Folke et al. 2002). The United Nations International Strategy for Disaster Reduction (UNISDR) defines resilience as "*an ability of a system, community or society exposed to hazards to resist, absorb, accommodate and recover from the effects of a hazard in a timely and efficient manner, including through the preservation and restoration of its essential basic structures and functions*" (https://www.unisdr.org/we/inform/terminology#letter-r).

In India, understanding of resilience is limited and now increasingly used by academia and agencies of the development sector. Especially, the term is frequently used post climate change and natural hazard studies carried out by national agencies MoEF and NDMA under resilience programs and with support from international donors (e.g., ACCCRN program by the Rockefeller Foundation[1]). During this research study, it was observed that Urban Local Body (ULB) officials/engineers are not aware about the word resilience and its true meaning. Understanding of officials is limited to engineering solutions (structural) as it is established, do not require experimenting, and practical in the Indian context.

This research article emphasis on financial aspects of urban flood resilience particularly for WSS infrastructure. WSS system has critical infrastructure and its flood resilience is to withstand changing conditions (environmental) and to provide continued services as designed before the changes. Thus, "flood resilient" infrastructure is one, which is able to survive flood event and to resume services during and after the flood event. However, in contrast to this "flood resilience" is response of system as a whole in city to floods including infrastructure, institutions that govern infrastructure, policies, and financial conditions that create and manages infrastructure in a sustainable manner. In most of Indian cities, these systems either function in isolation or even do not exist, and are major hurdle in achieving flood resilience. There are very few examples on urban flood resilience particularly for WSS sectors in India and, therefore, international literature was referred to understand urban flood resilience particularly for WSS sectors. Following section provides benchmark literature developed based on latest international experiences. It has been referred for this research and analysed from Indian context.

---

[1] Asian cities climate change resilience network (ACCCRN)- www.acccrn.net

49.3   Flood Resilience of WSS Sectors

The Construction Industry Research and Information Association (CIRIA), a UK based institution, has published a manual on flood resilience for critical urban infrastructure (McBain et al. 2010). According to it, flood resilience is designing or retrofitting an existing infrastructure in such a way that due to floods, no permanent structural damage occurs, or if at all it occurs then with minor repairs and minimum costs the services can be resumed. United States Environmental Protection Agency (USEPA) published in the year 2014 guidelines on flood resilience for water and wastewater utilities. This guideline discusses methodology to derive the financial costs of flood resilience for water and wastewater utilities. The method considers risk exposure, vulnerability of water assets like intake, treatment and distribution systems, and then lists activities to reduce flood impacts with required costs, which give the costs of resilience for water and wastewater utilities.

When present study cities were investigated in light of international guidelines, baseline data like historical floods information, inundation depths marking in different areas of city, extent of damage to critical infrastructure, etc. were not available. Particularly for the present study cities, only past two years' data on economic losses were available. Therefore, the first recommendation of this research article is to systematically record the baseline information in Indian cities after flood events.

The Zurich flood resilience alliance (2014) recommended four decision making tools, like cost–benefit analysis (CBA), cost-effective analysis (CEA), multi-criteria analysis (MCA) and robust-decision-making-approaches (RDMA) in order to make communities more flood resilient. Each of these tools has its own benefits and challenges. Out of all, CBA is considered more quantitative tool for flood resilience, however, data gaps remain an important challenge to apply this tool as it considers both tangible/measurable and non-tangible (indirect) losses during floods. In the Indian context, this tool is difficult to apply in practice as data scarcity is a main challenge. Whereas, tools like MCA are more realistic, but are applicable to specific region and non-replicable for other areas. Many risk assessments and vulnerability analysis studies conducted in India in past few years fit in MCA category.

The Collaborative Research for Floods in Urban Areas (CORFU), a European Union program proposes relief, resist, response, and reflect as the main elements of flood risk management cycle. It assumes urban areas as a complex system of built environment and society/community and, therefore, urban flood resilience is to accept flood disturbance, to maintain functions, to organize actions to reduce damages, and to recover from the floods with built environment and communities (Batica and Gourbesville 2014).

Indicator based approaches are commonly employed to study disaster resilience under broad dimensions of physical, ecological, social, economic, and institutional assessments. Flood Resilience Index (FRI) with five dimensions viz. natural, physical, economic, social, and institutional were analysed with indicators like sensitivity, availability, affordability, and relevance to determine FRI. Based on FRI at various scales, overall ranking of very low, low, medium, and high were assigned to various areas to know urban flood resilience.

Aerts et al. (2014) estimated the costs of flood resilience in the coastal megacity of New York for different strategies. Three strategies were proposed to make coastal areas flood resilient viz. (a) measures to enhance building codes in New York city, a non-structural approach, (b) different structural measures like levee barriers, beach nourishment in coastal areas of city, and (c) hybrid approach which combines building codes measures, and barrier construction only for very high risk areas as per probabilistic hazard estimation. Interesting policy recommendations like collection of US$10 resilience fees from tourists (50 million/year) visiting the city to recover an investment in building resilience measures are proposed in the study.

Based on international literature on flood resilience it emerges that the concept is gaining momentum post 2005. There is clear evidence from international literature that practical guidelines on flood resilience particularly on water and wastewater utilities are published in year 2014. Though the term resilience has been gaining attention since 2011 in India, these have been generic and not targeted specific sectors e.g. water and wastewater utilities.

49.4   Aim and Objectives

Studies/research in India on climate change impacts in urban areas are limited to risk profiling, vulnerability assessment of population and service sectors, and then recommend suitable adaptation/ resilience options. However, these attempts have lacked financial aspects of resilience options particularly for WSS sectors. Therefore, important aim of this article is to highlight status of urban flood resilience in general and of WSS sectors in two select cities in India namely Guwahati and Vishakhapatnam in particular. The specific objectives are,

1. To discuss gaps in WSS infrastructure, associated challenges, and how the gap influences flood resilience of the city
2. To appraise WSS infrastructure, institutional set up, and financial status in study cities from flood resilience perspective

Recently, Govt. of India has announced several national missions like smart cities,[2] Atal Mission for Rejuvenation and Urban Transformation (AMRUT),[3] and Swachh Bharat Mission (SBM)[4] to improve urban infrastructure and sanitation conditions in Indian cities. This research study aimed to propose a broad framework to measure the cost of urban flood resilience, which would be useful to city officials in preparing projects in WSS sectors under such missions. It is anticipated that by incorporating the resilience dimension for future WSS projects with some cost details; damages to

---

[2]https://smartcities.gov.in/

[3]https://amrut.gov.in/

[4]https://www.swachhbharaturban.in/sbm/home/

critical infrastructure would be minimized, and services will be less affected during urban floods by achieving some degree of urban flood resilience.

49.5 Material and Methods

Following sections covers systematic approach adopted to conduct this research study.

49.6 Study City Selection

This study aimed to understand status of urban flood resilience in select cities. The selection was done by considering (a) urban flood impacts in cities in general and on WSS sectors, (b) economic importance of cities, (c) performance of cities on infrastructure development, and (d) future prospectus' of cities in new missions. Baseline information of all important cities across India participated in previous and current national missions were scanned. As smaller cities lacked organized information on WSS sectors on infrastructure, institutions and finance (IIF) aspects; medium sized cities (<2 million populations) located in different geo-climatic zones of India and also flood prone were targeted.

In India, planning process starts with master/zonal plans at regional scale, and City Development Plans (CDPs) and City Sanitation Plans (CSPs) at local scale. Recently (2014) with launch of smart city mission by newly elected federal/central government, proposals were prepared to access grants from Ministry of Urban Development (MoUD), Govt. of India to improve infrastructure in cities. Thus such plans, documents of about 40 cities across India were referred to understand infrastructure, institutions and finance status and how WSS sectors have improved with time in the cities. Component of urban flood resilience for WSS sectors in projects submitted from cities was appraised to judge inclusiveness of resilience.

Two cities discussed in this research article are Guwahati in north-eastern state of Assam and the Vishakhapatnam in south-eastern coastal state of Andhra Pradesh. Both cities, which are part of previous and new missions, have different geo-climatic conditions but similar topographical conditions, which make them vulnerable to urban flooding. Communities and WSS infrastructure in both cities have been affected by urban flooding in recent past. WSS sectors have been severely damaged during urban floods.

49.7 Designing Survey Questionnaire

A questionnaire was designed to collect information on urban flood impacts on WSS sectors and flood resilience aspects. Personal discussions were held with city officials (administrative and technical) during city visits to know their understanding on resilience and status of WSS sectors as a system to be flood resilient. Raw water supply sources (intake points), water and sewage treatment plants, solid waste management system, disposal sites, and storm water drainage infrastructure were visited to understand their vulnerability to urban floods and any protections measures adopted.

49.8 Flood Resilience Appraisal

Finally, comparative appraisal of two study cities has been done for the urban flood resilience and a broad framework of flood resilience is proposed for small and medium size cities in India.

This article concludes that though not completely but two cities are certainly taking measures to achieve urban flood resilience. For example, the Guwahati city in north east has raised (retrofitted) electricity transformers to prevent their submergence in flood water, regularization of cleaning storm water drains with improved machinery to reduce water logging conditions, and prepared a standard operational procedure (SoP) to respond during flooding.

49.9 Results and Discussion

Main objectives of this article are elaborated in this section for study cities. Infrastructure gaps in study cities were analysed to understand their impact on making the city flood resilient. This gap is a key factor in achieving flood resilience. Challenges and issues of existing infrastructure gaps and how urban flooding further affects these challenges have been discussed. Three aspects, i.e. infrastructure, institution, and finance from study cities are compared. A need to first prepare a framework on urban flood resilience is found necessary for small and medium sized towns of India.

49.9.1 Geomorphological Setup and Impact on Urban Flooding

Geomorphology of an area is an expression of local landforms controlled by local topography. It is a net result of physiographic set up and geo-climatic conditions of the region. Geomorphology governs land use land cover (LULC) pattern of an area, which changes as urbanization takes place. LULC changes have significant influence on urban flooding, which usually goes unnoticed. Therefore, it was essential to understand local geomorphic conditions of study cities. To derive topography of study areas ASTER data was processed using open source GIS software. This was also useful to know slope conditions and local drainage systems of the study cities. Use of Google Earth imageries and base maps collected from city officials was made to understand regional geomorphological features. Following is a summary of geo-climatic conditions of two cities.

The Guwahati city receives approximately 1,600 mm annual rainfall. The regional slope of city is from south to north direction towards the Brahmaputra River (Fig. 49.1), the ultimate water discharge location. The plateau region in south of the city is the main catchment area for local rivers passing through the city. The city is surrounded by steep hills and, therefore, forms bowl shape topographic conditions. Two important local rivers named Bharalu and Bashishtha originate from southern plateau region. Both are important from urban flooding perspective in the city.

There are three important wetlands in and around city viz. Deepor, Borsola, and Silsako. Locally, they are termed as Beels *(e.g., Deepor Beel)*. Wetlands are very important ecosystems to accommodate large quantities of flood water coming from city side and thus act as buffer areas to prevent urban flooding. Size and shape of these wetlands are significantly reduced due to natural and anthropogenic reasons and are adversely affecting urban flooding in city. The city is facing severe urban flooding and related issues since the year 1980. Urban flooding is more severe when the Brahmaputra River is in the state and local rivers passing through the city cannot discharge water from city side as sluice gates are kept closed to prevent backflow. Raw water is extracted through floating barges in the Brahmaputra River, which are required to be shifted in protected areas when river is in the state. Water distribution lines in hills are damaged due to landslides, induced by heavy rainfall.

On the other hand, the Vishakhapatnam city receives 1,200 mm annual rainfall. The city is an important port city located on the south east coast of India (Fig. 49.1). The natural harbour conditions facilitated it to be an important naval dockyard of the country, and also is hub for many industries. Due to typical geomorphic set up of the city it has also a saucer shape with three steep hills in surrounding, vast low elevated tidal flats and linear coastal stretches. These conditions make the city particularly peri-urban areas vulnerable to urban as well as to coastal flooding. The regional slope in the city follows local topography and is towards Bay of Bengal in the east direction. The city is exposed to various hydro-meteorological hazards like cyclone, urban/coastal flooding, sea level rise, storm surges, and also possible tsunami waves.

Urban flooding is emerging as one of the challenges in city and recently experienced during the Hudhud cyclone in year 2014. The city experiences two monsoon seasons. At least one cyclone per year originating from the Bay of Bengal hits this region. During Hudhud cyclone in October 2014, one of the worst hits in the city resulted in very high economic losses. Low elevation coastal areas and tidal flats especially near airport areas and its surrounding were inundated due to storm surges during high cyclonic winds. Water supply system (both surface and groundwater) in

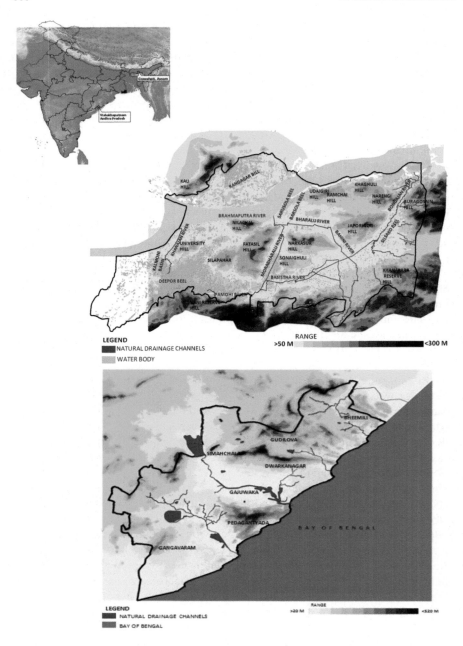

Fig. 49.1 Physiographic set up of two cities Guwahati and Vishakhapatnam *Source* ASTER data

the city was paralyzed and the administration had to import generators from neighbour states to run water treatment plants. In many areas water supply was regular after a week time.

49.10 Infrastructure and Institutional Appraisal (WSS Sectors)

The main aim of this research was to appraise study cities for urban flood resilience. This has been done by considering three important pillars of cities viz. infrastructure, institutions, and finance. To achieve urban flood resilience all the three pillars should be strong or be strengthened. Following paragraphs provides brief on these pillars particularly for two study cities. Since present article emphasise on financial aspects, it is discussed with more details.

a) Infrastructure Appraisal

It was observed that significant infrastructure improvements has occurred in both the study cities in last decade. This was essential with financial support from previous national mission Jawaharlal Nehru National Urban Renewal Mission (JnNURM) on urban infrastructure improvement. However, to fill existing gaps and to meet the increasing demands, new infrastructure projects are being implemented in cities, which utilize financial resources. One of the important reasons for additional infrastructure requirement is expansion of cities by merging peri-urban areas. In case of Vishakhapatnam city this has occurred repeatedly from the year 2005 to 2013. To achieve urban flood resilience of WSS sectors first these two cities will have to reduce the existing infrastructure gap. In case of Guwahati city this gap is wide compared to the Vishakhapatnam city, thus the costs of resilience will be over and above filling existing gaps, and remain higher in case of Guwahati. Moreover, the infrastructure is developed with the financial support from the international institutions like Japan International Cooperation Agency (JICA) and Asian Development Bank (ADB). In both study cities, conventional engineering infrastructure have been adopted which are costly, energy dependent, and require high operation and maintenance costs. Emerging concepts of green infrastructure are not given considerations.

b) Institutional Appraisal

This pillar indicates capacities of organizations or departments responsible in cities to manage WSS sectors and their understanding on the concept of flood resilience. Information on this part was collected during personal discussions with city officials and from web portals of respective municipal corporations. In both study cities, no single institution is responsible for urban flood management in terms of planning and designing resilient infrastructures, co-ordination of different agencies/departments, assessment for flood damages, and losses occurred to critical infrastructure. During discussions with city officials they indicated that in cities, where floods are recurrent phenomena, there is a need to form a dedicated cell. Capacity building activities of staffs are essential for making WSS sectors flood resilient. In case of flood event, how the critical infrastructure should be protected so that no or minimum damage occurs. Presently, training imparted is on SCADA system operations, construction of water supply systems, etc. The trainings are more on supply side management and not on demand side. The CEO of GMDA (Guwahati) informed that there is an urgent need to tie up with international institutions for capacity building, which have learned lessons from severe floods in the past, e.g., Bangkok in Thailand in the year 2011.

There is an important constitutional amendment done in 1992, known in country as 74th CAA. As per it, the state has to devolve power of functional responsibilities to urban local bodies. In case of Guwahati, many functions still remain with states, and the urban local body is not yet given responsibilities of all functions. Multiple agencies are involved in water supply (GMC, GMDA, PHED) and storm water drainage sector (PWD, T&CPD, WRD, GMC, GMD). Out of all these agencies, PHED, PWD and WRD are para-stately agencies not under urban local body. The sewerage infrastructure is completely absent in the city so no agency is responsible. Assam state has formed Guwahati Development Department (GDD), under which the Guwahati Municipal Corporation (GMC) and Guwahati Metropolitan Development Authority (GMDA) functions. Recently, the state has formed an agency named Guwahati Jal Board (GJB), which is supposed to be the umbrella agency for water and sanitation sectors in future. Presently, there are overlapping functions and responsibilities among different agencies, and no single agency is responsible to manage urban flooding. Inter-agency and even inter-departmental coordination is not happening that affects the systems. As informed during the discussions, two agencies, PWD and the GMC, do not coordinate to clean the city drains and does this in different months. This remains ineffective during monsoon despite of spending resources and high costs in cleaning the drains.

In second study city, the Greater Vishakhapatnam Municipal Corporation (GVMC), is empowered by the state to plan, design, and operation and maintenance (O&M) of WSS sectors. Regional urban planning is in the responsibility of Vishakhapatnam Urban Development Authority (VUDA). Thus, for various sectors (WSS) there are no multiple agencies involved, but there are different departments under an umbrella of GVMC. However, there is no accountability decided for urban flooding management. District collector office, which usually responds to natural hazards, has to coordinate for city floods. The area of municipal corporation limit has been substantially increased post 2005. Recently, in 2013, two nearby municipalities have also been merged. However, the infrastructure and institutional arrangements are not improved with respect to increase in the area.

c)   Financial Appraisal

Two types of cost scenarios were attempted to understand financial aspects of urban flood resilience. In the first scenario, the economic losses in the city due to floods were understood. This was essential to understand the costs of inaction. In the second scenario, to know the costs of resilience and the costs to fulfil the infrastructure gaps for WSS sector was appraised with HPEC recommendation on per capita expenditure in WSS sectors. HPEC estimated financial requirements in the country for capital and O&M for 20 years in urban areas of India. The current *appraisal indicates* that costs of resilience in cities with less infrastructure gap and favourable topographical *conditions* (flat terrain) will be less comparable to cities with wider infrastructure gaps and difficult topographical conditions (hilly terrains).

The following sections provides details on financial aspects of urban flood resilience from study cities.

49.11   Cost of Inaction-Financial Cost of Damage to Critical Infrastructure

Vishakhapatnam: In October 2014, the powerful cyclone the Hudhud made its landfall right in this city resulting into US$ 3.4 billion economic losses in the entire district being the most urbanized in the state. Urban flooding occurred in the low lying coastal areas of the city due to storm surges. Water supply was severely affected due to failure of electricity supply and non-availability of generators in water treatment plants. This has affected for a week time some parts of city. Groundwater extraction was also not possible due to failure of electricity supply. As per latest city development plan of 2015, within the GVMC jurisdiction about 393 km of drainage network and 750 km of water distribution network got damaged, 811 km of road surface got damaged as well; 3,830 trees and 60,000 street lights had fallen; 178 municipal buildings got damaged. As per status report prepared by Govt. of Andhra Pradesh overall, the estimated damage to municipal services was Rs.255.58 crore (US$39 million).

Guwahati: Urban flooding affected critical infrastructure (WSS sectors) in the city. Water distribution lines in the hills were damaged due to high rainfall resulting in landslides. Septic tanks, which are widespread in city due to lack of sewerage infrastructure, were choked due to silt load and overflows during floods, and imposed high risks of epidemic. The damage assessment particularly of utility assets was done by respective departments. State revenue department collated all damage details and estimated economic losses occurred due to floods for financial relief from central government. There are no systematic records on financial losses in the city due to urban flooding. As per the Assam State Disaster Management Authority (ASDMA), in past two years (2014 and 2015), total damage of Rs. 223 crore (US$ 34 million)[5] was reported in municipal corporation limits. This amount is much higher than annual budget of GMC. It is evident from expenditure data that water supply sector was affected due to urban floods. Expenditure was incurred on restoration of water supply system and de-silting of storm water drains after floods. Nevertheless, this partly estimated cost do not reveal the total damage cost to city in case of flooding or the inaction.

---

[5]*Conversion done @ Rs.65/US$ / one crore is—1,000,0000.*

49.12   Financial Costs of Resilience

This was very crucial part of study as measuring the costs of resilient city depends on its financial strength. To understand financial status, five years city budgets were analysed to see revenue receipts from different sources like own and external sources, like central, state and other grants, and loans from financial institutions. The revenue expenditure for WSS sectors were analysed for capital, O&M and other expenditure. An assessment of budgets was done from flood resilience view point. Infrastructure gaps in WSS sectors (critical infrastructure) are important in determining the costs of flood resilient WSS sectors. Costs of resilience are higher for those cities, which have wider infrastructure gaps because this will be over and above the infrastructure gap filling cost. In case of Guwahati, the gap is wider compared to the Vishakhapatnam. Thus, cost of resilience is higher in case of Guwahati compared to Vishakhapatnam.

In Guwahati, there are multiple agencies managing WSS sectors in the city. Budgets of all the agencies are separately done, and there are no collated data available with single agency on income and expenditure incurred by different agencies for new infrastructure projects, O&M, etc. NIUA team could only get budget data of GMC (urban local body) for past five years (2011–2012 to 2015–2016) from the office of chief accounts & audit officer. Actual budgets data of 2009–2010 to 2013–2014 were analysed. The data indicated that share of water sector for capital and O&M is quite negligible when compared to entire GMC budget. All new infrastructure projects are being constructed with international financial support on loan basis. GMC budget does not reflect these details. The revenue collected by GMC is not sufficient when compared to expenditure in water sector. The analysis indicates poor performance of WSS sector in terms of financial transactions. The city heavily depends on higher government fund transfers and on grants.

In Vishakhapatnam, analysis of budgets was easy as there are no multiple agencies doing expenditure. Past five years budget analysis indicates that per capita the annual capital expenditure in WSS sectors infrastructure projects is only 9% considering HPEC recommendations, whereas the per capita annual O&M costs are only 52% of recommendations by HPEC. This indicates that the O&M is comparatively better in the city and also validates benefit of having single agency in managing the city. Considering the HPEC recommendations, the capital expenditure requirements in Vishakhapatnam comes to Rs. 2,968 crore (US$457 million) for meeting 2021 population demand of city.

The appraisal of three pillars infrastructure, institutions and finance of study cities indicates that the financial cost to make WSS sector flood resilient will include costs for filling existing and future infrastructure gaps. In addition to this, the resilience costs will include the costs of innovative green infrastructure, institutional strengthening by capacity building to implement new cost effective options, and relying on own financial strength. City can only afford to spend additional infrastructure expenditure for resilience if its existing financial burden is low.

Acknowledgements   This study was part of the project supported by the Rockefeller Foundation.

Abbreviations

CAA	Constitutional Amendment
CEO	Chief Executive Officer
NATCOM	National Communication to United Nations Framework Convention on Climate Change
PHED	Public Health Engineering Department
PWD	Public Work Department
SCADA	Supervisory Control and Data Acquisition
T&CPD	Town and Country Planning Department
WRD	Water Resource Department

References

Aerts JCJH, Wouter Botzen W, Moel H, Bowman M (2013) Cost Estimates for Flood Resilience and Protection Strategies in New York City, Ann. Of NY Acad. of Science. Issue: Cost Estimates for Flood Resilience and Protection Strategies in New York City, vol 1294, pp 1–104

Batica J, Gourbesville P (2014) Flood resilience index - methodology and application. In International Conference on Hydroinformatics, City University of New York https://academicworks.cuny.edu/cc_conf_hic/433

CDP (2014) Revised City Development Plan for Vishakhapatnam – 2041, Capacity Building for Urban Development Project, Min. of Urban Dev. & World Bank

CWC (2013) Water and Related Statistics, Water Resources Information Directorate, Central Water Commission, 29

Folke C, Carpenter S, Elmqvist T, Gunderson L, Holling CS, Walker B (2002) Resilience and sustainable development: building adaptive capacity in a world of transformations. Ambio 31(5):437–440

Gosain A, Rao S, Arora A (2011) Climate change impact assessment of water resources of India. Curr Sci 101:356–371

HPEC Report (2011) Report on Indian Infrastructure and Services, Ministry of Urban Development, Govt. of India, High Powered Expert Committee Report

McBain W, Wilkes D, Retter M (2010) Flood Resilience and Resistance for Critical Infrastructure, Construction Industry Research and Information Association – CIRIA

Serre D, Barroca B (2013) Preface natural hazards and resilience cities. Nat. Hazards Earth Syst Sci 13:2675–2678

USEPA (2014) Flood Resilience- A Basic Guide for Water and Wastewater Utilities, Office of Water (4608T) – EPA 817–B–14 –006

Zurich Flood Resilience Alliance (2014) Making Communities More Flood Resilient The Role of Cost Benefit Analysis and Other Decision-Support Tools in Disaster Risk Reduction, 9

Chapter 50
Vulnerability to Disaster in a Multi-hazard Coastal Environment in Bangladesh

M. M. Islam, M. Mostafiz, P. Begum, A. Talukder, and S. Ahamed

Abstract Using the DPSIR (Drivers, Pressures, States, Impacts, and Responses) framework as an analytical lens, this study elucidates how climate-related hazards coupled with other drivers create disaster vulnerability to coastal communities in the southern region of Bangladesh. Primary data were collected through fieldwork from the five communities in the study area which includes individual interview, focus group discussion and key informant interview. Following this framework, the present study revealed that coastal communities in Bangladesh face recurrent hazards (e.g., cyclones) which coupled with social (e.g., poverty), demographic (e.g., migration) and economic drivers (e.g., poor economic system). These altogether transform the effects of hazards into disaster and make corresponding changes in human well-being and as well as in the environment. Also, environmental degradation (e.g., over-exploitation of common pool resources) seriously undermines the adaptive capacity of the population. As a result of a disaster, communities suffer: human causalities, food insecurity, and malnutrition. In response to the adverse impacts of disaster, affected communities adopt a variety of coping strategies, some of which led them to be worse off. For instance, taking loan for consumption needs, taking children out of school for child labor entrap them into long-term debt bondage and make further vulnerable to intergenerational poverty. To reverse this situation, the economic condition needs to be enhanced for overcoming poverty; disaster risk strategies need to address all factors related to water security of the coastal fishers.

Keywords Disaster · DPSIR framework · Bangladesh · Water security

50.1 Introduction

Bangladesh is ranked as the 5th most at risk country in the world in terms of disasters with a world risk index of 19.17% (World Risk Report 2016). The risk could be further

M. M. Islam (✉) · M. Mostafiz · P. Begum · A. Talukder · S. Ahamed
Department of Coastal and Marine Fisheries, Sylhet Agricultural University, Sylhet, Bangladesh
e-mail: mahmud.cmf@sau.ac.bd

M. Babel et al. (eds.), *Water Security in Asia*, Springer Water,
https://doi.org/10.1007/978-3-319-54612-4_50

675

aggravated with the anticipated impacts of climate change. This risk comprises exposure to natural hazards and the vulnerabilities of society. Particularly, coastal areas of Bangladesh are facing an increasing number of challenges including tropical cyclones, tidal surges, floods, drought, saline water intrusion, water logging, landslides, and arsenic contamination of groundwater which pose substantial threats to the livelihoods of the coastal inhabitants (Lazar et al. 2015). These frequent natural hazards account for significant losses in human lives and physical assets (Choudhury 2002; Khan 2008). The southern zone[1] of Bangladesh, exposing to the Bay of Bengal is geo-physiologically and ecologically diverse, and environmentally vulnerable. Of different hazards, the region is particularly vulnerable to cyclonic storm surge due to its location in the path of tropical cyclones, wide and shallow continental shelf and the funneling shape of the coast (Das 1972). Due to the shallow continental shelf, large tidal water influence and the vast amount of long and narrow shorelines or inlets from the land of the bay the surge amplifies to a considerable extent as it approaches low-lying and poorly protected land and causes disastrous floods along the coast (Shamsuddoha and Chowdhury 2007; Karim and Mimura 2008). From 1797 to 2009, a total of 65 devastating cyclones swept over Bangladesh and caused immense harm to the people (Rana et al. 2010). According to another estimate, since 1980 Bangladesh experienced over 200 natural disasters leaving a total death toll of approximately 200,000 people and causing economic loss worth nearly $17 billion. It is estimated that 14% of GDP Bangladesh is exposed to disasters. Each year the country incurred 1.8% of GDP loss due to natural disaster (CDMP II 2016).

Though the cyclones forming in the Bay of Bengal constitute only 5–6% of the global total, they are the deadliest of all the cyclones (Choudhury 2002). The severity of cyclones in Bangladesh is reflected in the fact that about 80–90% of global losses and 53% of global cyclone-related deaths occurred in Bangladesh (GoB 2008; Paul 2009). For instance, in April 1991, the cyclone Gorki caused material damage of about USD 2.4 billion and human casualties numbered around 140,000. On 15 November 2007, another cyclone Sidr struck the coastal region, which is the worst of its type since 1991, killed more than 3,300 people and created wide-scale damage and losses totaling over $1.5 billion (GoB 2008). On May 25, 2009, the cyclone Aila had hit the south-western part of Bangladesh affected the residents, homesteads, roads, and embankments. The cyclone Aila caused lesser fatalities than cyclone Sidr; however, the economic losses were much greater than cyclone Sidr.

In general hazards are potentially damaging events or phenomena and do not necessarily cause a disaster. Instead disasters occur by a mix of physical exposure and socio-economic pressures. Thus, hazards are placed within the broader context of society, and vulnerability is explained as a result of both biophysical dynamics and social, political and economic processes (Blaikie et al. 1994). Due to the geographical location of Bangladesh in low lying deltaic setting and funnel-shaped part of the Bay of Bengal make its population highly vulnerable to climate change impacts. The

---

[1]Following the gazette notification of the Bangladesh Government, the coastal zone comprises 19 districts. 14 districts that are bordering along the coastline and directly exposed to the Bay of Bengal are considered as the southern zone.

southern zone of Bangladesh faces multiple challenges that threaten its ecosystem and production process and dependent's livelihoods. Geographic setting, hydrological, morphological, geo-physical and bio-physical characteristics shape the conditions of challenges which compounded by widespread poverty and poor asset base of resident population (GoB/FAO 2013). The southern region of Bangladesh is socio-economically poor and disadvantaged. Given land is the most important natural asset; in the southern zone average land ownership per household is 0.29 ha and this is lower than the national average. The people in the southern region comprise different occupational group's viz. small agricultural farmers, agricultural labor, sharecropper, shrimp farmers, salt farmers, honey collectors, small-scale fishers; shrimp fry collector, boat building, net making etc. It is estimated that fishing is the predominant sources of livelihood for 14% households of 14 districts situated along the coastline of Bangladesh (GoB/FAO 2013). Thus, majority of the population in the southern zone are involved in occupation related to natural resources which are sensitive to climate changes and variability. Small-scale fishers are among the most vulnerable professional groups to waterborne disasters since they live close to coastal water and they are dependent on climate-sensitive coastal living resources. Using DPSIR as analytical tool, this study explores the mechanism of disaster vulner-abilities of coastal fishers in the southern Bangladesh. The findings could suggestions important lesson for ensuring water security of the coastal communities, particularly in the context of climate change impacts.

The DPSIR (Drivers-Pressures-State-Impacts-Responses) framework was devel-oped for assessing the causes, consequences, and responses to change in a holistic way (Atkins et al. 2011). In this framework, drivers refer as social, demographic and economic developments in societies and the corresponding changes in lifestyle, overall levels of consumption and production patterns (Gabrielsen and Bosch 2003). These Drivers creates several or many particular Pressures on the system (such as the exploitation of fisheries) As a result, a State of the environment changes and produces Impacts on both environment and society (e.g. degraded habitats, removal of species, loss of biodiversity, etc.), which through its links with human welfare can have positive and/or negative implications. Responses are the actions taken by society and government in efforts to minimize or mitigate the adverse impacts imposed on the environment and society and feedback to the drivers or pressures influencing anthropogenic developments (Gabrielsen and Bosch 2003; Hou et al. 2014). One of the important benefits of DPSIR is that the framework can clarify and disen-tangle complex multi-sectorial inter-relationship that makes it suitable for disaster vulnerability assessment.

50.2 Materials and Methods

This study was conducted in four study sites situated in Patuakhali and Borguna districts of the southern Bangladesh. The selection of study sites was intentional as

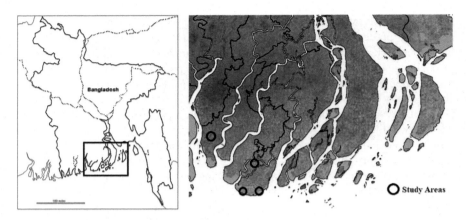

Fig. 50.1 Study areas in the coastal region of Bangladesh

these two districts are significantly vulnerable to a range of natural disasters and one of the hardest hit by the 2007 super cyclone Sidr. The study was conducted in four communities such as Hossain Pur, Char Gongamoti and Kuakata in Kalapara Upzilla (*sub-district*) of Patuakhali district and Padmapukur in Pathorgata upzilla of Borguna district (Fig. 50.1). The primary data were collected from the extensive fieldwork from July 2015 to December 2015 through a series of field visits using various data collection tools such as semi-structured interview, focus group discussion (FGD) and key informant interview. A total of 59 personal interviews were conducted with respondents who are mainly involved in the fishing occupation. Though the sample size is relatively smaller in comparison to community size, however level of saturation was followed and last few interviews any new findings. Additionally, small sample sizes are acceptable in qualitative research, since more data does not necessarily result in more information (Mason 2010). In this study, saturation was achieved, in that the last few interviews did not provide new information; indicating that the sample size was appropriate for this study. Since the study directed to a particular occupational group, so purposive sampling method was employed. Interviewees in the present study included fishermen, fisherwomen (wife of fisherman), *aratder* (fish entrepreneur and fish trader). Individual interviews were conducted in their house, working areas nearby river, on their fishing boat, or in the *arat* (business place) of *aratdar.* In the present study, four focus group discussions were conducted in study areas. Key informant interviews were conducted with knowledgeable persons. Key informant includes *majhi* (leader of the fishing team), fish trader as well as Upzilla Fisheries Officers, people involved in disaster management such as volunteers of Red Crescent Society. In total, 12 key informant interviews were conducted for this study. The collected data were stored, coded for thematic analysis.

50.3 Results and Discussion

50.3.1 Socio-Economic Profiles of the Respondents

The surveyed respondents were mainly from the age group of 18–40 years (63.4%) followed by 41–50 years (35.2%) and >50 years (1.4%). About 44.8% respondents were found illiterate where 42.9% completed their elementary education, 10.5% completed their secondary and only 1.4% completed their higher secondary. Occupationally the respondents were mainly fishermen (96.8%), and among them, 49.3% had secondary occupations. Average fishing experience of the respondents was found 10.5 (±4.1) years (Table 50.1). The study found that the average monthly income of the most of the respondents was Tk. 7,942 (±2,115) and only a small part of respondents (10.5%) have savings. Three types of house structures were found i.e., Tin-shed/Thatched (70.5%), semi-pucca building (23.8%) and building (5.7%). The

Table 50.1 Socio-economic profile of the respondents

Parameter	Description	Mean (±SD)	Frequency (%)
Age	<18 years	–	0
	18–40 years	–	63.4
	41–50 years	–	35.2
	>50 years	–	1.4
Education	Illiterate	–	44.8
	Primary	–	42.9
	Secondary	–	10.5
	Higher secondary	–	1.9
Main occupation (Fishing)	–	–	96.8
Secondary occupation	–	–	49.3
Fishing experience	–	10.5 (4.1)	–
Monthly income	–	7,942 (2,115)	–
House	Tin-shed/Thatched	–	70.5
	Semi-Building	–	23.8
	Building	–	5.7
Household member	–	6.25(1.775)	–
Electricity	–	–	45.2
Access to medical facilities	–	–	75.6
Member of any organization	–	–	65.7
Access to credit	–	–	89.5
Savings	–	–	10.5
Relief or Government assistance after disaster	–	–	86

average number of household members of the respondents was observed as 6.25 (±1.775) (Table 4.1). Most of the respondents (86%) got some kinds of relief materials or government assistance after disasters like dry foods, blanket, water purifying tablet, medicines, oral saline, etc. About 90% respondents have access to informal credit, mainly to *'dadon'* credit system (i.e., advance sale of catches) or micro-credit from NGOs. A large portion of the respondents (65.7%) were found to be a member of a social and professional organization cooperative society. In addition, 75.6% respondents have access to medical facilities, and 45.2% have electricity supply (Table 50.1).

50.3.2 Drivers

The study areas in Patuakhali and Borguna districts are situated in river mouth systems in the low-lying coastal ecosystem. This particular setting made this region vulnerable to different hydro-meteorological events, such as cyclone, storm surges, rising tidal waters from the Bay of Bengal and floods from the upstream. The respondents spoke about repeated exposure to cyclones originating in the Bay of Bengal. From 2007, seven cyclones made landfall on the Bangladesh coast such as the cyclone Sidr (15 November 2007 killed 3500 people), cyclone Rashmi (26–27 October 2008, killed 15 people), cyclone Bijli (19–21 April 2009), cyclone Aila (27–29 May 2009, killed 150 people), cyclone Viyaru (16–17 May 2013, killed 17 people), cyclone Komen (29 July 2015, killed 39 people), cyclone Roanu (21 May, 2016, killed 27 people). Of these extreme events, cyclones Sidr and Aila inflicted major damages in the region. Though other cyclones were less severe, however, several blows one after another created pressure in the region. A number of fishers reported that bad weather condition due to depression (due to low atmospheric pressure) in the bay is becoming more common in recent years. Increasing the numbers of low-pressure system means that increasing number of days of rough weather which hinders traditional fishing activities in the open sea. Analysis of the past depression data (1975–2015) shows that Bangladesh is experiencing with a number of depressions in the Bay of Bengal almost every year (Fig. 50.2).

The interruption of fishing activities due to bad weather condition could bring economic hardship when the majority of the households are already poor. As reflected in their socio-economic profile, for most of the respondents, monthly income is close to the poverty line, where majority of them have no savings. One respondent fisher said *"We have very limited income, but crises are numerous which comes one after another. Thus we could hardly think of any saving. If any disaster strikes, our priority is protecting our assets at any cost. I prefer to stay at home during cyclone because I have a fear of burglary of my belongings"*. Almost all fishers are found to have a contractual agreement with the middleman (local patron) to sell their fish catch as they gave advance money (locally known as *dadon*) to the fishers. Fishers have to pay a commission to the middleman for the sale price of their catch which already fixed at lower price due to contractual obligations. Concerning power relation, the

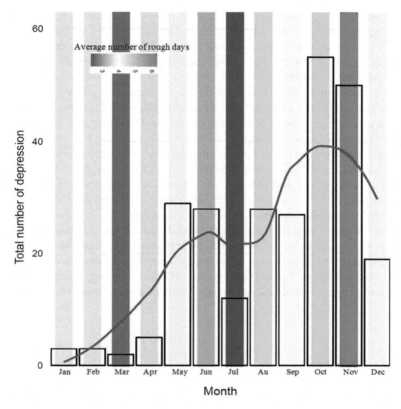

Fig. 50.2 Total number of depression events (left side) and number of total days of rough sea (right side) per month between 1975 and 2015. (Data collected from Meteorological Department)

patron-client relationship is most dominant in the communities that wielded most social power in the area. Socio-economic analysis of respondents identified that the educational level and other skills of fishers are substantially low. Given that scope for occupations other than fishing and agriculture is little thus fishers with few resources (agricultural land, capital) and little education, fishing is one of the few income options in the region they can avail.

Difficulties in communication are another driver identified by the respondents. The region is remotely located, far from the urban centers, thus non-farm economic opportunities are limited. A number of respondents also pointed out that spatial remoteness increased their vulnerabilities to disaster. A fisher elaborated as *"When any cyclone hits, we usually get warning, but getting into cyclone center is tough, particularly for older, women and children. Road connection to cyclone centers often not established. Local elite people usually force government authority to build cyclone center near their home, far from our home."* Majority of the poor people live in vulnerable place, close to or nearby of the coastal embankment which was built in 1960s to protect coastal settlement and crop lands from tidal inundation. A number of the respondent raised their concerns about the stability and effectiveness of the embankments in the face of repeated extreme events. Though, this embankment made crop cultivation, mainly rice possible in areas that would otherwise have been too salty, embankments often also restricted natural river flow patterns, resulting in heavy siltation of river channels, canals, and sluices, thereby adding to the problem of surface water salinization. This salinization is a major concern raised by majority of the respondents stating that it decreased agricultural productivity in this region drastically. One respondent stated "*When cyclone water flooded the agricultural land, it become too salty and cultivation of crops are not possible. Rainwater helps to reduce salt content of the land. However, before the removal of existing salt content through the natural process, another cyclone strikes with rising saline water. Now salinization became a permanent problem. We are struggling to cultivate crops. Few years back, I was sufficient in rice and vegetable production, now I buy a major portion of these crops to meet my family needs.*" Majority of the interviewed households are found functionally landless and a section them live on the government (*khas*) land in densely settlement. The government policy for infrastructural development and tourism expansion caused intense demand for land in the region. Meanwhile marginalized people living on the government-owned land are facing threats of displacement by powerful local elites.

50.3.3 Pressures

The exposure of extreme events coupled with socio-economic factors created pressure on the social and ecological system of the region. Besides human causalities, cyclones destroyed standing crops, productive assets for fishing, collapsed houses and other physical infrastructures that shattered local economy, food security and shelter as stated by a number of respondents. The tidal water washed away their households' assets, destroyed standing crops, flattened the road infrastructure and thatched houses with a heavy toll on livestock, poultry and pond aquaculture that further aggravated their poverty. Interviewed fishers reported that rough weather conditions due to frequent cyclones and depression (caused by lower atmospheric pressure) in the bay often force fishers to abandon their fishing trip and return to the coast. Incomplete fishing trips incur a substantial financial loss. They estimated that the loss is around 76,000 BDT (*950* US$) for a 10-day hilsa (*Tenualosa ilisha*) fishing trip. Respondent fishers mentioned that such income loss is quite common. They also estimated that when any depression occurs in the Bay of Bengal, it causes fishers to discontinue fishing for at least for a week. To avoid this pressure of economic loss and poverty, many fishers defy warnings and continue fishing, which results in fatalities and morbidities of fishers. As the majority of the fishers are waged labor, they do not want to miss their employment even if the weather is rough. One fisher said "*…we must obey the order of boat owner to fishing during rough weather. If we do not, he may punish us which includes, no wage for no fishing day, exclusion from fishing* *team in the following fishing season. Apart from these, physical and mental harass-ments are also done by them. We cannot afford to be unemployed for longer period".* When such natural disasters occur that create mounting pressure on the lives and livelihoods of fishers, while they already live on the margin of survival. Those who are dependent on agriculture also feel the pressure as the agricultural land is situated in cyclone prone areas. Loss of standing crops often pushes the marginal population to adopt fishing as a last resort activity. A number of respondents reported the trend of a crowded condition in the coastal fishery immediate aftermath of cyclone.

50.3.4 State

The majority of the respondent believed that overall environmental quality of their areas is getting worse. Due to saline water intrusion agricultural productivity decreased which reflected in decreased food production. Due to the intense compe-tition and reduced productivity of coastal fisheries, catch per unit effort (CPUE) also decreased, as a number of fishers reported. Since opportunity for alternative occu-pations are limited, and due to contractual obligations with middleman a number of fishers feel themselves trapped themselves in fishing occupation.

50.3.5 Impacts

When the respondent fishers were asked to compare their overall situation before and after a major cyclone Aila (in 2009), almost all respondent provided negative views regarding this cyclonic event. Majority of them also reported negative changes in the ecology of their environment. Many of the agricultural land became unsuitable for agriculture for prolonged period due to saline water intrusion. Crises for drinking water increased as salt water deteriorated the quality of fresh water. Almost all the respondents reported some sort of loss and damage of their assets. Disasters caused enormous destruction of local infrastructure and community assets such as housing, fishing boats and gears. That decimated the capacity to go for fishing and immediate survival. Moreover, as many failed to pay back previous loans, due to losses following the cyclone Aila, the microcredit providing NGOs had not provided further loans to support their livelihood re-establishment. These fall in agricultural and fishing related activities significantly affected the local labor markets and have led to decreased employment opportunities and income for agriculture wage laborers. A few fishers indicated that use destructive fishing practices rises significantly since majority of the fishers are indebted and are desperate to restore their livelihoods. A number of families reported that they have at least one member who migrated elsewhere to support the family. The financial capabilities of affected families were crippled. Some fishers left fishing after facing devastating cyclone during fishing in the sea. Majority of the households live in thatched house which was either destroyed or damaged. A section of the respondents said that they become indebted as they had to rebuild their housing and livelihoods after disasters. To cope with burden of load, fishers have to work more that negatively affect their health.

50.3.6 Responses

The immediate aftermath of any extreme event, the majority of respondents depended on their storage of food and own savings, also relied on relief distributed by the government and NGOs. After losing boats and fishing gears, a section of fisher left sea fishing and moved to riverine fishing. In other cases, man from fishing households migrated to nearby cities, leaving women and children behind who survive on small earning by catching shrimp seedlings. Further responses include reducing the frequency and quality of meals searching for wild foods from the forest instead of normal diet, using savings to meet basic needs, taking loans from NGOs, taking children out of school for employing in child labor. Different humanitarian agencies followed individual mixed approaches, but none of them could adopt a coordinated recovery approach, thus producing limited success towards the recovery of the affected communities. The overall relief and reconstruction work have not been sufficient to maintain the local people's lives, and so a great deal of people had to live in vulnerable condition for longer time. Key informant interview revealed that around 21 national and International NGOs participated in rehabilitation and livelihood recovery programs in the Aila affected areas. However, the coordination was weak, and most of the agencies implemented stand-alone program with very limited coordinating with other agency approaches which result in overlapping of services whereas some areas remained underserved. So far, there has been no formal evaluation of the responses.

River-mouth system in coast is particularly vulnerable to disaster (Newton et al. 2012). The present study also reflected this fact that river-mouth system is indeed vulnerable to climate-related hazards, coming from both sea and catchment areas. In the coastal zone of Bangladesh, hazards are recurring and cumulative, making coastal communities particularly vulnerable to disaster. Majority of hazards are related geo-morphological setting of the region. These hazards coupled with socioeconomic context created disastrous situation for the inhabitant in the southern zone of Bangladesh. Among all professional groups, the fishing communities, by their very nature are the first victim of cyclone and other water-borne hazards. Most of the fishery dependent communities are illiterate, poor, have limited economic capacity to generate savings for support in time of illness or hardship. They have limited room for maneuver due to the lack of alternative income generating activities, living far from main economic activities of urban centers. Limited ownership of different capitals restricts their livelihood space and reduces wellbeing and resilience. Thus fishers are more vulnerable to extreme events and disasters with less resilience. They are less resilient because they have limited capacity for surplus for accumulation and saving; the ability to set something aside for buffer against crises. Consequently, they are trapped in indebted. The provision of alternative livelihoods is very limited hence little scope to increase this leeway. The lack of alternative skills keeps fishers trapped in resource dependency with continuous exposure to hazards. Due to failure to secure a viable occupation on, some of them migrated to elsewhere. The population density (754/square km) of the southern zone is relatively lower compared to Bangladesh (964/square km) (BBS 2011), largely because of out-migration (GoB/FAO 2013).

The coastal population continuously attempts to adapt to the changing situation. However when communities suffer several shocks within a short period, and/or multiple simultaneous pressures, then the pace of change outstrips the adaptive capacity of the suffered population and local institutions. Consequently, vulnerability becomes starkly apparent and it becomes very hard for any individual to cope. In response to repeated water-related disaster in the coastal zone, fostering appropriate mitigation and adaptation using limited resources is a development challenge for Bangladesh. Such response requires being timely, effective and well co-ordinated with all stakeholders at multiple scales. Though some steps are taken by the affected communities and the government, however, the response to the change impacts was insufficient to make the system resilient to recurring changes. The respondents in the present study ask for creating economic opportunity in the region for income generation. They feel needs for better protection from hazards through technical measures such coastal afforestation, fortified coastal embankments. Better coordination, between GoB and NGOs and among the NGOs response and recovery programs could have produced better utilization of limited resources and resulted in the distribution of services and resources equally to all the affected parties that would ultimately benefit the vulnerable population in the coast.

50.4 Conclusion

Bangladesh made a commendable success in disaster risk reduction, yet steps for addressing the social vulnerability of communities needs further improvement. Thus a holistic approach is required to address all factors that create vulnerability, for successful disaster risk reduction. The coastal fishing communities of the Bangladesh are prone to water-related hazards, coupled with socio-economic drivers, thus created the space for the vulnerability to disaster. Hence achieving water security of coastal people is a major challenge. Though the present study reflected some positive indicators related to status of education, access to electricity, strong housing pattern etc. However water-related hazards and fishers livelihoods are twisted in the same rope. The anticipated impacts of climate change will further aggravate the conditions. Hence, coastal fishing communities need to achieve water security in several veins. While they need protection against water-related hazards, similarly sustainable development of water resources is also essential, as their livelihoods depends on water ecosystem services.

References

Atkins JP, Burdon D, Elliott M, Gregory AJ (2011) Management of the marine environment. Integrating ecosystem services and societal benefits with the DPSIR framework in a systems approach. Mar. Pol. Bull. 62:215–226

BBS (2011) Population & Housing Census 2011: Preliminary Result, Bangladesh Bureau of Statistics, Statistical Yearbook of Bangladesh, Planning Division, Ministry of Planning, Dhaka, Bangladesh

Blaikie P, Cannon T, Davis I, Wisner B (1994) At Risk – Natural Hazards, People's Vulnerability, and Disasters. Routledge, London

CDMP II (2016) Comprehensive Disaster Management Programme, Phase 2 (CDMP II), Ministry of Disaster Management and Relief, Bangladesh Government, Dhaka

Choudhury AM (2002) Managing natural disasters in Bangladesh. In: The Dhaka Meet on Sustainable Development in Bangladesh: Achievements, Opportunities and Challenges at Rio+10, Bangladesh Unnayan Parishad, pp 16–18

Das PK (1972) A prediction model for storm surges in the Bay of Bengal. Nature 239:211–213

Gabrielsen P, Bosch P (2003) Environmental Indicators: Typology and use in reporting, Copenhagen

GoB (Government of Bangladesh) (2008) Cyclone Sidr in Bangladesh: damage, loss, and needs assessment for disaster recovery and reconstruction. Dhaka: economics relations division, ministry of finance, Government of the People's Republic of Bangladesh

GoB/FAO (2013) Master plan for agricultural development in the southern region of Bangladesh. Ministry of Agriculture, The Government of the Peoples' Republic of Bangladesh and Food and Agricultural Organization of the United Nations

Hou Y, Zhou S, Burkhard B, Müller F (2014) Socioeconomic influences on biodiversity, ecosystem services and human well-being a quantitative application of the DPSIR model in Jiangsu, China. Sci Total Environ 490:1012–1028

Karim MF, Mimura N (2008) Impacts of climate change and sea-level rise on cyclonic storm surge floods in Bangladesh. Glo Environ Chan 18(30):490–500

Khan MSA (2008) Disaster preparedness for sustainable development in Bangladesh. Disaster Prev Manage 17(5):662–671

Lazar AN, Clarke D, Adams H, Akanda AR, Szabo S, Nicholls RJ, Matthews Z, Begum D, Saleh AFM, Abedin MA, Payo A, Streatfield PK, Hutton C, Mondal MS, Moslehuddin AZM (2015) Agricultural livelihoods in coastal Bangladesh under climate and environmental change e a model framework. Environ Sci Process Impacts 17:1018–1031

Mason M (2010) Sample size and saturation in PhD studies using qualitative interviews. Forum Qual Soc Res 11(3)

Newton A, Carruthers TJB, Icely J (2012) The coastal syndromes and hotspots on the coast. Estuar Coast Shelf Sci 96:39–47

Paul BK (2009) Why relatively fewer people died? The case of Bangladesh's cyclone Sidr. Nat Hazards 50(2):289–304

Rana MS, Gunasekara K, Hazarika MK, Samarakoon L, Siddiquee M (2010) Application of remote sensing and GIS for cyclone disaster management in coastal area: a case study at Barguna district Bangladesh. Int Arch Photogramm Remote Sens Spat Inf Sci XXXVIII(8):122–126

Regional Specialised Meteorological Centre. Cyclone Aila preliminary report, India meteorological department (2009). <https://www.imd.gov.in/section/nhac/dynamic/aila.pdf>. Accessed 5 Nov 2015

Shamsuddoha M, Chowdhury RK (2007) Climate change impact and disaster vulnerabilities in the coastal areas of Bangladesh, COAST Trust, Dhaka

World Risk Report 2016 Bündnis Entwicklung Hilft (Alliance Development Works) in cooperation with United Nations University. Institute for Environment and Human Security, Bonn (UNU-EHS)

Chapter 51
Analysis of Public Perceptions on Urban Flood in Phnom Penh, Cambodia

S. Heng, S. Ly, S. Chhem, and P. Kruy

Abstract A remarkable population and economic growth has been seen in Phnom Penh, a center of commerce, tourism and residence. Urbanization in this capital city of Cambodia, changes of land cover from pervious to impervious areas, and the climate change phenomenon introducing a more intense rainfall within a short time, have led to frequent flooding during the rainy season. Human activities might be a main contributing factor to urban floods in this area. To propose a sustainable measure for such water-related hazard, understanding on public perception is one among various important issues to be considered. This study is aimed at analyzing public perceptions on urban flood in Phnom Penh. A questionnaire survey on 100 samples was conducted within four dense districts of the city. Key findings of the survey are: (1) flood depth of 0.15–0.30 m and flood duration of less than 1 h was very common since it is a rainfall-flood phenomenon, no overflow from rivers; (2) the impacts were greatly on small businesses and health of citizens; (3) most of the people did nothing when there are flood occurrences because flood water does not flow into their house; (4) garbage in the drainage system reduced the flood flow capacity; and (5) education on water engineering and effective garbage management were recommended by citizens to solve flood problems on a long term basis. These results might urge the government to take immediate actions on this extreme event. They also provide essential information for proposing a sustainable flood management strategy.

Keywords Urban flood · Public perception · Phnom Penh

51.1 Introduction

Urban flood has become a major risk to many cities globally. A consistent result from many studies shows that damages caused by urban floods is projected to increase in the future (Hammond et al. 2015). In recent years, such disaster brought a significant disruption to the city service and wide negative effects on citizens. For instance,

S. Heng (✉) · S. Ly · S. Chhem · P. Kruy
Institute of Technology of Cambodia, Phnom Penh, Cambodia

© Springer Nature Switzerland AG 2021
M. Babel et al. (eds.), *Water Security in Asia*, Springer Water,
https://doi.org/10.1007/978-3-319-54612-4_51

in January 2011, the urban flood in Brisbane caused 23 people dead and 18,000 properties inundated (Honert and McAneney 2011). Bangkok was severely impacted by the 2011 flood, attributed to hundred dead and massive property damage (Komori et al. 2012). In July 2012, the urban flood in Beijing killed 79 people, affected over 1.6 million people, destroyed 8,200 houses and caused a direct economic loss of USD 2 billion (Huang et al. 2013).

Economic damages resulted from urban floods are of great concern once the city has been expanding rapidly (Inagaki et al. 2012). Urban flood in Phnom Penh, the capital of Cambodia, are very critical since the city is a center of commerce, tourism and residence. Recently, heavy rainfalls on 31 July and 1 August 2015 flooded a number of residential areas surrounding the city and its vicinity for several hours (Fig. 51.1). Roads were blocked, causing a great disturbance to Phnom Penh dweller while many vehicles have been damaged; moreover, these two large events of rainfall

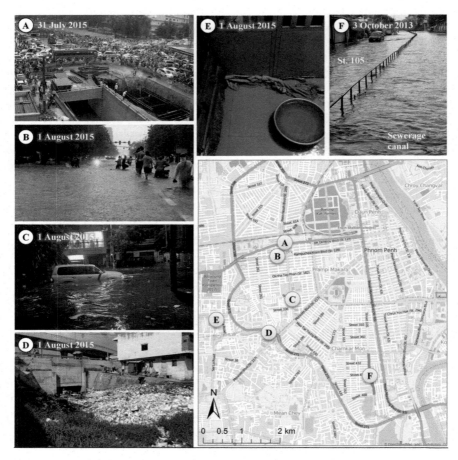

Fig. 51.1 Actual situations of urban floods in Phnom Penh

also inundated hundreds of houses. In Phnom Penh, the combined sewage system is used to drain both wastewater and rainwater; this contaminated water sometimes spills from sewerage canals and flows into household and non-household areas. For instance, black and smelly water in the open drainage canal along the Street No. 105 overflew during a heavy rainfall on 3 October 2013 (Fig. 51.1) causing a severe public health concern. This strongly confirms that urban flood in Phnom Penh has a wide impacts ranging from infrastructure to livelihood and environment. A sustainable flood management strategy is therefore important and urgent.

Many factors should be taken into account to propose an effective approach for flood management. The most important three and could not avoid are (1) stakeholder and legislation, (2) public perception and (3) flood/inundation modelling (Bonacci 2017). The first two factors are essential for setting up feasible and practical alternatives of the management method (scenario). Before selecting any alternative or the most suitable scenario for implementation, it must be evaluated, and this can be done through a modelling test. An analysis of stakeholder and legislation could identify relevant institutions together with scope of actions that should be involved in the decision making processes and implementation of the selected flood management strategy. Understanding on public perceptions is to define the actual needs of people and to ensure their acceptance on any proposed flood mitigation measures. This can be achieved via consultation workshops and/or interviews with a set of structured questions. A study on public perceptions could also provide in-depth information on local conditions, impacts and causes of flood occurrences, good data for the modelling examination. Up to now, there is no observation of flood characteristics (flood extent, depth and duration) in Phnom Penh, leading to a great difficulty in the development of a good flood/inundation model for this area. However, such important information could be collected by means of a questionnaire survey with local community.

Public awareness and attitudes towards urban flooding need to be understood for the flood management while this impact could be mitigated through the application of both structural and non-structural measures based on inputs from local residents. Therefore, the main objective of this research is to analyse public perceptions on urban flood in Phnom Penh, the capital of Cambodia. Key research questions are:

- What are the characteristics of urban flood in Phnom Penh?
- What are the major impacts of urban flood on the citizens' livelihood in Phnom Penh?
- What are the flood responses of people living in Phnom Penh?
- What are the main courses of flooding in Phnom Penh?
- What are the suggestions of Phnom Penh citizens to sustainably coping with floods?

51.2 Materials and Methods

51.2.1 Study Area

Phnom Penh is located on a flat alluvial plain at the western bank of the confluence of the Mekong, Tonle Sap and Bassac rivers, an intersection known as Chaktomuk (Yim et al. 2016). According to the population census 2008, the city has a total population of about 1.5 million people (JICA 2016). As a part of a tropical monsoon climate, Phnom Penh receives an average annual rainfall of approximately 1,500 mm, with a great variation ranging from less than 1,200 mm/year to more than 1,900 mm/year. A majority of annual precipitation are concentrated in rainy season (May–October). The average monthly minimum and maximum temperature is respectively 21.8 °C and 35.5 °C. The annual average humidity is around 77%. The drainage system is characterized by two main types: open channel (concrete and earth canal) and closed conduit (pipe and box culvert). Drainage facilities are constructed, operated and maintained by Department of Public Works and Transport (DPWT). The pipe culvert, the later called drainage pipe, is predominant and has a total length of 477 km (as in 2013). It is equivalent to a drainage density of 700 m/km^2. The size of drainage pipe ranges from 30 to 180 cm in diameter. It should be noted that this data is based on a record of DPWT between 2006 and 2013; there are still many drainage pipes in the city which have not been recorded yet (JICA 2016).

With an administrative area of 678.5 km^2, Phnom Penh is divided into 12 districts. Chamkarmon, Daun Penh, 7 Makara and Tuol Kouk are the top four districts in term of populated area; these four districts, the study area, has a population of 571,649 people (population census 2008) representing 38% of the city's population and they cover a territory of about 29 km^2 (Fig. 51.2). The area and population of each district is tabulated in Table 51.1. As the study area is the urban center of Phnom Penh, there are currently many high-rise buildings and commercial places (Fig. 51.3). The existing flood protection includes dikes (ring roads), pumping stations and maintenance/improvement of drainage system. Although flood countermeasures are in place, this urbanised area still experienced flooding during an intent rainfall within a short time (JICA 2016).

51.2.2 Questionnaire Survey

First of all, a set of questions was developed and pre-tested to make sure that it is practical. The questionnaire composes of five main sections:

– General information of respondents: age, duration of residence, education level and occupation
– Flood situations: depth and duration of flood
– Flood impacts and responses

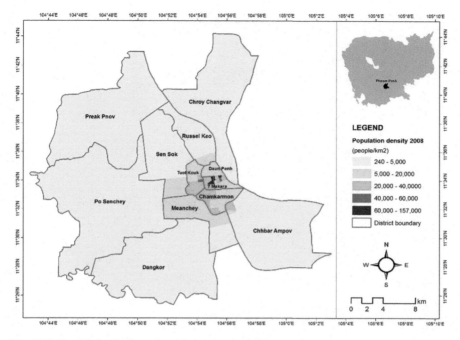

Fig. 51.2 Spatial distribution of population density in Phnom Penh

Table 51.1 Area, population and sample size of each district

District	Area (km^2)	Population (people)	Sample size (people)	Sample density (people/km^2)
Chamkarmon	11.15	182,004	38	3.41
Daun Penh	7.68	126,550	24	3.13
7 Makara	2.20	91,895	13	5.91
Tuol Kouk	7.97	171,200	25	3.14
Total	29.00	571,649	100	3.45

– Local causes of floods
– Recommendations of local people for solving flood problems.

After pre-testing of the questionnaire and its revision (if any), the survey by interview was carried out using a random sample method with some criteria as follow. To ensure a good information can be obtained, we considered only respondents with an age of 18 years old or more, living within the study area (Chamkarmon, Daun Penh, 7 Makara and Tuol Kouk district) for at least 1 year and finished primary school. Regarding the location of samples, we primarily selected areas where there were flood occurrences in the past; this was based on literature review and experiences of the research team. Spatial condition of samples (geographic distance between

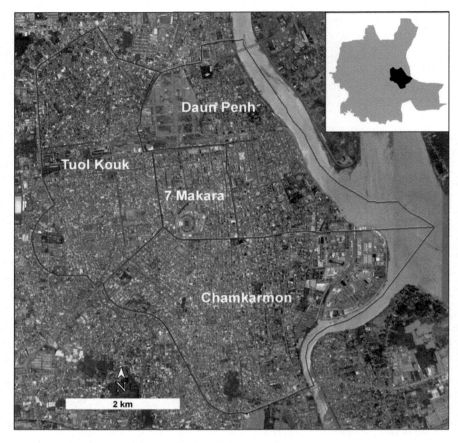

Fig. 51.3 Map of the study area (4 districts of Phnom Penh)

each sample) was also considered to avoid a redundancy of information provided by different respondents. Therefore, location of the next sample is dependent on that of the preceding sample. To achieve this, GPS was employed simultaneously to position the location of samples and analyse the spatial distribution during the survey. Although a great spatial distribution of samples within the study area is important, a distance between samples cannot be specified due to an uncertain participation of interviewees. A small souvenir was prepared for interviewees to obtain and acknowledge their active participation in the survey. Finally, the questionnaire survey was conducted from 19 March to 24 April 2016.

51.2.3 Verification of Flood Depth

Data of flood depth is a key information for various purposes such as flood management, economic analysis of flood damage and calibration/validation of the model simulation, etc. Data of flood depth obtained from the questionnaire survey was verified with another dataset resulted from analysis of flood marks on physical objects, the latter called photo survey (Fig. 51.4). The photo survey was carried out through a social media, Facebook. After receiving photo of flood, we identified the photo's location by using experiences of local research team or asking directly the person posting the photo. For the flood depth (D_{photo}), it can be approximated from the physical objects. It is understood that there are cutting-edge equipment that can be utilized to measure flood depth automatically. The photo survey via social media was selected since historical floods is the focus of this research. However, gauge measurements should be started from now on for future studies.

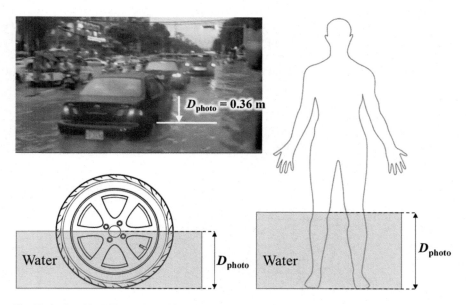

Fig. 51.4 Graphical illustration of flood depth observation via photo survey

51.3   Results and Discussion

51.3.1   Spatial Distribution of Samples and General Information of Respondents

Some questions were revised after pre-testing on a few samples. A common revision is the change from opened question to multiple choice one. Following a set of criteria mentioned above, a total number of 100 respondents were interviewed with the revised questionnaire; it is equivalent to 3.45 sample/km^2 which is reasonably enough for this kind of study. There are 38, 24, 13 and 25 samples located in Chamkarmon, Daun Penh, 7 Makara and Tuol Kouk district, respectively. The difference in sample sizes in different districts could be explained by the difference in area of each district (Table 51.1). The geographic location of the 100 samples is presented in Fig. 51.5. It can be seen that the sample distribution is relatively sparse in the North and the Southeast. These areas are the new reclamation of land where urbanization is ongoing and there is no people living there yet. The sample density in the central region is slightly higher than other places due to inaccessibility and non-participation of respondents at targeted areas. As a whole, the spatial distribution of samples is acceptable and could represent the entire area to some extent.

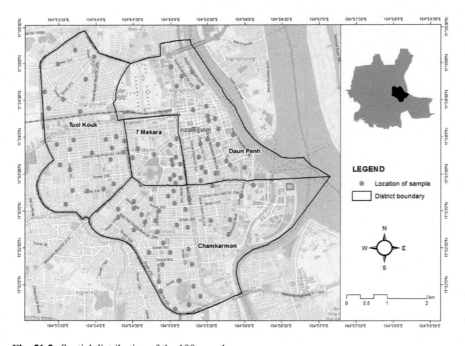

Fig. 51.5 Spatial distribution of the 100 samples

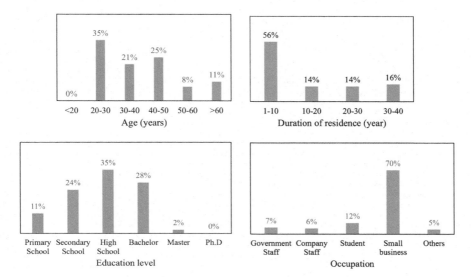

Fig. 51.6 General information of respondents

About 57% of the whole respondents are female. It is slightly higher than the rate of male respondent since women are responsible for taking care of family and/or do a small business at home. Figure 51.6 depicts the general information of respondents. All respondents are more than 20 years old while the majority (81%) are between 20 and 50 years old. Only 44% of respondents have been living in this area from 10 to 40 years and they might be a permanent resident; the remaining 56% have been living there from 1 to 10 years and this amount is probably a combination of permanent and temporary residents. Since this is a case study in urban area, many respondents (65%) finished their high school degree. The number of respondents having a primary school and secondary school education is 11% and 24%, correspondingly. Regarding their occupation, 70% of them do a small business, selling something, at home and in a small shop along the street. Only 13% of them are government and private company officer while 12% are students. Based on these results, it can be concluded that all respondents have enough ability to understand well questions and provide sophisticated answers.

51.3.2 Flood Situations

Figure 51.7 illustrates the variation of flood depth in the four busy districts of Phnom Penh. It is not a result of a rainfall event but an experience of the citizens. That is why information of flood depth is classified in different ranges. It is apparent that respondents experienced flooding up to 0.6 m deep. Floods also occurred even along the riverside and this could be due to the effect of high river water level together with

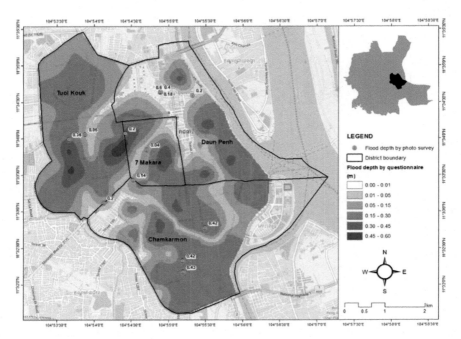

Fig. 51.7 Flood depth experienced by Phnom Penh citizens

the insufficient capacity of pumps and drainage system problem. In many parts of the flooded areas, the flood depth is 0.15–0.30 m. The spatial difference in flood depth could be explained by difference in topographic, drainage and rainfall conditions.

Regarding the photo survey, we obtained 13 reports (flood events in July–September 2015) and their approximated flood depth (D_{photo}) is presented in Fig. 51.7. The value of D_{photo} varies between 0.04 m and 0.60 m which is similar to information obtained from the questionnaire survey on 100 samples. The average D_{photo} of 0.28 m is between 0.15–0.30 m, the range of major flood depth by questionnaire. On the other hand, there are only three locations where flood depth by photo (D_{photo}) is within the range of flood depth by questionnaire. At two locations, D_{photo} is marginally below the range of flood depth by questionnaire. D_{photo} is above the range of flood depth by questionnaire for the remaining eight locations. There is of course an uncertainty in both datasets. Respondents require a good memory of flood occurrences to provide precise answer; information given to us might be one of an event that they remember, not information of an extreme flood; or it is their experience at another location, not the place where we interviewed them (mostly at home and shop). Moreover, information from photo just belongs to a specific event at a particular time. Overall, both results are comparable to some extent and contain a contradiction in term of geographic location.

We analyzed the flood duration in correspondence with three classes of flood depth (Table 51.2). As reported by 70% of respondents, the flood duration is less

Table 51.2 Cross correlation between flood depth and flood duration

Flood depth (m)	Percentage of respondent (%)			Total (%)
	Flood duration <1 h	Flood duration 1–6 h	Flood duration >6 h	
<0.40	66	22	3	91
0.40–0.50	1	1	0	2
0.50–0.60	3	4	0	7
Total	70	27	3	100

than 1 h. In some places, 27% of respondents said that it is more than 1 h and up to 6 h. Remarkably, flood events with a depth of less than 0.40 m happened for a period less than 1 h.

51.3.3 Flood Impacts and Responses

According to 31.62% of respondents, the major impact of urban flood in these four districts is on business. When floods occur, people face a great difficulty in travel (Fig. 51.8). In cases of a small business, floods sometimes caused a damage on the selling goods. The second major impact is on health as reported by 26.47% of respondents. Water supply and sanitation is important in this case. The third major impact is on traveling of citizens (23.53%). During the rainy season, heavy rainfalls commonly occur in the evening while people travel from works and schools to home. Thirty-nine percent of respondents told us that flood water broke their vehicles during their travel. The other impacts are on house and education sector. When there is a flood occurrence, 85% of respondents did nothing because the flood situation is not so serious and flood water does not flow inside their houses (Fig. 51.8). Also, they

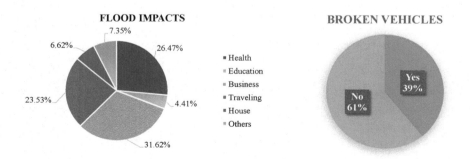

Fig. 51.8 Flood impacts

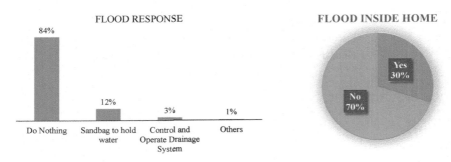

Fig. 51.9 Flood responses

might not know how to improve flood situation. Only 12% of respondents cope with flood by using sandbag or clothes to hold water in order to avoid water flowing into their houses (Fig. 51.9) and 3% go to check or control drainage system if it is blocked by something.

51.3.4 Causes of Flood

Many respondents, 35.59% and 32.43%, understand that urban flood in Phnom Penh is caused by heavy rainfall in a short time and insufficient capacity of existing drainage system, respectively (Fig. 51.10). About 24.77% of respondents answered that the

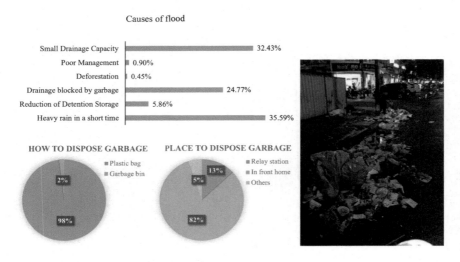

Fig. 51.10 Causes of flood and garbage disposal method

cause of flood is the blockage of drainage by garbage (Fig. 51.1). During the question-naire survey, we also checked the garbage disposal method of citizens. As a result, we found that most of the respondents use plastic bag to pack the garbage (98%) and dispose in front of their house (82%) for further collection by garbage trucks. The package is sometimes not well done and/or the plastic bag is broken, leading to a disorder of the disposed garbage (Fig. 51.10). In such condition, garbage can be transported into the drainage system easily by rainfall. Only 13% of respondents dispose their garbage at a relay station where there is a big garbage bin.

51.3.5   Suggestions of Local People to Cope with Floods

A large number of respondents (50.87%) suggested that an improvement of drainage capacity (bigger sewer) should be considered to reduce flood impacts (Fig. 51.11). As mentioned in the report of JICA (2016), the Royal Government of Cambodia is conducting a master plan study on drainage management in Phnom Penh which includes an improvement of drainage channels and pumping stations. Raising aware-ness on flood countermeasures is also a concern of 41.62% of respondents. For the time being, there is not much local academic institutions that provide training on flood management. According to the authors' knowledge, only the academic curriculum of Institute of Technology of Cambodia, Phnom Penh, contains some courses on water engineering and management which cover parts of flood issues like flood modeling. Nevertheless, budget limitation is commonly a big issue for a developing country like Cambodia, meaning that there is not enough budget for development of all sectors. Therefore, we also asked Phnom Penh citizens about their priority needs for improving their livelihood. As a result, their first priority need is a solution to garbage problem (32%), following by flood (30%), house fire (23%) and road (15%) (Fig. 51.11). Since this is a case study in the city, road infrastructures are mostly sufficient.

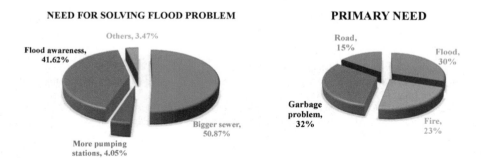

Fig. 51.11 Suggestions of Phnom Penh citizens to cope with floods

51.4 Conclusions

This study attempts to analyze the public perceptions on urban flood in Phnom Penh for a sustainable flood management in the city. A set of questions was developed and used to interview 100 citizens living in the most four populated districts, corresponding to about 3.45 sample/km². The questionnaire survey was conducted using a random sample method together with some criteria. The resulted flood depth was compared with that obtained from the photo survey via Facebook. Key findings of the research are: (1) flood depth of 0.15–0.30 m and flood duration of less than 1 h was very common since it is a rainfall-flood phenomenon, no overflow from rivers; (2) the impacts were greatly on small businesses and health of citizens; (3) most of the people did nothing when there are flood occurrences because flood water does not flow into their house; (4) garbage in the drainage system reduced the flood flow capacity; and (5) education on water engineering and effective garbage management were recommended by citizens to solve flood problems on a long term basis. These results might urge the government to take immediate actions on this extreme event. They also provide essential information for proposing a sustainable flood management strategy. The accuracy of this analysis could be further enhanced by increasing the sample size, especially to the northern and southeastern area. For future research, it is recommended to include the remaining eight districts so that the whole Phnom Penh city can be covered; the perception on less and highly urbanized areas could be different as well. The future development of infrastructure might alter the study results and therefore they should be used with caution.

Acknowledgements The authors would like to express high gratitude to JSPS Core-to-Core Program, B. Asia-Africa Science Platforms, for supporting this research study.

References

Bonacci O (2017) Floods: new concepts emerge-old problem remain. In: Bekic D, Carevic D, Vouk D (eds) Proceedings of the 15th international symposium on water management and hydraulics engineering, 6–8 September 2017, Primosten, Croatia, pp 19–26

Hammond MJ, Chen AS, Djordjevic S, Butler D, Mark O (2015) Urban flood impact assessment: a state-of-the-art review. Urban Water J 12(1):14–29

Honert RCVD, McAneney J (2011) The 2011 Brisbane floods: causes, impacts and implications. Water 3(4):1149–1173

Huang Y, Chen S, Cao Q, Hong Y, Wu B, Huang M, Qiao L, Zhang Z, Li Z, Li W (2013) Evaluation of version-7 TRMM multi-satellite precipitation analysis product during the Beijing extreme heavy rainfall event of 21 July 2012. Water 6(1):32–44

Inagaki I, Sisinggih D, Wahyuni S, Hapsari RI, Sunada K (2012) Assessment of public perceptions of urban flood as an effective approach for disaster mitigation: example from Tulungagung, Indonesia. J Jpn Soc Civil Eng Ser B1 (Hydraulic Engineering) **68**(4):I_103–I_108

JICA (Japan International Cooperation Agency) (2016) The study on drainage and sewerage improvement project in Phnom Penh Metropolitan area, JICA, Phnom Penh

Komori D, Nakamura S, Kiguchi M, Nishijima A, Yamazaki D, Suzuki S, Kawasaki A, Oki K, Oki T (2012) Characteristics of the 2011 Chao Phraya river flood in central Thailand. Hydrol Res Lett 6:41–46

Yim S, Aing C, Men S, Sovann C (2016) Applying PCSWMM for stormwater management in the Wat Phnom Sub Catchment, Phnom Penh, Cambodia. J Geogr Environ Earth Sci Int 5(3):1–11

Chapter 52
Assessment of Climate Change Impact on Drought in the Central Highlands of Vietnam

Dao Nguyen Khoi and Pham Thi Thao Nhi

Abstract In recent years, drought has been strongly affecting the Central Highlands of Vietnam and has resulted in crop damage, yield decline and serious water shortage. Understanding the characteristics of the drought will be helpful to decision makers in managing and planning water resources. The objective of the present study is to predict changes in droughts from agricultural and hydro-meteorological perspectives under the impact of climate change for the Srepok River Basin using three drought indices, such as standardized precipitation index (SPI), standardized runoff index (SRI), and standardized soil moisture index (SSWI). The climate data from the observational period and projected period from CMIP5 global climate models are used to calculate severity and duration of meteorological droughts, and soil water content and streamflow data from the well-calibrated SWAT hydrological model are used to examine the hydrological and agricultural droughts. Then, the droughts with severe, prolonged, and frequent characteristics were analysed. The results show that drought frequency is predicted to increase while drought duration and severity are projected to decrease in the future for the Srepok River Basin. The obtained results would help managers in sustainable water resources management and agricultural planning.

Keywords Climate change · Drought · Hydrology · SWAT model · Srepok River Basin

52.1 Introduction

Drought is a prolonged and abnormally dry period of water insufficiency for normal use. Drought normally starts with precipitation deficiency affecting hydrological processes, including evapotranspiration, runoff, and soil water content, and thus impacts on socio-economic system directly related to hydrological conditions (Duan and Mei 2014). According to Wilhite and Glantz (1985), drought is classified into

D. N. Khoi (✉) · P. T. T. Nhi
Faculty of Environment, University of Science, Ho Chi Minh City, Vietnam

Vietnam National University, Ho Chi Minh City, Vietnam

© Springer Nature Switzerland AG 2021
M. Babel et al. (eds.), *Water Security in Asia*, Springer Water,
https://doi.org/10.1007/978-3-319-54612-4_52

four basic types, including meteorological, agricultural, hydrological, and socio-economic droughts. In recent years, the frequency and severity of the flood and drought due to climate change have considerably increased (IPCC 2013). Under the impact of climate change, studies on monitoring and predicting droughts on a long-term scale are necessary to find countermeasures to cope with extreme drought conditions that may occur in the future (Kim et al. 2014). Generally, the method to monitor and predict drought is using observed hydro-meteorological data and projected data through outputs of the general circulation models (GCMs) and hydrological model. These hydro-meteorological data are converted into drought indices, such as the Standardized Precipitation Index (SPI), Palmer Drought Severity Index (PDSI), Standardized Soil Water Index (SSWI), Agricultural Rainfall Index (ARI), and Standardized Runoff Index (SRI), to predict drought severity. For example, Sayari et al. (2013) used the drought indices (SPI and ARI) to monitor and forecast the drought intensity and duration in the Kashafrood basin (Northeast Iran), and he reported that slight increases in the future precipitation and temperature would cause increase in drought frequency. Leng et al. (2015) investigated the potential impacts of climate change on drought characteristics in China, and he indicated that meteorological, agricultural, and hydrological droughts will become more severe, prolonged, and frequent under climate change scenarios.

Recently, Vietnam has faced severe and prolonged droughts, which cause water scarcity and loss of agricultural production with damage costs of hundreds of billion Vietnamese Dong (VND). Especially in the dry season of 2015–2016 with the effects of El Niño phenomenon, the Central Highlands region had faced the most severe droughts in the past 90 years, causing severe damage to agricultural production and farmer's income (FAO 2016). Moreover, Vietnam has experienced climate change. For the entire country, the annual temperature increased by 0.5 to 0.7 °C and annual precipitation decreased by 2% over the past 50 years (MONRE 2012). These changes have impacted significantly on water availability and droughts in Vietnam. In addition, impacts of climate change vary from region to region and need to be considered for a local scale.

The purpose of this study was to investigate the impacts of climate change on agricultural and hydro-meteorological droughts in the Srepok River Basin in the Central Highlands of Vietnam. The results obtained in this study are expected to help local water to managers understand more insight into the climate change impacts on the droughts.

52.2   Study Area

The Srepok River Basin, located in the Central Highlands of Vietnam, lies between latitudes 11°45′–13°15′N and longitudes 107°15′–109°E (Fig. 52.1). This river basin is an important sub-basin of the 3S (Sesan, Sekong, and Srepok) River Basin, which flow into the main channel of the Mekong River. The Srepok River is formed by two main tributaries, the Krong No and Krong Ana Rivers. The total area of this basin

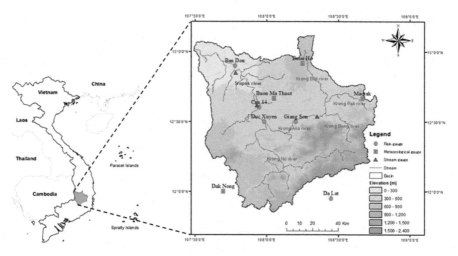

Fig. 52.1   Location of the Srepok River Basin

is approximately 12,000 km^2 with the population of approximately 2.4 million in 2014. The average altitude of the watershed varies from 100 m in the northwest to 2,400 m in the southeast. The climate in the area is very humid (78–83% annual average humidity) with annual rainfall varying from 1,700 to 2,300 mm and features two distinct wet and dry seasons. The wet season lasts from May to October and accounts for over 75–95% of the annual precipitation. The mean annual temperature is 23 °C. In this basin, there are two dominant soils: grey soils and red-brown basaltic soils. These soils are very consistent with agricultural development. Agriculture is the main economic activity in this basin.

52.3   Methodology

52.3.1   SWAT Hydrological Model

The SWAT model is a physically based, semi-distributed model used to predict the impact of land management practices on flow, sediment and agricultural chemical yield in watershed scale (Neitsch et al. 2011). This model has been promising in its application to various watersheds (see SWAT literature database: https://www.card.iastate.edu/swat_articles/). In SWAT, the catchment is firstly divided into sub-catchments and then further discretizing the sub-catchment into Hydrological Response Units (HRUs) based on soil, land use and slope classes that allows a high level of spatial detail simulation. Hydrological simulation is based on water balance equation of soil water, comprising precipitation, surface runoff, evapotranspiration,

Tab. 52.1 Input data for the Srepok River Basin

Data type	Description	Scale	Source
DEM	Topological features	90 m	U.S. Geological Survey (USGS)
Land-use	Land-use classification in 2003	1 km	Mekong River Commission (MRC)
Soil	Soil types and physical properties	10 km	Food and Agriculture Organization (FAO)
Weather	Precipitation and maximum/minimum temperature in the period 1981–2009 at three climate stations and six rain gauges	Daily	Hydro-Meteorological Data Center (HMDC)
Hydrology	Streamflow in the period 1981–2000 at the Ban Don station	Daily	HMDC

infiltration, and subsurface flow. Further details of hydrological processes can be found in the SWAT Theoretical Documentation (Neitsch et al. 2011).

SWAT model setup requires three spatial datasets, namely topography, land-use, and soil, and hydro-meteorological dataset. The data used in this study are shown in Table 52.1. ArcSWAT version 2012 was used in this study. The SWAT model set-up involved four steps: (1) data preparation, (2) sub-basin discretization, (3) HRU definition, and (4) calibration and validation. Hydrological parameters were calibrated and validated using sequential uncertainty fitting (SUFI-2) algorithm available in the SWAT-CUP version 2012.

The model performance was evaluated using statistical analysis to compare the quality and reliability of the simulated streamflow with the observed data. In this study, three statistical indicators, including Nash–Sutcliffe efficiency (NSE), percent bias (PBIAS), and ratio of root mean square error to the standard deviation of measured data (RSR), were used to evaluate the statistical significance between observed and simulated values. Generally, the model is considered to be satisfactory when the values of NSE greater than 0.5, the PBIAS values less than 25%, and the values of RSR less than 0.7 (Moriasi et al. 2007).

52.3.2 Drought Analysis Method

In this study, drought indices are used for quantitative analysis of drought. The widely used three drought indices are the standardized precipitation index (SPI), standardized soil moisture index (SSWI), and standardized runoff index (SRI) to characterize the hydro-meteorological and agricultural droughts respectively (e.g. Wang et al. 2011; Duan and Mei 2014; Leng et al. 2015). SPI is estimated based on probability distribution of precipitation using gamma density function for specified monthly time scales. The long-term precipitation data are fitted to the probability distribution, which is then transformed into a normal distribution (WMO 2012). The calculation procedure for SSWI and SRI is similar to SPI, but the lognormal

Tab. 52.2 Detailed information on the five GCMs used in this study

Code	Centre, country developed the GCMs	Centre abbreviation	Resolution	RCPs	Country used GCMs
1	Canadian Centre for Climate Modelling and Analysis, Canada	CanESM2	2.80 × 2.80°	RCPs 4.5 & 8.5	Malaysia
2	Centre national de Recherché Météorologiques, France	CNRM-CM5	1.40 × 1.40°	RCPs 4.5 & 8.5	Vietnam
3	Hadley Centre, UK Met Office	HadGEM2-AO	1.90 × 1.25°	RCP 4.5 & 8.5	South Korea
4	Institute Pierre-Simon Laplace, France	IPSL-CM5A-LR	3.75 × 1.90°	RCPs 4.5 & 8.5	Malaysia
5	Max Planck Institute for Meteorology	MPI-ESM-MR	1.90 × 1.90°	RCPs 4.5 & 8.5	Thailand

density function is used for SRI and the beta density function is used for SSWI. The calculation procedure for SPI, SSWI, and SRI can be found in McKee et al. (1993) and Duan and Mei (2014). SPI is calculated using monthly rainfall data, and SRI and SSWI are estimated using monthly runoff and soil water content data simulated from the SWAT model.

In this study, the 12-month drought indices (i.e. SPI 12, SSWI 12, and SRI 12) were calculated. The drought indices with annual timescale (12 month) were selected for the present study, because they are more suitable to describe the hydro-meteorological regimes than the drought indices with sub-seasonal and seasonal timescales (Spinoni et al 2014). A drought event was considered when the values of the drought indices are below -1.0. According to McKee et al. (1993), droughts can be classified into moderate ($-1.0 \leq$ SPI, SSWI, SRI < -1.5), severe (-1.5 $\leq$ SPI, SSWI, SRI < -2.0), and extreme ($-2.0 \leq$ SPI, SSWI, SRI) based on the normal distribution of SPI, SSWI, and SRI. In order to analyze drought characteristics (i.e. severity, duration, and frequency), severity-duration-frequency curves for the droughts were established. In terms of terminology, severity means the averaged SPI value for a drought event, duration refers to the length of each drought event, and frequency indicates the number of drought events over a defined period.

52.3.3 Climate Change Scenarios

Climate change scenarios were developed for future period 2045–2070 (2060s) using five CMIP5 GCM simulations (CanESM2, CNRM-CM5, HadGEM2-AO, ISPL-CM5A-LR, and MPI-ESM-MR, see Table 52.2 for details) driven by RCP 4.5 (medium emission) and RCP 8.5 (high emission) scenarios. The five GCM simulations were selected based on using GCM outputs of SEACLID/CORDEX member countries within Southeast Asia. In order to downscale these GCM simulations at global scale to regional scale, the perturbation (or delta change) method was selected (Sunyer et al. 2010), because this method has been widely used and it is quite simple in generating a large range of sensible climate scenarios in various hydrological studies under climate change impacts (Khoi and Suetsugi 2012).

52.4 Results and Discussion

52.4.1 Calibration and Validation of the SWAT Model

The SWAT hydrological parameters used for calibration and validation of the model were selected by referring the relevant study in the Be River Catchment (Khoi and Suetsugi 2012). The SWAT flow simulations were calibrated against monthly flow from 1981 to 1990 and validated from 1991 to 2000 at the Ban Don gauging station, as shown in Fig. 52.2. The observed and simulated monthly streamflow showed a good agreement with the NSE, PBIAS and RSR values of 0.72, –24%, and 0.54 in the calibration period respectively. For the validation period, the NSE, PBIAS, and RSR values were 0.85, –14%, and 0.39, respectively. General speaking, the result of the model calibration and validation reveals that hydrological processes in SWAT are

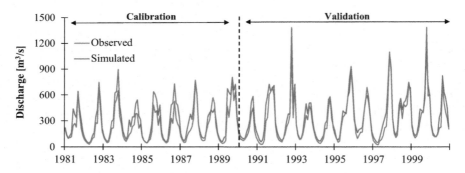

Fig. 52.2 Observed and simulated monthly flow hydrographs at the Ban Don station for the **a** calibration and **b** validation periods

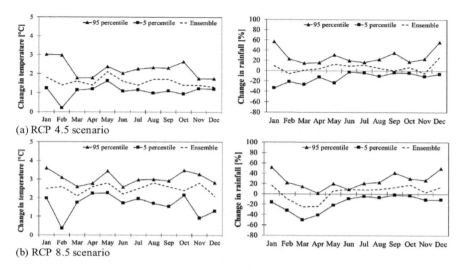

Fig. 52.3 Monthly changes in precipitation and temperature for the period 2060s

modelled realistically and the calibrated model could be used to examine the impacts of climate change on hydrology in the Srepok River Basin.

52.4.2   Climate Change Scenarios

Future climate conditions were determined using an ensemble of GCMs (CanESM2, CNRM-CM5, HadGEM2-AO, ISPL-CM5A-LR, and MPI-ESM-MR) driven by the RCP 4.5 and RCP 8.5 emission scenarios. Figure 52.3 illustrates the shift in annual distribution of temperature and precipitation with the uncertainty range of 5 and 95 percentile bounds for the 2060s (2045–2070) relative to the baseline period (1984 to 2009). Analysis of temperature change indicates an obvious increase in future temperatures. The annual average temperature has a mean shift of 1.6 °C (within the range of 1.1–2.2 °C) for the RCP 4.5 scenario and 2.5 °C (1.7–3.0 °C) for the RCP 8.5 scenario. Averaged over all GCMs ("ensemble average"), the annual precipitation increases 6.4% (within the range of −8.8–24.4%) in RCP 4.5 and 7.3% (−10.8–26.5%) in RCP 8.5. In the dry season (November to April), precipitation slightly changes within a range from −7.0–3.3% in the future. In the wet season (May to October), the precipitation significantly increases, ranging from 7.1 to 10.6%.

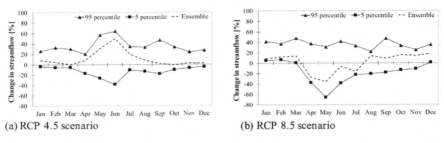

(a) RCP 4.5 scenario   (b) RCP 8.5 scenario

Fig. 52.4 Monthly changes in streamflow for the 2060s in **a** the RCP 4.5 scenario and **b** the RCP 8.5 scenario

52.4.3 Impacts of Climate Change on Streamflow

Figure 52.4 illustrate the changes in annual, seasonal, and monthly streamflow with the uncertainty range of 5 and 95 percentile bounds under the effects of projected climate change. Compared with the baseline period, the predicted annual streamflow change is expected to increase, with changes of 8.1% (within the range of −12.2–35.7%) in the RCP 4.5 scenario and 9% (−15.3–34.4%) in the RCP 8.5 scenario. The findings here are similar to the results of the similar study in the Srepok River Basin conducted by Kawasaki et al. (2010) and they indicated that the future streamflow will increase by 3–6%. The differences in terms of magnitude are understandable because the future climate change scenarios in those studies were generated based on CMIP3 GCM outputs. In the case of seasonal change, the average streamflow considerable increases from 8.3 to 9.2% in the wet season and 8.0–8.1% in the dry season for the RCP 4.5 and 8.5 scenarios.

52.4.4 Analysis of Changes in Severity, Duration, Frequency of the Hydro-Meteorological and Agricultural Droughts

The drought in the Srepok River Basin is becoming more frequent with adverse effects on social economy and agriculture. The calculation results for the period 1983–2009 by the SPI, SSWI, and SRI indices with 12 month scale are shown in Fig. 52.5. It is shown that the drought events were happened quite often in the dry season for the periods 1983, 1991, 1994–1995, 1997, and 2004–2005. The calculation results are similar to historical drought records. During the period 1983–2009, there were totally 55 meteorological drought events (58.2% for moderate drought, 29.1% for severe drought, and 12.7% for extreme drought) shown by SPI 12, 44 agricultural drought events (68.2% for moderate drought, 15.9% for severe drought, and 15.9% for extreme drought) shown by SSWI 12, and 53 hydrological drought events (58.5% for moderate drought, 39.6% for severe drought, and 1.9% for extreme drought). The

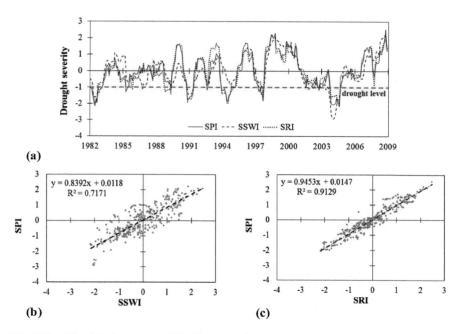

Fig. 52.5 **a** Monthly time series of SPI, SSWI, and SRI in the period 1982–2009, **b** Scatter plot of SPI vs SSWI, and **c** Scatter plot of SPI vs SRI

correlation coefficients of SPI vs SSWI and SPI vs SRI were quite high (Fig. 52.5). The results reveal a very strong linear relationship between SPI and SRI with R^2 of 0.91 and significance level of $p < 0.01$, but less strong with R^2 of 0.72 and significance level of $p < 0.01$ for the relationship of SPI and SSWI. In general, SSWI and SRI were quite similar to SPI in terms of severity and timing of droughts.

Under the impact of climate change, the agricultural and hydro-meteorological drought events increase 7–15% in the future for both the RCP 4.5 and 8.5 scenarios. The increases in the drought events are attributed to decrease in future rainfall in some dry-seasonal months and increase in future temperature. Figure 52.6 presents the severity-duration-frequency (SDF) curves for the droughts over the Srepok River Basin for the baseline and future periods. It indicated reductions in duration of agricultural and hydro-meteorological droughts throughout the SDF curves in the future comparing to the baseline period. In addition, a decrease in drought severity is projected in the future, but the most severe drought (i.e. the minimum negative drought severity) is predicted to slightly increase for hydro-meteorological and agricultural droughts. The finding here is similar to that of the study conducted by Vu et al. (2015) in the Dakbla River Basin in the Central Highlands of Vietnam. He reported an increase in future drought events but decreases in severity and frequency.

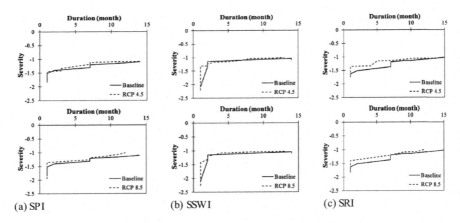

Fig. 52.6 Future changes in the severity, duration, and frequency of the droughts in the RCP 4.5 and 8.5 scenarios

52.5 Conclusions

This study examined the changes in agricultural, and hydro-meteorological droughts under climate change scenarios driven by the RCP 4.5 and 8.5 scenarios. The main findings can be summarized as follows: (1) future climate scenarios (precipitation and temperature) were generated for the period 2045–2070. It is indicated that the climate in the study area would generally warmer and wetter in the future; (2) under the possible climate change, the annual and seasonal streamflow would increase significantly in the future; and (3) the drought frequency would generally increase in the future, while the drought duration and severity would decrease. In general, the streamflow and droughts in the Srepok River Basin would not experience dramatic changes in the future according to this analysis. The results obtained in this study could be useful for planning and managing water resources in this region through enhancing the understanding of the impact of various climate change scenarios on droughts in terms of agricultural and hydro-meteorological perspectives.

Future work is in progress to quantify separate and combined impacts of climate change and land-use changes on meteorological, agricultural, and hydrological droughts. The related results will be given in a forthcoming paper.

Acknowledgements This research is funded by Vietnam National Foundation for Science and Technology Development (NAFOSTED) under grant number "105.06-2019.20".

References

Duan K, Mei Y (2014) Comparison of meteorological, hydrological and agricultural drought responses to climate change and uncertainty assessment. Water Resour Manage 28(14):5039–5054

FAO (2016) "El Nino" event in Vietnam: Agriculture food security and livelihood needs assessment in response to drought and salt water intrusion. Hanoi, FAO

IPCC (2013) The Physical Science Basis: Contribution of Working Group I to the Fifth Assessment Report of the Intergovernmental Panel on Climate Change. Cambridge University Press, Cambridge

Kawasaki A, Takamatsu M, He J, Roger P, Herath S (2010) An integrated approach to evaluate potential impact of precipitation and land-use change on streamflow in the Srepok River Basin. Theory and Application of GIS 18(2):9–20

Khoi DN, Suetsugi T (2012) Hydrologic response to climate change: a case study for the Be River Catchment Vietnam. J Water Climate Change 3(3):207–224

Kim CJ, Park MJ, Lee JH (2014) Analysis of climate change impacts on the spatial and frequency patters of drought using a potential drought hazard mapping approach. Int J Climatol 34:61–80

Leng G, Tang Q, Rayburg S (2015) Climate change impacts on meteorological, agricultural, and hydrological droughts in China. Global Planet Change 126:23–34

MONRE (2012) Climate change, sea level rise scenarios for Vietnam. Vietnam Ministry of Natural Resources and Environment: Hanoi

McKee TB, Doesken NJ, Kliest J (1993) The relationship of drought frequency and duration to time scales. Paper presented to the 8th Conference of Applied Climatology, American Meteorological Society, Anaheim, California

Moriasi DN, Arnold JG, van Liew MW, Bingner RL, Harmel RD, Veith TL (2007) Model evaluation guidelines for systematic quantification of accuracy in watershed simulations. Trans ASABE 50:885–900

Neitsch AL, Arnold JG, Kiniry JR, Williams JR (2011) Soil and Water Assessment Tool Theoretical Documentation Version 2009. Texas Water Resources Institute Technical Report No. 406: Texas A&M University, Texas

Sayari N, Bannayan M, Alizadeh A, Farid A (2013) Using drought indices to assess climate change impacts on drought conditions in the northeast of Iran (case study: Kasahfrood basin). Meteorol Appl 20:115–127

Spinoni J, Naumann G, Carrao H, Barbosa P, Vogt J (2014) World drought frequency, duration, and severity for 1951–2010. Int J Climatol 34:2792–2804

Sunyer MA, Henrik M, Keiko Y (2010) On the use of statistical downscaling for assessing climate change impacts on hydrology. In International Workshop Advances in Statistical Hydrology, Taormina, Italy

Vu MT, Raghavan VS, Liong S-Y (2015) Ensemble climate projection for hydro-meteorological drought over a river basin in Central Highland Vietnam. KSCE J Civil Eng 19(2):427–433

WMO (2012) Standardized precipitation index User guide (WMO-No.1090). World Meteorological Organization, Geneva, Switzerland

Wang D, Hejazi M, Cai X, Valocchi AJ (2011) Climate change impact on meteorological, agricultural, and hydrological drought in central Illinois. Water Resour Res 47:W09527

Wilhite DA, Glantz MH (1985) Understanding the drought phenomenon: the role of definitions. Water International 10(3):111–120

Chapter 53
Optimized Operation of Red-River Reservoirs System in the Context of Drought and Water Conflicts

L. X. Nguyen, T. D. Tran, S. T. Hoang, and P. T. Nguyen

Abstract Red-River System (RRS) is the second largest basin in Vietnam (second to Mekong river basin) consists of 25 provinces/cities, with population of 32 million people in 2014. The basin is also characterized with high hydropower potential due to generally steep topography and high rainfall in the mountain area. During recent years, the hydro-power reservoir system has been developed, nearly completed as planned and making up 24% of 48,573 MW total national electricity generation. The development of hydropower reservoir system has certain contribution to social economic status of the basin, however, since 1998, there have been about 8 drought events with increasing intensity in the basin. The impacts of drought would be minimal if the water conflicts do not exist between power generation and water supply. The conflicts often occur during peak demand of the two biggest water user groups, agriculture from January to March and hydropower from April to June. Operation optimization for 6 major reservoirs in RRS during dry season has been carried out using the Non-dominated Sorting Genetic Algorithm II (NSGA II) with 2 objective functions of total hydropower cost and Hanoi water level. The optimum parameter sets for reservoir operation policy were found using the parameterization simulation optimization (PSO) approach on the model of reservoir system connected with a downstream Artificial Neuron Network (ANN) emulator. The model has been validated for the year 2011. Optimized operation policy for dry season of 2015–2016 shows $2.67 \div 6.33\%$ increase in total relative hydropower cost and $1.18 \div 2.77\%$ increase in average water level at Hanoi. The model applied for reservoir system on RRS of Vietnam offers better option for water resources operators to reduce the water related conflict while increase the benefit of water resources of the basin.

L. X. Nguyen (✉)
Institute for Water and Environment, Vietnam Academy for Water Resources, Hanoi, Vietnam

T. D. Tran (✉) · P. T. Nguyen
Vietnam Academy for Water Resources, Hanoi, Vietnam
e-mail: trinhtd.rtc@vawr.org.vn

S. T. Hoang
Vietnam Academy of Science and Technology, Hanoi, Vietnam

© Springer Nature Switzerland AG 2021
M. Babel et al. (eds.), *Water Security in Asia*, Springer Water,
https://doi.org/10.1007/978-3-319-54612-4_53

Keywords Reservoir operation · Optimization · Genetic algorithm · Artificial
neural network

53.1 Introduction

Red River Basin in Vietnam plays an important role in water-food-energy security
for the country as it is the highest hydropower development potential and second
largest rice granary of Vietnam. The total hydro power potential of the Red River
Basin accounts to about 60% of 17,031 MW hydropower potential of the whole
country while its 1.8 million hectares of agriculture land producing about 6.7 million
tonnes of rice annually (Dang et al. 2010). In the basin, the main water user groups
are hydropower and agriculture even though the former is not consumptive user.
The differences in water demand schedule between the two main users group has
been shown as the main cause of the drought risk and constant water frictions,
which often need administrative procedures to arrange the water discharge protocol
of hydropower (Molle and Chu 2009). These procedures are still far from optimal
policy because of the constraints involve in the operation of the reservoir system and
the agricultural intake condition downstream has not been fully taken into account
(Castelletti et al. 2012a) (Fig. 53.1).

The situation is further complicated with many uncertainties and related hydropower dams regulation happening on the river basin. Firstly, since 1998, there have
been about 8 drought events (1998, 2002, 2004, 2005, 2009, 2010, 2015, 2016) with
the minimum dry flow reduced from 700–800 to 400 m^3s^{-1} (2010). Based on the CC
prediction published by Ministry of Natural Resources and Environment, it indicated

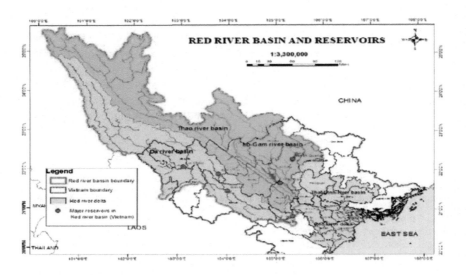

Fig. 53.1 Red River basin with its sub basins and reservoirs

Table 53.1 Spring releases and cropping areas over years in RRD

Year	Total Spring release (10^9 m^3)	Spring cultivation area (ha)	Remark
2010	2.782	627,401	3 reservoirs
2011	2.947	630,757	3 reservoirs
2012	3.967	635,117	Son La in operation
2013	4.556	634,275	
2014	5.77	636,275	

that the average flow in the basin will decrease in dry and flood season by 6.7–11.8% and 8.1–13.24% respectively (Schmidt-Thome et al. 2015). Secondly, the basin is shared mostly by Vietnam and China, accounted for more than 99% of the basin area (China 51%, the minor 1% Lao PDR, and 48% Vietnam at downstream), however the information from upstream activities in Chinese part is very limited, for example, up to now China has built dozens hydropower reservoirs and barrages, but their detail designs and operations are also restricted (Van Diep et al. 2007). The operation of those hydropower has changed the hydrological pattern in Vietnamese side, such as the dry season usually comes earlier, and during the hottest and driest period, there have been signs of flow interruption. Thirdly, the riverbed erosion due to the imbalance in the sediment trapping for reservoir system, thus every year the reservoirs system needs to discharge higher volume of water to ensure the gravitational flow into the intake structure of the irrigation system. Before 2003, the system just needs to discharge about 860 m^{3}s^{-1} but now it is 1,450 m^{3}s^{-1} in order to meet the water level requirement of 2.2 m at Hanoi. In which, the three main hydro-power reservoirs are Hoa Binh, Thac Ba and Tuyen Quang as main water contributors for 15 days meeting the requirements in of the transplanting period of spring rice season ("Operation rules for Red River reservoirs system" 2015). However statistics of 2010–2014 shows that the release has doubled whereas spring cropping area is still the same (Table 53.1 and Fig. 53.2).

The above difficulties and uncertainties in the operation of reservoir system in relation with water allocation for other economic sectors and electricity generation in Vietnam calls for the study regarding the optimization of reservoirs system. Operating procedure for reservoirs system often has to take into account conflicting objectives, between power generation, water storage versus water for agriculture, water for environment and services. Therefore, there have been a number of approaches developed in optimizing operation for reservoir system. To solve the linear equations of complex system involves in operating a reservoir system, implicitly stochastic optimization (ISO) has been developed to find optimal reservoir discharge. ISO normally used deterministic optimization to find the reservoir operation rules under several equally likely inflow scenarios and the reservoir capacity (Castelletti et al. 2012b; Karamouz and Houck 1982). In contrast to the ISO, the explicit stochastic optimization (ESO) method incorporates probabilistic inflow functions directly into the optimization problem (Goor et al. 2010; Powell 2007; Xu et al. 2014). Even though those approaches are very applicable and appealing for application, they still face with

Fig. 53.2 Riverbed exposed
in 2010 at Ha Noi

computational difficulties in solving conflicting objective function (Labadie 2004; Simonovic 1992; Wurbs 1993; Yeh 1985). In a different optimization scheme, parameterization simulation optimization (PSO) usually predefines the reservoir operation rule curve based on reservoir parameters and then applies nonlinear optimization procedure to find the parameters with best reservoir operation performance under different inflow scenarios. In this approach, most stochastic events of the problem such as inflow, spatial and temporal distribution parameters are introduced in the model implicitly, thus reducing the curse of dimensionality compare to ISO and ESO (Chang et al. 2005; Chen 2003; Momtahen and Dariane 2007).

With multi-objective operation optimization of reservoir system, there will not exist a single solution that improves all the objective function rather there will be a non-dominated set of solutions. The traditional approach to operate multi-objective reservoir system was often reduce the problem to single objective using weighted functions and consecutively select different weighted functions to find the optimal pareto front of solutions for the multi-objective problem (Shiau and Lee 2005; Srinivasan and Philipose 1998). Other method to solve the multi-objective optimization problem is the constraint method, in which all objectives except one are constrained to a specific value and the remaining objective is optimized. The solution then is an optimal Pareto front. This method is often referred as ε-constraint method (Ko et al. 1997; Mousavi and Ramamurthy 2000; Shirangi et al. 2008). However, with the increasingly number of objective functions of complex problem, the above approaches have been shown inefficient in solving the problem and often end up with same Pareto fronts for different combination of weights and often these approaches requires so many trials to get the non-dominated solutions.

In recent decades, multi-objective evolutionary algorithms (MOEAs) which uses a population-based search have been found very attractive as they find many Pareto optimal solutions in a single run (Nicklow et al. 2009). Therefore, MOEAs has been repeatedly applied to multi-purpose reservoir operation optimization. This is because the ability of the algorithm to connect with simulations models without simplification of some nonlinear relationships which cannot be implemented with the traditional approaches. In addition, the approach facilitates the parallel computation (Maier et al. 2014). In this study, the Non-dominated Sorting Genetic Algorithm-II (NSGA-II) (Deb et al. 2002) which is a subclass of MOEAs method has been employed to optimize reservoir operation policy, in conjunction with an artificial neural network model that emulates the 1-D hydraulic model downstream of Red River Delta (RRD).

53.2 Data and Methods

With the PSO adopted, all arrangements for models, policy operation function and optimization algorithm are briefed as followings:

53.2.1 Climate and Reservoir System

The climate of RRB is characterized of monsoon conditions, with the annual temperature spatially varies from 14 to 24 °C, humidity 80–87%, wind speed 0.8–4 ms^{-1}, evaporation (Piche measurement) 500 mm–1,000 mm. As for rainfall, in Chinese part, the annual rainfall varies from 600 to 3,000 mm, coming down to Vietnamese part, it ranges 1,100–5,000 mm. By calculation, total annual rainfall resource is 238.69 km^3 and Vietnamese amount makes up to 57.82% (Xuan 2012). However, due to the fact of highly uneven rainfall distribution within a year, flood season usually occurring in the period of May—Oct. The flood season accounts for 70–87% of total annual flow volume (121.8 km^3). The dry season (Nov–April) accounts for the rest of the total annual flow, with the driest months usually occurs in Jan—March or Feb–April.

At the end of 2015, in RRB, there are about 12 major and medium hydropower plants that have been in operation. By their active storage and installed power generation capacity, six reservoirs are considered the most important such as Ban Chat, Lai Chau Son La, Hoa Binh, Thac Ba and Tuyen Quang (Table 53.2). Among them, two largest reservoirs are Hoa Binh and Son La with about 6–6.5 bil m^3 of active storage and both of them belongs to Da river basin. Two smaller as Thac Ba and Tuyen Quang staying in Lo-Gam river basin with 1.5 and 1.5 bil m^3, however, together with Hoa Binh, they keep the role of direct supplying water for RRD (Fig. 53.3).

53.2.2 Objective Functions

Based on the fact that conflicts mainly are between hydropower and others such as agriculture, navigation, domestic water, environment, etc., which can be integrated as an indicator of Ha Noi water level. Two objective functions can be written as:

Table 53.2 Main parameters of 6 main hydropower reservoirs in Red River Basin

Reservoirs	Parameter					
	Normal water level (m)	Active storage (10^9 m^3)	Firm capacity (MW)	Installed capacity (MW)	Maximum turbine discharge (m$^{3-1}$)	Regulation type*
Hoa Binh	117	6,058	548 (671)	1,920	2,400	Year
Son La	215	6,504	522	2,400	3,460	Year
Tuyen Quang	120	1,079	83.3	342	750	Multi-year
Thac Ba	58	1,576	41.2	120	420	Multi-year
Ban Chat	475	1,702	74.7	220	273.3	Year
Lai Chau	295	0.7997	160	1200	1,665	Year
Total		16,919	1,552.2	6,202	8,695	

*Regulation type: based on the characteristic of the basin and the designed capacity of the reservoir to regulate flow on yearly basis or multi-year basis (discharge and storage plan need to closely regulate to meet the requirement of yearly or multi-year regulation)

Fig. 53.3 Diagram of simulated RRB system

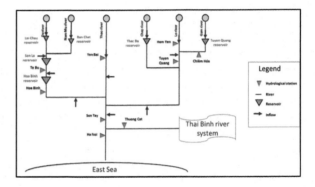

1. Total relative cost of hydro power generated during the dry season by the system over the horizontal time

$$max \sum_{t=1}^{T} \alpha_t \sum_{i=1}^{n} [r_{i,t}.g.\overline{h}_{i,t}.e_{i,t}.\Delta t] \qquad (53.1)$$

In which, T: number of time steps of simulation horizontal time; n: number of reservoirs; $\overline{h}_{i,t}$: average water head (m) equivalent to $\overline{h}_{i,t}^{up}$ - $h_{i,t}^{do}$, $\overline{h}_{i,t}^{up}$ average upstream reservoir water level and $h_{i,t}^{do}$, downstream water level of reservoir i during time step t; g: gravitational force = 9.81 (m/s^2); The power is expected in kWh, Δt will be number of hours per time step; $e_{i,t}$ is the efficiency of equipment should be equal to $e_{i,t}^{turbin} \times e_{i,t}^{plant}$, however, for simple, assume $e_{i,t}^{tu}$

and $e_{i,t}^{plant}$ be constant as 0.87 and 0.93 respectively; α_t is relative factor between the minimum and maximum price, and should be equivalent to $\alpha_t^{season} . \alpha_t^{peaktime}$.

2. Maximizing Ha Noi water level averaged over the horizontal time:

$$max \frac{1}{T} \sum_{t=1}^{T} f\left(r_{t-5}^{delta}, r_{t-5}^{nat}, Z_{t-3}^{BL}, WD_t^{delta}\right) \tag{53.2}$$

In which, r_{t-5}^{delta} as total releases from 3 reservoirs Hoa Binh, Thac Ba and Tuyen Quang: r_{t-5}^{nat} total inflow to the rivers from section from reservoirs to the delta; Z_{t-3}^{BL} is the water level at Ba Lat (tide level); WD_t^{delta}: water demand for the whole delta; $t-5$, $t-3$: lagging time as calculated from the hydrodynamic model.

53.2.3 Constraints

There are 5 constraints that govern the problem which need to be considered as.

1. System water balance equation:

$$s_{t+1}^i = s_t^i + C \cdot r_t^k + I_t^i + E_t^i - r_t^i, \quad \text{with } t = 1, \dots, T \text{ and } i = 1, \dots, 6 \tag{53.3}$$

In which, s_t^i, s_{t+1}^i, reservoir storage at the beginning of time step t and $t + 1$; I_t^i: unregulated or natural inflow to reservoir i during t; C: connectivity matrix of the reservoir system; r_t^k releases from upstream reservoirs (1..k) during t; E_t^i: loss due to evaporation to reservoir i during t and equal $e_t^i . S_t^i$, with e_t^i as evaporation rate unit and S_t^i surface area of reservoir i at the beginning of time step t; r_t^i: releases from reservoir i during t.

2. Lower and upper bounds of storages: Reservoir storages must stay within dead storage and maximum storage, and also need to take into account other purposes such as tourism, flood control, etc. For this problem, the lower and upper boundaries can be integrated as dead storage and normal storage during dry season, and the end storage of dry season must be under the flood-control thresholds for the early flood period, which is stipulated under the operation guideline No. 1622. Therefore, constraints on storage can be stated as:

$$s_{dead}^i \leq s_t^i \leq s_{normal}^i \quad and \quad s_{T+1}^i \leq S_{early \; flood \; control}^i \quad \text{with } t = 1 \dots T \tag{53.4}$$

3. Lower and upper bounds of storages: Similarly, reservoir release also depends on the capacity of the discharging structures such as turbine, spillway, culvert, and the policy that stimulating on the dam safety, flood control and environment flow. Moreover, with this problem, we intuitively understand optimization during dry season means that there should be no spillages and the release should be through

turbine as much as possible. In case of water level exceeding the normal water level that threatens the dam safety, the spillway can just be utilized to bring water level down under the normal level as soon as possible. Therefore, the constraints can be summarized as:

$$r_{t,i}^{min} \leq r_{i,t} \leq r_{t,i}^{max} \tag{53.5}$$

where:

$$r_{max} = \begin{cases} r_{turbin_m} + r_{spill\_max} \ if \ z_{up} > z_{normal\,WL} \\ r_{turbin\_max} \ if \ z_{up} \leq z_{normal\,WL} \ and \ z_{up} \geq z_d \\ 0 \ if \ z_{up} \leq z_d \end{cases} \tag{53.6}$$

$$r_{min} = \begin{cases} r_{turbin_m} + r_{spill\_max} \ if \ z_{up} > z_{normal\,WL} \\ r_{turbin\_max} \ if \ z_{up} = z_{normal\,WL} \\ r_{emf} \ if \ z_{tl} < z_{normal\,WL} \ and \ z_{up} \geq z_d \\ 0 \ if \ z_{up} \leq z_d \end{cases} \tag{53.7}$$

In formula (53.5), $r_{i,t}$ is the release of reservoir i during t; $r_{t,i}^{min}$, $r_{t,i}^{max}$ are minimum and maximum release of reservoir i during t. These bounds can be further expressed by formulas (53.6) and (53.7), in which, r_{turbin_max}, r_{spill_max} are maximum water volume released through turbine and spillway respectively; z_{up}, z_{normal_WL} and z_d are reservoir water level at time t, normal water level and dead water level; r_{emf} is the environmental flow that the reservoir hast to maintain all the time. Constraints except for (53.4) were directly embedded into the system model. Besides, the function of r_{max} and r_{min} also were modified by the Newton approximation:

$$V_t^{min|max} = \int_t^{t+\Delta t} \tilde{r}_t^{min|max} dt \tag{53.8}$$

$$r_t^{min|max} = V_t^{min|max} / k \tag{53.9}$$

4. Water level at Ha Noi: As we know, water level at Ha Noi is the control point of RRD, however, this point is not only affected by 3 releases of Hoa Binh, Thac Ba and Tuyen Quang, but also by the natural inflow at the downstream of those reservoirs, therefore, we put it out of the constraints above. In practice, to adapt with the changes, agriculture production now needs to have water supply at Ha Noi of more than 2.2 m during the 3 transplanting periods of intensive releases of (15 days). So, we set a constraint:

$$H_t^{HN} \geq 2.2 \tag{53.10}$$

Fig. 53.4 Methodology to develop emulator from Mike 11 data

53.2.4 Downstream Model

At downstream of RRB, there is a large irrigation and drainage system, which does not only provide services for navigation, domestic water supply, aqua-culture, environment services, etc., but also supplies for 650,000 ha of spring rice and a similar value for summer rice. The system comprises 54 sub-irrigation systems. The water intaking practice is largely dependent on water level on the mainstream, which affected by upstream inflow, tide regime and partly by salt-water intrusion.

In order to simulate water level at the control point at Ha Noi, the research used a one-dimension hydrodynamic model Mike 11 (DHI 2003). The domain of model ranges from upstream boundaries of 3 reservoirs Hoa Binh, Thac Ba and Tuyen Quang, and other natural boundaries to downstream boundaries of 9 estuaries. Besides, water intaking process of major works on the mainstream was simulated, and the models were calibrated and verified on several years. However, application of hydrodynamic model that integrated into an optimization framework on a 20-year time series, would take a lot of computation time. To replace this model, the research utilized Artificial Neural Network (ANN) with 2 layers and one hidden layer. The number of neurons were estimated through a trial process and evaluated with performance indicators.

Referring to sensitivity analysis, the research defined Ha Noi water level function:

$$h_{t+1}^{HN} = f\left(h_t^{HN}, q_t^{delt}, h_t^{BL}, WD_t\right) \tag{53.11}$$

In which, h_t^{HN}: water level at Ha Noi at the beginning of time step t, q_t^{delt}: total flow to the downstream delta, which includes reservoirs release and natural flows during time step t, h_t^{BL}: water level (tide) at Ba Lat station at the beginning of time step t; WD_t: total water demand during time step t. Also, lagging time of upstream inflow and tide to Ha Noi was analyzed and included. Hence, the Mike 11 model was run on period of 2001–2010 to provide output, later on combined with the input to form training and testing data for the ANN-based emulator (Fig. 53.4).

53.2.5 Operation Policy Function

Normally, reservoir policy function depends on time, physical and policy characteristic, current system state, inflow forecast, etc. However, in order to support real-time operation, simplified data requirements, reduced problem dimensions, resolve the conflicts between hydropower and other sectors, the research used a predefined function of operation policy:

$$u_t^k = Rmin_t^k + \theta_j^k.\alpha^k.\left(Rmax_t^k - Rmin_t^k\right) \qquad (53.12)$$

In which, u_t^k: decision variable of release for reservoir k and during time step of t; $Rmin_t^k$, $Rmax_t^k$: minimum and maximum release potential of reservoir k and during time step of t, these variables can be estimated by formulas (53.6, 53.7, 53.8, 53.9); α^k: coefficient of daily high peak, assuming 1.5. Therefore, the operation policy function was parameterized by θ_j^k: operation parameter of period j and reservoir k. Based on the electricity- demand variation and water-supply characteristics, the dry seasons was divided into 11 periods (Tuyen and Dung 2015).

53.2.6 Optimization Procedures

With such a division of eleven-period dry season and system of 6 reservoirs, the dimension of optimization problem grown up 6 × 11 dimensions, and the variables are real number, ranging from 0 to 1. The computation cost was expected to be extreme if rendering the classical methodologies such as dynamic programming (Bellman 1957). Therefore, the research used a MOEA algorithm as NSGA-II (Deb 2002) (Fig. 53.6), conducted optimization evaluation for dry season with time step of 4 h and horizontal time of 20 years. The calculation of flood season was used to initialize the beginning condition for dry model following Guideline 1622 ("Operation rules for Red River reservoirs system" 2015) (Fig. 53.5 and Fig. 53.7).

53.3 Results and Discussion

53.3.1 Downstream Model

The research has set up a 1-dimensional hydrodynamic that consists 31 rivers, 800 cross section, 27 natural inflow boundaries, 3 reservoir release boundaries, 9 water level (tide) boundaries and 201 control water intakes and hundreds of water intake in the form of "point sources". In total there are 8,920 computing nodes. The model has been validated by Nash–Sutcliffe (Nash and Sutcliffe 1970) model efficiency coefficient $NASH = 1 - \frac{\sum(X_{o,i}-X_{S,i})^2}{\sum(X_{o,i}-\overline{X_O})^2}$ (2–9) for 21 stations through calibration-verification process. The results showed Nash of calibration results achieves 89.39% in average in 2010 and 88.7% in 2011. The following shows the verification results of 9 most important stations (Fig. 53.8):

Fig. 53.5 Flowchart of
calculating objective
function and constraints and
initial conditions from flood
model for dry season

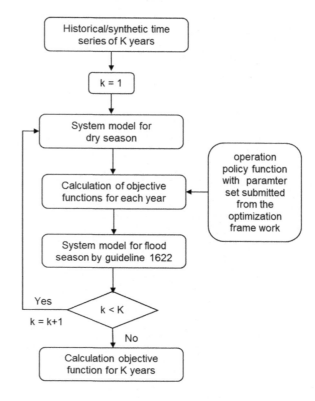

Due to the time consuming to simulate the hydraulic model for the whole system. It is unrealistic to couple hydraulic network simulation with optimization of the reservoirs policy function. Thus a library of hydraulic network simulation outputs has been prepared before running the optimization policy for function for the reservoir. The output from Mike 11 model application has provided training and testing data for developing ANN-based emulator. NETLAB toolbox developed (Nabney 2002) running on Matlab is utilized to define ANN network. Number of neuron is determined as 10 through a trial computing process with 9 performance indicators (Nabney 2004). Among them, coefficient of determination R^2 is calculated as 0.9925 (Fig. 53.9—top-left) (Fig. 53.10).

53.3.2 Optimal Operation Policy Function

The optimization framework has been written in Matlab programming language. In order to ensure the possibility of finding feasible solutions, a population of 5000 individuals was used, and the optimization evaluation was proceeded until 200th generation, at which, only minor improvements was observed. In total, 1,000,000

Fig. 53.6 Optimization framework of NSGA-II to develop non-dominated solutions pareto

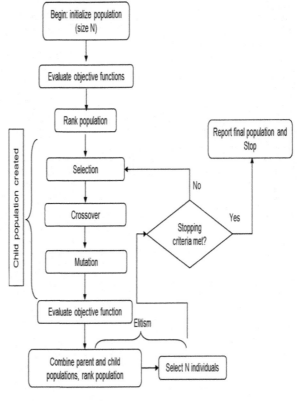

Fig. 53.7 Diagram of delta model on Mike 11 user interface

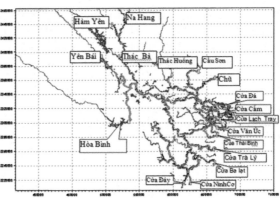

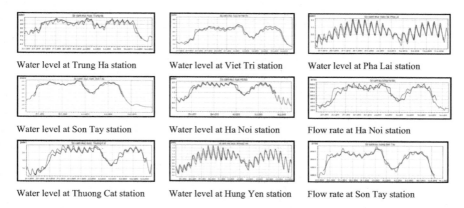

Water level at Trung Ha station	Water level at Viet Tri station	Water level at Pha Lai station
Water level at Son Tay station	Water level at Ha Noi station	Flow rate at Ha Noi station
Water level at Thuong Cat station	Water level at Hung Yen station	Flow rate at Son Tay station

Fig. 53.8 Verification of hydraulic model at main hydrological stations in 2011, simulated data (red), observed data (blue)

Fig. 53.9 Performance indicators of ANN-based emulator by number of neurons

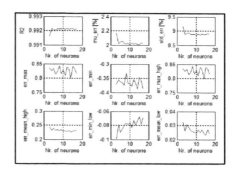

Fig. 53.10 Input and output of the RRD ANN-based emulator with 10 neurons

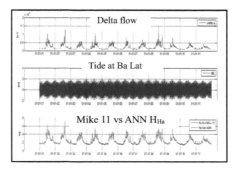

parameter sets were applied and evaluated on the policy function and a pareto of non-dominated alternatives was obtained (Fig. 53.11 and Fig. 53.12).

Data collection of inflow and reservoir operation was conducted for the period from Sept 2015 to June 2016, and then the optimal alternatives were applied and

Fig. 53.11 Pareto of 5000 optimal non-dominated alternatives

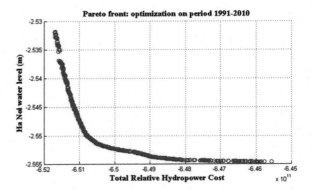

Fig. 53.12 Comparison between application of optimal alternatives and 2015 observed operation data

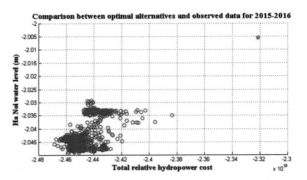

compared with the observation data on both objectives. As for the optimal alternatives, though a pareto could not be obtained (as the alternative is designated for long term), the results have shown an improvement compared to the observation, 2.67–6.33% of total relative hydropower cost and 1.18–2.77% of Ha Noi average water level (Table 53.3), which might be considered not to be significant due to the simplicity of the policy function with advantages of largely reducing the computing cost. However, there is no violation to the minimum threshold of navigation, $H_{HN\,min}$ = 1.213 m > 1.2 m (Guideline-1622) ("Operation rules for Red River reservoirs system" 2015) with the operation policy of the lowest performance in term of Ha Noi water level H1234 (Fig. 53.13), meanwhile, the observation data was shown as $H_{HN\,min}$ = 0.82 m < 1.2 m, and the violation occurred many times during April, May and June 2016 (Fig. 53.14).

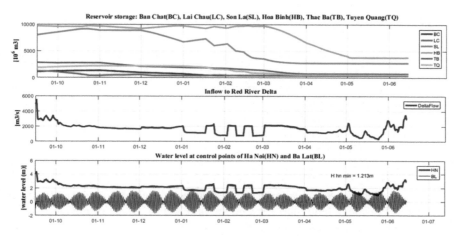

Fig. 53.13 Variation of reservoirs storages and water level of control points at Ha Noi and Ba Lat by alternative H1234 from 16 Sep 2015 to 14 June 2016

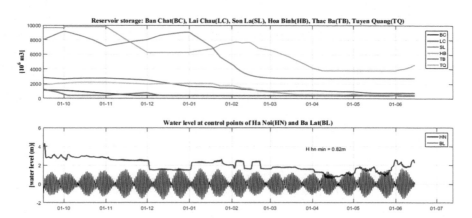

Fig. 53.14 Variation of observed reservoirs storages and water level of control points at Ha Noi and Ba Lat from 16 Sep 2015 to 14 June 2016

Table 53.3 Comparison between application of optimal alternatives and 2015 observed operation data

Objective function	5000 Optimal alternatives	2015 operation	Improvement (%)
Total relative hydropower cost	2.3837e10–2.4687e10	2.3217e10	2.67–6.33
Ha Noi water level (m)	2.0292–2.0491	2.0056	1.18–2.17

53.4 Conclusions

Red River reservoir system plays an important role in ensuring water, food and energy security for Vietnam. However, data analysis shows that there have been major conflicts, and these are being exaggerated due to increased drought risk in the context of Climate Change and changing river morphologies, socio-economic development coupled with a significant growth in energy demand. Based on that, the study has formulated an optimization framework to develop a real-time operation policy function that comprises 6 reservoirs which are Ban Chat, Lai Chau, Son La, Hoa Binh, Thac Ba and Tuyen Quang. The research has predefined the families of policy function with parameter sets. Through applying NSGA-II optimization algorithm on 20-year time-series of the coupled reservoir system model and ANN-based downstream emulator. The computation has handled the problem of "curse of dimensionality" and provided a 5,000-non-dominated-alternatives pareto. Theses optimal alternatives have been validated in dry season of 2015–2016 and indicated an improvement of 2.67–6.33% in total relative hydropower cost and 1.18–2.77% average Ha Noi water level. Moreover, there is no violation to the minimum threshold for navigation in the optimized policy while observed data show frequent violation of the threshold water level in that period. Nevertheless, more effort is needed on researching the policy function to explore more optimal solutions as well as algorithms in balance with computational cost.

Acknowledgements The authors would like to thank Vietnam Academy for Water Resources for facilitating the research. Besides, a special appreciation goes to Assoc. Prof. Hoang Minh Tuyen and Dr. Luong Huu Dung for their provision of data and valuable advices. Last but not least, a special thanks to the research group led by Prof. Rodolfo Soncini-Sessa who provided training on reservoirs operation course in the framework of IMRR project in April 2016.

References

Bellman R (1957) A Markovian decision process. J Math Mech 6:679–684
Castelletti A, Pianosi F, Quach X, Soncini-Sessa R (2012) Assessing water reservoirs management and development in Northern Vietnam. Hydrol Earth Syst Sci 16(1):189–199
Chang FJ, Chen L, Chang LC (2005) Optimizing the reservoir operating rule curves by genetic algorithms. Hydrol Process 19(11):2277–2289
Chen L (2003) Real coded genetic algorithm optimization of long term reservoir operation1. In: Wiley Online Library
Dang TH, Coynel A, Orange D, Blanc G, Etcheber H, Le LA (2010) Long-term monitoring (1960–2008) of the river-sediment transport in the red river watershed (Vietnam): temporal variability and dam-reservoir impact. Sci Total Environ 408(20):4654–4664
Deb K, Pratap A, Agarwal S, Meyarivan T (2002) A fast and elitist multiobjective genetic algorithm: NSGA-II. IEEE Trans Evol Comput 6(2):182–197
Dhi D (2003) Mike-11: a modelling system for rivers and channels, reference manual. DHI–Water and Development, Horsholm, Denmark

Goor Q, Halleux C, Mohamed Y, Tilmant A (2010) Optimal operation of a multipurpose multireservoir system in the Eastern Nile River Basin. Hydrol Earth Syst Sci 14(10):1895–1908

Karamouz M, Houck MH (1982) Annual and monthly reservoir operating rules generated by deterministic optimization. Water Resour Res 18(5):1337–1344

Ko SK, Oh MH, Fontane DG (1997) Multiobjective analysis of service-water-transmission systems. J Water Resour Plan Manag 123(2):78–83

Labadie JW (2004) Optimal operation of multireservoir systems: state-of-the-art review. J Water Resour Plan Manag 130(2):93–111

Maier HR, Kapelan Z, Kasprzyk J, Kollat J, Matott LS, Cunha MC, Dandy GC, Gibbs MS, Keedwell E, Marchi A (2014) Evolutionary algorithms and other metaheuristics in water resources: current status, research challenges and future directions. Environ Model Softw 62:271–299

Molle F, Chu TH (2009) Implementing integrated river basin management: lessons from the Red River Basin, Vietnam. IWMI

Momtahen S, Dariane AB (2007) Direct search approaches using genetic algorithms for optimization of water reservoir operating policies. J Water Resour Plan Manag 133(3):202–209

Mousavi H, Ramamurthy AS (2000) Optimal design of multi-reservoir systems for water supply. Adv Water Resour 23(6):613–624

Nabney I (2002) NETLAB: algorithms for pattern recognition. Springer

Nabney IT (2004) Efficient training of RBF networks for classification. Int J Neural Syst 14(03):201–208

Nash JE, Sutcliffe JV (1970) River flow forecasting through conceptual models part I - a discussion of principles. J Hydrol 10(3):282–290

Nicklow J, Reed P, Savic D, Dessalegne T, Harrell L, Chan-Hilton A, Karamouz M, Minsker B, Ostfeld A, Singh A (2009) State of the art for genetic algorithms and beyond in water resources planning and management. J Water Resour Plan Manag 136(4):412–432

Operation rules for Red River reservoirs system, guideline 1622 (2015) Ministry of Natural Resources and Environment, Vietnam Government

Powell WB (2007) Approximate dynamic programming: solving the curses of dimensionality, vol 703. Wiley, Hoboken

Schmidt-Thome P, Nguyen TH, Pham TL, Jarva J, Nuottimäki K (2015) Climate change in Vietnam. In: Climate change adaptation measures in Vietnam, pp 7–15. Springer

Shiau JT, Lee HC (2005) Derivation of optimal hedging rules for a water-supply reservoir through compromise programming. Water Resour Manag 19(2):111–132

Shirangi E, Kerachian R, Bajestan MS (2008) A simplified model for reservoir operation considering the water quality issues: application of the young conflict resolution theory. Environ Monit Assess 146(1–3):77–89

Simonovic SP (1992) Reservoir systems analysis: closing gap between theory and practice. J Water Resour Plan Manag 118(3):262–280

Srinivasan K, Philipose MC (1998) Effect of hedging on over-year reservoir performance. Water Resour Manag 12(2):95–120

Tuyen MH, Dung HL (2015) Developing guideline for operating Red River reservoir system (final report). Retrieved from Ministry of Natural Resources and Environment

Van Diep N, Khanh NH, Son NM, Van Hanh N, Huntjens P (2007) Integrated water resource management in the Red River basin-problems and cooperation opportunity. Paper presented at the proceedings of the CAIWA international conference on adaptive and integrated water management, Basel, Switzerland, pp 2–10

Wurbs RA (1993) Reservoir-system simulation and optimization models. J Water Resour Plan Manag 119(4):455–472

Xu W, Zhang C, Peng Y, Fu G, Zhou H (2014) A two stage Bayesian stochastic optimization model for cascaded hydropower systems considering varying uncertainty of flow forecasts. Water Resour Res 50(12):9267–9286

Xuan Thi T, Tuyen Minh H, Thuc T, Thai Hong T, Dung KN (2012) Water resources of main rivers in Vietnam. Vietnam Science Publishing House
Yeh WW-G (1985) Reservoir management and operations models: a state-of-the-art review. Water Resour Res 21(12):1797–1818

Part IX
Community Engagement for Water Security Enhancement

Chapter 54
Integrated Information Dissemination System for Coastal Agricultural Community

Aaron Firoz, Nazmul Huq, and Lars Ribbe

Abstract Smallholder farmers in the coastal zone are particularly vulnerable to the increasing impacts of salinity and salinity related problems stemming from climate change impacts. Availability of climatic information offers great potentials for informed decision making by farmers at the face of increasing uncertainty for their agricultural production system. This is also true for other related stakeholders such as policy makers. The key research question therefore addresses "how can we better inform different stakeholders through managing information to enhance the adaptive capacity of the community at the saline prone region?". Answering such question is not straight forward due to its multifaceted nature and the involvement of diverse knowledge architectures which prevail at different social and technical scales. While scientific models, data and results are a specialist concern with its own body of evidence within the scientific community, the interface to bring that scientific information to the local level arguably unlocks new potential to mitigate global food security challenges in a changing environment. Translating scientific information to the community level is an emerging area of research. We have developed a methodological framework which focuses mainly on how scientific data and information can be transcribed to the salinity affected community at multiple levels, so that all stakeholders can take optimal benefit from the information flow. This is a proposed research project, which has been submitted for obtaining funds from the German Ministry of Education and Research. The project will be implemented in coastal Bangladesh where salinity is the key issue for food production.

Keywords Coastal agriculture · Salinity · Adaptation · Information system · Bangladesh · INDICA

A. Firoz (✉) · N. Huq · L. Ribbe
Institute for Technology and Resources Management in Tropics and Subtropics (ITT), TH Köln, Cologne, Germany
e-mail: aaron.firoz@list.lu

A. Firoz
Environmental Research and Innovation Department (ERIN), Luxembourg Institute of Science and Technology (LIST), Esch-sur-Alzette, Luxembourg

© Springer Nature Switzerland AG 2021
M. Babel et al. (eds.), *Water Security in Asia*, Springer Water,
https://doi.org/10.1007/978-3-319-54612-4_54

54.1 Introduction

At the interface of land and sea, coastal areas in Bangladesh are environmentally fragile due to saline water intrusion (Dasgupta 2014; MoEF 2006), sea level rise (Rabbani et al. 2013) and tropical cyclones (Dasgupta 2014). The combined effects of these issues would limit the area of land and water availability for food production (Thomas et al. 2013). Solving these issues, improving the sustainability of the agricultural production system and facilitating their transition are some of the key agriculture objectives not only for Bangladesh but for throughout the world.

Agriculture has become more complex, resulting in an increased importance of farmers' access to reliable, timely, and relevant information. Farmers require access to more varied, multisource and context-specific information, related to not only the best practices against environmental problems such as salinity and crop production technologies, but also to information about post-harvest aspects. Information and communication technologies (ICT) (Ospina et al. 2012) have proven useful in tailoring responses to such situations faced by the agricultural community in a dynamic climatic situation (Ospina and Heeks 2010a). The types of ICT-enabled services for addressing such challenges are growing quickly, particularly in the field of monitoring and adaptation (Ospina and Heeks 2010a). ICT applications help to observe, describe, record and understand weather, climate, agricultural production and are pivotal in environmental research, real-time data capturing and analysis as well as visualization (Labelle et al. 2008).

The recent advancements of remote sensing, GIS applications, sensors for precision farming, and weather and soil information is significant and these advancements can be a vital tool for providing appropriate information to the farmers. Particularly important is the vast quantity of information that is available from the freely accessible remote sensing and GIS tools. These emerging technologies and products can be used to exhibit how ICT tools facilitate the capture, processing and modelling of the complex biophysical system (Labelle et al. 2008). In conjunction with these emerging technologies, ICT tools can also provide opportunities to further disseminate and broaden access to information towards the small holder farmers and decision makers (Ospina and Heeks 2010b). Furthermore, mobile phone technology which is a pioneer invention in the ICTs sector, allows the active engagement of communities in monitoring and disseminating all relevant information, ranging from environmental, agricultural and market based information to disaster or early warning.

Despite the rapid development and assessment of ICTs enabled services for climate change, food production and disaster management at different sectors, the major challenge which remains for most projects and literature is the transformation of this information so it is relevant and accessible to local actors (Ospina et al. 2012). Very often it was found that there is a significant information and knowledge gap at the grassroots level, for instance when it comes to climate information, a tension exists between localisation of climate data and reality (Ospina and Heeks 2010a). Similarly, when considering early warning systems, key challenges remain at the local level, particularly when it comes to last-mile technology (Ospina et al. 2012).

The critical challenge of upscaling the use of ICTs from field based to community level to regional level also remains.

In a coastal zone, key information related to the level of salinity of soil and water, its spatial distribution and extent, suitable crops which can grow in the zone, and specific agricultural input at near real time would certainly provide more valuable information. Recent advancement of the information flow, particularly freely available earth observation data, and the continued increase of spatial and temporal resolution of these remotely sensed data has made it possible to monitor the agricultural and related information at near real time from the local to global level (Scudiero et al. 2015; Taghizadeh-Mehrjardi et al. 2014; Garnett et al. 2013; Whitecraft et al., 2015; Justice et al. 1998). Combining this information with local in situ measurements of various land-surface-atmospheric parameters can increase the robustness of the information for application at field level farming activities. Purposeful and wide use of these subsystems requires a higher level of interdisciplinary collaboration and would require engaging with experts who have advanced knowledge of mainstream products, application and engagement of local knowledge and adaptation strategy.

Research, capacity development, collaboration and innovative activities should play a key role in efforts to move towards a more resilient production system. Often this is undermined by a lack of efficient approaches to bring the appropriate information to the local community and demonstrate the validity of such innovative knowledge. There is a need to explore how the information flow, its dissemination and how the demonstration of that information could deliver mutually beneficial impacts for communities in the coastal areas.

Development of a methodological framework addressing the key information demand along with appropriate means of dissemination have a major role to play in the application of adaptive capacity development of small-holder farmers and the national agencies. Efforts are needed to develop their potential and prepare for scientific connectivity.

The overall goal of this paper is to propose a conceptual framework to integrate ICT in a coastal land use system to enhance the adaptive capacity of the local small holders. The framework will provide integrated solutions of information flow to enhance the adaptive capacity of the stakeholders (smallholder farmers, local and national agencies) through- a) local adaptation strategies and options, b) Integration of scientific data, information and technologies including remote sensing information & low cost environmental sensor technology c) Dissemination of information to the different layer of the actors in the coastal agricultural sectors.

54.2   Theoretical Framework and Literature Review

54.2.1   ICT-Enabled Services and Projects

The types of ICT-enabled services that are useful for improving the capacity and livelihoods of poor smallholders are growing quickly. One specialized and arguably the most practiced service is agricultural information using mobile phones. Mobile phones are used as a platform to exchange and access price and market information, coordinate input/output resources and production techniques (Parikh et al. 2007; Tickner 2009; Baumuller 2012). Some examples of this in action are Reuters Market Light (RML) or the subsidiary group of the Indian Farmer's Fertilizer Cooperative, Kisan Sanchar Limited, who both offer information to farmers on crops, diseases, and market prices. Similar kinds of information were provided in Kenya by KACE (Kenya Agricultural Commodity Exchange, ltd) (https://www.kacekenya. co.ke/), which provides full market information with access through information centres. In India MCX (Milti-commodity exchange) (https://www.mcxindia.com) has partnered with rural postal offices to display prices on an electronic display.

One of the major areas of innovation is mobile based value-chain development of agricultural products in order to ensure a fair price to the producers and improving overall economic strength of all stakeholders. Prominent examples include the Agricultural Market Information Systems in Bangladesh, Farmer's Friend (a Google product) in Uganda, and Ovi Life Tools by Nokia. Having the benefit of low cost internet services, Web-based initiatives are currently being developed in India (https://www.e-krishi.org/; https://www.hatbazaar.com); Bangladesh (https://www.ais.gov.bd/; https://www.isapindia.org/) Central and South America (https://www.radajamaica.com.jm/; https://www.infoagro.com/), Africa (https://cgs pace.cgiar.org/handle/10568/57546) for information about market prices and access to the latest agricultural practices.

Another type of specialized application is providing weather information to the farming community through mobile and Internet technology. aWhere Inc. (www. awhere.com/) has developed an Agronomics-Weather Index using a combination of high resolution weather data and knowledge about farming practices from all over the world. The index looks at variations from average weather to determine where a farmer's crops and therefore yields are most likely flattering (visit aWhere website). A similar approach to inform farmers regarding weather information is offered by ESOKO (https://esoko.com/). ESOKO offers a mobile agricultural and market information platform, which disseminates seasonal weather forecasts and gives climate-smart agricultural tips to farmers in the Lawra and Jirapa districts in the Upper West Region of Ghana. Early warning information (e.g., flood, drought, or other disaster events) through mobile phone is being quite widely used in developing countries (Ospina et al. 2012).

Since the beginning of the new millennium, new ICTs have given farmers and stakeholders better opportunities to manage location oriented climate risk (Finlay and Adera 2012). Low cost environmental monitoring sensors, freely available remote sensing technology and climatic data transformed into information have been useful to agricultural stakeholders, helping them make informed decisions about choosing crops, planting time, applying fertilizers or other inputs. Until now, this information is mostly capitalized on by multinational or largescale commercial farming companies (e.g., Bayer Crop Science- Digital Farming, Monsanto- The Climate Corporation, BASF- Schlagkartei). However, availability and use of such information and technology can also greatly assist subsistence farmers to make informed decisions on their production methods and to increase their resilience against climatic extreme events. For example, utilizing these tools can help farmers in the coastal zone to monitor soil salinization and take anticipatory resilience measures.

The advent of high resolution (ranging from 10 to 250 m) remote sensing products (e.g., Centinal-2, Landsat -7, MODIS) along with the capability of near real time data availability, has made using these products the most cost-effective method for monitoring soil salinization in the coastal zone, where salinity is the major drawback for agricultural production (Casas 1995). Soil salinity can be determined using the remote sensing indicator soil salinity index (SI), which maps the spatially distribution of soil salinity based on vegetation growing conditions (Tilley et al. 2007). With the help of in situ measurement coupling with remote sensing information, SI can detect the salinity ranges that are most relevant for agricultural productivity (less than 20 ds m^{-1}) (Scudiero et al. 2015). Certain features of crop stress, dehydration, and senescence have been examined using vegetation indices to determine their response in different spectra (Carter 1993; Psilvoiks and Elhag 2013). Additionally, low cost sensor technology (https://www.libelium.com/, www.freeduino.org/), has been used widely for community based monitoring activities and thereby providing information about the water salinity. The Common-Sense Net project (Panchard et al. 2007) implemented in rural Karnataka, India is an integrated sensor network system aimed at improving farming strategies in the face of highly variable conditions, in particular for risk management strategies.

Previously, the accessibility of this vast potential has been mostly limited to developed countries, e.g. precision farming. However, the usefulness and accessibility of the current open source software, technology and data (remote sensing products) can be considered as second generation features of ICTs for agriculture due to the opportunity it has given to disseminate this information to the wider community at minimal or no cost.

Global agriculture development on the other hand currently struggles with the significant gaps between the information and technology utilized by its most sophisticated practitioners, and the overwhelming majority of smallholder farmers practicing worldwide (World Bank 2011). The data and information flow which has been generated by the scientific community needs to be translated to the agricultural community in order for the full utilization of the scientific advances to be put into practice (Huq et al. 2014). There are encouraging signs that agricultural systems have started taking interest in database management, development and deployment of crop models, GIS systems and knowledge-based systems (Finlay and Adera 2012). To be functional and useful these applications require greater collaboration and multi- or cross-disciplinary partnerships which may span several units, departments and institutions. The building and use of these applications indicate a trend towards digital tools and technologies and knowledge sharing between partners at the project, institute and system level.

54.2.2   ICT Model and Solutions to Combat Climate Change

In order to analyze the potential role of ICTs in marginalized communities, the "Information Chain" model (Heeks 2005) identifies the key elements that need to be in place for data resources to translate into development results. The major component of the model is the inter linkages of knowledge and data resources to the information domain, which guide the necessary steps needs for economic and social resources for obtaining the development goals. The model indicates that the mere presence of data is not enough, but that communities need to be able to access the data, assess its qualities and apply it to their own local needs (Ospina et al. 2010a).

The dynamic nature of information and its role as a cross-cutting factor within livelihoods that are vulnerable to climate change, can be represented by integrating the Information Chain into a traditional Sustainable Livelihood Approach (SLA) framework (Duncombe 2006). The SLA approach presents interconnected and complex set of livelihood options and strategies where ICT and information tools can contribute to reduce vulnerability and build resilience. But while the SLA provides the basis for a more system based approach (linking vulnerability, adaptation and development outcome), it does not identify any specific role for ICTs. More specifically, it does not recognize the roles of "digital capital" (May et al. 2011; Ospina et al. 2012). The "Information Chain" models opens up the pathways for seamless integration of information and information flows to empower communities at micro, meso, and macro level.

ICTs, climate change and development overview model (Heeks and Molla 2009), addressed the multidimensional issues of livelihoods, resilience and climate change and how each of this can be linked with the ICT. This model identified four main areas in which ICTs relate to climate change; namely mitigation, monitoring, adaptation and strategy. Given the fact that the intersection between ICTs, climate change and development is a new field of enquiry, the model should be seen as providing a broad overview of key issues and links rather than as an exhaustive account of topics.

An updated methodological framework ICT for development (ICT4D) (Ospina and Heeks 2010a), was presented in which the potential of ICT to contribute to water-related adaptations is analyzed. The framework suggests a chain of linkages that exist with short- and long-term climate change impacting the six water-related vulnerability dimensions (livelihood, socio-political condition, health, habitats, food security, biodiversity and ecosystem services) of households, communities, regions, etc. These impacts demand adaptive actions which are shaped by the vulnerabilities, but which in turn reshape those vulnerabilities, ultimately leading to outcomes in terms of broader development goals. But this current model does not offer a conceptual foundation for those seeking to understand how ICTs make this contribution (Ospina et al. 2012).

An improved version of the conceptual model has been proposed (Ospina et al. 2012) to contribute to the implementation of ICT-enabled adaptation projects in the water sector, aiding practitioners to identify the key elements that need to be considered while designing a project. One of the key features of this framework that it identifies systematic prerequisites of ICT deliverables (e.g., GIS, Mobile phone, Web based software program etc.) for any ICT4D initiatives as well as the adoption of this ICT deliverables. However, this model cannot be automatically transferred without necessary customization according to local socio-ecological conditions. It is important that the model is adequately narrowed down to focus on the key areas and sectors that are impacted by climate change, where ICT can be most influential.

Each of the model described in this chapter, one of the missing elements which is perhaps the most important is the linkages between stakeholder within the socio-political context and how the information flow is moving from one to another group. Identifying this gap will establish the linkages between different stakeholder and their information flow which help to inform the different decision making levels to achieve their target goal and most importantly it make sure that vast knowledge generated by the scientific and technical community is accessible by the local small holder farmers.

Lessons from the models and case studies shaped to design and propose an updated framework presented in this chapter named *'Integrated Information Dissemination System for Coastal Agricultural Community'* (INDICA) framework. The INDICA framework can be considered as a conceptual development mainly aimed to integrate all aspects of agricultural value chain, stakeholders, communication medium to capture the ICT benefits. However, the INDICA framework should also be considered as a living framework which should be further validated and modified according to local socio-ecological context and climatic hazards. A detail description of the INDICA framework is provided in the following section.

54.3   Result and Discussion

54.3.1   The INDICA Framework

The proposed INDICA framework will be based on four different levels progressive stages such as adaptation assessment, information need assessment, ICT measures and strategies and piloting and dissemination. The first level contextualizes the main linkages that exist between salinity, vulnerability, and food production. This includes the vulnerability dimensions that are exacerbated by acute climatic shocks and slow-changing trends on water salinity. The second level of the framework introduces the dimension of the local adaptation context, identifying the adaption deficits and

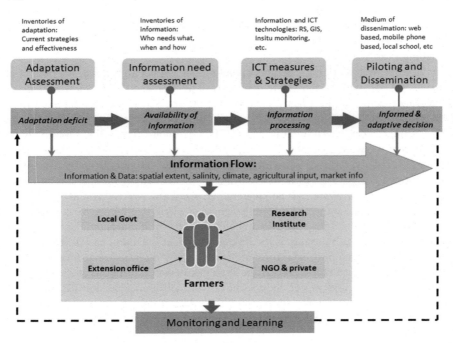

Fig. 54.1 Proposed information dissemination framework for coastal agricultural community

assessing ICT's potential to mitigate such deficits. An information need assessment at different stakeholder levels will be done along with identifying appropriate mediums of dissemination. These will lead to developing the framework for data management, and will help the entire production system, including the value chain. The third level will analyse processes and mediate scientific data and information (using RS and Sensors) and link these information hubs at multiple levels. This will be done through the developed framework to ensure that appropriate information reaches the right place at the right time in the best format and medium. This level focuses on the key factors that need to be considered in order to effectively integrate ICTs into the design, operation and evaluation of the framework including involvement of different stakeholders and their level of intervention. The final level will be the piloting of the developed framework in a field site to evaluate the applicability of such a framework in a real world situation (Fig. 54.1).

54.3.2 Adaptation Assessment

The central purpose of ICT is to connect stakeholders through passing necessary and important information using different communication mediums to empower relevant stakeholders. In this direction, the first stage of the proposed methodology is assessing the adaptation deficit which will reveal what kind of adaptation in farming sectors communities need. The purpose of the assessment is to identify adaptation needs or shortages in qualitative or quantitative form, for example, effectiveness of current approaches in preventing salinization. This assessment provides the general basis of subsequent phases of the methodology to take place.

54.3.3   Information Need Assessment (INA)

After a detail assessment of adaptation needs and deficits, an inventory would be proposed which precisely points out "who needs what"? The inventory will identify how information can be best available and affordable to each of the relevant stakeholders. As a precedent, a stakeholder identification process will be conducted. The whole process can be done using both qualitative and quantitative methods. The major advantage of developing such an inventory to guide the policy makers and scientists to customize the messages and information technology (IT) applications according to user needs. As a baseline of INA, current advances on INA from practical case studies related to ICT applications mostly identified various types of information needs of the stakeholders and major communication channels for disseminating the information. Studies on developing countries identified major information needs at different scales such as (i) facilitating access to land records/ online registration, (ii) information about rural development programmes and subsidies (iii) latest (best) packages of practices, (iv) post-harvest technology, (v) general agricultural news, (vi) information on crop insurance, (vii) farm business and management information, (viii) input prices and availability, (ix) early warning and management of diseases and pests, (x) dairying and marketing of milk and milk product, (xi) accounting and payment, (xii) soil testing and soil sampling information (Infocom Technologies 2014).

54.3.4   ICT Measures and Strategies

The information medium is also spreading its breadth from traditional methods, such as radio, to state of the art technologies such as application and smartphones. Mobile phones are the most preferred medium for communication, however, radio and farm radio, SMS and interactive voice response system, web pages, internet, internet, optical media, protocol TV, and video, telecentres and computers are all also used. This stage will see various development, piloting and reproduction of ICT technologies according to user needs. Various kinds of "hardware" and "software" applications will be developed and mainly stem from INA. For each the user groups and stakeholders, technologies will be first produced within the scientific research community and then shared with local agricultural extensions officers and selected farmers for further improvement. It is envisaged that different stages of training will be necessary to deliver the messages and technologies to the target groups. Therefore, a modularized training infrastructure needs to be developed at this stage. In this methodology, emerging technologies are proposed apart from traditional approaches such as low-cost sensors, participatory GIS at farmer level and use of airborne images and open source data infrastructure at meso level such as extension offices.

54.3.5   Piloting and Dissemination

The final phase "piloting and dissemination" will determine whether the proposed method is effectively providing necessary ICT support to different levels of stakeholders, such as farmers, research and scientific communities, NGOs, extensions services, and local government. There will be monitoring mechanisms to measure how each stakeholder community's needs are being targeted and information is delivered. One particular inclusion of the proposed methodology is to include local government, which is not very highlighted in exiting IT case studies in agriculture. Given the close links of farmers and local consumers with the local tier of governance, their inclusion not only widens the path of greater stakeholder inclusion but also supports to establish IT infrastructure at local level. Periodic monitoring is suggested using apps based monitoring system developed as part of the system. Another proposed addition of the model is converting local schools into information centres and school-goers as trained agents for spreading technologies. Developing countries like Bangladesh are currently investing substantial efforts for IT education at school levels. The usefulness of such infrastructure can be multiplied by using it as the local information centre where students can be trained as local monitors and future change makers.

54.3.6   Information Channeling Through Stakeholders

Enabling ICT for improving food security, climate resilience and agricultural production is commonly a participatory process (Smit and Skinner 2002; Brown et al. 2012). Similar to other resilience and adaptation philosophies, success depends on active involvement of stakeholders in defining and operating ICT at the user level (Huq 2016; Chen et al. 2010). Reviews of existing case studies identified different levels of stakeholders at different scales such as (i) input services (government agencies, extension services, farm stores, research institutions), (ii) productions (farmers), (iii) post-productions (processors, intermediaries, transporters), (iv) marketing (whole sellers and retailers) and (v) consumers (individuals and corporations) (Infocom Technologies 2014). In most of the development countries, farmer or producer communities often become the central consumer communities, therefore, effectively shorten the value chain (Gatzweiler and Braun 2016). Given that situation, enabling ICT for agricultural production can serve dual purposes, connecting both producers and consumers. On the other hand, the role of local government is not largely mentioned as a potential agent of accelerating ICT at the local level. Local governments have key roles in connecting different policy making levels with other actor's, e.g. primary producers and consumers, therefore their active inclusion can fundamentally benefit the ICT applicability at all levels of value chains (Huq and Stubbings 2015; Morrison and Pickering 2013; Gatzweiler and Braun 2016). Each group of stakeholders again is proposed to carry-out certain roles in regular monitoring of the progress so that the framework can be regularly modified and updated based of stakeholder's feedback.

54.4   Limitations of the Framework

The INDICA framework is a conceptual approach which needs to be validated though field testing and piloting. At the same time, it should be considered that the AISP framework is a general framework which mainly focusing on stakeholder integration, nonetheless, institutional policies and environment are not considered with greater scope. Creating ICT enabling policies and environment is the forefront of successful demonstration of such a framework. The INDICA framework is on a favouring policy environment, however, contexts can be different, in practice. In spite of such drawbacks, the embedded flexibilities of the framework allow to integrate policy changes and uncertainties, if necessary.

54.5   Conclusion

The proposed method seeks to support the agricultural sectors of the salinity affected area through enhanced information flow, accessing scientific information and data and disseminating information in the most appropriate way. Despite the fact that much remains to be explored in terms of the role and potential of ICTs within the agricultural field, the methodology proposed here sheds light on key conceptual foundations that help better understand the complex linkages that exist within the overall lifecycle of the agricultural systems. The methodology will also ultimately determine the role of information flows among the multilevel stakeholders and sectors in achieving resilience against an uncertain climatic future. The framework developed integrates the key concepts that mediate the information flow in agricultural production system through four key components: adaptation assessment, information need assessment, strategy and dissemination. It provides a basis to analyse how the dynamic interaction of each of the processes are linked with the information flow system and how this dynamic information is mediated to the end users, here in our case the small holders' farmers. The outcomes not only benefit the coastal zone in Bangladesh, but also increases the potential for implementation in similar regions.

References

Baumuller H (2012) Facilitating Agricultural Technology Adoption among the Poor: The Role of Service Delivery through Mobile Phones, ZEF Working Paper Series 93, Centre for Development Research, Bonn

Boschetti L, Roy DP, Justice CO, Humber ML (2015) MODIS–Landsat fusion for large area 30 m burned area mapping. Remote Sens Environ 161:27–42. https://doi.org/10.1016/j.rse.2015.01.022

Brown G, Montag JM, Lyon K (2012) Public participation GIS: a method for identifying ecosystem services. Soc Nat Resour 25(7):633–651

Carter GA (1993) Responses of leaf spectral reflectance to plant stress. Am J Bot 80(3):239–243

Casas S (1995) Salinity assessment based on combined use of remote sensing and GIS. In: Use of Remote Sensing Techniques in Irrigation and Drainage. FAO, Rome, Italy, pp 141–150

Chen L, Zuo T, Rabina GR (2010) Farmer's adaptation to climate risk in the context of China: a research on Jianghan Plain of Yangtze River Basin. In: Agriculture and Agricultural Science Procedia, pp 116–125

Dasgupta S, Hossain Md.M, Huq, M, Wheeler D (2014) Climate Change, Soil Salinity, and the Economics of High-Yield Rice Production in Coastal Bangladesh: Policy research Working Paper no 7140. The World Bank

Duncombe R (2006) Analysing ICT Applications for Poverty Reduction via Microenterprise Using the Livelihoods Framework. IDPM Development Informatics Working Paper no. 27. University of Manchester. https://www.sed.manchester.ac.uk/idpm/research/publications/wp/di/documents/DIWkPpr27.pdf

Finlay A, Adera E (2012) Application of ICTs for climate change adaptation in the water sector: developing country experiences and emerging research priorities. APC, IDRC

Garnett T, Appleby MC, Balmford A, Bateman IJ, Benton TG, Bloomer P et al. (2013) Agriculture. Sustainable intensification in agriculture: premises and policies. Science 341(6141): 33–34. https://doi.org/10.1126/science.1234485

Gatzweiler FW, Von Braun J (2016) Technological and Institutional Innovations for Marginalized Smallholders in Agricultural Development

Feugard DC, Abner R, Smith P, Wayland D (1994) Modelling daylight illuminance. J. Climate Appl. Meterol. 23–109

Heeks R (2005) Foundations of ICTs in Development: The Information Chain. DIG eDevelopment Briefings, No. 3/2005. Manchester: Institute for Development Policy and Management. https://www.sed.manchester.ac.uk/idpm/publications/wp/di/short/DIGBriefing3Chain.doc. Accessed 24 May 2006

Heeks R, Molla A (2009) Impact Assessment of ICT-for-Development Projects: A Compendium of Approaches, IDPM Development Informatics Working Paper No. 36, University of Manchester, Manchester, UK. www.sed.manchester.ac.uk/idpm/research/publications/wp/di/di_wp36.htm

Huq S, Rai N Huq M (2014) Climate resilient planning in Bangladesh: a review of progress and early experiences of moving from planning to implementation, 2014, Development in Practice

Huq N (2016) Institutional adaptive capacities to promote Ecosystem-based Adaptation (EbA) to flooding in England. Int J Clim Change Strat Manag 8(2):212–235

Huq N, Stubbings A (2015) How is the role of ecosystem services considered in local level flood management policies: case study in Cumbria, England. JEAPM 17(4):1550032

Infocom Technologies (2014) ICTs in Agricultural Value Chains in the Caribbean – Study and Promotion, Kingston

Justice CO, Vermote E, Townshend JRG, DeFries R, Roy DP, Hall DK, Salomonson VV, Privette JL, Riggs G, Strahler A (1998) The moderate resolution imaging spectroradiometer (MODIS): land remote sensing for global change research. IEEE Trans Geosci Remote Sens 36:1228–1249

Labelle R, Rodschat R, Vetter T (2008) ICTs for eEnvironment: Guidelines for Developing Countries with a Focus on Climate Change. International Telecommunication Union (ITU), Geneva. https://www.itu.int/ITUD/cyb/app/docs/ituictsforeenvironment.Pdf

May J, Dutton V Munyakazi L (2011) Poverty and Information and Communications Technology in Urban and Rural Eastern (PICTURE) Africa, International Development Research Centre (IDRC), Kenya

MoEF (2006) Bangladesh Climate Change Strategy and Action Plan 2008. Ministry of Environment and Forests, Government of the People's Republic of Bangladesh, Dhaka, Bangladesh

Morrison C, Pickering CM (2013) Perceptions of climate change impacts, adaptation and limits to adaption in the Australian Alps: The ski-tourism industry and key stakeholders. J Sustain Tour 21(2):173–191

Ospina AV Heeks R (2010a) Linking ICTs and Climate Change Adaptation: A Conceptual Framework for e-Resilience and e-Adaptation, Centre for Development Informatics, Institute for Development Policy and Planning (IDPM), University of Manchester, UK. www.niccd.org/Conceptua lPaper.pdf

Ospina AV Heeks R (2010b) Unveiling the Links between ICTs & Climate Change in Developing Countries: A scoping Study. Centre for Development Informatics, Institute for Development Policy and Planning (IDPM), University of Manchester, UK. www.niccd.org/ConceptualPaper. pdf

Ospina AV, Heeks R, Adera E (2012) The ICTs, climate change adaptation and water project value chain: a conceptual tool for practitioners. In: Alan F, Edith A (eds.) Application of ICT s for Climate Change Adaptation in the Water Sector. Developing country experiences and emerging research priorities. South Africa: Association for Progressive Communications (APC) and the International Development Research Centre (IDRC), pp 17–31

Parikh TS, Patel N, Schwartzman Y (2007) A survey of information systems reaching small producers in global agricultural value chains, Berkeley, UC: School of Information. https://hci. stanford.edu/neilp/pubs/ictd2007.pdf. Accessed 12 Oct 2013

Panchard J, Prabhakar TV, Hubaux JP, Jamadagni HS (2007) COMMONSense Net: a wireless sensor network for resource-poor agriculture in the semiarid areas of developing countries. Inf Technol Int Dev 4(1): 51–67. https://itidjournal.org/index.php/itid/article/download/244/114

Psilovikos and Elhag (2013) Forecasting of remotely sensed daily evapotranspiration data over Nile Delta region. Egypt Water Resour Manag 27(12):4115–4130

Rabbani G, Rahman A, Mainuddin k. (2013) Salinity-induced loss and damages to farming households in coastal Bangladesh. Int J Global Warm 5(4):400–415

Scudiero E, Skaggs TH, Corwin DL (2015) Regional-scale soil salinity assessment using Landsat ETM+ canopy reflectance. Remote Sens Environ 169:335–343. https://doi.org/10.1016/j.rse. 2015.08.026

Smit B, Skinner MW (2002) Adaptation options in Agriculture to climate change a typology. Mitig Adapt Strat Glob Change 7:85–114

Taghizadeh-Mehrjardi R, Minasny B, Sarmadian F, Malone BP (2014) Digital mapping of soil salinity in Ardakan region, central Iran. Geoderma 213:15–28. https://doi.org/10.1016/j.geo derma.2013.07.020

Thomas T, Mainuddin K, Chiang C, Rahman A, Haque A, Islam N, Quasem S, Sun Y (2013), Agriculture and Adaptation in Bangladesh: Current and Projected Impacts of Climate Change. IFPRI Discussion Paper 01281 (July). Washington, DC: IFPRI. https://www.ifpri.org/sites/def ault/files/publications/ifpridp01281.pdf

Tilley DR, Ahmed M, Son JH, Badrinarayanan H (2007) Hyperspectral reflectance response of freshwater macrophytes to salinity in a brackish subtropical marsh. J Environ Qual 36(3):780–789

Whitecraft A, Becker-Reshef I, Killough B, Justice C (2015) Meeting earth observation requirements for global agricultural monitoring. An evaluation of the revisit capabilities of current and planned moderate resolution optical earth observing missions. Remote Sens 7(2):1482–1503. https://doi. org/10.3390/rs70201482

World Bank (2011) ICT in Agriculture -Connecting Smallholders to Knowledge, Networks, and Institutions, e-Sourcebook.m, Report Number 64605. The World Bank

Chapter 55
Citizen Science on Water Resources Monitoring in the Nhue River, Vietnam

N. H. Tran, T. H. Nguyen, T. H. Luu, M. M. Rutten, and Q. N. Pham

Abstract The explosion of citizen science (CS) in data collecting, including hydrology and water resources management, results from information and communication technology development. This approach is still a new topic in Vietnam, while CS development stages are not popular in the previous works. This paper demonstrates how a CS project can be developed, a pilot area at Lien Mac 2 sluice, Noi Bridge, and Dong Bong 1 pumping station on the Nhue River, Vietnam. There are seven main stages to implement a CS project, in which monitosring observation choose is a crucial factor to attract participants and obtain long- recorded observations. Social network and smartphone are a tool to boost CS development and the young generation is interested in this approach. Citizen-based water level monitoring was conducted by images that will be uploaded on social networks. The water levels obtained from the CS on the Nhue River are as good as the conventional approach. The step-by-step CS approach can be applied in other aspects of water resources management such as land use, water quality monitoring to promote the CS development.

Keywords Citizen science · Citizen science approach · Water resources monitoring · Water level · Nhue River

N. H. Tran (✉) · Q. N. Pham
Faculty of Water Resources, Hanoi University of Natural Resources and Environment (HUNRE), Hanoi, Vietnam
e-mail: tnhuan.tnn@hunre.edu.vn

N. H. Tran
Faculty of Agriculture and Environmental Sciences, University of
Rostock, Rostock, Germany

T. H. Nguyen
Institute of Mechanics, Vietnam Academy of Science and Technology, Hanoi, Vietnam

T. H. Luu
College of Hydrology and Water Resources, Hohai University, Hohai, China

M. M. Rutten
Delta Futures Lab,
Delft University of Technology, Delft, The Netherlands

© Springer Nature Switzerland AG 2021
M. Babel et al. (eds.), *Water Security in Asia*, Springer Water,
https://doi.org/10.1007/978-3-319-54612-4_55

55.1 Introduction

Citizen Science (CS) is defined as the participant of local people in projects organized by government institutions, researchers to collect data (Walker et al. 2016), while Buytaert et al. (2014) illustrated CS as an involvement of the general public in creating new knowledge, which can help to address local problems. CS is a new and innovative approach to collect and gather data in the all field, including hydrology and water resources (Buytaert et al. 2014), (Cohn 2008), (Beza 2017). Data amassed by communities can be at low cost, more spatially distributed, and relatively accurate (Buytaert et al. 2014; Gardiner et al. 2012; Zeng et al. 2020; de Bruijn et al. 2019). With the explosive development of Internet and Information Technology, CS has been stimulated by using social media like Facebook, Tweeter, or Web-based platforms to attract and retain local people's participation in collecting data and information (Davids et al. 2019).

There are several CS projects to collect water-related data over the globe over the past 50 years. In the 1970s, Minnesota Pollution Control Agency implemented citizen-based water quality observation to protect the environment, which helps people change participation attitudes from curiosity at the starting time to self-responsibility about environmental protection after 30 years (MPCA 2020). Peters-Guarin (2008) researched on applying the local knowledge in flood risk assessment for urban areas in the Philippines. The work emphasized that local knowledge had a crucial contribution to flood risk assessment. The author concentrated on exploiting information and data from communities through stories, oral descriptions and narratives. In order to improve hydro-meteorological monitoring networks in Ethiopia, a community-based monitoring program from 2014 to 2015 was established to monitor daily rainfall, surface, and groundwater level to maintain the conventional observations stopped due to lack of funding (Walker et al. 2016). Recently, Davids et al. (2019) developed CS in precipitation monitoring in Nepal, known as Smartphone4Water.

Apart from mining data from local communities, CS is considered non-structural measures for solving water resources (Buytaert et al. 2014) or flood risk management problem (Ferri et al. 2019). Some researchers presented that CS projects help local people raise their awareness about water resources and strengthen their capacity to react to natural catastrophes (Cheung and Feldman 2019; Starkey et al. 2017). For example, local people involved in flood observation of flood risk projects can reduce annual flood damage, while the cost of CS approach is less expensive than structure measures (Ferri et al. 2019). The citizen participation can create transparent information in order to promote democratization, which stimulates debate and supports the decision-making process between stakeholders (Buytaert et al. 2014) and policy building (Minkman et al. 2017).

Despite the significant development of CS on water resource monitoring globally, it is still an alien approach in Vietnam. Since 2009, CS has been applied in transportation to inform the traffic situation in the Vietnam capital (VOV Giao thông 2019), Hanoi, which can help listeners choose an appropriate road in rush hours. This model could be considered the first CS approach in Vietnam. Le (2015) and Đinh Hoe et al. (2011) highly recommended the contribution of local people on water resources management in general and reservoir operation in particular. However, these previous research on water resources have just been the ideas. It is necessary to implement a CS project in water quality, quantity, and morphology in Vietnam (Nhan et al. 2015).

The water level is one of the primary parameters in water resources information, while it does not require some special equipment and physical installation like discharge measurement or rainfall (Etter et al. 2020). This paper presented the stages to implement a CS project on water level monitoring on the Nhue River, Vietnam, which can be scaled up in other basins or other objects.

55.2   Material and Methods

55.2.1   Nhue River Overview

The Nhue River is a distributary of the Red River, a primary river flowing through Hanoi. It transfers on water from the Red River at Lien Mac sluice to irrigate for the Dan Phuong - Hoai Duc irrigation area and surrounding area. In addition, The Nhue River is responsible for draining off water for the Western districts of Hanoi city, water generating from To Lich, Cau Nga, and Pheo (other name Dam) River (Tran et al. 2016). The Nhue River flows into the Day river at Phu Ly city, Ha Nam province, a 74 km length (Fig. 55.1). In recent years, the water level of the Nhue River did not change much except during rainstorms in the flood season. The Nhue River receives the wastewaters without treatment from households, the craft villages, and the small to medium-sized industrial zones leading to the serious-polluted river. The mean annual water level is around 3.06 m for Ha Dong.

The pilot river is of 18 km length, limited from Lien Mac sluice to Ha Dong sluice, far Ha Dong Bridge about 2 km toward the downstream river. There are some key water-work along the section, such as Lien Mac, Lien Mac No.2 and Ha Dong sluices, Thuy Phuong, and Dong Bong pumping stations (Tran et al. 2016). A manual staff gauge was erected on these positions to monitor the water level. The monitored daily water level at 7 am and 1 pm at Lien Mac, and Ha Dong are updated on the website of Day River Irrigation Development Investment Co., Ltd (shorturl.at/acgw7).

There are several universities located on the basin, including HUNRE. It might be a favorable condition to organize field trips and recruit participants.

Fig. 55.1 The diagram of
the study area and SC
monitoring locations

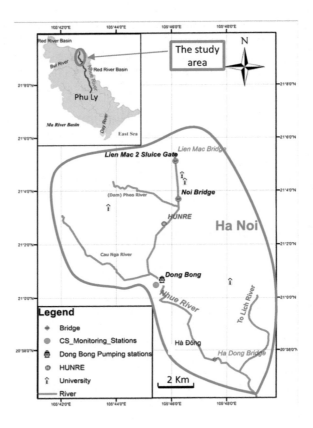

55.2.2   Research Method

The approach to developing a CS model in the water level monitoring of the Nhue
River consists of seven main steps, are shown as Fig. 55.2.

The first step needs to have an overview of research on CS and features of the
study area. Following this, the research conducts surveys of monitoring works in the
study area. Afterward, we select the monitoring locations with existing works such as
pumping stations, bridges, culverts, and locating areas convenient for data collection.
In the following stages, after selecting the monitoring locations, the research proceeds
to establish the measuring structures by painting water level gauges on the works and
assuming the presumptive elevations for each location or using the available water
level gauges at the waterworks.

Simultaneously, the research team will invite participants. At this stage, we
develop a questionnaire to identify the local people's concerns about water resources
and their expectations and motivation to participate in the project. The survey participants' personal information will also be processed and analyzed by age, education
level to determine the potential participants. Interviewing and selecting participants

Fig. 55.2 The key stages of
SC project on water level
monitoring on the Nhue
River

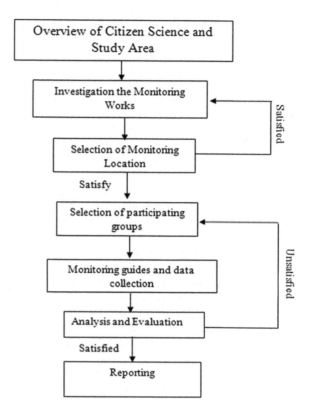

is conducted in three periods (December 10, 2015, March 12, 2016, and April 24, 2016) to attract more participants in the project and improve the suitable participants' approach.

In the following stage, we guide participants on observing and sending data at the monitoring stations. The study also depends on the actual situation to choose an appropriate communication method with the participants via SMS messages, social networks, etc.

Afterward, water level obtained from participants will be compared with the water level at some existing monitoring stations on the river section. Thereby, it helps check and detect erroneous data and promptly respond to the participants to improve in the next observations.

The final step is after verifying and assessing the suitability of the data, the water level data will be published widespread so that all participants can follow, or other peoples can access the data for the research, assessment, or management of water resources.

55.3   Results and Discussion

55.3.1   Selecting and Constructing CS Observation Points

To choose the appropriate monitoring stations, research was conducted to investigate existing monitoring stations and constructions along the Nhue River to install staff gauges to reduce tasks. Monitoring stations should be accessible to everyone and must not be harmful to participants during the observing process. Several water level monitoring works were currently installed by Nhue River Irrigation Development Investment Co., Ltd to operate drainage and irrigation systems on the Nhue River, such as Lien Mac 1, 2 sluices, Dong Bong 1, 2 pumping stations, and Ha Dong sluice, etc. The Nhue River separates urban and sub-urban areas of Hanoi. There are, therefore, many bridges across the river like Noi, Dien, Nhue, Ha Dong bridges. Three stations are chosen: Lien Mac sluice No.2, Noi bridge, Dong Bong No.1 pumping stations to implement the CS project. The locations of CS observations in water resource monitoring is shown in Fig. 55.1 and Fig. 55.3.

These selected stations are located on a straight branch, which meets the station designing standards, while these areas are of crowded residents and can avoid tracks and busy streets at highways and main avenues. Lien Mac No. 2 sluice is a control structure integrated traffic bridge, which combines with Lien Mac No 1 sluice to control flood from Red River for Ha Noi. Lien Mac No 2 sluice gates normally open to transfer water from Red River to Nhue River for irrigation purposes. A small bridge was designed to operate and maintain structures on the top of the sluice gate where participants can stand there to monitor the water level. Noi Bridge is far from Lien Mac No sluice more than 2 km toward downstream, with two bridges to serve traveling, a smaller and lower one for pedestrians, bicycles, and scooters. We installed a staff gauge on a pier of the remaining bridge, while a vertical datum was assumed at pile cap level (Fig. 55.3.a, b). Dong Bong No1 pumping station is an outlet of My Dinh urban area's combined drainage system, consisting of complex constructions to drain into Nhue River such as pumping and culverts. Staff gauges were installed

| (a) Constructing a staff gauge on the Noi Bridge pier | (b) Noi Bridge | (c) Lien Mac No. 2 sluice | (d) Dong Bong No.1 pumping station |

Fig. 55.3 The views and staff gauges of pilot areas on the Nhue River

on the culvert, hidden when the valve gate is heightened to drain water into the Nhue River (Fig. 55.3.d).

55.3.2 Participants in CS

The interview took place in three phases in the monitoring stations and surrounding areas, which lasted one day for each period (December 10, 2015; March 3 and April 24, 2016). The first three authors interviewed 69 people, and the number of respondents per phase was 19, 9, and 45, respectively. After finishing each phase, the research reformulated the questionnaire form to reduce redundant questions and match the actual situations. The information of the interviewer was announced to be anonymous to get honest responses.

Third-four of the interviewers agreed to join the CS project with some appealing reasons such as environment protection, research supporting, and their curiosity about CS. The 25% of people interviewed disagree to join the project; most of them are the elders being afraid of health conditions and technical skills to join projects. However, one of them is keen to share their knowledge of the river, which they witnessed. There are three age groups taking part in interviews, including 18–34, 35–60, and 60–75 year olds. The 69% of interviewers were 18–34 year-olds, mainly students studying at the study area universities. Two remaining groups only occupied one-third of the response, sharing 22% and 9%, respectively (Fig. 55.4).

The participants agreeing to join the project graduated with bachelor's degrees and high school with 59%, 33%, respectively, and less than 15% only study secondary and primary school.

The participation recruitment processes were divided into different periods. Figure 55.5 demonstrated the percentage of the number of interviewees joining the

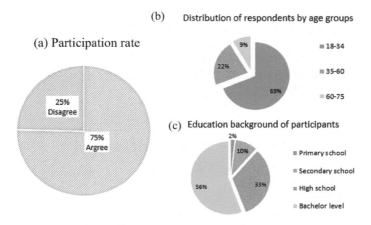

Fig. 55.4 The agreement rate and demographic of the participant in CS research

Table 55.1 The rate of participants preferring data collection methods and frequent monitoring contribution

Commitment of monitoring frequency	Percentage (%)	Data collecting method	Percentage (%)
Twice day	20.0	Photo	81.0
Once a day	35.0	SMS	2.4
2–4 time/ week	37.5	Email	4.8
Once a week	7.5	Notebook	11.9

project gradually increased. These results showed the efficiency of improvement of questionnaires and chose more appropriate persons.

55.3.3 Data Collecting and Sending Methods

The survey results about the data collecting method and participant's commitment on monitoring frequency are given in Table 55.1. A monitoring frequency question was performed to ask the commitment of participation spending their time on monitoring activities. During surveying periods, most participants (more than 70%) believed that they can monitor water level every day; 20% of participants committed to monitor twice a day. Regarding the data collecting method, 81.0% were photo-taking, 11.9% used hard report, and 7.2% constituted both Email and SMS. Currently, Smart-phones are commonly for people. Therefore, most participants preferred taking the photo to collect water level values instead of using notes. Besides, they can take advantage of social networks such as (Facebook, Zalo a social media platform developed by a Vietnamese technology company) to share and send data.

The research used the social network, a public Facebook page (the source of the Facebook page: https://www.facebook.com/groups/233323730344680/) as a data collection aggregation platform. This approach can avoid service fees to maintain like a private website. Besides, it can access many people and is convenient to use. After collecting water level by taking the photo, the participant can upload its photo in the group with some short information, including location, time and value of measurement, and notes.

55.3.4 Evaluating Water Level Obtained from CS

The Lien Mac No. 2 sluice, Noi Bridge, and Dong Bong No. 1 pumping station were chosen to implement citizen-based observations.

At Noi bridge point, thanks to convenient transportation and a high student density; there was more than 104 observations from March to November 2016. The community- monitored water level at Noi Bridge is shown in Fig. 55.7a reflecting flow regime in flood and dry seasons with a flood season from May to October. The dry season was frequently monitored because this time coincided with the surveying campaign, accounting for 76% of the recorded value. Recorded water level ranging from −2 m to 1 m, is relative water level compared with the Noi Bridge's pile cap level. The water level at Noi Bridge has been updated in the group since then, but occasionally.

The two remaining stations are in remote areas. There were about ten citizen-based observations for each Lien Mac 2 sluice and Dong Bong 1 pumping station in 2016. The later station's staff gauge will be hidden by the sluice gate when it is opened to drain water for the urban area. The participant can use an alternative way to estimate the water level position with some stable points there such as the embankment toe, the embankment peak, etc.

To estimate the accuracy of water level from CS, water level obtained from CS compared to formal observations of existing monitoring works such as Lien Mac 2 sluice and Dong Bong 1 pumping station. Due to the time limitation, research collected only water level of Lien Mac 2 sluice published on the Website of Day River Irrigation Development Investment Co., Ltd, is considered reference values. In the paper's scope, the water level obtained from CS at Lien Mac 2 sluice and Noi Bridge was compared with reference values. The community-based versus formal water level on the Nhue River is shown in Fig. 55.8.

At the Lien Mac 2 sluice, although there were nine recorded values from CS from 24th April to 5th May 2016, the water level trend of the two methods is the same. The error ranges from 0.02 m to 0.61 m, the mean error was 0.15 m. Unfortunately, there is only nine values overlapping between the formal and the community water level data. Therefore, research will not consider the correlations with formal sources (Fig. 55.8).

In general, the water level at Noi bridge has the same trend as Lien Mac 2 sluice, especially in the dry season. The correlation coefficient (R) is 0.95, which indicates two datasets are closely relative. The flow significantly fluctuates in flood season, while citizen-based observations were intermittent. Therefore, some recorded values

Fig. 55.5 The agreement rate of interviewees joining CS project

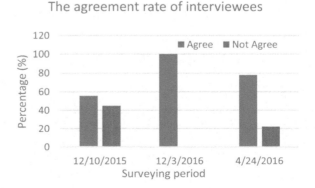

| Collecting data | Uploading data on the social network |

Fig. 55.6 Photographs of (left to right) the water level monitoring by photo-taking (a) and uploading data on the social network (b)

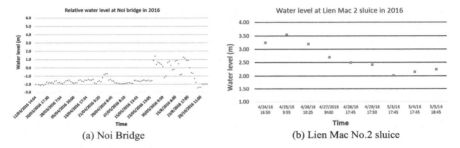

(a) Noi Bridge   (b) Lien Mac No.2 sluice

Fig. 55.7 Water level from community-based monitoring on the Nhue River

did not match the reference value trend (for instance: flood in the end of May). The datum level of the Noi Bridge station has not been transmitted according to the national datum level (VN 2000). Therefore, there is a large gap between a citizen-based station and a traditional station although two stations are located in a short river section (Fig. 55.8). The research will not consider the error between data pairs of two sets of data.

55.3.5 Reporting and Exploiting Water Level Values from CS

Monitoring data from the community is aggregated and displayed in the process line for the whole monitoring period. The 2016 Noi Bridge water level monitoring results were aggregated and shared on the group's fan page. Some members are interested in the result and asked to have access to data for research.

In 2016, Tran et al. used water level obtained from CS in this project to explore the contribution of CS and modeling in water resources monitoring and controlling.

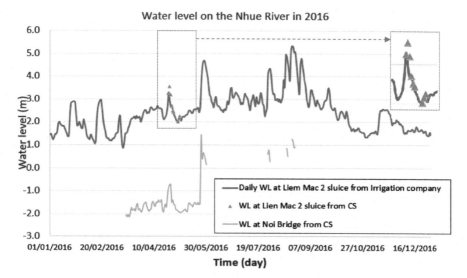

Fig. 55.8 The community-based versus formal water level on the Nhue River

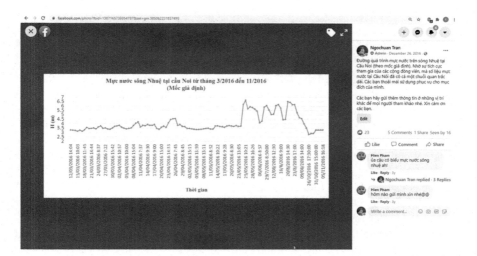

Fig. 55.9 The water level from CS model displaying on group's fan page

The work highlights that citizens can provide timely observation, which reflects natural phenomena such as storms, wastewater discharge, etc. At the same time, modeling can estimate water level along the river, from that datum level of citizen-based monitoring can be determined (Tran et al. 2016). This issue is still a limitation of CS on water resource monitoring when researchers only consider relative water level (Etter et al. 2020; Starkey et al. 2017; Walker et al. 2016), while water level relative to national datum reference can help stakeholders to use it easily.

55.4 Conclusions

The CS on water resources monitoring in the Nhue River was demonstrated and water level was collected by the local community at Lien Mac No.2 sluice, Noi Bridge, and Dong Bong No.1 pumping station. The major conclusions are mentioned below:

1. CS project on water resource monitoring underwent seven main stages: investigating, surveying, recruiting participants, monitoring guidance, analysing data to publish, and using data for different purposes. The measurement location selection was a crucial factor in the project's success. In addition, the station should be located in a population density area.
2. Most of the participants are young generation aging from 18 to 35. They are familiar with smartphone and social networks usages, a pillow in developing citizen projects to collect and transmit data. They are enough time and health to join the CS project, and pioneers to disseminate CS to other classes.
3. Citizen-based water level monitoring was conducted by images that will be uploaded on social networks. The water levels obtained from the CS on the Nhue River are as good as the conventional approach. The correlation coefficient is relatively high.

Our research organized that CS projects need to be maintained regularly to attract the attention of people. The motivation of enthusiasm of participants has been decreasing over time. Although they committed to collect data every day or every two days, the community-based water level is intermittent for flood reasons after finishing survey campaigns. In order to motivate the participants, a clear and simple data collection method is necessary, like AKVO flow, ODK Collect apps (Davids et al. 2019). The correlation equation of water level among measurement locations can be considered in the future to extend missed data.

In developing countries citizen science is quite alien. It can be applied in other aspects of water resources management such as land use, and water quality monitoring to promote the development of CS. The citizens can understand the role and significance of water resources monitoring and supervision, especially areas that are faced with the problems related to water recourse when they join CS projects.

Acknowledgements This work was supported by the MK27 project under CGIAR Research Program on Water, Land and Ecosystems in the Greater Mekong, funded in part by the Australian Government. The authors are thankful to Mariette van Tilburg and Maurits Ertsen of Civil Engineering and Geosciences Faculty of TUDelft and this book editor's reviewer for their comments and suggestions. We sincerely thank all citizen scientists who volunteered in this project for their whole-hearted support and devotion. The first author gratefully acknowledges a scholarship grant of Catholic Academic Exchange Service (KAAD) during doctoral research period at University of Rostock.

References

Beza EA (2017) Citizen science and remote sensing for crop yield gap analysis. https://doi.org/10. 18174/420049

Buytaert W, Zulkafli Z, Grainger S, Acosta L, Alemie TC, Bastiaensen J, De Bièvre B, Bhusal J, Clark J, Dewulf A, Foggin M, Hannah DM, Hergarten C, Isaeva A, Karpouzoglou T, Pandeya B, Paudel D, Sharma K, Steenhuis T, Tilahun S, Van Hecken G, Zhumanova M (2014) Citizen science in hydrology and water resources: opportunities for knowledge generation, ecosystem service management, and sustainable development. Front Earth Sci 2:1–21. https://doi.org/10. 3389/feart.2014.00026

Cheung W, Feldman D (2019) Can citizen science promote flood risk communication? Water 11(10):1–9. https://doi.org/10.3390/w11101961

Cohn JP (2008) Citizen science: can volunteers do real research? Bioscience 58(3):192–197. https:// doi.org/10.1641/b580303

Davids JC, Devkota N, Pandey A, Prajapati R, Ertis, BA, Rutten MM, Lyon SW, Bogaard TA, van de Giesen N (2019) Soda bottle science—citizen science monsoon precipitation monitoring in Nepal Front Earth Sci 7. https://doi.org/10.3389/feart.2019.00046

de Bruijn JA, de Moel H, Jongman B, de Ruiter MC, Wagemaker J, Aerts JCJH (2019) A global database of historic and real-time flood events based on social media. Sci Data 6(1):311. https:// doi.org/10.1038/s41597-019-0326-9

Đinh Hoe N, Bac Giang N (2011) The participation of community in Huong and Bo watersheet in building and operating hydropower dams in Thua Thien Hue province (Sự tham gia của cộng đồng lưu vuệc sông Hương, sông Bồ trong xây dựng và vận hành hồ đập thủy điện ở Thừa Thiên Huế). VNU J Sci Nat Sci Technol. https://js.vnu.edu.vn/NST/article/view/1316/1280

Etter S, Strobl B, Seibert J, van Meerveld HJI (2020) Value of crowd-based water level class observations for hydrological model calibration. Water Resour Res 56(2):1–17. https://doi.org/ 10.1029/2019WR026108

Ferri M, When U, See L, Fritz S (2019) The value of citizen science for flood risk reduction: cost-benefit analysis of a citizen observatory in the Brenta-Bacchiglione catchment. Hydrol Earth Syst Sci Discuss. 0.8: 1–27 (2019)

Gardiner MM, Allee LL, Brown PMJ, Losey JE, Roy HE, Smyth RR (2012) Lessons from lady beetles: accuracy of monitoring data from US and UK citizenscience programs. Front Ecol Environ 10(9):471–476. https://doi.org/10.1890/110185

Le AT (2015) Supervising of citizen on water resources management in general and reservoir operation in particular, Can Tho University

Minkman E, Van Der Sanden M, Rutten M (2017) Practitioners' viewpoints on citizen science in water management: a case study in dutch regional water resource management. Hydrol Earth Syst Sci 21(1):153–167. https://doi.org/10.5194/hess-21-153-2017

MPCA (2020) About the Minnesota Pollution Control Agency (MPCA). Minnesota Pollution Control Agency. https://www.pca.state.mn.us/about

Nhan QP, Huan NT, Khoa VLT, Anh BN, Rutten M (2015) Ecosystem services monitoring using remote sensing, citizen science and other ground observations and current practices in Vietnam. In: The 4th international symposium and exhibition of the vietnam cooperation initiative (VACI) water security in a changing area. https://repository.tudelft.nl/islandora/object/uuid%3A031d ab85-cd80-462d-9b7c-4a5da79a8c64

Peters-Guarin G (2015) Integrating local knowledge into GIS-based flood risk assessment (ITC Ph.D.) [International Institute for Geo-Information Science and Earth observation (ITC)]. https:// www.itc.nl/library/papers_2008/phd/peters.pdf

Starkey E, Parkin G, Birkinshaw S, Large A, Quinn P, Gibson C (2017) Demonstrating the value of community-based ('citizen science') observations for catchment modelling and characterisation. J Hydrol 548:801–817. https://doi.org/10.1016/j.jhydrol.2017.03.019

Tran NH, Nguyen TH, Rutten M, Nguyen TT, Pham QN (2016) The comparison of citizen science and numerical modeling in water resources monitoring and control: a case study in the Nhue river. In: Proceedings of international conferences of VACI 2016 (Vietnam WAter Cooperation Initiative)

VOV Giao thông: Marked 10 year anneversary of VOV tranportation channel (Dấu ấn 10 năm - Kênh phát thanh VOV Giao thông) (2019). https://vovgiaothong.vn/dau-an-10-nam-kenh-phat-thanh-vov-giao-thong. Accessed 19 Jan 2021

Walker D, Forsythe N, Parkin G, Gowing J (2016) Filling the observational void: scientific value and quantitative validation of hydrometeorological data from a community-based monitoring programme. J Hydrol 538:713–725. https://doi.org/10.1016/j.jhydrol.2016.04.062

Zeng Z, Lan J, Hamidi AR, Zou S (2020) Integrating Internet media into urban flooding susceptibility assessment: a case study in China. Cities 101(1037):102697. https://doi.org/10.1016/j.cities.2020.102697

Chapter 56
Remuneration for Conservation: An Ecosystem Service Changing Lives in the Hills of Bangladesh

F. K. Pushpa

Abstract A major concern regarding the effects of climate change is fresh-water scarcity. Even though a key source of sustainable fresh water is the world's remaining natural forests, research on fresh water is not being prioritized in forest conservation projects. Understanding the link between fresh water and forest conservation, few organizations in Bangladesh have moved to conserve the natural forests of the Chittagong Hill Tracts, also known as the Village Common Forests (VCFs), in order to safeguard a sustainable safe water supply for communities using the Gravity Flow Water System (GFWS). With this system, communities living around the VCFs have gained access to clean drinking water and come to view the forests around them as a clean water source. Accordingly, their dependency on the unsustainable extraction of forest resources has decreased. This work reports on a case study that drew on the experiences of communities and organizations in the Chittagong Hill Tracts. The case study explored how, with limited support, initiatives for community forest conservation could ensure a clean and sustainable drinking water source. This study also reports on an analysis of publications on adaptation practices. Findings indicate that communities using GFWS have gained access to a safe drinking water source and relief to women and children, who had previously spent hours collecting drinking water. Study findings have implications for countries concerned with the relationship between fresh-water scarcity and forest conservation efforts amidst climate change effects.

Keywords Fresh water · Ecosystem · Forest conservation

56.1 Introduction

Forests' most important role to the hydrological balance of watershed ecosystems is to preserve high-quality water (Dale et al. 2001). Ample amount of clean water is a valuable resource and one of the most valued products provided by public and private forestlands. One of most direct links between people and the valuable services that

F. K. Pushpa (✉)
Arannayk Foundation, Dhaka, Bangladesh

© Springer Nature Switzerland AG 2021
M. Babel et al. (eds.), *Water Security in Asia*, Springer Water,
https://doi.org/10.1007/978-3-319-54612-4_56

forests provide is drinking water (Todd and Weidner 2010). Forest cover influences groundwater levels, wells and springs, as well as safeguard water quality. The safest protection for groundwater is forest cover on its sources, and water may be forests' one of the most valuable and essential products (Hamilton 2008). Forest cover may have direct positive impact on available water resources in regions, where water from other sources is in shortage. Forests, which are in a healthy condition, support watersheds. Such watersheds provide important water related ecosystem services and could work as atmospheric 'water pumps' that prevent many areas not to turn into deserts. However, forests are under huge pressure from climate change. In recent years, climate change has been considered as the greatest threats to mankind in the twenty-first century (Alauddin and Rahman 2013). Changes due to climate change are expected to aggravate water-related hazards and water scarcity. Only 3% of total global water is available as freshwater (UNFCCC 2011). Further, climate change is projected to affect water quantity and quality, and has a significant impact on hydrology and water resources (Hamilton 2008). Finally, deforestation is expected to further contribute to overall declines in water quality (UNFCCC 2011).

Bangladesh is a forest poor country. Official record of forest area is 2.5 mill. ha, or 17% of the total land area of the country, but virtually the amount is less than that (Daily News 2016). In spite of limited forest resources, Chittagong Hill Tracts (CHT) is considered as one of the forest richest areas of Bangladesh. This area is different from other areas of Bangladesh due to its unique geographic and social structures, characterized by hilly topography and inhabited by people from 12 indigenous groups (Jashimuddin and Inoue 2012). The indigenous culture, lifestyle, and livelihood are mostly related to forest and forest resources. Unfortunately, over the past several decades, because of unsustainable use of these resources, forest resource oriented indigenous communities faced several crises for their subsistence necessities. With the objective of maintaining tree cover and biodiversity to protect the environment facing the deforestation, indigenous communities of CHT are maintaining small patches of natural forests around their villages, which are also known as Village Common Forests (VCFs) (Quddus et al. 2009). Such natural forests like VCFs provide a buffer against climate change and serve to maintain the integrity of watersheds and freshwater supplies (Foster-Turley et al. 2016). Water is another key motivation for them to conserve the VCFs. These VCFs are mostly small, averaging 20–120 ha in size, which are maintained and utilized collectively by the indigenous communities under the leadership of the head of the mouza ('Headman') or village ('Karbari') or by educational or religious institutions or a committee formed by the community leaders. These leaders govern the management according to customary rules (Quddus et al. 2015). Though shifting cultivation is strictly prohibited in the VCFs, the people living around the VCFs do it in the lands surrounding the VCFs, and shifting cultivation in these areas are continuously expanding keeping the VCFs under increasing threat of further degradation. In addition, many VCFs are already degraded due to over extraction and selling of the mother trees. VCFs are degrading both in quantity (number and size) and quality (Jashimuddin and Inoue 2012). There is no reliable statistics regarding the number of VCFs, but it is assumed that it may be around 700–800, and many of them might have been lost due to encroachment, unmanaged use and conversion to other land-uses (Quddus et al. 2015). Aside from deforestation and land-use change, now boulders are removed by people from channels, streams and canal beds, which are sold to traders and used for making stone chips for construction works. This practice aggravates soil erosion and landslide, and reduces water holding ability of the soil, causing drying out of water streams in the dry season (Quddus et al. 2009). This has been a serious threat to the indigenous communities for their water source (Miah 2012) as they rely mostly on these water bodies for drinking water. Such local anthropogenic causes to deforestation along with the global effect of warming have made this situation worse off (Miah 2012). Various non-governmental organizations (NGOs) and other stakeholders have implemented many development projects to combat the forest loss and land degradation, and to improve the livelihoods of the CHT people (Jashimuddin and Inoue 2012). Projects addressing need of water in the CHT has been implemented as well. However, not all initiatives have been successful in achieving their target objectives due to several reasons, of which most important are lack of good governance and rejection of the approach by the tribal people (Jashimuddin and Inoue 2012).

In this context, from 2009, Arannayk Foundation (AF), a civil society organization took an initiative of conserving some of the VCFs of CHT through various projects involving some NGOs. It was identified that improvement of livelihood can significantly contribute in forest conservation by the indigenous communities. Hence, their initiative included a Revolving Loan Fund (RLF). In each selected VCF community, with savings of the selected VCF community members and matching grant from the AF, the RLF supported the forest dependent community members taking loans for improved Alternative Income Generating Activities (AIGA) rather than forest resource extraction and traditional shifting cultivation. A committee consisting of the VCF community members themselves manages RLF in each community. This initiative also included following mechanisms: a) establishing community based co-management system for maintaining VCFs, b) developing institutional mechanism for conservation and sustainability of the VCFs, c) capacity building and organizing indigenous communities to participate in community based conservation, d) mobilizing communities for conservation and restoration of degraded and threatened VCFs, e) strengthening management capacities of the VCF management committees, and f) watershed and its catchments area protection (Quddus et al. 2009).

Though a relation between improved livelihood and forest conservation was identified, the main motivation for the indigenous communities to conserve the VCFs was the sustained supply of water in the streams (Quddus et al. 2009). The indigenous communities living around the degraded VCFs were facing water scarcity. Women and children needed to travel long distances to streams and rivers to collect water, which is often polluted and causes water-borne diseases.

To motivate the selected VCF communities for conserving the VCF and to reduce their stress through ensuring an easily accessible and consistent water supply, AF along with a local NGO 'Tahzingdong' supported the selected VCF communities. They established Gravitational Flow Water System (GFWS) of water collection from the VCF under a project titled 'Community based conservation of village common forest in Rowangchari, Bandarban'. Such GFWS brought water to their doorsteps and reduced burden of women for water collection (Quddus et al. 2015).

It has been found that in countries like Bangladesh, people of the climate change affected areas have been practicing different options to adapt to climate change. However, in many cases such adaptive measures are not adequate to meet their safe water demand (Alauddin and Rahman 2013). Installations of different water technologies like deep tube-well and deep-set pump are the common and useful options for the hilly areas. Rainwater harvesting (Quddus et al. 2009), collection of water with pipes from fountains and conservation in big reservoir in the hilly areas, mainly in the CHT is an important adaptation practice. People collect water from this reservoir with hand tube-well installed beside (Alauddin and Rahman 2013). Community based seepage water harvesting, hand pump, ring-wells, dug well, etc. are some of the options practiced in the CHTs. The existing adaptation options for safe water for drinking, domestic use and agricultural purpose are not sufficient in any ecosystem compared to their demand, and most are short and midterm options, which need to be more improved. Therefore, innovations of more improved yet low cost scientific safe water options are necessary for each ecosystem (Alauddin and Rahman 2013). In Chittagong hill tracts, Gravitational Flow System (GFS) also known as Gravitational Flow Water Supply System or Gravitational Flow Water System (GFWS) is an important adaptation practice for drinking water, small-scale agriculture and horticulture, and daily household purposes (Sutradhar et al. 2015). In the hilly areas, where other adaptation technologies are difficult to install, GFWS could provide the forest adjacent communities with a supply of water directly from the forests that too within a very low cost. This study reports on a case study that drew on the experiences of the communities and organizations to explore how, with limited financial and technical support, the community based forest conservation initiatives could ensure a clean, sustainable drinking water source. With this system, communities living around the VCFs have gained access to clean drinking water and come to view the forests around them as a clean water source. Accordingly, their dependency on the unsustainable extraction of forest resources has decreased.

56.2   Material and Methods

56.2.1   Study Area

The studied sites Tulachari Para (Sadar Union, Rowangchari Upazilla), Plede Para (Taracha Union, Bandarban Sadar Upazilla), and Rinikkhyon Bagan Para (Swowalokh union, Bandarban Sadar Upazilla) are located in Bandarban, Chittagong Hill Tracts (Bangladesh). At Tulachari Para, 48 'Marma' indigenous community families (240 members) are living, who are maintaining their adjacent Tulachari VCF for around 68 years. The forest is almost of 24 ha. 22 families (217 members) of 'Mro' indigenous community living in Plede Para are conserving their adjacent

VCF for almost 100 years. The size of the VCF is approximately 12 ha. Other 'Mro' community members have conserved Rinikkhyon Bagan Para VCF for almost 22 years. 31 families (217 members) are conserving this 16 ha VCF.

GFWS were established in these communities in three different time phases. Tulachari para's GFWS was established on October 2009, Plede Para's GFWS on February 2016, and at Rinikkhyon Bagan Para on April 2016.

56.2.2 Research Methods

The study is based on both primary and secondary data. Primary data were collected through household surveys, FDGs, KIIs, informal group interviews, and personal conversations to collect preliminary information regarding communities' GFWS previous condition, current usage status of GFWS, overview of livelihood and forest conservation, etc. Annual reports of the donor organization, project's baseline study, and evaluation reports were reviewed as well. Secondary literature was reviewed, related to climate change, fresh-water scarcity, forest and environment, and adaptation practices from different journal articles, books, reports, and related web information.

56.3 Results and Discussion

56.3.1 Situation Prior to Establishment of GFWS

With the financial and technical support from the AF, Tahzingdong helped the VCF communities establishing Gravitational Flow Water System (GFWS) of water collection from the VCF that brought water to their doorstep. These organizations started working with the communities for improvement of livelihood aiming to significantly contribute in forest conservation and to build institutional capacity of the forest based communities. Watershed and its catchment area conservation was one of the main objectives of the project. While implementing the project, it was realized that for the indigenous communities to conserve the VCFs, consistent supply of water is the main motivation (Quddus et al. 2009). Collection of water is one of the toughest tasks that CHT communities have to accomplish. In these communities, the water collection responsibility for daily usage is generally on the women and children of each family, who need to travel long distances to streams and rivers for collecting water (Quddus et al. 2015). In the case of the studied areas, women and children used to face the same situation before installation of GFWS. On an average, a 3–5 members' family requires 25 L of water every day for household chores. The traditional water collection pots they used could hold 1–5 L of water at a time depending on the size of the pot. Normally, women could carry 5–8 pots while travelling almost

1.5 kms to the nearest water source in the hills. In case of children, who mostly accompanied their mothers or sisters, they could carry 1–3 smaller pots at a time. Except for the traditional pots, they also used large pitcher pots, which could hold 10–15 L of water at a time, depending on their sizes. They usually walked to the water collection point, and it used to take them around 1.5 h to walk there and the time extended while coming back with water-filled pots or pitchers. On an average, women and children spent 3–4 h of their time daily for this task. Mostly, they used to put up all the pots together in bamboo basket and carry the baskets on their backs. The distance and travelling time used to vary by season. During rainy season, two of the studied communities had to look for other water collection points as the regular points used to become inaccessible due to inundated muddy hilly roads. In winter, because of low water amount in the collection points, water-collecting time used to get prolonged. The families could collect water only one time per day due to the long distance and lengthy travelling time to and from the water collection point. Depending only on these water sources, the community people frequently fell ill. Water borne diseases like diarrhoea, cholera, and jaundice were regular in the rainy season, mostly.

To reduce such stress through ensuring an easily accessible and consistent water supply, and to motivate them to conserve their VCFs, the GFWSs were established in the studied sites (Quddus et al. 2015).

56.3.2 Mechanism of GFWS

The GFWS, established in the three communities, used the action of gravity to move the water downhill from a water source to the village. This water distribution system does not require any generator or electricity. In the first part of the system, water is collected from upstream natural water sources like springs and stored through a check dam in an up-hill position within the forest. The stored water is then channelled through pipeline along hill slope to another constructed reservoir in the village using the natural force of gravity. The reservoir is specially constructed for water treatment consisting of boulders. It has three chambers. First water collection point chamber is filled with large boulders and sand, from that chamber water is transported to second chamber having small sized boulders, and finally the third chamber contained the filtered water. Water passed from first to third chamber through the boulders and sand to remove dirt and sediments, and dispensed either directly through taps installed with the third chamber or to another collection point in the community through pipeline using gravitational force. The reservoirs are covered to save the water from detritus and dust.

The maximum water holding capacity of GFWS at Tulachari Para is 12,000 L. The capacity is same for Rinikkhyon Bagan Para. Plede Para GFWS water reservoir can hold up to 10,000 L of water at a time. The community people regularly collect sufficient amount of water from GFWS reservoirs. Even after that, often the surplus amount of water is dispensed from the reservoirs when over-flown. There is no key

Fig. 56.1 Water collected from upstream natural resources and stored through a check dam

Fig. 56.2 A pipeline is connected to the check dam for channelling water to the reservoir

or any such device for managing the over-flown water. However, the reservoirs either have a small hole or a pipe for dispensing the surplus amount of water. Tulachari Para community people connect a rubber-pipe with the reservoir and use the surplus water for watering their paddy fields, vegetables, and fruit gardens nearby. However, the other two communities' have not yet utilized the excess water in any of such work, yet.

Tulachari Para community people collect the water from a collection point within their community. Plede Para and Rinikkhon Bagan Para community regularly collect the water directly from the reservoir (Figs. 56.1, 56.2, 56.3, 56.4 and 56.5).

The chronological order of the following photos describes the general mechanism of the GFWS:

Fig. 56.3 The water is channelled to the reservoir using the natural force of gravity

Fig. 56.4 Two of the communities directly collect the water from the reservoir

Fig. 56.5 Tulachari Para community collect the water from a collection point in their community. Photo source: (Arannayk Foundation 2015)

56.3.3 Maintenance of GFWS

In all three of the studied sites, VCF communities have chosen three representatives among them, who monitor the GFWS of their respective community. Inspection of the check dam, water reservoir and collection point takes place weekly and cleaning of water reservoir is done in every three months interval. In case of major maintenance problem, there is probation of collecting fund from the communities but no such incident has happened yet.

56.3.4 Supports of GFWS

After installation of these GFWSs, the communities gained access to a safe and secured supply of drinking water at their doorstep. The purified water collection point was placed within their communities saving a massive amount of their time. Now it takes 10–15 min from their houses to reach to the collection points, collect water, and get back, which previously was around 3–4 h. Seasonal variation does not hamper the water collection duration any more. They can collect water anytime during the day or night depending on their requirement. It has been a huge relief to the women and children of the communities. It was found that previously 68% of the women, who collected the water used to have severe back and/or neck pain as a result of walking while carrying 20–25 L of water at a time. No such physical problems are faced now as they can collect whatever amount they want all through the day. 25% of the women and children had experienced minor physical injuries because of slipping down while carrying the water pots. The percentage is zero now. Several times, women had encounter with snakes while travelling to long and remote distances in search of water collection points. Such incidents were never experienced after they started collecting water from the GFWS. As the water collection point now has been located within their communities, women and children get a very good amount of time. Children support their families in household chores, spend in studying or playing, and women spend in various activities. 56% of them stay engaged with household chores and/or take rest, 25% dedicate their time in their vegetable-fruit gardens while 19% spend nurturing their livestock and poultry.

The water the communities are getting through the GFWS is filtered in different sections. Previously, the water they used to directly collect from the streams or rivers contained detritus and were highly contaminated of germs. Such water was directly consumed without any sort of filtration. As a result, they used to fell ill. Previously, 66% of the community members suffered from water borne diseases, but it has reduced to 8% after they started using water from GFWS. There were incidents of cutting trees from the VCFs for immediate income for treatment of the community members previously, but none of such incidents took place after installation of the GFWS. The overall sanitation quality of the communities has improved as well. 100% of the community people think they are living in a better situation in terms of health and hygiene because of using water from the GFWS.

The communities are concern about the natural forest with fresh water. The degraded areas of their VCFs have been reforested through enrichment plantation, and illegal extraction of VCF resources has become nil. However, with the permission of the Karbari of the community, people sustainably harvest bamboo and collect other forest products and fruits. Though sustained supply of water in the streams is the main motivation for the communities to conserve the VCFs, improved livelihood significantly contributed to it too. The RLF support for adopting various AIGAs, technical guidance for skill development has played a key role in supporting the

VCF communities conserving the VCFs (Quddus et al. 2015). Before the intervention of the project, 54% of the community households depended on shifting cultivation as their major income generating source, 20% were depended on fruit-vegetable gardening, 14% were day labourers, 10% used to only extract forest resources and sell for earning money, and the remaining 2% were service holders. Now, the households have taken loan from the RLF of respective VCF communities and adopted various AIGAs. Currently, as their major income generating activity, 64% of the community households have chosen vegetable cultivation and fruit gardening, 7% depend on day labouring, and 5% depend on pig and poultry rearing. However, 24% still depend on shifting cultivation. Apart from these major income sources, the community people also depend on other options for income. As their secondary source of income, 52% rely on pig and poultry rearing, 10% on homestead vegetable cultivation, and the rest on other options. While selecting their income sources, availability of water played a key role. As they usually get a consistent supply of water through GFWS, they do not have to take mental and/or physical stress of carrying the extra amount of water for vegetable or fruit gardening of pig or poultry rearing. The overall improvement in the average household income has been noticed in all three of the studied sites. Their monthly income has increased up to 36%, approximately. The improvement has been possible because of the financial, technical, and managerial support of the AF and Tahzingdong. Such support along with regular awareness raising and visible impact of forest conservation for water has encouraged the communities for conservation of their adjacent VCFs.

56.4 Conclusions

Natural forests like VCFs provide a buffer against climate change and serve to maintain the integrity of watersheds and freshwater supplies. However, these VCFs are degrading or shrinking in size and number gradually due to anthropogenic and natural causes, which is ultimately impacting on the water availability and quality. In these locations, women and children face difficulties collecting water on a regular basis. To reduce their stress through ensuring an easily accessible and consistent water supply and to motivate them to conserve their VCF, the communities were supported to establish GFWS of water collection from the VCF. The interventions and continuous backstopping for improving livelihood of the community people has visibly worked well and this has played a key role in conserving the VCFs, but sustained supply of water in the streams has been the main motivation for the communities for conserving the natural forests. Through ensuring this water supply from the VCFs to their communities via GFWS, a positive impact in their lives has been noticed. VCF communities are exploring more options of utilizing the water in their lives. Tulachari community people are using the excess water of the reservoir for irrigation, nurturing vegetable and fruit gardens, which have contributed to their increased income. However, the other two communities are yet to plan anything for utilizing such water from their particular GFWS's reservoirs. In all three the communities,

GFWS users were found very much satisfied with the water quality and water services in general. It has been noticed at Tulachari community, where the GFWS has been established before than other two sites, the health of the VCF (vegetation cover, biodiversity, growth of trees and bamboos) improved and flow of water in the springs and canals in the villages during the dry season increased significantly. If the current condition is sustained, the same output is expected from the other two sites as well. All the communities are getting access to clean drinking water at their doorsteps. The communities are getting this benefit because of conserving the VCFs.

Acknowledgements The author would like to thank 'Arannayk Foundation' for providing with the opportunity to explore and learn about the GFWS and their various interventions for conserving the VCFs and ensuring sustained water to the communities. The author is equally thankful to Mr. Paiching U Marma of Tahzindong for helping throughout the time with information from the field. And finally, author's family and friends for their consistent support, throughout.

References

Ahmed FU, Mannan A, Pushpa FK (2014) Report on Training of Stakeholders on Adaptation to Change including Climate Change Issues. Arannayk Foundation, Dhaka

Alauddin S, Rahman KF (2013) Vulnerability to climate change and adaptation vulnerability to climate change and adaptation. J. SUB 4(2):25–42

Arannayk Foundation: Access to clean water can change lives. https://www.facebook.com/pg/arannaykfoundation/photos/?tab=album&album_id=468365510021391

Daily News. https://en.dailynews.com.bd/archives/18593

Dale VH, Joyce LA, McNulty S, Neilson RP, Ayres MP, Flannigan MD, Wotton BM (2001) Climate change and forest disturbances. Bioscience 51(9):723–734

Foster-Turley P, Das R, Hasan MK, Hossain PR (2001) Bangladesh Tropical Forests and Biodiversity Assessment. USAID, Bangladesh

Hamilton L (2008) A Thematic Study Prepared in the Framework of the Global Forest Resources Assessment 2005. Food and Agricultural Organization (FAO), Rome

Jashimuddin M, Inoue M (2012) Management of village common forests in the Chittagong hill tracts of Bangladesh: historical background and current issues in terms of sustainability. Open J. Forestry 2(3):121–137

Miah MD (2012) Evaluation report on Community based conservation of village common forest in Rowangchari Bandarban. Arannayk Foundation, Dhaka

Quddus MA, Chowdhury AH, Mannan A, Pushpa FK, Daru JN, Ashraf A (2015) Seeding Hopes in the Coast Hills and Haors. Arannayk Foundation, Dhaka

Quddus MA, Hossain MS, Chowdhury AH (2009) Conserving Forests for the Future. Arannayk Foundation, Dhaka

Quddus MA, Hossain MS, Chowdhury AH (2010) Promoting Alternative Livelihood for Forest Conservation. Arannayk Foundation, Dhaka

Sutradhar, L., Bala, S., Islam, A., Hasan, M., Paul, S., Rhaman, M., Billah, M.: A review of good adaptation practices on climate change in Bangladesh. In: International Conference on Water and Flood Management, pp. 607–614 (2015)

Todd, A.H., Weidner, E.: Valuing drinking water as an ecosystem service. Pinchot Lett. (2010)

UNFCCC: Climate Change and Freshwater Resources, United Nations Framework Convention on Climate Change, Bonn (2011)

Chapter 57
Role of Citizen Science in Safe Drinking Water in Nepal: Lessons on Water Quality Monitoring from Brazil

A. Gautam, J. Ramirez, L. Ribbe, K. Schneider, S. Panthi, and M. Bhattarai

Abstract Availability and accessibility of safe drinking water remain some of the main challenges in many parts of the world, especially in developing countries. Brazil and Nepal are two developing countries from two different continents, and both are considered water resourceful countries, but they still face significant water and sanitation problems. An experience of a research work in the rural area of Rio de Janeiro, Brazil *'Design of a Community-Based Water Quality Monitoring (CBWQM) Strategy'* implies a huge potential of involving active stakeholders in assessing and monitoring water quality to ensure the improved health of the citizens. In the case of Nepal, about 80% of prevalent communicable diseases are due to poor sanitation and lack of access to quality water. There is a big gap between the coverage and functionality of the water supply system owing to the degraded quality of water. In fact, there is no data available on water quality. The Water Safety Plan (WSP) tool has been helping the government strategy, but the data accuracy of Citizen Based Monitoring (CBM) needs to be checked. In order to provide reliable baseline of water quality data and institute a monitoring mechanism with proper SDG indicators, a specific methodological approach has been designed for the selected area of Nepal (Pokhara Metropolitan City). The approach focuses mainly on the selection of appropriate tools (Information and Communication Technologies (ICTs) like mobile applications, sensors) implemented with stakeholders' participation (through water user's committees, local schools and authorities), which in conjunction will provide a suitable data management system for the proper collection, storage, analysis, verification, and dissemination of acquired data.

A. Gautam (✉) · J. Ramirez · L. Ribbe
Technical University of Cologne (Technische Hochschule Köln), Cologne, Germany

A. Gautam · K. Schneider
University of Cologne (Universität zu Köln), Cologne, Germany

S. Panthi
World Health Organization (WHO), Kathmandu, Nepal

A. Gautam · M. Bhattarai
Institute of Engineering (IOE), Tribhuvan University (TU), Kathmandu, Nepal

© Springer Nature Switzerland AG 2021
M. Babel et al. (eds.), *Water Security in Asia*, Springer Water,
https://doi.org/10.1007/978-3-319-54612-4_57

Keywords Safe drinking water · Citizen science · ICTs · Water supply · Water quality monitoring

57.1   Introduction

Safe drinking water and basic sanitation facilities are an indispensable need for human civilization. Access to these facilities plays a crucial role in the overall social and economic development of a community and a nation. The facts about Global Water, Sanitation, and Hygiene (WASH) indicate that the sector has the potential to prevent at least 9.1% of the global disease burden and 6.3% of all deaths (NEWAH 2011). WHO/UNICEF Joint Monitoring Programme (JMP) for water supply and sanitation also refers to the development of an innovative and cost-effective approach to testing water quality implying access and quality monitoring possibilities as well (WHO 2015).

Sustainable Development Goal[1] no. 6 aims to "ensure availability and sustainable management of water and sanitation for all," and comprises of six technical categories (UNDP 2015); in order to accelerate the development sustainable ideas, it proposes several indicators and priorities to meet by 2030. The first two targets (6.1 and 6.2) hold continuation of MDG 7.C in expanded and refined forms of definitions and scope (UN Water 2015; GEMI 2015). Issues like water resources, wastewater and water quality that were left out of the MDG period, hindering the global attention, commitment and investment, are now embedded into the SDGs (6.3 and 6.5) to ensure a coherent monitoring approach for the sector. Goal No. 6b calls for projects to "support and strengthen the participation of local communities in improving water and sanitation management" (UN-Water 2015).

About 13% of the world's surface water resources are in Brazil. Therefore, it is known as a country in which water is plentiful, with the highest total renewable fresh water supply of the planet (Gleick 1998; Tereza and Alves 2014). The main challenges of water resources management in Brazil are unreliable access to water with a strong adverse impact on the living and health standards of rural populations, water pollution in and near large urban centres, which compromises poor populations' health, and increases the cost of water treatment for downstream users (Gondim et al. 2006; Porto and Kelman 2000; Tundisi 2011). Due to high population density and industrialization, Rio de Janeiro has considerable water quantity and quality issues. A symposium held in Rio de Janeiro in 1991, dedicated to water resources and environment, proposed correcting drastic water pollution as the main national priority (Dominguez 2015).

---

[1] Sustainable Development Knowledge Platform. United Nations. Available at: https://sustainabled evelopment.un.org/.

Nepal is a landlocked country, where planners and policy makers propagandize it as a water resourceful country,[2] but in reality, safe water scarcity is a significant problem both in cities and villages (Suwal 2015). Only around 25.4% of piped water supply systems are well functioning (NMIP/DWSS 2014). Communities throughout the country are exposed to several major health threats resulting from water contamination from sewage, agriculture, and industry Warner et al. (2008). The nation has already faced several tragic water epidemics in places like Nepalgunj, Jajarkot, which almost killed all of the inhabitants of the particular area (WASH-RCNN 2016; Suwal 2015). The WASH report and initiative sector status for Nepal stated that "all" agencies should begin complying with national standards in developing new and rehabilitated services. They should increase efforts to implement water safety plans to ensure delivery of safe drinking water" (SEIU/MoUD 2011), referring more to the need of functional, coordinated, information-based planning and monitoring systems to have better water and sanitation sector performance. Engaging most trusted members of the community, known as "gatekeepers," in the monitoring activities from the beginning, can help steer monitoring ideas into goal achievements (Conrad and Hilchey 2011). Without the clear linkage between emerging technologies for monitoring water quality issues, and a policy of cooperation with local stakeholders from the outset (Nare et al. 2011), any desired water monitoring networks in water supply schemes that are based on results will be ineffective (CAO 2008; Pratihast 2015). An adequate information enhances water monitoring approaches helping in decision-making process for water resources management (Ribbe et al. 2008; Hernández 2010).

The proposed research work in Nepal aims to develop a systematic approach to assess and monitor rural water quality by proposing relevant technical solutions (computer and geo-information science) and incorporating the needs and capabilities of local communities for a safe drinking-water monitoring mechanism. This refers to the idea of conceptualizing the role of information in water security concerning quality factor with a Community-Based Participatory Approach (CBPA) for managing quality in water supply schemes. It allows checking the reliability of data obtained by CBPA. Identifying potentials of emerging ICTs to facilitate a Water Quality Monitoring (WQM) mechanism is vital to accelerate the process of establishing a meaningful flow of information of the monitoring system, accessing suitability, and sustainability. The appropriate use of the web or smart-phone applications and their acceptance by the local communities will be analysed. Specific strategies in the selected area, the Pokhara Metropolitan City of Nepal will be applied considering lessons learned from the research carried out in Brazil on the "Community-Based Water Quality Monitoring (CBWQM) strategy" to produce reliable SDG indicators. Finally, the effective evaluation of the applied strategies and the recommendations for scaling in different regions will help to support the sustainable development of ideas beyond the theoretical world.

---

[2]Nepal Environmental and Scientific Services. Consultancy/ Environmental Laboratory (NESS).Available at: https://nesspltd.com/info/2012/03/water-quality-monitoring-and-assessment/.

57.2 Theoretical Framework

The experience of the research work in Brazil on CBWQM in the micro-basins Barracão dos Mendes, Santa Cruz, and São Lourenço, Rio de Janeiro points out the potentials to use local communities in water monitoring activities. The research was conducted to design and to develop a strategy for monitoring water quality at the community level with the direct, indirect, and active participation of various stakeholders. At first, the main concept of the topic and the area were reviewed together with distinctive internet research for selecting field tools, equipment, and kits. The initial idea of selecting monitoring points by criteria such as stream order, land use, point source pollution, and accessibility was also defined (Fig. 57.1).

Likewise, the demands, purpose, and the knowledge of the selected subject of various stakeholders were analysed through participatory workshops, followed by an intensive field data collection (water tests), a survey for identifying monitoring points, and its verification via interviews. The mobile application (ODK) was introduced and tested for collecting, storing, transferring and sharing of data. Major results of the study include the identification of thirteen monitoring points, determination of appropriate water test kits, formation of an active working team (students, teachers, and community representatives) dubbed the "Water Club," defining the roles of main stakeholders, cost estimation of materials and the development of an online platform (website) for data sharing and dissemination. Lack of proper coordination and

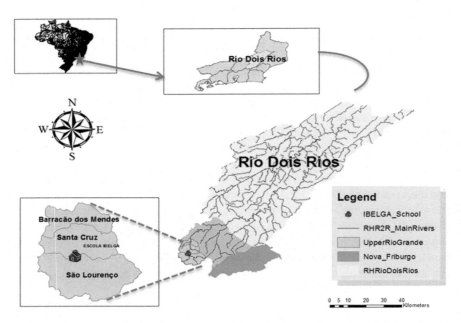

Fig. 57.1 Case study area in Rio de Janeiro, Brazil (data *source* INTECRAL 2014-Geodata; Gautam 2015-GPS data)

accountability should be handled with better cooperation and understanding between stakeholders with an adequate balance between data collection, data storage, and data use to ensure a sensible decision-making process and continuous monitoring activities (Gautam 2015).

57.2.1   Identifying Research Gaps in Nepal

Nepal, being a country with about 42% of the population below the poverty line, also represents various issues in the context of safe drinking water and sanitation facilities.

57.2.1.1   Theoretical Approach

Adequate knowledge of the key components is important for analysing the necessary scope of any research. Hence, relevant literature was reviewed to clarify concepts of basic terminology like water quality, monitoring, WSP, CBPA, ICTs, etc., and also for understanding the existing situation regarding water supply schemes, safe drinking water, availability of water quality data, and use of tools and techniques in the context of Nepal (Fig. 57.2).

57.2.1.2   Research Demand and the Specific Study Area

Nepal shows the low status of sanitation services with only 27% of citizens having access to improved sanitation. This leads to a number of issues and challenges in both rural and urban areas of the country in relation to water quality and monitoring aspects, and the degree of compliance with the country's own water quality standards is weak (UN-HABITAT 2011; NMIP/DWSS 2014). The public lacks awareness and education on proper sanitation issues, and more domestic and industrial wastewater treatment plants needed to provide coverage of the population. Nepal struggles to overcome these obstacles and needs solutions to eradicate this problem so that the poor citizens can live healthier lives. Through access to community friendly, low cost, and sustainable technology (Suwal 2015; Bdour et al. 2015); (WASH-RCNN 2016). The Global MDG Assessment Report points out the progress on Sanitation and Drinking Water and indicates monitoring process of MDG 7C in Nepal by WHO/UNICEF Joint Monitoring Programme (JMP) (WHO/UNICEF 2015). It implies that the estimations are based on national surveys and censuses, and states that official definitions of urban and rural vary across countries and may not be directly comparable. Furthermore, national data has differed from JMP in the case of Nepal. There remains a gap between coverage (quantity) and quality assurance of extant drinking water supply projects with no updated sector assessment (NMIP/DWSS 2014). WSP tool has been useful for the government in developing a strategy but it is

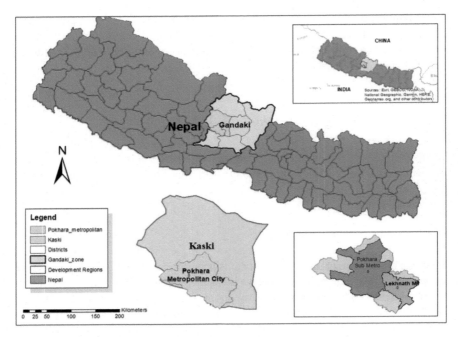

Fig. 57.2 Location map for the study area in Nepal (Data *source* DIVA-GIS 2016)

not enough yet. There should be a mechanism to check the reliability of data obtained by CBM (Barrington et al. 2013; Ballard et al. 2016). Though, water supply target for MDG has been met, that of sanitation was not. The actual scenario is different (NMIP/DWSS 2014; WASH-RCNN 2016). The SDG on water and sanitation has been introduced with more indicators, so there is a need for reliable baseline data and monitoring mechanisms to show the effectivity of merged information products in describing the current water supply situation. It is important to analyse the acceptance of mobile or web technologies by local communities in order to provide flexible and sustainable monitoring network (REACH 2016).

Nepal's Kaski district is one of the most vulnerable areas[3] in the context of this subject. The district headquarter Pokhara is one of the best tourist destinations in the world, covers parts of the Annapurna mountain range, and is full of rivers such as the Seti-Gandaki, Modi, and Madi along with other rivulets. This district in total has about 1002 water supply schemes (Gravity flow: 994, surface pumping: 5 and schemes older than 20 years: 292) reflecting water supply and sanitation coverage

---

[3]Lack of data availability of water quality. Information received during a distant interview (Online-email and telephone) with National Professional Officer, Environment and Health Division, WHO—Dr. Sudan Raj Panthi.

of about 90.6 and 53.5% respectively. Among these water supply schemes,[4] only around 59.8% provide supply throughout the year, though only 18.5% are well-functioning, which reveals a large gap between coverage and functionality in the supply system owing to the degraded quality of the water. In fact, there is no data available on water quality (NMIP/DWSS 2014). These gaps symbolize the lack of proper system assessment, the flow of information, understanding responsibilities, suitable instruments, and tools to monitor and to evaluate the water quality factors and linkages. Hence, the research proposed for Nepal will help to develop the described mechanism with the proper use of appropriate technologies assigning high priority to supporting relevant and concerned indicators of SDG in a sustainable and socially acceptable manner.

57.2.2   Translating Conceptual Idea from Brazil to Nepal

The process of transferring ideas holds both benefits and challenges. Certain aspects can be translated with less effort and some ideas have to be largely modified. Depending upon various circumstances, different concepts and strategies may need to be redefined and recreated. In this particular case, two main components can be considered in transcribing ideas to conduct the proposed research in Nepal. They are stakeholders' integration and the use of mobile application from data collection to dissemination. Some other steps and strategies applied during field research in Brazil are somewhat relevant to the context of Nepal but cannot be directly translated. For instance, in the case of Brazil, three micro-sheds were analysed, whereas different water supply schemes of the Kaski District will be investigated in Nepal.

57.2.2.1   Stakeholders' Integration

The meaningful integration of concerned stakeholders is a key to the success of any project, especially when involving a participatory approach requiring the collection of primary data from the field. During initial meetings with the stakeholders in Brazil, one school showed great interest in this concept and helped to organize a workshop in their facilities. This workshop served to elucidate the school's points of view and to explain the basic idea of CBWQM. An extensive stakeholder analysis helped to identify the demand for CBWQM at the institutional level and the educational possibilities for the students carrying out the monitoring. During this cooperation with the local school and the accompanying field surveys, the idea of starting a regional water club was raised. This club would play a vital role in a proper implementation of monitoring activities in the long term and engage students in the ongoing monitoring

---

[4]Schemes-Need repair: 65%, Need reconstruction: 6.3%, Have Water Supply and Sanitation Technician: 29.5%, Adequate tools: 29.4%, Water and Sanitation Users' Committee: 26.3%, No operation and maintenance fund.

of their local river. Accordingly, the water club 'CLUBE DA ÁGUA' was formed in 'IBELGA' school (located in Santa Cruz) consisting of active and interested students (higher level), concerned teachers, and community representatives (Gautam 2015). For the case of Nepal, this serves as a good example for involving school students and community representatives to ensure effective monitoring to provide safe drinking water to citizens.

57.2.2.2   Use of Mobile Application

The chosen mobile application for this monitoring activity (data collection, storage, transfer, and share) in Brazil was Open Data Kit (ODK). ODK is a free and open-source set of tools, which helps to collect data using mobile phones and to transfer them to an online server (Hartung et al. 2010). The usage of a mobile application for data collection and transfer was perceived by the participants of the monitoring as easy, convenient, user-friendly, and less time consuming than paperwork. The utilization of ODK offered the possibility to design separate forms specially designed for each of the water quality test kits in English and Portuguese. After entering, the data was transferred to a Google Engine server and visualised on a website by ODK. This website is publicly accessible and aimed to further involve the public in waterway issues. Other municipalities and organizations in Rio de Janeiro have already stated their interest in CBWQM, and follow up projects are being discussed (Gautam 2015). This concept can be widely utilized in the research in Nepal. A similar application or a new application can be identified or developed considering the local situation and research facilities.

57.3   Conceptual Framework

This section describes the conceptual structure of the research as a hypothetical result. The methodological approach has been designed for the proposed research and the specific case study in Nepal (Fig. 57.3).

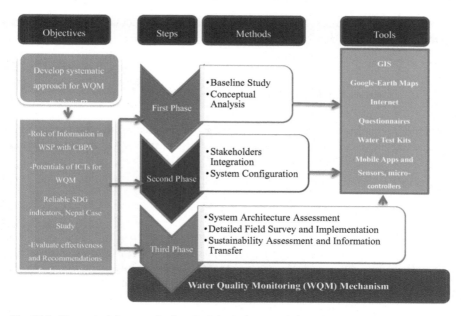

Fig. 57.3 Conceptual framework of methodological approach for proposed research in Nepal

57.3.1 Baseline Study

Bibliographic review will identify and evaluate present situation of data available, information, and management strategies. Open access data will be given priority.

57.3.2 Conceptual Preparation and Analysis

Preparation prior to the field visit will include analysis of information and planning (GIS tools, land use maps, satellite images for water resources-catchment context), and selection of appropriate tools and criteria for identifying suitability of different technological approaches (mobile applications, sensors, remote sensing, etc.)

57.3.3 Stakeholders' Integration

This step includes organizing stakeholders and workshops/trainings for assessing and evaluating the interest and requirements in the local level and informing local communities and authorities about the subject and the benefits of their participation. Water users' committees, local schools and authorities, and local educational

centres like Karkhana[5] and The GLOBE program[6] will be encouraged to support the
activities.

57.3.4 System Configuration

Possible approaches to configure mobile applications (including algorithms), sensors,
microcontrollers, etc. will be developed. Local organizations will be considered as
well.

57.3.5 System Structure Assessment

In order to understand each system better, some specific steps include water resources
estimation, observation of different components of water supply, identifying sources
of pollution, selection of water quality parameters to be tested, etc.

57.3.6 Detailed Field Survey and Implementation (From Data Collection to Dissemination)

A suitable database management system will be applied for the proper collection,
storage, analysis and verification of different forms of acquired data (primary and
secondary) according to the system's configuration and strategies developed.

57.3.7 Sustainability Assessment and Information Transfer

Finally, the results will be conveyed and the indicators/strategies that best assess the
effective monitoring will be presented and recommended for possible replication.

---

[5] An education company and makerspace with a unique approach to learning. https://www.karkhana.
asia/-advance technical and logical support during field visit.

[6] The GLOBE program. A worldwide science and education program. https://www.globe.gov/.
Citizen science approach.

57.4 Conclusions

The conceptual framework is drawn following the analysis and evaluation of previous works and methods in the similar context of Brazil, along with the review of relevant literature. The resulting methodological approach will be applied in the proposed research work in Nepal. Systematic planning of the fieldwork involves analysis of ICTs—appropriate tools/techniques, related information such as GIS tools and applications, smart phones and sensors, etc. The concept of community participation (mainly students and community representatives with meaningful interest from active stakeholders' coordination with local bodies and related organizations) is considered as one of the key factors in ensuring systematic monitoring mechanism of drinking water quality. Main challenges can occur during this phase. Therefore, sufficient consultation with local people, organization, and experts should be done to avoid probable difficulties in coordination. Required tools and equipment should be configured before and during fieldwork for applying in the selected area (Pokhara Metropolitan City, Nepal) for efficient flow of water quality information from data collection to dissemination after understanding different components of the water supply system. Then, evaluation of all these steps has to be verified and explained so that similar strategies can be followed for possible replication and best practices.

References

Ballard HL et al (2016) Youth-focused citizen science: examining the role of environmental science learning and agency for conservation. BiolConserv 2–6

Barrington D et al (2013) Water safety planning: adapting the existing approach to community-managed systems in rural Nepal. Water Sanit Hyg Dev 3:3

Bdour AN et al (2015) Real-time remote monitoring (RTRM) of selected water quality parameters in marine ecosystem using wireless sensor networks. In: 14th international conference on environmental science and technology, CEST, Rhodes, Greece, pp 3–5

CAO (2008) Participatory water monitoring. A Guide for Preventing and Managing Conflict. The Office of the Compliance/Ombudsman

Conrad CC, Hilchey KG (2011) A review of citizen science and community-based environmental monitoring: issues and opportunities. Environ Monit Assess 176(1–4):273–291

DIVA-GIS (2016) Download Data by Country. https://www.diva-gis.org/gdata

Dominguez AF (2015) Case study: Brazil national water agency capacity development. In: 2015 UN-water annual international Zaragoza conference

Gautam A (2015) Design of a community based water quality monitoring (CBWQM) strategy in the micro-basins Barracão dos Mendes, Santa Cruz and São Lourenço, RJ, Brazil

GEMI (2015) Monitoring Waste Water, Water Quality and Water Resources Management: Options for Indicators and Monitoring Mechanisms for the Post-2015 Period. Global Expanded Water Monitoring Initiative

Gleick PH (1998) The World's Water 1998–1999: The Biennial Report on Freshwater Resources. Island Press, Washington DC

Gondim et al (2006) Brazil Water Resources Management: Evolution and Challenges. American Society of Civil Engineers, pp 1–11

Hartung C et al (2010) Open data kit: tools to build information services for developing regions. In: Proceedings of international conference on information and communication technologies and development, pp 1–11

Hernández JHA (2010) Water quality monitoring system approach to support Guapi-Macacu river basin management, Rio de Janeiro, Brazil

INTECRAL (2014) INTECRAL RBIS. Integrated Eco Technologies and Services for a Sustainable Rural Rio de Janeiro. River Basin Inventory System. https://leutra.geogr.uni-jena.de/intecralR BIS/metadata/login.php?url=start.php

Nare L et al (2011) Framework for effective community participation in water quality management in Luvuvhu Catchment of South Africa. Phys Chem Earth Parts A/B/C 36(14–15):1063–1070

NEWAH (2011) Water, Sanitation and Hygiene in Nepal. Nepal Water for Health

NMIP/DWSS (2014) Nationwide Coverage and Functionality Status of Water Supply and Sanitation in Nepal. National Management Information Project (NMIP). Department of Water Supply and Sewerage (DWSS), Panipokhari, Kathmandu

Porto M, Kelman J (2000) Water resources policy in Brazil. Rivers 7(3):250–257

Pratihast AK (2015) PhD Thesis on "interactive community-based tropical forest monitoring using emerging technologies". Wageningen University, Netherlands

REACH (2016) Establishing a water quality monitoring network in mid-western Nepal. REACH: Improving water security for the poor. https://reachwater.org.uk/funding/catalyst-projects-call-1/establishing-a-water-quality-monitoring-network-in-mid-western-nepal/

Ribbe L et al (2008) Monitoring to support water quality management in north-central Chile. In: IWRA world water congress 13

SEIU/MoUD (2011) Water Supply, Sanitation and Hygiene Sector Status Report May 2011. Sector Efficiency Improvement Unit. Ministry of Urban Development

Suwal S (2015) Water in Crisis – Nepal. https://thewaterproject.org/water-in-crisis-nepal

Tereza M, Alves R (2014) A global scientific literature of research on water quality indices: trends, biases and future directions. Acta Limnol Bras 26(3):245–253

Tundisi JG (2011) Water Policy in Brazil. International Institute of Ecology, São Carlos, Brazil, National Institute of Science and Technology in Mineral Resources, Water Resources and Biodiversity. (Ministry of Science and Technology). Feevale University, RS. Brazilian Ac

UN-HABITAT (2011) Nepal Country Impact Study Document 3. UN-HABITAT, Nairobi

UN Water (2015) Consolidated technical input from UN agencies on water and sanitation related indicators. https://sustainabledevelopment.un.org/index.php?page=view&type=400&nr=2076&menu=35

UNDP (2015) Sustainable Development Goals. https://www.undp.org/content/undp/en/home/librar ypage/corporate/sustainable-development-goals-booklet.htm

Warner NR et al (2008) Drinking water quality in Nepal's Kathmandu valley: a survey and assessment of selected controlling site characteristics. Hydrogeol J 16(2):321–334

WASH-RCNN (2016) Drinking Water Policies and Quality issues in Nepal. https://www.wash-rcnn.net.np/drinking-water-policies-and-quality-issues-in-nepal

WHO (2015) Key Facts from JMP 2015 Report. https://www.who.int/water_sanitation_health/mon itoring/jmp-2015-key-facts/en/

WHO/UNICEF (2015) Progress on Sanitation and Drinking Water. 2015 Update and MDG Assessment. https://www.unicef.org/publications/index_82419.html

Chapter 58
Community Mitigation Approaches to Combat Safe Water Scarcity in the Context of Salinity Intrusion in Coastal Bangladesh

M. Tauhid Ur Rahman, Md. Rasheduzzaman, Md. Arman Habib, Afzal Ahmed, Syed M. Tareq, and S. Md. Muniruzzaman

Abstract Scarcity of safe drinking water is now a global problem. Today, nearly 1.2 billion people in the developing world lack access to safe drinking water. The crisis of safe drinking water is increasing in Bangladesh with the increasing population. Mainly, the coastal regions of Bangladesh experience severe safe drinking water scarcity due to salinity intrusion. In this context, this research is undertaken in Satkhira district, which is one of the severe climate hit water stressed regions of Bangladesh. The study was conducted using water quality parameters related to salinity, household questionnaire survey and focus group discussions (FDGs) to investigate how local people's cope with safe drinking water scarcity. Water quality parameters show the severe conditions of the drinking water sources in the study area. The average value of total dissolved solids (TDS), electrical conductivity (EC) and chloride concentration (Cl^-) was found in the study area 3,178.86 mg/L, 5,788.5 μS/cm, and 2,272.85 mg/L, respectively. Though there are many socioeconomic and locational factors increasing the drinking water vulnerability, local peoples have their own mitigation approaches to cope with the crisis. The study efforts to devise an integrated community based model by highlighting both present community mitigation methods and efforts of the stakeholders, which would be effective for reducing the drinking water crisis.

Keywords Salinity intrusion · Community adaptation · Water scarcity

58.1 Introduction

Safe drinking water scarcity in south-western coastal regions of Bangladesh is not a recent phenomenon. According to the coastal zone policy (2005), out of 64 districts of Bangladesh, 19 are bordered as coastal areas, which are affected directly or indirectly by tidal surges, salinity intrusion, drought and storm surges (Tareq et al. 2018).

M. T. Ur Rahman (✉) · Md. Rasheduzzaman · Md. A. Habib · A. Ahmed · S. M. Tareq · S. Md. Muniruzzaman
Military Institute of Science and Technology (MIST), Dhaka, Bangladesh

© Springer Nature Switzerland AG 2021
M. Babel et al. (eds.), *Water Security in Asia*, Springer Water,
https://doi.org/10.1007/978-3-319-54612-4_58

Coastal regions of Bangladesh have been experiencing acute shortage of safe drinking water over the past few years, and this problem is becoming more acute and fatal as to the increase in salinity intrusion in surface and ground water sources. Groundwater is unsuitable for drinking or irrigation in the coastal regions as the deep groundwater in the coastal area is relatively vulnerable to the contamination of saline water intrusion (Kim et al. 2006). Also, excessive use of groundwater near the coast increases the salinity intrusion into the aquifer (Abedin and Shaw 2013). In this region, near 15 million people are forced to drink saline water, and more than 30 million people are facing difficulties to collect safe drinking water because of unavailability of safe water sources (Hoque 2009).

Over the years, the people of coastal regions of Bangladesh are coping with different coastal hazards and disasters, and have implemented different mitigation approaches to combat the safe water scarcity. Providing access to safe drinking water is very important for public health and sustainable socioeconomic development in these areas. Various government organizations; national, international and local NGOs; the private sector; and the communities are working to resolve the safe drinking water crisis. Without the participation of local communities in the planning and development processes, it is not possible to achieve sustainable safe drinking water supply and access. Since communities respond first to any kind of natural disaster, which happens in their locality, local community participation and ownership are key to any successful program (Habiba and Shaw 2012). For this reason, it is a must to find out the existing mitigation and coping approaches on drinking water scarcity.

Several researchers have conducted diverse studies the coastal threats and fresh water paucity of coastal regions in Bangladesh. Abedin et al. (2014) studied the community perception and adaptation to safe drinking water scarcity, which is triggered by the combined effects of salinity, arsenic, and drought. This study reveals that salinity is the major cause of fresh water scarcity and the present adaptation and coping mechanisms will not be adequate to cope with the future challenges of fresh drinking water supply (Fig. 58.1).

Parvin et al. (2008) also studied the coastal people's perception to the hazards, their vulnerabilities to these hazards, and their adaptation methods for different hazards. This research indicates that both the frequency of coastal hazards and community's vulnerabilities are increasing with time.

Though there are many studies about the coastal regions of Bangladesh, only few have focused on severe water stressed regions. Moreover, there is barely any research that focused predominantly on "Shyamnagar" and "Tala", which are the hardest climate hit water stressed regions of Bangladesh (Abedin et al. 2014; Rahman et al. 2017). This study provides a brief overview of safe drinking water scarcity caused by salinity at the villages of the south-western coast of Bangladesh. It also examines current water quality conditions and the community adaptation measures to safe drinking water scarcity. This research suggests a community based adaptability action plan at the local level to combat drinking water scarcity.

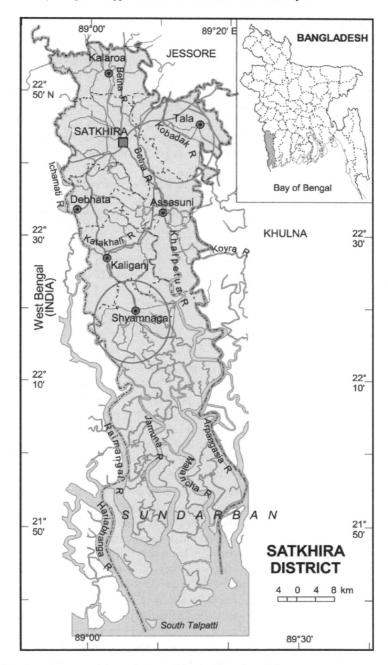

Fig. 58.1 Location map of the study areas (Memim Encyclopedia)

58.2 Study Area

The study was undertaken in one of the severe salinity prone districts namely Satkhira, which is located in the coastal region of Bangladesh. Two Upazila (sub-district) namely Shyamnagar and Tala were studied of Satkhira district. Salinity intrusion has been striking great danger to the drinking water sources of these areas. The study area is geographically situated between 21°36′ N and 22°54′ N latitude and 88°54′ E and 89°20′ E longitude. Satkhira district is part of a barren delta of large Himalayan Rivers, and the Sundarban mangrove forests protect this region from tidal surges (Mahmud and Barbier 2010).

With regards to the climate conditions, Satkhira shows variation in temperature and rainfall patterns over the past few decades (Miah 2010). The highest average maximum temperature is 33 °C and above during the month of March to May, and the lowest average minimum temperature is about 18.9 °C in December and January (UNICEF, 2014). On the other hand, the coastal region of Bangladesh (mainly Satkhira district) receives an average precipitation of about 1710 mm per annum, much of which falls within the month of May to October (UNICEF, 2014).

58.3 Methodology

This research is based on the primary water sample data and preliminary field investigation using a semi structured questionnaire. The goal of this study is to investigate existing water quality parameters related to salinity, to analyse the impacts of drinking water scarcity, and to identify the coping and adaptation practices of local communities for safe drinking water scarcity. This research also investigates the contribution of different organizations (Government, Private) and institutions, and practices in facilitating local adaptation to fresh drinking water crisis caused by salinity.

Water samples were collected in 500 ml preconditioned high density polyethylene bottles. Mainly water from, tubewell, reserved ponds, river and treatment plants were collected. Prior to collect sample, all bottles were cleaned with five percent (5%) nitric acid and finally washed with distilled water several times. This was carried out to ensure that the sampling bottles were free from contaminant. Before sampling, the bottles were rinsed again three times with the water which was going to be sampled for investigation. After sample collection, the bottles were sealed immediately to avoid exposure to air. Samples were kept in an ice contained field sampling box to keep the temperature below at a lower level (<4 °C) to avoid the alteration of the water quality. Plastic bottles were labeled separately with a unique identification number. Such data as date of collection, location, time, etc. were recorded in the note book to provide necessary information for each sample. The bottles were kept in a clean, cool and dry place in the laboratory, before analyzed. The onsite parameter tests such as electrical conductivity, EC, total dissolved solids, TDS and salinity were measured onsite by employing a handheld multi-parameter instrument. The chloride, Cl content was measured by using Mohr's titration method.

A total of 150 questionnaires were circulated and collected at these Upazila (75 questionnaires from each Upazila). 95% confidence interval and ±5% level of accuracy were used for calculating the sample size of the study areas. The questionnaire was designed in three parts. In the first part, the questions were designed to collect the socioeconomic information of the respondents, and in the second part, the questions highlighted their perception to the condition of the drinking water and its availability. In the last part, the questions looked into the different impacts of safe drinking water scarcity on a respondent's daily life and their various adaptation measures towards salinity intrusion. Four group discussions (FDGs) were conducted to cross check and to validate the responses from respondents.

58.4 Results and Discussion

After collecting data through field visit and questionnaire survey, the data are analysed. The results of water quality and survey are discussed in the following.

58.4.1 Socio-economic Profile

Socioeconomic features such as gender, age group, education level, occupation, and income level help to recognize the vulnerable portion of the community to hazards and crisis. A total of 150 households was included in the study. Our survey results reveal that almost all households are dominated by male persons, which is more than 65%. The percent of illiterate people in the study areas are more than 40%, and a very few people have completed their college education. It is observed that the peoples of the study areas are engaged in a different set of financial activities. According to the survey results, the status of primary occupation of local peoples greatly varies in two ways viz. agriculture and labour (average of 42.5% and 23.5%, respectively). Most of the people in the study area have low income (BDT 46,000 to BDT 60,000 per year = $US 550-720). Thus, these peoples are more vulnerable to safe drinking water crisis because of low literacy rates and low income level.

58.4.2 Present Water Quality of the Study Areas

Salinity intrusion in drinking water sources in this region is increasing day by day. All the ground and surface water sources in almost all the Upazila of Khulna and Satkhira districts exceed the threshold values (<600 ppm) as given by the Bangladesh drinking water standard, except for some locations of Tala Upazila (Table 58.1).

Table 58.1 Drinking water quality standards (*Source* WHO 2011.)

Water quality parameters	Unit	Bangladesh standards	WHO guideline values
TDS	mg/L	1,000	1,000
EC	μS/cm	1,563	1,563
Chloride	mg/L	150–1,000	250
Salinity	mg/L	<600	–

Salinity of drinking water sources (surface and groundwater) in most of the Upazila lies between 600 and 1,500 ppm. After testing the water samples, the EC value in Shyamnagar and Tala Upazila was found ranging from 129.4 μS/cm to 29,600 μS/cm with an average of 5,788 μS/cm. According to WHO and the Bangladesh standards, the threshold value of EC is 1,563 mg/L for drinking water, which indicates that all drinking water sources do not fall within the drinking water quality standards. Presence of high EC in the drinking water sources indicates that the water of the study areas is highly affected. TDS of the water samples was also investigated in order to determine the salinity in the drinking water sources. According to WHO and Bangladesh standards, the threshold value of TDS is 1,000 mg/L for drinking water. The TDS of the Shyamnager and Tala Upazila was found at ranges from 59 mg/L to 17,600 mg/L with an average of 3,179 mg/L. Almost all water samples exceeded the TDS limit, which is indicating a threat to safe drinking water in the study areas. As chloride concentration is mainly responsible for salinity, the concentration of chloride ions was also measured. The average value of chloride was found 2,273 mg/L. The threshold value of chloride is 150–1,000 mg/L for drinking water in the coastal regions of Bangladesh. Most of the drinking water sources of Shyamnagar and Tala exceeded this limit, but comparatively a large number of drinking water sources of Tala were found within permissible limits. The average values of Salinity at Shyamnagar and Tala were found 4,236 and 2,370 mg/L, respectively (Rahman et al. 2017).

58.4.3 Impact of Salinity on Drinking Water and Environment

Saline water intrusion in the study areas increases the scarcity of fresh drinking water for the rural people. All drinking water sources in these areas are somehow affected. Especially the poor peoples in these areas face acute drinking water problems because of salinity problems. In some areas, neighbourhood drinking water sources are affected by severe salinity, so the women and adolescent girls must travel distances to collect safe salinity free drinking water, which is economically infeasible and time consuming. The findings of our field visits and questionnaire survey revealed that about 100% respondents face drinking water problem because of salinity problem. The increased unavailability of fresh water, costly and time

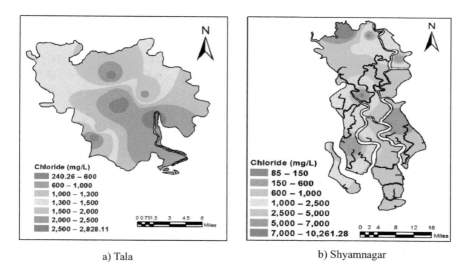

a) Tala   b) Shyamnagar

Fig. 58.2   Spatial distribution of chloride concentrations in the study areas

consuming treatment facilities of salt water force people to drink contaminated saline water containing harmful trace elements, which can lead to various water borne diseases like diarrhoea, indigestion, fever and other intestinal diseases. Respondents, who suffered from diarrhoea, were found in more than 85% in the study locations. Indigestion, skin diseases, and high blood pressure were also found in significant percentages in the study areas (Fig. 58.2).

A gynaecological anomaly was observed in several women in the study area in the sense that they were found to suffer from the lowering of the uterus, having pain in the lower abdomen, caused by regular carrying of water jars on their hips. Salinity has also a direct effect on stroke, left ventricular mass, cancer of the stomach, and many more diseases (He and MacGregor 2008).

Salinity is greatly responsible for the loss of ecosystem and loss of bio-diversity. The world largest mangrove forest, the Sundarbans is located in the south-western part of Bangladesh covering parts of the Khulna, Satkhira, and Bagerhat districts. Increased amount of salinity has changed the habitat pattern of the mangrove forest. The trees of the Sundarbans are suffering from salinity-induced top-dying disease. More than 50% respondents in the study areas perceived that they noticed changes in the environmental patterns as salinity levels went higher over the years. Shrimp cultures in these areas also threaten the existence of the fresh water fish species and lead to their extinction (Mustari and Karim 2014). More than 30% respondents in the study areas told that their livestock have been affected by high salinity levels (Rahman et al. 2017).

58.4.4 Current Mitigation Approaches at Different Levels

Despite the different impacts of salinity directly affect livelihoods, drinking water sources, public health and the total environment of this region, however, the local people, government, and different local and international organizations are trying to do their best to overcome these negative situations.

People are practicing different adaptation measures to mitigate the impacts of increased salinity. The adaptation measures at the household level are usually carried out by a particular household to meet their household demands. To meet the scarcity of safe drinking water, household practice measures such as rainwater harvesting, conservation of pond waters, and harvesting drinking water through ring wells. As Rainwater harvesting is a relatively costlier option, conservation of ponds and pond water is more popular among local peoples. More than 65% peoples utilizes pond water for drinking purposes. At community level, digging of artificial ponds and using pond sand filters (PSF) are additional techniques above the individual level. Rainwater harvesting gets preferred at the community level as there are more hands to join together as far as the cost is concerned. Use of pond sand filter is the most popular practice in the community level (48%) (Fig. 58.3).

Institutions play a leading role in supplying safe drinking water in the whole country. Different national, government, NGO, and International organizations like Sushilan, Uttaran, Action Aid, Caritas, Concern Worldwide, UNICEF, USAID, Department of Public Health (DPHE), and Comprehensive Disaster Management Program (CDMP) are working relentlessly to ensure safe drinking water to the people of the zone hit by water scarcity. NGOs are implementing different projects like pond re-excavation, deep tube well with overhead tank to combat drinking water scarcity in the study areas. Another national NGO, The Bangladesh Centre for Advanced

Pond Sand filters (PSF)   Deep tube well with overhead tank

Fig. 58.3 Various adaptation measures to combat safe drinking water crisis

Studies (BCAS), also help people in their own communities run water organizations, called *Pani Parishad*, which bring people together to identify the best water supply options for the households and the communities in their own villages.

58.5 Conclusions

Bangladesh, particularly at the south-western coastal regions faces severe drinking water scarcity due to increasing salinity intrusion. According to IPCC (2007), surface and groundwater resources in Bangladesh will be affected by the effect of climate change. Therefore, low to medium saline areas in the coastal regions will become highly to extremely saline areas, and will cause severe drinking water scarcity. This research identifies and reveals people's adaptation and coping measures along with water quality parameters related to salinity intrusion and its imposed threat to the drinking water sources and the environment. Water quality parameters (TDS, EC and Cl^-) in most of the locations in the study areas were found to be exceeding the allowable limit by a significantly greater margin.

Though local peoples have own adaptation measures to cope with drinking water scarcity, the current level of adaptation measures is not sufficient to cope with the future challenges of safe drinking water scarcity. This research presented the efforts of government organizations and NGOs for overcoming safe drinking water crisis. Though salinity intrusion is becoming more alarming in the context of fresh drinking water, remaining stakeholders and constitutional management operators are not giving appropriate thought to it. Local government organizations and various national and international organizations are providing support to the local communities besides their own adaptation measures, but these practices and supports are not enough to combat safe drinking water crisis in the near future. Therefore, it is a must to take an integrated approach with including local communities in that area. It is suggested that the understanding of the community's perceptions and evaluation of their adaptive and proactive capacities is vital for creating successful safe water adaptability programs. There are many government organizations and NGOs working to solve this problem, but the main problem is harmonization between them.

Based on the results of the study, Fig. 58.4 proposes a community based model to integrate an operational coping mechanism at the community level. This model recommends that for a functional, proficient and environmentally friendly coping approach, the communities, government, NGOs and international organizations need to function as a single unit. In place of disconnected activities, integration is needed between different stakeholders and constitutional management operators to resolve the safe drinking water crisis and to ease the community's adaptation.

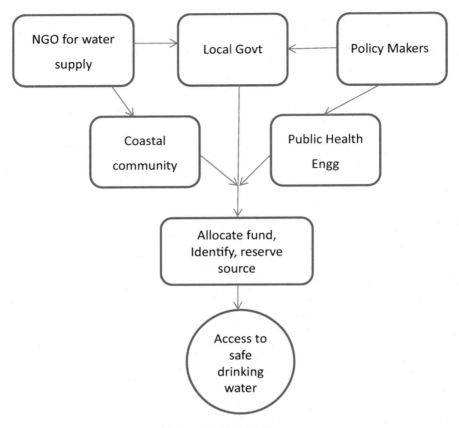

Fig. 58.4 Community based model to combat safe drinking water scarcity

Acknowledgements This research received funding from the Higher Education Quality Enhancement Project (HEQEP, CP-3143) jointly funded by the Government of Bangladesh (GoB) and the World Bank, and implemented by the University Grants Commission (UGC) of Bangladesh. The authors gratefully acknowledge the supports of concerned authorities and the staffs of the Climate Change Laboratory, and Environmental Engineering Laboratory of Military Institute of Science and Technology (MIST), Dhaka. Help and sincere cooperation of the local communities, local government representatives, NGOs of Tala and Shyamnagar Upazila of Satkhira district are greatly appreciated as well.

References

Abedin MA, Shaw R (2013) Safe water adaptability for salinity, arsenic and drought risks in southwest of Bangladesh. Risk Hazards Crisis Publ Pol 4(2):62–82

Abedin MA, Habiba U, Shaw R (2014) Community perception and adaptation to safe drinking water scarcity: salinity. Arsenic Drought Risks Coast Bangladesh Int J Disaster Risk Sci 5:110–124

Habiba, U, Shaw R (2012) Bangladesh experiences of community based disaster disk reduction. In: Community, environment and disaster risk management, pp 91–111. Emerald Publishers, Bingley

He FJ, MacGregor GA (2008) A comprehensive review on salt and health and current experience of worldwide salt reduction programs. J Hum Hypertens 23:363–384

Hoque RM (2009) Access to safe drinking water in rural Bangladesh: Water governance by DPHE. BRAC University, Dhaka, Institute of Governance Studies

IPCC (2007) Summary for policymakers. In: Parry, ML, et al (eds) Climate Change: Impacts, Adaptation and Vulnerability. Contribution of Working Group II to the Fourth Assessment Report of the Intergovernmental Panel on Climate Change, vol 1000. Cambridge University Press, Cambridge

Kim RH, Kim JH, Ryu JS, Chang HW (2006) Salinization properties of a shallow groundwater in a coastal reclaimed area, Yeonggwang, Korea. Environ Geol 49:1180–1194

Mahmud S, Barbier EB (2010) Are private defensive expenditures against storm damages affected by public programs and natural barriers? Evidence from the aoastal areas of Bangladesh. In: South Asian Network for Development and Environmental Economics (SANDEE), Kathmandu, Nepal. no 54-10

MEMIM Encyclopedia: https://memim.com/satkhira-district.html

Miah MMU (2010) Assessing long-term impacts of vulnerabilities on crop production due to climate change in the coastal areas of Bangladesh. Bangladesh center for Advance Studies, Bangladesh

Mustari S, Karim AHMZ (2014) Impact of salinity on the socio-environmental life of coastal people of Bangladesh. Asian J Soc Sci Humanit 3(1):12–18

Parvin RH, Takahashi F, Shaw R (2008) Coastal hazards and community-coping methods in Bangladesh. J Coast Conserv 12(4):181–193

Rahman MTU, Rasheduzzamana Md, Habib MA, Afzal A, Tareq SM, Muniruzzaman SM (2017) Assessment of fresh water security in coastal Bangladesh: an insight from salinity, community perception and adaptation. Ocean Coast Manag 137:68–81

Tareq SM, Rahman MTU, Islam AZMZ, Baddruzzaman ABM, Ali MA (2018) Evaluation of climate-induced waterlogging hazards in the south-west coast of Bangladesh using Geoinformatics. Environ Monit Assess 190(4):1–14

UNICEF (2014) District Equity Profile-Satkhira, Local Capacity Building and Community Empowerment (LCBCE) Program, Bangladesh,

WHO (2011) Guidelines for Drinking Water quality, 4th edn. WHO Press, Switzerland